Viral Therapy OF Human Cancers

Viral Therapy OF Human Cancers

EDITED BY

Joseph G. Sinkovics
University of South Florida College of Medicine
Tampa, Florida, U.S.A.
University of Texas MD Anderson Cancer Center
Houston, Texas, U.S.A.

Joseph C. Horvath
St. Joseph's Hospital
Cancer Institute Biotherapy Research Laboratory
University of South Florida College of Medicine
Tampa, Florida, U.S.A.

CRC Press is an imprint of the
Taylor & Francis Group, an **informa** business

CRC Press
Taylor & Francis Group
6000 Broken Sound Parkway NW, Suite 300
Boca Raton, FL 33487-2742

First issued in paperback 2019

ISBN-13: 978-0-8247-5913-1 (hbk)
ISBN-13: 978-0-367-39334-2 (pbk)

Visit the Taylor & Francis Web site at
http://www.taylorandfrancis.com

and the CRC Press Web site at
http://www.crcpress.com

Preface

Could viruses contribute to the cure of cancer? Viruses certainly contributed to the cause of cancer. Retroviruses transduced cellular proto-oncogenes and converted them into oncogenes. Certain DNA viruses knock out the tumor suppressor gene product proteins of p53 and Rb and counteract early apoptotic death of the infected cell, thus eliminating a self-defensive reaction of the infected host, in which these virions or at least their genomes will persist. The infected cells assume transformed or malignant pheno- and genotypes, gain "immortality" and grow as cancers. On the other hand, some naturally acquired virus infections induced remissions in tumor-bearing patients. There are rare but well documented and published records of various human malignancies, more leukemias and lymphomas than other types of cancers, yielding to a natural viral infection and regressing without relapse. In the laboratory, mice and other experimental animals are readily cured of their cancers, ascitic tumors in particular, by direct inoculation of these tumors, or by infection of the animals with some selected viruses that are not pathogenic in these hosts. These "oncolytic viruses" readily kill tissue cultures or xenografts of human cancers. In clinical trials, be as simplistic as letting patients repeatedly drink or intranasally inhale or receive as an enema untitrated veterinary virus preparations, or as sophisticated as the intratumoral injections of a genetically engineered virus, tantalizing, rare, temporary remissions of cancers occur and are lost to relapses. Long stabilizations of the disease are more commonly recorded and are readily attributed to the viral intervention, disregarding that some advanced tumors can assume a Gompertzian growth curve and slow down their growth rate on their own. Decades go by, but not one particular virus emerges as the ideal oncolytic agent for human tumors.

In melanoma, there is a strong claim that oncolysates prepared with Newcastle disease virus eliminated subclinical microscopic disease left behind after surgical removal of gross tumors. Because the Cassel-Murray clinical trial was evaluated against historical controls and was not prospectively randomized, it is not readily accepted as valid. Even though the trial was impeccably carried out for almost two decades, the academic institution (Emory University, Atlanta, GA) keeps accurate records of its patients and tumor-free patients are alive, this claim of successful virotherapy of a highly malignant human cancer is often ignored and dismissed. However, we accepted the curative value of the MOPP regimen for Hodgkin's disease and that of the first cis-platin, bleomycin and vinblastine protocols for testicular carcinoma without prospectively randomized trials. If it is true that microscopic subclinical cancer can be eradicated by a virus, then it is also possible that resident apathogenic viral flora of the human host (parvoviruses; reoviruses; and as yet undiscovered other viruses) can eliminate incipient newly emerging tumors. Further, if this really could happen, some anti-tumoral immunity should remain in these individuals. We cite not only the surveillance by innate NK cells that kill virally infected cells and tumor cells and respond first to immunizations of patients with viral oncolysates, but also the immune T cells, which arise after the inception of a tumor. If one of us could document immune T memory cells specifically reactive to tumor-associated antigens in a healthy and tumor-free individual who never had any documented cancer before, would this finding be acceptable as a proof for the rejection of an incipient tumor in that individual? It would require a series of additional documentation to establish just how that incipient tumor was rejected; viral attack on that tumor would be one of the possible means of its rejection.

In this volume, virologists of the old school meet with the new generation of molecular virologists. Together we try to decipher the controversies embedded in the intricate processes of viral oncolysis, as this discipline of treatment for cancer is applied in preclinical and clinical trials. We are comparing the efficacy of naturally oncolytic and genetically engineered viruses. Gene therapy of cancer has not been reviewed, however, it has not been ignored! It is frequently referred to in this volume. The new generation of virologists persuing the virotherapy of human cancer forms a nicely knit cooperative group. They meet, confer and publish together. Here we list just a few of their recent conferences to guide the reader to early abstracts and full reports. A. Cid-Arregui (Heidelberg, Germany) and A. García-Carrancá (Ciudad Mexico, Mexico) edited the volume "Viral Vectors: Basic Science and Gene Therapy" published by Bio-Techniques/Eaton (Natick, MA) in 2000. In that volume from adenoviruses to vesicular stomatitis virus all genetically engineered viral agents useful for the viral therapy of cancer are elaborated on. A conference on "Replicating Vectors for Gene Therapy" took place in Rochester, Minnesota on October 5-7, 2001. This conference was organized by the Mayo Clinic and its abstracts are printed in Gene Therapy volume 8, Supplement 1, 2001. The abstracts of the "Eleventh International Conference on Gene Therapy of Cancer" was held on 12-14 December, 2002 in San Diego, California; its abstracts are printed in a special extra volume of Cancer Gene Therapy 2003; 10: S1-S44. Papers on naturally oncolytic and genetically engineered viral vectors were presented at these conferences. Finally, the conference entitled "Oncolytic Viruses as Cancer Therapeutics" was held on March 26-30, 2003 in Banff, Canada, organized by the Ottawa Regional Cancer Centre (www.orcc.on.ca). There, a most spectacular array of presentations took place with the take home message that we do have a new and working weapon against cancer in the form of naturally oncolytic and/or genetically engineered viruses. Will viral therapy of human cancers match or surpass the efficacy of chemoradiotherapy or that of the new biological therapeuticals: gene therapy (T. Wasil and A. Buchbinder), ribozymes (Louise Wright and P. Kearney), or antisense oligonucleotides (B.P. Monia, J. Holmlund and F.A. Dorr) as reviewed in Cancer Investigation (Marcel Dekker) volumes 18-9, 2000-1. Is viral therapy of human cancers a collaborator or a competitor of tyrosine kinase inhibitors targeting oncogene cascades or that of cancer vaccines or monoclonal antibodies/immunotoxins? Would sensational announcements of success with virotherapy of cancer (T. Pawlik and K. Tanabe inducing death of colon carcinoma cells with Herpes simplex virus mutant rRp450 and using it with cyclophosphamide and ganciclovir for the treatment of liver metastases) translate for therapy and withstand the tear and wear of late relapses (Oncology News International, June, 2000).

It is a privilege for us to serve as editors of a volume on virotherapy of cancer. In this volume, we present virotherapy of cancer in context, that is, how it is related to work done along these lines decades ago and how it is comparable with other biological therapeuticals now in clinical trials. The abundance of choices is overwhelming. Close to innumerable adenoviral vectors populate the field. We could have catalogued or tabulated them giving each item one line. Instead, one of us (JGS) condensed each of them in a sentence or two with a reference. Enough information is given to the reader to locate the viral vector of interest in the computer or in the library. There is a saying in medical oncology: "if a tumor category is being treated with more than five regimens, that means none of them is working". Hopefully, this dictum is not applicable to the rapidly proliferating adenoviral vectors. Each of them is the result of superb basic knowledge of tumor cell biology and congenial thinking; each of them deserved to be cited.

One of us (JGS) expresses his gratitude to Dr. Robert W. Simpson, emeritus professor of virology, The State University, Rutgers, The Waksman Institute, New Brunswick and Piscataway, New Jersey, and to Dr. H. David Kay, virologist, immunologist and patent attorney formerly affiliated with The University of Texas M. D. Anderson Hospital, Houston, Texas, and with The University Medical School, Omaha, Nebraska for reading and criticizing the JGS manuscripts.

While this volume was edited, The U.S.A. Department of Agriculture launched an investigation on all Newcastle disease virus strains used for research and vaccination to find out if these virus strains posed any biohazard to wild and domesticated birds, especially poultry, and to human beings. One of us (JCH) had to turn his full attention toward cloning and sequencing part of the genome of the 73T Newcastle disease virus strain with which we prepare our viral oncolysates for immunization of patients. However, the virus turned out to be not viscerotropic and it poses no biohazards in the human community or in the ecosystems in which we live. Our work at Emory University (Atlanta, GA) with this virus may continue unimpeded.

We regret the delays this publication had to sustain. Nevertheless, we proudly present to the readers some magnificent articles on viral oncolysis and are very grateful to the authors for their contributions. In the chapter on biological therapy of cancer, one of us (JGS) frequently mentions (in parentheses) drug companies and laboratories sponsoring research on this subject matter. This is done to assist the interested reader to locate better the source of information and facilitate making contact with the manufacturer of a new investigational product. Here we, the editors of this volume, explicitly declare that neither one of us received any monetary rewards or presents from any one of the cited (or non-cited) drug companies or research laboratories.

The contributing authors were given unrestricted liberty as to the construction of their chapters and as to their choices for literature citations. Since we could not engage Dr. Patrick Lee for a reovirus chapter, we asked Drs. M. Bergmann and T. Muster to include a review of oncolytic reoviruses into their chapter. Who could have contributed better than the P.W.K Lee team [1]? After CRC Press and Marcel Dekker publishers united, we had the good fortune to have the assignment of Ms Kathryn Phillips of Maryland Composition as project manager to take charge of the accelerated publication of this volume. Ms Phillips removed all elements of stagnation and acted with care and expertise. We wish to express our gratitude for her efficient interventions. We hope that we have rendered a useful service for those who work in the laboratories or conduct clinical research on the virotherapy of human cancers; to the medical oncologists in practice; and to those patients who are in need of innovative new treatment modalities.

Joseph C. Horvath, M.D and Joseph G. Sinkovics M.D.

1. Norman KL, Farassati F, Lee FWK. Oncolytic viruses and cancer therapy. Cytokine Growth Factor Rev 2001;12:271–282.

Contents

Contributors

John C. Bell Ottawa Regional Cancer Centre Research Laboratories, Ottawa, Ontario, Canada

Michael Bergmann Department of Surgery, University of Vienna Medical School, Vienna, Austria

William A. Cassel Emory University School of Medicine, Atlanta, Georgia

Roberto Cattaneo Molecular Medicine Program, Mayo Clinic Rochester, Rochester, Minnesota

Jan J. Cornelis Applied Tumor Virology Program (Abteilung F0100 and Institut National de la Santé et de la Recherche Médicale U375), Deutsches Krebsforschungszentrum, Heidelberg, Germany

Celina Cziepluch Applied Tumor Virology Program (Abteilung F0100 and Institut National de la Santé et de la Recherche Médicale U375), Deutsches Krebsforschungszentrum, Heidelberg, Germany

Adele Fielding Molecular Medicine Program, Mayo Clinic Rochester, Rochester, Minnesota

Xinping Fu Center for Cell and Gene Therapy, Departments of Pediatrics, Molecular Virology and Microbiology, Baylor College of Medicine, Houston, Texas

Eva Galanis Molecular Medicine Program, Mayo Clinic Rochester, Rochester, Minnesota

Nathalia Giese Applied Tumor Virology Program (Abteilung F0100 and Institut National de la Santé et de la Recherche Médicale U375), Deutsches Krebsforschungszentrum, Heidelberg, Germany

Matthias Gromeier Department of Molecular Genetics & Microbiology, Duke University Medical Center, Durham, North Carolina

Peter Hersey Oncology & Immunology Unit, Newcastle, NSW, Australia

Joseph C. Horvath St. Joseph's Hospital Cancer Institute Biotherapy Research Laboratory, University of South Florida College of Medicine, Department of Medical Microbiology & Immunology, Tampa, Florida

Frank McCormick University of California-San Francisco Cancer Research Institute, San Francisco, California

Melinda Merrill Department of Molecular Genetics & Microbiology, Duke University Medical Center, Durham, North Carolina

Douglas R. Murray Emory University School of Medicine, Atlanta, Georgia

Thomas Muster Department of Dermatology, University of Vienna Medical School, Vienna, Austria

Mikihito Nakamori Center for Cell and Gene Therapy, Departments of Pediatrics, Molecular Virology and Microbiology, Baylor College of Medicine, Houston, Texas

Zbigniew L. Olkowski† Emory University School of Medicine, Atlanta, Georgia

---

†Deceased.

Jennifer M. Paterson Ottawa Regional Cancer Centre Research Laboratories, Ottawa, Ontario, Canada

Kah-Whye Peng Molecular Medicine Program, Mayo Clinic Rochester, Rochester, Minnesota

Jean Rommelaere Applied Tumor Virology Program (Abteilung F010C and Institut National de la Santé et de la Recherche Médicale U375), Deutsches Krebsforschungszentrum, Heidelberg, Germany

Stephen J. Russell Molecular Medicine Program, Mayo Clinic Rochester, Rochester, Minnesota

Volker Schirrmacher Division of Cellular Immunology, Tumor Immunology Program, German Cancer Research Center (DKFZ), Heidelberg, Germany

Joseph G. Sinkovics Cancer Institute, St. Joseph's Hospital; Departments of Medicine and Medical Microbiology-Immunology, The University of South Florida College of Medicine; The H. Lee Moffitt Cancer Center, Tampa, Florida

David Solecki Department of Developmental Neurobiology, Rockefeller University, New York, New York

Christoph Springfeld Molecular Medicine Program, Mayo Clinic Rochester, Rochester, Minnesota

Rebecca Ann C. Taylor Ottawa Regional Cancer Centre Research Laboratories, Ottawa, Ontario, Canada

Xiaoliu Zhang Center for Cell and Gene Therapy, Departments of Pediatrics, Molecular Virology and Microbiology, Baylor College of Medicine, Houston, Texas

1

Progressive Development of Viral Therapy of Human Cancers

A PERSONAL NARRATIVE ACCOUNT

JOSEPH G. SINKOVICS

Cancer Institute, St. Joseph's Hospital;
Departments of Medicine and Medical
Microbiology–Immunology, The University of
South Florida College of Medicine; and The H.
Lee Moffitt Cancer Center
Tampa, FL

EARLY HISTORY

In 1893, the Hungarian physician F. Kovács, practicing in Vienna, Austria [1], observed remissions in leukemic patients after natural infections. Viral infections? Perhaps, but only by exclusion. The diagnosis of bacterial infections must have been fairly well established by then and in the absence of a bacterial pathogen, viral infection could have been suspected. Kovács mentions “Influenza,” “Herpes,” and “Morbilli” in his paper, but these were clinical diagnoses at the best; those viruses were not isolated and identified until many decades later. Indeed he does not mention the word *Virusarten* because that word probably did not exist in the medical vocabulary at that time. Leukemia as *weißes Blut* was well described by Virchow in the mid-1800s and Löffler & Frosch discovered the causative virus of foot and mouth disease in 1898. Therefore, it must not have been possible to readily diagnose and prove human viral diseases as early as in the last decade of the nineteenth century.

G. Dock is credited as being the first to realize that coincidental viral infections can ameliorate the clinical course of human leukemia [2]: A woman with “myelogenous leukemia” and hepatosplenomegaly contracted “influenza” with a reduction of her leukocyte count from 367,070 to 4,775 and diminishing hepatosplenomegaly; her remission lasted more than 6 months.

Subsequently, in 1964, E. F. Wheelock* and J. H. Dingle, working at the Cleveland Research Center of the University Hospital, Cleveland, OH, inoculated influenza A and B, Newcastle disease and arboviruses into a patient with acute myelogenous leukemia and recorded repeated partial remissions of the disease that eventually terminated in death [3].

In 1910, an Italian woman was suffering from carcinoma of the uterine cervix. Her case history was presented that year at the International Cancer Conference in Paris, France. After an accidental exposure to lyssa, she received the attenuated live rabies (Pasteur-Roux) vaccine. Her enormously large vegetating tumors unexpectedly regressed [4]. The author, N. G. De Pace, gave the patient 30 injections of *emulsione di midollo rabido attenuato secondo il metodo di Pasteur* and clearly recognized that *il virus rabido* eliminated all gross remnants of

* Dr. E. F. Wheelock is the discoverer of gamma interferon (IFNγ); when he moved to The University of Texas at Galveston to chair the Microbiology Department, this author had the privilege to meet him on several occasions for discussions on matters of mutual interest.

the tumors (*l'eliminazione di grossi pezzi di tumore*) due to cytolysis *un'azione citolitica sulle cellule neoplastiche*. One wonders whether the regressing tumor contained bona fide Negri inclusions?

Thereafter, in the 1940s, G. T. Pack in New York City used the live attenuated rabies vaccine to treat patients with metastatic melanoma [5,6]. He reported the first remissions ever induced in that malignancy! After these spectacular remissions received extraordinary publicity and posed unrealistic demands for this modality of treatment, most of the patients relapsed. Unfortunately, virus therapy of melanoma was then abandoned.

Virus therapy of human tumors was resumed again worldwide in the 1950s and 1960s—in the United States, England, Latvia, the Soviet Union, and Japan; in the late 1970s in China; and in the late 1980s in Germany and Hungary (literature cited) [7–14]. However, this modality of treatment could not advance to the accepted rank of a proven, effective, and established therapeutic modality. None of the real pioneers took out a patent on the virus therapy of human cancers (as Albert Sabin did not patent his polio vaccine and Cesar Milstein and Georg Köhler did not patent monoclonal antibodies).

There are fascinating similarities regarding the histories for virus therapy of human cancers and virus therapy of bacterial infections. Bacteriophages have evolved by recombination and chimerism. Thus, derivatives of some of their ancient genes have reappeared in adeno- or reoviruses. Phage Φ6 may be a possible ancestor of retroviruses. The latent state of lysogenic prophages resembles that of the retro-provirus, which is reversely transcribed and vertically transmitted as retroviral RNA → DNA sequences inserted in the genome of their host cells. Phages may acquire host cell genes that enable them to encode bacterial toxins (example: the 933W λ coliphage encoding Shiga toxin), whereas retroviruses can transduce cellular proto-oncogenes. The estimated biomass of phages is 10% or more of that of the prokaryotes populating the Earth [15].

After the discovery of bacteriophages at the Pasteur Institute in Paris in 1917, Felix d'Herelle and his colleagues in

France successfully treated bacterial infections in children and adults using phage preparations. Eli Lilly in the United States produced therapeutic "phage preparations" in the 1940s. The Eliava Institute in Tbilisi, Georgia, in the Soviet Union, mass produced and distributed phage preparations in alliance with d'Herelle.

Then disastrous events occurred: These efforts received adverse criticism. Two articles in the United States described bacteriophages as "inanimate enzymes" and the work with them was declared to be unscientific, with grossly exaggerated results. There was a lack of control groups (no prospectively randomized trials; "low priority," some wise project site visitors would say today); Giorgi Eliava was executed by the KGB; d'Herelle died; and the original crude phage preparations were found to have contained bacterial toxins [16] (maybe inadvertently combining phage therapy with Coleys 's bacterial toxin therapy). Thereafter, the discovery of antibiotics rendered "phage therapy" unnecessary.

However, multiple drug-resistant microbes emerged and perpetuated their antibiotic-resistance plasmids to each other. Suddenly it was rediscovered (and publicized by the media) that the Eliava Institute still manufactures therapeutic phage preparations and that in Wroclav, Poland (formerly Breslau, Germany) at the Hirszfeld Institute of Immunology, a branch of the Polish Academy of Sciences, there still is an active pursuit of clinical research in "phage therapy" for veterinary and human bacterial infections. Currently there is access to phage preparations against all common and uncommon bacterial infections, including the Mycobacteria of tuberculosis, vancomycin-resistant *Enterococcus faecium,* and methicillin-resistant *Staphylococcus aureus*; *Serratia* endocarditis; and for infectious complications (including Salmonella and Bacteroides septicemias) in cancer patients [17–32]. There is even "a bacteriolytic agent that detects and kills *Bacillus anthracis*" and *Nature* publishes an editorial entitled "Virus Deals Anthrax a Killer Blow" [33,34].

There are no harmful side effects of properly purified new phage preparations, but antibodies of the host may neutralize the virus after repeated inoculations of the same phage preparation. However, by serial passages of virulent phages, subclones of the virus may be selected out that are less immunogenic [35,36]. In bacteria, natural killer genes exist that can eliminate plasmid-free individual cells in a process resembling programmed cell death. Genetically engineered non-lytic phages could deliver such killer genes (the *gef* genes in the M13 phagemid system) into bacterial cells. *Pseudomonas aeruginosa,* a major pathogen in cancer patients, is very sensitive to *gef* gene-mediated killing [37].

In comparison, virus therapy of cancer is an approach that was also repeatedly initiated and abandoned. Its value has been (and sometimes still is) grossly exaggerated (and publicized in the media) or completely denied. It received no academic credentials. It was not supported by the National Institutes of Health/National Cancer Institute (NIH/NCI). Antibody production of the host was believed to have hampered its efficacy. Cancer chemotherapy rendered it unnecessary, but as cancer cells gained multiple drug-resistance, it has been re-enlisted to take its place among the new biological treatment modalities. A possible caveat is forming: Because P-glycoprotein-overexpressing multidrug-resistant tumor cells do not admit hydrophobic peptides and eject drugs through efflux mechanisms, they also resist infection with enveloped viruses (influenza virus fusion protein) [38]. Nevertheless, oncolytic viruses re-emerge now as their potential efficacy is supported by evidence gathered from molecular virology–immunology and from new controlled clinical trials.

In addition to naturally oncolytic viruses, genetically engineered viruses can deliver apoptosis-inducer suicide genes into cancer cells [39–43]. The transduction efficiency of filamentous phagemid vectors recently developed for gene delivery to mammalian cells now reaches 45%. These vectors of low immunogenicity operate with a mammalian reporter gene and express a cell-targeting ligand that confers assured specificity for a cell surface receptor [44].

By the 1950s, the diagnosis of natural human viral infections had become greatly improved. Case reports of children with natural varicella or measles infections whose documented coexisting malignancies (acute leukemia, Hodgkin's disease, Burkitt's lymphoma) rapidly entered remissions appeared in succession in the literature [45–52]. Adult patients with chronic lymphocytic leukemia received vaccinia virus for vaccination against small pox (variola) (a procedure forbidden in this clinical setting) and remission of the leukemia ensued [53,54]. Other patients with Hodgkin's disease receiving live attenuated yellow fever (Theiler's) vaccine experienced remissions. In patients with Hodgkin's disease, hepatitis A viral infections coincided with some temporary remissions and an amelioration of the clinical course for these neoplasms [55]. Patients with acute leukemia acquiring infection with hepatitis A virus were observed to experience higher chemotherapy-induced remission rates and longer remission durations than similarly treated leukemic patients who were exempt from this viral infection [56,57].

All these remissions are known from brief but documented and published case reports. However, the case of the most popular and most frequently cited Hungarian poultry farmer remains (in the new document-based medicine) an anecdote: He has no name, no hospital chart, no pathological diagnosis, and no tissue sections on which a diagnosis of cancer could be based. Even the year of his alleged recovery from alleged disseminated stomach cancer during an alleged epidemic of fowl plague (Newcastle disease virus; NDV) outbreak in Hungary remains uncertain (unless his factual case history could be located) [58–61].

VIRUSES INTERACT

In 2001, it was observed that in human immunodeficiency virus (HIV)-1–infected patients harboring hepatitis G viral coinfection, acquired immunodeficiency syndrome (AIDS) ran a milder and slower course [62,63]. Viruses obviously interact with each other through the modulation of their hosts. Some unconventional practitioners, in order to induce interferon production, treated genital herpes viral infections in homosexual men using live attenuated poliovirus vaccines grown in monkey kidney cell cultures (prior to the availability of interferon-a or acyclovir). These practitioners were soon falsely accused to have initiated the AIDS epidemic in the Western hemisphere by inadvertently infecting their patients with a simian immunodeficiency virus thought to be the predecessor of HIV-1. Subsequent work established that both Koprowski's and Sabin's polio vaccines were free of (did not contain) a pre-HIV-1 simian virus [64,65].

Some aspersions are still cast on these vaccines (that saved so many lives) because the early Salk and Sabin polio vaccines harbored simian vacuolating virus-40 (SV40), a recognized oncogenic DNA virus. The presence of SV40 DNA sequences or structural proteins in some human tumors (mesothelioma, osteogenic sarcoma, medulloblastoma, thyroid and other carcinomas) strengthens the argument for a viral cause of at least some human tumors, even though the prevalence of SV40 infection in the human population is not known, except that geographic differences that exist are attributed to the type of polio vaccines used [66,67].

Human T cell leukemia virus-I (HTLV-I) causes human lymphoma-leukemia [68], and its xenografted tumors in SCID mice are inhibited by Bay 11-7082 (Calbiochem, La Jolla, CA), an inhibitor of IκBα phosphorylation and translocation of NF-κB into the nucleus, whereby treated tumor cells die apoptotic deaths [69]. Mouse mammary tumor-related retroviral agent is detectable in some cases of human breast carcinoma [70–72]. Epstein-Barr virus (EBV) and human herpes virus type 8 (HHV-8) certainly contribute to the malignant transformation of human lymphoid or endothelial cells, respectively [73–76]. One can also ask whether the Jamestown Canyon virus is a passenger or an etiologically involved agent in human colon carcinomas [77]. These are examples of tumors which, when superinfected with another virus (harmless in the host), might regress owing to viral interference or other related mechanisms, such as certain virally induced cytokines.

Three years after its first isolation (1953), an adenovirus strain was already shown to replicate and cause cytopathic effects with tumor regression in uterine cervical carcinoma [78,79]. Newcastle disease virus achieved the same effect [10]. Human papilloma virus (HPV)-induced squamous cell carcinomas of the uterine cervix may be unusually sensitive to another viral infection. Instead of viral interference, the doubly infected cancer cells die. This is the tumor that regressed in the Italian woman after vaccination with live attenuated lyssa virus (*vide supra*). In this entity, HPV E6 and E7 genes surrender the $p110^{RB}$ and the p53 proteins to the ubiquitin-proteasome system for degradation (anti-apoptotic events) to secure tumor cell survival. In contrast, viral gene product structural proteins expressed in the tumor cells induce antibody- and T cell-mediated immune reactions against the HPV-transformed cells [80,81].

The alphavirus Semliki Forest virus (SFV) can be genetically engineered to express a fusion protein from HPV gene product proteins E6 and E7. The SFV-enhE6, 7 construct induces strong antibody and immune T cell responses against the fusion oncoprotein in immunized hosts, whereas in mice it protects against xenografts of human squamous cell carcinoma of the uterine cervix [82]. Nonstructural regions of the viral genome induce apoptosis in human cancer cells even when p53 is deleted. Recombinant replication-defective SFV strains can deliver the IL-12 gene, or the GM-CSF gene or the endostatin gene, into human tumor cells to attract dendritic cells to, and to inhibit neovascularization within, the tumor bed, respectively [83–87]. Another alphavirus of the *Togaviridae* family, the Sindbis virus, attaches to the 67-kDa laminin receptors of its host cells. Tumor cells overexpress laminins and are readily targeted by the virus and succumb to apoptotic death upon viral entry. This virus has no nuclear phase in its replicative cycle and cannot integrate into the genome of the infected cells [88]. Another alphavirus, Venezuelan equine encephalitis virus, when used as a viral replicon, carrying the HPV E7 gene, induces strong CD8 T cell-mediated immunity against HPV-induced tumors [89].

Additionally, humans harbor highly prevalent, nonenveloped, small circular ssDNA TT miniviruses, members of the families *Circo-*, *Circino-* or *Para-circinoviridae*, that coevolved with their hosts through primate speciation in Africa [90]. It is entirely unknown what interaction, if any, these resident viral populations have with naturally infectious or therapeutically inoculated oncolytic viruses. The prevalence of TT viruses ranges from 18% in Europe to 50% in Asia and 70% in Africa and South America [91]. TT viruses are subject to exacerbations in patients with organ transplants, hepatitis, leukemias, and other neoplasms. Are we transferring TT viruses into our tumor tissue cultures or xenografts when we test them for viral oncolysis?

FROM MICE TO MAN

Returning to the past, in the 1920s, laboratory models of viral oncolysis were set in motion in Paris, France, reaching rapidly spectacular conclusions. Levaditi and Nicolau cured malignant tumors in mice by infecting the tumor-bearing animals or the tumors directly with viruses having no, or limited, pathogenicity in these hosts. Vaccinia, ectromelia and fowl plague (probably "classical" fowl plague and not the later-discovered Newcastle disease) viruses excelled most in their oncolytic activities. Levaditi exclaimed: *"le tumeur fait fonction d'eponge!"* [92–94]. Alice Moore* at the Sloan-Kettering Institute in New York City confirmed and extended these experimental studies on viral oncolysis [95,96]. Her work gave the impetus for Chester Southam's* clinical trials for the "virus therapy of human cancers" conducted in the 1950s with arboviruses of African origin (West Nile; Egypt 101 viruses) [97,98]. When their clinical results were presented in 1957 at M. D. Anderson Hospital in Houston, TX, Albert Sabin* pressed for the details. The investigators had to admit that the tumors regrew after the initial virally induced partial remissions were obtained. Sabin concluded:

"Dr. Moore mentioned some very disappointing aspects of the possible use of viruses as oncolytic agents in human

* The author had the privilege to repeatedly visit with the Sloan-Kettering team of virologists, including Dr. Charlotte Friend in 1957 in New York City, and Dr. A. Sabin in Cincinnati, OH, in 1960–1961.

beings. The most disappointing aspect is the fact that even when a virus is oncolytic and it punches a hole in a tumor the immune response of the individual to the virus occurs so fast that the effects are quickly wiped out and the tumor continues to grow" [99].

In this early period, other disquieting reports emerged: One observer described a phenomenon as if rabies and yellow fever viruses enhanced the growth of squamous cell carcinoma explants deriving from the hard palate of a patient, whereas influenza A, St. Louis and Western equine encephalitis (WEE) viruses destroyed these explants. However, the influenza A and WEE viruses were ineffective in halting tumor growth after intratumoral injections in the patient [100,101]. Thus, there was no impetus or convincing evidence to support expections of cures from the virus therapy of human cancers at that time.

POSTONCOLYTIC ANTITUMOR IMMUNITY

Still in the 1950s, J. Lindenmann and P. Klein [102,103], W. W. Ackerman and H. Kurtz [104], A. D. Flanagan et al [105], and W. Cassel and R. E. Garrett [106] cured ascitic mouse tumors with orthomyxo- and paramyxo- (influenza and Newcastle disease) viruses. At this time, working at the State Institute of Public Health in Budapest, Hungary, I gained my first experience with the oncostatic–oncolytic effects of NDV on Ehrlich ascites carcinoma cells in mice. These NDV strains isolated by us from cases of human oculoglandular illness (*vide infra*) were adapted to newborn mice.

Professors R. Doerr and C. Hallauer in Switzerland welcomed the manuscript and subsequently published it in the *Archiv für die gesamte Virusforschung* in 1957 (with more than 1-year delay due to the events of the heroic Hungarian uprising of 1956) [107]. In the meantime, in my textbook of virology, *Die Grundlagen der Virusforschung* (Hungarian Academy of Sciences, Budapest, 1956), I reviewed all known examples of viral interference and viral oncolysis [9,108]. The work on the full or incomplete replicative cycles of NDV in the mouse brain or in mouse tumors and the relationship (or differences) between mumps virus and NDV constituted the subject matter of my first lectures in the United States in 1957, at Vincent Groupé's Department of Virology, Waksman Institute, Rutgers University, New Brunswick, NJ; at the Veterinary Medical School, University of Illinois, Urbana, IL; at Jerome Syverton's Department of Virology at the University of Minnesota's medical school in Minneapolis; and at the annual meeting of the American Microbiological Society (as published in abstract form in the *Bacteriological Proceedings* in 1957) (references cited) [109,110].

Much later, in our (Harris JE, Sinkovics JG, authors) monograph: "*The Immunology of Malignant Disease,*" Second Edition, published in 1976 by C.V. Mosby, St. Louis, MO, I listed many more examples of viral oncolysis [111]. Especially impressive was the oncolytic efficacy of bovine enterovirus-1, regarding which three articles are quoted in that chapter; it remains entirely unexplained why no reports on human clinical trials followed.

Regression of virally infected ascitic murine tumors was not a simple matter of viral oncolysis. When mice rendered tumor-free by viral oncolysis were challenged with inocula of the same strain of tumor cells but without further admixture or administration of the virus, there were no takes! These animals had acquired solid antitumor immunity following viral oncolysis (Lindenmann et al; H. Koprowski et al) [102,103,112]. In those years of labile and rapidly and repeatedly revised tenets of tumor immunology, we misinterpreted a single similar experiment. We attributed the "no takes" response to challenge with tumor cells to the effect of a second episode of viral oncolysis by the virus (NDV) latently persisting in the host. In overlooking antitumor host immunity developing consequentially to viral oncolysis, we missed the great opportunity to contribute to the discovery of the phenomenon now known as *postoncolytic antitumor immunity*. Postoncolytic antitumor immunity is not limited to oncolysis mediated by RNA viruses. The HF10 derivative of herpes simplex virus-1 (HSV-1) with deletions of a 3.9 kilobase pair (kbp) at the right end and a 2.3 kbp at the left end of its UL and UL/IRL junction in its genome destroys intraperitoneal neoplasms of immunocompetent mice. When these animals are challenged with the same but virus-free tumor cells, they reject these tumor cells [113].

Studies on viral oncolysis of ascitic tumors have produced some discrepancies concerning the results of laboratory assays and clinical trials. According to one statement, "the most effective viruses were Bunyamwera and Egypt 101, which caused all tumors to regress with survival of the animals" [95]. In another laboratory, Bunyamwera virus killed the ascites tumor-bearing mice [112] and Egypt 101 (and other African viruses; for example, the highly oncolytic Rift Valley fever virus) [114] failed to induce durable remissions when administered to tumor-bearing patients [7]. The dsRNA Bunyamwera virus activates protein kinase R (PKR), but it is its nonstructural (NS) protein acting as a virulence factor that blocks interferon production [115]. The NDV strain used at Sloan-Kettering inhibited the growth of murine ascites tumors, but without viral replication occurring in the tumor cells [94,95], whereas other NDV strains replicate in, and kill, Ehrlich ascites carcinoma cells [105,107].

VIRAL ONCOLYSATES

More than 10 years later, this notable phenomenon induced us and William Cassel to be the first to prepare "viral oncolysates" (VO) for the active tumor-specific immunotherapy of human tumors. Sinkovics used the PR8 influenza A virus for the treatment of patients with sarcoma and melanoma at M. D. Anderson Hospital in Houston, TX, in 1972–1973 and thereafter (references cited) [116–120]; W. Cassel and D. R. Murray used the live 73T strain of NDV for the treatment of patients with melanoma at Emory University in Atlanta, GA, in 1973 and thereafter [121,122]. Later, M.K. Wallack and H. Koprowski used vaccinia virus for the treatment of patients with melanoma (and patented the procedure; something that neither Cassel nor I ventured to do) at the Wistar Institute in Philadelphia, PA, in 1977 and thereafter [123]; and V. Schirrmacher used the Ulster strain of live NDV for the treatment of patients with a large variety of tumors at the Krebsforschungszentrum in Heidelberg, Germany, from the mid-1980s [124] until the present time (see V. Schirrmacher, in this volume).

Active tumor-specific immunotherapy with VO for gynecologic tumors continued until recently at the M.D. Anderson Cancer Center in Houston, TX, by R. Freedman, J. Bowen, and C. Ioannides (references cited) [119]. W. Cassel, Z. I. Olkowski, and D. R. Murray have submitted a report for this volume (see Chapter 9) on their 20 years follow-up of their patients with melanomas metastatic to regional lymph nodes and immunized (after surgical removal of gross tumors) with live 73T NDV oncolysates; at 10 years after surgery, more than 60% remained tumor free, in contrast to 15% tumor-free survival of surgically treated contemporary (not prospectively randomized) historical control patients [122].

The Surveillance Committee at M. D. Anderson Hospital in the early 1960s was more lenient than the Institutional Review Boards and the regulatory committees for investigational protocols of the NCI and the Food and Drug Administration are now. I was allowed to register, among others, the following human clinical research projects at M. D. Anderson Hospital: J. G. Sinkovics: *Project M27 gm/13*: Effect of Oncolytic Viruses in Selected Cases of Human Malignant Diseases (1966-) and *Project M27/gm 15*: Immunotherapy of Cancer by Immunization with Cultured Tumor Cells (1967-). Beginning in the early 1970s, these two projects continued, but in a united form, as active immunotherapy with viral oncolysates for sarcoma and melanoma (and kidney carcinoma to be added later) (Figures 1A,1B).

By then, the Institutional Surveillance Committee required that we inactivate (or "attenuate") the live PR8 influenza virus in our VO preparations. We complied, but by titrations we found some residual low titers of live influenza virus remaining in these lysates (which were used for human immunotherapy) [125]. The live PR8 virus exerted oncolysis in cultured cells of many human tumors, especially sarcomas, melanoma, kidney and ovarian carcinomas; we tested it on a large "bank" of human tumor cell lines established in our laboratory

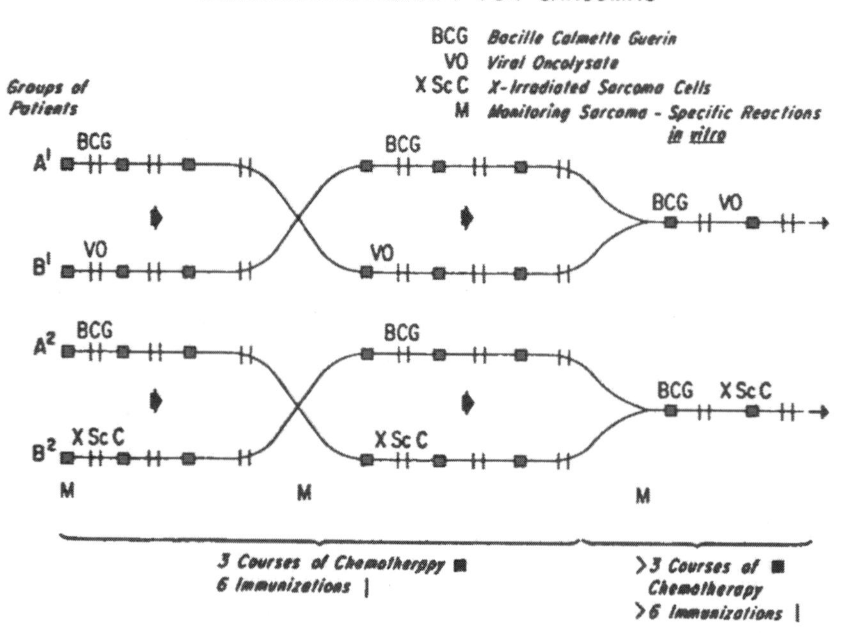

Figure 1 1A. The first chemo–immunotherapy protocol designed for the treatment of patients with metastatic sarcomas was approved in 1971–1972 at The University of Texas M. D. Anderson Hospital, Houston, TX. The protocol was designed to determine whether immunotherapy increased sarcoma-specific immune reactions in vitro. These reactions were [1] titrated antibodies for cytolysis with complement and [2] tested in indirect immunofluorescence assays for cell surface, cytoplasmic, or intranuclear bindings with autologous or allogeneic sarcoma cells, and for [3] lymphocytes extracted with ficoll-hypaque from the buffy coats and tested in a chamber-slide cytotoxicity assay against autologous and allogeneic primary and established tumor cell cultures, [4] with or without antibodies added. The project was "approved but without funding" by the National Cancer Institute, National Institutes of Health, Bethesda, MD [109,117–120,130–134,146,147].

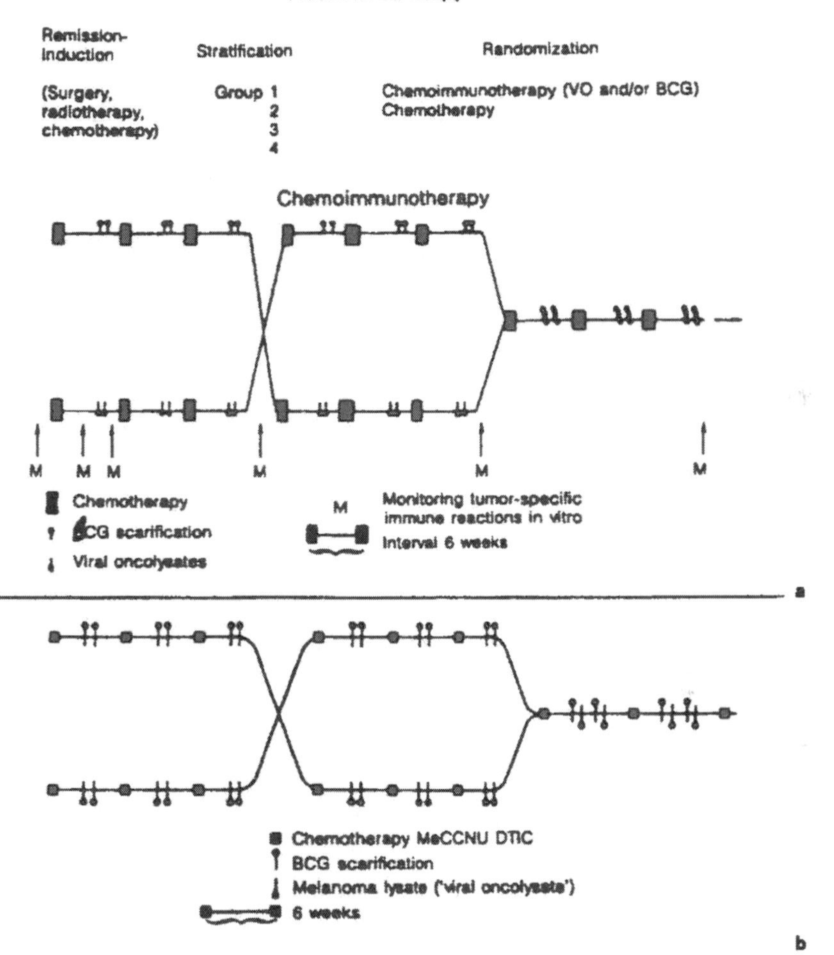

Figure 1 1B. Thereafter, immunization with X-irradiated sarcoma cells was omitted and only viral oncolysates (VO) were used for "active tumor-specific" immunizations. The PR8 influenza A virus was used to prepare VO. Over the site of intra- or subcutaneous VO inoculations, live Chicago BCG was scarified. Later, patients with malignant melanomas of adverse prognostic factors (deep or ulcerated primary tumors; satellitosis or in-transit metastases; regional lymph node metastases, visceral metastases with or without surgical removals of gross disease) were included. These protocols were active from 1972 to 1976. These and related protocols were reactivated at the Department of Gynecologic Oncology of the same institution in the early 1980s as "immunotherapy for gynecological tumors with virus-modified tumor cells" [109,118,182–186].

[126]. Even in vivo, the PR8 virus on occasions performed quite well: We observed regression of directly injected Kaposi's sarcomas (in the pre-AIDS era). In one case, in a female patient, a hemangiosarcoma of the scalp sloughed off after direct inoculation of the PR8 virus; however it regrew and resisted repeated viral inoculations while the patient developed high titers of antibodies, inhibiting viral hemagglutination and neutralizing the virus. The case of another patient (hospital chart on microfilm), whose biopsy-proven hemangiosarcoma regressed during an influenza-like upper respiratory tract infection, remains an anecdote, inasmuch as laboratory proof for the viral infection has not been provided [127]. However, because influenza viruses readily recombine with each other, there was a possibility that the PR8 virus used might regain its virulence by recombination with one of the newer-generation influenza viruses circulating in the human populations. As a case in point, it was shown that a Russian "inactivated" influenza virus vaccine acquired new virulence through recombination ("multiplicity reactivation") in nature, and caused fatal illnesses in Mongolian camels [128].

We reported and published the results of our clinical trials with viral oncolysates at the then famous and well-attended Chicago Symposia (identified in the references). Once published there, the work lost its qualification for publication in journals that insisted on receiving manuscripts not published anywhere else prior to their submission; we conscientiously observed this rule. Therefore our work received no wide recognition: Only investigators who actually worked along the same lines referred to them (very favorably so), and now our work of the 1970s has fallen into oblivion. In recapitulating some of our clinical trials, we treated with VO adult patients with metastatic sarcomas at the Section of Clinical Tumor Virology and Immunology and at the clinical Melanoma-Sarcoma Service, Department of Medicine of The University of Texas M.D. Anderson Hospital in Houston, TX. The chemotherapy regimens in the pre-doxorubicin era consisted of vincristine, cyclophosphamide, actinomycin D and methotrexate (often in high dose with leucovorin rescue); after adriamycin (doxorubicin) became available, it was the

CyVADIC regimen in which DTIC (dacarbazine) was believed to have acted additively to adriamycin, and chemotherapy courses were administered at 28-day intervals. Of 49 patients treated with chemotherapy only, 36 (72%) progressed over a 6-month observation period eventually to death, occurring within the observation period or later. Of 19 patients receiving CyVADIC chemotherapy with BCG scarifications, 10 (53%) progressed within the same period of time, eventually to death. Of 19 patients treated with CyVADIC chemotherapy and PR8 VO prepared from autologous (preferred) or allogeneic cultured sarcoma cells and administered intra- or subcutaneously with BCG scarified over the inoculation sites of the VO on the fourteenth postchemotherapy day, when the patients' white blood cell counts were on the rise, 6 (32%) progressed in the same observation period, eventually to death [129,130]. The NCI considered our grant applications of "low priority; approved without funding," even though we requested support for a prospectively randomized clinical trial with stratification of subgroups of patients.

Our clinical trials with melanoma included postoperative patients with deep primary melanomas; patients whose regional lymph node metastases were surgically removed; or patients with gross metastatic disease in various viscera. This was the era when adjuvant chemotherapy and BCG scarifications were extensively tried to suppress micrometastases left behind after surgical removal of primary tumors or tumors metastasizing to regional lymph nodes or to treat gross metastatic disease. Dacarbazine, actinomycin D, and the nitrosourea semustine constituted the chemotherapeutic agents used. Of 27 patients of all stages and with adverse prognosis receiving chemotherapy and BCG, 14 (52%) progressed in a 6-month observation period and eventually died. Of 36 patients of all stages and with adverse prognosis receiving chemotherapy, BCG, and PR8 VO prepared from cultured autologous (preferred) or allogeneic melanoma cells between courses of chemotherapy with BCG scarified over the sites of inoculation of the VO, 14 (39%) progressed and eventually died. Patients with Stage IV disease formed three groups: 19 patients received chemotherapy only with zero complete remissions (CR) and 86% progression rate; 24 patients received chemotherapy and BCG with two CR and a progression rate of 73%; and 11 patients received chemotherapy, BCG, and PR8 VO with one CR and a high progression rate: six of eight adequately treated and three inadequately treated patients progressed. In each of these three groups of very ill patients, there were several who could not tolerate the treatment (especially semustine, which was not licensed) and were "inadequately" treated. However, when survival times were compared, patients with Stage IV disease who received chemotherapy, BCG, and PR8 VO survived a mean of 12.8 months vs. 5.1 months for patients receiving chemotherapy only and 6.25 months for chemotherapy and BCG [131–133].

When these results were reported at the 12th International Cancer Congress in Buenos Aires, Argentina in 1978, the trend favoring patients who also received melanoma VO was ignored; statistically highly significant results of tumor regressions were expected and the consecutively occurring minor responses (rare partial remissions, more frequent stabilization of disease, prolonged survival) were dismissed. Patients with metastatic tumors evaluated today are counted as responding if the disease stabilizes and no progression occurs for several months; these results were below our expectations in the 1970s. Stratification of these patients into comparable subgroups posed close to insurmountable difficulties. The NCI again considered the work "of low priority; approved without funding"; the approval was of a small consolation without the funds of the NCI, which then was operating with the increased budget of President Nixon's "Conquer of Cancer" program. BCG itself and *Corynebacteria* as immunostimulants for cancer patients were of "high priority" at that time. These simplistic and now abandoned measures received extraordinary NCI support and funding, whereas more sophisticated interventions decades ahead of their time were not understood or supported by that noble institution.

These and other patients overlapping with these groups, or independent from these groups, were reported elsewhere, always showing a trend favoring patients who received VO in addition to standard therapy, but not quite reaching an overwhelmingly significant advantage at the end of prolonged observation periods [134]. We recently reported the case of a patient exemplifying those who fail despite chemobiotherapy (including interferon-α2b and interleukin-2 [IL-2]), NDV oncolysate, direct intratumoral injections of NDV, GM-CSF (sargramostim), and immune lymphocyte infusions [135]. Metastatic melanoma remains an incurable disease.

Our present NDV oncolysate (autologous VO preferred) protocol for adjuvant therapy aimed at the eradication of melanoma micrometastases is combined with subcutaneous injections of either interferon-α2b or IL-2. For stage IV disease, we alternate biochemotherapy with immune lymphocytes: lymphokine-activated killer (LAK) cells, a subpopulation of NK cells obtained from the blood or CD4 and CD8 immune lymphocytes extracted from autologous tumors (tumor-infiltrating lymphocytes: TIL), both administered with low-dose (5 million units) subcutaneous IL-2 [136,137]. We can measure quantitatively antibody- and lymphocyte-mediated cytotoxicity aimed at the patient's tumor cells before and after immunotherapy (but must use the same target cell population for valid evaluation).

IMMUNE REACTIONS

Incorporated in our projects of the 1970s were tests for autologous and allogeneic antitumor immune reactions of patients quantitatively evaluated before and after active immunizations with VO. We measured humoral and cell-mediated immune reactions at a time when monoclonal antibodies were yet to be discovered. Our early observation (1969–1970) that antibody-producing lymphoid cells can fuse with lymphoma cells and thereby continue the secretion of a specific antibody was not translated into practical therapeutic use of the antibody produced by the fused ("chimeric") cells (Figures 2–4A,4B) [138–142].

At that time, the separation of B, T, and NK lineages and their subgroups of lymphocytes was not yet clearly defined

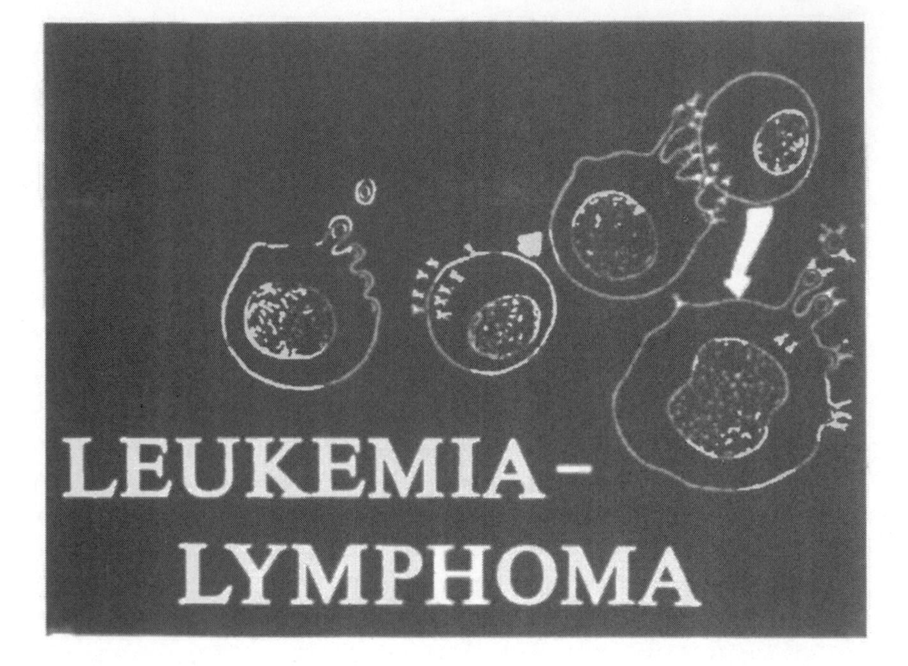

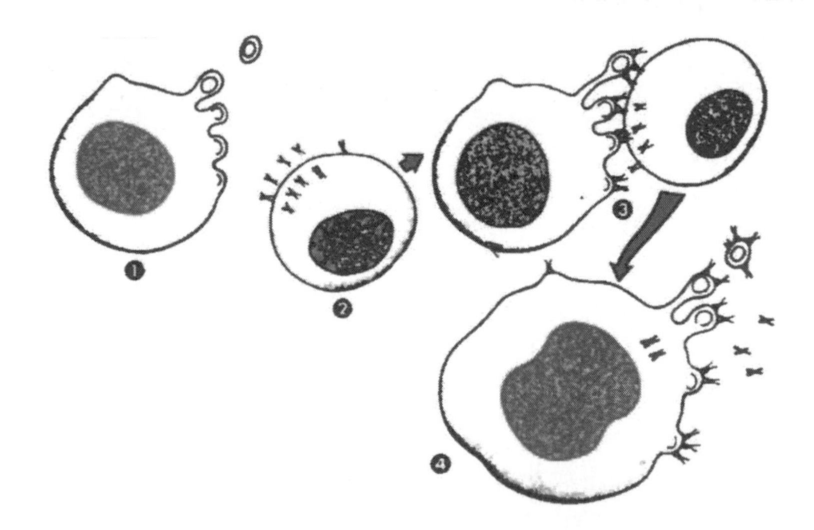

Figure 2 The first concept of the "natural hybridoma" formation as depicted and presented in 1969 [138]. Expressing budding retrovirus particles, a diploid murine lymphoma cell fuses with an autologous plasma cell secreting retrovirus-neutralizing antibodies. It was clearly recognized what the fused cells were: "··· tetraploid immunoresistant lymphoma cells in the mouse emerge by fusion of the diploid virus-producing lymphoma cell with a plasma-cell producing virus-

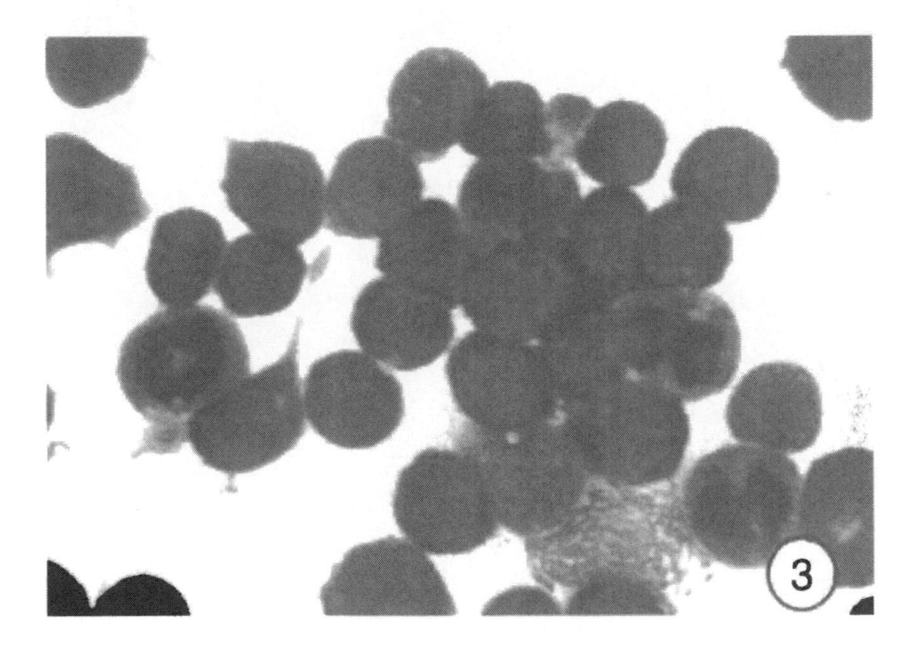

Figure 3 Suspension culture cell line #818 consisted of immortalized specific antibody-producing poly- and tetraploid lymphoid cells that grew as highly malignant ascitic tumors in mice (see Fig. 4A,4B). The immune globulins extracted from the culture fluid (by Dr. Roman Pienta) exerted minimal retardation of growth or no effect at all (Dr. Eiichi Shirato) on the pathogenicity of the lymphoma cells (for references, see Chapter 2, "New Biological Therapeutics," Part IV, Ref. 78–82,84,85 in this volume). (From J Med 1985; 16:509–524.)

specific globulins··· The resulting tetraploid cell will retain malignant growth potential and the genetically determined committedness of both parent cells—to produce leukemia virus, as coded for by the viral genome within the neoplastic cell, and to synthesize virus-specific globulins, as coded for by the genome of the plasma-cell" [139]. The tetraploid fusion product cells remained alive in culture and in ascites tumors of mice and continued secreting the antibody (under the care of Dr. Jose Trujillo†) [140]. (From the volume Leukemia-Lymphoma, A Collection of Papers Presented at the Fourteenth Annual Clinical Conference on Cancer, 1969, at The University of Texas M. D. Anderson Hospital and Tumor Institute at Houston, Texas. Chicago: Year Book Medical Publishers, Year Book Medical Publishers, 1970.)

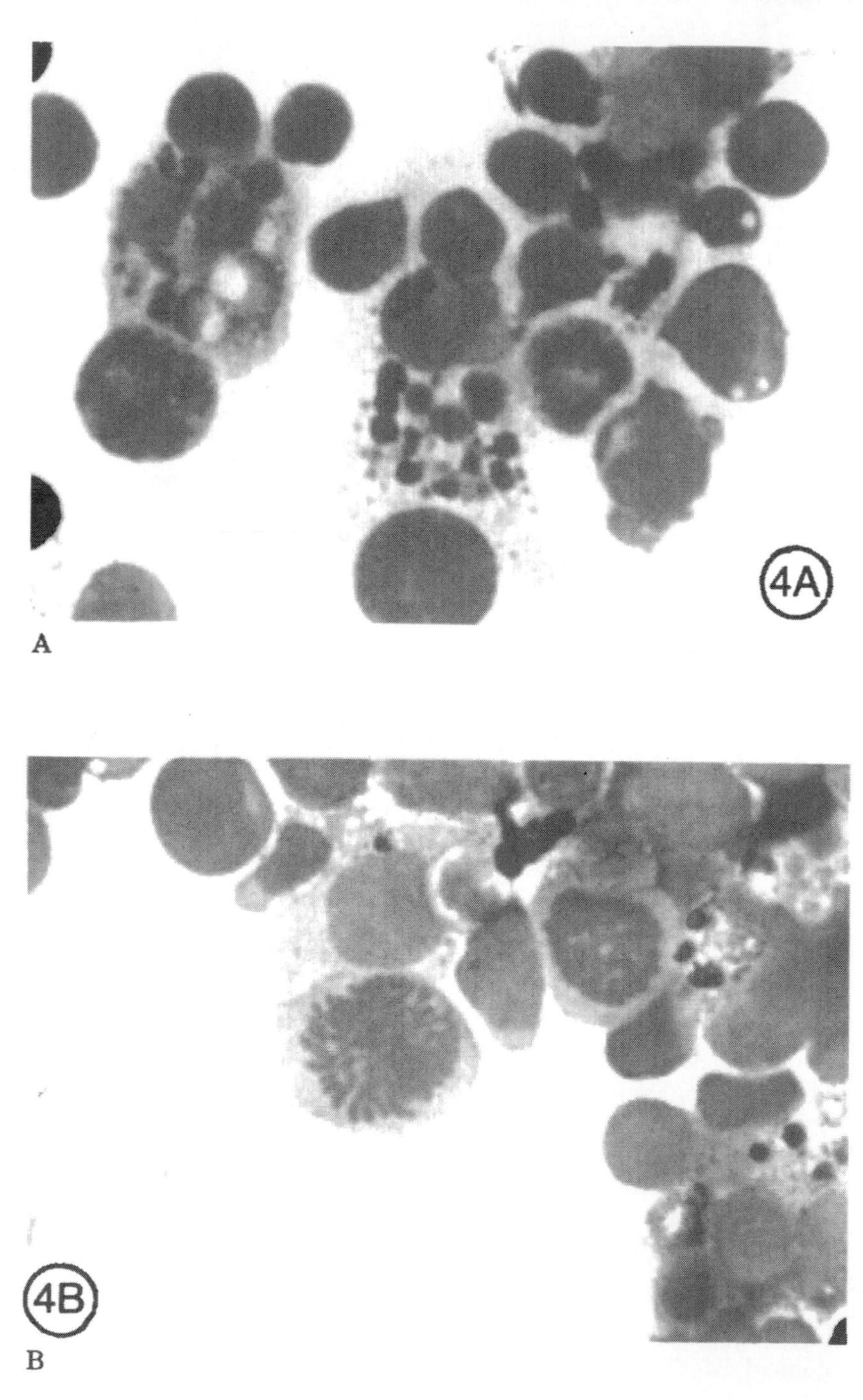

A

B

either. We observed and photographed in the early 1970s spectacular encounters between tumor cells and autologous and allogeneic host lymphocytes (Figures 5–8) [143–150]. Large lymphocytes with granular cytoplasm frequently killed allogeneic tumor cells by lysing the cytoplasm of the tumor cells. Blood samples of healthy (tumor-free) individuals (first among them this author) yielded such cells. Smaller lymphocytes with more compact nuclei frequently killed autologous tumor cells by clumping the nuclei of the tumor cells. The phenomenon of apoptosis was not yet recognized as "programmed cell death," as it is known today. After immunization with VO, the numbers of antitumor cytotoxic lymphocytes increased.

Regarding serology, there existed antibodies blocking lymphocyte-mediated cytotoxicity; after immunization with VO, these antibodies were replaced by antibodies that intensified lymphocyte-mediated cytotoxicity [151–154]. Blocking "serum factors" worked against the small lymphocytes in autologous settings; serum factors intensifying cytotoxicity worked best with the large lymphocytes in allogeneic settings. It was not known to us in 1969 and the early 1970s that what we called *small lymphocytes* were immune T cells. The "large lymphocytes" were designated later to the class of NK cells. With their Fc receptors, NK cells could practice the as yet

Figure 4 A, 4B Reproducing natural hybridoma formation in 1969 by injecting intraperitoneally antibody-producing normal diploid plasma cells and retrovirus antigen- expressing diploid lymphoma cells and recovering from the peritoneal cavity large poly- and tetraploid fused cells that grew in vitro or in the mouse in vivo in an immortalized fashion and continued the expression of deformed retrovirus particles and the secretion of antibodies specifically neutralizing the retrovirus [138,139]. These experiments were successfully repeated by Dr. H. David Kay (M.D. Anderson Hospital Reseach Report, 1970, pp 331–336; 1972, pp 476–479; 1974, pp 454–455). In 1969–1970, this phenomenon was not referred to as "natural hybridoma formation" (for further references, see Chapter 2, "New Biological Therapeutics," Part IV, Refs. 78–82,84,85, in this volume). (From J Med 1985; 16:509–524.)

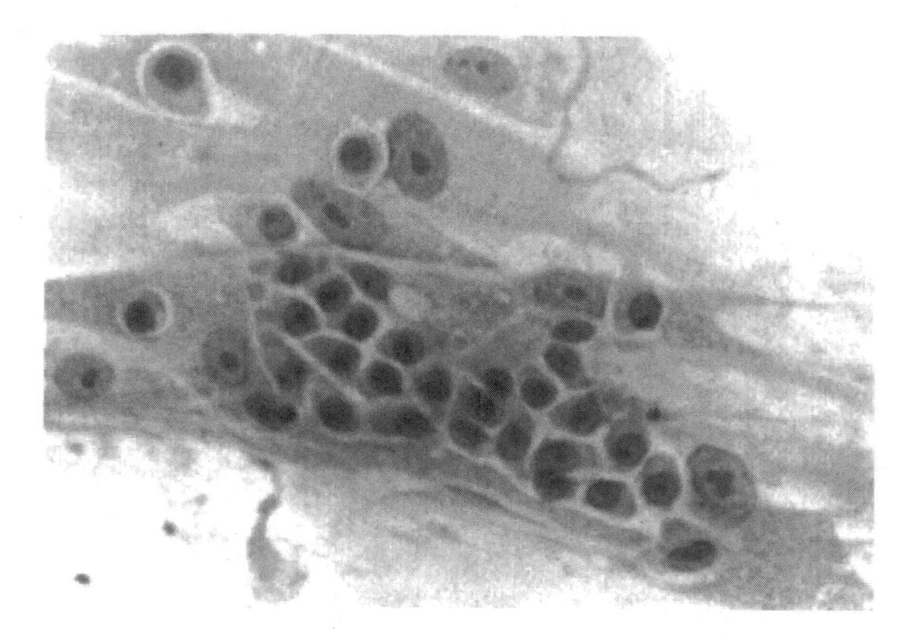

Figure 5 Relationship of lymphocytes with host stromal cells and with tumor cells. B-CLL cells rest and multiply within large autologous fibroblast-like cells after explantation: the "feeder layer principle." Otherwise the leukemic cells would die. Both participants of this union survive. On occasions, cytoplasmic bridges connect the lymphocytes with the host cells [148,149] (for further reference, see Chapter 2, "New Biological Therapeutics," Part I, Ref. 57).

undiscovered ADCC reaction (hence the explanation for "unblocking serum factors" intensifying lymphocyte-mediated cytotoxicity). These were the lymphocytes whose numbers increased first after the administration of VO in tumor-bearing patients. Indeed, one of the physiological functions of the in-

Figure 6 Plate reproduced with original legends showing microphotographs from the early 1970s. "Small compact lymphocytes" later recognized to be immune T cells and "large granular lymphocytes" later recognized to be natural killer cells kill autologous and allogeneic tumor cells either by cytoplasmic lysis or by nuclear clumping, much later recognized as "programmed cell death or apoptosis." (From Acta Microbiologica Hungarica 1991; 38:321–334.)

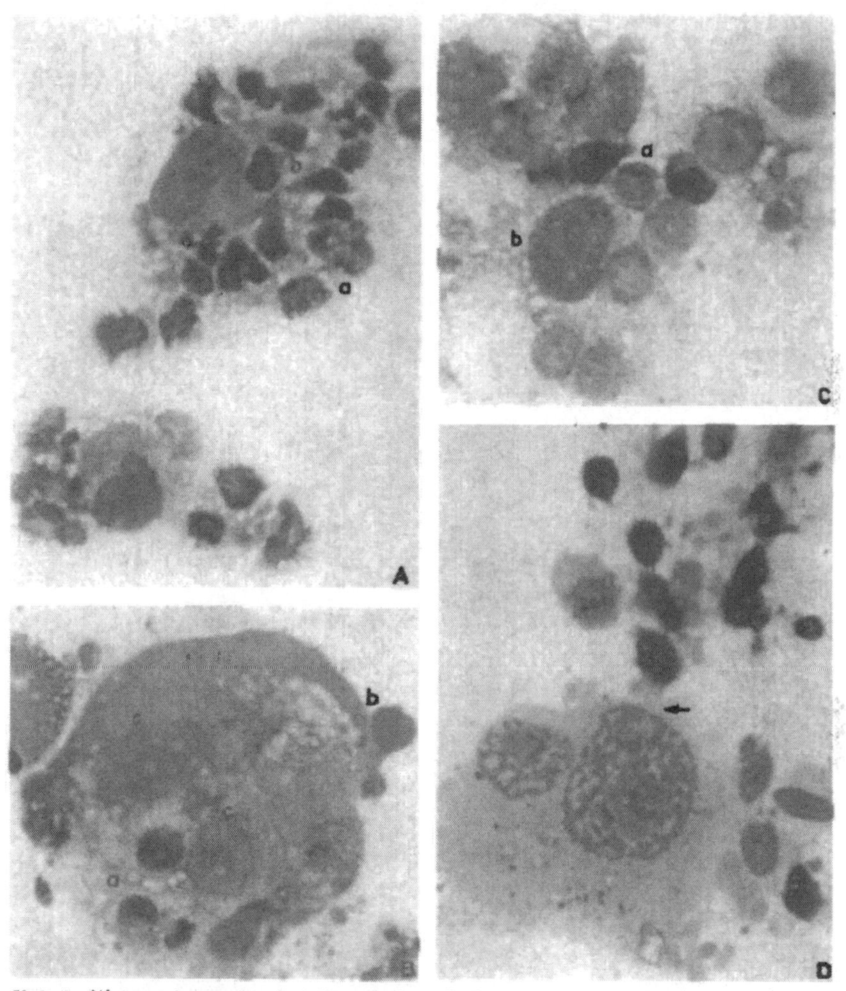

Plate 1. (A) Large lymphocytes lyse the cytoplasm of allogeneic tumour (sarcoma) cell while the nucleus of the target cell remains intact: a — cell debris and denuded nuclei; b — lymphocytes lysing cytoplasm

(B) Allogeneic tumour (sarcoma) cell undergoes cytoplasmic lysis: a — three large lymphocytes destroy cytoplasm; b — as one lymphocyte attaches to cell membrane, major vacuolization of target cell cytoplasm takes place beneath site of attachment; c — target cell nuclei remain intact

(C) After cytoplasmic lysis of numerous target tumour (melanoma) cells: a — three attacker lymphocytes remain tinctorially intact; would commit further acts of cytolysis when in native state were re-exposed to target cells (not shown); b — some denuded target cell nuclei and nucleoli remain intact

(D) After destroying a target tumour cell (melanoma) above, one small autologous lymphocyte makes cytoplasm to cell membrane contact (arrow) with a new target cell which undergoes rapid nuclear lysis

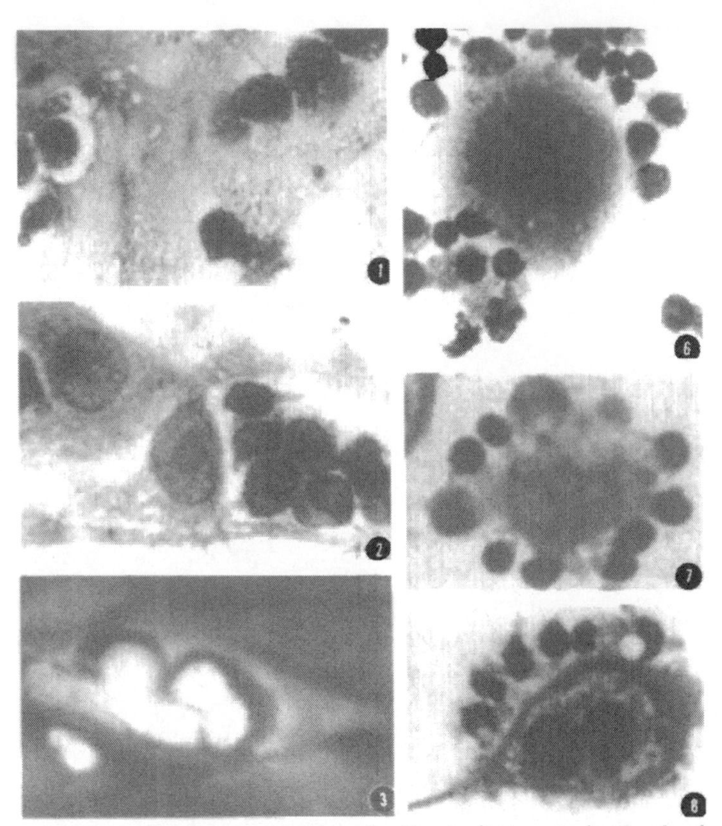

FIGURE 1. Small lymphocytes in perinuclear vacuole and large lymphocytes moving along the surface of autologous fibroblast (Wright's stain)

FIGURE 2. Lymphocytes coexisting with autologous fibroblast (Wright's stain)

FIGURE 3. Lymphocytes occupying cytoplasmic vacuole in autologous fibroblast in lymph node culture from patient with Hodgkin's disease. Orange-staining granules travel through cytoplasmic bridges from fibroblast to lymphocyte (acridine-orange stain)

FIGURE 6. Small lymphocytes taken from blood immediately surround (and destroy in 12–24 h, not shown) autologous chondrosarcoma cell (Wright's stain)

FIGURE 7. Small lymphocytes of healthy twin surround allogeneic rhabdomyosarcoma cells. Lymphocytes of the tumor-bearing twin were devoid of this reactivity. Fusion of cytoplasm between lymphocytes and tumor cell occurs[88] (Wright's stain)

FIGURE 8. Small lymphocytes attach to and vacuolize autologous sarcoma cell (Wright's stain)

Figure 7 A & B Plates reproduced with original legends showing microphotographs from the early 1970s, in which lymphocytes coexist with host fibroblasts; "small compact lymphocytes" surround autologous sarcoma cells and appear as if injecting substances into the cytoplasm of the tumor cells with cytoplasmic lysis ensuing; a large lymphocyte vacuolizes the cytoplasm of an allogeneic sarcoma cell; and large and small lymphocytes destroy tumor cells, causing nuclear clumping, while the lymphocytes preserve their integrity as judged by their intact morphology and tinctorial charactersitics. These lymphocytes remained viable and, when transferred into new tumor cell cultures, they could kill tumor cells again. (From Annales of Clinical and Laboratory Science 1986; 16:488–496).

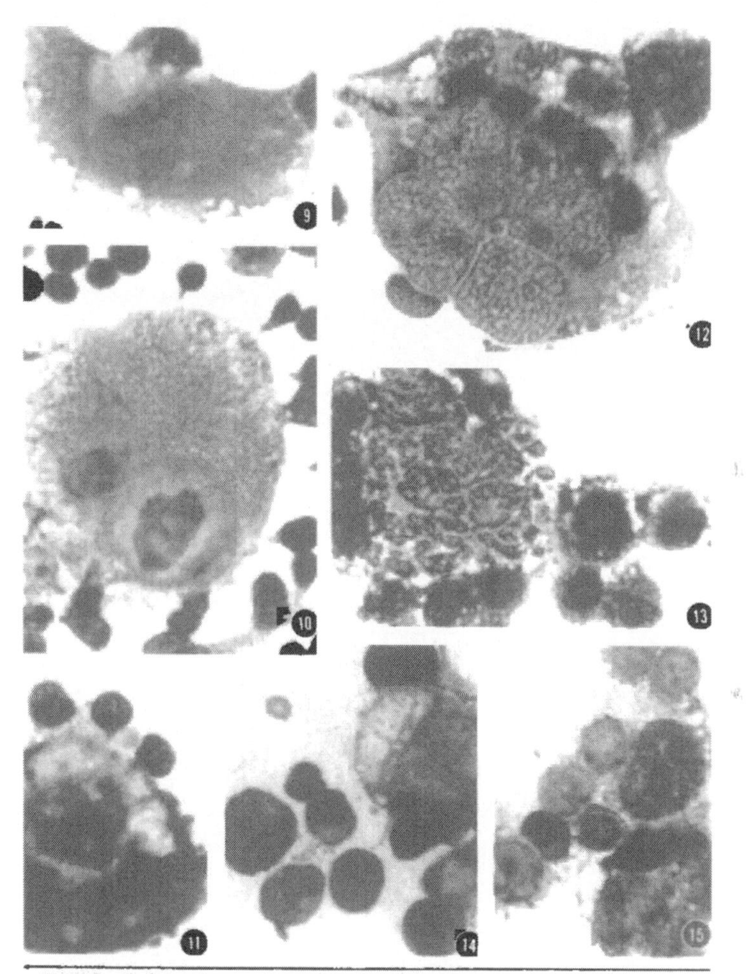

FIGURE 9. Single large lymphocyte resembling NK cell attaches to and vacuolizes the cytoplasm of allogeneic sarcoma cell without any immediate damage to target cell nucleus (Wright's stain).

FIGURE 10. Autologous small lymphocytes surround large tumor cell and emit cytoplasmic protrusions into the cytoplasm of the tumor cell. Vacuolization of the cytoplasm and nuclear damage in the tumor cell are evident (Wright's stain).

FIGURE 11. Three small lymphocytes attach to autologous tumor cell and emit protrusions into the tumor cell's cytoplasm resulting in vacuolization of the tumor cell's cytoplasm (Wright's stain).

FIGURE 12. Multinucleated kidney carcinoma cell is penetrated by 5 small autologous lymphocytes as the tumor cell's nuclei undergo chromatinic disintegration (Wright's stain).

FIGURES 13, 14, AND 15. Large and small lymphocytes remain tinctorially intact after complete lysis and disintegration of tumor target cells (Wright's stain).

Figure 7 (Cont)

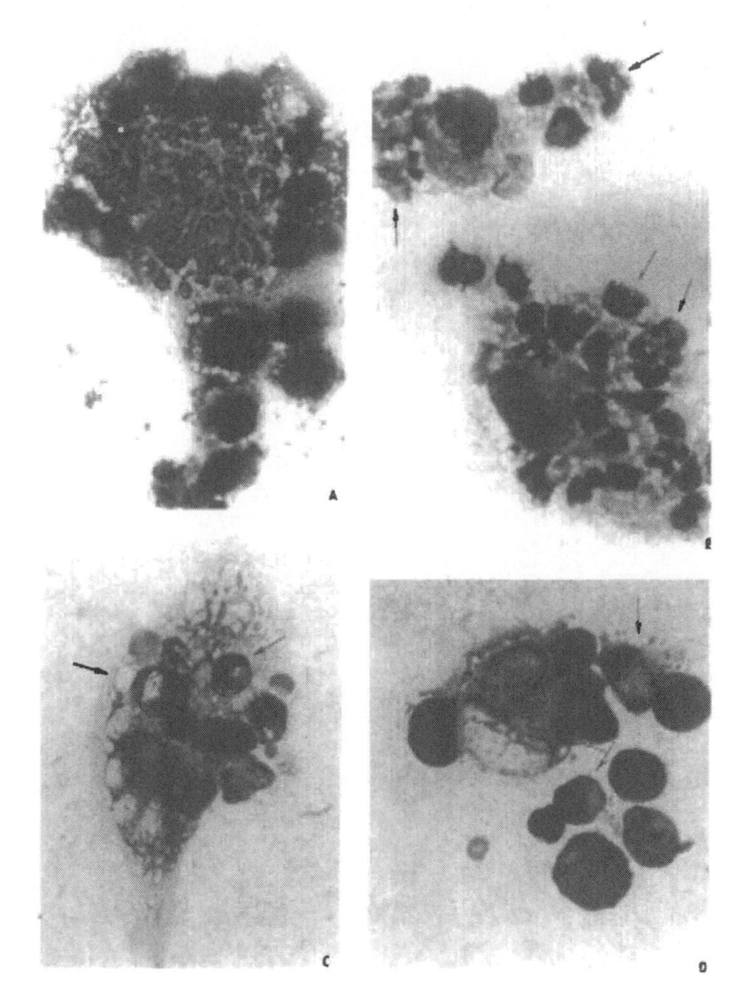

Fig. 2. Review of 25 year old scenes of human tumour cell destruction by autologous and allogeneic lymphocytes. Observation: Established cell lines in allogeneic setting are attacked by large granular lymphocytes (now known as NK cells); tumour cells in primary cultures in autologous setting are attacked by smaller lymphocytes without granular cytoplasm (now known as immune T cells) [31, 32]. All preparations shown were stained by Wright:

A. Embryonal rhabdomyosarcoma cell line (McAllister) #2089 (received from American Type Culture Collection on 12•14•1970) is destroyed by allogeneic lymphocytes of patient RA (MDAH #85779) with rhabdomyosarcoma. Lymphocytes retain good tinctorial characteristics and viability (Experiment #2875 3•8•1972).

B. Fibrosarcoma cell line T2 (from Dr. Jose Trujillo) carried by us as culture #2117 resists destruction by allogeneic lymphocytes of patients JS (MDAH #82358) with neurofibrosarcoma: nuclei intact; cytoplasmic lysis questionable. Lymphocytes approaching sarcoma cells disintegrate some with nuclear clumping (arrows). (Experiment #2950 4•3•1972)

C. Primary culture of rhabdomyosarcoma cell of patient MM (MDAH #79161) undergoes some mild vacuolization while autologous lymphocytes show semilunar nuclear clumping and cytoplasmic vacuolization (arrows). (Experiment #2494 10•2•1971)

D. Primary culture of melanoma cell of patient KC (MDAH #90640) exposed to autologous lymphocytes: target cell dies with nuclear clumping and cytoplasmic vacuolization; some lymphocytes show nuclear clumping and disintegrate (arrows). (Experiment #3024 4•24•1972) nate NK cells is to protect the host against virally infected cells. Immune T cells appeared later, when viral antigens were processed and presented by the as yet undiscovered dendritic cells. These sequences of the immune reaction were not well known in the early 1970s.

For metastatic disease, we had to incorporate the active immunization with VO (and live BCG scarified over the site of the intra- or subcutaneous inoculations of the VO) into standard schedules of chemotherapy regimens because it would have been ethically unacceptable to withhold standard therapy in favor of an entirely investigational treatment regimen from these patients (Figures 1A,1B). It was observed *in vitro* that tumor cells that survived exposure to cytotoxic lymphocytes not only resisted, but actually could kill, the lymphocytes (H. D. Kay and J. G. Sinkovics: *Project M30/gm23*. Self-defensive Mechanisms of Tumor Cells, 1976).* Only much later could we connect the killing of lymphocytes by tumor cells with the expression of *Fas* ligand by the tumor cells [155,156] (Figures 8–10A,10B).

It was in 1968–1969 that I observed that my own lymphocytes appearing in stained preparations as large cells with granular cytoplasm killed a variety of allogeneic tumor cells; we referred to the phenomenon as "Burnet's immune surveillance at work" and did not call those lymphocytes "natural killer cells: NK" (H.D. Kay and J.G. Sinkovics: *Project M30/gm21*. An Investigation into Antitumor Surveillance of Normal Individuals, 1976)* [138,143,148,157].

Our pioneering (ahead of its time) and impeccably conducted projects had to be abandoned at M. D. Anderson Hospi-

* All *Projects* cited in this chapter are listed in the bi-annual *Research Reports* of The University of Texas M.D. Anderson Hospital, Printing Division, Austin, TX, 1962–1980.

Figure 8 Plate reproduced with original legend showing microphotographs from the early 1970s. Some of the lymphocytes attacking tumor cells also die, with nuclear clumping resembling apoptotic death (which was not recognized as such in the early 1970s). (From Acta Microbiologica et Immunologica Hungarica 1997; 44:295–307.)

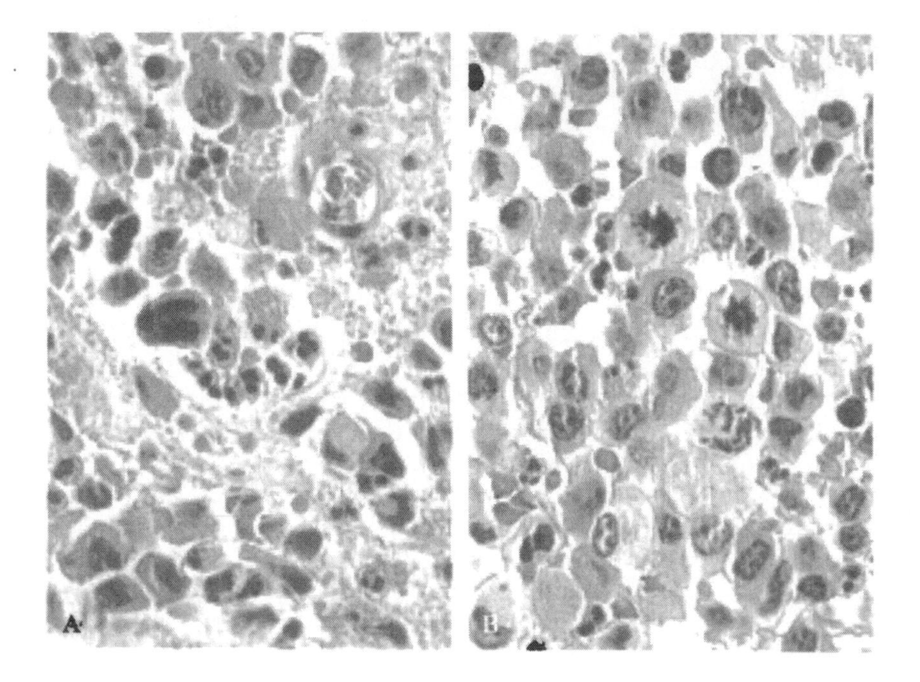

Figure 9 Apoptotic deaths commonly occur in large cell diffuse malignant lymphomas. Both large lymphoma cells and small, healthy-appearing host cells die. (From the clinical material of the author, St. Joseph's Hospital, Tampa, FL.)

tal in the late 1970s because of the "approved without funding" ratings that our grant applications received. Much later, Joseph Horvath and I resumed work along these lines at the Cancer Institute of St. Joseph's Hospital in affiliation with the University of South Florida College of Medicine in Tampa, FL. In the early 1970s, W. Cassel received approval from Emory University's Surveillance Committee to use live attenuated NDV (passaged in murine ascites carcinoma). The generous gift of his 73T NDV strain allowed us to switch from PR8 VO to the use of NDV VO (J. C. Horvath, Chapter 5 in this volume). After lymphocyte-mediated killing of human tumor cells was rediscovered at the NCI, where major projects for the therapeutic use of LAK cells and TIL driven by "T cell growth factor," IL-2, were launched in the mid-1980s, we could incorpo-

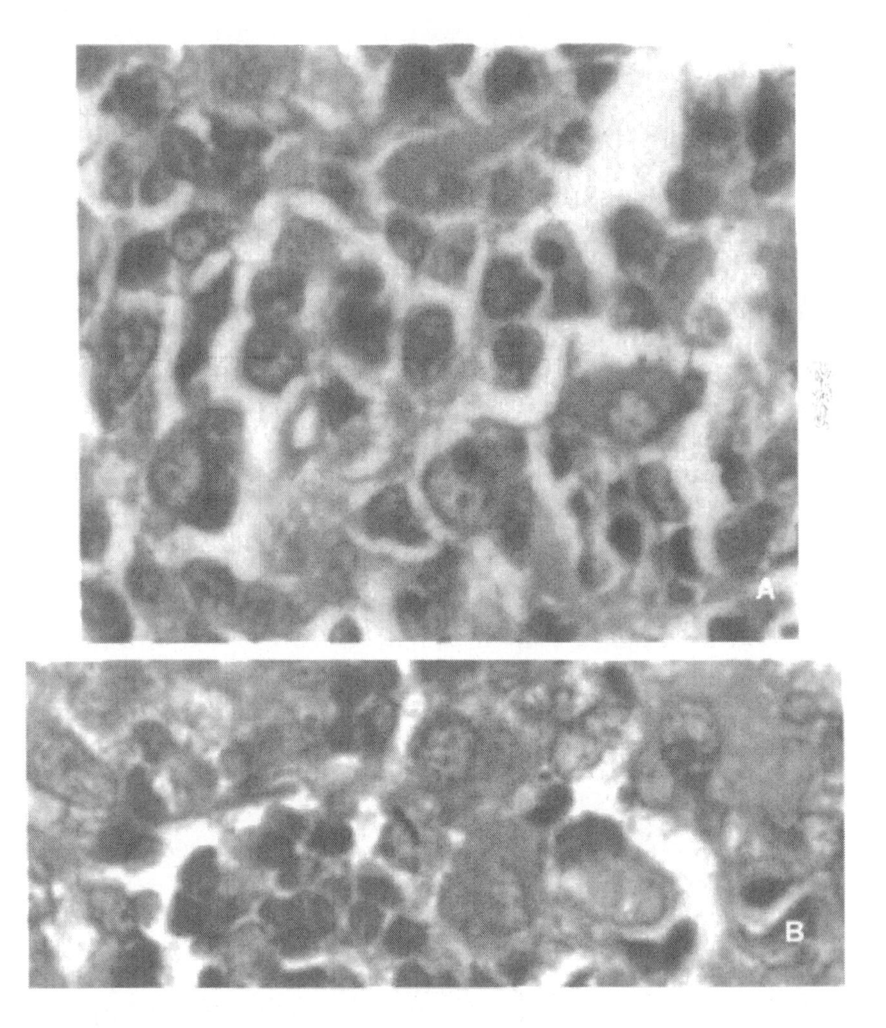

Figure 10 A, 10B Fas-receptor-positive large lymphoma cells (dark red cytoplasm) are attacked by presumably FasL^{+} host lymphocytes (A), while Fas-ligand-expressing lymphoma cells (faint red cytoplasm) escape immune attack by host lymphocytes, whereas some presumably Fas-receptor-expressing lymphocytes appear to die, with nuclear clumping. Histochemical stains with commercial Fas-receptor and Fas-ligand antibodies, horseradish peroxidase technique, and counterstain with H&E (with the help of Professor Bela Bodey, Los Angeles, CA, 2000). (From the clinical material of the author, St. Joseph's Hospital, Tampa, FL.)

rate this modality for the treatment of metastatic melanoma and kidney carcinoma in our program [136,137].

MORE RECENT HISTORY

In 1962 and thereafter, at the Roswell Park Memorial Institute in Buffalo, NY, veterinary viruses apathogenic in the human host were adapted to tissue cultures of human tumors and then used for the therapy of patients with various cancers [158,159]. Again, remissions so induced were lost to relapses and the project was abandoned. In a personal note, David Yohn explained to this author that fear of possible recombination of the veterinary virus with latent related or unrelated human viruses contributed to the denial of licensure for the clinical use of this procedure. The clinical trials with a veterinary NDV vaccine (renamed MTH-68N virus) conducted (but now discontinued) in Hungary have been referred to extensively elsewhere [61].

In the meantime, in Japan in the 1970s, live attenuated mumps virus was found to induce remissions of human ovarian and other carcinomas and glioblastomas [160–162]. Teruo Asada, in a letter addressed to this author, stated that public health authorities of that country disallowed the use of live mumps virus for cancer therapy; only investigational trials supervised in qualified institutions may be continued. Yoshio Shimizu also abandoned the use of mumps virus for the treatment of ovarian carcinomas [162]; we invited his contribution to this volume but he declined. In 1978, the Hokkaido University published Hiroshi Kobayashi's monograph on the artificial xenogenization of tumor cells. Tumor cells infected with the Friend erythroleukemia virus of mice suffered no direct oncolytic effect from the virus, but evoked strong antitumor immune reactions from the host [163,164].

Later, in 1989, at a M. D. Anderson Hospital symposium, these authors presented work on tumor cell xenogenization with viral gene transfers (Friend virus *env* gene; influenza A virus hemagglutinin gene) into tumor cells, rendering the tumor cells more antigenic in their hosts. At the European Association for Cancer Research symposium in Athens, Greece, in 1989, I reviewed this work in comparison with the direct effect of oncolytic viruses [165]. Unfortunately, the Moloney mouse leukemia virus envelope protein, when expressed in the membranes of tumor cells, elicited the opposite effect in both allogeneic and syngeneic systems, whereby the tumor cells became immunosuppressive and evaded rejection [166]. Curiously, this article, edited by George Klein, did not mention tumor cell xenogenization by Kobayashi, or the controversy regarding results obtained with the Friend virus previously and differing from those obtained with the Moloney virus, as both retroviral envelope proteins are expressed by tumor cells.

The human genome harbors large numbers of nonfunctional retrotransposon-like elements. These include mobile insertional DNA sequences; short and long interspersed elements; ψg pseudogenes deriving from retrotranscribed mRNA; long terminal repeats and endogenous proviral DNA sequences existing as remnants of prior exogenous retroviral infections poised with the potential to spring into action as new retroviruses or evolve into useful genes. The RAG1, 2 system, telomerase, syncytin in the mammalian placenta, or centromers directing cell divisions and many others take their origin from genomic retrotransposons. Endogenous retroviruses, by gaining *env* sequences ("natural xenogenization") appear as budding particles in syncytiotrophoblasts of the placenta, teratocarcinomas, certain breast cancers, lupus erythematosus and multiple sclerosis, and in lymphoma and leukemia cells, especially when lymphocytes undergo somatic hypermutations of theVDJ sequences initiated by the RAG1, 2 enzymes. Endogenous retroviruses also exist in Kaposi's sarcoma and other sarcoma cells. A xenografted human tumor may pick up endogenous retroviruses of its new host (Figures 11–13). These *env* proteins and other retroviral structural proteins may induce autoimmune or antitumor immune reactions, or they may immunologically compromise the host. However, their exact role remains quite underestimated [167*,

* In this article [167], because of regretful mislabeling of two electron microscopic prints, the retroviral particles shown in Figure 12 are now identified to be not in Reed-Sternberg cells, but in Kaposi's sarcoma cells, as shown in a new reference [171].

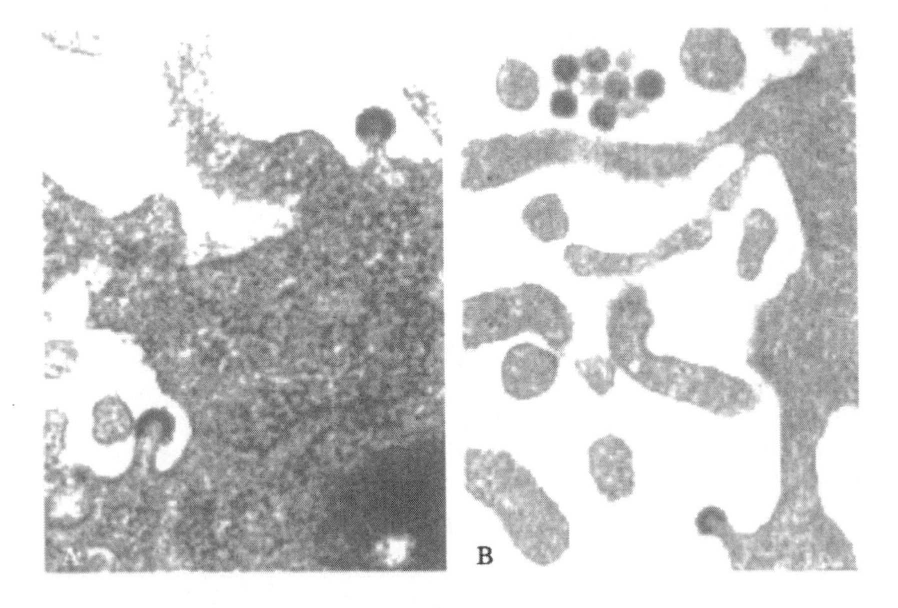

Figure 11 Plate showing budding endogenous retrovirus particles in human B lineage lymphoma cells presumably activated by acquiring new *env* gene-coding sequences during somatic hypermutation [167–170]. Transmission electron microscopy, original magnifications 40,000 and 30,000. (From Orvosi Hetilap [Hungarian Weekly Medical Journal] 2001; 142:1352–1353.)

168–170] and it is entirely unknown how resident cellular retrotransposons react to oncolytic viruses.

In the 1980s, W. Cassel's 73T NDV strain was used again (as a gift of Dr. Cassel to a Chicago team) to kill xenografted human neuroblastoma and fibrosarcoma and various carcinoma cells without harming the tumor-bearing host [172]. This work was editorialized in the *Journal of the National*

Figure 12 A, and B Herpesviruses (CMV; HHV-8) and unidentified, presumably endogenous, retroviruses (Fig 12C) share a Kaposi's sarcoma tumor [167–171]. Transmission electron microscopy, original magnifications 120,000 & 50,000 (†Ferenc & †Phyllis Gyorkey, Joseph Sinkovics, VA Medical Center, Houston, TX, 1983). (A: From J.G. Sinkovics, J.E. Harris, in: J.E. Harris, J.G. Sinkovics, authors. The Immunology of Malignant Disease, second edition. St. Louis: C.V. Mosby, 1976: fig. 5.5, p. 432.)

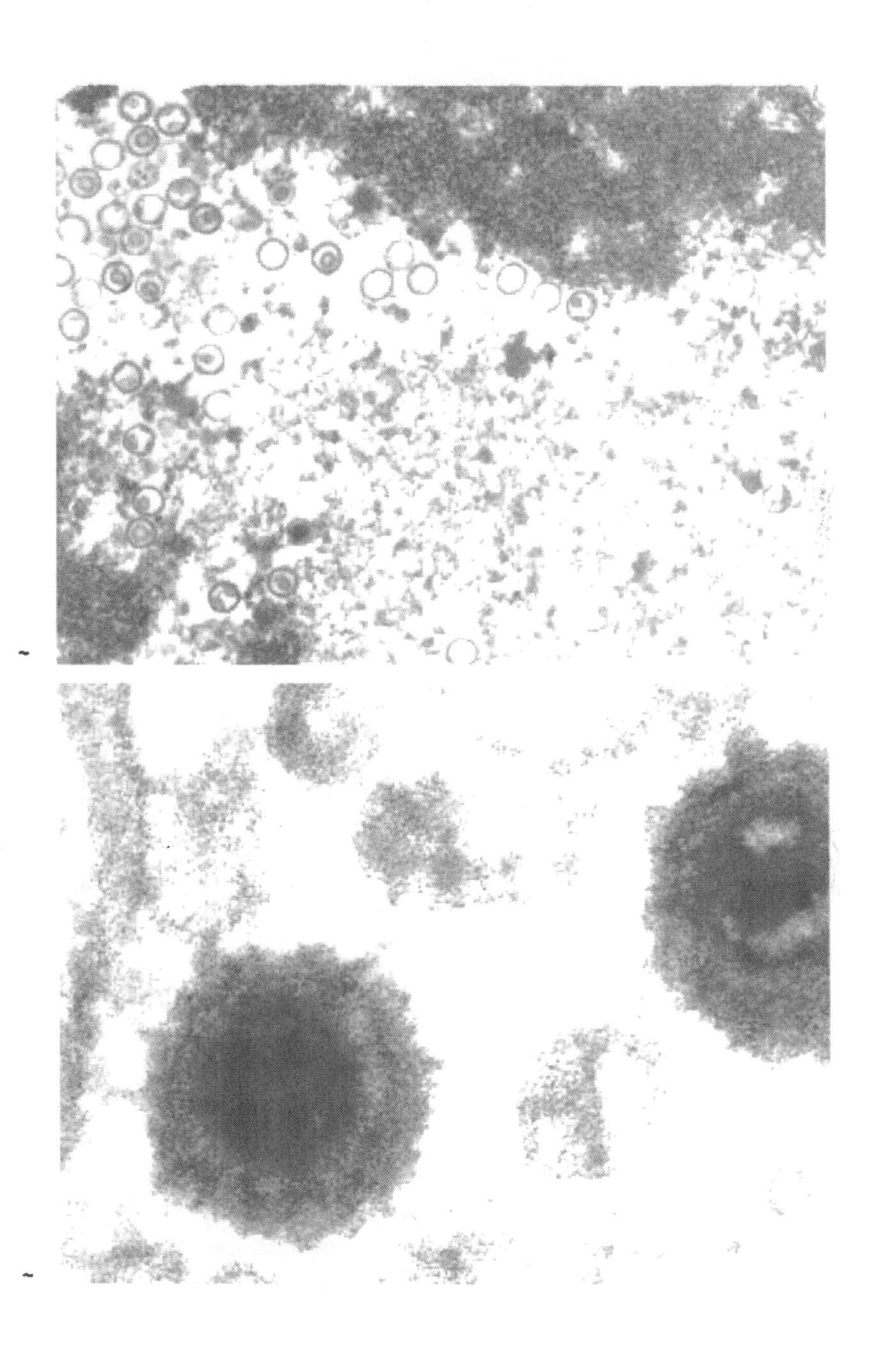

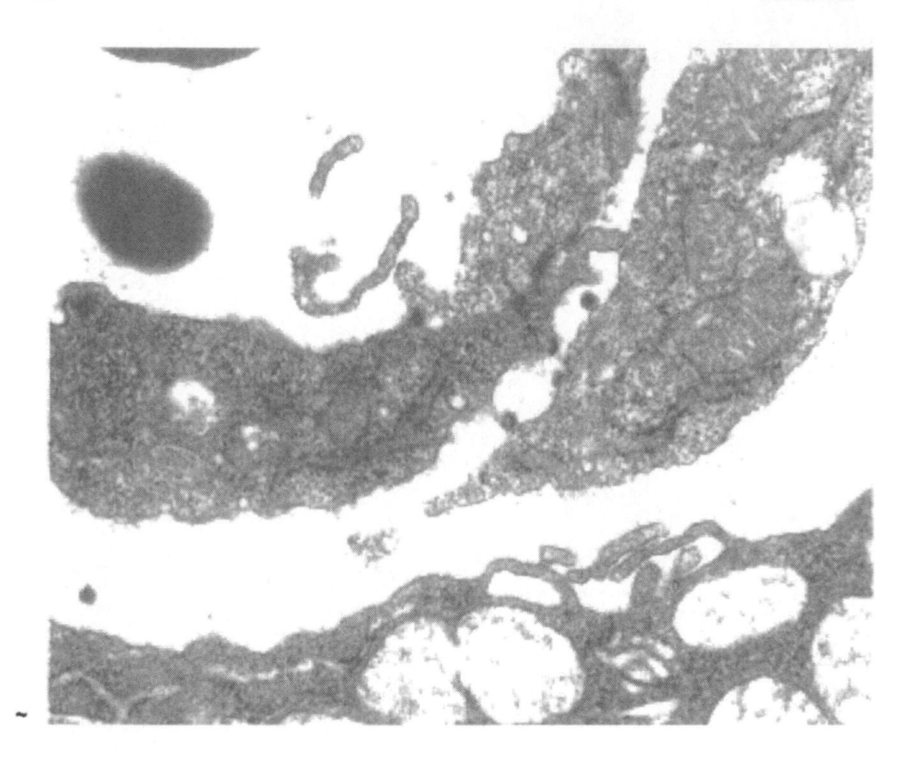

Figure12C

Cancer Institute [173] announcing that a "new era" for the virus therapy of cancer had begun. The origin and relationship of NDV 73T and a new NDV strain—P701—now used for clinical trials by the same investigators [61,174] who revitalized the field of the virus therapy for human cancers in eradicating human tumor xenografts with W. Cassel's 73T NDV have not been clearly explained, however. What is the origin and derivation of the new P701 NDV? Is it related to 73T NDV or is it entirely different from it and, if so, are the results of the preclinical work done with 73T NDV applicable to P701 NDV?

VIRAL ONCOLYSIS TODAY

Influenza Virus

The negative-sense ssRNA influenza orthomyxovirus, with its 13.6 kb fragmented genome, is re-emerging as an oncolytic

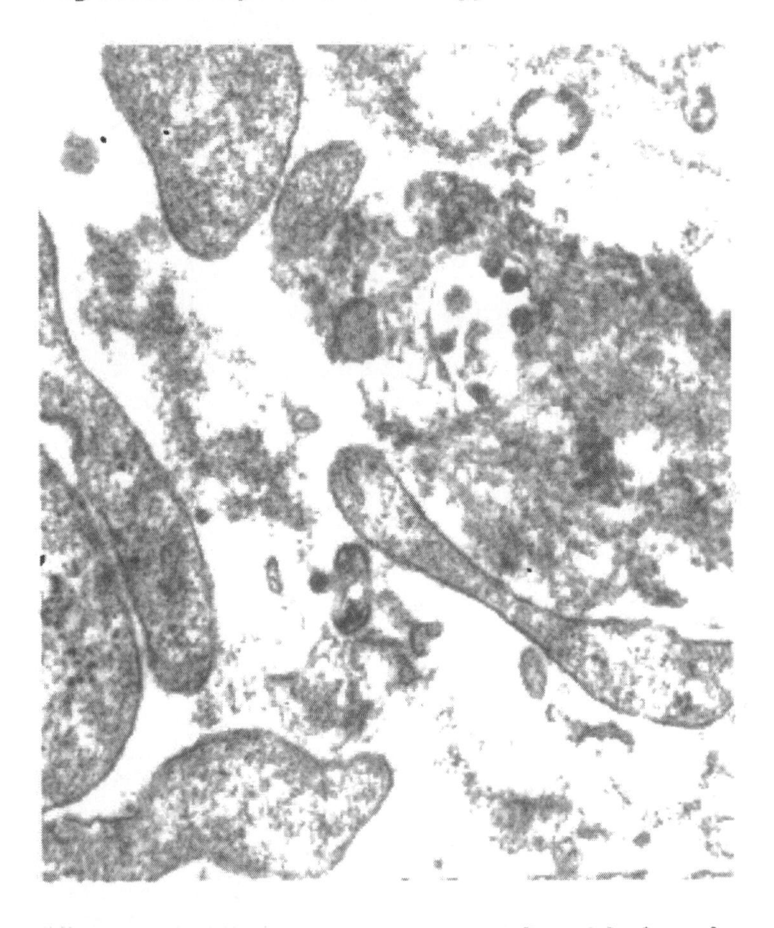

Figure 13 Murine retrovirus particles picked up by a human tumor xenograft [126]. Transmission electron microscopy, original magnification 35,000 (†Ferenc & †Phyllis Gyorkey, Alex Sandor Gergely, Joseph Sinkovics, VA Medical Center, Houston, TX, 1977).

agent. Its genome is transcribed within the nucleus of the host cell under the direction of its nucleoprotein (NP) [175]. This NP is a multifunctional RNA-binding protein of 498 amino acids with kinase activity. It interacts with the viral RNA-dependent RNA polymerase, the viral matrix protein, importins/exportins, export receptor CRM-l, F-actin, and the enzyme helicase. Dendritic cells of the infected host produce interferons, phagocytize apoptotic influenza virus-infected cells, and express the viral antigens to CD4 and CD8 T cells, eliciting a Th1 response in which CD8 T cells secrete IFNγ and inhibit a Th2 response. Interleukin-5 levels drop but eotaxin/CC chemokine ligand 11 and chemokine ligand 2 (monocyte chemoattractant protein-1) levels rise. Influenza virus-infected cells elicit a strong MHC class I-restricted CD8 T cell-mediated immune response [176,177]. The nonstructural viral protein NS1A suppresses the formation of anti-viral mRNAs and inactivates the cellular enzyme PKR, the dimerizing and autophosphorylating serine/threonine protein kinase, which is activated by interferon. The activated (phosphorylated) PKR blocks protein synthesis, including that of viral structural proteins. The PR8 influenza A virus with deleted NS1A gene cannot inactivate PKR; therefore, infected healthy cells allow its replication. However, in cells of STAT knockout mice (no signal transducer and activator transcription), or in tumor cells harboring a mutated *ras* gene, cellular inhibitors have already incapacitated PKR. In these cells, the PR8/34delNS1 virus replicates uninhibited, whereas healthy cells of the same host suppress its replication. Herein lies the selective oncolytic potency of the genetically engineered influenza A virus strain [178,179] (M. Bergmann and T. Muster, Chapter 6 in this volume).

Attenuated temperature sensitive ("cold-adapted") live influenza viral vaccines are being developed for intranasal instillation and show a reduced ability to recombine [180,181]. Mucosal immunity is primarily mediated by secretory inmmunoglobulin A and, when generated by the vaccine, may prevent the initiation of infection with wild influenza virus. These influenza virus strains may replicate in cancers of the oral or nasal cavities (not yet tested) or possibly produce viral oncolysates safer than those prepared from the PR8 virus in the early 1970s (Figure 14) [129–132].

However, these earlier PR8 influenza A viral oncolysates remained in use for the treatment of gynecological cancers (ovarian adeno- and vulvar squamous cell carcinomas) at M.D. Anderson Hospital in Houston, TX until recently [182–186]. Apparently, after being reclassified as "virus-modified autologous or homologous extracts" and administered to gynecologi-

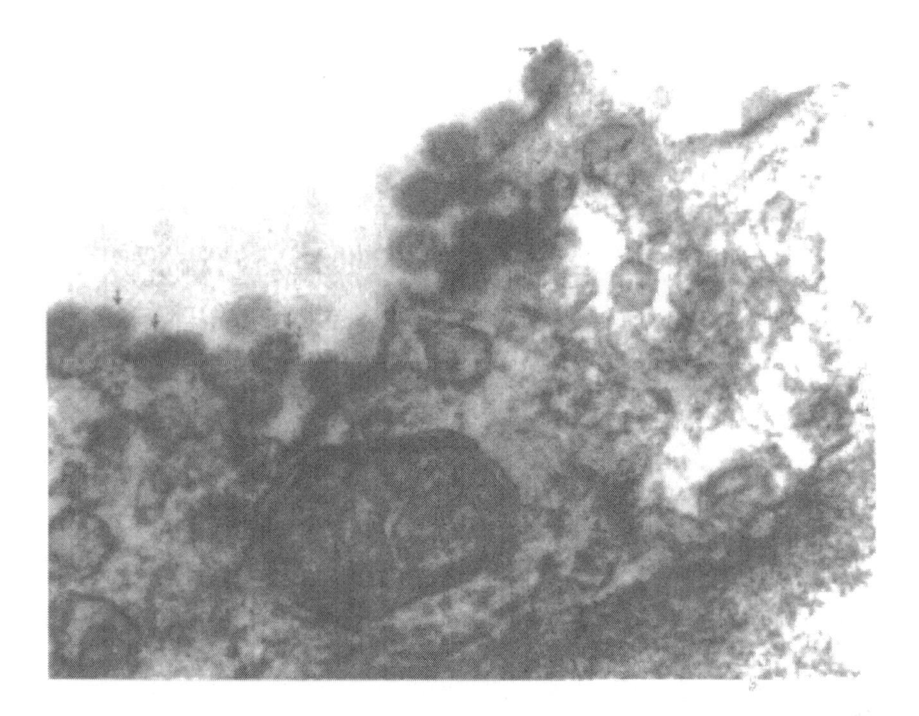

Figure 14 Ewing's sarcoma cell line #1846 [126] infected with PR8 influenza A virus in preparation of viral oncolysate (1972). Transmission electron microscopy, original magnification 40,000. (From Intern Rev Immunol 1991; 7:259–287).

cal patients, funding (that was not provided for melanoma and sarcoma patients in the 1970s) became available in the 1980s at the same institution, but at a different department, where the preventive therapeutic and antitumor immune stimulatory value of influenza A viral oncolysates were appreciated [119,184,185].

Any influenza viral strain used for oncolysis must be deprived of its ability to recombine with a wild strain. For example, H1N2 influenza A viruses (Egypt/96/02; Saudi Arabia/2231/01) emerging in 2001 were shown to be reassortants between H1N1 (New Caledonia/20/99; Madagascar/55/94/00) and H3N2 (Moscow/10/99; Panama2007/99) viruses [187]. Superficially reviewed (not checking directly for virally induced remissions, if any), natural influenza viral infections in patients of a cancer hospital were followed by life-threatening bacterial pneumonias [188,189].

Measles Virus

One of the most vicious and notoriously immunosuppressive viruses, the enveloped wild type measles virus (MV), with its 16–20 kb helical RNA genome, is a member of the *Paramyxoviridae* family within the genus *Morbillivirus*. Measles remains endemic [190]. The virus attacks dendritic cells and T lymphocytes and induces their apoptotic death, but its attenuated derivative strains replicate in, and kill, human myeloma or glioblastoma cells (S. J. Russell and associates, Chapter 3 in this volume).

The Janus protein kinases (JAK) are phosphorylated when the cytoplasmic domains of INFαβ type I receptors are phosphorylated, which occurs after the extracellular domain captures its ligand. Activated JAK phosphorylates the signal transducer and activator of transciption (STAT) system, which in turn activates IFN-responsive promoter elements. Members of the *Paramyxoviridae* inactivate the IFN signaling pathway at many levels. Rubulaviruses (NDV) and mumps virus deplete STAT proteins. MV suppresses JAK phosphorylation and downregulates IL-12 production by monocytes–macrophages. The MV nucleocapsid protein binds to FcγRIII and inhibits the generation of humoral immunity. When the virus infects dendritic cells (DC) and epidermal Langerhans cells, inducing cell fusion and abrogating DC and CD4 T cell interactions, ultimately both DC and T cells die apoptotic deaths [191–198].

In a Chinese clinical trial [199], when its immunosuppressive effects might have been counterbalanced with the coadministration of BCG (an extremely precarious balance it must have been!), MV added to chemotherapy for acute leukemias induced the highest remission rates. Remissions were in the 80% range in the MV-treated group of patients vs. only 31% remission rates in chemotherapy-only control patients; remission rose to 45% when only BCG was added to chemotherapy. Standard MV strains (Edmonston; Halle) enter cells through the CD46 receptor. Another MV receptor, the signaling lymphocyte activation molecule (SLAM; CD150) is expressed on lymphocytes, including memory cells, CD4 cells, Th1 cells, and EBV-transformed B cells [200]. This latter event may explain the regressions of African Burkitt's lymphoma or Hodgkin's disease in children experiencing natural infection with MV [52]. The former event (loss of memory T cells) explains death caused by natural measles infection in patients with non-Hodgkin's lymphoma [201]. Measles virus and antibodies may form large amounts of immune complexes ("atypical measles") [202] and MV may cause acute sclerosing panencephalitis. Hamster brain tumor cells persistently infected with MV and exposed to antiviral antibodies gained an enhanced growth rate [203].

Attenuation of the wild Edmonston MV [204] was achieved by passages of the virus through human kidney and amnion cell cultures (Edmonston-Enders strain) and through chicken embryo amnion and fibroblast cells (Moraten vaccine strain). The Zagreb strain was passaged in Wistar Institute human diploid fibroblast cultures (WI-38). To direct the attenuated vaccine strain of MV exclusively into myeloma cells, the virus particle must express a single-chain antibody directed against the myeloma cell receptor CD38. When used against ovarian carcinoma or glioblastoma cells, the Edmonston attenuated MV enters through the CD46 receptor, which is overexpressed in the tumor cells (Russell and Associates, in this volume). In orthotopically transplanted human glioblastoma xenografts, the genetically engineered Edmonston vaccine measles virus (MV-CEA) revealed its presence by releasing CEA from the infected tumor cells before complete destruction of the tumors occurred, leaving glial scars and some inflammatory cells behind. The Edmonston MV-CEA completely destroyed xenografts of the $CD46^+$ SKOV3ip.1 human ovarian carcinoma cell line in athymic mice [205–208]. Thus, MV closes ranks with NDV and vesicular stomatitis virus (VSV), considered to be "naturally oncolytic viruses."

Rubella virus also can attack selectively astrocytes but its apoptosis-inducing activities were studied in Vero cells possessing intact wild p53 [209–211]. Possibly, the attenuated live rubella virus (Meruvax, Wistar Institute RA 27/3 strain) would attack human tumor cells with deleted (neutralized by MDM2 protein) or mutated p53. Also, this virus, when combined with the Jeryl Lynn strain of attenuated mumps virus (Biavax, Merck & Co., Whitehouse Station, NJ) might exhibit a increased (or lost) oncolytic potency (if any).

Newcastle Disease Virus

The enveloped, negative-stranded helical RNA NDV belongs in the genus *Rubulavirus* within the family of *Paramyxoviridae*. NDV is the causative agent of fowl plague, a highly contagious disease of poultry. Virulent strains could be recovered from its cloned cDNA [212] (J. C. Horvath, Chapter 5 in this volume). NDV can cause conjunctivitis, with enlargement of the preauricular lymph node, in human beings ("the oculoglandular syndrome" described by M. Radnót in Hungary, who initiated in 1948 our studies of virus isolation from human conjunctival washings) [213,214].

Rubulaviruses evade the host cells' interferon response by V protein production that targets STATs (see *New Biological Therapeutics* by Sinkovics) for proteasome-mediated degradation [215] and thereforce can replicate their progeny to full maturation, resulting in cell death. When the host cell dies from apoptosis early during the viral cycle, however, no infectious viral particles are formed and there is no further spread of infectious virus within the tumor mass. Even when the viral replication cycle is incomplete in a given tumor, such as with the Ulster strain of NDV (V. Schirrmacher, Chapter 9 in this volume), host antibodies, activated macrophages, antigen-loaded dendritic cells, NK cells, immune T cells, and lympho- and cytokines produced by them mediate destruction of the infected cell [216–221]. It is debated whether fully replicative, early apoptosis-inducing, or incompletely replicating but cytokine- and immune reactions-evoking NDV strains exert the best oncolytic effect. If a transfected NDV gene encodes viral hemagglutinin-neuramidase antigenic proteins in a tumor cell (like in a murine myeloma cell), the host's tolerance will be overcome and host-vs-tumor immune responses will be evoked

[222,223]. Could virus-neutralizing antibodies curtail cell-to-cell viral spread in a tumor mass? Indeed, the possession of hydrophobic fusion peptides within the viral envelope promote syncytium formation of infected tumor cells, whereby the virus spreads without an extracellular phase and renders the syncytiosome highly immunogenic (as in the case of vesicular stomatitis virus, *vide infra*).

The early apoptotic death of NDV-infected rat pheochromocytoma cells [224,225] is probably not a valid model of NDV oncolysis for the tumor-bearing human host. Despite of the high publicity it receives [59–61,226], the claim of the sponsor of this report, that 20 CR and 9 PR in 29 patients with "advanced cancers" occurred in response to NDV (MTH-68/H) treatment [224], should be viewed with great reservations (unless confirmed convincingly and independently) [156].

W. Cassel, in the early 1960s, induced the death of human adenocarcinoma cells implanted in hamster cheek pouches by direct intratumoral injections of NDV [10]. The same virus induced temporary regression of uterine cancer in a patient [10]. W. Cassel was first to declare NDV to be an antineoplastic agent [10] and translated the outcome of animal experiments into a successful clinical therapeutic trial (W. Cassel, Chapter 9 in this volume). It is not yet proven whether oncolytic viruses that perform so well against cultured or xenografted human tumors, such as W. Cassel's 73-T NDV [172], will be able to sustain this level of antitumor activity in the tumor-bearing human host.

Whether NDV oncolysates would perform better than direct viral inoculations is unknown because no comparative clinical trial exists. However, W. Cassel's and V. Schirrmacher's trials, taken together [122,227–230], indicate that NDV oncolysates as "preventive vaccines" perform very well against micrometastases left behind after surgical removal of gross tumors (melanoma and various adenocarcinomas). A NDV melanoma oncolysate prepared at M.D. Anderson Hospital but using techniques different from those of W. Cassel, induced melanoma cell-specific antibodies in patients, but provided no significant protection against relapses; W. Cassel clarified the issue [231–233]. Newcastle disease virus oncolysates may not work as therapeutic vaccines against gross metastatic disease. If it were possible to incorporate NDV in a dendritic cell vaccine [125,234] to be co-administered with IFNs, IL-2, or IL-12, its performance might rise to the level of a therapeutic vaccine (V. Schirrmacher; J. C. Horvath; Chapter 4 and 5, respectively, in this volume).

In a recent professionally conducted clinical trial [174], using the newly introduced attenuated PV701 NDV strain (without publishing its preclinical performance against xenografted human tumors), some 80 patients with advanced cancers were treated. A few partial remissions (PR), minor responses and stabilizations of disease were recorded but the single complete remission (CR) was lost to relapse. A second Phase I trial using "desensitization" and a very high dose of PV701 NDV (24–120 billion PFU/m^2) and prolonged infusion time produced three partial remissions (PR), two minor responses (MR) and four stabilizations of disease in 11 patients with metastatic cancers refractory to standard therapy [235]. Still, no CR could be observed, in contrast to complete eradication by the 73T NDV in preclinical experiments with human tumor xenografts in mice.

In a Chinese clinical trial [236] for the treatment of cancers of the digestive tract, 257 patients were treated only surgically; 310 patients received an autologous tumor vaccine and NDV (LaSota strain) therapy in addition to surgical resection. Overall survival was more than 7 years in the immunized group vs. 4.46 years in the surgically treated group. The mean survival time was 5.13 years for the immunized group and 4.15 years for the surgically treated group. A stratification of these patients will be necessary for valid evaluation of these results.

V. Schirrmacher (Chapter 4 in this volume) has used the incompletely replicating but cytokine- and immune- reactions–evoking Ulster strain of NDV for human tumor immunotherapy. Dendritic cells pulsed with Ulster NDV oncolysates induce Matzinger's danger signals, re-enlist memory cells resting in the bone marrow, and react to tumor (breast cancer) antigens. Another German group used the Ulster strain of NDV to prepare glioblastoma vaccines for active tumor-specific immunotherapy given postresection and postradiotherapy [237]. Median patient survival was 46 weeks; mean, 60 weeks, ranging from 31 to 105 weeks; only one patient survived (more than 88 weeks), even though the patients appeared to be immunocompetent to the extent that they produced skin reactions to the autologous vaccine inoculations. Relapsing and re-resected tumors were infiltrated by CD4/CD8 lymphocytes.

NDV oncolysates are in use as preventive vaccines administered after tumor removal (melanoma; kidney carcinoma) when clinical criteria are such that persistence of micrometastases is expected (Horvath, in this volume). However, in Germany, kidney carcinoma cells infected with NDV 73T and irradiated with 100 Gy thereafter induced no antitumor antibodies or immune T cells against tumor cells not virally infected. Anti-NDV antibodies exerted ADCC reaction against virally infected tumor cells only [238,239].

Cassel's 73-T NDV strain was investigated by the U.S. Department of Agriculture in 2003. It was declared to be not a viscerotropic virus. In the laboratory, it may kill newly hatched chickens after cerebral inoculation, but it is not considered to be pathogenic to birds or to human beings through any natural routes of infection. It is safe to use it as live virus in VO preparations.

Vesicular Stomatitis Virus

The enveloped negative-stranded helical *Rhabdoviridae* RNA virus with a 13–16 kb genome is oncolytic [240] (R. A. C. Taylor, J. M. Paterson, and J. C. Bell, Chapter 7 in this volume). Vesicular stomatitis virus was shown already in the mid-1960s to kill mouse tumors [241] and have an unusual relationship with IFN [242]. In Finland, it was used for viral oncolysate production to treat mouse tumors in 1971 [243]. In the 1970s, VSV was found to inhibit the growth of, or render murine mastocytoma and L1210 murine leukemia cells more immunogenic in their hosts [244,245].

It is conceivable that *Rhabdoviridae* and *Paramyxoviridae* descend from an ancient common ancestor that has conferred natural oncolytic properties to some members of these families and that these viruses exercise natural oncolysis since their remote past. When VSV was passaged for short, intermediate, or late intervals, viral populations passaged for intermediate intervals were the most stable [246]. It recently was discovered that VSV induces apoptosis in p53-defective human tumor cell lines in vitro or in their xenografts in vivo (G.N. Barber and associates). However, VSV is known to induce apoptosis very early after its entry into the cytoplasm. This effect is not inhibited by cycloheximid. Ultraviolet-irradiated virus particles also elicit apoptosis. The matrix (M) protein of VSV in itself elicits the apoptotic response of the cell and suppresses host cell gene expressions, such as JAK/STAT and IL-6. Tumor cells (HeLa) overexpressing Bcl-2, resist M-induced cell-rounding and apoptosis [247–249]. The VSV fusogenic membrane glycoprotein (G protein) induces multinucleated syncytia of infected human tumor fibro- and osteosarcoma, melanoma, and colon and lung cancer) cells, resulting in cell death caused by premature chromosome condensation and adenosine triphosphate depletion. Dying syncytiosomes load dendritic cells with their surface antigens. A plasmid expression system encoding VSV G protein induced murine colon and kidney carcinoma cells to form syncytia that were highly immunogenic, especially in the presence of IL-12 [250–253]. Although VSV-induced syncytia were described much earlier [242], it was recently shown that these syncytia are highly immunogenic and attract NK cells, macrophages, and dendritic cells. Vesicular stomatitis virus antigen processing and presentation occur via MHC class I and class II pathways [254].

In early clinical trials, a VSV human melanoma oncolysate failed to induce the production of melanoma cell-specific antibodies in patients [255]. Tumors with deleted p53, mutated *ras* or *c-myc* and glioblastoma cells are susceptible to VSV oncolysis in vitro or as xenografts [256,257]. Human hepatocellular carcinoma cells orthotopically xenografted into immunocompetent rats completely regressed after one intratumoral injection of VSV, although the virus was not able to replicate in healthy hepatocytes. It was not explicitly stated whether the virus destroyed the tumor cells or virally infected tumor cells were immunologically attacked by the host, or if tumor cell syncytia were formed. Tumor cell death was described as "large necrotic areas within the tumor" but not in normal liver parenchyma [258]. Vesicular stomatitis virus vectors carrying the herpes viral gene for thymidine kinase (TK) or the gene for IL-4 gain oncolytic potency and evoke granulocytic and T cell infiltrations into the murine tumors against which they were tested [259]. Moloney murine leukemia retroviral vectors pseudotyped for VSV G protein expression and armed with an insulin-like growth factor receptor antisense sequence or with the herpes viral TK sequence inhibited tumor (colon and pancreatic carcinomas) growth and metastasis formation in xenografts. In these xenografts, ganciclovir exerted enhanced bystander effect [260,261]. Indomethacin suppresses the replication of VSV [262], so this medication should be withheld from patients entering clinical trials to receive oncolytic VSV.

Reovirus

The respiratory–enteric orphan (Sabin) icosahedral naked reoviruses with a 22–27 kb RNA genome cannot readily infect healthy cells because of the cells' vigorous IFN response, whereas tumor cells with defective machinery for IFN production (especially in *ras*-mutated tumor cells) are exquisitely susceptible to this virus and die apoptotic death upon viral entry [263]. However, tumor cells inherently resist programmed cell death caused by tumor suppressor gene mutations or deletions. Instead, tumor cells may allow the virus to complete its full replicative cycle, leading to cell death by burst as the new viral progeny exits.

Reovirus-induced apoptotic cell death involves either or both the extrinsic (through cell surface death receptors) and intrinsic (through cytochrome C release from mitochondria) pathways [264]. From modest beginnings—that is, immunologically mediated rejection of murine leukemia–lymphoma cells and tumors by reovirus, as reported in 1976 and 1988 (especially in combination with the nitrosourea BCNU)

[265–267]—reoviruses have now advanced to occupy a prominent place among the most respected oncolytic viruses ready for Phase I and II clinical trials.

The reoviral gene S1 is bicistronic and, as such, encodes the viral protein σ1 that attaches to sialic acid receptors and the nonvirion-associated protein σ1s that induces cell cycle arrest at G_2-M [268]. Some reoviruses may expropriate the EGF-R as their entry site into the cytosol [269]. The reoviral gene M2 encodes the outer capsid protein (μ1c) of the virus particle, which enables the virus to transgress the cell membrane. Proteins encoded by reoviral genes were characterized with the use of monoclonal antibodies from cloned hybridomas [270]. When the nonenveloped dsRNA reovirus (type 3 Abney T3A or Dearing T3D strains) induces apoptosis, both the extrinsic (through cell surface receptor activation involving the tumor necrosis factor or TNF-related apoptosis-inducing ligand released from infected cells) and intrinsic (through cytochrome C release from mitochondria and caspase 7 and 8 activation) pathways are activated [264].

Activation of c-*jun*–encoded c-Jun kinase occurs in type 3 but not in type 1 (Lang T1L) reovirus infections. The c-Jun proteins either homodimerize or heterodimerize with c-Fos protein and activate nuclear transcription factor AP1 (or the c-*jun* gene actually encodes AP1). Counterparts of the genes c-*jun* (1p31–33) and c-*fos* (14q24.3–22) were discovered as v-*jun* and v-*fos* in acutely transforming retroviruses (avian sarcoma virus 17 and murine osteosarcoma virus FBJ-MuSV, respectively) and are considered to be proto-oncogenes when activated in adult life. In normal fetal life, however, c-*fos* and c-*fms* are strongly expressed in trophoblasts. The c-Jun protein also activates NFκB—an anti-apoptotic event; further, it is essential for the innate IFNα and IFNβ responses [271,272].

However, reoviruses are not oncogenic; they induce or allow pre-existing cell survival mechanisms (as in cells already transformed, especially by *ras* oncogenes) to function and cause lytic death of cells through full maturation of their progeny, at which time the infected cell dies by burst. Normal cells produce IFNα and IFNβ and disallow full maturation of the viral progeny.

Although healthy cells are not readily susceptible to reoviral infection, *ras*-mutated transformed cells are fully susceptible, presenting an environment in which the dsRNA-activated PKR is inactive. This IFN-activated enzyme shuts down protein synthesis in normal cells, including the synthesis of structural proteins of a viral usurper [273]. Without functional (dephosphorylated) PKR, *ras*-mutated cells succumb to full replicative cycles of reoviruses.

Reovirus serotype 3 (Dearing) produced extensive oncolysis of human cancer cells in vitro (in tissue cultures) and in vivo (in xenografts). The most sensitive human tumors were glioblastoma, breast, ovarian, colon and pancreatic carcinomas, and malignant lymphomas [274–278]. In a bilateral xenograft model, whereby the pancreatic cancer cell xenograft was injected in one side, the uninfected xenograft on the other side also picked up and replicated the virus and attracted an NK cell-mediated host reaction [279]. Orthotopically transplanted human brain tumor xenografts (glioblastoma; medulloblastoma) injected intratumorally with reovirus either regressed or showed extremely retarded growth rates and failed to metastasize [280]. In co-cultured spheroids of human transitional carcinoma cells of the urinary bladder and their fibroblasts, reovirus destroyed the cancer cells and selectively spared the fibroblasts [281].

A substantial segment of the human populations have antibodies against reoviruses. It is not known whether the virus will perform as well in clinical trials as it does in its pre-clinical assays or whether pre-existing or newly acquired antiviral immune reactions (antibodies; NK and immune T cells) would counteract or promote viral oncolysis.

Vaccinia Virus

Vaccinia and ectromelia viruses already excelled in the 1920s as oncolytic agents against murine malignant tumors [94,281]. The large, enveloped vaccinia virus (VV), with its linear DNA genome of over 220 kb, undergoes the following serial stages of maturation within the cytoplasm of its host cell: intracellular mature virus (IMV), intracellular enveloped virus (IEV), cell-associated enveloped virus (CEV), and release of extracellular enveloped virus. Vacinia virus replicates exclusively in the cytoplasm. Viral genes B5R and F13L provide packaging for IMV to become IEV. Gene A36R forms the actin tail. Gene A56R forms the hemagglutinin; A56R$^-$ particles produce syncytial plaques in monolayers. Gene F12L product protein transports the virus particle through microtubules to the cell surface. Genes A33R and A34R render the virus particle infectious. The viral gene B8R encodes a 43kDa glycoprotein, which is secreted from infected cells and displays homology for amino acid sequence in a manner similar to the extracellular domain of the INFγ-receptor. It binds INFγ and neutralizes it, thus depriving the host of its molecular mediator that would have initiated Th1-type immune reactions. The C12L viral gene product protein binds IL-18, which would have otherwise interacted with IL-12 to further initiate Th1-type immune reactions. Viral gene N1L encodes the 14kDa virulence factor. Virus particles lacking N1L genes are fully attenuated. Viral genes B13R and B22R encode serpins that inhibit the IL-1β–converting enzyme and prevent the cell from committing TNF- or Fas-induced programmed death. Viral genes E3L and K3L encode proteins that suppress IFN-induced antiviral cellular proteins. Otherwise, the antiviral state of the cell is set by the IFN-induced serin-threonine kinase PKR and the 2′–5′ oligoadenylate synthetase. Viral gene A41L encodes a chemokine-binding protein. Early gene expression of VV induces apoptotic death of macrophages and dendritic cells by suppressing the anti-apoptotic protein Bcl_{XL}. Interferon-γ does not, but Bcl-2 can protect against VV-induced apoptosis. The serine protease inhibitor serpin crmA is a prominent NFκB-activator and anti-apoptotic protein produced by poxviruses [282–288]. Monoclonal antibody C3, aimed at the 14kDa trimeric VV membrane protein, neutralizes the virus as it emerges from the cytoplasm of its host cell [289]. Vaccinia virus and other pox viruses nonpathogenic in the human host are being considered for gene delivery systems for gene therapy of human cancers [290–294].

Vaccinia virus elicits strong oncolytic activity or augments in syngeneic hosts the immunogenicity of tumor cells expressing both tumor-associated transplantation antigens and VV antigens [281]. Replication-selective oncolytic VV strains are deprived of their thymidine kinase and vaccinia growth factor genes [294]. Genetically engineered VV contain IL-4 and IL-10 genes dictating that the host develop not Th1- but Th2-type immune reaction with much delayed viral clearance [295]. The GM-CSF gene-containing VV induced remissions of melanoma skin nodules after direct intratumoral injections [296]. The TA-HPV recombinant VV contains the E6 and E7 genes of HPV-18 and immunizes (mice) against HPV infection [297]. The gene of the overexpressed tumor-associated antigen 5T4 could be inserted into the genome of the attenuated Ankara strain of VV. Mice immunized with this virus rejected murine and human tumor xenografts [298].

When tested in Australia, the Wistar Institute VV melanoma oncolysate showed only a favorable trend in prolonging survival of patients, but without significance (P. Hersey, in this volume). Overall survival was 151 months in treated patients and 88 months in the control patient group. The 5- and 10-year survival rates for treated and control patients were 60% and 53% vs. 55% and 41%, respectively. Median relapse-free survival was 83 months in treated and 43 months in control patients. By the fifth year, these figures fell to 51% for the treated and 47% for the control group [299–301].

Herpes Viruses

The icosahedral, enveloped, linear dsDNA-containing herpes viruses with a genome size of 150–200 kb are widely distributed in nature exhibit both oncogenic (EBV and HHV-8) [73–76,302–305] and oncolytic [306] potentials, and evade defensive reactions of their hosts at many levels. Two strains of herpes viruses, rendered avirulent through genetic exchange and fusion, may create virulent recombinants [307]. The coevolution and coexistence of herpes and retro(lenti)viruses of African origin are most remarkable and the presence of both retroproviral DNA sequences, in addition to HHV-8, in Kaposi's sarcoma cells has been postulated [308] (Figures 12A,12B,12C).

Herpes viruses can persist in a latent form. Herpes viruses abrogate or evade immune reactions of the host at many levels. Virally encoded complement and Fc decoy receptors, downregulated major histocompatibility complex (MHC) class I and II antigens, and cell surface adhesion molecules (ICAM-1 and LFA-3) appear, and suppressed viral antigen processing and presentation to immune T cells are readily practiced by herpes virions. Infected cell protein ICP47 disrupts normal peptide loading on MHC class I molecules, thus averting attacks by cytotoxic T cells. The synthesis of MHC class II molecules is suppressed by viral gene product proteins Vhs and ICP34.5 from genes U_L41 and $\gamma_1 34.5$. Herpes viruses express glycoproteins having homology to host cell glycoproteins, including those that imitate G protein-coupled receptors. There is a virally encoded tolerogen and immunosuppressive cytokine—IL-10.

The U_L40 gene in cytomegalovirus (CMV) codes for a glycoprotein that imitates the structure of the cell surface antigen human leukocyte antigen (HLA)-E, allowing the virus to incapacitate NK cell receptors CD94/NKG2A and thereby avert NK cell attack on virally infected cells. Cytomegalovirus also downregulates epidermal growth factor (EGF) receptor expression and signal conduction (autophosphorylation) of the tyrosine kinase, possibly a pro-apoptotic event.

Herpes simplex virus types 1 and 2 block apoptotic response of infected cells, even under conditions of experimental induction by TNFα, anti-Fas antibody, C_2-ceramide or thermal shock. Herpes simplex virus constructs with deleted genes $\alpha 4$ and U_s3 do not protect infected cells against apoptosis. Apoptosis triggered by the $\alpha 4$-defective herpes virus can be blocked by Bcl-2. The virally encoded protein U_s3 neutralizes the proapoptotic Bcl-2-associated cellular protein BAD by forming a complex with it, thus preventing programmed cell death. The ICP27 and ICP4 viral proteins of the wild-type HSV-1 are necessary to maintain intracellular levels of bcl-2 mRNA and protein. In an attempt to block apoptosis, deletion mutants $ICP27^-$ and $ICP4^-$ HSV-1 viruses fail to counteract the host cell's apoptotic response to viral entry.

Human herpes virus-6, the causative agent of roseola infantum and exanthema subitum, encodes counterfeit and secreted (solubilized) receptors for chemokines RANTES (regulated activation normal T cell-expressed and secreted) and eotaxin-β. In this case, the infected cell will not bind chemokines and will not attract immune T cells, NK cells, or macrophages. The viral *lat* gene codes for the anti-apoptotic Lat protein, permitting survival of the infected cell until completion of the viral replicative circle and full maturation of the new viral progeny [309–323]. In mice, HSV-2 acts as a double agent. Replicative viral infection exerts an oncolytic effect on cellular transplants of a syngeneic tumor, whereas enhanced tumor growth occurs in mice harboring latent, nonreplicating HSV-2 infection [324].

Of the 152 kb herpes virus genome that codes for some 89 proteins, 30 kb is replaceable (dispensible for viral replication). Thymidine kinase-defective HSV *dl*sptk replicates well in rapidly growing glioma cells that abundantly express cellular *tk* gene, whereas resting brain cells resist the virus, although the virus retains its neurotoxicity and acyclovir fails to inhibit its replication. By deleting the $\gamma_1$34,5 viral gene, HSV neurovirulence decreases, but the virus retains its selectivity to replicate in malignantly transformed nerve cells. On the other hand, glioma cells (and many other cancer cells) transduced with HSV *tk* gene by retroviral vectors may be killed by acyclovir-ganciclovir, ionizing radiation, or bromovinyldeoxyuridine [325–335]. In addition to directly attacked tumor cells, "bystander tumor cells" may also succumb [336–338]. This system worked well against HSV *tk* gene-transduced lung metastases of murine melanoma, or glioma grafts, or Lewis lung carcinoma in rats [339–341].

The G207 mutant of the parent HSV R3616 (MediGene, Inc., San Diego, CA) is defective for the $\gamma_1$34.5 (neurovirulence) genes and contains an inactivated viral ribonucleotide reductase (ICP6) subunit. The cellular counterpart of this enzyme is overexpressed in rapidly replicating tumor cells, where G207 replicates with preference over resting cells and elicits immune T cell reactions of the murine host to infected tumor cells. When the cellular enzyme is inhibited with hydroxyurea, the virus loses its oncolytic potential and ionizing radiation fails to enhance it. Ionizing radiation or cisplatin promote viral oncolysis in xenografts of human uterine cervical carcinomas [342,343]. The G207 mutant virus cannot replicate in the brain of owl (*Aotus*) monkeys, which are highly susceptible to wild type HSV. G207 is in clinical trials for human malignant gliomas with or without adjuvant radiotherapy and for colorectal, breast, ovarian, and prostate cancers. In a group of patients with high-grade gliomas who were treated intratumorally, two survived 4 years and by autopsy a patient who died of cerebral infarction showed no residual glioma. The virus did not cause encephalitis in any of the patients (additional literature cited: [331–335,344–346]).

Cytomegalovirus replicates in human rhabdomyosarcoma cells. Both in Europe and in San Diego, CA, G207 also destroyed childhood rhabdomyosarcoma xenografts [347–349] (but when we applied for its use in one patient, our request was not granted). The fusogenic G207, Fu-10, spreads through syncytia formed by coalescing tumor cells and escapes antiviral antibodies, while showing an increased oncolytic potency [350]. The Synco-2D herpes virus is doubly fusogenic in that it encodes and produces the hyperfusogenic glycoprotein of the gibbon ape leukemia virus, enabling it to eradicate human ovarian carcinoma xenografts in athymic *nu/nu* mice [351] (Fu, in this volume).

The NV1020 HSV construct (R7020, MediGene, Inc., San Diego, CA) with major (700 bp) genomic deletions lacks its endogenous *tk* gene, one copy of its two $\gamma_1 34.5$ genes and the joint region of the long and short components of its genome, while retaining its ICP6 region. An exogenous copy of viral *tk* gene was inserted under the control the viral $\alpha 4$ gene acting as a promoter. So constructed, NV1020 destroys xenografted human gastric, pancreatic, prostatic, and squamous cell carcinomas and synergizes with ionizing radiation. In investigational protocols, the engineered virus is being injected intra-arterially into liver metastases of colorectal carcinomas.

The ICP4 gene of HSV mutant G92a is under the transcriptional control of the albumin gene. This gene is overexpressed in hepatoma cells, where G29a replicates to 10^4-fold higher titers than in healthy hepatocytes [331–335]. Cyclophosphamide (CTX) synergizes with oncolytic HSV in more than one way. Herpes simplex virus particles containing the cytochrome p450 oxidase gene activate intracellular CTX for antitumor effects, and CTX exerts an immunosuppressive effect on the host, resulting in suppressed innate immune reactions and low virus-neutralizing antibody titers, allowing for cell-to-cell spread of the oncolytic virus. Complement deprivation achieves similar effects [352,353]. HSV1716 with deleted ICP34.5 protein selectively grows in rapidly dividing tumor cells (melanoma) [354]. Upon direct injection into high-grade gliomas, this deletion mutant induced tumor regressions with greater than 3-year survivals of three of nine patients so treated (at the University of Glasgow, UK) [355,356]. When the two replication-competent multimutants G207 and NV1020 herpes viruses were compared in intraperitoneally growing human gastric cancer xenografts, NV1020 performed better. Against prostatic adenocarcinoma xenografts, both G207 and NV1020 performed equally well [346]. NV1020 kills human head and neck squamous carcinoma cell lines in vitro and the xenografts of these tumors in vivo [357].

Herpes viruses enlisted for gene therapy of human cancers deliver genes into tumor cells—CEA, GM-CSF, IL-2, and the cyclophosphamide-activator transgene CYP2B1. Incorporated are in amplicon plasmids, endostatin or secondary lymphoid tissue chemokine, and CD ligand for the generation of augmented Th1-type antitumor immunity [358–364]. In Hungary, a ribonucleotide reductase null mutant of the pseudorabies (Aujeszky) virus replicated in, and destroyed, human neuroblastoma, glioblastoma, and hepatoma cells and it is considered to be a gene delivery vector for the gene therapy of human cancers [365]. Replication-competent herpes viruses are oncolytic and spread within the tumor. For gene therapy, replication-defective mutant herpes viruses are used as delivery vectors.

Adenoviruses

The naked icosahedral adenoviruses possess linear dsDNA of 38 kb and are capable of killing cells by lysis, or malignantly transforming them, although adenoviruses are not known to cause human cancers. Early after viral entry through the coxsackie-adenovirus receptor (CAR) capturing the fiber knob of, and internalizing, the adenovirus type 5 particle, apoptotic cell death is triggered by viral protein E1A. However, the virus has evolved anti-apoptotic and anti-IFN strategies to spare the host cell until the new mature viral progeny exits.

In human cancer cells of the urinary bladder [366], CAR functions as an intercellular adhesion molecule and elicits a growth inhibitory cascade. Tumor cells with low CAR expression grew faster than tumor cells with high CAR expression. Low-grade prostatic carcinoma cells underexpress CAR, whereas its metastasizing cells display strong CAR expression [367]. Like cadherins, CAR may be tumor suppressing in the organ of origin for the tumor, but tumor promoting in an environment distant from the organ in which the tumor originated. In astrocytomas [368], high-grade tumors expressed less CAR than low-grade tumors.

In addition to CAR, adenovirus particles also attach to MHC class II molecules and α-integrins (the penton base of Ad2). Although human prostate cancer cells express CAR, CAR may not be overexpressed in many other human cancer cells. Solubilized sCAR may inhibit adenovirus particles extracellularly [369]. Adenovirus 5 carrying a gene of adenovirus 3 attaches to a receptor different from CAR and can enter human ovarian carcinoma cells with downregulated CAR expression [370]. As a conditionally replicating adenovirus, AdsCAR-EGF enters human tumor cells through the overexpressed EGF-R [371]. Adenovirus armed with the sCARfC6.5 protein enters cancer cells through the erb2 (HER2/neu) receptor [372]. Coxsackie-adenovirus receptor -independent cellular entry mechanisms through the α-integrins facilitate the infection of human kidney carcinoma cells with retargeted adenoviruses [373]. Fibroblast growth factor (FGF)-Ad containing the HSV*tk* gene targets the basic FGF receptor (bFGF-R) [374], which is overexpressed in proliferating endothelial cells in tumor-induced neoangiogenesis and in certain tumor cells. Adenovirus constructs with an integrin-binding motif (RGD peptide in the fiber knob) enter low CAR-expressor squamous carcinoma cells of the oral cavity [333,361,375–380]. Retention of the E3 region in Ad5-Δ24RGD increased its oncogenic potency [381].

Adenoviral replication-defective ($E1^-$ and $E3^-$) vectors induce immune reactions. For example, the proinflammatory intercellular adhesion molecule ICAM-1 is overexpressed in the infected tissues through activation of mitogen-activated protein kinases (MAP kinases) and (nuclear factor kappa B) NFκB [382]. The Hungarian adenovirus research team has proposed that viral capsid alterations reduce the antigenicity of the virus, thereby delaying the elimination of adenoviral vectors by immune reactions through virus-neutralizing antibodies of the host [383]. Immunostimulatory adenoviral vectors have also been constructed [384]. The $E1^-/E3^-$ Ad4-1BBL construct carrying the gene for the TNF family ligand 4-1BBL, infects human adenocarcinoma cells, which attract T-LAK cells that are highly cytotoxic for the tumor cells and eradicate tumor xenografts. In addition to NK cells [385,386], immune T cells cytotoxic for adenovirus-infected cells are also mobilized [387,388].

The relationship of adenovirus-infected tumor cells and macrophages is very complicated. Adenoviruses induce dendritic cell maturation [389,390] whereas viral E1A protein sensitizes tumor cells to apoptosis by TNFα of macrophage origin. However, E1A also activates heat shock proteins that antagonize TNFα-mediated apoptosis [391,392]. It is possible to generate cytotoxic T cells reacting specifically with E1A proteins, therefore attacking and destroying adenovirally induced tumors [387]. The adenovirus IFNγ gene construct TG1041 is expected to upregulate MHC class I and II expression in melanoma cells, rendering the infected cells more immunogenic [393]. E1A proteins also block IFN-induced gene expressions [394]. Although IFNs promote NK cell-mediated lysis of adenovirus-infected cells, they protect noninfected cells from NK cell-mediated cytotoxicity. This is the phenomenon of IFN-mediated resistance to cytolysis or cytoprotection (IFN-MCP). E1A (expression of its first exon) counteract IFN-MCP [386].

Soon after viral entry, the tumor suppressor gene for the Rb protein is inactivated [395–397]. As the normal cell cycle approaches S phase, the Rb protein is in the nucleus bound to cellular transcription activator family proteins E2F. Adenoviral E1A protein binds to the Rb protein that releases E2F, making it available to induce transcription of cellular genes for DNA synthesis initiation—especially that of the viral DNA. The construct AdE2F is a carrier of the E2f gene. The virus is highly cytotoxic for human non–small-cell lung cancer cells. In Ad(E2F)-1^{RC}, the E1A gene is under the control of the E2F-l promoter element. Quiescent normal cells do not express E2F, whereas tumor cells overexpress E2F and support viral replication with high efficiency [398–400]. The fusion protein encoded by a gene located in the viral genome between the E2F-binding Rb protein and the DNA-binding domain of E2F inhibits DNA replication.

The gene for the fusion protein was placed under the promoter of p53 in the genome of the 01/PEME adenovirus construct (Canji) with deletion 1101 from its E1A sequence. Its inhibitory effect on p53 is therefore disabled [401]. As expected, this virus construct will not grow in normal cells expressing wild-type p53.

The gene of the adenovirus protein E3–11.6K was placed under the control of the major late viral promoter to achieve an intensified late cytotoxic effect in tumor cells. 01/PEME *in vitro* and in xenografted human tumors surpasses the cytotoxicity of *E1B*- deleted *dl* 1520 viruses. CyclinD/CDK4 and CDK6 (cyclin-dependent kinases) phosphorylate the Rb protein, rendering it unable to arrest the cell cycle [396]. Both mutated cyclins and CDK can act as oncogenes, whereas CDK inhibitors such as p16 can act as tumor suppressors. When such CDK inhibitors are mutated (independently from adenoviral infection, as in many human tumors, such as melanoma, lymphoma, breast cancer, etc.), oncogenesis is promoted. The Ad-PML construct delivers an intact tumor suppressor gene, the one that is translocated in acute promyelocytic leukemia—t(15;17)—to breast carcinoma cells. These tumor cells failed to grow as xenografts in nude mice and showed cell cycle arrest at G1, decreased levels of cyclin D1 and CDK2, and increased p21 CDK-inhibitor activities. The Rb protein became dephosphorylated, making it able again to stop the cell cycle [402].

The adenovirus receptor internalization and degradation protein [403] is anti-apoptotic. It antagonizes TNF-induced apoptosis and internalizes the Fas receptor, surrendering it to degradation within lysosomes. However, it also internalizes EGF-R, depriving a tumor cell of EGF-induced mitotic stimuli. The adenovirus proteins E1B (19K, 55K, 14.7K0) act as Bcl-2 homologues and inhibit extrinsic apoptosis initiated from receptors Fas or TNFα, activators of FADD (Fas-associating protein with death domain) but not of FLICE (FADD-like IL-1β converting enzyme) [404]. E1B 14.7 antagonizes TNFα. E1B55K inactivates p53 protein encoded by gene on chromosome 17p13. Whereas E1A increases the activity of p53 (pro-apoptotic events), the 55 kDa E1B and E4ORF6 proteins bind p53 at two sites, remove it from the nucleus and degrade it (anti-apoptotic events). One of the viral E4 gene product proteins, E4ORF4, causes apoptotic cell death independently from E1A or p53, unless counteracted by Bcl-2 or Bcl_{XL}.

Delivery of Fas ligand (L) or FADD protein genes by adenovirus constructs (Adeno-FL; Adeno-FADD) into human glioblastoma cells expressing Fas but resisting anti-Fas antibody-induced apoptosis rendered these cells highly prone to programmed cell death [405,406]. Many human tumors downregulate Fas receptor expression but express FasL and kill Fas CD95 receptor-positive defensive host immune T cells. Colon carcinoma cells expressing Fas receptor succumbed to programmed cell death upon intracellular expression of adenovirus-encoded FasL and attracted inflammatory cell infiltrations. Adenovirus-mediated transfer of FasL expressed internally, induced apoptotic death of prostatic carcinoma cells [407]. When transferred into endothelial cells by adenovirus, FasL inhibited neovascularization [408]. Adenovirus type 2 early transcription unit 3 protein 14.7ORF downregulates Fas receptor expression at the tumor cell surface, exerting its anti-apoptotic effect [409,410].

The death domain containing protein p84N5 triggers apoptosis from within the nucleus; its gene N5, delivered by adenovirus AdN5, induced apoptosis in tissue cultures and xenografts of human tumors, but spared normal diploid fibroblasts [411,412].

The Bcl-associated BAX is a pro-apoptotic member of the Bcl-2 family. Its gene is driven by vascular endothel growth factor (VEGF) overexpressed in lung cancer cells. The *bax* gene is incorporated into the adenoviral genome together with the VEGF promoter element to form the AdVEGFBAX construct, which induces apoptotic death of human lung cancer cells, whereas it spares normal bronchial epithelial cells that do not overexpress the VEGF gene [413].

Recombinant adenoviruses (rAd-EGFP) infecting dendritic cells elicit rapid antibody, NK, and immune T cell defensive reactions in the murine host, and MHC and costimulatory molecules also incease. Levels of cytokines, IL-6, IL-12p40, IL-15, IFNγ and TNFα rise, whereas IL-10, IL-12p35, and TGFβ tend to decrease. Cytoplasmic NF-κB rapidly translocates to the nucleus, a reaction inhibited by an adenovirus encoding a mutant IκBα gene [389]. The adenovirus Ad.IkappaBM triggers apoptosis of hepatocytes, which is counteracted by Bcl-2, allowing for viral persistence through suppression of innate defensive reactions. Adenoviruses produce VA-RNA I to block PKR and escape the antiviral effect of IFNα [414,415] (similarly to NS1 of influenza virus, σ3 protein of reovirus, and E3L and K3L proteins of VV). The E1A gene product protein suppresses malignant transformation mediated by the *neu* oncogene [416]; additional literature cited [417–419].

Of innumerable adenoviral constructs created for oncolytic activity, none emerges as being exclusively the best; the ONYX construct is a prototype that has been clinically tested most frequently. The dl1520 adenovirus (ONYX-015; CI-1042) (McCormick, in this volume; and extensively reviewed elsewhere [333,361,420–427]) replicates without its *E1B* gene and cannot produce its E1B55K protein (which would bind p53). Its *E1B19K* gene and its Bcl-2 homologue gene-product protein remain intact. In human malignant glioma xenografts ONYX is oncolytic independently from the p53 status of the tumors [428]. In head and neck squamous cell carcinomas, it induced major tumor regressions (additively to or synergistically with cisplatin and radiotherapy) [429,430]. Intraarterial administration of ONYX for the treatment of gastrointestinal adenocarcinomas metastatic to the liver in addition to 5-FU and leucovorin induced 11% PR and 37% MR with stabilization of disease [431]. Intratumoral injection of ONYX into hepatobiliary tumors induced 6 % PR and 56 % MR and stabilization of disease [432]. In a patient with adenocarcinoma of the gall bladder, it was proven by RT-PCR (real time polymerase chain reaction) aimed at viral hexon expression that the virus replicated in the injected host both in tumor and stromal cells [433]. Another E1B-55kDa-deleted adenovirus (YKL-1) killed human gastric carcinoma cells in a p53-independent manner; that is, even cancer cells with wild p53 still functional failed to resist viral replication. Possibly, loss of the alternative reading frame protein $p14^{ARF}$ also allows E1B 55 kDa-deleted adenoviruses to replicate, inasmuch as $p14^{ARF}$ interacts with Mdm2, the major antagonist of the p53 protein. E1B 19 kDa protein inhibits E1B 55 kDa protein-induced and p53-induced apoptosis. E1B 55 kDa-deleted adenoviruses (Ad-ΔE1B19; Ad-ΔE1B19/55) exhibit greatly increased apoptosis-inducing efficacy and are awaiting clinical trials [434,435].

E1A exerts tumor-suppressive effects on the HER2/neu proto-oncogene with differentiation–induction, reduced metastatic potential, and increased susceptibility to TNFα-induced apoptosis [436,437]. While E1A may act as an oncogene product protein in rodent cells, in contrast, in human cells it exerts tumor suppressive activity. Human rhabdo- and osteosarcoma cells, and lung and breast cancer cells transfected with retrovirally inserted *E1a* gene showed attenuation of their malignant phenotype and their growth rate was suppressed. Ad5*dl*01/07 produces a defective E1A protein and triggers a strong apoptotic response in infected tumor cells, thereby inhibiting the growth of xenografted human tumors.

When the human tyrosinase promoter is used to replace the adenoviral E1A promoter in the viral genome, the genetically engineered adenovirus (AdTyrΔ24 and AdTyrΔ2Δ24) becomes highly cytotoxic for melanoma cells with a high yield of viral progeny in melanoma cells and low yield, if any, in other (control) cells [438–440]. Estrogen- and hypoxia-responsive promoters and telomerase transcriptase promoters of the *E1a* gene drive adenoviruses AdEHT2 and AdEHE2F in ER^+ human breast cancer cells [400]. The *E1a* gene of the Delta24 adenovirus with a 24 bp deletion codes for a protein that cannot inactivate the Rb protein in normal cells, disallowing viral replication. In Rb-defective cells, such as high-grade gliomas, the virus replicates without restraint.

Ad5-Delta24 destroyed human glioma cells in the brains of nude mice without causing encephalitis. The Delta24 virus expresses arginine-glycine-aspartic acid (RGD) peptides for high affinity to α-integrins that are overexpressed in glioma cells. Delta24 is in clinical trial at the M.D. Anderson Cancer Center [441]. Ad E3 sequence encodes the adenovirus death protein activated at the end of the viral cycle to dissolve the cell and allow lateral dispersion of the new viral progeny [381]. The $E3^+$ Ad5-Δ24RGD with deletion of base pairs 923 to 946 of the E1A sequence was much more cytotoxic for human cancer cells than its $E3^-$ variant. Ad5-Δ24RGD suppresses the growth of human glioma xenografts in nude mice and synergizes with radiotherapy. Adenoviruses with mutated *E1a* gene replicate only in tumor cells with E1A-like activity, such as in HeLa cells, undifferentiated embryonic carcinoma cells, or hepatocellular carcinoma cells. As shown at Baylor in Houston, TX, *E1a*-mutated adenoviruses destroyed hepatocellular carcinoma xenografts [442]. At this stage, it is not possible to omit and ignore any of the adenoviral constructs: One may score extraordinarily well in a future clinical trial.

Intravenous (i.v.) infusion of ONYX-015 in combination with chemotherapy (irinotecan and 5-FU) and IL-2 induced both stabilization of disease and PR in patients with advanced malignancies as well as virus-neutralizing antibody production; however, virus particles accessed tumor masses. For patients with chemotherapy-refractory metastatic colorectal cancers, ONYX-015 stabilized the course of disease for 11 to18 weeks [443].

Of the attenuated replication-competent adenoviruses, agents CV706 $E3^-$ and CV787 $E3^+$ destroy xenografted human prostatic carcinoma cells [444]. In these viruses, encoding of the E1A protein is under the direction of a PSA gene promoter-enhancer. Agent CV787 eradicated prostate cancer xenografts after one i.v. injection. Unlike Ad-sPSA-E1, replicating only in PSA^+ tumor cells, Ad-hOC-E1 is under the promoter osteocalcin and it replicates also in PSA^- or PSA^+ androgen-dependent and independent prostate cancer cells [445]. The replication-competent adenovirus Ad5-CD/TK*rep* inserts a fused HSV *tk* gene and a cytosine deaminase (CD) gene into the genome of transfected prostate cancer cells after direct injection of the virus into the tumors. Thereafter, patients receiving 5-fluorocytosine (5-FC) and ganciclovir experienced reduction of tumor size and a drop of PSA [446]. Post-treatment prostatic tumor biopsies showed extensive coagulation necrosis and two patients remained in CR 1 year after treatment. Metastatic androgen-resistant prostate cancer cells show extensive invaginations of their cell membrane, in contrast to healthy prostatic cells, which are free of this development. These tumor cells overproduce PSA, resist apoptosis, gain androgen-independence, or continue to use testosterone as a growth hormone and overexpress the gene *cav*-1 that encodes the proteins caveolae [447].

Tumors other than those of the prostate (breast, colon, and ovarian carcinomas) also overexpress caveolae and gain multiple drug resistance. Endothelial cells undergoing neoangiogenesis also overexpress caveolae and show extensive invaginations of their cell membranes. A murine promoter of the *cav*-1 gene was inserted into the genome of an adenovirus upstream to the HSV *tk* gene. Adcav-1*tk* was thus created. Upon treatment with ganciclovir, in these hosts orthotopically grafted with prostate cancer cells that were transfected with Adcav-1*tk*, massive necrosis of the tumor cells occurred. Human tumor cells not in S phase may escape cytotoxicity of the HSV *tk* gene-ganciclovir (TK-GCV) system; when overexpressed, however, adenoviral E1A protein drives the cell toward S phase and cell-killing efficacy of the TK-GCV system increases [448].

Ovarian carcinoma cells are often deficient in the expression of CAR. The Ad5/3luc1 virus possesses a chimeric Ad5 fiber protein with an Ad3 knob domain directed at the Ad3 integrin receptors, allowing it to enter CAR^- ovarian carcinoma cells [449]. Another agent, Ad-Δ24RGD, readily infects the CAR^- ovarian carcinoma cell through integrins. AdCMVHSV-TK delivers the *tk* gene (driven by CMV or Rous sarcoma virus promoters) into ovarian carcinoma cells and renders the tumor cells highly susceptible to ganciclovir [450]. The transmembrane protein CD40 appears overexpressed in many tumors, including ovarian carcinomas. Its natural ligand, CD154 of the TNF family, may induce a growth inhibitory effect [451]. The fusion protein CAR/G28 ligand targets CD40 and, when expressed in the construct Ad5luc1-CAR/G28, it facilitates viral infection of ovarian carcinoma cells, which undergo apoptosis and gain increased susceptibility to cisplatin [452,453]. Good basic knowledge of cancer cell biology and congenial thinking led to the creation of these adenoviral constructs, which are poised to spring into wide clinical application.

Adenoviral transfer of the HSV *tk* gene into anaplastic lymphoma cells renders these cells ganciclovir-sensitive [454]. Epstein-Barr virus immediate early protein encoding genes BZLF1 and BRLF1 convert the latent EBV state into lytic infection. Adenoviral vectors delivering these genes into EBV^+ gastric carcinoma cells induced tumor cell lysis [455]. This type of intervention may possibly work in EBV^+ lymphomas (African Burkitt's lymphoma, Hodgkin's disease, primary brain lymphoma in AIDS). Recombinant adenoviral vectors delivering HSV*tk* gene into human lung and other solid tumors, including adrenocortical carcinoma [456–458], sensitize the targeted tumor cells and some bystander cells to ganciclovir-mediated lysis. The aziridine-cyclobenzamide (CB1954) is converted by the *Escherichia coli* enzyme nitroimidazole reductase (NTR) into a cytotoxic alkylator. The delivery of the NTR gene by adenovirus into primary or metastatic liver tumors is expected to render these cells highly susceptible to i.v. (or intraarterial) administration of CB1954 [459].

Rabbit carboxylesterase (rCE) converts the camptothecin prodrug, irinotecan, into its cytotoxic metabolite. The rCE gene under the influence of the Rous sarcoma virus (RSV) promoter is delivered by AdRSVrCE into neuroblastoma cell, with tumor cell death occurring upon treatment with irinotecan [460]. This treatment spares CD34 stem cells and purges stem cell preparations (collected before high-dose myeloablative chemotherapy) of tumor cells before its reinfusion into patients for rescue.

Ad-CMV-CD contains the cytosine deaminase (CD) transgene driven by a cytomegalovirus promoter; specifically, its protein product enzyme CD sensitizes breast cancer cells to 5-FC. This strategy has been proposed for purging bone marrow stem cell preparations, which are to be reinfused into patients after myeloablative high-dose chemotherapy [461]. Ad-CD vectors infect pancreatic and prostatic carcinoma cells. As few as 400 vector particles per cell render these cells highly susceptible to 5-FC [462,463].

Replication-competent adenoviral vectors carrying either L-plastin or CMV promoters for an inserted CD gene and for the viral E1A were designed to contain also an intraribosomal entry site (IRES) sequence (AdCMVCDIRESE1A; dLpCDIRESE1A). These constructs were cytotoxic for infected and for bystander human breast, ovarian, and colon cancer cells and FC increased cytotoxicity in an additive fashion, whereas normal parenchymal and stromal cells were spared [464].

Adenovirus 5 variant CV890 (Calydon, Sunnyvale, CA) is allowed to express both E1A and E1B from a single transcription response element whilst an encephalomyocarditis virus element regulates its internal ribosome entry site. An α-fetoprotein (AFP) enhancer-promoter element amplified from human genomic DNA drives its replication. In non-AFP-producer cells, there is minimal viral replication, but in AFP-producer cells, viral replication increases up to 10^5-fold. CV890 replicates to cytotoxicity in human hepatocellular carcinoma cells and synergizes with doxorubicin [465]. Adenovirus 5 CG8840 (Cell Genesys, San Francisco, CA) is attenuated and replication competent. Its E1A and E1B genes are driven by an uroplakin promoter deriving from the uroplakin gene *UPII*, which is specifically expressed only in cells of the urothelium. CG8840 replicates to cytotoxicity 10^4-fold better in transitional carcinoma cells of the urinary bladder than in glioma, melanoma, prostatic or ovarian carcinoma cells, or normal fibroblasts. The virus is effective in combination with docetaxel [466]. The clinical performance of these adenovirus constructs is expected to be highly additive, if not synergistic, to current treatment modalities for cancers of the genitourinary tract.

Adenovirus vectors are constructed for the gene therapy of human cancers (mentioned frequently, but not reviewed in this volume; extensively reviewed elsewhere-[39-43, 333,361, 410,417-419]). Wild-type p53 gene adenovirally transferred into human tumors in vitro or in xenografts suppressed tumor growth or even induced apoptotic deaths of infected tumor cells, but in Phase I clinical trials, apart from inducing remissions in rare individual cases, no cures resulted. Stagnation (stabilization) of disease progression is the best result so far recorded.

In glioblastoma, p53 mutations are not essential for the malignant transformation [467,468]. In head and neck squamous cell or nasopharyngeal carcinomas [469–471] and in lung cancers [472,473], synergism of adenoviral vectors carrying wild-type p53 with chemoradiotherapy is evident in xenografts [472] but not in clinical trials [473]. In breast cancer [474], reinsertion of wild-type p53 by adenoviral vector sensitized the tumor cells to TNF-induced apoptotic death. In prostate or urinary bladder carcinomas [475–479], intratumoral injections or intravesical instillations of adenoviral vectors armed with wild-type p53 show favorable clinical results [479]. Controversies between preclinical assays showing efficacy of viral vectors against human tumor xenografts, but limited performance of the same agents in subsequent clinical trials, call for further efforts to explain the reason for the discrepancy and to reconcile it in favor of successsful clinical trials [480,481]. For example, in a comparison of wild-type p53 and adenoviral death protein expression for killing lung cancer cells, the replicating Adp53rc performed better [482]. Therefore, replication-competent vectors should be favored, however, these vectors must remain harmless in normal tissues of the host.

The actin-regulatory protein gelsolin is present in normal, but it is absent in cancerous epithelial cells. When the gelsolin gene was inserted by adenoviral vector Ad-GSN into the genome of human bladder cancer cells orthotopically xenografted into the bladders of nude mice, tumor growth was suppressed [483]. The construct Ad-RB94 inserts the gene of the truncated (short of 112 amino acids at its NH_2 terminal) retinoblastoma protein p94 kDa into squamous cell carcinomas of the head and neck. Xenografts of these tumors failed to grow [484]. The PTEN (phosphatase on chromosome ten) tumor suppressor gene, when reinduced into prostatic cancinoma cells by an adenoviral vector, induced no tumor cell death, but blocked the metastatic potential of the tumor cells [485]. The gene of the IFN-inducible protein p202, when inserted into breast cancer cells by an adenoviral vector, rendered the tumor cells highly susceptible to TNF-induced apoptosis [486]. The cDNA of the oncoprotein p14ARF incorporated into an adenoviral vector infecting MCF-7 breast cancer cells sensitized the tumor cells to cisplatin [487].

Further efforts concern the delivery of lymphokine (IFNs and ILs) and cytokine (hematopoietic growth factors, such as GM-CSF) genes into tumor cells. Anti-neoangiogenesis (Ad-mEndo for endostatin) [488] genes, immunostimulatory co-factor (B7s, heat shock proteins; MHC) genes, cell suicide genes, or these genes combined in the same adenoviral vector (such as endostatin and *tk* genes for the treatment of renal cancer) and delivered into human tumors of all cell types follow suit. In many cases, the genes were shown to resume their normal function, directing the tumor cells toward differentiation or apoptotic death and increasing the tumor cells' susceptibility to ionizing irradiation, to cytotoxic chemotherapy, or to immune attack by host antibodies, NK cells, and immune T cells [489]. Antigens released from apoptotic (but not from necrotic) tumor cells undergoing programmed cell death are re-expressed not in a tolerogenic but in an immunogenic manner by phagocytic macrophages and mature dendritic cells, thereby inducing a form of postoncolytic antitumor immunity [490–494].

What dangers and side effects do these new therapeutic measures entail, such as insertion of viral genes that may transactivate proto-oncogenes in the host cell? Pathogenicity of replication-competent viral agents may be regained due to recombination with a latent virus harbored in the host. What possible countermeasures might the tumor be able to mobilize to escape destruction by these new therapeutic interventions? It is apparent that no single intervention will kill out the entire tumor cell population in its natural environment (the tumor-bearing host) and that combination therapy involving viral vectors with antitumor immunological attacks or chemoradiotherapy will be necessary for the induction of durable CR that may eventually lead to cure.

Parvoviruses

Parvoviruses are small, icosahedral particles containing a ssDNA linear genome of 5 kb with terminal palindromes. Whereas adenoviruses infect quiescent cells, parvoviruses require cells in the S phase of their cycle for replication. Second-strand synthesis of the viral DNA and completion of a lytic infection of the cell are accomplished. The left viral gene driven by promoter P4 encodes nonstructural proteins NS1 and NS2 acting as nuclear phosphoproteins and regulators of viral capsid assembly; the right viral gene driven by promoter P38 encodes structural proteins VP1 and VP 2, forming the viral capsid.

Autonomous parvoviruses do not require a helper virus for their replication. Dependoviruses, like the adenoassociated virus (AAV), integrate their genomes into host cell DNA sequences and lay dormant without causing disease until an unrelated helper virus (adeno- or herpes virus) superinfects the cell and induces the parvoviral genome to replicate itself [495–499]. Heather Mayor at the Melnick Department of Virology, Baylor University, Houston, TX, concluded that harboring some parvoviral genomes was "good for your health" [500], inasmuch as these agents suppress oncogenesis. But certainly not that of parvovirus B19: This is the causative agent of erythema infectiosum in children and induces hemolytic anemia with aplastic crisis in the bone marrow [501–505]. The possible association of parvoviruses as causally involved agents in rheumatoid arthritis [506] remains uninvestigated.

Minute virus of mice (MVM) existing as prototype and immunosuppressive agents (MVMp and MVMi) are able to infect and kill human tumor cells. The MVMp autonomous parvovirus destroys xenografts of human tumors. Armed with the anti-angiogenic chemokine, IP-10, this virus was shown to destroy human tumors formed by hemangiosarcoma or Kaposi's sarcoma cells [507] (Rommelaere, in this volume). MVMp was able to transfect human tumor cells with the HSV *tk* gene and render them highly susceptible to ganciclovir [508]. The feline panleukopenia autonomous parvovirus retargeted by insertion of an integrin-binding peptide could infect the human rhabdomyosarcoma cell line Rh18A [509]. Mink plasmacytosis with hyper-gammaglobulinemia or Aleutian mink disease is caused by an autonomous parvovirus.

Viral vaccines containing parvoviral capsid proteins VP1 or VP2 did not prevent but, rather, enhanced disease progression. Antibody-virus immune complexes enter macrophages through the Fc receptor; the virus is not neutralized and infects the macrophage. Vaccines incorporating the virally encoded nonstructural protein NS1 induced CD8 lymphocytes reactive with infected cells and provided partial protection [510–512].

Canine, feline, and porcine parvovirus vaccines are in use in veterinary medicine. In beagle dogs, venereal sarcomas regressed when the animals became infected with the canine parvovirus. Live attenuated feline panleukemia parvovirus vaccine could prevent the outgrowth of canine sarcomas but it was ineffective against established tumors [513]. The chicken anemia parvovirus encodes an apoptosis-inducing 121 amino acid protein, apoptin. This protein elicits p53-independent apoptosis and kills human osteogenic sarcoma and other human tumor cells in vitro [514]. A recombinant H1 parvovirus vector that encodes apoptin protein, a p53-independent and Bcl-2-insensitive apoptotic effector, was shown to act as an apoptosis-inducer but only in tumor cells and not in normal fibroblasts [515]. It remains to be determined whether certain myxo- and paramyxoviruses (influenza A virus; NDV), grown in embryonated chicken eggs that may harbor the chicken anemia parvovirus, exerted unusually fast apoptotic death of tumor cells instead of full replicative cycles because of this undetected contaminant [516].

In the initial clinical tests carried out at Sloan-Kettering, the human autonomous parvovirus H1 was shown to cause viremia but was ineffective against disseminated sarcomas [517]. Later, it was shown that it destroyed human breast cancer [518] or melanoma cells and rendered infected melanoma cells more immunogenic through release of heat shock proteins (HSP72) [519]. Lymphocytes, monocytes, and dendritic cells resisted infection with H1. Hepatocellular carcinoma cells are susceptible to H1, whereas resting hepatocytes resisted infection [520]. The chemotactic CC chemokine protein-3 of monocytic origin (MCP-3), when expressed in tumor cells (osteosarcoma, mastocytoma), elicits Th1-type antitumor immune reactions in immunocompetent hosts. When H1 was used as a vector to deposit the MCP-3 gene in HeLa cells, xenografts of these MCP-3-transduced HeLa cells elicited NK cell- and macrophage-mediated immune reactions in nude mice, leading to rejection of the xenografts [521]. Attempts at the construction of an adenoviral E1A gene-carrier H1 parvovirus vector (H1Ad5E1A) for oncolytic virus therapy of human cancers have not been abandoned [516]. Autonomous parvovirus vectors possess a packaging capacity for foreign DNA sequences not much more than 4.8 kb [522].

The carrier state of AAV may result in suppression of naturally occurring tumors. This concept is deduced from serological data showing low titers of AAV antibodies in 15% of cancer patients, and higher titers in 34% of cancer-free individuals [523]. Women with or without squamous cell carcinoma of the uterine cervix, especially, displayed such differences. In this entity, there is an interplay between AAV and HPV. Human papilloma virus-16 E2 protein can activate latent AAV [524–527]. Therefore, if uterine cervical cells are undergoing malignant transformation by certain anogenital HPV strains and if these cells harbored latent AAV genomes, activation of AAV could result in the suppression of HPV-induced oncogenesis. However, when the dsDNA HPV genome integrates itself into host cell chromosomes, its E2 function is lost and AAV will not be activated. The malignant transformation of the cell will proceed unopposed. If the AAV carrier state involves the testicles, oligospermia and male infertility may result. Adenoassociated virus infection of the female genitalia may contribute to spontaneous abortions [528].

Adenoassociated virus possesses 4700 to 5000 nucleotide bases and integrates preferentially into locus 19q13 [529,530]. Its genome accepts inserted foreign DNA sequences (clusters of differentiation CD antigens; neomycin phosphotransferase gene; cystic fibrosis transmembrane conductance regulator gene, etc.) elevating AAV to the rank of a very promising vector for gene therapy.

Although adenoviral genomic sequences may persist in human lymphocytes, these cells, with or without prior adenoviral exposure, accept genomic insertions by an AAV [531–534]. Angiostatin gene carrier AAV was injected intramuscularly (one single injection) into nude mice harboring growing human glioblastoma xenografts in their brains. Mortality was 100% for untreated mice, but 40% of treated mice survived more than 10 months. Surviving mice secreted high levels of angiostatin in their blood over 250 days. Sacrificed animals showed inhibition of neoangiogenesis in the regressing brain tumors [535]. When activated, the AAV *rep* gene encodes multifunctional Rep proteins, including Rep68 and Rep78, that suppress mitoses of the cell. Rb proteins p107 and p130 also suppress cell divisions. Adenoviruses inactivate these proteins to initiate cell divisions. Cellular transcription factor YY1 [536], an antagonist of TGFβ, suppresses AAV Rep proteins and no transcription for AAV reproduction occurs. Adenovirus E1A interacts with YY1, and liberates Rep.Transcription of AAV for its reproduction is initiated. Human papilloma virus E2 protein also interacts with cellular YY1 and with Rep. Rep proteins also bind to E1A, therefore preventing phosphorylation (inactivation) of Rb proteins, which therefore do not dissociate from E2F, causing the cell cycle to remain arrested. Rep78 interacts with p53 protein and protects it from degradation. Rep proteins 52, 68 and 78 bind E1A. Intact E1A phosphorylates Rb and separates it from E2F. Cell divisions proceed under these conditions (dephosphorylated Rb bound to E2F would block cell divisions) [537].

Transformed cells with intact DNA repair mechanisms and with wild type p53 are resisant to parvovirus infection, whereas DNA repair-defective and p53-deficient cells are susceptible in accepting parvoviral entry and replication. Whereas H1 and AAV-5 do not increase cisplatin sensitivity of tumor cells, AAV-2 does; together they induce apoptosis, but through mitochondrial (endogenous) mechanisms (except in small-cell undifferentiated carcinoma of the lung) rather than through the FasL → Fas system [538].

HeLa cells harboring integrated AAV become more susceptible to gamma rays [539] and undergo apoptosis when treated with anti-Fas immune serum. During this process, the latent AAV is activated and its genomic sequences replicate [540]. Saos p53-null and Rb-null osteosarcoma cells die apoptotic death upon entry of AAV. In contrast, normal osteoblasts with wild p53 and intact Rb pause in G_2 and resist AAV-induced death. When HPV-16 E6 protein degrades p53, AAV infection kills the cell. Apparently, the entry of a special DNA, like the AAV DNA with hairpin structures at both of its ends, induces cell death in tumor suppressor gene-defective cells but not in healthy cells [541]. The "Achilles heel of cancer" has purportedly been found [542].

Polio–Rhinovirus Recombinants

The naked icosahedral ssRNA poliovirus (PV) enters through its receptor CD155 (PVR) by eliminating viral protein VP4 and exposing the hydrophobic viral structural protein VP1 to extrude from the virion's surface as the virion passes through the cell membranes. Poliovirus receptor-related molecules include PRR-1 nectin-1, lymphocyte homing receptor CD44 (only when associated with PVR), and a heparan sulfate. The viral RNA of 7–8 kb is either extracted from the viral capsid in endosomes or is ejected by VP1 through the cell membrane into the cytosol. As a positive strand viral RNA, the poliovirus genome has dual functions. It is either copied into negative strands that serve as templates for the synthesis of positive-strand genomic RNAs and it enters the translational system to act as mRNA and encodes the formation of structural protein capsid pentameres that encapsidate the positive-strand genomic RNA. Poliovirus mRNA is uncapped and its translation occurs on internal ribosomes through IRES (internal ribosome entry sites), which are 400-nucleotide-long segments within the 5′nontranslated region of the viral genome. Translation occurs by binding of a 40S ribosomal subunit to IRES in a 5′-independent and cap-independent manner. Internal ribosome entry sites also have the capacity to encode cell type-specific restrictions regulating viral propagation to occur [543–545].

Meanwhile, a cellular inhibitor of the RNA-activated serine-threonine PKR blocks IFN production. The PKR N-terminal is the regulatory domain activated by dsRNA. Its C-terminus is the catalytic domain. Poliovirus degrades PKR. Poliovirus protein 3A is anti-apoptotic, inasmuch as it inhibits TNF-induced apoptosis. Poliovirus proteases 2A and 3C, on the other hand, may trigger apoptotic death of the infected cell as they interact with the anti-apoptotic Bcl_{XL} or pro-apoptotic Bcl_{XS}.

Within the cell (anterior horn motor cells of the spinal cord), PV proteins 2B, 2BC and 3A inhibit vesicular transport from the endoplasmic reticulum and disassemble the Golgi apparatus. Host cell protein translation is blocked and is replaced by viral protein translation. The N-S dipeptide is cleaved to liberate viral proteins VP2 and VP4 from VP0 and the viral proteases will not be incorporated into the mature virus particles. After VP0 cleavage, the viral genome is “spring-loaded” so as to be readily ejectable from the virions upon their entry into a new host cell. Infectious virions exit through the cell membrane within 8 hours after viral entry, resulting in cytoplasmic lysis of the cell [543].

Mutations in the 5′noncoding (untranslated) region (5′-UTR) of the viral genome deprives the virus of its neurovirulence. P1Sabin was mutated in nucleotide 480, changing amino acid sequences in capsid proteins VP1, VP3, and VP4. P2Sabin was mutated in nucleotide 481 and VP1. These mutations revealed that domain V within the IRES of PV was the residence of the neurovirulence gene; later it was found that domains V and VI are coordinately responsible for the neurovirulence of the wild PV. P3Sabin was mutated in nucleotide 472 and VP3. Poliovirus P3/119 isolated from a case of vaccine-associated poliomyelitis was a revertant from P3Sabin to P3/Leon [543].

The family *Picornaviridae* incorporates the genera *Enterovirus* (with PV included) and *Rhinovirus*. Minor and major rhinoviruses enter host cells through Ig-like ICAM and low-density lipoprotein and sialic acid receptors. Polio- and rhinoviruses exhibit canyons in their capsids. In the case of rhinoviruses, ICAM domains penetrate deep into these canyons [543]. Virus-neutralizing antibodies also intrude into the canyons. Internal ribosome entry site elements of these two genera are closely related, form type 1 IRES elements, and are interchangeable between members of the two genera. Human rhinovirus type 2 (HRV2) IRES is devoid of neuroxicity-encoding elements and it could replace PV type 1 (Mahoney) IRES elements; thus the chimeric virus PV1(RIPO) was created (see in this volume). Poliovirus 1(RIPO) does not replicate in neuroblastoma cells or in cells of neuronal origin, but grows well in HeLa cells. It infects the skeletal musculature of CD155 transgenic mice but exerts no neurotoxicity in them; it is not neurotoxic in *Cynomolgus* monkeys upon intraspinal injection. If HRV2 receives IRES domains V and VI from PV1Mahoney, the parent virus of PV1(RIPO), however, it starts replicating in neuroblastoma cells and gains neurovirulence in CD155 transgenic mice. Human glioma and neuroectodermal tumor cells express CD155. Human glioblastoma xenografts in athymic mice were completely eradicated after a single intratumoral injection of PV1(RIPO) (Melinda Merrill, D. Solecki and M. Gromeier, in this volume) [546–550].

SAFETY MEASURES AND REGULATIONS

The ideal oncolytic virus should not cause disease in the tumor-bearing host. It should not integrate in the host cell genome (unless it is a retroviral vector intentionally delivering a gene). Preferably, it should not be excreted in urine and saliva, so it should not spread in the patient's environment. It should be genetically stable, to avoid recombination, multiplicity reactivation, and other means for regaining virulence upon encounter with a latent, related or unrelated virus carried by the tumor-bearing host or existing in the patient's domicile. If the oncolytic virus is excreted, it should not be readily infectious in healthy human beings. It would be most advantageous if a genetic switch-off function could be incorporated into the viral genome to curtail unwanted replication of the oncolytic virus in the patients. Ideally, the virus should be highly tumor-selective, sparing normal healthy cells and replicating exclusively in tumor cells. The oncolytic virus should not be highly immunogenic (rather, it should be "tolerogenic").

It would not be without additional risk-taking if oncolytic virus therapy were to be combined with immunosuppressive measures.

The candidate virus for virotherapy should derive from safe sources. Tissue cultures, serum preparations, and other ingredients can be contaminated with exo- or endogenous avian or mammalian retroviruses; potentially oncogenic simian viruses (SV40), such as in the first preparations of Salk and Sabin poliovirus vaccines and in an adenovirus vaccine; latent herpes viruses; porcine parvoviruses (in trypsin preparations); or hepatitis A or B viruses (in an early formulation of yellow fever vaccine stabilized with human sera). Bacterial contaminants must be excluded; especially frequent are mycoplasma contaminations of tissue culture preparations. Finally, the viral preparations should be endotoxin-free. Some cartilage samples taken from cadavers and transplanted into joints were found to be contaminated with *Clostridia* and caused fatal septicemias in the recipients.

The facility where the virus preparations are produced should be licensed, properly staffed with qualified and experienced personnel, and equipped with a filtered air supply. Special attention should be given to the use of high-quality and controlled ingredients. The indiscriminate use of human sera (except the patients' own blood products) should be prohibited to avoid use of oncolytic viral preparations containing hepatitis viruses, HIV-1, prions of Creutzfeld-Jakob disease, and other contaminants. The facility should follow approved manufacturing instructions. The premises should be supervised and periodically inspected by the licensing authority to assure that the required and agreed upon manufacturing procedures resulting in a clinically administered oncolytic virus preparation are meticulously observed. Record keeping should be thoroughly accurate so that batches issued would be clearly identifiable (for urgent withdrawal from use in case of trouble). The product should be distributed in a safe manner, so that the quality of the preparation would not diminish because of inappropriate transportation. Good guidelines for manufacturing practices are available [551,552].

The risk–benefit ratio must be clearly explained to the patient in the approved informed consent forms. Patients with treatment-refractory advanced cancers may accept higher risks than patients with cancers potentially curable by means other than oncolytic virus therapy. The risk–benefit ratio of competing treatment modalities should be presented to the patient in a balanced and honest manner before selecting an oncolytic virus for therapy. Qualified staff of physicians and nurses, preferably specialty board-certified in hematology-oncology and infectious diseases, should attend the patients. Clinical records and documents should be kept and preserved according to guidelines of good clinical practice.

The approved investigational protocols should not be administered in private offices. Preferably, academic institutions or well-administered non-private, non-profit hospitals should take charge of the clinical testing of oncolytic viruses. The Institutional Review Board should exercise rigorous supervision over the conduct of these clinical trials.

Recitations of anecdotal case histories and inviting patient testimonials should be discouraged. Premature media publicity should be scrupulously avoided. Media publicity should be allowed only for proven results and always with moderation. The principal investigator's announcement should be followed by comments from independent experts (in good taste and in good faith). Verily, gross exaggerations and distortion of facts should not be tolerated or awarded by casual and indiscriminate citations in the literature without criticism being exercised. Financial involvement of the principal investigators should be discouraged or declared and promotional maneuvers for the sale of the product during its investigational period of work (Phase I and II clinical trials) should be prohibited.

THE FUTURE

Will naturally oncolytic or genetically engineered viruses be the best clinically useful oncolytic agents? Will antiviral host immunity be detrimental (antibodies neutralizing the oncolytic virus) or promotional, with NK cells practicing antibody-directed cellular cytotoxicity (ADCC) against virally infected tumor cells? Can oncolytic viruses be combined with each other simultaneously or sequentially, or with other therapeutic interventions, such as chemotherapy, monoclonal antibodies and immunotoxins, inhibitors of oncogene- or fusion oncoprotein-associated tyrosine kinases, or antisense oligonucleotides (see in this volume)? Will viruses contribute further to the efficacy of cancer vaccines—to viral oncolysates made with reo- or vesicular stomatitis (VS) viruses, or to dendritic cell vaccines in which x-ray–inactivated tumor cells are fused with mature live autologous dendritic cells and the fusion is mediated by a viral fusion protein (NDV, Sendai, VSV)? Would lipid-encapsulated nonpathogenic but oncolytic viruses [553–557] be taken up preferentially by voraciously feeding cancer cells, whereas the encapsulated virus would not infect normal cells through the customary viral envelope or capsid to cell receptor pathway? Polymer-coated [558,559], cationic liposome-conjugated [555,556], or microbead-conjugated adenovirus preparations [560] are already being tested for oncolytic viral therapy. Target tumor cells of low viral permissivity yield to microbead-conjugated adenoviruses, accepting them readily [560]. The M. D. Anderson Cancer Center tested a cationic liposome-encapsulated E1A gene preparation for the treatment of breast and ovarian carcinomas [556]. Perhaps the antiviral antibody response, which may be detrimental to oncolysis in cases of repeated viral inoculations, may be suppressed or averted in this way? Are three-dimensional organotypic multicellular spheroids more reliable models than refined tissue cultures (deprived of stromal cells) or xenografts for preclinical testing of oncolytic viruses [561]?

AFTERTHOUGHTS

Viral invasion of the host elicits prompt innate and subsequent adaptive immune reactions. These immune reactions differ, if the virus is a virulent one, from those mobilized against an attenuated live viral vaccine. In the first case, the virus overcomes IFN production [562,563]; neutralizes other lympho-, cyto- and chemokines by encoding soluble decoy receptors for them [64,564–567]; escapes neutralization by antibodies in producing *quasi species* particles [568]; kills dendritic cells, macrophages, and immune T cells [569]; evades NK cells [570]; and suppresses the presentation of its antigens in an immunogenic fashion [571–573]. Many viruses are able to prevent immediate apoptotic death of infected host cells to secure the full maturation of their entire progeny [574,575]. Alternatively, these virulent pathogens may kill the host or evoke tolerance to establish a carrier state of viral latency without causing any morbidity (HSV-1, HSV-2; zoster virus; EBV). In contrast, the attenuated virus induces the production of virus-neutralizing antibodies and immune T cells with the persistence of appropriate memory cells to effectively overcome subsequent infections by the virulent virus of the same species. The relation of a host with, or its immune reactions to, replication-competent or genetically modified oncolytic viruses is unnatural, very complex and the least well understood. Indeed, both attenuated (rabies) and virulent viruses (measles) have induced rare remissions of malignant tumors in patients. Although viral infection of patients with cancer is not unusual, the rarity of this phenomenon—that is, remission of cancer coincidentally with a natural viral infection—is conspicuous.

Although oncolytic viruses have the capacity to prolong or spare the lives of cancer patients, these attenuated or genetically modified viruses may be less endowed for the evasion of host immunity and succumb to it rapidly, allowing partially lysed tumors to re-grow. For example, pre-existing immunity to adenovirus type 5 (Ad5) hampers its usefulness in protocols for viral oncolysis or gene therapy, whereas viral toxicity may prevail even in immune individuals. In contrast, Ad35 is much less immunogenic in human communities, pre-existing immunity to Ad5 does not neutralize Ad35, and Ad35 constructs readily insert transgenes into human cells [576].

The situation with retro(lenti)viral vectors is even more complex. The DTA lentiviral vector delivers the diphtheria toxin A gene into human prostatic carcinoma cells and kills the tumor cells (in xenografts) [577]. However, work with replication-competent retro(lenti)viral vectors is subject to the infidelity of reverse transcription, mistakes at the transgene insertion, suppressed or altered transgene expression, and difficulties in tagging gene-modified cells, especially when nerve growth factor receptor technology is used for this purpose. The host's immune surveillance system is not overlooking the genetically engineered retro(lenti)virus; innate and adaptive immune reactions may eliminate the virus upon its repeated inoculations. In the murine or human hematopoietic systems, retroviral insertion may unexpectedly induce apoptosis, differentiation or clonal expansion of the infected cell population. Manipulation of the NGFR during retroviral insertion induced clonal expansion of monocytes, culminating in monocytic leukemia, in the mouse. It is not yet known what leukemogenic genes might have been inadvertently transactivated in a child receiving successful gene therapy for X-linked SCID by retroviral gene transfer in France [578]. Thus, the ideal oncolytic virus or viral vector has not as yet been identified.

Work with enteroviruses at the August Kirchenstein Institute in Riga, Latvia and in the former Soviet Union more than 30 years ago for the virotherapy of human cancers [12,13] now reveals that these viruses induced apoptosis in the infected cells [579]. This event would result in incomplete oncolysis and explains the rapid regrowth of most virally infected tumors. If the cell dies immediately after viral entry, there will be no mature virions produced to propagate viral oncolysis within the tumor. However, tumor apoptosomes could induce antitumor immunity mediated by DC.

Mediators of this "postoncolytic immunity" could attack virally not-infected tumor cells as well [580]. First, the innate NK cells and, later, the immune T cells are mobilized to kill tumor cells by cytoplasmic lysis with perforins or antibodies (in an ADCC reaction with NK cells) or by nuclear clumping (programmed cell death; apoptosis) through cascades emanating from "death receptors and ligands" from the cell membrane. Conventional immune surveillance predating the appearance of the tumor is attributed to patrolling NK cells [581]. This is the cell population that responded first in patients to immunization with VOs. Natural T cell immune responses to tumor-associated antigens arise after the inception of the tumor and appear to co-exist in vivo with advancing tumors, contrary to the case with in vitro assays, which show that these T cell clones are able to expand and to kill autologous tumor cells [582]. Large, granular and smaller, more compact lymphocytes were shown to kill allogeneic and autologous tumor cells more than 30 years ago (J. G. Sinkovics, Chapter 2 in this volume) [149], but their characterization, the identification of the antigens (TAA) with which they react, and their use for adoptive immunotherapy did not materialize until recently. Preventive (VO) [134] and therapeutic (DC) cancer vaccines induce clonal expansion of these T cell populations. Even vaccination against growth factors (VEGF and VEGF-R; bFGF) that induce neoangiogenesis could result in tumor regression [583].

Subverted macrophages are frequently enlisted to produce cytokines or chemokines that act as paracrine growth factors for certain tumor cells [584]. Some viruses (influenza, measles, vaccinia) infect and may kill macrophages; others (NDV) activate macrophages (for the advantage or disadvantage of the host, as the individual case may be) (V. Schirrmacher, Chapter 4 in this volume). However, in case of intact macrophages, the ADCC reaction may be mobilized, as macrophages, through their FcR, capture antibody-coated tumor cells with cognate immunoglobulins attached to tumor antigen epitopes, and engulf and digest them.

What immunomodulators (BCG scarified over the site of VO injections 30 years ago; GM-CSF administered in the past few years; lipid encapsulated viral constructs engulfed by macrophages in 2004) would activate the macrophage to kill the tumor cell and avoid being subverted by it? Viruses evoke cytokines [585] that may act against the tumor (TNFα; IFNαβγ IL-2, IL-12; Fas ligand) or may induce cytokines (IL-10), rendering the host tolerant toward the same tumor antigens that the virus was supposed to co-immunize against. Certain virally induced lympho- and cytokines may be those that the tumor had already expropriated for its own auto- or paracrine growth loops. Some melanoma cells use the IL-2 → IL-2R loop and even FasL→Fas a circuit to induce mitosis; some neuroblastoma cells are driven by the TNFα→TNFαR loop. Interleukin-6 is a growth factor for multiple myeloma, Kaposi's sarcoma, and kidney and some prostate carcinoma cells; IFNγ is a growth factor for Kaposi's sarcoma and certain osteogenic sarcoma cells (reviewed by Sinkovics in this volume). Paracrine (of keratinocyte origin) IL-7 drives the growth of Sèzary cells in mycosis fungoides [586–588]. Adenovirally (AdCAIL-2) mediated delivery of IL-2 into prostatic carcinoma cells has been accomplished. Interferon-γ- and IL-4-secreting T cells invaded the transfected tumors and induced tumor cell death, while PSA levels dropped in the patients' blood [589]. The IL-2 gene-carrier vaccinia virus VVIL-2, deriving from xenogeneic culture, infected immature dendritic cells, which matured, expressed CD80, CD86, and MHC class II antigens and started to secrete IL-12. When VVIL-2 derived from syngeneic cultures, the infected DC did not produce IL-12; instead they secreted IL-10 and TNFα [590]. Thus, DC can induce, alternatingly, immunity or tolerance. The major suppressor of DC maturation and activation is IL-21, a relative of IL-4 and IL-15 [591]. A refined technology can be envisioned in which the oncolytic virus would be given with IL-21 and the host would tolerate it, whereas TAA would be processed by mature DC for IL-2 and IL-12 production and for immunogenic presentation to T cells. In mice, the genetically engineered (rVSV)-IFNβ acted as an attenuated, tolerated, and highly oncolytic agent [592], but it has not as yet been tested against human tumors. The effect of virally induced chemokines on tumor growth is still a *terra incognita*. Antiviral or other antibodies, such as those resulting from Th2-type immune response, may complex with the antigenic epitopes by coating them. These concealed antigens, that would be otherwise targeted by cytotoxic T cells, may escape this type of immune attack. Some lymphocyte subclones of tumor-bearing mice could accelerate tumor growth [241,593–598]. No tumor enhancement has been as yet reported in patients as a consequence of oncolytic virus therapy; did it occur and remained unreported?

The task remains to find systems in which oncolytic viruses persist until total oncolysis is achieved, regardless of the mechanism—apoptosis; cytolysis caused by host immunity directed at immature viral particles and tumor antigens co-expressed; or by the outburst of a fully mature viral progeny, thus lysing the tumor cell, but without morbidity [599–601]. Promising systems to achieve these effects are those in which fusogenic viruses induce tumor cell syncytia (NDV, VSV, HSV, in this volume) and spread within them without being exposed to virus-neutralizing antibodies; or those of dendritic cells, which phagocytize apoptotic tumor cells and co-express viral and tumoral epitopes to immune T cells (as in viral oncolysates). Influenza virus-infected macrophages are engulfed by dendritic cells that induce T cell-mediated immunity directed against virally infected cells [602]. If the virally infected apoptotic cell is a tumor cell, this mechanism of antigen presentation may explain the antitumor efficacy of the first influenza viral oncolysates (Sinkovics, in this volume). The crystal structure of the human class II MHC protein HLA-DR1 complexed with an influenza virus peptide is a phenomenal sight [603] and the structure is of high immunogenicity.

Could we envision that viral and tumoral antigenic epitopes are co-expressed in the same fashion in viral oncolysates? In patients immunized with viral oncolysates, we should be able to visualize the "immunological synapse" between professional antigen-presenting cells (mature dendritic cells), as antigenic peptide epitopes of viral and tumoral origin are co-expressed in the grove of MHC molecules for presentation to T cell receptors next to co-stimulatory mediators (B7.2, IL-2, and IL-12) [604], which are necessary for breaking the tolerance to, and for the initiation of, a sustained Th1-type antitumor immune reaction of therapeutic strength. However, immunological synapses can be formed between dendritic cells and B lymhocytes, which are representatives of Th2-type immune reactions. Inhibitory NK cells also form their immunological synapses [605], in which case the target will be spared.

Further, NKT cells arise in the tumor-bearing host and are guided by their receptors (KIRs; NKp30,44, 46; NKG2D) to transformed cells to act as agents of antitumor surveillance [606]. When the placenta established itself in the ancestors of mammals some 300 million years ago, it must have compromised the, by then fully operational, adaptive immune system. The VDJ hypermutations and their initiators, the retrotransposon-derived RAG1 and RAG2 systems, originated in the ancestors of cartilaginous fishes (*Chondrichthyes*) some 450 millions years ago [607–609]. This compromise consists of the forced acceptance of paternal antigens expressed by the semi-allograft fetus. The trophoblast, but especially the syncytiotrophoblast, is an invasive and metastasis-prone entity. Its stem cells proliferate under the effect of fibroblast growth factors [610] (frequently used also by tumor cells in their autocrine or paracrine growth loops). The syncytiotrophoblast expresses syncytin [611], a HER-W retrovirally encoded fusion protein, in addition to hCG and TGFβ (a Th3-type reaction) and interacts with temporarily activated *c-myc* and *jun* protooncogenes. It suppresses expanding maternal T cell clones, and expresses FasL to kill Fas receptor-positive maternal lymphocytes [612,613].

The phylogenetic origin of NK cells descends way back to the *Botryllus* urochordates [614]. It is surmised that ancient NK cells exerted antiviral defense, but it is proven that urochordate NK cells prevent incompatible fusion of host cell colonies. Some of their ancestral receptors appear to remain preserved throughout evolution, up to the species *Homo*. Maternal NKT cells emerged for the containment of the fetal trophoblast [615–617]. Since then, closely related NKT cell populations operate in the tumor-bearing host for the containment of invasive and metastatic tumor cells [618–620]. The placenta temporarily activates the proto-oncogenes and evasive reactions that tumor cells display constitutively. Placental proto-oncogenes are not mutated, whereas in the tumor, mutated, truncated, translocated, and fused oncogenes operate. Tumor-specific T cell memory can be generated after NK cell-mediated tumor rejection [621]. A virally infected tumor cell would be a primary target of NKT cells of the host.

Preparations of LAK cells consist of subclasses of NK cells. Could LAK cells be enriched of NKT cells, if collected after immunization of the donor with oncolytic viruses or with viral oncolysates? The "immunological synapse" and NKT cells were the subject matters of discussions at the 56th Annual Symposium on Fundamental Cancer Research, "Cancer Immunity: Challenges for the Next Decade," at the University of Texas M.D. Anderson Cancer Center, Houston, TX, October 14–17, 2003) [622].

In the pregnant woman T and/or NK cell, clones emerge to suppress maternal immune reactions to paternal antigens expressed by the fetus [623,624]; if they fail, the fetus is lost [625,626]. Pregnancy carried to term is supported by Th2-type immune reactions of the mother; Th1-type reactions lead to fetal loss [625]. In analogy, Th2-type reactions in a tumor-bearing host promote the tumor, whereas the induction of Th1-type reactions could lead to tumor rejection. When tumor cells secrete fetal antigens (AFP; hCG; CEA; CA-15-3; CA-125, CA-19-9, etc.), would these substances activate similar classes of suppressor cells for the protection of the tumor (as another analogy of immune evasion as practiced both by the placenta and malignant tumors)? Would multiple pregnancies alter the woman's susceptibility to subsequent neoplasms [627]? The rapidly expanding clone of the newly emerging suppressor cells could be decimated by low-dose cyclophosphamide [109] co-administered with an oncolytic virus. If molecular virology and molecular immunology unite, this unison may overcome expanding tumor cell populations.

REFERENCES

1. Kovács F. Zur Frage der Beeinflussung des leukämischen Krankheitsbildes durch complicirende Infectionskrankheiten Wien Klin Wochenschr 1893; 6:701–704.

2. Dock G. The influence of complicating diseases upon leukæmia Am J Med Sci 1904; 127:563–592.

3. Wheelock EF, Dingle JH. Observations on the repeated administration of viruses to a patient with acute leukemia N Engl J Med 1964; 27:645–651.
4. De Pace NG. Sulla scomparsa di un enorme cancro vegetante del collo dell'utero senza cura chirurgica Ginecologia 1912; 9: 82–88.
5. Pack GT. Note on the experimental use of rabies vaccine for melanomatosis Arch Dermatol Syphil 1950; 62:694–695.
6. Higgins GK, Pack GT. Virus therapy in the treatment of tumors Bull Hosp Joint Dis 1951; 12:379–382.
7. Southam CM, Moore AE. Clinical studies of viruses as antineoplastic agents with particular reference to Egypt 101 virus Cancer 1952; 5:1025–1034.
8. Newman W, Southam CM. Virus treatment in advanced cancer Cancer 1954; 7:106–118.
9. Sinkovics J. Virusvermehrung in Tumorzellen. In: Sinkovics J, Author. Die Grundlagen der Virusforschung. Budapest: Verlag der Ungarischen Akademie der Wissenschaften, 1956: 98–103.
10. Cassel WA, Garrett RE. Newcastle virus as an antineoplastic agent Cancer 1965; 18:863–868.
11. Webb HE, Smith CEG. Viruses in the treatment of cancer Lancet 1970; i:1206–1209.
12. Mutseniyetse AY. Onkotropism Virusov i Problema Viroterapii Zlokachestvennih Opuholei (Oncotropism of Viruses and the Problem of Virus Therapy of Malignant Tumors). Riga. Latvia: Zinatne, 1972.
13. Voroshilova MK. Potential use of nonpathogenic enteroviruses for control of human disease Prog Med Virol 1989; 36:191–202.
14. Sinkovics J, Horvath J. New developments in the virus therapy of cancer: a historical review Intervirology 1993; 36: 193–214.
15. Campbell J. The future of bacteriophage biology Nature Rev Genetics 2003; 4:471–477.
16. Sulakvelidze A, Alavidze Z, Morris JG. Bacteriophage therapy Antimicrob Agents Chemother 2001; 45:649–659.

17. Alisky J, Iczkowski K, Papoport A, Troitsky N. Bacteriophages show promise as antimicrobial agents J Infect 1998; 36:5–15.

18. Carlton RM. Phage therapy: past history and future prospects Arch Immunol Ther Exp (Warsz) 1999; 47:267–274.

19. Weber-Dabrowska B, Mulczyk M, Gorski A. Bacteriophage therapy of bacterial infections: an update of our institute's experience Arch Immunol Ther Exp (Warsz) 2000; 48: 547–551.

20. Summers WC. Bacteriophage therapy Annu Rev Microbiol 2001; 55:437–451.

21. Ho K. Bacteriophage therapy for bacterial infections. Rekindling a memory from the pre-antibiotics era Perspect Biol Med 2001; 44:1–16.

22. Pirisi A. Phage therapy – advantages over antibiotics? Lancet 2000; 356:1418.

23. Sinkovics JG, Smith JP. Salmonellosis complicating malignant disease Cancer 1969; 24:631–636.

24. Akimkin VG, Bondarenko VM, Voroshilova NN, Darbeeva OS, Baiguzina FA. Practical use of adapted Salmonella bacteriophage for the treatment and prevention of nosocomial infections Zh Mikrobiol Epidemiol Immunobiol 1998; 6:85–86.

25. Sinkovics JG, Smith JP. Septicemia with Bacteroides in patients with malignant disease Cancer 1970; 25:663–671.

26. Kochetkova VA, Mamontov AS, Moskovtseva RL, Erastova EI, Trofimov EI, Popov MI, Dzhubalieva SK. Phagotherapy of postoperative suppurative-inflammatory complications in patients with neoplasms Sov Med 1989; 6:23–26.

27. Zemskova ZS, Dorozhkova IR. Pathomorphological assessment of the therapeutic effect of mycobacteriophages in tuberculosis Probl Tuberk 1991; 11:63–66.

28. Locatcher-Khorazo D, Sullivan N, Gutierrez E. *Staphylococcus aureus* isolated from normal and infected eyes. Phage types and sensitivity to antibacterial agents Arch Ophthalmol 1967; 77:370–377.

29. Slopek S, Kucharewicz-Krukowska A, Weber-Dabrowska B, Dabrowski M. Results of bacteriophage treatment of suppura-

tive bacterial infections. VI. Analysis of treatment of suppurative staphylococcal infections Arch Immunol Ther Exp (Warsz) 1985:261–273.

30. Slopek S, Weber-Dabrowska B, Dabrowski M, Kucharewicz-Krukowska A. Results of bacteriophage treatment of suppurative bacterial infections in the years 1981-1986 Arch Immunol Ther Exp (Warsz) 1987; 35:569–583.

31. Biswas B, Adhya S, Washart P, Paul B, Trostel AN, Powell B, Carlton R, Merril CR. Bacteriophage therapy rescues mice bacteremic from a clinical isolate of vancomycin-resistant *Enterococcus faecium* Infect Immun 2002; 70:204–210.

32. Grimont PA, Grimont F, Lacut JY, Issanchou AM, Aubertin J. Treatment of a case of endocarditis caused by Serratia with bacteriophages Nouv Presse Med 1978; 24:2251.

33. Schuch R, Nelson D, Fischetti VA. A bacteriolytic agent that detects and kills *Bacillus anthracis* Nature 2002; 418: 884–888.

34. Rosovitz MJ, Leppia SH. Virus deals anthrax a killer blow Nature 2002; 418:825–826.

35. Payne RJ, Jansen VA. Phage therapy: the peculiar kinetics of self-replicating pharmaceuticals Clin Pharmacol Ther 2000; 68:225–230.

36. Merril CR, Biswas B, Carlton R, Jensen NC, Creed GJ, Zullo S, Adhya S. Long-circulating bacteriophage as antibacterial agents Proc Nat Acad Sci USA 1996; 93:3188–3192.

37. Westwater C, Kasman LM, Schofield DA, Werner PA, Dolan JW, Schmidt MG, Norris JS. Use of genetically engineered phage to deliver antimicrobial agents to bacteria: an alternative therapy for treatment of bacterial infections Antimicrob Agents Chemother 2003; 47:1301–1307.

38. Raviv Y, Puri A, Blumenthal R. P-glycoprotein-overexpressing multidrug-resistant cells are resistant to infection by enveloped viruses that enter via the plasma membrane FASEB 2000; 14:511–515.

39. Kirn D, Martuza RL, Zwiebel J. Replication-selective virotherapy for cancer: biological principles, risk management and future directions Nature Med 2001; 7:781–787.

40. Mah C, Byrne BJ, Flotte TR. Virus-based gene delivery systems Clin Pharmacokinet 2002; 41:901–911.

41. Connell PP, Weichselbaum RR. Gene therapy: the challenge of translating laboratory research into clinical practice J Clin Oncol 2003; 21:2230–2231.

42. Dobbelstein M. Viruses in therapy – royal road or dead end? Virus Res 2003; 92:219–221.

43. Kay MA, Glorioso JC, Naldini L. Viral vectors for gene therapy: the art of turning infectious agents into vehicles of therapeutics Nature Med 2001; 7:33–40.

44. Larocca D, Burg MA, Jensen-Pergakes K, Ravey EP, Gonzalez AM, Baird A. Evolving phage vectors for cell targeted gene delivery Curr Pharmacol Biotechnol 2002; 3:45–57.

45. Bierman HR, Crile DM, Dod KS, Kelly KH, Petrakis NL, White LP, Shimkin MB. Remissions in leukemia of childhood following acute infectious disease Cancer 1953; 6:591–605.

46. Pasquinucci G. Possible effect of measles on leukæmia Lancet 1971; i:136.

47. Gross S. Measles and leukæmia Lancet 1971; i:397.

48. Medical News. Can some viral infections protect against leukemia? JAMA 1973; 225:1303.

49. Taqi AM, Abdurrahman MB, Yakubu AM, Fleming AF. Regression of Hodgkin's disease after measles Lancet 1981; ii:1112.

50. Schattner A. Therapeutic role of measles vaccine in Hodgkin's disease Lancet 1984; i:171.

51. Greentree LB. Hodgkin's disease: therapeutic role of measles vaccine Am J Med 1983; 75:928.

52. Bluming AZ, Ziegler JL. Regression of Burkitt's lymphoma in association with measles infection Lancet 1971; ii:105.

53. Hansen RM, Libnoch JA. Remission in chronic lymphocytic leukemia after smallpox vaccination Arch Intern Med 1978; 138:1137–1138.

54. DiStefano AD, Buzdar AU. Viral-induced remission in chronic lymphocytic leukemia? Arch Intern Med 1979; 139:946.

55. Hoster HA, Zanes RP, Von Haam E. Studies in Hodgkin's syndrome. IX. The association of "viral" hepatitis and Hodgkin's disease Cancer Res 1949; 9:473–480.

56. Weintraub LR. Lymphosarcoma JAMA 1969; 210:1590–1591.

57. Wade JC, Gaffey M, Wiernick PH, Schimpf SC, Schiffer CA, Wesley M, Hoofnagle JH. Hepatitis in patients with acute nonlymphocytic leukemia Am J Med 1983; 75:413–422.

58. Csatary LK. Viruses in the treatment of cancer Lancet 1971; ii:825.

59. Moss RW. From the Cancer Chronicles #23. http:/ ralphmoss.com/newcastle.html.

60. Russ M. Questioning conventional oncology. An interview with cancer activist Ralph W Moss PhD Altern Complement Ther 2001; 7:21–26.

61. Lorence RM, Roberts MS, Groene WS, Rabin H. Replication-competent oncolytic Newcastle disease virus for cancer therapy. In: Hernáiz Driever P, Rabkin SD, Eds. Replication-Competent Viruses for Cancer Therapy Monographs in Virology, 2001.

62. Xiang J, Wünschmann S, Diekema DJ, Klinzman D, Patrick KD, George SL, Stapleton JT. Effect of coinfection with GB virus C on survival among patients with HIV infection N Engl J Med 2001; 345:707–714.

63. Tillman HL, Heiken H, Knapik-Botor A, Heringlake S, Ockenga J, Wilber JC, Goergen B, Detmer J, McMorrow M, Stoll M, Schmidt RE, Manns MP. Infection with GB virus C and reduced mortality among HIV infected patients N Engl J Med 2001; 345:715–724.

64. Sinkovics J, Horvath J, Horak A. The origin and evolution of viruses Acta Microbiol Immunol Hung 1998; 45:349–390.

65. Sinkovics JG, Horvath JC. Viral contaminants of poliomyelitis vaccines Acta Microbiol Immunol Hung 2000; 47:471–476.

66. Vilchez RA, Kozinetz CA, Butel JS. Conventional epidemiology and the link between SV40 and human cancers Lancet Oncol 2003; 4:188–191.

67. Garcea RL, Imperiale MJ. Simian virus 40 infection of humans J Virol 2003; 77:5039–5045.

68. Franchini G. Molecular mechanisms of human T-cell leukemia/lymphotropic virus type I infection Blood 1995; 86: 3619–3639.

69. Dewan MZ, Terashima K, Taruishi M, Hasegawa H, Ito M, Tanaka Y, Mori N, Sata T, Koyanagi Y, Maeda M, Kobuki Y, Okayama A, Fujii M, Yamamoto N. Rapid tumor formation of human T-cell leukemia virus type 1-infected cell lines in novel NOD-SCID/γc^{null} mice: suppression by an inhibitor against NF-κB J Virol 2003; 77:5286–5294.

70. Szakacs JG, Moscinski LC. Sequence homology of deoxyribonucleic acid to mouse mammary tumor virus genome in human breast tumors Ann Clin Lab Sci 1991; 21:402–412.

71. Wang Y, Pelisson I, Melana SM, Go V, Holland JF, Pogo BG-T. MMTV-like *env* gene sequences in human breast cancer Arch Virol 2001; 146:171–180.

72. Melana SM, Wang Y, Dales S, Holland JF, Pogo BG-T. Characterization of retroviral particles from human breast cancer, 92nd Annual Meeting Am Assoc Cancer Res, March 24-28, New Orleans, LA. Proceedings 2001; 42:115 621.

73. Cohen I. Benign and malignant Epstein-Barr virus-associated B-cell lymphoproliferative diseases Semin Hematol 2003; 40: 116–123.

74. Yachie A, Kanegane H, Kasahara Y. Epstein-Barr virus-associated T-/natural killer cell lymphoproliferative disease Semin Hematol 2003; 40:124–132.

75. Swaminathan S. Molecular biology of Epstein-Barr virus and Kaposi's sarcoma-associated herpesvirus Semin Hematol 2003; 40:107–115.

76. Aoki Y, Tosato G. Pathogenesis and manifestations of human herpesvirus 8-associated disorders Semin Hematol 2003; 40: 143–153.

77. Enam S, Del Valle L, Lara C, Gan DD, Ortiz-Hidalgo C, Palazzo JP, Khalili K. Association of human polyomavirus JCV with colon cancer: evidence for interaction of viral T-antigen and beta-catenin Cancer Res 2002; 62:7093–7101.

78. Rowe WP, Huebner RJ, Gilmore LK, Parrott RH, Ward TG. Isolation of a cytopathogenic agent from human adenoids undergoing spontaneous degeneration in tissue culture Proc Soc Exper Biol Med 1953; 84:570–573.

79. Smith RR, Huebner RJ, Rowe WOP, Schatten WE, Thomas LB. Studies on the use of viruses in the treatment of carcinoma of the cervix Cancer 1956; 9:1211–1218.

80. Scheffner M, Whitaker NJ. Human papillomavirus-induced carcinogenesis and the ubiquitin-proteasome system Semin Cancer Biol 2003; 13:59–67.

81. Kast WM, Feltkamp MCW, Ressing ME, Vierboom MPM, Brandt RMP, Melief CJM. Cellular immunity against human papillomavirus associated cervical cancer Virology 1996; 7: 117–123.

82. Daemen T, Regts J, Holtrop M, Wilschut J. Immunization strategy against cervical cancer involving an alphavirus vector expressing high levels of a stable fusion protein of human papillomavirus 16 E6 and E7 Gene Ther 2002; 9:85–94.

83. Murphy AM, Morris-Downes MM, Sheahan BJ, Atkins GJ. Inhibition of human lung carcinoma cell growth by apoptosis induction using Semliki Forest virus recombinant particles Gene Ther 2000; 7:1477–1482.

84. Zhang J, Asselin-Paturel C, Bex F, Bernard J, Chehimi J, Willems F, Caignard A, Berglund P, Liljeström P, Burny A, Chouaib S. Cloning of human IL-12 and p35 DNA into the Semliki Forest virus vector: expression of IL-12 in human tumor cells Gene Ther 1997; 4:367–374.

85. Yamanaka R, Zullo SA, Ramsey J, Yajma N, Tsuchiya N, Tanaka R, Blaese M, Xanthopoulos KG. Marked enhancement of antitumor immune responses in mouse brain tumor models by genetically modified dendritic cells producing Semliki Forest virus-mediated interleukin-12 J Neurosurg 2002; 97: 611–618.

86. Yamanaka R, Zullo SA, Ramsey J, Chodera M, Tanaka R, Blaese M, Xanthopoulos KG. Induction of therapeutic antitumor antiangiogenesis by intratumoral injection of genetically engineered endostatin-producing Semliki Forest virus Cancer Gene Ther 2001; 10:796–802.

87. Withoff S, Glazenburg KL, van Veen ML, Kraak MM, Hospers GA, Storkel S, de Vries EG, Wilschut J, Daemen T. Replication-defective recombinant Semliki Forest virus encoding GM-CSF as a vector system for rapid and facile generation of autologous tumor cell vaccines Gene Ther 2001; 8:1511–1523.

88. Tseng J-C, Levin B, Hirano T, Yee H, Pampeno C, Meruelo D. *In vivo* antitumor activity of Sindbis viral vectors J Natl Cancer Inst 2002; 94:1790–1802.

89. Kast WM. Venezuelan equine encephalitis virus for treatment of cervical cancer. In: Cancer Immunology Immunotherapy. The Albert B. Sabin Vaccine Institute 4th Annual Colloqium on Cancer Vaccines and Immunotherapy, Walker's Cay, March 6-10, Proceedings 2002, S17 Abaco, Bahamas.

90. Thom K, Morrison C, Lewis JCM, Simmonds P. Distribution of TT virus (TTV), TTV-like minivirus and related viruses in human and nonhuman primates Virology 2003; 306:324–333.

91. Balogh ZS, Takács M. Prevalence of TT virus in the Hungarian population, Jubilee Meeting Hungarian Society Microbiology, Oct 10-12, 2001, Balatonfüred, Hungary. Acta Microbiol Immunol Hung 2003; 50:256–257.

92. Levaditi C, Nicolau S. Sur le culture du virus vaccinal dans les néoplasmes épithelieux CR Soc Biol 1922; 86:928.

93. Levaditi C, Nicolau S. Vaccine et néoplasmes Ann Inst Pasteur 1923; 37:1–106.

94. Levaditi C, Haber S. Affinité du virus de la peste aviare pour les cellules néoplastiques (épithelioma) de la souris CR Soc Biol 1936; 202:2018–2020.

95. Moore AE. The oncolytic viruses Prog Exp Tumor Res 1960; 1:411–439.

96. Moore AE. Oncolytic properties of viruses. In: 11th Annual Symposium on Fundamental Cancer Research, The University of Texas M.D. Anderson Hospital and Tumor Institute, March 7–9, 1957, Houston, TX. Texas Rep Biol Med 1957; 3: 140–154.

97. Southam CM, Moore AE. Induced virus infections in man by the Egypt isolates of West Nile virus Am J Trop Med Hyg 1954; 3:19–50.

98. Southam CM, Noyes WF, Mellors R. Virus in human cancer cells *in vivo* Virology 1958; 5:395–400.

99. Sabin A. Discussion Texas Rep Biol Med 1957; 3:151–153.

100. Pollard M, Bussell RH. Oncolytic effect of viruses in tissue cultures Proc Soc Exp Biol Med 192[??]; 80:574–578.

101. Pollard M, Snyder CC. Treatment of a human cancer case with viruses Texas Rep Biol Med 1954; 12:341–344.

102. Lindenmann J, Klein PA. Viral oncolysis: increased immunogenicity of host cell antigen associated with influenza virus J Exp Med 1967; 126:93–108.

103. Lindenmann J, Klein PA. Immunological Aspects of Viral Oncolysis. Recent Results Cancer Research. Vol. 9. New York: Springer Verlag, 1967:1–84.

104. Ackermann WW, Kurtz H. A new host-virus system Proc Soc Exp Biol 1952; 81:421–423.

105. Flanagan AD, Love R, Tesar W. Propagation of Newcastle disease virus in Ehrlich ascites cells *in vitro* and *in vivo* Proc Soc Biol Med 1955; 90:82–86.

106. Cassel WA, Garrett RE. Tumor immunity after viral oncolysis J Bacteriol 1966; 92:792.

107. Sinkovics J. Studies on the biological characteristics of the Newcastle disease virus (NDV) adapted to the brain of newborn mice Arch Ges Virusforsch 1957; 7:403–411.

108. Sinkovics J. Virusinterferenzen. In: Sinkovics J, Author. Die Grundlagen der Virusforschung. Budapest: Verlag der Ungarischen Akademie der Wissenschaften, 1956:235–247.

109. Sinkovics JG. On the threshold of the door of "no admittance". In: Friedman H, Szentivanyi A, Eds. The Immunologic Revolution: Facts and Witnesses. Boca Raton: CRC Press, 1994: 241–286.

110. Sinkovics JG, Horvath JC. Newcastle disease virus (NDV): brief history of its oncolytic strains J Clin Virol 2000; 16:1–15.

111. Sinkovics JG. Immunology of tumors in experimental animals. Oncolytic viruses. In: Harris JE, Sinkovics JG, Authors. The

Immunology of Malignant Disease. 2nd edition. Saint Louis: C.V. Mosby, 1976:93–282.

112. Koprowski H, Love R, Koprowska I. Enhancement of susceptibility to viruses in neoplastic tissues Texas Rep Biol Med 1957; 15:11–128.

113. Takakuwa H, Goshima F, Nozawa N, Yoshikawa T, Kimata H, Nakao A, Nawa A, Kurata T, Sata T, Nishiyama Y. Oncolytic viral therapy using a spontaneously generated herpes simplex virus type 1 variant for disseminated peritoneal tumor in immunocompetent mice Arch Virol 2003; 148: 813–825.

114. Takemori N, Nakano M, Hemmi M, Ikeda H, Yanacida S, Kitaoka M. Destruction of tumour cells by Rift Valley fever virus Nature 1954; 174:698–700.

115. Streitenfeld H, Boyd A, Fazakerley JK, Bridgen A, Elliott RM, Weber F. Activation of PKR by Bunyamwera virus is independent of the viral interferon antagonist NSs J Virol 2003; 77: 5507–5511.

116. Austin FC, Boone CW. Virus augmentation of the antigenicity of tumor cell extracts Adv Cancer Res 1979; 30:301–345.

117. Sinkovics JG, Loh KK, Shullenberger CC. Use of viral oncolysates for tumor-specific immunotherapy in man. In: Chirigos MA, Regelson W, Wheelock EF, Eds. Modulation of Host Immune Resistance in the Prevention or Treatment of Induced Neoplasia. Fogarty International Center Proceedings (NIH Publication # 77-893). Vol. 28, 1974:235–236.

118. Sinkovics JG. Viral oncolysates as human tumor vaccines Int Rev Immunol 1991; 7:259–287.

119. Ioannides CG, Platsoucas CD, O'Brian CA, Patenia R, Bowen JM, Taylor Wharton J, Freedman RS. Viral oncolysates in cancer treatment: immunological mechanisms of action Anticancer Res 1989; 9:535–544.

120. Sivanandham M, Wallack MK. Viral oncolysates. In: DeVita VT, Hellman S, Rosenberg SA, Eds. Biologic Therapy of Cancer. 2nd edition. Philadelphia: JB Lippincott, 1995:659–667.

121. Murray DR, Cassel WA, Torbin AH, Olkowski ZI, Moore ME. Viral oncolysate in the management of malignant melanoma. II. Clinical studies Cancer 1977; 40:680–686.

122. Cassel WA, Murray DR. A ten-year follow-up on stage II malignant melanoma patients treated postsurgically with Newcastle disease virus oncolysate Med Oncol 1992; 9:169–171.

123. Wallack MK, Steplewski Z, Koprowski H, Rosato E, George J, Hulihan B, Johnson J. A new approach in specific, active immunotherapy Cancer 1977; 39:560–564.

124. Heicappel R, Schirrmacher V, von Hoegen P, Ahlert T, Appelhans B. Prevention of metastatic spread by postoperative immunotherapy with virally modified autologous tumor cells. I. Parameters for optimal therapeutic effect Int J Cancer 1986; 37:569–577.

125. Horvath JC, Sinkovics JG. Human tumor antigens released in viral oncolysates (VO) are loaded onto dendritic cells (DC), 94th Annual Meeting American Association for Cancer Research, April 5–9, 2003, Toronto, Canada. Proceedings 2003; 44:1091–1092 abstract #4762.

126. Sinkovics JG, Györkey F, Kusyk C, Siciliano MJ. Growth of human tumor cells in established cultures. In: Harris Busch, Ed. Methods Cancer Research. Vol. 14, 1978:243–323.

127. Sinkovics JG. Oncolytic viruses and viral oncolysates Ann Immunol Hung 1986; 26:271–290.

128. Yamnikowa SS, Mandler J, Beck-Ocher ZH, Dachtzeren P, Ludwig S, Lvov DK, Scholtissek C. A reassortant H1N1 influenza A virus caused fatal epizootics among camels in Mongolia Virology 1993; 197:558–564.

129. Sinkovics JG, Plager C, Romero J. Immunology and immunotherapy of patients with sarcomas. In: Crispen RG, Ed. Neoplasm Immunity: Solid Tumor Therapy, Franklin Institute Press. 1977:211–219.

130. Sinkovics JG. Immunotherapy with viral oncolysates for sarcoma JAMA 1977; 237:869.

131. Sinkovics JG, Campos LT, Loh KK, Cormia F, Velasquez W, Shullenberger CC. Chemoimmunotherapy for three categories of solid tumors (sarcoma, melanoma, lymphoma): the problem of immunoresistant tumors. In: Crispen RG, Ed. Neoplasm Immunity: Mechanisms: Chicago Symposium at the University of Illinois Medical Center, 1775. Proceedings 1976: 193–212.

132. Sinkovics JG, Plager C, McMurtrey M, Papadopoulos NE, Waldinger R, Combs S, Romero J, Romsdahl MM. Adjuvant chemoimmunotherapy for malignant melanoma. In: Crispen RG, Ed. Neoplasm Immunity: Experimental and Clinical. Amsterdam. Oxford: Elsevier/North Holland, 1980:481–519.

133. Horvath JC, Horak A, Sinkovics JG, Pritchard M, Pendleton S, Horvath E. Cancer vaccines with emphasis on a viral oncolysate melanoma vaccine Acta Microbiol Immunol Hung 1999; 46:1–20.

134. Sinkovics JG, Horvath JC. Vaccination against human cancers Int J Oncol 2000; 16:81–96.

135. Sinkovics JG, Horvath JC. Virus therapy of human cancers Melanoma Res 2003; 13:431–432.

136. Horvath J, Szabo-Szabari M, Sinkovics JG. Autologous activated lymphocyte therapy in a community hospital Acta Microbiol Hung 1994; 41:205–214.

137. Horvath J, Sinkovics JG. Adoptive immunotherapy with peripheral blood lymphocytes Leukemia 1004; 8:S121–126.

138. Sinkovics JG, Shirato E, Gyorkey F, Cabiness JR, Howe CD. Relationship between lymphoid neoplasms and immunologic functions. In: Shullenberger CC, Ed. Leukemia-Lymphoma. Chicago: Year Book Medical Publishers, 1970:53–92.

139. Sinkovics JG, Drewinko B, Thornell E. Immunoresistant tetraploid lymphoma cells Lancet 1970; i:139–140.

140. Sinkovics JG. Early history of specific antibody-producing lymphocyte hybridomas Cancer Res 1981; 41:1246–1247.

141. Yelton DE, Scharff MD. Monoclonal antibodies: powerful new tool in biology and medicine Ann Rev Biochem 1981; 50: 657–680.

142. Wainwright M. The Sinkovics hybridoma – The discovery of the first "natural hybridoma" Perspect Biol Med 1992; 35: 372–379.

143. Sinkovics JG, Cabiness JR, Shullenberger CC. Monitoring *in vitro* immune reactions to solid tumors Front Radiat Ther Oncol 1972; 7:141–154.

144. Sinkovics JG, Tebbi K, Cabiness JR. Cytotoxicity of lymphocytes to established cultures of human tumors: evidences for specificity National Cancer Institute Monograph 1973; 37: 9–18.

145. Sinkovics JG. Monitoring in vitro of cell-mediated immune reactions to tumors. In: Harris Busch, Ed. Methods Cancer Research. 1973; 8:107–175.

146. Sinkovics JG, Reeves WJ, Cabiness JR. Cell- and antibody-mediated immune reactions of patients to cultured cells of breast carcinoma J Natl Cancer Inst 1972; 48:1145–1149.

147. Sinkovics JG, Kay HD, Thota H. The evaluation of chemoimmunotherapy regimens by *in vitro* lymphocyte cytotoxicity directed to cultured human tumor cells Bibliotheca Haematologica 1976; 43:281–384.

148. Sinkovics JG. Cytotoxic lymphocytes Ann Clin Lab Med 1986; 16:488–496.

149. Sinkovics JG. A reappraisal of cytotoxic lymphocytes in human tumor immunology. In: Cory JG, Szentivanyi A, Eds. Cancer Biology and Therapeutics. London: Plenum Press, 1987:225–253.

150. Sinkovics JG. Programmed cell death (apoptosis): its virological and immunological connections Acta Microbiol Hung 1991; 38:321–334.

151. Sinkovics JG. Malignant lymphoma arising from natural killer cells: report of the first case in 1970 and newer developments in the FasL → FasR (CD95) system Acta Microbiol Immunol Hung 1997; 44:295–307.

152. Sinkovics JG, Thota H, Kay HD, Loh KK, Williams DE, Howe CD, Shullenberger CC. Intensification of immune reactions by immunotherapy: attempts at measuring sarcoma-specific reactions *in vitro*. In: Crispen RG, Ed. Neoplasm Immunity: Theory and Application. Proceedings Chicago Symposium 1974 Published by ITR 904 W Adams. Chicago. IL 60607 (Library of Congress LC75-18796), 1975:137–152.

153. Sinkovics JG, Williams DE, Campos LT, Kay HD, Romero JJ. Intensification of immune reactions of patients to cultured sarcoma cells: attempts at monitored immunotherapy Semin Oncol 1974; 1:351–365.

154. Sinkovics JG, Thota H, Loh KK, Gonzalez F, Campos LT, Romero JJ, Kay HD, King D. Prospectives for immunotherapy of human sarcomas. In:. Cancer Chemotherapy: Fundamental Concepts and Recent Advances. Chicago: Year Book Medical Publisher, 1975:417–443.

155. Horvath JC, Horvath E, Sinkovics JG, Horak A, Pendleton S, Mallah J. Human melanoma cells (HMC) eliminate autologous host lymphocytes (Ly^{FasR+}) and escape apoptotic death by utilizing the FasL→FasR system as an autocrine growth loop ($HMC^{FasL \rightarrow FasR+}$), 89th Annual Meeting American Association Cancer Research, March 28-Apr 1, 1998, New Orleans, LA, Proceedings 1998; 39:584 3971.

156. Sinkovics JG, Horvath JC. Virological and immunological connotations of apoptotic and anti-apoptotic forces in neoplasia Int J Oncol 2001; 19:473–488.

157. Kay HD, Thota H, Sinkovics JG. A comparative study on *in vitro* cytotoxic reactions of lymphocytes from normal donors and patients with sarcomas to cultured tumor cells Clin Immunol Immunopathol 1976; 5:218–234.

158. Hammon WMCD, Yohn DS, Casto BC, Atchison RW. Oncolytic potentials of nonhuman viruses for human cancer. I. Effects of twenty-four viruses on human cancer cell lines J Natl Cancer Inst 1963; 31:329–345.

159. Yohn DS, Hammon WMCD, Atchison RW, Casto BC. Oncolytic potentials of nonhuman viruses for human cancer. II. Effects of five viruses on heterotransplantable human tumors J Natl Cancer Inst 1968; 41:523–529.

160. Asada T. Treatment of human cancer with mumps virus Cancer 1974; 34:1907–1928.

161. Yumitori K, Handa H, Yamashita J, Suda K, Otsuka S, Shimizu Y. Treatment of malignant glioma with mumps virus No Shinkei Geka 1982; 10:143–147.

162. Shimizu Y, Hasumi K, Okudaira Y, Yamanishi K, Tanakahashi M. Immunotherapy of advanced gynecologic cancer patients utilizing mumps virus Cancer Detect Prevent 1988; 12: 487–495.

163. Kobayashi H, Takeichi N, Kuzumaki N. Xenogenization of Lymphocytes, Erythroblasts and Tumor Cells. Sapporo. Japan: Hokkaido University School of Medicine, 1978.

164. Kobayashi H. Viral xenogenization of intact tumor cells Adv Cancer Res 1979; 30:279–299.

165. Sinkovics JG. Oncogenes-antioncogenes and virus therapy of cancer Anticancer Res 1989; 9:1281–1290.

166. Mangeney M, Heidman T. Tumor cells expressing a retroviral envelope escape immune rejection *in vivo* Proc Natl Acad Sci USA 1998; 95:14920–14925.

167. Sinkovics JG, Horvath JC. Kaposi's sarcoma: breeding ground of Herpesviridae. A tour de force over viral evolution Int J Oncol 1999; 14:615–646.

168. Sinkovics JG, Horvath JC. Acquisition of new genomic sequences by retro(lenti)viruses and by advanced mutated transposons-retrotransposons, 92nd Annual Meeting American Association Cancer Research, March 24-28, 2001, New Orleans, LA. Proceedings 2001; 42:115 abstract #622.

169. Sinkovics JG, Horvath JC. Endogenous retroviruses in human lymphoma cells, 38th Annual Meeting American Society Clinical Oncology, May 18-21, 2002, Orlando, FL. Proceedings 2002; 21:288a abstract #1151.

170. Sinkovics JG, Horvath JC. Transposons→retrotransposons→retro(lenti)viruses: their role in genomics and proteomics during evolution, ontogenesis and morbidity, 18th Clinical Virology Symposium & Annual Meeting PanAmerican Society for Clinical Virology, Apr 28-May 1, 2002, Clearwater, FL. Abstracts 2002; abstract M17.

171. Sinkovics JG. Comments on cultured human sarcoma cells Sarcoma 2003; 7:75–77.

172. Phuangsab A, Lorence RM, Reichard KW, Peeples ME, Walter RJ. Newcastle disease virus therapy of human tumor xenografts: antitumor effects of local or systemic administration Cancer Lett 2001; 22:27–36.

173. Kenney S, Pagano JS. Viruses as oncolytic agents: a new age for "therapeutic" viruses J Natl Cancer Inst 1994; 36: 1185–1186.

174. Pecora AL, Rizvi N, Cohen GI, Meropol NJ, Sterman D, Marshall JL, Goldberg S, Gross P, O'Neil JD, Groene WS, Roberts MS, Rabin H, Bamat MK, Lorence RM. Phase I trial of intravenous administration of PV701, an oncolytic virus in patients with advanced solid cancers J Clin Oncol 2002; 20:2251–2266.

175. Portela A, Digard P. The influenza virus nucleoprotein: a multifunctional RNA-binding protein pivotal to virus replication J Gen Virol 2002; 83:723–734.

176. Parker CE, Gould KG. Influenza A virus — a model for viral antigen presentation to cytotoxic T lymphocytes Semin Virol 1996; 7:61–73.

177. Wohlleben G, Müller J, Tatsch U, Hambrecht C, Herz U, Renz H, Schmitt E, Moll H, Erb KJ. Influenza A virus infection inhibits the efficient recruitment of Th2 cells into the airways and the development of airway eosinophilia J Immunol 2003; 170:4601–4611.

178. Bergmann M, Romirer I, Sachet M, Fleischhacker R, García-Sastre A, Palese P, Wolff K, Pehamberger H, Jakesz R, Muster T. A genetically engineered influenza A virus with *ras*- dependent oncolytic properties Cancer Res 2001; 61:8188–8193.

179. Salvatore M, Basler CF, Parisien J-P, Horvath CM, Bourmakina S, Zheng H, Muster T, Palese P, García-Sastre A. Effects of influenza A virus NS1 protein on protein expression: the NS1 protein enhances translation and is not required for shut-off of host protein synthesis J Virol 2002; 76:1206–1212.

180. Belshe RB, Mendelman PM, Treanor J, King J, Gruber WC, Piedra P, Bernstein DI, Hayden FG, Kotloff K, Zangwill K, Iacuzio D, Wolff M. The efficacy of live attenuated, cold adapted, trivalent, intranasal influenzavirus vaccine in children N Engl J Med 1998; 338:1405–1412.

181. Zangwill KM. Cold-adapted live attenuated intranasal influenza virus vaccine Pediatr Infect Dis 2003; 22:273–274.

182. Freedman RS, Bowen JM, Herson H, Wharton JT, Edwards CL, Rutledge F. Immunotherapy for vulvar carcinoma with virus-modified homologous extracts Obstet Gynecol 1983; 62: 707–714.

183. Freedman RS, Edwards CL, Bowen JM, Lotzova E, Katz R, Lewis E, Atkinson N, Carsetti R. Viral oncolysates in patients

with advanced ovarian cancers Gynecol Oncol 1988; 29: 337–347.

184. Ioannides CG, Platsoucas CD, Patenia R, Kim YP, Bowen JM, Morris M, Edwards C, Wharton JT, Freedman RS. T-cell functions in ovarian cancer patients treated with viral oncolysates. I. Increased helper activity to immunoglobulin production Anticancer Res 1990; 10:645–654.

185. Ioannides CG, Platsoucas CD, Freedman RS. Immunological effects of tumor vaccines. II. T cell responses directed against cellular antigens in the viral oncolysates In Vivo 1990; 4: 17–24.

186. Freedman RS, Edwards CL, Bowen JM, Tomasovic B, Patenia R, Scott W. 2nd European Congress Gynecologic Endoscopy New Surgical Techniques. 21-23 October 1993. Bologna. Italy: Monduzzi Editore, 1993:35–42.

187. Gregory V, Bennett M, Orkhan MH, Al Hajjar S, Varsano N, Mendelson E, Zambon M, Ellis J, Hay A, Lin YP. Emergence of influenza A H1N2 reassortant viruses in the human population during 2001 Virology 2002; 300:1–7.

188. Yousuf HM, Englund J, Couch R, Rolston K, Luna M, Goodrich J, Lewis V, Mirza NQ, Andreeff M, Koller C, Elting L, Bodey GP, Whimbey E. Influenza among hospitalized adults with leukemia Clin Infect Dis 1997; 24:1095–1099.

189. Hicks KL, Chemaly RF, Kontoyiannis DP. Common community respiratory viruses in patients with cancer Cancer 2003; 97:2576–2587.

190. Santibanez S, Tischer A, Heider A, Siedler A, Hengel H. Rapid replacement of endemic measles virus genotypes J Gen Virol 2002; 83:2699–2708.

191. Karp CL, Wysocka M, Wahl LM, Ahearn JM, Cuomo PJ, Sherry B, Trinchieri G, Griffin DE. Mechanism of suppression of cell-mediated immunity by measles virus Science 1996; 273: 228–230.

192. Fugier-Vivier I, Servet-Delprat C, Rivailler P, Rissoan MC, Liu YJ, Rabourdin-Combe C. Measles virus suppresses cell-mediated immunity by interfering with the survival and function of dendritic and T cells J Exp Med 1997; 186:813–823.

193. Kaiserlian D, Grosjean I, Caux C. Infection of human dendritic cells by measles virus induces immune suppression Adv Exp Med Biol 1997; 417:421–423.

194. Grosjean I, Caux C, Bella C, Berger I, Fabian W, Banchereau J, Kaiserlian D. Measles virus infects human dendritic cells and blocks their allostimulatory properties for $CD4^+$ T cells J Exp Med 1997; 186:801–812.

195. Schnorr J-J, Xanthacos S, Keikavoussi P, Kämpgen E, Meulen V, Schneider-Schaulies S. Induction of maturation of human blood dendritic cell precursors by measles virus is associated with immunosuppression Proc Nat Acad Sci USA 1997; 94: 5326–5331.

196. Ravanel K, Castelle C, Defrance T, Wild TF, Charrion D, Lotteau V, Rabourdin-Combe C. Measles virus nucleocapsid protein binds to FcγRII and inhibits human B cell antibody production J Exp Med 1997; 186:269–278.

197. Steineur M-P, Grosjean I, Bella C, Kaiserlian D. Langerhans cells are susceptible to measles virus infection and actively suppress T cell proliferation Eur J Dermatol 1998; 8:413–420.

198. Yokota S-I, Saito H, Kubota T, Yokosawa N, Amano K-I, Fujii N. Measles virus suppresses interferon-α signaling pathway: suppression of Jak1 phosphorylation and association of viral accessory proteins, C and V, with interferon-α receptor complex Virology 2003; 306:135–146.

199. Zhifei Y, Maofang L, Bichang W, Xiuji L, Xiuhui L. Chemoimmunotherapy vs. chemotherapy in acute leukemia remission induction Chinese Med J 1981; 94:31–34.

200. Ynagi Y, Ono N, Tatsuo H, Hashimoto K, Minagawa H. Measles virus receptor SLAM (CD150) Virology 2002; 299: 155–161.

201. Klimkiewicz A, Muller-Schulz M, Gerigk C, Neumann U, Ostendorf P. Fatal course of measles infection in a patient with a low-grade malignant non-Hodgkin lymphoma Dtsch Med Wochenschr 1998; 123:901–904.

202. Polack FP, Auwaeter PG, Lee S-H, Nousari HC, Valsamakis A, Leiferman KM, Diwan A, Adams RJ, Griffin DE. Production of atypical measles in rhesus macaques: evidence for disease

mediated by immune complex formation and eosinophils in the presence of fusion-inhibiting antibody Nature Med 1999; 5:629–634.

203. Evermann JF, Burnstein T. Immune enhancement of the tumorigenicity of hamster brain tumor cells persistently infected with measles virus Int J Cancer 1975; 16:861–869.

204. Takeuchi K, Takeda M, Miyajima N, Kobune F, Tanabayashi K, Tashiro M. Recombinant wild-type and Edmonston strain measles viruses bearing heterologous H proteins: role of H protein in cell fusion and host cell specificity J Virol 2002; 76: 4891–4900.

205. Peng K-W, Ahmann GJ, Greipp PR, Cattaneo R, Russell SJ. Systemic therapy of myeloma xenografts by an attenuated measles virus Blood 2001; 98:2002–2006.

206. Peng KW, Donovan KA, Schneider U, Cattaneo R, Lust JA, Russell SJ. Oncolytic measles viruses displaying a single chain antibody against CD38, a myeloma cell marker Blood 2003; 101:2557–2562.

207. Peng K-W, TenEyck CJ, Galanis E, Kalli KR, Hartmann LC, Russell SJ. Intraperitoneal therapy of ovarian cancer using an engineered measles virus Cancer Res 2002; 62:4656–4662.

208. Phuong LK, Allen C, Peng K-W, Gianni C, Greiner S, TenEyck CJ, Mishra PK, Macura SI, Russell SJ, Galanis EC. Use of a vaccine strain of measles virus genetically engineered to produce carcinoembryonic antigen as a novel therapeutic agent against glioblastoma multiforme Cancer Res 2003; 63: 2462–2469.

209. Chantler JK, Smyrnis L, Tai G. Selective infection of astrocytes in human glial cell cultures by rubella virus Lab Invest 1995; 72:334–340.

210. Natale VAI, Caffrey JF, Sheahan BJ, Atkins GJ. Effect of infection with rubella virus on the development of rat cerebellar cells in culture Neuropathol Appl Neurobiol 1993; 19:530–534.

211. Megyeri K, Berencsi K, Halazonetis TD, Prendergast GC, Gri G, Plotkin SA, Rovera G, Gönczöl E. Involvement of a p53-dependent pathway in rubella virus-induced apoptosis Virology 1999; 259:74–84.

212. Krishnamurthy S, Huang Z, Samal S. Recovery of a virulent strain of Newcastle disease virus from cloned cDNA. Expression of a foreign gene results in growth retardation and attenuation Virology 2000; 278:168–182.

213. Radnót M. Maladie oculo-glandulaire jusqu'á présent inconnue Ophthalmologica 1949; 113:106–108.

214. Sinkovics J. The isolation of Newcastle-virus from human oculoglandular disease Kisérl Orvostud (Budapest, Hungary) 1949; 1:34–43.

215. Gotoh B, Komatsu T, Takeuchi K, Yohoo J. Paramyxovirus strategies for evading the interferon response Rev Med Virol 2002; 12:337–357.

216. Cassel WA, Weidenheim KM, Campbell WG, Murray DR. Malignant melanoma: inflammatory mononuclear cell infiltrates in cerebral metastases during concurrent therapy with viral oncolysate Cancer 1986; 57:1302–1312.

217. Zorn U, Dallmann I, Groe J, Kirchner H, Poliwoda H, Atzpodien J. Induction of cytokines and cytotoxicity against tumor cells by Newcastle disease virus Cancer Biother 1994; 9: 225–235.

218. Schirrmacher V, Haas C, Bonifer R, Ertel C. Virus potentiation of tumor vaccine T-cell stimulatory capacity requires cell surface binding but not infection Clin Cancer Res 1997; 3: 1135–1148.

219. Schirrmacher V, Haas C, Bonifer R, Ahlert T, Gerhards R, Ertel C. Human tumor cell modification by virus infection: an efficient and safe way to produce cancer vaccine with pleiotropic immune stimulatory properties when using Newcastle disease virus Gene Ther 1999; 6:63–73.

220. Schirrmacher V, Bai L, Umansky V, Yu L, Xing Y, Qian Z. Newcastle disease virus activates macrophages for anti-tumor activity Int J Oncol 2000; 16:363–373.

221. Zeng J, Fournier P, Schirrmacher V. Stimulation of human natural interferon-α response via paramyxovirus hemagglutinin lectin-cell interaction J Mol Med 2002; 80:443–451.

222. Risinskaya NV, Vasilenko OV, Fegeding KV, Sudarikov AB. Transfection of the Newcastle virus hemagglutinin-neurami-

dase gene into murine myeloma cells for induction of hot-vs.-tumor immune response Dokl Biochem Biophys 2001; 378: 217–220.

223. Horvath J, Horak A, Sinkovics JG. Comparison of oncolytic Newcastle virus strains, 86th Annual Meeting American Association Cancer Research, March 18-22, 1995, Toronto, Canada, Proceedings 1995; 36:439 abstract #2619.

224. Fábián ZS, Töröcsik B, Kiss K, Csatary L, Bodey B, Tigyi J, Csatary C, Szeberényi J. Induction of apoptosis by a Newcastle disease virus vaccine (MTH-68/H) in PC12 rat phaeochromocytoma cells Anticancer Res 2001; 21:125–136.

225. Szeberényi J, Fábián ZS, Töröcsik B, Kiss K, Csatary LK. Newcastle disease virus-induced apoptosis in PC12 phaeochromocytoma cells Am J Ther 2003; 10:282–288.

226. San Diego Clinic. Therapies: Newcastle virus disease. http:/ www.sdclinic.com/therapies/cancer.

227. Cassel WA, Murray DR, Phillips HS. A Phase II study on the postsurgical management of stage II malignant melanoma with a Newcastle disease virus oncolysate Cancer 1983; 52: 856–860.

228. Batliwalla FM, Bateman BA, Serrano D, Murray D, Macphail S, Maino VC, Ansel JC, Gregersen PK, Armstrong CA. A 15-year follow-up of AJCC stage III malignant melanoma patients treated postsurgically with Newcastle virus (NDV) oncolysate and determination of alterations in the CD8 T cell repertoire Mol Med 1998; 4:783–794.

229. Ahlert T, Schirrmacher V. Isolation of a human melanoma adapted Newcastle disease virus mutant with highly selective replication patterns Cancer Res 1990; 50:5962–5968.

230. Plaksin D, Porgador A, Vadai E, Feldman M, Schirrmacher V, Eisenbach L. Effective anti-metastatic melanoma vaccination with tumor cells transfected with MHC genes and/or infected with Newcastle disease virus (NDV) Int J Cancer 1994; 59: 796–801.

231. Savage HE, Rossen RD, Hersh EM, Freedman RS, Bowen JM, Plager C. Antibody development to viral and allogeneic tumor cell-associated antigens in patients with malignant melanoma

and ovarian carcinoma treated with lysates of virus-infected tumor cells Cancer Res 1986; 46:2127–2133.

232. Plager C, Bowen JM, Fenoglio C, Papadopoulos NEJ, Murray L, Chawla SP, Benjamin RS, Hersh EM. Adjuvant immunotherapy with Newcastle disease virus oncolysate at MDAH stage III malignant melanoma, 22nd Annual Meeting American Association Cancer Research, Spring, 1986, Los Angeles, CA, Proceedings 1986; 5:137 abstract #534.

233. Cassel WA, Olkowski ZI, Murray DR. Immunotherapy in malignant melanoma J Clin Oncol 1999; 17:1963.

234. Bai L, Koopmann J, Fiola C, Fournier P, Schirrmacher V. Dendritic cells pulsed with viral oncolysates potently stimulate autologous T cells from cancer patients Int J Oncol 2002; 21: 685–694.

235. Hotte SJ, Major PP, Hirte HW, Polawski S, Bamat MK, Rheaume N, Buasen PT, Groene WS, Roberts MS, Lorence PM. Prolonged intravenous infusion of PV701: safety profile and objective tumor responses in a Phase I study of advanced cancer patients Bell J, Ed. Oncolytic Viruses Cancer Therapeutics. Banff Alberta Canada Scientific Program 2003, March 26-30 2003:23.

236. Liang W, Wang H, Sun T-M, Yao W-Q, Chen L-L, Jin Y, Li C-L, Meng F-J. Application of autologous tumor cell vaccine and NDV vaccine in treatment of tumors of digestive tract World J Gastroenterol 2003; 9:495–498.

237. Schneider T, Gerhards R, Kirches E, Firsching R. Preliminary results of active specific immunization with modified tumor cell vaccine in glioblastoma multiforme J Neuro-Oncol 2001; 53:39–46.

238. Kirchner HH, Anton P, Atzpodien J. Adjuvant treatment of locally advanced renal cancer with autologous virus-modified tumor vaccines World J Urol 1995; 13:171–173.

239. Zorn U, Duensing S, Langkopf F, Anastassiou G, Kirchner H, Hadam M, Knüver-Hopf J, Atzpodien J. Active specific immunotherapy of renal cell carcinoma: cellular and humoral immune responses Cancer Biother Radiopharmaceut 1997; 12: 157–165.

240. Stojdl DF, Lichty BL, Knowles S, Marius R, Atkins H, Sonenberg N, Bell JC. Exploiting tumor-specific defects in the interferon pathway with a previously unknown oncolytic virus Nat Med 2000; 6:821–825.

241. Sinkovics JG, Howe CD. Superinfection of tumors with viruses Experientia 1969; 25:733–734.

242. Sinkovics JG, Groves GF, Howe CD. Actions of interferon in tissue cultures harboring mouse leukemia virus Experientia 1968; 24:927–928.

243. Hakkinen I, Halonen P. Induction of tumor immunity in mice with antigens prepared from influenza and vesicular stomatitis virus grown in suspension culture of Ehrlich ascites cells J Natl Cancer Inst 1971; 46:1161–1167.

244. Bandlow G, Schlemminger B. Growth of a transplantable mast-cell neoplasm in the mouse inhibited by vesicular stomatitis virus Z Krebsforsch 1974; 81:151–159.

245. Wise KS. Vesicular stomatitis virus-infected L1210 murine leukemia cells: increased immunogenicity and altered surface antigens J Natl Cancer Inst 1977; 58:83–90.

246. Elena SF. Evolutionary history conditions the timing of transmission in vesicular stomatitis virus Infect Genet Evol 2001; 1:151–159.

247. Gadaleta P, Vacotto M, Coulombié F. Vesicular stomatitis virus induces apoptosis at early stages in the viral cycle and does not depend on virus replication Virus Res 2002; 86:87–92.

248. Kopecky SA, Lyles DS. The cell-rounding activity of the vesicular stomatitis virus matrix protein is due to the induction of cell death J Virol 2003; 77:5524–5528.

249. Terstegen L, Gatsios P, Ludwig S, Pleschka S, Jahnen-Dechent W, Heinrich PC, Graeve L. The vesicular stomatitis virus matrix protein inhibits glycoprotein 130-dependent STAT activation J Immunol 2001; 167:5209–5216.

250. Linardakis E, Bateman A, Phan V, Ahmed A, Gough M, Olivier K, Kennedy R, Errington F, Harrington KJ, Melcher A, Vile R. Enhancing the efficacy of a weak allogeneic melanoma vaccine by viral fusogenic membrane glycoprotein-mediated tumor cell-tumor cell fusion Cancer Res 2002; 62:5495–5505.

251. Bateman AR, Harrington KJ, Kottke T, Ahmed A, Melcher AA, Gough MJ, Linardakis E, Riddle D, Dietz A, Lohse CM, Strome S, Peterson T, Simari R, Vile RG. Viral fusogenic membrane glycoproteins kill solid tumor cells by nonapoptotic mechanisms that promote cross presentation of tumor antigens by dendritic cells Cancer Res 2002; 62:6566–6578.

252. Eslahi NK, Muller S, Nguyen S, Wilson E, Thull N, Rolland A, Pericle F. Fusogenic activity of vesicular stomatitis virus glycoprotein plasmid in tumors as an enhancer of IL-12 gene therapy Cancer Gene Ther 2001; 8:55–62.

253. Peisajovich SG, Shai Y. New insights into the mechanism of virus-induced membrane fusion Trends Biochem Sci 2002; 27: 183–190.

254. Reiss CS, Gapud CP, Keil W. Newly synthesized class II MHC chains are required for VSV presentation to CTL clones Cell Immunol 1992; 139:229–238.

255. Livingston PO, Albino AP, Chung TJ, Real FX, Houghton AN, Oettgen HF, Old LJ. Serological response of melanoma patients to vaccine prepared from VSV lysates of autologous and allogeneic cultured melanoma cells Cancer 1985; 135: 713–720.

256. Balachandran S, Porosnicu M, Barber GN. Oncolytic activity of vesicular stomatitis virus is effective against tumors exhibiting aberrant p53, Ras or Myc function and involves the induction of apoptosis J Virol 2001; 75:3474–3479.

257. Barber GN. Host defense, viruses and apoptosis Cell Death Differ 2001; 8:113–126.

258. Ebert O, Shinozaki K, Huang T-G, Savontaus MJ, García-Sastre A, Woo SLC. Oncolytic vesicular stomatitis virus for treatment of orthotopic hepatocellular carcinoma in immune-competent rats Cancer Res 2003; 63:3605–3611.

259. Fernandez M, Porosnicu M, Markovic D, Barber GN. Genetically engineered vesicular stomatitis virus in gene therapy: application for treatment of malignant disease J Virol 2002; 76:895–904.

260. Samani AA, Fallavollita L, Jaalouk DE, Galipeau J, Brodt P. Inhibition of carcinoma cell growth and metastasis by a

vesicular stomatitis virus G-pseudotyped retrovector expressing type I insulinlike growth factor receptor antisense Human Gene Ther 2001; 12:1969–1977.

261. Howard BD, Boenicke L, Schniewind B, Henne-Bruns D, Kalthoff H. Transduction of human pancreatic tumor cells with vesicular stomatitis virus G-pseudotyped retroviral vectors containing herpes simplex virus thymidine kinase mutant gene enhances bystander effects and sensitivity to ganciclovir Cancer Gene Ther 2000; 7:927–938.
262. Mukherjee PK, Simpson RW. Indomethacin inhibits viral RNA and protein synthesis in cells infected with vesicular stomatitis virus Virology 1985; 140:188–191.
263. Coffey MC, Strong JE, Forsyth PA, Lee PWK. Reovirus therapy of tumors with activated Ras pathway Science 1998; 282: 1332–1334.
264. Kominsky DJ, Bickel RJ, Tyler KL. Reovirus-induced apoptosis requires both death receptor- and mitochondrial-mediated caspase-dependent pathways of cell death Cell Death Differ 2002; 9:926–933.
265. Kollmorgen GM, Cox DC, Killion JJ, Cantrell JL, Sansing WA. Immunotherapy of EL-4 lymphoma with reovirus Cancer Immunol Immunother 1976; 1:239–244.
266. Bryson JS, Cox DC. Characteristics of reovirus-mediated chemoimmunotherapy of murine L1210 leukemia Cancer Immunol Immunother 1988; 26:132–138.
267. Steele TA, Cox DC. Reovirus type 3 chemoimmunotherapy of murine lymphoma is abrogated by cyclosporine Cancer Biother 1995; 10:307–315.
268. Lee PWK. Reovirus protein σ1: from cell attachment to protein oligomerization and folding mechanisms Bioessays 1994; 16: 199–206.
269. Strong JE, Lee PW. The v-erbB oncogene confers enhanced cellular susceptibility to reovirus infection J Virol 1996; 70: 612–616.
270. Lee PWK, Hates EC, Joklik WK. Characterization of anti-reovirus immunoglobulins secreted by cloned hybridoma cell lines Virology 1981; 108:134–146.

271. Vogt PK. Fortuitous convergences: the beginnings of JUN Nature Rev Cancer 2002; 2:465–468.

272. Clarke P, Meintzer SM, Widmann C, Johnson GL, Tyler KI. Reovirus infection activates JNK and the JNK-dependent transcription factor c-Jun J Virol 2001; 75:11275–11283.

273. Norman KL, Farassati F, Lee PWK. Oncolytic viruses and cancer therapy Cytokine Growth Factor Rev 2001; 12: 271–282.

274. Wilcox ME, Yang WQ, Senger D, Rewcastle NB, Morris DG, Brasher PMA, Shi ZQ, Johnston RN, Nishikawa S, Lee PWK, Forsyth PA. Reovirus as an oncolytic agent against experimental human malignant gliomas J Natl Cancer Inst 2001; 93:903–912.

275. Norman KL, Coffey MC, Hirasawa K, Demetrick DJ, Nishikawa SG, DiFrancesco LM, Strong JE, Lee PWK. Reovirus oncolysis of human breast cancer Human Gene Ther 2002; 13: 641–652.

276. Alain T, Hirasawa K, Pon KJ, Nishikawa SG, Urbanski SJ, Auer Y, Luider J, Martin A, Johnston RN, Janowska-Wieczorek A, Lee PWK, Kossakowska AE. Reovirus therapy of lymphoid malignancies Blood 2002; 100:4146–4153.

277. Hirasawa K, Nishikawa SG, Norman KL, Alain T, Kossakowska A, Lee PWK. Oncolytic reovirus against ovarian and colon cancer Cancer Res 2002; 62:1696–1701.

278. Etoh T, Himeno Y, Matdumoto T, Aramaki M, Kawano K, Nishizono A, Kitano S. Oncolytic viral therapy for human pancreatic cancer cells by reovirus Clin Cancer Res 2003; 9: 1218–1223.

279. Yang WQ, Senger D, Muzik H, Shi ZQ, Johnson D, Brasher PMA, Rewcastle NB, Hamilton M, Rutka J, Wolff J, Wetmore C, Curran T, Lee PWK, Forsyth PA. Reovirus prolongs survival and reduces the frequency of spinal and leptomeningeal metastases from medulloblastoma Cancer Res 2003; 63: 3162–3172.

280. Kilani RT, Tamini Y, Hanel EG, Wong KK, Karmali S, Lee PWK, Moore RB. Selective reovirus killing of bladder cancer in a co-culture spheroid model Virus Res 2003; 93:1–12.

281. Ito T, Wang D-Q, Maru M, Nakajima K, Kato S, Kurimura T, Wakamiya N. Antitumor efficacy of vaccinia virus-modified tumor cell vaccine Cancer Res 1990; 50:6915–6918.

282. Krauss O, Hollinshead R, Hollinshead M, Smith GL. An investigation of incorporation of cellular antigens into vaccinia virus particles J Gen Virol 2002; 83:2347–2359.

283. Xiang Y, Condit RC, Vijaysri S, Jacobs B, Williams BRG, Silverman RH. Blockade of interferon induction and action by the E3L double-stranded RNA binding protein of vaccinia virus J Virol 2002; 76:5251–5259.

284. Symons JA, Tscharke DC, Price N, Smith GL. A study of the vaccinia virus interferon-γ receptor and its contribution to virus virulence J Gen Virol 2002; 83:1953–1964.

285. Symons JA, Adams E, Tscharke DC, Reading PC, Waldmann H, Smith GL. The vaccinia virus C12L protein inhibits mouse IL-18 and promotes virus virulence in murine intranasal model J Gen Virol 2002; 83:2833–2844.

286. Kettle S, Alcami A, Khanna A, Ehret R, Jassoy C, Smith GL. Vaccinia virus serpin B13R (SPI-2) inhibits interleukin-1β-converting enzyme and protects virus-infected cells from TNF- and Fas-mediated apoptosis, but does not prevent IL-1β-induced fever J Gen Virol 1997; 78:677–685.

287. Barlett N, Symons JA, Tscarke DC, Smith GL. The vaccinia virus N1L protein is an intracellular homodimer that promotes virulence J Gen Virol 2002; 83:1965–1976.

288. Engelmayer J, Larsson M, Subklewe M, Chahroudi A, Cox WI, Steinman RM, Bhardwaj N. Vaccinia virus inhibits the maturation of human dendritic cells: a novel mechanism of immune evasion J Immunol 1999; 163:6762–6768.

289. Ramirez JC, Tapia E, Esteban M. Administration to mice of a monoclonal antibody that neutralizes the intracellular mature virus form of vaccinia virus limits virus replication efficiently under prophylactic and therapeutic conditions J Gen Virol 2002; 83:1059–1067.

290. Wang M, Bronte V, Chen PW, Gritz L, Panicali D, Rosenberg SA, Restifo NP. Active immunotherapy of cancer with a non-replicating recombinant fowlpox virus encoding a model tumor-associated antigen J Immunol 1995; 154:4685–4692.

291. Timiryasova TM, Li J, Chen B, Chong D, Langridge WH, Gridley DS, Fodor I. Antitumor effect of vaccinia virus in glioma model Oncol Res 1999; 11:133–144.

292. Bartlett DL. Vaccinia virus. In: Hernáiz Driever P, Rabkin SD, Eds. Replication Competent Viruses for Cancer Therapy. Vol. 22: Karger, Basel Monogr Virol, 2001:130–159.

293. Zeh HJ, Bartlett DL. Development of a replication-selective, oncolytic poxvirus Cancer Gene Ther 2002; 9:1001–1012.

294. MCCart JA, Ward JM, Lee J, Hu Y, Alexander HR, Libutti SK, Moss B, Bartlett DL. Systemic cancer therapy with a tumor-selective vaccinia virus mutant lacking thymidine kinase and vaccinia growth factor genes Cancer Res 2001; 61:8751–8757.

295. Ramshaw I, Ruby J, Ramsay A, Ada G, Karupiah G. Expression of cytokines by recombinant vaccinia viruses: a model for studying cytokines in virus infections *in vivo* Immunol Rev 1992; 127:157–182.

296. Mastrangelo MJ, Lattime EC. Virotherapy clinical trials for regional disease: *in situ* immune modulation using recombinant poxvirus vectors Cancer Gene Ther 2002; 9:1013–1021.

297. Kaufmann AM, Stern PL, Rankin EM, Sommer H, Nuessler V, Schneider A, Adams M, Onon TS, Bauknecht T, Wagner U, Kroon K, Hickling J, Boswell CM, Stacey SN, Kitchener HC, Gillard J, Wanders J, Roberts JSTC, Zwierzina H. Safety and immunogenicity of TA-HPV, a recombinant vaccinia virus expressing modified human papillomavirus (HPV-) 16 and HOPV-18 *E6* and *E7* genes, in women with progressive cervical disease Clin Cancer Res 2002; 8:3676–3685.

298. Mulryan K, Ryan MG, Myers KA, Shaw D, Wang W, Kingsman SM, Stern PL, Carroll MW. Attenuated recombinant vaccinia virus expressing oncofetal antigen (tumor-associated antigen) 5T4 induces active therapy of established tumors Molec Cancer Ther 2002; 1:1129–1137.

299. Wallack MK, Bash JA, Leftheriotis E, Seigler H, Bland K, Wanebo H, Balch C, Bartolucci AA. Positive relationship of clinical and serologic responses to vaccinia melanoma oncolysate Arch Surg 1987; 122:1460–1463.

300. Hersey P, Edwards A, Contes A, Shaw H, McCarty W, Milton G. Evidence that treatment with vaccinia melanoma lysates

(VMCL) may improve survival of patients with stage II melanoma Cancer Immunol Immunother 1987; 254:257–265.

301. Hersey P, Coates AS, McCarty WH, Thompson JF, Sillar RW, McLeod R, Gill PG, Coventry BJ, McMullen A, Dillon H, Simes RJ. Adjuvant immunotherapy of patients with high-risk melanoma using vaccinia viral lysates of melanoma; results of a randomized trial J Clin Oncol 2002; 29:4181–4190.
302. Nemerow GR, Cooper NR. CR2 (CD21) mediated infection of B lymphocytes by Epstein-Barr virus Semin Virol 1992; 3: 117–124.
303. Jenner RG, Boshoff C. The molecular pathology of Kaposi's sarcoma-associated herpesvirus Biochim Biophys Acta 2002; 1602:1–22.
304. An J, Sun Y, Sun R, Rettig MB. Kaposi's sarcoma-associated herpesvirus encoded vFLIP induces cellular IL-6 expression: the role of the NF-κB and JNK/AP1 pathways Oncogene 2003; 22:3371–3385.
305. Dourmishev LA, Dourmishev AL, Palmeri D, Schwartz RA, Lukac DM. Molecular genetics of Kaposi's sarcoma-associated herpesvirus (human herpesvirus 8) epidemiology and pathogenesis Microbiol Mol Biol Rev 2003; 67:175–212.
306. Varghese S, Rabkin SD. Oncolytic herpes simplex virus vectors for cancer virotherapy Cancer Gene Ther 2002; 9: 967–978.
307. Javier RT, Sederati F, Stevens JG. Two avirulent herpes simplex viruses generate lethal recombinants *in vivo* Science 1986; 234:746–748.
308. Sinkovics JG, Horvath JC, †Györkey F. From evolutionary coexistence to criminal collusion of herpes- and retro(lenti)viruses in human pathological entities, 15th Annual Clinical Virology Symposium and Annual Meeting Pan American Society Clinical Virology, May 9–12, 1999 abstract T32, Clearwater, FL.
309. Confer DL, Vercellotti GM, Kotasek D, Goodman JL, Ochoa A, Jacob HS. Herpes simplex virus-infected cells disarm killer lymphocytes Proc Nat Acad Sci USA 1990; 87:3609–3613.
310. Banks TA, Rouse BT. Herpesviruses—escape artists Clin Infect Dis 1992; 14:933–941.

311. Chou J, Roizman B. Herpes simplex virus 1 $\gamma_1$34.5 gene function, which blocks the host response to infection, maps in the homologous domain of the genes expressed during growth arrest and DNA damage Proc Nat Acad Sci USA 1994; 91: 5247–5251.

312. York IA, Roop C, Andrews DW, Riddell SR, Graham FL, Johnson DC. A cytosolic herpes simplex virus protein inhibits antigen presentation to $CD8^+$ T lymphocytes Cell 1994; 77: 525–535.

313. Galvan V, Roizman B. Herpes simplex virus 1 induces and blocks apoptosis at multiple steps during infection and protects cells from exogenous inducers in a cell-type-dependent manner Proc Nat Acad Sci USA 1998; 95:3931–3936.

314. Raftery MJ, Behrens CK, Muller A, Krammer PH, Walczak H, Schonrich G. Herpes simplex virus type 1 infection of activated cytotoxic T cells: induction of fratricide as a mechanism of viral immune evasion J Exp Med 1999; 190:1103–1114.

315. Liu T, Khanna KM, Chen X, Fink DJ, Hendricks RL. CD8(+) T cells can block herpes simplex virus type 1 (HSV-1) reactivation from latency in sensory neurons J Exp Med 2000; 191: 1459–1466.

316. Milne RS, Mattick C, Nicholson L, Devaraj P, Alcami A, Gompels UA. RANTES binding and down-regulation by a novel human herpesvirus-6 beta chemokine receptor J Immunol 2000; 164:2396–2404.

317. Perng GC, Jones C, Ciacci-Zanella J, Stone M, Henderson G, Yukht A, Slanina SM, Hofman FM, Ghiasi H, Nesburn AB, Wechsler SL. Virus-induced neuronal apoptosis blocked by the herpes simplex virus latency-associated transcript Science 2000; 287:1500–1503.

318. Ulbrecht M, Martinozzi S, Grzeschik M, Hengel H, Ellwart JW, Pla M, Weiss EH. Cutting edge: the human cytomegalovirus UL40 gene product contains a ligand for HLA-E and prevents NK cell-mediated lysis J Immunol 2000; 164:5019–5022.

319. Munger J, Roizman B. The U_s3 protein kinase of herpes simplex virus 1 mediates the posttranslational modification of BAD and prevents BAD-induced programmed cell death in

the absence of other viral proteins Proc Nat Acad Sci USA 2001; 98:10410–10415.

320. Zachos G, Koffa M, Preston CM, Clements B, Conner J. Herpes simplex virus type 1 blocks the apoptotic host cell defense mechanisms that target Bcl-2 and manipulates activation of p38 mitogen-activated protein kinase to improve viral replication J Virol 2001; 75:2710–2728.

321. Fairley JA, Baillie J, Bain M, Sinclair JH. Human cytomegalovirus infection inhibits epidermal growth factor (EGF) signaling by targeting EGF receptors J Gen Virol 2002; 83: 2803–2810.

322. Trgovcich J, Johnson D, Roizman B. Cell surface major histocompatibility complex class II proteins are regulated by the product of the $\gamma_1$34.5 and U_L41 genes of herpes simplex virus 1 J Virol 2002; 76:6974–6986.

323. Wang J, Fu Y-X. LIGHT (a cellular ligand for herpes virus entry mediator and lymphotoxin receptor)-mediated thymocyte deletion is dependent on the interaction between TCR and MHC/self peptide J Immunol 2003; 170:3986–3993.

324. Reiss-Gutfreund RJ, Dostal V, Binder M, Letnansky K. The tumor enhancing property of herpes simplex virus type-2 (HSV-2) Österreich Kneipp Magaz 1977; 3:148–154.

325. Martuza RL, Malick A, Markert JM, Ruffner KL, Coen DM. Experimental therapy of human glioma by means of a genetically engineered virus mutant Science 1991; 252:854–855.

326. Klatzmann D, Valery CA, Bensimon G, Marro B, Boyer O, Mokhtary K, Diquet B, Salzman JL, Philippon J. A Phase I/II study of herpes simplex virus type I thymidine kinase "suicide" gene therapy for recurrent glioblastoma Human Gene Ther 1998; 9:2595–2604.

327. Mineta T, Rabkin SD, Martuza RL. Treatment of malignant gliomas using ganciclovir-hypersensitive, ribonucleotide reductase-deficient herpes simplex virus mutant Cancer Res 1994; 54:3963–3966.

328. Toda M, Rabkin SD, Kojima H, Martuza RL. Herpes simplex virus as an *in situ* cancer vaccine for the induction of specific anti-tumor immunity Human Gene Ther 1999; 10:385–393.

329. Kim JH, Kim SH, Brown SL, Freytag SO. Selective enhancement by an antiviral agent of the radiation-induced killing of human glioma cells transduced with HSV-*tk* gene Cancer Res 1994; 54:6053–6056.

330. Izquierdo M, Martin V, deFelipe P, Izquierdo JM, Pérez-Higueras A, Cortés ML, Paz JF, Isla A, Blázquez MG. Human malignant brain tumor response to herpes simplex thymidine kinase (HSVtk)/ganciclovir gene therapy Gene Ther 1996; 3: 491–495.

331. Advani SJ, Weichselbaum RR, Whitley RJ, Roizman B. Friendly fire: redirecting herpes simplex virus-1 for therapeutic applications Clin Microbiol Infect 2002; 8:551–563.

332. Rabkin SD, Hernáiz Driever P. Replication-competent herpes simplex virus vectors for cancer therapy. In: Hernáiz Driever P, Rabkin SD, Eds. Replication-Competent Viruses for Cancer Therapy. Basel: Karger. Monogr Virol 2001; 22:1–45.

333. Mullen JT, Tanabe KK. Viral oncolysis The Oncologist 2002; 7:106–119.

334. Kokoris MS, Black ME. Characterization of herpes simplex virus type 1 thymidine kinase mutants engineered for improved ganciclovir or acyclovir activity Protein Sci 2002; 11: 2267–2272.

335. Carew JF, Kooby DA, Halterman MW, Kim SH, Federoff HJ, Fong Y. A novel approach to cancer therapy using oncolytic herpes virus to package amplicons containing cytokine genes Mol Ther 2001; 4:250–256.

336. Hamel W, Magnelli L, Chiraugi VP, Israel MA. Herpes simplex virus thymidine kinase/ganciclovir-mediated apoptotic death of bystander cells Cancer Res 1996; 56:2697–2702.

337. Touraine RI, Ishii-Morita H, Ramsey WJ, Blaese RM. The bystander effect in the HSVtk/ganciclovir system and its relationship to gap junctional communications Gene Ther 1998; 5:1705–1711.

338. van Dillen IJ, Mulder NH, Vaalburg W, de Vries EF, Hospers GA. Influence of the bystander effect on HSV-tk/GCV gene therapy Cancer Gene Ther 2002; 2:307–322.

339. Vile RG, Nelson JA, Castleden S, Chong H, Hart IR. Systemic gene therapy of murine melanoma using tissue specific expres-

sion of the HSVtk gene involves an immune component Cancer Res 1994; 54:6228–6234.

340. Park K-H, Kim G, Jang SH, Kim CH, Kwon S-Y, Yoo C-G, Kim YW, Kwon HC, Kim C-M, Han SK, Shim Y-S, Lee C-T. Gene therapy with GM-CSF, inteerleukin-4 and herpes simplex virus thymidine kinase shows strong antitumor effect on lung cancer Anticancer Res 2003; 23:1559–1564.

341. Jia WW-G, McDermott M, Goldie J, Cynader M, Tan J, Tufaro F. Selective destruction of gliomas in immunocompetent rats by thymidine kinase-defective herpes simplex virus type 1 J Natl Cancer Inst 1994; 86:1209–1215.

342. Stanziale SF, Petrowsky H, Joe JK, Roberts GD, Zager JS, Gusani NJ, Ben-Porat L, Gonen M, Fong Y. Ionizing radiation potentiates the antitumor efficacy of oncolytic herpes simplex virus 207 by upregulating ribonucleotide reductase Surgery 2002; 132:353–359.

343. Blank SV, Rubin SC, Coukos G, Amin KM, Albelda SM, Molnar-Kimber KL. Replication-selective herpes simplex virus type 1 mutant therapy of cervical cancer is enhanced by low-dose radiation Human Gene Ther 2002; 13:627–639.

344. Kooby DA, Carew JF, Halterman MW, Mack JE, Bertino JR, Blumgart LH, Federoff HJ, Fong Y. Oncolytic viral therapy for human colorectal cancer and liver metastases using a multi-mutated herpes simplex virus type-1 (G207) FASEB 1999; 13: 1325–1334.

345. Cozzi PJ, Burke PB, Bhargav A, Heston WD, Huryk B, Scardino PT, Fong Y. Oncolytic viral gene therapy for prostate cancer using two attenuated replication-competent, genetically engineered herpes simplex viruses Prostate 2002; 53: 95–100.

346. Bennett JJ, Delman KA, Burt BM, Mariotti A, Malhorta S, Zager J, Petrowski H, Mastorides S, Federoff H, Fong Y. Comparison of safety, delivery and efficacy of two oncolytic herpes viruses (G207 and NV1020) for peritoneal cancer Cancer Gene Ther 2002; 9:935–945.

347. Cinatl J, Cinatl J, Radsak K, Rabenau H, Weber B, Novak M, Benda R, Kornhuber B, Doerr HW. Replication of human cytomegalovirus in a rhabdomyosarcoma cell line depends on

the state of differentiation of the cells Arch Virol 1994; 138: 391–401.

348. Cinatl J, Cinatl J, Michaelis M, Kabickova H, Kotchetkov R, Vogel J-U, Doerr HW, Klingebiel T, Hernáiz Driever P. Potent oncolytic activity of multimutated herpes simplex virus G207 in combination with vincristine against human rhabdomyosarcoma Cancer Res 2003; 63:1508–1514.

349. Currier MA, Adams LC, Horsburgh B, Tufaro F, Cripe TP. Local control of large human rhabdomyosarcoma xenograft tumors by fractionated multiple direct intratumoral injections of oncolytic herpes simplex viruses Bell J, Ed. Oncolytic Viruses as Cancer Therapeutics Banff Canada Scientific Program 2003, March 26–30, 2003:41.

350. Fu X, Zhang X. Potent systemic antitumor activity from an oncolytic herpes simplex virus of syncytial phenotype Cancer Res 2002; 62:2306–2312.

351. Nakamori M, Fu X, Meng F, Jin A, Tao L, Bast RC, Zhang X. Effective therapy of metastatic ovarian cancer with an oncolytic herpes simplex virus incorporating two membrane fusion mechanisms Clin Cancer Res 2003; 9:2727–2733.

352. Ikeda K, Ichikawa T, Wakimoto H, Silver JS, Deisboeck TS, Finkelstein D, Harsh GR, Louis DN, Bartus RT, Hochberg FH, Chiocca EA. Oncolytic virus therapy of multiple tumors in the brain requires suppression of innate and elicited antiviral responses Nat Med 1999; 5:881–887.

353. Ikeda K, Wakimoto H, Ichikawa T, Jhung S, Hochberg FH, Louis DN, Chiocca EA. Complement depletion facilitates the infection of multiple brain tumors by an intravascular, replication-conditional herpes simplex virus mutant J Virol 2000; 74: 4765–4775.

354. MacKie RM, Stewart B, Brown SM. Intralesional injection of herpes simplex virus 1716 in metastatic melanoma Lancet 2001; 357:525–526.

355. Papanastassiou V, Rampling R, Fraser M, Petty R, Hadley D, Nicoll J, Harland J, Mabbs R, Brown M. The potential for efficacy of the modified (ICP 34.5(−)) herpes simplex virus HSV1716 following intratumoral injection into human malignant glioma Gene Ther 2002; 9:398–406.

356. Harland J, Papanastassiou V, Brown SM. HSV1716 persistence in primary glioma cells in vitro Gene Ther 2002; 9: 1194–1198.

357. Wong RJ, Kim SH, Joe JK, Shah JP, Johnson PA, Fong Y. Effective treatment of head and neck squamous cell carcinoma by an oncolytic herpes simplex virus J Am Coll Surg 2001; 193:12–21.

358. Liu B, Robinson M, Han Z, McGrath Y, Branston R, Reay P, Coffin R. Optimized oncolytic herpes simplex virus for cancer treatment, 11th International Conference on Gene Therapy Cancer, Dec 12-14, 2002, San Diego, CA. Cancer Gene Ther 2003; 10:S1–S42 abstract #066.

359. Hu JCC, Han Z, Liu B, Robinson M, Branston R, Coombes RC, Coffin RS. Combination of OncoVEX with chemotherapy for cancer treatment, 11th International Conference on Gene Therapy Cancer, Dec 12-14, 2002, San Diego, CA. Cancer Gene Ther 2003; 10:S1–S42 abstract #067.

360. Coukos G, Makrigiannakis A, Kang EH, Caparelli D, Benjamin I, Kaiser LR, Rubin SC, Albelda SM, Molnar-Kimber KL. Use of carrier cells to deliver a replication-selective herpes simplex virus-1 mutant for the intraperitoneal therapy of epithelial ovarian cancer Clin Cancer Res 1999; 5:1523–1537.

361. Chiocca EA. Oncolytic viruses Nature Rev Cancer 2002; 2: 938–950.

362. Tolba KA, Bowers WJ, Muller J, Houseknecht V, Giuliano RE, Federoff HJ, Rosenblatt JD. Herpes simplex virus (HSV) amplicon-mediated codelivery of secondary lymphoid tissue chemokine and CD40L results in augmented antitumor activity Cancer Res 2002; 62:6545–6551.

363. Delman KA, Zager JS, Bennett JJ, Malhotra S, Ebright MI, McAuliffe PF, Halterman MW, Federoff HJ, Fong Y. Efficacy of multiagent herpes simplex virus amplicon-mediated immunotherapy as adjuvant treatment for experimental hepatic cancer Ann Surg. 2002; 236:337–342; discussion 342–3.

364. Pawlik TM, Nakamura H, Mullen JT, Kasuya H, Yoon SS, Chandrasekhar S, Chiocca EA, Tanabe KK. Prodrug bioactivation and oncolysis of diffuse liver metastases by herpes sim-

plex virus 1 mutant that expresses the CYP2B1 transgene Cancer 2002; 95:1171–1181.

365. Boldogkoi Z, Bratincsak A, Fodor I. Evaluation of pseudorabies virus as a gene transfer vector and an oncolytic agent for human tumor cells Anticancer Res 2002; 22:2153–2160.

366. Okegawa T, Pong R-C, Li Y, Bergelson JM, Sagalowsky AI, Hsieh J-T. The mechanism of the growth-inhibitory effect of coxsackie and adenovirus receptor (CAR) on human bladder cancer: a functional analysis of CAR protein structure Cancer Res 2001; 61:6592–6600.

367. Rauen KA, Sudilovsky D, Le JL, CHew KL, Hann B, Weinberg V, Scmitt LD, McCormick F. Expression of the Coxsackie adenovirus receptor in normal prostate and in primary and metastatic prostate carcinoma: potential relevance to gene therapy Cancer Res 2002; 62:3812–3818.

368. Fuxe J, Liu L, Malin S, Philipson L, Collins VP, Petterson RF. Expression of the Coxsackie and adenovirus receptor in human astrocytic tumors and xenografts Int J Cancer 2003; 103:723–729.

369. Bernal RM, Sharma S, Gardner BK, Douglas JT, Bergelson JM, Dubinett SM, Batra RK. Soluble coxsackievirus adenovirus receptor is a putative inhibitor of adenoviral gene transfer in the tumor milieu Clin Cancer Res 2002; 8:1915–1923.

370. Kanerva A, Mikheeva GV, Krasnykh V, Coolidge CJ, Lam JT, Mahasreshti PJ, Barker SD, Straughn M, Barnes MN, Alvarez RD, Hemminki A, Curiel DT. Targeting adenovirus to the serotype 3 receptor increases gene transfer to ovarian cancer cells Clin Cancer Res 2002; 8:275–280.

371. Hemminki A, Dmitriev I, Liu B, Desmond RA, Alemany R, Curiel DT. Targeting oncolytic adenoviral agents to the epidermal growth factor pathway with a secretory fusion molecule Cancer Res 2001; 61:6377–6381.

372. Kashentseva EA, Seki T, Curiel DT, Dmitriev IP. Adenovirus targeting to c-*erb*B-2 oncoprotein by single-chain antibody fused to trimeric form of adenovirus receptor ectodomain Cancer Res 2002; 62:609–616.

373. Haviv YS, Blackwell JL, Kanerva A, Nagi P, Krasnykh V, Dmitriev I, Wang M, Naito S, Lei X, Hemminki A, Carey D,

Curiel DT. Adenoviral gene therapy for renal cancer requires retargeting to alternative cellular receptors Cancer Res 2001; 62:4273–4281.

374. Gu D-L, Gonzalez AM, Printz MA, Doukas J, Ying W, D'Andrea M, Hoganson DK, Curiel DT, Douglas JT, Sosnowski BA, Baird A, Aukerman SL, Pierce GF. Fibroblast growth factor 2 retargeted adenovirus has redirected cellular tropism: evidence for reduced toxicity and enhanced antitumor activity in mice Cancer Res 1999; 59:2608–2614.

375. Dehari H, Ito Y, Nakamura T, Kobune M, Sasaki K, Yonekura N, Kohama G, Hamada H. Enhanced antitumor effect of RGD fiber-modified adenovirus for gene therapy of oral cancer Cancer Gene Ther 2003; 10:75–85.

376. Hunt KK, Vorburger SA. Hurdles and hopes for cancer treatment Science 2002; 297:415–416.

377. Yu DC, Working P, Ando D. Selectively replicating oncolytic adenoviruses as cancer therapeutics Curr Opin Mol Ther 2002; 4:435–443.

378. Bauerschmitz GJ, Barker SD, Hemminki A. Adenoviral gene therapy for cancer: from vectors to targeted and replication competent agents Int J Oncol 2002; 21:1161–1174.

379. Davison E, Kirby I, Whitehouse J, Hart I, Marshall JF, Santis G. Adenovirus type 5 uptake by lung adenocarcinoma cells in culture correlates with Ad5 fibre binding is mediated by $\alpha_v\beta_1$ integrin and can be modulated by changes in β_1 integrin function J Gene Med 2001; 3:550–559.

380. Dirven CM, Grill J, Lamfers ML, van der Valk P, Leonhart AM, van Beusechem VW, Haisma HJ, Pinedo HM, Curiel DT, Vandertop WP, Gerritsen WR. Gene therapy for meningioma: improved gene delivery with targeted adenoviruses J Neurosurg 2002; 97:441–449.

381. Suzuki K, Alemany R, Yamamoto M, Curiel DT. The presence of the adenovirus E3 region improves the oncolytic potency of conditionally replicative adenoviruses Clin Cancer Res 2002; 8:3348–3359.

382. Tamanini A, Rolfini R, Nicolis E, Melotti P, Cabrini G. MAP kinases and NF-κB collaborate in gene expression in the early phase of adenovirus infection Virology 2003; 307:228–242.

383. Nász I, Ádám É, Lengyel A. Alternate adenovirus type-pairs for a possible circumvention of host immune response to recombinant adenovirus vectors Acta Microbiol Immunol Hung 2001; 48:141–146.

384. Yoshida H, Katayose Y, Unno M, Suzuki M, Kodama H, Takemura S-I, Asano R, Hayashi H, Yamamoto K, Matsuno S, Kudo T. A novel adenovirus expressing human 4-1BB ligand enhances antitumor immunity Cancer Immunol Immunother 2003; 52:97–106.

385. Ruzek MC, Kavanagh BF, Scaria A, Richards SM, Garman RD. Adenoviral vectors stimulate murine natural killer cell responses and demonstrate antitumor activities in the absence of transgene expression Mol Ther 2002; 5:115–124.

386. Routes JM. Adenovirus E1A inhibits IFN-induced resistance to cytolysis by natural killer cells J Immunol 1993; 150: 4315–4322.

387. Yang Y, Nunes FA, Berencsi K, Furth EE, Gönczöl E, Wilson JM. Cellular immunity to viral antigens limits E1-deleted adenoviruses for gene therapy Proc Nat Acad Sci USA 1994; 91: 4407–4411.

388. Kast WM, Offringa R, Peters PJ, Voordouw AC, Meloen JM, van der Erb AJ, Melief CJM. Eradication of adenovirus E1-induced tumors by E1A-specific cytotoxic T lymphocytes Cell 1989; 59:603–614.

389. Morelli AE, Larregina AT, Ganster RW, Zahorchak AF, Plowey JM, Takayama T, Logar AJ, Robbins PD, Falo LD, Thomson AW. Recombinant adenovirus induces maturation of dendritic cells via an NF-kB-dependent pathway J Virol 2000; 74: 9617–9628.

390. Miller G, Lahrs S, Pillarisetty VG, Shah AB, DeMatteo RP. Adenovirus infection enhances dendritic cell immunostimulatory properties and induces natural killer and T-cell-mediated tumor protection Cancer Res 2002; 62:5260–5266.

391. Miura TA, Morris K, Ryan S, Cook JL, Routes JM. Adenovirus E1A, not human papillomavirus E7, sensitizes tumor cells to lysis by macrophages through nitric oxide- and TNF-α-dependent mechanisms despite up-regulation of 70-kDa heat shock protein J Immunol 2003; 170:4119–4126.

392. Haviv YS, Blackwell JL, Li H, Wang M, Lei W, Curiel DT. Heat shock and heat shock protein 70i enhance the oncolytic effect of replicative adenovirus Cancer Res 2001; 61: 8361–8365.

393. Khorana AA, Rosenblatt JD, Sahasrabudhe DM, Evans T, Ladrigan M, Marquis D, Rosell K, Whiteside T, Phillippe S, Acres B, Slos P, Squiban P, Ross M, Kendra K. A Phase I trial of immunotherapy with intratumoral adenovirus-interferon-gamma (TG1041) in patients with malignant melanoma Cancer Gene Ther 2003; 10:251–259.

394. Kalvakolanu DVR, Bandyopadhyay SK, Harter ML, Sen GC. Inhibition of interferon-inducible gene expression by adenovirus E1A proteins: block in transcriptional complex formation Proc Nat Acad Sci USA 1991; 88:7459–7463.

395. Paggi MG, Martelli F, Fanciulli M, Felsani A, Sciacchitano S, Varmi M, Bruno T, Carapella CM, Floridi A. Defective human retinoblastoma protein identified by lack of interaction with the E1A oncoprotein Cancer Res 1994; 54:1098–1104.

396. Harlow E. A research shortcut from a common cold virus to human cancer Cancer 1996; 78:558–565.

397. Li D, Day KV, Yu S, Shi G, Liu S, Guo M, Xu Y, Sreedharan S, O'Malley BW. The role of adenovirus-mediated retinoblastoma 94 in the treatment of head and neck cancer Cancer Res 2002; 62:4637–4644.

398. Tsukuda K, Wiewrodt R, Molnar-Kimber K, Jovanovic VP, Amin KM. An E2F-responsive replication-selective adenovirus targeted to the defective cell cycle in cancer cells: potent antitumoral efficacy but no toxicity to normal cells Cancer Res 2002; 62:3438–3447.

399. Kuhn H, Liebers U, Gessner C, Schumacher A, Witt C, Schauer J, Kovesdi I, Wolff G. Adenovirus-mediated E2F-1 gene transfer in nonsmall-cell lung cancer induces cell growth arrest an apoptosis Eur Respir J 2002; 20:703–709.

400. Hernandez-Alcoceba R, Pihalja M, Qian D, Clarke MF. New oncolytic adenoviruses with hypoxia- and estrogen-receptor-regulated replication Human Gene Ther 2002; 13:1737–1750.

401. Ramachandra M, Rahman A, Zou A, Vaillancourt M, Howe JA, Antelman D, Sugarman B, Demers GW, Engler H, Johnson D,

Shrabram P. Re-engineering adenovirus regulatory pathways to enhance oncolytic specificity and efficacy Nature Biotech 2001; 19:1035–1041.

402. Le XF, Vallian S, Mu ZM, Hung MC, Chang KS. Recombinant PML adenovirus suppresses growth and tumorigenicity of human breast cancer cells by inducing G_1 cell cycle arrest and apoptosis Oncogene 1998; 16:1839–1849.

403. Tollefson AE, Hermiston TW, Lichtenstein DL, Colle CF, Tripp RA, Dimitrov T, Toth K, Wells CE, Doherty PC, Wold WSM. Force degradation of Fas inhibits apoptosis in adenovirus-infected cells Nature 1998; 392:726–730.

404. Perez D, White E. ER1B 19K inhibits Fas-mediated apoptosis through FADD-dependent sequestration of FLICE J Cell Biol 1998; 141:1255–1266.

405. Shinoura N, Yoshida Y, Sadata A, Hanada KI, Yamamoto S, Kirino T, Asai A, Hamada H. Apoptosis by retrovirus- and adenovirus-mediated gene transfer of Fas ligand to glioma cells: implications for gene therapy Human Gene Ther 1998; 9:1983–1993.

406. Arai H, Gordon D, Nabel EG, Nabel GJ. Gene transfer of Fas ligand induces tumor regression *in vivo* Proc Nat Acad Sci USA 1997; 94:13862–13867.

407. Hedlund TE, Meech SJ, Srikanth S, Kraft AS, Miller GJ, Schaack JB, Duke RC. Adenovirus-mediated expression of Fas ligand induces apoptosis of human prostate cancer cells Cell Death Differ 1999; 6:175–182.

408. Sata M, Perlman H, Muruve DA, Silver M, Ikebe M, Libermann TA, Oettgen P, Walsh K. Fas ligand gene transfer to the vessel wall inhibits neointima formation and overrides the adenovirus-mediated T cell response Proc Nat Acad Sci USA 1998; 95:1213–1217.

409. Elsing A, Burgert HG. The adenovirus E3/10.4K–14.5K proteins down-modulate the apoptosis receptor Fas/Apo-1 by inducing its internalization Proc Nat Acad Sci USA 1998; 95: 10072–10077.

410. Burgert HG, Ruzsics Z, Obermeier S, Hilgendorf S, Windheim M, Elsing A. Subversion of host defense mechanisms by adenoviruses Curr Top Microbiol Immunol 2002; 269:272–318.

411. Doostzadeh-Cizeron J, Terry NH, Goodrich DW. The nuclear death domain protein p84N5 activates a G_2/M cell cycle checkpoint prior to the onset of apoptosis J Biol Chem 2001; 276: 1127–1132.

412. Yin S, Bailiang W, Xie K, Goodrich DW. Adenovirus-mediated N5 gene transfer inhibits tumor growth and metastasis of human carcinoma in nude mice Cancer Gene Ther 2002; 9: 665–672.

413. Kaliberov SA, Buchsbaum DJ, Gillespie GY, Curiel DT, Arafat WO, Carpenter M, Stackhouse MA. Adenovirus-mediated transfer of BAX driven by vascular endothel growth factor promoter induces apoptosis in lung cancer cells Mol Ther 2002; 6:190–198.

414. Desai SY, Patel RC, Sen GC, Malhorta P, Ghadge GD, Thimmapaya B. Activation of interferon-inducible 2′–5′ oligoadenylate synthetase by adenoviral VAI RNA J Biol Chem 1995; 270:3454–3461.

415. Lei M, Liu Y, Samuel CE. Adenovirus VAI RNA antagonizes the RNA-editing of the ADAR adenosine deaminase Virology 1998; 245:188–196.

416. Yu D, Scorsone K, Hung M-C. Adenovirus type 5 E1A gene products act as transformation suppressors of the *neu* oncogene Molec Cell Biol 1991; 11:1745–1750.

417. Yu D, Working P, Ando D. Selectively replicating oncolytic adenoviruses as cancer therapeutics Curr Opin Mol Ther 2002; 4:435–443.

418. Ring CJ. Cytolytic viruses as potential anti-cancer agents J Gen Virol 2002; 83:491–502.

419. McCormick F. dl1520 (Onyx-015) as an antitumor agent. In: Hernáiz Driever P, Rabkin SD, Eds. Replication-Competent Viruses for Cancer Therapy. Basel: Karger. Monogr Virol 2001; 22:46–55.

420. Bonnet MC, Tartaglia J, Verdier F, Kourilsky P, Lindberg A, Klein M, Moingeon P. Recombinant viruses as a tool for therapeutic vaccination against cancers Immunol Lett 2000; 74: 11–25.

421. Howe JA, Demers GW, Johnson DE, Neugebauer SEA, Perry ST, Vaillancourt MT, Faha B. Evaluation of E1-mutant adenoviruses as conditionally replicating agents for cancer therapy Mol Therap 2000; 2:485–495.

422. Steele T. Recent developments in the virus therapy of cancer Proc Soc Exp Biol Med 2000; 223:118–125.

423. Wodarz D. Viruses as antitumor weapons: defining conditions for tumor remission Cancer Res 2001; 61:3501–3507.

424. Nemunaitis J, Edelman J. Selectively replicating viral vectors Cancer Gene Ther 2002; 9:987–1000.

425. Green NK, Seymour LW. Adenoviral vectors: systemic delivery and tumor targeting Cancer Gene Ther 2002; 9: 1036–1042.

426. Reid T, Warren R, Kirn D. Intravascular adenoviral agents in cancer patients: lessons from clinical trials Cancer Gene Ther 2002; 9:979–986.

427. Biederer C, Ries S, Brandts CH, McCormick F. Replication-selective viruses for cancer therapy J Mol Med 2002; 80: 163–175.

428. Geoerger B, Grill J, Opolon P, Morizet J, Aubert G, Terrier-Lacombe M-J, Bressac-de Paillerets B, Barrois M, Feunteun J, Kirn DH, Vassal G. Oncolytic activity of the E1B-55 kDa-deleted adenovirus ONYX-015 is independent of cellular p53 status in human glioma xenografts Cancer Res 2002; 62: 764–772.

429. Nemunaitis J, Khury F, Ganly I, Arseneau J, Posner M, Vokes E, Kuhn J, McCarty T, Landers S, Blackburn A, Romel L, Randlev B, Kaye S, Kirn D. A Phase II trial of intratumoral injection of ONYX-015 in patients with refractory head and neck cancer J Clin Oncol 2001; 19:289–298.

430. Nemunaitis J, Cunningham C, Tong AW, Post L, Netto G, Paulson AS, Rich D, Blackburn A, Sands B, Gibson B, Randlev B, Freeman S. Pilot trial of intravenous infusion of a replication-selective adenovirus (ONYX-915) in combination with chemotherapy or LI-2 treatment in refractory cancer patients Cancer Gene Ther 2003; 10:341–352.

431. Reid T, Galanis E, Abbruzzese J, Sze D, Wein LM, Andrews J, Randlev B, Heise C, Uprichard M, Hatfield M, Rome L,

Rubin J, Kirn D. Hepatic arterial infusion of a replication-selective oncolytic adenovirus (*dl*1520): Phase II viral, immunologic and clinical endpoints Cancer Res 2002; 62: 6070–6079.

432. Makower DM, Rozenblitt A, Kaufman H, Edelman M, Lane ME, Zwiebel J, Haynes H, Wadler S. Phase II clinical trial of intralesional administration of the oncolytic adenovirus ONYX-015 in patients with biliary tumors with correlative p53 studies Clin Cancer Res 2003; 9:693–702.

433. Wadler S, Yu B, Tan J-Y, Kaleya R, Rozenblit A, Makower D, Edelman M, Lane M, Hyjek E, Horwitz M. Persistent replication of the modified chimeric adenovirus ONYX-015 in both tumor and stromal cells from a patient with gall bladder cancer implants Clin Cancer Res 2003; 9:33–43.

434. Lee B, Choi J, Kim J, Kim J-H, Joo CH, Cho YK, Kim YK, Lee H. Oncolysis of human gastric cancers by an E1B 55kDa-deleted YKL-1 adenovirus Cancer Lett 2002; 185:225–322.

435. Kim J, Cho JY, Kim J-H, Jung KC, Yun C-O. Evaluation of E1B gene-attenuated replicating adenoviruses for cancer gene therapy Cancer Gene Ther 2002; 9:725–736.

436. Deng J, Kloosterbooer F, Xia W, Hung M-C. The NH_2-terminal and conserved region 2 domains of adenovirus E1A mediate two distinct mechanisms of tumor suppression Cancer Res 2002; 62:346–350.

437. Frisch SM, Dolter KE. Adenovirus *E1a*-mediated tumor suppression by a c-*erb*B-2/*neu*-independent mechanism Cancer Res 1995; 55:5551–5555.

438. Siders WM, Halloran PJ, Fenton FG. Transcriptional targeting of recombinant adenoviruses to human and murine melanoma cells Cancer Res 1996; 56:5638–5646.

439. McCarty J, Wang Z, Xu H, Hu Y, Park B, Alexander H, Bartlett D. Development of melanoma-specific adenovirus Mol Ther 2002; 6:471.

440. Nettelbeck DM, Rivera AA, Balagué C, Alemany R, Curiel DT. Novel oncolytic adenoviruses targeted to melanoma: specific viral replication and cytolysis by expression of E1A mutants from the tyrosinase enhancer/promoter Cancer Res 2002; 62: 4663–4670.

441. Lamfers MLM, Grill J, Dirven CMF, van Beusechem VW, Geoerger B, van den Berg J, Alemany R, Fueyo J, Curiel DT, Vassal G, Pinedo HM, Vandertop WP, Gerritsen WR. Potential of the conditionally replicative adenovirus Ad3-Δ24RGD in the treatment of malignant gliomas and its enhanced effect with radiotherapy Cancer Res 2002; 62:5736–5742.

442. Zhao T, Rao X-M, Xie X, Li L, Thompson TC, McMasters KM, Zhou HS. Adenovirus with insertion-mutated E1A selectively propagates in liver cancer cells and destroys tumors *in vivo* Cancer Res 2003; 63:3073–3078.

443. Hamid O, Varterasian ML, Wadler S, Hecht JR, Benson A, Galanis E, Uprichard M, Omer C, Bycott P, Hackman RC, Shields A. Phase II trial of intravenous CI-1042 in patients with metastatic colorectal cancer J Clin Oncol 2003; 21: 1498–1504.

444. Henderson DR, Chen Y, Yu D-C. Attenuated replication-competent adenoviruses (ARCA©) for prostate cancer: CV706 to CV787. In: Hernáiz Driever P, Rabkin SD, Eds. Replication-Competent Viruses for Cancer Therapy. Basel: Karger. Monogr Virol 2001; 22:56–80.

445. Hsieh C-L, Yang L, Miao L, Yeung F, Kao C, Yang H, Zhau HE, Chung LWK. A novel targeting modality to enhance adenoviral replication by vitamin D3 in androgen-independent human prostate cancer cells and tumors Cancer Res 2003; 62: 3084–3092.

446. Freytag SO, Khil M, Stricker H, Peabody J, Menon M, DePeralta-Venturina M, Nafziger D, Pegg J, Paielli D, Brown S, Barton K, Lu M, Aguilar-Cordova E, Kim JH. Phase I study of replication-competent adenovirus-mediated double suicide gene therapy for the treatment of locally recurrent prostate cancer Cancer Res 2002; 62:4968–4976.

447. Pramudji C, Shimura S, Ebara S, Yang G, Wang J, Ren C, Yuan Y, Tahir SA, Timme TL, Thompson TC. In situ prostate cancer gene therapy using a novel adenoviral vector regulated by the caveolin-1 promoter Clin Cancer Res 2001; 7: 4272–4279.

448. Parada C, Hernández Losa J, Guinea J, Sanchez-Arévalo V, Fernández Soria V, Alvarez-Vallina L, Sánchez-Prieto R,

Ramón Y, Cajal S. Adenovirus E1a protein enhances the cytotoxic effects of the herpes thymidine kinase-ganciclovir system Cancer Gene Ther 2003; 10:152–160.

449. Bauerschmitz GJ, Lam JT, Kanerva A, Suzuki K, Nettelbeck DM, Dmitriev I, Krasnykh V, Mikheeva GV, Barnes MN, Alvarez RD, Dall P, Alemany R, Curiel DT, Hemminki A. Treatment of ovarian cancer with tropism modified oncolytic adenovirus Cancer Res 2002; 62:1266–1270.

450. Rosenfeld ME, Feng M, Michael SI, Siegal GP, Alvarez RD, Curiel DT. Adenoviral-mediated delivery of herpes simplex virus thymidine kinase gene selectively sensitizes human ovarian carcinoma cells to ganciclovir Clin Cancer Res 1995; 1:1571–1580.

451. Hakkarainen T, Hemminki A, Pereboev AV, Barker SD, Asiedu CK, Strong TV, Kanerva A, Wahlfors J, Curiel DT. CD40 is expressed on ovarian cancer cells and can be utilized for targeting adenoviruses Clin Cancer Res 2003; 9:619–624.

452. Barnes MN, Coolidge CJ, Hemminki A, Alvarez RD, Curiel DT. Conditionally replicative adenoviruses for ovarian cancer therapy Mol Cancer Therap 2002; 1:435–439.

453. Russell W. Adenovirus gene therapy for ovarian cancer J Natl Cancer Inst 2002; 94:706–707.

454. Meeker TC, Lay LT, Wroblewski JM, Turturro F, Li Z, Seth P. Adenoviral vector efficiently target cell lines derived from selected lymphocytic malignancies including anaplastic large cell lymphoma and Hodgkin's disease Clin Cancer Res 1997; 3:357–364.

455. Feng WH, Westphal E, Mauser A, Raab-Traub N, Gulley ML, Bussin P, Kenney SC. Use of adenovirus vectors expressing Epstein-Barr virus (EBV) immediate-early protein BZLF1 or BRLF1 to treat EBV-positive tumors J Virol 2002; 76: 10951–10959.

456. Smythe WR, Hwang HC, Amin KM, Eck SL, Davidson BL, Wilson JM, Kaiser LR, Albelda SM. Use of recombinant adenovirus to transfer the herpes simplex virus thymidine kinase (HSV*tk*) gene to thoracic neoplasms: an effective *in vitro* drug sensitization system Cancer Res 1994; 54:2055–2059.

457. Wierwrodt R, Amin K, Kiefer M, Jovanovic VP, Kapoor V, Force S, Chang M, Lanuti M, Black ME, Kaiser LR, Albelda SM. Adenovirus-mediated gene transfer of enhanced herpes simplex virus thymidine kinase mutants improves prodrug-mediated tumor cell killing Cancer Gene Ther 2003; 10: 353–364.

458. Wolkersdofer GW, Bornstein SR, Higginbotham JN, Hiroi N, Vaquero JJ, Green MV, Blaese RM, Aguilera G, Chrousos GP, Ramsey WJ. A novel approach using transcomplementing adenoviral vectors for gene therapy of adrenocortical cancer Horm Metab Res 2002; 34:279–287.

459. Chung-Faye G, Palmer D, Anderson D, Clark J, Downes M, Baddeley J, Hussain S, Murray PI, Searle P, Seymour L, Harris PA, Ferry D, Kerr DJ. Virus-directed, enzyme prodrug therapy with nitroimidazole reductase: a Phase I and pharmacokinetic study of its prodrug, CB1954 Clin Cancer Res 2001; 7:2662–2668.

460. Wagner LM, Guichard SM, Burger RA, Morton CL, Straign CM, Ashmun RA, Harris LC, Houghton PJ, Potter PM, Danks MK. Efficacy an toxicity of a virus-directed enzyme prodrug therapy purging method: preclinical assessment and application to bone marrow samples from neuroblastoma patients Cancer Res 2002; 62:5001–5007.

461. Lillo R, Ramírez M, Álvarez Á, Santos S, García-Castro J, Fernández de Velasco J, Avilés MJ, Gómez-Pineda A, Díez JL, Balas A, Vicario JL, Bueren JA, García-Sánchez F. Efficient and nontoxic adenoviral purging method for autologous transplantation in breast cancer Cancer Res 2002; 62:5013–5018.

462. Pan X, Li Z, Xu G, Et A. Adenoviral mediated suicide gene transfer in the treatment of pancreatic cancer Chin Med J 2002; 115:1205–1208.

463. Yin L, Fu S, Wang X, Nanakorn T, Won J, Deisseroth A. Sensitization of prostate cancer cell lines to 5-fluorocytosine induced by adenoviral vector carrying a CD transcription unit Chin Med J 2001; 114:972–975.

464. Akbulut H, Zhang L, Tang Y, Deisseroth A. Cytotoxic effect of replication-competent adenoviral vectors carrying L-plastin promoter regulated E1A and cytosine deaminase genes in can-

cers of the breast, ovary and colon Cancer Gene Ther 2003; 10:388–395.

465. Li Y, Yu D-C, Chen Y, Amin P, Zhang H, Nguyen N, Hengerson DR. A hepatocellular carcinoma-specific adenovirus variant, CV890, eliminates distant human liver tumors in combination with doxorubicin Cancer Res 2001; 61:6428–6436.

466. Zhang DR, Ramesh N, Chen Y, Li Y, Dilley J, Working P, Y D-C. Identification of human uroplakin II promoter, a urothelium-specific adenovirus variant that eliminates established bladder tumors in combination with docetaxel Cancer Res 2002; 62:3743–3750.

467. Van Meir EG, Kikuchi T, Tada M, Li H, Diserens A-C, Wojcik BE, Huang H-JS, Friedmann T, de Tribolet N, Cavenee WK. Analysis of the p53 gene and its expression in glioblastoma cells Cancer Res 1994; 54:649–652.

468. Lang FL, Bruner JM, Fuller GN, Aldape K, Prados MD, Chang S, Berger MS, McDermott MW, Kunwar SM, Junck LR, Chandler W, Zwiebel JA, Kaplan RS, Yung WKA. Phase I trial of adenovirus-mediated p53 gene therapy for recurrent glioma: biological and clinical results J Clin Oncol 2003; 21: 2506–2518.

469. Clayman GL, El-Naggar AK, Lippman SM, Henderson YC, Frederick M, Merritt JA, Zumstein LA, Timmons TM, Liu T-J, Ginsberg L, Roth JA, Hong WK, Bruso P, Goepfert H. Adenovirus-mediated p53 gene transfer in patients with advanced recurrent head and neck squamous cell carcinona J Clin Oncol 1998; 16:2221–2232.

470. Higuchi Y, Asaumi J-I, Murakami J, Wakasa T, Inoue T, Kuroda M, Shibuya K, Shigehara H, Kawasaki S, Fukui K, Kishi K, Hiraki Y. Adenoviral p53 gene therapy in head and neck squamous cell carcinoma cell lines Oncol Rep 2002; 9: 1233–1236.

471. Chen W, Lee Y, Wang H, Yu G-G, Jiao W, Zhou W, Zeng Y. Suppression of human nasopharyngeal carcinoma cell growth in nude mice by the wild-type p53 gene J Cancer Res Clin Oncol 1992; 119:46–48.

472. Nishizaki M, Meyn RE, Levy LB, Atkinson EN, Roth JA, Ji L. Synergistic inhibition of human lung cancer cell growth by

adenovirus-mediated wild-type *p53* gene transfer in combination with docetaxel and radiation therapeutics *in vitro* and *in vivo* Clin Cancer Res 2001; 7:2887–2897.

473. Schuler M, Herrman R, De Greve JLP, Stewart AK, Gatzemeier U, Stewart DJ, Laufman L, Gralla R, Kuball J, Buhl R, Heussel CP, Kommoss F, Perruchaud AP, Shepherd FA, Fritz MA, Horowitz JA, Huber C, Rochlitz C. Adenovirus-mediated wild-type p53 transfer in patients receiving chemotherapy for advanced non-small-cell lung cancer: results of a multicenter Phase I study J Clin Oncol 2001; 19:1750–1758.

474. Ameyar M, Shatrov V, Bouquet C, Capoulade C, Cai Z, Stancou R, Bdie C, Haddada H, Chouaib S. Adenovirus-mediated transfer of wild-type p53 gene sensitizes TNF resistant MCF7 derivatives to the cytotoxic effects of this cytokine: relationship with c-*myc* and Rb Oncogene 1999; 18:5464–5472.

475. Eastham JA, Hall SJ, Sehgal I, Wang J, Timme TL, Yang G, Connell-Crowley L, Elledge SJ, Zhang W-W, Harper JW, Thompson TC. In vivo gene therapy with *p53* or *p21* adenovirus for prostate cancer Cancer Res 1995; 55:5151–5155.

476. Yang C, Cirielli C, Capogrossi MC, Passaniti A. Adenovirus-mediated wild-type *p53* expression induces apoptosis and suppresses tumorigenesis of prostatic tumor cells Cancer Res 1995; 55:4210–4213.

477. Wada Y, Gotoh A, Shirakawa T, Hamada K, Kamidono S. Gene therapy for bladder cancer using adenoviral vector Mol Biol 2001; 5:47–52.

478. Kuball J, Wen SF, Leissner J, Atkins D, Meinhardt P, Quijano E, Engler H, Hutchins B, Maneval DC, Grace MJ, Fritz MA, Störkel S, Thüroff JW, Huber C, Schuler M. Successful adenovirus-mediated wild-type p53 gene transfer in patients with bladder cancer by intravesical vector instillation J Clin Oncol 2002; 20:957–965.

479. Pagliaro LC, Keyhani A, Williams D, Woods D, Liu B, Perrotte P, Slatan JW, Merritt JA, Grossman HB, Dinney CP. Repeated intravesical instillation of an adenoviral vector in patients with locally advanced bladder cancer: a Phase I study of p53 gene therapy J Clin Oncol 2003; 21:2247–2253.

480. van Beusechem VW, van den Doel PB, Grill J, Oinedo HM, Gerritsen WR. Conditionally replicative adenovirus express-

ing p53 exhibits enhanced oncolytic potency Cancer Res 2002; 62:6155–6171.

481. Haviv YS, Takatyama K, Glasgow JN, Blackwell JL, Wang M, Lei X, Curiel DT. A model system for the design of armed replicating adenoviruses using p53 as a candidate transgene Mol Cancer Ther 2002; 1:321–328.

482. Sauthoff H, Pipiya T, Heitner S, Chen S, Norman RG, Rom WN, Hay JG. Late expression of p53 from a replicating adenovirus improves tumor cell killing and is more tumor cell specific than expression of the adenovirus death protein Human Gene Ther 2002; 13:1859–1871.

483. Sazawa A, Watanabe T, Tanaka M, Haga K, Fujita H, Harabayashi T, Shinohara N, Koyanagi T, Kuzumaki N. Adenovirus-mediated gelsolin gene therapy for orthotopic human bladder cancer in nude mice J Urol 2002; 168:1182–1187.

484. Li D, Day KV, Yu S, Shi G, Liu S, Guo M, Xu Y, Sreedharan S, O'Malley BW. The role of adenovirus-mediated retinoblastoma 94 in the treatment of head and neck cancer Cancer Res 2002; 62:4637–4644.

485. Davies MA, Kim SJ, Parikh NU, Dong Z, Bucana CD, Gallick GE. Adenoviral-mediated expression of MMAC/PTEN inhibits proliferation and metastasis of human prostate cancer cells Clin Cancer Res 2002; 8:1904–1914.

486. Ding Y, Wen Y, Spohn B, Wang L, Xia W, Kwong KY, Shao R, L Z, Hortobagyi GN, Hung M-C, Yan D-H. Proapoptotic and antitumor activities of adenovirus-mediated p202 gene transfer Clin Cancer Res 2002; 8:3290–3297.

487. Deng X, Kim M, Vandier D, Jung YJ, Rikiyama T, Sgagias MK, Goldsmith M, Cowan KH. Recombinant adenovirus-mediated p14(ARF) overexpression sensitizes human breast cancer cells to cisplatin Biochem Biophys Res Comm 2002; 296:792–798.

488. Calvo A, Feldman AL, Libutti SK, Green JE. Adenovirus-mediated endostatin delivery results in inhibition of mammary gland tumor growth in C3(1)/SV40 T-antigen transgenic mice Cancer Res 2002; 62:3934–3938.

489. Mogi S, Ebata T, Setoguchi Y, Fujime M, Heike Y, Kohsaka T, Yagita H, Okumura K, Azuma M. Efficient generation of

autologous peripheral blood-derived cytotoxic T lymphocytes against poorly immunogenic human tumors using recombinant CD80-adenovirus together with interleukin 12 and interleukin 2 Clin Cancer Res 1998; 4:713–720.

490. Buschle M, Schmidt W, Zauner W, Mechtler K, Trska B, Kirlappos H, Birnstiel ML. Translocating of tumor antigen-derived peptides into antigen-presenting cells Proc Nat Acad Sci USA 1997; 94:3256–3261.

491. Bellone M, Iezzi G, Rovere P, Galati G, Ronchetti A, Protti MP, Davoust J, Rugarli C, Manfredi AA. Processing of engulfed apoptotic bodies yields T cell epitopes J Immunol 1997; 159: 5391–5399.

492. Kotera Y, Shimizu K, Mule JJ. Comparative analysis of necrotic and apoptotic tumor cells as a source of antigen(s) in dendritic cell-based immunization Cancer Res 2001; 61: 8105–8109.

493. Strome SE, Voss S, Wilcox R, Wakefield TL, Tamada K, Flies D, Chapoval A, Lu J, Kasperbauer JL, Padley D, Vile R, Gastineau D, Wettstein P, Chen L. Strategies for antigen loading of dendritic cells to enhance the antitumor immune response Cancer Res 2002; 62:1884–12889.

494. Scheffer SR, Nave H, Korangy F, Schlote K, Pabst R, Jaffeee EM, Manns M, Greten TF. Apoptotic, but not necrotic, tumor cell vaccines induce a potent immune response *in vivo* Int J Cancer 2003; 103:205–211.

495. Rubio M-P, Guerra S, Almendral JM. Genome replication and postencapsidation functions mapping to the nonstructural gene restrict the host range of a murine parvovirus in human cells J Virol 2001; 75:11573–11582.

496. Davis C, Segev-Amzaleg N, Rotem I, Mincberg M, Amir N, Sivan S, Gitelman I, Tal J. The P4 promoter of the parvovirus minute virus of mice developmentally regulated in trandgenic P4-*LacZ* mice Virology 2003; 306:268–279.

497. Anderson MJ, Pattison JR. The human parvoviruses Arch Virol 1984; 82:137–148.

498. Cotmore SF, Tattersall P. The autonomously replicating parvoviruses of vertebrates Adv Virus Res 1987; 33:91–173.

499. Rommelaere J, Cornelis JJ. Autonomous parvoviruses. In: Hernáiz Driever P, Rabkin SD, Eds. Replication-Competent Viruses for Cancer Therapy. Basel: Karger. Monograph Virol 2001; 22:100–129.

500. Mayor HD. Defective parvoviruses may be good for your health! In: Melnick JL, Ed. Basel: Karger. Progr Med Virol 1993; 40:193–205.

501. Kurtzman G, Frickhofen N, Kimball J, Jenkins DW, Nienhuis AW, Young NS. Pure red-cell aplasia due to persistent parvovirus B19 infection and its cure with immunoglobulin therapy N Engl J Med 1989; 321:5219–523.

502. Nerlich A, Schwarz TF, Roggendorf M, Roggendorf H, Ostermeyer E, Schramm T, Gloning K-P. Parvovirus B19-infected erythroblats in fetal cord cord blood Lancet 1991; 337:310.

503. Brown KE, Hibbs JR, Gallinella G, Anderson SM, Lehman ED, McCarthy P, Young NS. Resistance to parvovirus B19 infection due to lack of virus receptor (erythrocyte antigen) N Engl J Med 1994; 330:1192–1196.

504. Bell LM, Naides SJ, Stoffman P, Hodinka RL, Plotkin SA. Human parvovirus B19 infection among hospital staff members after contact with infected patients N Engl J Med 1989; 321:485–491.

505. Schild RL, Bald R, Plath H, Eis-Hübinger AM, Enders G, Hansmann M. Intrauterine management of fetal parvovirus B_{19} infection Ultrasound Obstet Gynecol 1999; 13:161–166.

506. Simpson RW, McGinty L, Simon L, Smith CA, Godzeski CW, Boyd RJ. Association of parvoviruses with rheumatoid arthritis of humans Science 1984; 223:1425–1428.

507. Giese NA, Raykov Z, DeMartino L, Vecchi A, Sozzani S, Dinart C, Cornelis JJ, Rommelarere J. Suppression of metastatic hemangiosarcoma by parvovirus MVMp vector transducing the IP-10 chemokine into immunocompetent mice Cancer Gene Ther 2002; 9:432–442.

508. Dupont F, Avalosse B, Karim A, Nine N, Bosseler M, Maron A, van den Broeke AV, Ghanem GE, Burny A, Zeicher M. Tumor-selective gene transduction and cell killing with an oncotropic parvovirus-based vector Gene Ther 2000; 7:790–796.

509. Maxwell IH, Chapman JT, Scherrer LC, Spitzer AL, Leptihn S, Maxwell F, Corsini JA. Expansion of tropism of a feline parvovirus to target a human tumor cell line by display of an alpha(v) integrin binding peptide on the capsid Gene Ther 2001; 8:324–331.

510. Dworak LJ, Wolfinbarger JB, Bloom ME. Aleutian mink disease parvoviruas infection of K562 cells is antibody-dependent and is mediated via an Fc(gamma)RII receptor Arch Virol 1997; 142:363–373.

511. Aasted B, Alexandersen S, Christensen J. Vaccination with Aleutian mink disease parvovirus (AMDV) capsid proteins enhances disease, while vaccination with the major non-structural AMDV protein causes partial protection from disease Vaccine 1998; 16:1158–1165.

512. Best SM, Shelton JF, Pompey JM, Wolfinbarger JB, Bloom ME. Caspase cleavage of the nonstructural protein NS1 mediated replication of Aleutian mink disease parvovirus J Virol 2003; 77:5305–5312.

513. Yang TJ. Parvovirus-induced regression of canine transmissible venereal sarcoma Am J Vet Res 1987; 48:799–800.

514. Zhuang S-M, Shvarts A, van Ormondt H, Jockemsen AJ, van der Eb AJ, Noteborn MHM. Apoptin, a protein derived from chicken anemia virus, induces p53-independent apoptosis in human osteosarcoma cells Cancer Res 1995; 55:486–489.

515. Olijslagers S, Dege AY, Dinsart C, Voorhoeve M, Rommelaere J, Noteborn MH, Cornelis JJ. Potentiation of a recombinant oncolytic parvovirus by expression of apoptin Cancer Gene Ther 2001; 8:958–965.

516. Sinkovics JG, Horvath J. Can virus therapy of human cancer be improved by apoptosis induction? Med Hypotheses 1995; 44:359–368.

517. Toolan HW, Saunders EL, Southam CM, Moore AE, Levin AG. H-1 virus viremia in the human Proc Soc Exp Biol Med 1965; 119:711–715.

518. Dupressoir T, Vanacker J-M, Cornelis JJ, Duponchel N, Rommelaere J. Inhibition by parvovirus H-1 of the formation of tumors in nude mice and colonies *in vitro* by transformed

human mammary epithelial cells Cancer Res 1998; 49: 3203–3208.

519. Moehler M, Zeidler M, Schede J, Rommelaere J, Galle PR, Cornelis JJ, Heike M. Oncolytic parvovirus H1 induces release of heat-shock protein HSP72 in susceptible human tumor cells but may not affect primary tumor cells Cancer Gene Ther 2003; 10:477–480.

520. Moehler M, Blechacz B, Wiskopf N, Zeidler M, Stremmel W, Rommelarer J, Galle PR, Cornelis JJ. Effective infection, apoptotic cell killing and gene transfer of humas hepatoma cells but not primary hepatocytes by parvovirus H1 and derived vectors Cancer Gene Ther 2001; 8:158–167.

521. Wetzel K, Menten P, Opdenakker G, Van Damme J, Grone HJ, Giese N, Vecchi A, Sozzani S, Cornelis JJ, Rommelaere J, Dinsart C. Transduction of human MCP-3 by a parvoviral vector induces leukocyte infiltration and reduces growth of human cervical carcinoma cell xenografts J Gene Med 2001; 3:326–337.

522. Maxwell IH, Terrell KL, Maxwell F. Autonomous parvovirus vectors Methods 2002; 28:168–181.

523. Mayor HD, Drake S, Stahmann J, Mumford DM. Antibodies to adeno-associated satellite virus and herpes simplex in sera from cancer patients and normal adults Am J Obstet Gynecol 1976; 126:100–104.

524. Shadan FF, Villareal LP. Parvovirus-mediated antineoplastic activity exploits genome instability Med Hypotheses 2000; 55: 1–4.

525. Coker AL, Russell RB, Bond SM, Pirisi L, Liu Y, Mane M, Kokorina N, Gerasimova T, Hermonat PL. Adeno-associated virus is associated with a lower risk of high grade cervical neoplasia Exp Mol Pathol 2001; 70:83–89.

526. Hermonat PL. Adeno-associated virus inhibits human papillomavirus type 16: a viral interaction implicated in cervical cancer Cancer Res 1994; 54:2278–2281.

527. Ogston P, Raj K, Beard P. Productive replication of adeno-associated virus can occur in human papillomavirus type 16 (HPV-16) episome-containing keratinocytes and is augmented by the HPV-16 E2 protein J Virol 2000; 74:3494–3504.

528. Erles K, Rohde V, Thaele M, Roth S, Edler L, Schlehofer JR. DNA of adeno-associated virus in testicular tissue and in abnormal semen samples Hum Reprod 2001; 16:2333–2337.

529. Berns KI, Hauswirth WW. Adeno-associated viruses Adv Virus Res 1079; 25:407–449.

530. Walz C, Schlehofer JR. Modification of some biological properties of HeLa cells containing adeno-associated virus DNA integrated into chromosome 17 J Virol 1992; 66:2990–3002.

531. Muzyczka N. Adeno-associated virus (AAV) vectors: will they work? J Clin Invest 1994; 94:1351.

532. Kotin RM. Prospects for the use of adeno-associated virus as a vector for human gene therapy Hum Gene Ther 1994; 5: 793–801.

533. Horvath J, Palkonyai L, Weber J. Group C adenovirus DNA sequences in human lymphoid cells J Virol 1986; 59:189–192.

534. Muro-Cacho CA, Samulski RJ, Kaplan D. Gene transfer in human lymphocytes using a vector based on adeno-associated virus J Immunother 1992; 11:231–237.

535. Ma H-I, Guo P, Li J, Lin S-Z, Chiang Y-H, Xiao X, Cheng S-Y. Suppression of intracranial human glioma growth after intramuscular administration of an adeno-associated viral vector expressing angiostatin Cancer Res 2002; 62:756–763.

536. Kurisaki K, Kurisaki A, Valcourt U, Terentiev AA, Pardali K, ten Dijke P, Heldin C-H, Ericsson J, Moustakas A. Nuclear factor YY1 inhibits transforming growth factor β- and bone morphogenic protein-induced cell differentiation Mol Cell Biol 2003; 23:4494–4510.

537. Batchu RB, Shammas MA, Wang JY, Freeman J, Rosen N, Munshi NC. Adeno-associated virus protects the retinoblastoma family of proteins from adenoviral-induced functional inactivation Cancer Res 2002; 62:2982–2985.

538. Duverger V, Sartorius U, Klein-Bauernschmitt P, Krammer PH, Schlehofer JR. enhancement of cisplatin-induced apoptosis by infection with adeno-associated virus type 2 Int J Cancer 2002; 97:706–712.

539. Walz C, Schlehofer JR, Flentje M, Rudat V, zur Hausen H. Adeno-associated virus sensitizes HeLa cell tumors to gamma rays J Virol 1992; 66:5651–5657.

540. Mori S, Murakami M, Takeuchi T, Kozuka T, Kanda T. Rescue of AAV by antibody-induced Fas-mediated apoptosis from viral DNA integrated in HeLa chromosome Virology 2002; 301:90–98.

541. Raj K, Ogston P, Beard P. Virus-mediated killing of cells that lack p53 activity Nature 2001; 412:914–916.

542. Vogelstein B, Kinzler KW. Achilles' heel of cancer? Nature 2001; 412:865–866.

543. Flint SJ, Enquist LW, Krug RM, Racaniello VR, Skalka AM. Principles of Virology, Molecular Biology, Pathogenesis, and Control. Washington DC: Am Soc Microbiol Press, 2000: 117–118; 148–149; 396; 444–449; 596–598; 670–673; 750–751.

544. Ida-Hosonuma M, Sasaki Y, Toyoda H, Nomoto A, Gotoh O, Yonekawa H, Koike S. Host range of poliovirus is restricted to simians because of a rapid sequence change of the poliovirus receptor gene during evolution Arch Virol 2003; 148:29–44.

545. Calandria C, López-Guerrero JA. Poliovirus modulates Bcl-xl expression in the human U937 promonocytic cell line Arch Virol 2002; 147:2445–2452.

546. Gromeier M, Alexander L, Wimmer E. Internal ribosomal entry site substitution eliminates neurovirulence in intergeneric poliovirus recombinants Proc Nat Acad Sci USA 1996; 93:2370–2375.

547. Gromeier M, Bossert B, Arita M, Nomoto A, Wimmer E. Dual stem loops within the poliovirus internal ribosomal entry site controls neurovirulence J Virol 1999; 73:958–964.

548. Gromeier M, Lachman S, Sosenfeld MR, Gutin PH, Wimmer E. Intergeneric poliovirus recombinants for the treatment of malignant glioma Proc Nat Acad Sci USA 2000; 97:6803–6808.

549. Dufresne AT, Dobrikova EY, Schmidt S, Gromeier M. Genetically stable picornavirus expression vectors with recombinant internal ribosomal entry sites J Virol 2002; 76:8966–8972.

550. Lee SG, Kim DY, Hyun BN, Bae YS. Novel design architecture for genetic stability of recombinant poliovirus: the manipulation of G/C contents and their distribution patterns increases

the genetic stability of inserts in a poliovirus-based RPS-Vax vector system J Virol 2002; 76:1649–1662.

551. Vile R, Ando D, Kirn D. The oncolytic virotherapy treatment platform for cancer: unique biological and biosafety points to consider Cancer Gene Ther 2002; 9:1062–1067.

552. Wisher M. Biosafety and product release testing issues relevant to replication-competent oncolytic viruses Cancer Gene Ther 2002; 9:1056–1061.

553. Hodgson CP, Solaiman F. Virosomes: cationic liposomes enhance retroviral transduction Nat Biotech 1996; 14:339–342.

554. Kaneda Y. Virosomes: evolution of the liposome as a targeted drug delivery system Adv Drug Deliv Rev 2000; 43:197–205.

555. Mizuno M, Ryuke Y, Yoshida J. Cationic liposomes conjugation to recombinant adenoviral vectors containing herpes simplex virus thymidine kinase gene followed by ganciclovir treatment reduces viral antigenicity and maintains antitumor activity in mouse experiemental glioma model Cancer Gene Ther 2002; 9:825–829.

556. Hortobagyi G, Ueno NT, Xia W, Zhang S, Wolf JK, Putnam JB, Weiden PL, Willey JS, Carey M, Branham DL, Payne JY, Tucker SD, Bartholomeusz C, Kilbourn RG, De Jager RL, Sneige N, Katz RL, Anklesaria P, Ibrahim NK, Murray JL, Theriault RL, Valero V, Gershenson DM, Bevers MW, Huang L, Lopez-Berestein G, Hung MC. Cationic liposome-mediated E1A gene transfer to human breast and ovarian cancer cells and its biologic effects: a Phase I clinical trial J Clin Oncol 2001; 19:3422–3433.

557. Hubberstey AV, Pavliv M, Parks RJ. Cancer therapy utilizing an adenoviral vector expressing only E1A Cancer Gene Ther 2002; 9:321–329.

558. Fisher KD, Stallwood Y, Green NK, Ulbrich K, Mautner V, Seymour LW. Polymer-coated adenovirus permits efficient retargeting and evades neutralizing antibodies Gene Ther 2001; 8:341–348.

559. Fisher KD. Polymer-coated adenovirus can be programmed to infect cells using selected ligands and shows extended plasma circulation following intravenous injection. In: Bell JC, Ed.

Oncolytic Viruses as Cancer Therapeutics, March 26-30, 2003, Scientific Program 2003:17.

560. Pandori M, Hobson D, Sano T. Adenovirus-microbead conjugates possess enhanced infectivity: a new strategy for localized gene delivery Virology 2002; 299:204–212.

561. Grill J, Lamfers ML, van Beusechem VW, Dirven CM, Pherai DS, Kater M, van der Valk P, Vogels R, Vandertop WP, Pinedo HM, Curiel DT, Gerritsen WR. The organotypic multicellular spheroid is a relevant three-dimensional model to study adenovirus replication and penetration in human tumors *in vitro* Mol Ther 2002; 6:609–614.

562. Katze MG, He Y, Gale M. Viruses and interferon: a fight for supremacy Nature Rev Immunol 2002:675–687.

563. Kim M-J, Latham AG, Krug RM. Human influenza viruses activate an interferon-independent transcription of cellular antiviral genes: outcome with influenza A virus is unique Proc Nat Acad Sci USA 2002; 99:10096–10191.

564. Reiss CS, Rouse BT. The current status of viral immunology Immunology Today 1993; 14:333–335.

565. Zinkernagel RM, Hengartner H. Antiviral immunity Immunology Today 1997; 18:258–259.

566. Bachmann MF, Zinkernagel RM. Neutralizing antiviral B cell responses Annu Rev Immunol 1997; 15:235–270.

567. Seiler P, Bründler M-A, Zimmermann C, Weibel D, Bruns M, Hengartner H, Zinkernagel RM. Induction of protective cytotoxic T cell responses in the presence of high titers of virus-neutralizing antibodies: implications for passive and active immunization J Exp Med 1998; 187:649–654.

568. Eigen M. Error catastrophe and antiviral strategy Proc Nat Acad Sci USA 2002; 99:13374–13376.

569. Razvi ES, Welsh RM. Programmed cell death of T lymphocytes during acute viral infection: a mechanism for virus-induced immune deficiency J Virol 1993; 67:5754–5765.

570. Orange JS, Fassett MS, Koopman LA, Boyson JE, Strominger JL. Viral evasion of natural killer cells Nat Immunol 2002; 3: 1006–1012.

571. Rinaldo CR. Modulation of major histocompatibility complex antigen expression by viral infection Am J Pathol 1994; 144: 637–650.

572. McMichael AW, Klenerman P, Rowland-Jones S, Gotch F, Moss P. Recognition of viral antigens at the cell surface Mol Mechan Immune Respir 1995; 22:51–62.

573. Yewdell JW, Hill AB. Viral interference with antigen presentation Nat Immunol 2002; 3:1019–1025.

574. Benedict CA, Norris PS, Ware CF. To kill or to be killed: viral evasion of apoptosis Nat Immunol 2002; 3:1013–1018.

575. Hay S, Kannourakis G. A time to kill: viral manipulation of the cell death program J Gen Virol 2002; 83:1547–1564.

576. Vogels R, Zuijdgeest D, van Rijnsoever R, Hartkoorn E, Damen I, de Béthune M-P, Kostense S, Pendrs G, Helmus N, Koudstaal W, Cecchini M, Wetterwald A, Sprangers M, Lemckert A, Ophorts O, Koel B, van Meerendonk M, Quax P, Panitti L, Grimbergen J, Bout A, Goudsmit J, Havenga M. Replication-deficient human adenovirus type 35 vectors for gene transfer and vaccination: efficient human cell infection and bypass of preexisting adenovirus immunity J Virol 2003; 77:8263–8271.

577. Zheng J-Y, Chen D, Chan J, Yu D, Ko E, Pang S. Regression of prostate cancer xenografts by a lentiral vector specifically expressing diphtheria toxin A Cancer Gene Ther 2003; 10: 764–770.

578. Baum C, Düllmann J, Li Z, Fehse B, Meyer J, Williams DA, von Kalle C. Side effects of retroviral gene transfer into hematopoietic stem cells Blood 2003; 101:2099–2114.

579. Kuo R-L, Kung S-H, Hsu Y-Y, Liu W-T. Infection with enterovirus 71 or expression of its 2A protease induces apoptotic death J Gen Virol 2002; 83:1367–1376.

580. Ferdat AK, Bruvere RZH, Vitolin LA, Petrovska RG. Immunomodulation mechanisms in the anti-tumor effect of the ECHO-7 enterovirus Eksp Onkol 1989; 11:43–48.

581. Koh CY, Ortaldo JR, Blazar BR, Bennett M, Murphy WJ. NK-cell purging of leukemia: superior antitumor effects of NK cells

H2 allogeneic to the tumor and augmentation with inhibitory receptor blockade Blood 2003; 192:4067–4075.

582. Nagorsen D, Scheibenbogen C, Marincola FM, Letsch A, Keilholz U. Natural T cell immunity against cancer Clin Cancer Res 2003; 9:4296–4303.

583. Nair S, Boczkowski D, Moeller B, Dewhirst M, Vieweg J, Gilboa E. Synergy between tumor immunotherapy and antiangiogenic therapy Blood 2003; 102:964–971.

584. Hanahan D, Lanzavecchia A, Mihics E. Fourteenth Annual Pezcoller Symposium: the novel dichotomy of immune interactions with tumors Cancer Res 2003; 63:3005–3008.

585. Mogensen TH, Paludan SR. Molecular pathways in virus-induced cytokine production Microbiol Molec Biol Rev 2001; 65: 131–150.

586. Dalloul A, Laroche L, Bagot M, Mossalayi MD, Fourcade C, Thacker DJ, Hogge DE, Merle-Beral H, Schmitt C. Interleukin-7 is a growth factor for Sézary lymphoma cells J Clin Invest 1992; 90:1056–1060.

587. Asadullah K, Haeussler A, Friedrich M, Siegling A, Olaizola-Horn S, Trefzer U, Volk HD, Sterry W. IL-7 mRNA is not overexpressed in mycosis fungoides and pleomorphic T-cell lymphoma and is unlikely to be an autocrine growth factor *in vivo* Arch Dermatol Res 1996; 289:9–13.

588. Qin JZ, Kamarashev J, Zhang CL, Dummer R, Burg G, Dobbeling U. Constitutive and interleuken 7- and interleukin 15-stimulated DNA binding of STAT and novel factors in cutaneous T cell lymphoma cells J Invest Dermatol 2001; 117: 583–589.

589. Trudel S, Trachtenberg J, Toi A, Sweet J, Hua Z, Jewett M, Tshilias J, Zhuang LH, Hitt M, Wan Y, Gauldie J, Graham FL, Dancey J, Stewart K. A Phase I trial of adenovector-mediated delivery of interleukin-2 (AdIL-2) in high risk localized prostate cancer Cancer Gene Ther 2003; 10:755–763.

590. Mukherjee S, Upham JW, Ramshaw I, Bundell C, van Bruggen I, Robinson BWS, Nelson DJ. Dendritic cells infected with a vaccinia virus interleukin-2 vector secrete high levels of IL-12 and can become efficient antigen presenting cells that se-

crete high levels of immunoistimulatory cytokine Il-12 Cancer Gene Ther 2003; 10:591–602.

591. Brandt K, Bulfone-Paus S, Foster DC, Rückert R. Interleukin-21 inhibits dendritic cell activation and maturation Blood 2003; 102:4090–4098.

592. Obuchi M, Fernandez M, Barber GN. Development of recombinant vesicular stomatitis viruses that exploit defects in host defense to augment specific oncolytic activity J Virol 2003; 77: 8843–8856.

593. Rao VS, Bonavida B. Specific enhancement of tumor growth and depression of cell-mediated immunity following sensitization to soluble tumor antigens Cancer Res 1976; 36: 1384–1391.

594. Rumi L, Comerauer ME, Pasqualini CD. Conditions favoring immunological enhancement of a murine allogeneic lymphoma Arch Geschwulstforsch 1976; 46:457–460.

595. Bauer H, Hayami M, Stehfen-Gervinus JC. Influence of different routes of anti-tumor immunization: alternative induction of tumor immunity and tumor enhancement J Immunol 1979; 122:806–812.

596. Dei T, Tachibana T. Induced resistance in normal mice and growth enhancement in tumor-bearing mice by inoculation with syngeneic hybrid cells J Natl Cancer Inst 1980; 65: 739–749.

597. Bubenik J, Turano A. Enhancing effect on tumour growth of humoral antibodies against tumour specific transplantation antigens induced by murine sarcoma virus (Harvey) Nature 1986; 30:928–930.

598. Ershler WB, Tuck D, Moore AL, Klopp RG, Kramer KE. Immunologic enhancement of B16 melanoma growth Cancer 1988; 61:1792–1797.

599. Jolly D. Viral vector systems for gene therapy Cancer Gene Ther 1994; 1:51–64.

600. Bell JC, Garson KA, Lichty BD, Stojdl DF. Oncolytic viruses: programmable tumour hunters Curr Gene Ther 2002; 2: 243–254.

601. Wodarz D, Christensen JP, Thomsen AR. The importance of lytic and nonlytic immune responses in viral infections Trends Immunol 2002; 23:194–200.

602. Jenne L, Arrighi J-F, Jonuleit H, Saurat J-H, Hauser C. Dendritic cells containing apoptotic melanoma cells prime human $CD8^+$ T cells for efficient tumor cell lysis Cancer Res 2000; 60:4446–4452.

603. Stern LJ, Brown JH, Jardetzky TS, Gorga JC, Urban RG, Strominger JL, Willey DC. Crystal structure of the human class II MHC protein HLA-DR1 complexed with an influenza virus peptide Nature 1994; 368:215–221.

604. Trinchieri G. Interleukin-12 and its role in the generation of T_H1 cells Immunology Today 1993; 7:335–337.

605. Davis DM. Assembly of the immunological synapse for T cells and NK cells Trends Immunol 2002; 23:356–363.

606. Farag SS, Fehniger TA, Ruggeri L, Velardi A, Caligiuri MA. Natural killer cell receptors: new biology and insights into the graft-vs.-leukemia effect Blood 2002; 100:1935–1947.

607. Du Pasquier L, Litman G, Eds Origin and Evolution of the Vertebrate Immune System. New York: Springer Verlag, 2000.

608. Flajnik MF, Muller K, Du Pasquier L. Evolution of the immune system. In: Paul W, Ed. Fundamental Immunology. 5th edition. Philadelphia: Lippincott Williams & Wilkins, 2003: 519–570.

609. Kaufman J. The origins of the adaptive immune system: whatever next? Nature Immunol 2002; 3:1124–1125.

610. Tanaka S, Kunath T, Hadjantonakis A-K, Nagy A, Rossant J. Promotion of trophoblast stem cell proliferation by FGF4 Science 1998; 282:2072–2075.

611. Mi S, Lee X, Li X, Veldman GM, Finnerty H, Racie L, LaVallie E, Tang XY, Edouard P, Howes S, Keith JC, McCoy JM. Syncytin is a captive retroviral envelope protein involved in human placental morphogenesis Nature 2000; 403:785–799.

612. Zentella A, Weis FMB, Ralph DA, Laiho M, Massagué M. Early gene responses to transforming growth factor-β in cells

lacking growth-suppressive RB function Mol Cell Biol 1991; 11:4952–4958.

613. Uckan D, Steele A, Cherry X, Wang BY, Chamizo W, Koutsonikolis A, Gilbert-Barnes E, Good RA. Trophoblasts express Fas ligand: a proposed mechanism for immune privilege in placenta and maternal invasion Mol Hum Reprod 1997; 3: 655–652.

614. Khalturin K, Becker M, Rinkevich B, Bosch TCG. Urochordates and the origin of natural killer cells: identification of a CD94/NKR-P1-related receptor in blood cells of *Botryllus.* Proc Nat Acad Sci USA 2003; 100:622–627.

615. Cross JC, Baczyk D, Dobric N, Hemberger M, Hughes M, Simmons DG, Yamamoto H, Kingdom CP. Genes, development and evolution of the placenta Placenta 2003; 24:123–130.

616. Moffett-King A. Natural killer cells and pregnancy Nat Rev Immunol 2002; 2:656–663.

617. Boyson JE, Rybalov B, Koopman LA, Exley M, Balk SP, Racke FK, Schatz F, Masch R, Wilson SB, Strominger JL. CD1d and invariant NKT cells at the human maternal-fetal interface Proc Nat Acad Sci USA 2002; 99:13741–13746.

618. Smyth MJ, Hayakawa Y, Takeda K, Yagita H. New aspects of natural-killer cell surveillance and therapy of cancer Nat Rev Cancer 2002; 2:850–861.

619. Smyth MJ, Crowe NY, Hayakawa Y, Takeda K, Yagita H, Godfrey DI. NKT cells - conductors of tumor immunity? Curr Opin Immunol 2002; 14:165–171.

620. Brutkiewitz RR, Sriram V. Natural killer T (NKT) cells and their role in antitumor immunity Crit Rev Oncol Hematol 2002; 41:287–298.

621. Kelly JM, Darcy PK, Markby JL, Godfrey DI, Takeda K, Yagita H, Smyth MJ. Induction of tumor-specific T cell memory by NK cell-mediated tumor rejection Nat Immunol 2002; 3: 83–90.

622. Sinkovics JG, Horvath JC, Kay HD. NKT cells emerge to restrain the trophoblast: tumor cells imitating the trophoblast evoke NKT cell reactions, 56th Annual Symposium on Fundamental Cancer Research, The University of Texas MD Ander-

son Cancer Center, October 14-17, 2003; abstract I33: 117–118, Houston, TX.

623. Clark DA, Slapsys R, Croy BA, Kreek J, Rossant J. Local active suppression by suppressor cells in the decidua: a review Am J Reprod Immuno 1984; 5:78–83.

624. Szekeres-Bartho J, Barakonyi A, Miko E, Polgar B, Palkovics T. The role of γ/δ T cells in the feto-maternal relationship Semin Immunol 2001; 13:229–233.

625. Ragupathy R. Pregnancy: success and failure within the Th1/Th2/Th3 paradigm Semin Immunol 2001; 13:219–227.

626. Piccini M-P, Scaletti C, Vultaggio A, Maggi E, Romagnani S. Defective production of LIF, M-CSF and Th2-tyep cytokines by T cells at fetomaternal interface is associated with pregnancy loss J Reprod Immunol 2001; 52:35–43.

627. Agrawal B, Reddish MA, Krantz MJ, Longenecker BM. Does pregnancy immunize against breast cancer? Cancer Res 1995; 55:2257–2261.

2

New Biological Therapeutics

COMPETITORS OR COLLABORATORS OF VIRAL THERAPY FOR HUMAN CANCERS

JOSEPH G. SINKOVICS

Cancer Institute, St. Joseph's Hospital;
Departments of Medicine and Medical
Microbiology-Immunology, The University of
South Florida College of Medicine; The H. Lee
Moffitt Cancer Center
Tampa, FL

INTRODUCTION

There is an explosive and close to overwhelming array of preclinical and actual Phase I and Phase II clinical trials of biological agents for the therapy of human cancers. A new era of cancer therapeutics is taking its long-awaited place. At the cutting edge are naturally occurring attenuated and genetically engineered oncolytic viruses and viral vectors for gene therapy. Put in context, viral therapy of human cancers, tried and abandoned decades ago, re-emerges now to occupy a respectable position among these trials, even when the most sophisticated protocols of gene therapy with viral vectors [especially those with retro(lenti)viral vectors] are dealt with separately (regretfully not detailed in this volume, although we mention in this text some of the viral vectors that insert genes into the genome of cancer cells).

Pre-clinical testing for viral therapy of human cancers is quite similar to procedures in use for all other biotherapeutical agents: tumor cell killing in vitro in tissue cultures and in vivo in xenografted human tumors. Even the results are quite similar: extraordinary success for high levels of tumor cell killing in these pre-clinical tests; disappointingly weak results in Phase I and Phase II clinical trials; long stabilizations of disease commonly achieved. However, there are most favorable exceptions of success: imatinib mesylate (Gleevec, Novartis Pharmaceuticals, East Hanover, NJ) for chronic myelogenous leukemia (CML); bortezomib (Velcade, Millennium Pharmaceuticals, Cambridge, MA) for myeloma; erlotinib (Tarceva, Genentech, South San Francisco, CA; OSI Pharmaceuticals; and Roche) and bevacizumab (Avastin; Genentech) for non–small-cell lung cancers [Eastern Cooperative Oncology Group (ECOG) and M.D. Anderson Cancer Center protocols] [American Society of Clinical Oncologists (ASCO) #2521, 2003; Physicians Desk Reference, 2004]. The Discussion includes comments on several treatment modalities (monoclonal antibodies and immunotoxins, preventive and therapeutic cancer vaccines, adoptive immune lymphocyte therapeutics, etc.) that may compete or co-operate with the virus therapy of cancer. The scope of this volume is not to review tumor immunology and immunotherapy, but because these disciplines overlap with the virus therapy of cancer, it is incumbent upon the contributing authors not to leave these connections unnoticed.

This chapter provides a much-abbreviated cross section of these new biological therapeutic trials and seeks out the place where viral therapy of human cancers could fit in, either in competition, or in collaboration with, other new promising biotherapeutical agents. Of many pathological entities under investigation for the possible application of new biological treatment modalities, two—multiple myeloma and glioblastoma multiforme—have been selected to be reviewed here, inasmuch as they remain incurable and are targeted for oncolytic viral therapy.

FARNESYL TRANSFERASE INHIBITORS

c-*ras*[*]

The *ras* family of proto-oncogenes first revealed itself when Kirsten[†] and Harvey[**] **r**at **s**arcoma (*ras*) retroviruses yielded these genes transduced by these viruses from their host cells [1–3]. Later, point-mutated human *ras* genes were found in transitional cell carcinoma of the urinary bladder and, thereafter, in a very large number of diverse human cancers. In an early analysis of malignant human tumors, the incidences of mutations of p53 and the K-*ras* gene were 50% and 41%, respectively. The specific target of tobacco smoke for mutation induction is codon 12 of the K-*ras* proto-oncogene. This mutation is more common in lung adeno- than in lung squamous cell carcinomas. In lung adenocarcinomas, mutations in codon 12 K-*ras* exceed 50% [4–7]. In colon adenocarcinomagenesis, K-*ras* mutations occur very early and are followed by $5q^-$ (the locus of the adenomatous polyposis coli APC gene at 5q21), $18q^-$ (the locus of the deletion in the colorectal carcinoma DCC gene), and 17^- (the locus of p53) deletion. After K-*ras* and p53 mutations (or p53 deletion) occur, the patient's prognosis worsens [7–11].

---

[*] Italics indicate the gene (*bcl*-2); capital letters (Bcl-2; BCL-2) designate the gene product protein.

[**] The author had the privilege to become a friend of Dr. Werner Kirsten† in Chicago, IL in the late 1950s and worked at the same Section of Virology at M.D. Anderson Hospital in Houston, TX, where Dr. Jennifer Harvey spent a year of her fellowship in the late 1960s.

[†] Deceased

Co-operating with mutated K-*ras* in colorectal carcinoma cells, β-catenin translocates into the nucleus, initiates its own oncogenic cascade, and aggravates the prognosis [12]. K-*ras*-mutated colon carcinoma cells circulate in the blood and reach the bone marrow [13]. Co-mutations of K-*ras* and APC genes are frequent (27%), whereas co-mutations of K-*ras* and p53 genes are rare (6.6%); single mutations of one of these genes are the most common (38.7%) [10,11].The autosomal dominant hereditary nonpolyposis colorectal cancer (Lynch syndrome) arises because of the failure of DNA mismatch repair genes (hMLH1,2,3,6) [14,15]. In pancreatic ductal adenocarcinomas, K-*ras* mutations are frequent [16,17]. Mutated K-*ras* gene product proteins elicit CD4 T lymphocytic infiltrates in the periductal pancreatic tissues [18]. In head and neck squamous cell carcinomas mutated K-*ras* is rare, but amplified wild K-*ras* is overexpressed [19].

R-*ras* (a subclass in the H-*ras* family) is active in squamous carcinoma cells of the uterine cervix and promotes tumor cell proliferation and migration but is sensitive to inhibition by LY294002 (Calbiochem) [20]. The Ras association domain family 1A gene (RASSF1A) exerts suppressor effects on mutated *ras*; its locus is at 3p21, from which it is frequently eliminated (3 short arm deleted) in uterine cervical carcinomas. Human papilloma virus (HPV)-16 and HPV-18 possess transforming genes: E6 degrades p53; E7 inactivates Rb. In squamous cell carcinomas of the uterine cervix, HPVs are present in 90% or more of the cases; in adenocarcinomas of the uterine cervix, HPV presence stands at 50% [21,22].

The tumor suppressor gene RASSF1A (locus at 3p21.3) may be inactivated by promoter hypermethylation or by the loss of its hetero- or homozygosity (deletions). The intact RASSF1A protein binds GTP-*Ras* but hypermethylated RASSF1A is inactive. In 24% of cervical adenocarcinomas and 10% of cervical squamous cell carcinomas, RASSF1A is hypermethylated. Correlation of RASSF1A methylation and the presence of human papilloma virus (HPV) is uncertain, as if HPV-infected cervical epithelial cells advanced to malignant transformation without need for RASSF1A inactivation [22].

Whereas MAPK activation by *Ras* correlates with the progression of hormone refractory disease in prostate cancer, the anti-androgen bicalutamide (Casodex, AstraZeneca, Wilmington, DE) inhibits anchorage-dependent cell growth and *Ras* signaling [23,24].

Experimental introduction of N-*ras* into CD34 stem cells resulted in the clinical picture of myelodysplasia. H-*ras*, K-*ras*, and N-*ras* mutations occur in patients with acute myelogenous leukemia (AML) infrequently (15%–30%) and mainly in patients who were occupationally exposed to potentially carcinogenic chemicals; involvement of H-*ras* in the cause of chronic leukemias (CLL; CML) appears to be insignificant [25–27].

N-*ras* codon 61 mutations occur early and persist in melanoma cells and result in deeper invasions. In familial melanoma, germline mutations of the CDKN2A/ARF (cyclinedependent kinase/alternative reading frame) and CDK4 chromosomal regions (at 9p21 and 12q13) are passed on; the very same melanoma cells also display somatic (ultraviolet light-induced) N-*ras* and BRAF mutations. The cascade emanating from a mutated N-*ras* gene involves the proto-oncogene *raf* and terminates through serine-threonine kinase activity in MAPK (mitogene activated protein kinase) cascade: the RAS-RAF-MAPK pathway. In melanoma cells, the BRAF protein exhibits amino acid substitutions (V599E or V599R or V599K) consequentially to somatic mutations in exon 15 of the gene [28–33]. When the RAF cascade goes toward the MEK/ERK (mitogen activated protein/extracellular signal-regulated kinase) pathway, the cell will upregulate its adenoviral receptor CAR, presumably rendering these tumor cells more susceptible to adenoviral oncolysis [34–36]. The *raf* kinase inhibitor BAY43-9006 is in clinical trials for the treatment of melanoma; early reports indicate that partial remissions (PR) and disease stabilizations were induced.

The wild-type K- or H-*ras* genomic sequences encode their protein products that anchor themselves at the inner aspect of the cell membrane and on intracellular compartment membranes and are recruited into action when certain growth factor receptors (GF-R) capture their cognate ligands: epidermal (EGF-R); platelet-derived (PDGF-R); and HER2neu/Erb2

(HER for human epidermal receptor). In healthy cells, the guanine diphosphate (GDP)-Ras protein complex is inactive and in "off" position and is active in the "on" position. Guanine triphosphate (GTP)-Ras dephosphorylates a double-stranded RNA-activated protein kinase (PKR), thus creating an interferon (IFN)-free environment favorable for the full replication of certain viruses (*vide infra*) in the cell so afflicted. In healthy cells, Ras and extracellular signal-regulated kinases (ERK) and protein kinase C (PKC) cooperate in immune cell and mast cell activation [37].

The *ras*→Ras cascade activates a number of proto-oncogenes and neoangiogenesis factors downstream and suppresses Fas receptor (cluster of differentiation, CD95) expression, thereby providing exemption from one form of apoptosis for the cell. It would be of considerable interest if Fas ligand (FasL) overexpression also occurred in these cells, in that, armed with FasL, these *ras*-mutated cells could kill CD95-positive defensive host lymphocytes (unpublished). Among the proto-oncogenes activated by GTP-Ras is c-*raf*, encoding a serine-threonine kinase acting in unison with the mitogen-activated protein kinase (MAPK) system. Mutated B-Raf proteins are frequently detectable in melanoma and colon carcinoma cells. Physiological activation of Ras is reversible, but when point-mutated Ras and Raf-1 are constitutively active, the cells harboring them assume malignant phenotype. Ras mediates radioresistance through phosphatidylinositol 3-kinase (PI3K) and initiates Raf-dependent activities working within the MAPK pathway. Raf-1 may be activated independently from Ras by Bcl-2 and by protein kinase C (*vide infra*). Further, activated Ras acts upon the cytoskeleton's relationship with the intercellular matrix and its cell adhesion molecules and integrins; consequentially, *ras*-mutated cells exercise increased locomotion and invasiveness. Wild-type neuroblastoma N-*ras* may exert a tumor suppressor-effect against murine thymic lymphomas, but mutated N-*ras* becomes oncogenic. N-*ras* regulates lymphoid and myeloid cell homeostasis, but N-*ras* knockout mouse embryos develop normally, whereas K-*ras* knockout mouse embryos fail to reach normal maturation [38]. Trans-farnesylthiosalicylic acid (FTS) is a N-*ras* antagonist.

Under physiological conditions, GTP-Ras converts into the inactive GDP-Ras by hydrolysis because of interaction with guanine nucleotide exchange factor (GEF). For point-mutated (most frequently at codons 12, 13, 61) *ras* gene product proteins, hydrolysis does not take effect and GTP-Ras remains constitutively activated in the "on" position [1–3].

Up-regulated peroxisome proliferator-activated nuclear receptors (PPARs) exert anti-inflammatory effects in macrophages [39–41]; suppress tumor growth, especially PPARγ in K-*ras*–driven tumor cells by inhibiting PI3K/Akt and cyclinD1 expression; and may induce terminal differentiation of human tumor (liposarcoma) cells: Troglitazone (TGZ) attachment to PPARγ induced this differentiation process. When PPARs capture their ligand, they emit pro-apoptotic signals in human breast and prostate carcinoma cells. A PPAR-agonist γ-ligand (such as pioglitazone, sigma) inhibited *ras*-mediated malignant transformation and induced apoptosis in numerous cultured human cancer cell lines and synergized with histone deacetylase inhibitors. Ligands (TGZ) of PPARγ are in use to treat Type II diabetes, to induce regression of precancerous lesions in the upper digestive tract, and increase the efficacy of combination chemotherapy for metastatic cancers (AACR 2003 Symposium I by Eva Szabo of the National Cancer Institute). HER2 upregulates PPARγ in breast cancer cells but reduces its responsiveness to its ligands; trastuzumab (Herceptin, Genentech) neutralizes HER2 and synergizes with TZG in inbiting growth or inducing apoptosis of breast cancer cells [41]. PPARδ is upregulated by K-*ras* and activated by prostaglandins, products of cyclooxygenase-2 (see Part II).

In the multistep model of colonic adenocarcinomagenesis, the adenomatous polyposis coli APC (5q21), K-*ras* (12p), p53 (17p13), and the human homologue of the murine double minute *mdm* (12q13-14) genes mutate sequentially, and E3 ubiquitin ligase MDM2 degrades the p53 protein (the "guardian of the genome"). An antisense oligonucleotide (see Part III) directed to neutralize translation of the *mdm*-2 gene spares the integrity of p53 [42]. Mutated K-*ras* (glycine to valine substitution at codon 12) is overrepresented in Duke's C tumors, suggesting that K-*ras* contributes to tumor progression through lymph node metastases [4].

Viral Attack; Host Counterattack

It now appears that *ras*-mutated malignant cells support the replication of certain viruses [influenza virus, Newcastle disease virus (NDV), reovirus, vesicular stomatitis virus (VSV)] to the full maturation of the viral progeny, resulting in bursting death attributable to lysis of the tumor cells. Nonmalignant counterparts of these cells protect themselves and neighboring cells with IFN production and resist the replication of these viruses. Whereas replication-competent wild or attenuated NDV strains accomplish the full maturation of their progeny by inhibiting innate IFN production, the incompletely replicating Ulster strain NDV induces IFN production [43]; replication-competent NDV strains are oncolytic by apoptosis induction or by bursting tumor cells, whereas replication-incompetent NDV strains are oncolytic by eliciting innate and adaptive humoral and cell-mediated immune reactions of the host directed against infected cells expressing viral and tumoral antigens co-jointly. These two processes of oncolysis may overlap; for example, when epitopes of apoptosomes of tumor cells are re-expressed by host dendritic cells (DC), or when the replication-competent virus somehow fails to suppress innate host reactions, such as IFN production.

These viruses therefore emerge as selectively oncolytic agents ready for clinical trials. In human cell lines, genetically engineered so that they express the simian paramyxovirus 5 V protein, the STAT system (Figures 1 and 2) is subjected to proteasome-mediated degradation. In these IFN-free cell lines, viruses replicate uninhibited [44]. Wild-type influenza A viruses abound in nature, especially in avian reservoirs [45] and readily recombine with each other, so an attenuated live influenza virus may re-emerge as a pathogenic recombinant.

Influenza A and B strains initially overcome host defenses by their NS1A and NS1B proteins [46]. Viral neuraminidase (HN) activates NK cell receptors NKp44 and NKp46 and

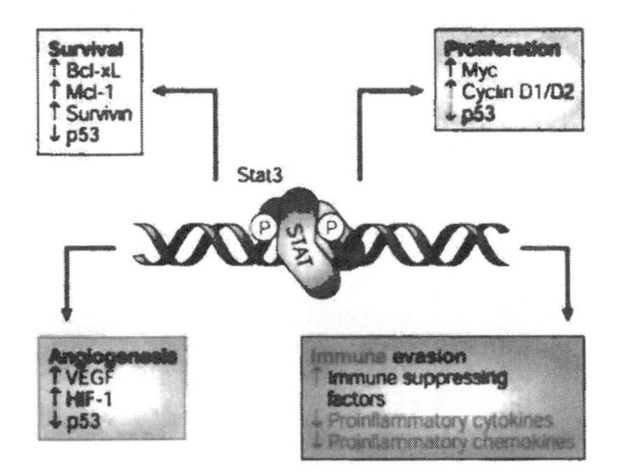

Figure 1 The activated STAT system decisively contributes, if not to the initiation, certainly to the sustained maintenance of the malignant geno- and phenotype of the cancer cell. Until now, targeting STATs for cancer therapy received lower priority than other targets, such as the EGF-R/HER2 or the Bcl-2 anti-apoptotic systems. However, at the University of South Florida H. L. Moffitt Cancer Center, Tampa, FL, STAT receives the attention it deserves. (These illustrations are gratefully acknowledged as the courtesy of Drs. Richard Jove and Hua Yu. Reprinted with permission *from Nature Reviews Cancer*, February 2004.)

induces cytotoxicity against infected cells; but when new class I MHC molecules are generated in the infected cells and bind the killer cell immunoglobulin (Ig)-like receptor 2 domain long tail 1 of NK cells (KIR2DL1), cell killing is inhibited and the viral progeny can mature [47]. From the point of view of influenza viral oncolysates (VO), it is of paramount importance that nonimmunogenic and even tolerogenic proteins (ovalbumin) generate antibody and immune CD4 and CD8 responses in hosts infected with influenza viruses [48]; probably in the tumor-bearing host, tolerogenic tumor epitopes become immunogenic when presented within influenza VO.

The oncolytic virus elicits antiviral immune reactions in the inoculated host. A three-dimensional mathematical model of a tumor infected by a replication-competent virus indicates

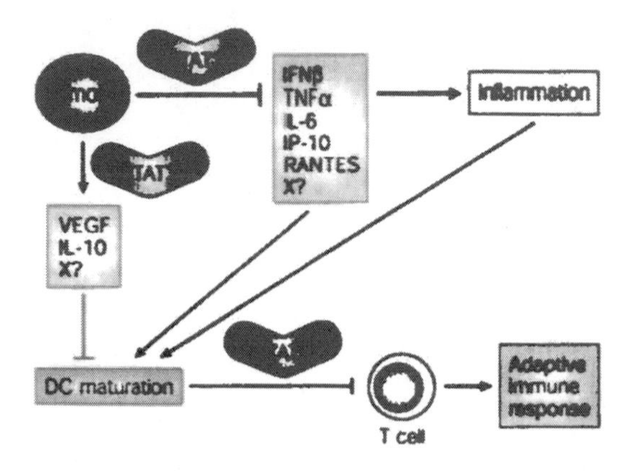

Figure 2 The activated STAT system decisively contributes, if not to the initiation, certainly to the sustained maintenance of the malignant geno- and phenotype of the cancer cell. Until now, targeting STATs for cancer therapy received lower priority than other targets, such as the EGF-R/HER2 or the Bcl-2 anti-apoptotic systems. However, at the University of South Florida H. L. Moffitt Cancer Center, Tampa, FL, STAT receives the attention it deserves. (These illustrations are gratefully acknowledged as the courtesy of Drs. Richard Jove and Hua Yu. Reprinted with permission *from Nature Reviews Cancer*, February 2004.)

high chances for tumor escape because of immediately mobilized innate and subsequent adaptive immune reactions [49]. Some early cytokine reactions may promote tumor cell growth. The CXC chemokine IL-8 is pro-angiogenic and is prominently expressed in transitional cells of the urinary bladder; it may even act in an autocrine growth loop in these tumors. When NFκB is liberated from its inhibitor IκB, it translocates to the nucleus, where it transactivates the IL-8 gene [50].

Virus-neutralizing antibodies may eliminate the virus (an unfavorable event for oncolysis), or the replicating virus may produce "quasi-species" new progeny [51,52] and escape neutralization. Some viruses may exist complexed with antibodies; these immune complexes may be deposited in synovial membranes and kidney glomeruli. Antiviral antibodies reacting with tumor cells expressing viral antigens may evoke the

ADCC reaction by enlisting NK cells or macrophages in reacting with their FcRs, and thereby inducing tumor cell lysis. Virally infected cells induce NK cell and immune T cell reactions (a most favorable event for oncolysis), but viruses may escape immune T cells [53,54] and viral peptide antigen may become tolerogenic [55]. When immunological synapses are formed between DC and lymphocytes, a subclass of NK cells or T cells may misconstrue instructions so received and become tolerogenic or immune-suppressive agents [56], an unfavorable event for oncolysis. Under such circumstances, immune reactions engendered by viral peptide epitopes may promote tumor growth (such as "enhancing antibodies"). T cell interactions with virally infected or transformed cells are very complex, inasmuch as viral peptides may become tolerogenic; or tumor cells armed with FasL may kill immune T cells [57]. Regulatory and CD4 CD25 suppressor T cells (Tr; Ts) rule over tolerance vs. immune reactions [58–61]. Viruses switching on the "burglar alarm" upon entering a cell induce programmed cell death, often mediated by TNFα, whereby no new mature viral progeny can be formed; in response, viruses evolved or expropriated genes encoding proteins inhibitory to apoptosis. Cells dying apoptotic death are engulfed by macrophages or load their antigens onto DC MHC type II molecules [62,63] to present them to CD4 helper lymphocytes. Endogenously processed epitopes are presented by MHC type I molecules to CD8 cytotoxic lymphocytes. If these antigens are expressed by mature DC, not in a tolerogenic (with IL-10), but in an immunogenic fashion (with co-stimulatory molecules B7.2, IL-2, IL-12), they may elicit extremely strong immune reactions capable of lysing gross established tumors (a most favorable event).

Viruses potentially oncogenic in the human host [human papilloma virus (HPV), Epstein-Barr virus (EBV), hepatitis B virus (HBV)] excel in overcoming host defense reactions at many levels and prevail [64], but malignant transformation by adenoviral E1a and *ras*, simian virus 40 (SV40), and HPV was inhibited by adeno-associated virus (AAV) Rep78-expressing plasmids, which preserve phosphorylated Rb and inhibit protein kinase A (PKA) when they arrest the cell cycle [65]. Protein kinase A is activated by luteinizing hormone in endometrial cancer cells and increases tumor cell invasiveness; in breast cancer cells, PKA and p38 intensified paclitaxel-induced apoptosis [66,67]. The dephosphorylated and inactivated nonstructural protein NS1 of minute virus of mice regains its helicase activity when co-incubated with protein kinase C, which is overactive in transformed cells, allowing parvovirus replication and genomic integration to take place preferentially in malignant cells. Adeno-associated viruses integrate their genomes into host cell chromosomes (AAV-2 to locus 19q13.3), inhibit mRNA and transcription of adenoviral gene E2a, readily penetrate solid tumors, inhibit neoangiogenesis by transferring the endostatin gene (rAA-HuEndo) into tumor-bearing hosts, and are elevated to the rank of good candidates for cancer gene therapy [68–74].

When an attenuated and replicating or, for gene therapy, genetically engineered, replication-competent or replication-restricted (to selected cells, such as organ-specifically targeted tumor cells) virus infects a tumor-bearing host, it is quite unpredictable whether the host could eliminate it or the virus manages to evade its disposal. Mixtures of replication-competent and incompetent viral particles in the inoculum (for example VSV inocula) profoundly influence the host's immune response [75]. Would suppression of antibody-mediated immune reactions (cytarabine; anti-B cell monoclonal antibodies) and promotion of Th1-type cytotoxic T cell reactions (IL-2, IL-12) directed against virally infected host cells tilt the balance in favor of viral oncolysis? A Sindbis virus replicon containing herpesviral gene VP22 and HPV-16 E7 gene induces E7-protein-specific CD8 cytotoxic lymphocyte-mediated immune reactions against HPV-infected cells [76]. Although human reoviruses are quite apathogenic in adult tumor-bearing and immunocompetent mice [77], infected newborn mice succumb to encephalitis with apoptotic death of nerve cells [78]. Hopefully immunosuppressed patients with cancer will not suffer any pathogenic effect of oncolytic reoviruses.

Prior to the mobilization of elements of adaptive immunity in response to a viral infection, innate immune reactions are immediately activated, with an upsurge of lympho-, cyto-, and chemokines. The mammalian CXC chemokine family consists of 16 ligands and six receptors [79]. Some chemokines are inhibitory to tumor cells: Macrophage inflammatory protein-1α/CCL3 and macrophage-derived chemokine CCL22 inhibit cell divisions and invite infiltration of macrophages, neutrophil leukocytes, T and NK lymphocytes and DCs into tumors that express these chemokines. Whereas the macrophage-inhibitory cytokine-1 (MIC-1) expressed by gastric cancer cells increased the invasiveness of the tumor cells, it made them susceptible to the MAPK inhibitor PD98059 (Calbiochem). However, tumor cells recruit many defensive and other cytokines (IFNs, ILs, TNFs, FasL, G- and GM-CSF, stromal-derived GF, PDGF, ligands for EGF or HER2/neu, etc.) to promote their mitoses in autocrine and paracrine growth loops (Sinkovics, in this volume). In their native microenvironment, some tumor cell subpopulations may undergo rapid replication driven by the cytokines of the innate immune response; other tumor cell populations, especially those exposed to the oncolytic virus after their extrication from their microenvironment (in tissue cultures; in xenografts), succumb to the shower of cytokines or die because of the absence of some growth-supportive cytokines (see Part IV). Some viruses encode immunosuppressive cytokines (decoy IL-10) or soluble receptors for immunostimulatory cytokines (IFN-γ; anti-apoptotic serpins) and thereby neutralize them [80–89]. These extremely complicated interactions between virus and host are known at least in the case of oncolytic viruses. Attenuated measles virus-specific lymphocytes persist longer than antibody production after vaccination [90]; if the oncolytic measles virus elicited a similar reaction that would favor the tumor-bearing host and delay elimination of the oncolytic virus by antibody. Wild measles virus suppresses Janus kinase 1 (JAK1) phosphorylation induced by IFN-α and abolishes the IFN-induced antiviral state; IFN-γ is not inhibited [91]. Although the influenza A viral NS1 protein is essential for the inhibition of IFN response of the infected cell, it is instrumental for the induction of immune T cells [92]. Influenza A virus could activate tumor antigen-specific (HER2/neu) cytotoxic lymphocytes, supporting the rationale for viral oncolysates as cancer vaccines (Sinkovics, in this volume). In mice devoid of mature B cells and antibodies, immune T cells could clear influenza A viral infection [93], an event most favorable for the tumor-bearing host when its tumor cells are infected with an oncolytic influenza A virus.

When oncolytic viruses (influenza, NDV, reovirus, VSV) multiply to their full maturation in host cells with constitutively activated GTP Ras and with defective IFN response because if inactivated PKR [94,96], their oncolytic efficacy should match or surpass that of farnesyl transferase (FT) inhibitors for the selection of these viruses in clinical trials. The inhibitors have an advantage over the virus in that they do not induce an immune reaction that would neutralize them in the host. However, these enzyme inhibitors now also in clinical trials do suppress the malignant phenotype of *ras*-mutated cells. Would this intervention restore resistance of these cells toward the oncolytic virus or, to the contrary, if the two interventions were combined (enzyme inhibitors and oncolytic viruses administered simultaneously or sequentially), could they act additively or synergistically? The FT inhibitor R115777 (*vide infra*) failed to exert any antitumor activity in patients with pancreatic adenocarcinoma. In contrast, reovirus serotype 3 readily destroys cell cultures and xenografts of human pancreatic adenocarcinoma cells; here the oncolytic virus promises better results. A Raf kinase inhibitor is undergoing clinical trials in patients with metastatic colorectal carinoma; BAY43-9006 also inhibits ERK, induces long stabilizations of disease, and lends itself to combination with standard chemotherapy. In three clinical trials, orally administered BAY43-9006 induced 2 partial remissions (PR) in 114 patients with metastatic cancers (2%); stopped disease progression in 42 patients (stabilization of disease, 37%), but allowed disease progression in 37 patients (32%); 33 patients remain on treatment with stable disease for further evaluation; of 60 patients with metastatic colorectal cancers, disease stabilization occurred in 25 (42%) [97]. Could reovirus match these results; or could it improve these results if administered in combination with BAY43-9006?

Adenoviral Oncolysis

Some adenoviruses borrow their oncolytic potency from herpesviruses. Herpesviral replicons (G207, HE7*tk*) either render infected cells (melanoma; hepatocellular carcinoma) antigenic in the host of origin and elicit T cell-mediated immune reactions or insert the *tk* gene into tumors of the brain or into tumors of other organs, sensitizing the tumor cells (and some bystander cells) to ganciclovir [98–101]. Adenoviruses are used as gene vectors (not discussed but occasionally mentioned in this volume; well reviewed elsewhere) [102–109]. The transfer of wild-type p53 by retro- or adenoviral vectors by direct repeated injections into lung cancers revealed minor and partial responses occurring despite antiviral antibody production in the patients; on rare occasions (in one patient with bronchioalveolar carcinoma) intratumoral injections of viral vector of p53 resulted in partial regression of noninjected distant (brain, liver) metastases (ASCO #487a 2000).

Adenoviruses are subjected to gene deletions; ONYX-015 or dl1520 has its E1B55 deleted and is supposed to favor for its replication p53-defective tumor (pancreatic and hepatocellular carcinoma) cells [110–112]. In 27 patients with metastatic colon cancer, dl1520 given through the hepatic artery with 5FU given intravenously (i.v.) induced seven responses (PR, MR) and nine stabilizations of disease occurred, whereas 11 patients continued to progress [103]. Adenoviral vectors Ar6-pAE2fE3F or Ar6pAE2fF (with or without E3 region) are expected to attack Rb-defective (tumor) cells, where the viral E1A depends on overexpressed cellular E2F1 promoter [113–115]. The entry of Ad5 into tumor cells underexpressing the entry-site receptor CAR could be promoted if Ad5 possessed serotype 3 knobs [116,117]. Adenoviral E1A protein sensitizes cells to TNFα, but AdTRAIL expressing a ligand kills not only hepatocellular carcinoma cells, but also normal hepatocytes [118,119]. CV890 AvE1a04i replicates only in hepatocellular carcinoma cells that are overexpressing the αFP promoter; CN706 and CV767 exercise prostate-specific expression of their E3 region and are aimed to grow in prostatic carcinoma cells [120–122]. Adenoviruses with deleted E1b genes (Ad-ΔE1B19.55) do not express the 19kDa or 55kDa proteins that protect infected cells from apoptosis and are highly apoptotic for infected tumor cells. Ewing sarcoma xenografts are susceptible to Ad-E1A gene therapy which increases tumor cell sensitivity to VP16 (etoposide) [123]. Liposome-encapsulated E1A gene therapy is in clinical trials for lung, breast, and ovarian carcinoma [124]. AdmRTVP-1 carries the proapoptotic mouse gene RTVP-1; it inhibits neoangiogenesis and elicits macrophage, DC, and immune CD8 T cell infiltrations into infected tumors. Adenoviruses can deliver tumor suppressor genes into tumor (colon carcinoma) cells: p16INK4/CDKN2p16, the cyclin-dependent kinase inhibitor, or p53 into bladder or lung cancer cells and SCH58500 into ovarian carcinoma cells or BRCA1 into breast and ovarian carcinoma cell lines [125–129]. The N5 gene encodes the death domain protein p84N5, which triggers apoptosis from within the nucleus even when the cell is resistant to p53 therapy [130]. The protein NK4 is a structurally related antagonist of hepatocyte growth factor and, when expressed in AdNK4, it suppresses the growth of xenografted human pancreatic adenocarcinoma cells in the peritoneal cavity of mice [131,132]. Ad-IGF-IR/950 and /482 introduce an insulinlike growth factor antisense sequence into tumor cells and inhibit tumor cell divisions; tumor cells so affected may undergo apoptotic death [133]. Preoperative intratumoral (prostate cancer) injection of AdIL-2 vector resulted in inflammatory reactions, tumor necrosis, and falling PSA levels [134]. Adenovirally delivered IL-12 gene encodes the cytokine and promotes the development of Th1-type immune reactions [135]. When adenoviruses deliver the herpesviral *tk* and endostatin (ES) genes into brain tumor cells or into xenografts of human kidney cancer cells, the tumor cells become vulnerable to ganciclovir and bystander cells may also be killed [136]. The IL-11 receptor α-locus chemokine CCL27 gene inserted into the genome of ovarian carcinoma cells by Ad-RGD-mILC worked to induce immune T and NK cell reactions, rejecting the tumor xenografts in immunocompetent mice [137]. AxdAdB-3 has mutated E1A and deleted E1B, and it is unable to react with p53. E1A-mutated or deleted adenoviruses are unable to bind Rb protein; AxdAdB-3 selectively replicated in, and was cytotoxic to, gall bladder carcinoma xenografts [138]. Ad-TRAIL, carrier of the gene encoding the TNF-related apoptosis-inducing ligand, induced apoptotic death of the infected cancer cells and exerted a bystander effect in tumor cells in the vicinity of infected cells [118,119]. Malignant glioma cells with homozygous p16 gene deletion accepted infection with Ad-p16, experienced cell cycle arrest, and underwent senescence. The gene product protein $p16^{INK4a}$ is a cyclin D1-dependent kinase inhibitor, preventing phosphorylation of the Rb protein (the cell cycle stands still in G_1); the $p14^{ARF}$ gene product protein counteracts the effects of the proto-oncogene MDM2: p53 escapes MDM2 binding and is not degraded by the ubiquitin-proteasome pathway. In human tumors, the gene for INK4a/ARF (alternative reading frame) at locus 9p21 is frequently mutated or deleted, removing the brake from the cell cycle and exempting the cell from p53-induced apoptosis. The p53 gene inserted into the genome of tumor cells by Ad-p53 is protected from degradation by co-infection of the same cell by Ad-ARF; human lung cancer cell xenografts doubly infected in *nu/nu* mice die programmed cell death as induced by p53 [139,140]. In hormone- and multiple–drug-resistant prostate cancer cells, NF-κB is constitutively activated and translocates to the nucleus unopposed, an anti-apoptotic event; adenovirus pAxCAmIκB-M expressing the negative dominant IκB super-repressor abrogates NF-κB activation and restores the tumor cell's susceptibility to apoptosis induction by chemotherapeuticals (paclitaxel) in both androgen-sensitive and androgen-insensitive prostatic carcinoma cell lines [141]. Canji's adenovirus 01PEME has its E1a gene and its E3 region deleted and has a p53-responsive promoter (to drive E2F antagonist E2F-Rb) and a late promoter inserted. It is unable to replicate in normal cells but reaches very high titers in human tumor cells, even after i.v. inoculations [142].

Cancer cells operate with reactivated telomerase reverse transcriptase (hTERT). Adenovirus hTERT-Ad expresses the gene of E1A protein under the control of the hTERT promoter and replicates best in hTERT-overexpresing tumor cells; it is more cytopathic in xenografted human tumor cells than ONYX-015, even when p53 is intact [143]. The histone deacetylase inhibitor FR901228 induces cell-surface overexpression of CAR and α_v integrins; in these cells, transgene insertion by AdCMVβgal was most efficient [144]. Cyclooxygenase (COX)-

2 gene expression is very low in normal cells but is increased in tumor cells. COX-2 gene-driven adenoviral E1A gene facilitates viral replication best in *ras*-mutated and MAPK-overactive tumor cells [145,146]. Replication-defective and survivin gene-carrier adenoviral vector (Ad-survivin) and a phosphorylation-defective $Thr^{34} \rightarrow Ala$ survivin gene-carrier adenoviral vector (pAd-T34A) protected and destroyed capillary networks of human breast cancer xenografts, respectively; pAd-T34A induced mitochondrial apoptosis with cytochrome C release, resulting in apoptotic death of both endothelial and tumor cells [147,148]. The CD154→CD40 activation stimulates antigen presentation; adenovirally (Ad-CD154) inserted CD154 gene upregulates α_v integrin expression in murine lymphoma cells and augments their immunogenicity [149]. ONYX-411 replicates in tumor cells in which the Rb protein is inactivated, E2F is deregulated, and the cell cycle rolls forward uninhibited [152]. Tumor necrosis factor (TNF)-α is antagonized by NF-κB and c-FLIP, the inhibitor of caspase 8. Adenoviral protein E1A sensitizes cells to TNFα and antagonizes both c- and v-FLIPs, sending them to ubiquitin/proteasome-mediated degradation [151]. An immunoconjugate-encoding adenoviral construct induces tumor cells to overproduce the immunoconjugate directed at factor VIII of endothelial cells of the tumor bed; as neovasculature is destroyed, the tumor regresses [152]. The combination of replication-defective (without E1A, E1B, and E3 transcription units) and replication-competent (such as ONYX-051 without E1B55K protein) adenoviral vectors, which overexpress adenoviral death protein, synergize in killing human lung and liver cancer cell xenografts [153] (Table 1).

Adenoviral vectors induce humoral and cell-mediated immune reactions in the infected hosts and are quickly eliminated, or cannot be repeatedly and effectively used. Polyethylene glycol-coated or liposome-encapsulated viral vectors are less immunogenic; corticosteroid- or budesonide-treated hosts are less immunoreactive. Some host tissues can be rendered tolerant to the viral vector (whereas other tissues of the same host remain immunoreactive); the vector may carry immunosuppressive genes prohibiting antigenic expression by MHC

Table 1 Principles of Viral Engineering for Virotherapy

Virus	Effect
Ad.CAR-EGFΔ24	EGF-R-overexpressing tumor cells are attacked
Ad5Δ24RGD	Viral entry into CAR tumor cells through integrins
AdenoCV706	Replicates in prostate cancer cells under the effect of PSA promoter
AdenoCV890	Replicates in αFP (fetoprotein)-expressing hepatocellular carcinoma cells
AdenoCG884	Replicates under the uroplakin promoter expressed by transitional cancer cells of the urinary bladder
AdTyrΔ24	Tyrosinase promoter in melanoma cells drives viral replication to cytotoxicity
AdEHT2	Hypoxia-responsive/estrogen responsive promoters drive viral replication in breast cancer cells
ONYX-015	*E1b* gene-deleted, E1B55K protein-deprived adenovirus replicates in p53-deficient tumor cells
Adeno01/PEME	Attacks cancer cells with mutated/lost p53
Ad-ARF	Protects wild-type or virally inserted p53 from degradation in proteasomes by antagonizing MDM2 through insertion of gene for $p14^{ARF}$
Adenovirus Delta24	Deprived of gene product protein that inactivates Rb in normal cells; it cannot grow in normal cells; replicates in RB-deficient (glioma) cells
Ad(E2F)-1RC	E2F tumor cells support viral replication; no viral replication in E2F quiescent cells
Adeno-FL; Adeno-FADD	Delivers Fas ligand into tumor cells expressing CD95 Fas receptor, inducing apoptosis
Adeno VEGF BAX	VEGF promoter drives expression of pro-apoptotic BAX in lung cancer cells
Adenovirus hTERT	Its E1A protein is under the control of the hTERT promoter in telomerase-preserving tumor cells
Ad-CMV-CD	Inserts cytosine deaminase transgene under CMV (cytomegalovirus) promoter into breast cancer cells, where the encoded enzyme renders the prodrug 5-fluorocytosine cytotoxic
HSV1 G92A	Replicates under albumin promoter in hepatocellular carcinoma cells
HSV1DF 3834.5	Replicates preferentially in MUC-expressing pancreatic carcinoma cells

class I molecules to CD8 lymphocytes; not adeno- but less antigenic adeno-associated or retrolentiviral vectors may be used [154–156]; or anti-B or -T cell monoclonal antodies are to be co-administered.

Farnsyl Transferase Inhibitors

One major enzyme family responsible for Ras activation is that of the FT; inhibitors of these enzymes defuse the cascade. For example, the very high incidence of breast and salivary gland carcinomas in the *ras*-transgenic "oncomice" is reduced to a minimal occurrence by FT inhibitors [157]. Posttranslational modifications of the Ras protein consist of farnesylation (addition of farnesyl lipid group to Ras isoform proteins by FT, a zinc metalloenzyme heterodimer); phrenylation (covalent addition by thioether linkage of 15-C farnesyl or 20-C geranylgeranyl group to C-terminal cysteine residues); and geranylgeranylation [158–163]. These events are now targeted by enzyme inhibitors that have already passed pre-clinical testing in vitro (in tissue cultures) and in vivo (in xenografted human tumors) and are in Phase I or Phase II clinical trials. The chemical structures of FT inhibitors are well depicted in a particular article from the University of Florida at Gainesville [164].

Among the competitors or collaborators of *ras*-mutation–dependent oncolytic viruses, FT inhibitors occupy an important position. SCH66336 (lonafarnib, Sarasar, Schering-Plough) is a non-peptidomimetic CAAX (C = cysteine; AA = aliphatic amino acids leucine, isoleucine, valine; X = methionine, serine, glutamine, leucine) inhibitor of FT. This compound inhibited in vitro growth of H*ras*-mutated human tumor cells (especially when the wild-type p53 was preserved) and suppressed the growth of xenografted *bcl2-abl*-mutated leukemic cells. In Phase I clinical trials, it induced only 1 PR in 44 patients with non–small-cell lung cancer (NSCLC) and other solid tumors; provided no survival benefit to patients with colorectal or pancreatic carcinomas; and was toxic when 400 mg twice daily dosage was exceeded, but in lower nontoxic dosage it synergized with taxanes and cisplatin. Tested against human melanoma cell lines in vitro, lonafarnib exerted antiproliferative and pro-apoptotic effects and synergized with cisplatin. Tested against Ph^+ CML cell lines in vitro, lonafarnib acted additively with imatinib or cytosine arabinoside but daunorubicin or etoposide antagonized its efficacy. In squamous carcinoma cells, SCH66336 inhibited AKT; adenoviral vector expressing constitutively active *act* reversed this effect [165–169].

R115777 (tipifarnib; Zarnestra, Janssen Pharmaceutica Products, Titusville, NJ; Ortho Biotech, Bridgewater, NJ) is a nonpeptidomimetic quinolone-derived inhibitor of FT on CAAX. It prevents benzopyrene-induced lung cancer development in mice [170]. When tested in patients with acute myelogenous leukemia (AML), 8 clinical responses occurred in 25 patients (32%) and two responses were complete remissions (CR). More recently, 10 of 30 elderly patients with AML responded to R115777 and 8 of these responses were CR; there was hematologic toxicity associated with remission induction. Of 36 patients with acute leukemias, 10 clinical responses occurred and 2 were CR. In myelodysplasia, 2 of 16 patients experienced CR. Chronic myelogenous leukemic (CML) patients in blast crisis also responded. This compound could clear leukemic blast cells from bone marrow in vitro before its reinfusion into patients, who received myeloablative high-dose chemotherapy [171–175]. Some patients with colon cancer responded with a drop of CEA levels. Human colon cancer cell lines may acquire resistance to tipifarnib; the resistant cell line remained sensitive to chemotherapeuticals, the phosphatidylinositol 3′-kinase (PI3K) inhibitor LY294002, and the EGF-R tyrosine kinase inhibitor PD153035, which inhibit xenografted human cancer cell growth, whereas R115777 is quite inactive against pancreatic cancer cells. R115777-resistant colon cancer cells remain sensitive to etoposide, platinol, paclitaxel, and MAPK/ERK inhibitor UO126 or PI3K inhibitor LY294002 [176]. In comparison, 18 patients with metastatic colorectal adenocarcinoma received i.v. ONYX-015 E1B55kD gene-deleted replication-selective adenovirus; only in 4 patients was disease stabilization observed for 2 to 4 months and eventually all patients progressed [102]. In 30 patients with advanced cancers treated with tipifarnib, 1 complete remission (CR) and 8 PR were recorded, at the price of hematologic, gastrointestinal, and dermatologic toxicities. Of 28 patients with advanced solid tumors, stabilization of disease occurred in 3 cases. This compound in patients with pancreatic adenocarcinoma or non–small-cell lung cancer induces rare PR or no remissions (NSCLC), but it is being re-tested in combination with standard chemotherapy. With gemcitabine, R115777 induced myelosuppression and other toxicities in 30 patients with advanced solid tumors, but 8 PR were recorded. The PR rate in patients with metastatic breast cancer was 14%. Intermittent administration (ID) of tipifarnib was better tolerated than the continuously administered (CD) drug and tipifarnib showed weak activity against advanced breast cancer: 10% PR and 15% stabilization of disease occurred in the CD cohort, and 14% PR and 9% strabilization of disease occurred in the ID cohort [177–186]. Patients with AML or with advanced breast cancer are being recruited for clinical trials with R11577 (*http://www.centerwatch.com/trials/*).

The summary from the literature of NSCLC treatment data shows the peformance of tipifarnib (with docetaxel and/or gemcitabine) to be 1 CR and 5 PR in 20 patients, and lonafarnib (with paclitaxel) to be 12 PR in 45 patients and 14 minor response (MR) or stabilization of disease (SD) in 33 patients. These are weak agents; viral therapy could match or surpass their efficacy [187].

L778123 (Eli Lilly, Indianapolis, IN; Merck, Whitehouse Station, NJ) is a peptidomimetic FT inhibitor active on CAAX. It is a dual inhibitor of FT and geranylgeranyl protein transferase (GGPT) [188–190]. It caused thrombocytopenia and partial heart block (prolonged QT), but it induced CR in some exceptional patients with NSCLC and head and neck carcinomas, especially when given in combination with radiotherapy. Farnesyl transferase inhibitors now in clinical trials (R11577, BMS-214662, BMS-186511, L-774123, L-778123, SCH44342, PD083176, PD98059) have been repeatedly reviewed [187,191]. Extracellular signal-regulated kinases are overactive in AML cells but can be inhibited by PD98059 (New England BioLabs, Beverly, NJ); downmodulation of ERK resulted in cell-cycle arrest and apotosis of the leukemic cells [190].

Rho GTPses bind GTP and hydrolyze it to GDT. Rho GTPase-activating proteins (Rho-GAPs) organize eukaryotic cytoskeleton and gene transcription. Cells of inflammatory carcinoma of the breast overexpress RhoC GTPase and treatment with L774123 or L774832 suppressed the expression of this enzyme and the malignant phenotype (decrease of anchorage-independent growth; locomotion and invasiveness) of these cells [193].

In summary, FT inhibitors (R115777, BMS214662, L778,123) used singly in patients with metastatic disease induced only occasional PR or MR, but more frequently caused stabilization of disease. In patients with pancreatic or lung cancers, there were no real responders; in metastatic breast carcinoma, 10% PR and 16% stabilization of disease occurred; occasional PR are recorded in patients with recurrent glioma, AML, or myelodysplasia. In combination with gemcitabine, paclitaxel, or platinols, or with radiotherapy, the PR rate rises and occasional CR are also recorded, with a trend suggestive of additive effects.

Ras antagonists transfarnesylthiosalicylic acid (FTS) or SCH66336 block the growth of human melanoma xenografts in SCID mice [166,194]. Ras-transformed tumor cells do not express Fas (CD95) and therefore escape externally induced apoptosis. Farnsyl transferase inhibitor LB42722 upregulates Fas (CD95) expression and restores susceptibility of the cell to apoptosis; the same compound inhibits VEGF expression [195]. N-ras–transformed neuroblastoma cells utilize insulinlike growth factor (IGF)-II, nerve growth factor (NGF) and brain-derived neurotrophic factor in their autocrine growth loops. The FT inhibitor FTI-277 disrupts these autocrine growth loops [196]. The peptidomimetic FT inhibitors B956 and B1086 suppress colony formation in soft agar, or growth in xenografts of human fibrosarcoma, urinary bladder carcinoma, and colon carcinoma cells [197]. The enzyme nucleoside diphosphate kinase (Ndk) attacks the mutated Ras protein RasD12, but not the wild-type Ras protein; it may act in vivo as a metastasis suppressor [198]. The small molecular compound HA14-1 binds and inhibits Bcl-2 and increases cytochrome C release from mitochondria. These are pro-apoptotic events. The isoquinolone carboxamide derivative PK11195 is an antagonist of the mitochondrial benzodiazepine receptor (PBR) and thereby reinforces HA14-1-induced apoptosis, whereas Ro-4864, an agonist of PBR, protects tumor cells against TNFα-induced apoptosis [199]. The HA14-1 and PK11195 combination is awaiting clinical trials. PK11195 eliminates Bcl_{XL}^{+} and $Mcl\text{-}1^{+}$ cholangiocarcinoma xenografts [200]. PK11195 sensitizes AML cells to daunomycin and cytarabine [201]. The lactone of *Aspergillus terreus*, lovastatin, depletes intracellular mevalonates and synergizes with interferon (IFN-α2b) in suppressing the growth of, and inducing apoptosis in, the CML cell line K562 [202]. SU11248 [Sugen (now Pfizer, New York, NY)] inhibits PDGF-R, VEGF-R, c-KIT (cyclin-dependent kinase inhibitor tumor), and FLT3 (the *fms*-like tyrosine kinase) and suppresses the growth of M.D. Anderson Hospital breast cancer xenografts [203].

The PI3K inhibitor LY294002 restores cancer cells' susceptibility to ionizing radiation. LY294002 renders pancreatic cancer cells susceptible to sulindac. LY294002 and the mTOR inhibitor rapamycin induce the expression of cycline kinase inhibitor p21 in prostate cancer cells but Bcl_{XL} protects prostate cancer cells against LY294002-induced apoptosis (204–207).

Tyrphostin AG490, the JAK inhibitor in pancreatic carcinoma cells, downregulated cyclin D1, Bcl_{XL}, and VEGF mRNA. In anaplastic lymphoma cells, it induced expression of the proapoptotic caspase 3 and suppressed Bcl-2 and BCL_{XL}. Tyrphostins arrest the cell cycle in ovarian carcinoma cells [208–210]. These tumors are targeted for reoviral therapy: There appears to be a rationale for a combined attack on these tumor cells.

TYROSINE KINASE INHIBITORS

Imatinib Mesylate

The reciprocal translocation t(9;22)(q34;q11) linking the gene c-*abl* to the 5′ segment of gene *bcr* forms the fusion gene *bcr-abl*. The $p210^{bcr\text{-}abl}$ oncoprotein product of the fused gene functions as a tyrosine kinase in Ph^{+} CML cells. It is recognized by CD4 and CD8 T cells of the host. The implications of graft-vs leukemia reaction and immunotherapy directed against this target have been explored [211,212]. The phenylamino pyrimidine STI-571 (imatinib mesylate, Gleevec, Novartis) emerges as the most efficient inhibitor of the phosphorylated gene product protein. It induces 94% CR in CML with 91% cytogenetic response; freedom from progression and survival at 18 months are 93% and 96%. This drug is now fully licensed for the primary treatment of CML, even in the accelerated or blast crisis stages of the disease. In the accelerated phase of CML, complete cytogenetic response rate is 24%, extending 18-month survival up to 73% [213–223]. Imatinib may control hypereosinophilic syndromes with or without the translocation t(5; 12)(q33;p13) [222]. The FT inhibitors SCH66336 or R115777 are additive to imatinib in suppressing Ph^+ CML cells [225,226].

The pyrido-pyrimidine compounds PD166326, PD173955, and PD180970 inhibit both *abl* and *src* kinases and suppress the growth of imatinib-resistant CML cells [227,228]. CML K562 xenografts regress in nude mice treated with the perorally administered anilino-quinoline-carbonitrile SKI-606 (Wyeth), a dual inhibitor of *src* and *abl* tyrosine kinases [229,230]. A 10-fold increase in the BCR-ABL fusion transcript (caused by mutations in the *bcr-abl* genes) is the cause of imatinib resistance [231,232]. When imatinib destroyed mast cells in the tumor bed of human mammary carcinoma xenografts, accelerated growth of the tumors occurred [233]. This compound may be active in the treatment of systemic mastocytosis. However, in an actual assay, STI571 failed to inhibit the growth of human leukemic mast cell lines [234].

The Hardy-Zuckerman (HZ4) feline sarcoma retrovirus harbors the oncogene v-*kit*, which derives from the host cell gene c-*kit*. It serves as the receptor (CD117) for its cognate ligand stem cell factor (SCF). When phosphorylated on Tyr^{567} and Tyr^{719}, the gene product protein of this gene induces the JAK/STAT (Janus family kinases/signal transducers and activators of transcription), the Ras-Raf protein kinases, the PI3K, the MAPK, and the PCL (phosphokinase C) systems, leading to mitoses and apoptosis-free survival of the cell. If the signal transduction emanates from a point-mutated c-*kit* (exons 9,

11, or 13), the system remains constitutively activated and the cell exhibits all criteria of the malignant geno- phenotype, especially those of invasiveness and neoangiogenesis induction. STAT3 is constitutively activated, but JAK2 can be blocked with AG490 (Tyrphostin); MAPK are blocked by PD98059; PI3K is blocked by LY294002, but only STI571 (imatinib) or AG490 could inhibit proliferation of malignant gastrointestinal stromal tumor (GIST) cells [235–245]. Malignant cells expressing the receptor c-*kit* and its ligand SCF utilize the system as an autocrine growth loop. Prominently expressed in the notoriously chemotherapy-resistant GIST, c-*kit* signal transduction is inhibited by imatinib mesylate. However, c-*kit* is frequently mutated and constitutively activated without capturing its ligand. Occasionally, not c-*kit* but the PDGF-R gene is mutated. When imatinib mesylate inhibits c-*kit* signal transduction, the cell becomes susceptible to programmed death; when Bcl-2 is amplified (not translocated), GIST cells become apoptosis-resistant. Gastrointestinal stromal tumor cells are usually positive for vimentin and CD34 and could express a-smooth muscle actin and S-100. Bcl-2, c-Myc, and Ki-67 overexpression correlates with a highly malignant course. Prognosis worsens with tumors being aneuploid [237] and harboring a mutated $p16^{INK4}$ gene at 9p21 [239]. This *c-kit* suppressor gene is frequently mutated in a number of human malignant tumors. Imatinib induces up to 40% PR rates in metastatic disease and is now fully licensed for the treatment of GIST [240]. Dosage exceeding 500 mg/day induces toxicities (edema, nausea, diarrhea, myalgia, skin rash, myelosuppression with bleeding). Overexpressed and/or mutated c-*kit* is operational in colorectal adenocarcinomas or in glioblastoma cells; cultured or xenografted cells of these and other (small-cell undifferentiated carcinoma of the lung) tumors are also inhibited by imatinib mesylate [246,247].

In small-cell carcinoma of the lung (SCLC), the tumor cells use c-*kit* and its ligand SCF in autocrine and paracrine growth loops [248,249]. Anaplastic thyroid carcinoma cells also expressing CXCR4, the chemokine receptor for stromal cell-derived factor-1α (SDF-1α), continue to proliferate in the presence of imatinib; the PI3K-inhibitor LY294002 inactivated these cells [250], calling for clinical trials in which imatinib and LY294002 are combined. Imatinib potentiated the efficacy of cis-platin against lung cancer cells and clinical trials of carboplatin with paclitaxel plus imatinib are being initiated. Recently, c-*kit* expression was detected in choroidal melanoma cells [251]. Pediatric sarcoma, synovial sarcoma, osteosarcoma, and Ewing sarcoma cells are frequently positive for c-*kit* [236,252]. Nonmutated c-*kit* is overexpressed in the majority of postradiation sarcomas [253], which are usually chemotherapy-resistant and are awaiting clinical trials with imatinib.

Some sarcomas (osteosarcoma, dermatofibrosarcoma protuberans, and others) utilize platelet-derived growth factor and its receptor (PDGF;R) and its ligand as an autocrine growth loop. Nonmesenchymal tumors with an operational PDGF→R system include melanoma, neuroendocrine carcinoids and carcinomas, and many adenocarcinomas. Imatinib mesylate suppressed PDGF-R signaling in cultured and xenografted human osteosarcoma cells and induced remissions in patients with DFSP. In this latter tumor, the PDGF gene (encoding for the ligand) undergoes translocation t(17;22)(q22; q13), in which the collagen type 1 promoter gene drives the PDGF β-chain gene, resulting in an autocrine growth circuit. The retrovirally transduced counterpart of this gene v-*sis* (c-*sis*: 22q12.3-12.1 in the human genome) was discovered in the simian sarcoma retrovirus. The gene c-*sis* is active in wound healing, atheroscleotic plaques, first trimester cytotrophoblasts, transformed fibroblasts, Kaposi's sarcoma cells, other sarcomas, and in several carcinomas, especially in desmoplastic reactions induced by neuroendocrine tumors, certain melanomas, and breast carcinomas. Imatinib mesylate induces dramatic remissions in cases of DFSP by suppressing signal transduction from PDGF-R (failures of treatment are also on record) [254–257]. Expectations have been raised as to the treatment of myelofibrosis, desmoid tumors, and some sarcomas with imatinib mesylate.

New phenylamino-methyl-pyrido-pyrimidine compounds (PD173955; PD166326; PD180970) unrelated to, but potentially cooperative with, imatinib emerge as most potent inhibitors of the Bcr-Abl oncoprotein autophosphorylation. The *src* kinase (the Rous sarcoma virus oncogene) inhibitor PD180970 suppressed the growth of CML blast crisis-derived leukemic cells (K562). Mutated Abl kinase (threonine for methionine at amino acid 351—M351T) gains resistance to imatinib mesylate, but remains susceptible to PD180970; this compound also inhibits signal transduction from nonmutated (wild-type) c-*kit* [258–261]. Bcr-Abl$^+$ CML cells exhibit increased susceptibility to STI571 combined with histone deacetylase inhibitors [aphicidin, depsipeptide, soberoylanilide hydroxamide acid (SAHA), sodium butyrate, oxamflatin, trichostatin, FR901228, MS-275]; SAHA in itself could induce either differentiation or apoptosis of leukemic or solid tumor (pancreatic, small-cell undifferentiated lung carcinoma) cells [262–268]. Raf inhibitor L-779,450 and PI3K inhibitor LY294002 act additively against leukemic cells [269]. HT29 human colon carcinoma cells overexpress c-*kit* and respond to SCF, while remaining susceptible to inhibition by PI3K antagonist LY294002, to Rho GTPases (*Clostridium botulinum* exoenzyme transferase), and to inhibitors of Rho kinases (L27632, STI571, Y27632) [270]. The ROCK (Rho-associated kinase) inhibitor Y-27632 and *C. botulinum* C3 exoenzyme inhibit lysophosphatic acid expression in ovarian carcinoma cells that is essential for tumor cell migration [271]. SU11248 (Sugen) inhibits signal transduction from both EGF-R and PDGF-R and is being tested in patients with SCLC. EGF-R and HER2/neu may form homo- or heterodimers. SU11248 inhibits c-*kit* and PDGF-Rβ in SCLC; GW2016 inhibits both EGF and Erb2 tyrosine kinases [272–274].

The *fms*-like tyrosine kinases (FLT) and their ligands drive hematopoietic cell maturation from their loci at chromosomes 5q31 and 13q12. The gene was discovered in the McDonough feline sarcoma retrovirus (v-*fms*); *c-fms* encodes the receptor for M-CSF. This receptor family includes stem (steel) factor receptor Kit and platelet-derived growth factor receptors A and B (PDGF-R). When c-*fms* undergoes internal tandem duplication (ITD) mutation, it contributes to the generation of AML cells. Tyrosine kinase inhibitors (TKI) aimed at FLT include indolo-carbazoles and the herbimycins and geldanamycins, which inhibit heat shock protein 90, the chaperone that protects oncoproteins from degradation in proteasomes. Sugen product SU5416, quinazolines (MLN518), and staurosporine derivatives (PKC412) inhibit FLT signaling pathways. Patients with AML receiving treatment with one of these TKI may achieve PR. The preparations GTP14564 and SU11248 also inhibit FLT3 [275–277]. SU6656 (Sugen) and/or PD180970 inhibit *src*-kinase and the STAT pathway in lung cancer cells, eliciting apoptosis. Antisense STAT3 oligonucleotides or adenoviral vector expressing dominant negative STAT3 isoforms also induced tumor cell death [278]. CP-65477 is a selective Erb2 kinase inhibitor in breast cancer cells [279]. The monoclonal antibody 2C4 also targets Erb2 because it is expressed in breast and prostate cancer cells [280]. The growth of a renal cancer cell line is stimulated by EGF and TGFα; the EGF-R tyrosine kinase inhibitor PK11166 and paclitaxel cooperate in apoptosis induction of tumor xenografts and endothelial cells supporting the tumor [281]. The isoquinoline ligand of the mitochondrial benzodiazepine receptor ihibitor PK11195 counteracts multiple drug resistance of some leukemic and ovarian carcinoma cells [282].

Epstein-Barr virus carrier malignant lymphomas utilize IL-6 and IL-10 in their autocrine growth loops. The microbial macrolide (macrocyclic lactone) rapamycin (sirolimus; Rapamune, Wyeth-Ayerst Laboratories, St. Davids, PA) is immunosuppressive, arrests cells in G1, and kills lymphocytes, so it emerges as a therapeutic modality for those lymphomas, which appear in organ transplant recipients. Its analogues CC1-779 (Wyeth-Ayerst) and RAD001 (everolimus; Novartis) act upon the serine-threonine kinase receptor mTOR, which bears homology with PI3K. Combining rapamycin with LY294002 increased its growth-inhibitory effect on prostate cancer cells because PI3K, AKT, and mTOR signaling drives the growth of prostatic cancers. The growth of osteo- and rhabdomyosarcoma cells; medulloblastoma, neuroblastoma, and glioblastoma cells; lung cancer cells (SCLC); and some adenocarcinoma (breast and prostate) cells was inhibited by rapamycin [283–292].

ZD1839 Gefitinib

Epidermal growth factor receptor is commonly constitutively overexpressed in human cancer cells and it is targeted by inhibitors of its tyrosine kinase and by monoclonal antibodies [293–301]. The anilino-quinazoline derivative ZD1839 (Iressa, AstraZeneca) selectively and reversibly blocks the EGF-R tyrosine kinases. The EGF-R/ErbB1 (7p11-13 in the human genome) and its family (HER2, 3, 4 for human epidermal receptor) with their ligands, heregulin for HER3 and neuregulin for HER4, are proto-oncogenes whose ancestors were discovered in the genome of the avian erythroblastosis retroviruses as v-*erb*2, 3, 4. EGF-R captures EGF, transforming growth factor alpha (TGFα), amphiregulin, and heparin-binding growth factor as its ligands. Truncated EGF-RvIII without its extracellular domain is constitutively active in glioblastoma and in some breast carcinoma cells. However, in some human tumor cell lines (breast cancer, lung cancer) ZD1839 acted independently from EGF-R expression [302,303]. ZD1839 in low dose suppressed the growth of cultured or xenografted human carcinoma cells; in high dose it induced apoptotic deaths. In treated tumor cells, the cell cycle stops at G_1 and the p27 cyclin-dependent kinase inhibitor levels rise. Apoptosis induction is enhanced by platinols or radiation. In xenografted tumors, VEGF, bFGF (*vide infra*), and TGFα levels drop: Iressa exerts anti-angiogenesis. Signal transduction in breast carcinoma cells from the HER2/neu/ErbB2 receptor is also suppressed; this effect is intensified by the co-administration of trastuzumab (Herceptin, Genentech). The growth rate of xenografted ovarian carcinoma cells was significantly retarded by ZD1839 in SCID mice. When tested against xenografted prostate carcinoma cells, Iressa synergized with bicalutamide [304–306].

In an AstraZeneca trial (reported by R.S. Herbst from M.D. Anderson Cancer Center, Houston, TX) of 71 patients with NSCLC, only one experienced PR, but in most of the patients, the disease stabilized for 3 (45%) to 12 months (7%) [307,308].

In Phase II clinical (IDEAL—Iressa dose evaluation in advanced lung cancer) trials, chemotherapy (taxanes and platinols)-resistant NSCLC appeared to be sensitive to Iressa monotherapy: The orally administered drug induced 8% to 18% durable PR rates; with stabilization of disease, the response rate rose to 53%, but 34% of patients progressed after a stagnation of tumor growth for 4 months. When combined with carboplatin or paclitaxel for 24 untreated patients, one CR, five PR, and eight stabilizations of disease occurred. Clinical trials of Iressa in combination with 5FU or gemcitabine with or without platinols or taxanes continue. Tolerated dosage is 250 to 500 mg/day. Higher dosages cause diarrhea and acneiform skin eruptions. The Phase III (INTACT) clinical trials failed to prove improved survival rates when Iressa was added to chemotherapy for the treatment of NSCLC because the control group in this trial achieved an unexpectedly high survival rate. The addition of gefitinib to the standard paclitaxel and carboplatin regimen in the treatment of advanced NSCLC did not prolong survival (INTACT 2 trial). There is good agreement among clinical trials on Iressa's tolerable toxicity profile and on its efficacy in inducing responses (PR; stabilization of disease) even in chemotherapy-refractory patients [309–317]. According to case reports, bronchioalveolar carcinoma appears to respond better to Iressa than other histological types of lung cancer [318,319] due to EGF-R mutations. In pre- or early clinical studies, Iressa suppressed the growth of some selected solid tumors (colorectal adenocarcinoma, squamous cell carcinoma of the head and neck) and synergized with radiotherapy (oral squamous cell carcinoma) and chemotherapy (5FU, capecitabine, and the platinols) [320–330].

In Japan, tens of thousands of patients received Iressa but 80 cases of interstitial pneumonitis, sometimes advancing into fatal pulmonary fibrosis, emerged [331]. Bronchial alveolar cell damage [332] might have occurred prior to pulmonary fibrosis.

The Oncologic Drug Advisory Committee (ODAC) approved the licensure of Iressa [333] for the treatment of patients with NSCLC who failed chemotherapy with platinols and taxanes.

OSI-774 Erlotinib

The quinazoline-derivative CP-358,774 or OSI-774 (erlotinib, Tarceva, OSI Pharmaceuticals; Genentech) inhibits EGF-R tyrosine kinases. The drug gained fame when it induced CR in a patient with metastatic kidney carcinoma (whose primary tumor was then resected) for more than 20 months. Other patients with a variety of metastatic carcinomas experienced PR (11% to 13%) and long stabilizations of disease (34% to 58%). Of bronchogenic carcinomas, bronchioalveolar carcinoma appears to respond best to erlotinib (ASCO #2491 2003). Some of these patients developed severe but reversible papulopustular acneiform skin eruptions. Both Iressa and OSI-774 act on the normal epidermis of cancer patients in upregulating cell-cycle inhibitor p27 and decreasing autophosphorylation of the EGF-R on Tyr1173. Of 34 patients with ovarian carcinoma, 3 PR and 15 stabilization of disease occurred. Clinical trials with erlotinib and docetaxel for advanced NSCLC are not without side effects: Febrile neutropenia, diarrhea, stomatitis, and acneiform skin rash occur, and the maximum tolerated dosages (100 mg/day and 75 mg/m^2 every 3 weeks) are not to be exceeded. [334–337].

PART I: RECENT DEVELOPMENTS IN VIRAL ONCOLYSIS

The Chinese HOT vaccine consists of a recombinant oncolytic adenovirus (Ad-sp/E1A) expressing heat shock protein (HSP70); HSP-complexed antigenic peptides from tumors, when presented to DCs, elicit rejection-strength immune reactions of the host [338]. Viral vectors (AdCMVIl-12) injected intratumorally may disseminate in the host [339]. The histone deacetylase inhibitor depsipeptide (FR901228; FK228, NCI, Bethesda, MD) is in Phase II clinical trial; it enhances adenoviral receptor CAR expression and thereby promotes adenovirally inserted transgene expression in targeted cells [340]. The efficiency of adenoviral gene transfer depends on the extent of CAR expression in NSCLC xenografts [341]. The current status of gene therapy for lung and head and neck cancers is dominated by adenovirally inserted wild-type p53 gene into tumor cells [342]. Although success rate is limited, adenovirally inserted wild-type p53 increases the radiosensitivity and decreases the DNA repair capacity of the tumor cells [343]. The R1 subunit of the enzyme ribonucleotide reductase (RNR) acts as a tumor suppressor; intratumorally injected rAd5-R1 inhibits the growth of human colon carcinoma xenografts [344]. Both Ras and the adenovirally produced small VA (virus-associated) RNA block PKR; otherwise, adenovirally produced dsRNA would activate PKR, viral protein translation would be blocked, and IFN would abolish viral replication, as it does in healthy cells. In Ras-mutated tumor cells, adenoviral infection does not depend on its VA small RNAs; there PKR is already blocked. Therefore, VA gene-defective adenovirus could not replicate in normal cells but would selectively grow in Ras-mutated tumor cells. The dl330 (29 bp-deleted VAI RNA-coding region) Ad5 VAI mutant is therefore a new candidate for virotherapy of Ras-mutated cancers [345]. Human cancer cells often overexpress MMAC (mutated in multiple advanced cancers) and PI3K/Akt and eliminate the tumor-suppressor gene PTEN (phosphatase and tensin homologue deleted on chromosome ten). The "kinase dead mutant of Akt" replication-defective Ad-Akt-DN E1/E3-deleted adenoviral vector selectively induces apoptosis in tumor cells [346]. In kidney carcinoma cells with mutated von Hippel-Lindau (VHL) gene, the hypoxia-inducible factor (HIF) is overexpressed and escapes proteasomal degradation because of faulty ubiquination. The replication of adenovirus construct Ad9xHRE1A is HIF-dependent and, therefore, it selectively kills VHL gene-mutated tumor cells (pheochromocytoma, renal cell carcinoma) [347]. A replication-competent adenovirus contains two cell suicide genes fused: those of cytosine deaminase (CD) and herpes simplex virus (HSV) *tk*. Prostate cancer cells infected with AD5-CD/TKrep cannot rapidly eliminate 5-fluorocytosine and die upon exposure to valganciclovir; these tumor cells show increased sensitivity to radiation therapy [348]. Genetically engineered adenoviruses facilitate pro-drug therapy by sensitizing tumor cells (hepatocellular carcinoma, human tumor xenograft 293/E4, breast carcinoma) to gemcitabine, irinotecan, or mitomycin C 349–351 (Table 1). Vaccinia and canarypox (ALVAC, Aventis Pasteur, Swiftwater, PA) viruses delivering genes for HPV16, HPV18E6, and HPV18E7; GM-CSF; or gp100, respectively, effectively immunize against malignant transformation of vulvovaginal and uterine cervical intraepithelial neoplasia, overcome tumor-related immunological ignorance, and, with high-dose IFN, recall previously activated immune T cells to react with and kill melanoma cells [352–354].

REFERENCES FOR INTRODUCTION AND PART I

1. Bos JL. Ras oncogenes in human cancer: a review Cancer Res 1989; 49:4682–4689.
2. Bos JL. Ras-like GTPases Biochim Biophys Acta 1997; 1333: M19–31.
3. Malumbres M, Barbacid M. RAS oncogenes: the first 30 years Nat Rev Cancer 2003; 3:459–465.
4. Graziano SL, Gamble GP, Newman NB, Abbott LZ, Rooney M, Mookherjee S, Lamb ML, Kohman LJ, Poiesz BJ. Prognostic significance of K-*ras* codon 12 mutations in patients with resected stage I and II non-small-cell lung cancer J Clin Oncol 1999; 17:668–675.
5. Westra WH, Slebos RJ, Offerhaus GJ, Goodman SN, Evers SG, Kensler TW, Askin FB, Rodenhuis S, Hruban RH. Activation of the K-ras oncogene activation in lung adenocarcinomas from former smokers. Evidence that K-*ras* mutations are an early and irreversible event in the development of adenocarcinoma of the lung Cancer 1993; 72:432–438.
6. Mills NE, Fishman CL, Rom WN, Dubin N, Jacobson DR. Increased prevalence of K-ras oncogene mutations in lung adenocarcinoma Cancer Res 1995; 55:1444–1447.
7. Tórtola S, Marcuello E, González I, Reyes G, Arribas R, Aiza G, Sancho FJ, Peinado MA, Capella G. p53 and K-ras gene mutations correlate with tumor aggressiveness but are not of routine prognostic value in colorectal cancer J Clin Oncol 1999; 17:1375–1381.
8. Kane K. The genetics of colorectal carcinoma Ann Clin Lab Sci 1994; 24:287–293.

9. Markowitz S, Hines JD, Lutterbaugh J, Myeroff L, Mackay W, Gordon N, Rustum Y, Luna E, Kleinerman J. Mutant K-ras oncogenes in colon cancers do not predict patient's chemotherapy response or survival Clin Cancer Res 1995; 1: 441–445.

10. Smith G, Carey FA, Beattie J, Wilkie MJ, Lightfoot TJ, Coxhead J, Garner RC, Steele RJ, Wolf CR. Mutations in APC, Kirsten-ras, and p53--alternative genetic pathways to colorectal cancer Proc Natl Acad Sci U S A 2002; 99:9433–9438.

11. Leslie A, Pratt NR, Gillespie K, Sales M, Kernohan NM, Smith G, Wolf CR, Carey FA, Steele RJ. Mutations of APC, K-ras, and p53 are associated with specific chromosomal aberrations in colorectal adenocarcinomas Cancer Res 2003; 63:4656–45.

12. Zhang B, Ougolkov A, Yamashita K, Takahashi Y, Mai M, Minamoto T. β-Catenin and ras oncogenes detect most human colorectal cancer Clin Cancer Res 2003; 9:3073–3079.

13. Tórtola S, Steinert R, Hantschick M, Peinado MA, Gastinger I, Stosiek P, Lippert H, Schlegel W, Reymond MA. Discordance between K-ras mutations in bone marrow micrometastases and the primary tumor in colorectal cancer J Clin Oncol 2001; 19:2837–2843.

14. Bertario L, Russo A, Sala P, Varesco L, Giarola M, Mondini P, Pierotti M, Spinelli P, Radice P. Multiple approach to the exploration of genotype-phenotype correlations in familial adenomatous polyposis J Clin Oncol 2003; 21:1698–1707.

15. Liu HX, Zhou XL, Liu T, Werelius B, Lindmark G, Dahl N, Lindblom A. The role of hMLH3 in familial colorectal cancer Cancer Res 2003; 63:1894–1899.

16. Kondo H, Sugano K, Fukayama N, Hosokawa K, Ohkura H, Ohtsu A, Mukai K, Yoshida S. Detection of K-ras gene mutations at codon 12 in the pancreatic juice of patients with intraductal papillary mucinous tumors of the pancreas Cancer 1997; 79:900–905.

17. Matsubayashi H, Watanabe H, Nishikura K, Ajioka Y, Kijima H, Saito T. Determination of pancreatic ductal carcinoma histogenesis by analysis of mucous quality and K-ras mutation Cancer 1998; 82:651–660.

18. Brembeck FH, Schreiber FS, Deramaudt TB, Craig L, Rhoades B, Swain G, Grippo P, Stoffers DA, Silberg DG, Rustgi AK. The mutant K-ras oncogene causes pancreatic periductal lymphocytic infiltration and gastric mucous neck cell hyperplasia in transgenic mice Cancer Res 2003; 63:2005–2009.

19. Hoa M, Davis SL, Ames SJ, Spanjaard RA. Amplification of wild-type K-ras promotes growth of head and neck squamous cell carcinoma Cancer Res 2002; 62:7154–7156.

20. Rincón-Arano H, Rosales R, Mora N, Rodriguez-Castaneda A, Rosales C. R-Ras promotes tumor growth of cervical epithelial cells Cancer 2003; 97:575–585.

21. Kuzmin I, Liu L, Dammann R, Geil L, Stanbridge EJ, Wilczynski SP, Lerman MI, Pfeifer GP. Inactivation of RAS association domain family 1A gene in cervical carcinomas and the role of human papillomavirus infection Cancer Res 2003; 63: 1888–1893.

22. Cohen Y, Singer G, Lavie O, Dong SM, Beller U, Sidransky D. The RASSF1A tumor suppressor gene is commonly inactivated in adenocarcinoma of the uterine cervix Clin Cancer Res 2003; 9:2981–2984.

23. Bakin RE, Gioeli D, Sikes RA, Bissonette EA, Weber MJ. Constitutive activation of the Ras/mitogen-activated protein kinase signaling pathway promotes androgen hypersensitivity in LNCaP prostate cancer cells Cancer Res 2003; 63: 1981–1989.

24. Bakin RE, Gioeli D, Bissonette EA, Weber MJ. Attenuation of Ras signaling restores androgen sensitivity to hormone-refractory C4–2 prostate cancer cells Cancer Res 2003; 63: 1975–1980.

25. Knauf WU, Ho AD. Polymorphism of the human Ha-ras oncogene locus in chronic lymphocytic and chronic myelogenous leukemia Hematol Oncol. 1991; 9:157–162.

26. Taylor JA, Sandler DP, Bloomfield CD, Shore DL, Ball ED, Neubauer A, McIntyre OR, Liu E. ras oncogene activation and occupational exposures in acute myeloid leukemia J Natl Cancer Inst 1992; 84:1626–1632.

27. Beaupre DM, Kurzrock R. RAS and leukemia: from basic mechanisms to gene-directed therapy J Clin Oncol 1999; 17: 1071–1079.

28. Omholt K, Karsberg S, Platz A, Kanter L, Ringborg U, Hansson J. Screening of N-ras codon 61 mutations in paired primary and metastatic cutaneous melanomas: mutations occur early and persist throughout tumor progression Clin Cancer Res 2002; 8:3468–3474.

29. Kraemer KH. NRAS hypermutability in familial melanoma with CDKN2A mutations: cause and effect J Natl Cancer Inst 2003; 95:768–769.

30. Eskandarpour M, Hashemi J, Kanter L, Ringborg U, Platz A, Hansson J. Frequency of UV-inducible NRAS mutations in melanomas of patients with germline CDKN2A mutations J Natl Cancer Inst 2003; 95:790–798.

31. Gorden A, Osman I, Gai W, He D, Huang W, Davidson A, Houghton AN, Busam K, Polsky D. Analysis of BRAF and N-RAS mutations in metastatic melanoma tissues Cancer Res 2003; 63:3955–3957.

32. Laud K, Kannengiesser C, Avril MF, Chompret A, Stoppa-Lyonnet D, Desjardins L, Eychene A, Demenais F. French Herediatary Melanoma Study Group, Lenoir GM Paillerets BB-de. BRAF as a melanoma susceptibility candidate gene Cancer Res 2003; 63:3061–3065.

33. Meyer P, Klaes R, Schmitt C, Boettger MB, Garbe C. Exclusion of BRAFV^{599E} as a melanoma susceptibility mutation Int J Cancer 2003; 106:78–80.

34. Mercer KE, Pritchard CA. Raf proteins and cancer: B-Raf is identified as a mutational target Biochim Biophys Acta 2003; 1653:25–40.

35. Anders M, Christian C, McMahon M, McCormick F, Korn WM. Inhibition of the Raf/MEK/ERK pathway up-regulates expression of the coxsackievirus and adenovirus receptor in cancer cells Cancer Res 2003; 63:2088–2095.

36. Tarutani M, Cai T, Dajee M, Khavari PA. Inducible activation of Ras and Raf in adult epidermis Cancer Res 2003; 63: 319–323.

37. Kawakami Y, Kitaura J, Yao L, McHenry RW, Kawakami Y, Newton AC, Kang S, Kato RM, Leitges M, Rawlings DJ, Kawakami T. A Ras activation pathway dependent on Syk phosphorylation of protein kinase C Proc Natl Acad Sci U S A 2003; 100:9470–9475.

38. de Castro IP, Diaz R, Malumbres M, Hernández MI, Jagirdar J, Jiménez M, Ahn D, Pellicer A. Mice deficient for N-ras: impaired antiviral immune response and T-cell function Cancer Res 2003; 63:1615–1622.

39. Chang TH, Szabo E. Enhanced growth inhibition by combination differentiation therapy with ligands of peroxisome proliferator-activated receptor-γ and inhibitors of histone deacetylase in adenocarcinoma of the lung Clin Cancer Res 2002; 8: 1206–1212.

40. Henson P. Suppression of macrophage inflammatory responses by PPARs Proc Natl Acad Sci U S A 2003; 100: 6295–6296.

41. Yang Z, Bagheri-Yarmand R, Balasenthil S, Hortobagyi G, Sahin AA, Barnes CJ, Kumar R. HER2 regulation of peroxisome proliferator-activated receptorγ (PPARγ) expression and sensitivity of breast cancer cells to PPARγ ligand therapy Clin Cancer Res 2003; 9:3198–3203.

42. Wang H, Yu D, Agrawal S, Zhang R. Experimental therapy of human prostate cancer by inhibiting MDM2 expression with novel mixed-backbone antisense oligonucleotides: in vitro and in vivo activities and mechanisms Prostate 2003; 54:194–205.

43. Fournier P, Zeng J, Schirrmacher V. Two ways to induce innate immune responses in human PBMCs: paracrine stimulation of IFN-α responses by viral protein or dsRNA Int J Oncol 2003; 23:673–680.

44. Young DF, Andrejeva L, Livingstone A, Goodbourn S, Lamb RA, Collins PL, Elliott RM, Randall RE. Virus replication in engineered human cells that do not respond to interferons J Virol. 2003; 77:2174–2181.

45. Baigent SJ, McCauley JW. Influenza type A in humans, mammals and birds: determinants of virus virulence, host-range and interspecies transmission Bioessays 2003; 25:657–671.

46. Krug RM, Yuan W, Noah DL, Latham AG. Intracellular warfare between human influenza viruses and human cells: the roles of the viral NS1 protein Virology 2003; 309:181–189.

47. Achdout H, Arnon TI, Markel G, Gonen-Gross T, Katz G, Lieberman N, Gazit R, Joseph A, Kedar E, Mandelboim O. Enhanced recognition of human NK receptors after influenza virus infection J Immunol 2003; 171:915–923.

48. Brimnes MK, Bonifaz L, Steinman RM, Moran TM. Influenza virus-induced dendritic cell maturation is associated with the induction of strong T cell immunity to a coadministered, normally nonimmunogenic protein J Exp Med 2003; 198:133–144.

49. Wein LM, Wu JT, Kirn DH. Validation and analysis of a mathematical model of a replication-competent oncolytic virus for cancer treatment: implications for virus design and delivery Cancer Res 2003; 63:1317–1324.

50. Karashima T, Sweeney P, Kamat A, Huang S, Kim SJ, Bar-eli M, McConkey DJ, Dinney CP. Nuclear factor-κB mrdiates angiogenesis and metastasis of human bladdr cancer through the regulation of interleukin-8 Clin Cancer Res 2003; 9: 2786–2797.

51. Graci JD, Cameron CE. Quasispecies, error catastrophe, and the antiviral activity of ribavirin Virology 2002; 298:175–180.

52. Eigen M. Error catastrophe and antiviral strategy Proc Natl Acad Sci U S A 2002; 99:13374–13376.

53. Koup RA. Virus escape from CTL recognition J Exp Med. 1994; 180:779–782.

54. Moskophidis D, Zinkernagel RM. Immunobiology of cytotoxic T-cell resistant virus variants: studies on lymphocytic choriomeningitis virus LCMV) Semin Virol 1996; 7:3–11.

55. Toes RE, Blom RJ, Offringa R, Kast WM, Melief CJ. Enhanced tumor outgrowth after peptide vaccination. Functional deletion of tumor-specific CTL induced by peptide vaccination can lead to the inability to reject tumors J Immunol 1996; 156: 3911–3918.

56. Bluestone JA, Abbas AK. Natural versus adaptive regulatory T cells Nat Rev Immunol. 2003; 3:253–257.

57. Sinkovics JG, Horvath JC. Virological and immunological connotations of apoptotic and anti-apoptotic forces in neoplasia Internat J Oncol 2001; 19:473–488.

58. Benedict CA. Viruses and the TNF-related cytokines, an evolving battle Cytokine Growth Factor Rev 2003; 14: 349–357.

59. Bachmann MF, Kundig TM, Freer G, Li Y, Kang CY, Bishop DH, Hengartner H, Zinkernagel RM. Induction of protective cytotoxic T cells with viral proteins Eur J Immunol 1994; 24: 2228–2236.

60. Guidotti LG, Chisari FV. To kill or to cure: options in host defense against viral infection Curr Opin Immunol 1996; 8: 478–483.

61. Vanlandschoot P, Leroux-Roels G. Viral apoptotic mimicry: an immune evasion strategy developed by the hepatitis B virus Trends Immunol 2003; 24:144–147.

62. Sauter B, Albert ML, Francisco L, Larsson M, Somersan S, Bhardwaj N. Consequences of cell death: exposure to necrotic tumor cells, but not primary tissue cells or apoptotic cells, induces the maturation of immunostimulatory dendritic cells J Exp Med 2000; 191:423–434.

63. Scheffer SR, Nave H, Korangy F, Schlote K, Pabst R, Jaffee EM, Manns MP, Greten TF. Apoptotic, but not necrotic, tumor cell vaccines induce a potent immune response in vivo Int J Cancer 2003; 103:205–211.

64. Tindle RW. Immune evasion in human papillomavirus-associated cervical cancer Nat Rev Cancer 2002; 2:59–65.

65. Di Pasquale G, Stacey SN. Adeno-associated virus Rep78 protein interacts with protein kinase A and its homolog PRKX and inhibits CREB-dependent transcriptional activation J Virol 1998; 72:7916–7925.

66. Dabizzi S, Noci I, Borri P, Borrani E, Giachi M, Balzi M, Taddei GL, Marchionni M, Scarselli GF, Arcangeli A. Luteinizing hormone increases human endometrial cancer cells invasiveness through activation of protein kinase A. Cancer Res 2003; 63:4281–4286.

67. Reshkin SJ, Bellizzi A, Cardone RA, Tommasino M, Casavola V, Paradiso A. Paclitaxel induces apoptosis via protein kinase

A- and p38 mitogen-activated protein-dependent inhibition of the Na(+)/H(+) exchanger (NHE) NHE isoform 1 in human breast cancer cells Clin Cancer Res 2003; 9:2366–2373.

68. Nüesch JP, Corbau R, Tattersall P, Rommelaere J. Biochemical activities of minute virus of mice nonstructural protein NS1 are modulated in vitro by the phosphorylation state of the polypeptide J Virol 1998; 72:8002–8012.

69. Hüser D, Heilbronn R. Adeno-associated virus integrates site-specifically into human chromosome 19 in either orientation and with equal kinetics and frequency J Gen Virol 2003; 84: 133–137.

70. Nada S, Trempe JP. Characterization of adeno-associated virus rep protein inhibition of adenovirus E2a gene expression Virology 2002; 293:345–355.

71. Musatov SA, Dudus L, Parrish CM, Scully TA, Fisher KJ. Spontaneous mobilization of integrated recombinant adenoassociated virus in a cell culture model of virus latency Virology 2002; 294:151–169.

72. Ponnazhagan S, Curiel DT, Shaw DR, Alvarez RD, Siegal GP. Adeno-associated virus for cancer gene therapy Cancer Res 2001; 61:6313–6321.

73. Enger PO, Thorsen F, Lonning PE, Bjerkvig R, Hoover F. Adeno-associated viral vectors penetrate human solid tumor tissue in vivo more effectively than adenoviral vectors Hum Gene Ther 2002; 13:1115–1125.

74. Shi W, Teschendorf C, Muzyczka N, Siemann DW. Adeno-associated virus-mediated gene transfer of endostatin inhibits angiogenesis and tumor growth in vivo Cancer Gene Ther 2002; 9:513–521.

75. Browning MJ, Huneycutt BS, Huang AS, Reiss CS. Replication-defective viruses modulate immune responses J Immunol 1991; 147:2685–2691.

76. Cheng WF, Hung CF, Hsu KF, Chai CY, He L, Polo JM, Slater LA, Ling M, Wu TC. Cancer immunotherapy using Sindbis virus replicon particles encoding a VP22-antigen fusion Hum Gene Ther 2002; 13:553–568.

77. Hirasawa K, Nishikawa SG, Norman KL, Coffey MC, Thompson BG, Yoon CS, Waisman DM, Lee PW. Systemic reovirus

therapy of metastatic cancer in immune-competent mice Cancer Res 2003; 63:348–353.

78. Richardson-Burns SM, Kominsky DJ, Tyler KL. Reovirus-induced neuronal apoptosis is mediated by caspase 3 and is associated with the activation of death receptors J Neurovirol 2002; 8:365–380.

79. Huising MO, Stet RJ, Kruiswijk CP, Savelkoul HF, Lidy Verburg-van Kemenade BM. Molecular evolution of CXC chemokines: extant CXC chemokines originate from the CNS Trends Immunol 2003; 24:307–313.

80. Lu P, Nakamoto Y, Nemoto-Sasaki Y, Fujii C, Wang H, Hashii M, Ohmoto Y, Kaneko S, Kobayashi K, Mukaida N. Potential interaction between CCR1 and its ligand, CCL3, induced by endogenously produced interleukin-1 in human hepatomas Am J Pathol 2003; 162:1249–1258.

81. Wolf M, Clark-Lewis I, Buri C, Langen H, Lis M, Mazzucchelli L. Cathepsin D specifically cleaves the chemokines macrophage inflammatory protein-1 α, macrophage inflammatory protein-1 β, and SLC that are expressed in human breast cancer Am J Pathol 2003; 162:1183–1190.

82. Gear AR, Camerini D. Platelet chemokines and chemokine receptors: linking hemostasis, inflammation, and host defense Microcirculation 2003; 10:335–350.

83. Yi F, Jaffe R, Prochownik EV. The CCL6 chemokine is differentially regulated by c-Myc and L-Myc, and promotes tumorigenesis and metastasis Cancer Res 2003; 63:2923–2932.

84. Homey B, Müller A, Zlotnik A. Chemokines: agents for the immunotherapy of cancer Nat Rev Immunol 2002; 2:175–184.

85. Flanagan K, Kaufman HL. Chemokines and cancer Cancer Invest 2002; 20:825–834.

86. Lee DH, Yang Y, Lee SJ, Kim KY, Koo TH, Shin SM, Song KS, Lee YH, Kim YJ, Lee JJ, Choi I, Lee JH. Macrophage inhibitory cytokine-1 induces the invasiveness of gastric cancer cells by up-regulating the urokinase-type plasminogen activator system Cancer Res 2003; 63:4648–4655.

87. Spriggs MK. Cytokine and cytokine receptor genes "captured" by viruses Curr Opin Immunol 1994; 6:526–529.

88. Biron CA. Cytokines in the generation of immune responses to, and resolution of, virus infection Curr Opin Immunol 1994; 6:530–538.

89. Wilson J, Balkwill F. The role of cytokines in the epithelial cancer microenvironment Semin Cancer Biol 2002; 12: 113–120.

90. Ward BJ, Boulianne N, Ratnam S, Guiot MC, Couillard M, De Serres G. Cellular immunity in measles vaccine failure: demonstration of measles antigen-specific lymphoproliferative responses despite limited serum antibody production after revaccination J Infect Dis 1995; 172:1591–1595.

91. Yokota S, Saito H, Kubota T, Yokosawa N, Amano K, Fujii N. Measles virus suppresses interferon-α signaling pathway: suppression of Jak1 phosphorylation and association of viral accessory proteins, C and V, with interferon-α receptor complex Virology 2003; 306:135–146.

92. Efferson CL, Schickli J, Ko BK, Kawano K, Mouzi S, Palese P, García-Sastre A, Ioannides CG. Activation of tumor antigen-specific cytotoxic T lymphocytes (CTLs) by human dendritic cells infected with an attenuated influenza A virus expressing a CTL epitope derived from the HER-2/neu proto-oncogene J Virol 2003; 77:7411–7424.

93. Epstein SL, Lo CY, Misplon JA, Bennink JR. Mechanism of protective immunity against influenza virus infection in mice without antibodies J Immunol 1998; 160:322–327.

94. Clemens MJ. Interferons and apoptosis J Interferon Cytokine Res. 2003; 23:277–292.

95. Blackman MA, Rouse BT, Chisari FV, Woodland DL. Viral immunology: challenges associated with the progression from bench to clinic Trends Immunol 2002; 23:565–567.

96. Chatterjee M, Osborne J, Bestetti G, Chang Y, Moore PS. Viral IL-6-induced cell proliferation and immune evasion of interferon activity Science 2002; 298:1432–1435.

97. DeGrendele H, Chu E, Marshall J. Activity of the raf kinase inhibitor BAY 43–9006 in patients with advanced solid tumors Clin Colorect Cancer 2003; May:16–18.

98. Toda M, Iizuka Y, Kawase T, Uyemura K, Kawakami Y. Immuno-viral therapy of brain tumors by combination

of viral therapy with cancer vaccination using a replication-conditional HSV Cancer Gene Ther 2002; 9:356–364.

99. Endo T, Toda M, Watanabe M, Iizuka Y, Kubota T, Kitajima M, Kawakami Y. In situ cancer vaccination with a replication-conditional HSV for the treatment of liver metastasis of colon cancer Cancer Gene Ther 2002; 9:142–148.

100. Wang S, Qi J, Smith M, Link CJ. Antitumor effects on human melanoma xenografts of an amplicon vector transducing the herpes thymidine kinase gene followed by ganciclovir Cancer Gene Ther 2002; 9:1–8.

101. Latchman DS. Gene delivery and gene therapy with herpes simplex virus-based vectors Gene 2001; 264:1–9.

102. Harrington KJ, Melcher AA, Bateman AR, Ahmed A, Vile RG. Cancer gene therapy: Part 2. Candidate transgenes and their clinical development Clin Oncol 2002; 14:148–169.

103. Kerr D. Clinical development of gene therapy for colorectal cancer Nat Rev Cancer 2003; 3:615–22.

104. Renaut L, Bernard C, D'Halluin JC. A rapid and easy method for production and selection of recombinant adenovirus genomes J Virol Methods 2002; 100:121–131.

105. Lotze MT, Kost TA. Viruses as gene delivery vectors: application to gene function, target validation, and assay development Cancer Gene Ther 2002; 9:692–699.

106. Yan W, Kitzes G, Dormishian F, Hawkins L, Sampson-Johannes A, Watanabe J, Holt J, Lee V, Dubensky T, Fattaey A, Hermiston T, Balmain A, Shen Y. Developing novel oncolytic adenoviruses through bioselection J Virol 2003; 77: 2640–2650.

107. Demers GW, Johnson DE, Tsai V, Wen SF, Quijano E, Machemer T, Philopena J, Ramachandra M, Howe JA, Shabram P, Ralston R, Engler H. Pharmacologic indicators of antitumor efficacy for oncolytic virotherapy Cancer Res 2003; 63: 4003–4008.

108. Gottesman MM. Cancer gene therapy: an awkward adolescence Cancer Gene Ther 2003; 10:501–508.

109. Post DE, Khuri FR, Simons JW, Van Meir EG. Replicative oncolytic adenoviruses in multimodal cancer regimens Hum Gene Ther 2003; 14:933–946.

110. Hecht JR, Bedford R, Abbruzzese JL, Lahoti S, Reid TR, Soetikno RM, Kirn DH, Freeman SM. A phase I/II trial of intratumoral endoscopic ultrasound injection of ONYX-015 with intravenous gemcitabine in unresectable pancreatic carcinoma Clin Cancer Res 2003; 9:555–561.

111. Habib N, Salama H, Abd El Latif Abu Median A, Isac Anis I, Abd Al Aziz RA, Sarraf C, Mitry R, Havlik R, Seth P, Hartwigsen J, Bhushan R, Nicholls J, Jensen S. Clinical trial of E1B-deleted adenovirus (dl1520) gene therapy for hepatocellular carcinoma Cancer Gene Ther 2002; 9:254–259.

112. Habib NA, Mitry RR, Sarraf CE, Jiao LR, Havlik R, Nicholls J, Jensen SL. Assessment of growth inhibition and morphological changes in in vitro and in vivo hepatocellular carcinoma models post treatment with dl1520 adenovirus Cancer Gene Ther 2002; 9:414–420.

113. Cook JL, Miura TA, Ikle DN, Lewis AM, Routes JM. E1A oncogene-induced sensitization of human tumor cells to innate immune defenses and chemotherapy-induced apoptosis in vitro and in vivo Cancer Res 2003; 63:3435–3443.

114. Jakubczak JL, Ryan P, Gorziglia M, Clarke L, Hawkins LK, Hay C, Huang Y, Kaloss M, Marinov A, Phipps S, Pinkstaff A, Shirley P, Skripchenko Y, Stewart D, Forry-Schaudies S, Hallenbeck PL. An oncolytic adenovirus selective for retinoblastoma tumor suppressor protein pathway-defective tumors: dependence on E1A, the E2F-1 promoter, and viral replication for selectivity and efficacy Cancer Res 2003; 63: 1490–1499.

115. Hubberstey AV, Pavliv M, Parks RJ. Cancer therapy utilizing an adenoviral vector expressing only E1A Cancer Gene Ther 2002; 9:321–329.

116. Kawakami Y, Li H, Lam JT, Krasnykh V, Curiel DT, Blackwell JL. Substitution of the adenovirus serotype 5 knob with a serotype 3 knob enhances multiple steps in virus replication Cancer Res 2003; 63:1262–1269.

117. Campbell M, Qu S, Wells S, Sugandha H, Jensen RA. An adenoviral vector containing an arg-gly-asp (RGD) motif in the fiber knob enhances protein product levels from transgenes refractory to expression Cancer Gene Ther 2003; 10:559–570.

118. Seol JY, Park KH, Hwang CI, Park WY, Yoo CG, Kim YW, Han SK, Shim YS, Lee CT. Adenovirus-TRAIL can overcome TRAIL resistance and induce a bystander effect Cancer Gene Ther 2003; 10:540–548.

119. Armeanu S, Lauer UM, Smirnow I, Schenk M, Weiss TS, Gregor M, Bitzer M. Adenoviral gene transfer of tumor necrosis factor-related apoptosis-inducing ligand overcomes an impaired response of hepatoma cells but causes severe apoptosis in primary human hepatocytes Cancer Res 2003; 63: 2369–2372.

120. Galanis E, Vile R, Russell SJ. Delivery systems intended for in vivo gene therapy of cancer: targeting and replication competent viral vectors Crit Rev Oncol Hematol 2001; 38:177–192.

121. Kim J, Cho JY, Kim JH, Jung KC, Yun CO. Evaluation of E1B gene-attenuated replicating adenoviruses for cancer gene therapy Cancer Gene Ther 2002; 9:725–736.

122. Satoh T, Timme TL, Saika T, Ebara S, Yang G, Wang J, Ren C, Kusaka N, Mouraviev V, Thompson TC. Adenoviral vector-mediated mRTVP-1 gene therapy for prostate cancer Hum Gene Ther 2003; 14:91–101.

123. Zhou RR, Jia SF, Zhou Z, Wang Y, Bucana CD, Kleinerman ES. Adenovirus-E1A gene therapy enhances the in vivo sensitivity of Ewing's sarcoma to VP-16 Cancer Gene Ther 2002; 9:407–413.

124. Ueno NT, Bartholomeusz C, Xia W, Anklesaria P, Bruckheimer EM, Mebel E, Paul R, Li S, Yo GH, Huang L, Hung MC. Systemic gene therapy in human xenograft tumor models by liposomal delivery of the E1A gene Cancer Res 2002; 62: 6712–6716.

125. Kim SK, Wang KC, Cho BK, Lim SY, Kim YY, Oh CW, Chung YN, Kim CY, Lee CT, Kim HJ. Adenoviral p16/CDKN2 gene transfer to malignant glioma: role of p16 in growth, invasion, and senescence Oncol Rep 2003; 10:1121–1126.

126. Tamm I, Schumacher A, Karawajew L, Ruppert V, Arnold W, Nussler AK, Neuhaus P, Dorken B, Wolff G. Adenovirus-mediated gene transfer of P16INK4/CDKN2 into bax-negative colon cancer cells induces apoptosis and tumor regression in vivo Cancer Gene Ther 2002; 9:641–650.

127. Swisher SG, Roth JA. p53 Gene therapy for lung cancer Curr Oncol Rep 2002; 4:334–340.

128. Kuball J, Wen SF, Leissner J, Atkins D, Meinhardt P, Quijano E, Engler H, Hutchins B, Maneval DC, Grace MJ, Fritz MA, Störkel S, Thüroff JW, Huber C, Schuler M. Successful adenovirus-mediated wild-type p53 gene transfer in patients with bladder cancer by intravesical vector instillation J Clin Oncol 2002; 20:957–965.

129. Buller RE, Shahin MS, Horowitz JA, Runnebaum IB, Mahavni V, Petrauskas S, Kreienberg R, Karlan B, Slamon D, Pegram M. Long term follow-up of patients with recurrent ovarian cancer after Ad p53 gene replacement with SCH 58500 Cancer Gene Ther 2002; 9:567–572.

130. Yin S, Bailiang W, Xie K, Goodrich DW. Adenovirus-mediated N5 gene transfer inhibits tumor growth and metastasis of human carcinoma in nude mice Cancer Gene Ther 2002; 9: 665–672.

131. Manabe T, Mizumoto K, Nagai E, Matsumoto K, Nakamura T, Nukiwa T, Tanaka M, Matsuda T. Cell-based protein delivery system for the inhibition of the growth of pancreatic cancer: NK4 gene-transduced oral mucosal epithelial cell sheet Clin Cancer Res 2003; 9:3158–3166.

132. Saimura M, Nagai E, Mizumoto K, Maehara N, Okino H, Katano M, Matsumoto K, Nakamura T, Narumi K, Nukiwa T, Tanaka M. Intraperitoneal injection of adenovirus-mediated NK4 gene suppresses peritoneal dissemination of pancreatic cancer cell line AsPC-1 in nude mice Cancer Gene Ther 2002; 9:799–806.

133. Lee CT, Park KH, Adachi Y, Seol JY, Yoo CG, Kim YW, Han SK, Shim YS, Coffee K, Dikov MM, Carbone DP. Recombinant adenoviruses expressing dominant negative insulin-like growth factor-I receptor demonstrate antitumor effects on lung cancer Cancer Gene Ther 2003; 10:57–63.

134. Trudel S, Trachtenberg J, Toi A, Sweet J, Li ZH, Jewett M, Tshilas J, Zhuang LH, Hitt M, Wan Y, Gauldie J, Graham FL, Dancey J, Stewart AK. A phase I trial of adenovector-mediated delivery of interleukin-2 (AdIL-2) in high risk localized prostate cancer Cancer Gen Ther 2033; 10:755–763.

135. Liu Y, Ehtesham M, Samoto K, Wheeler CJ, Thompson RC, Villarreal LP, Black KL, Yu JS. In situ adenoviral interleukin 12 gene transfer confers potent and long-lasting cytotoxic immunity in glioma Cancer Gene Ther 2002; 9:9–15.

136. Pulkkanen KJ, Laukkanen JM, Fuxe J, Kettunen MI, Rehn M, Kannasto JM, Parkkinen JJ, Kauppinen RA, Pettersson RF, Yla-Herttuala S. The combination of HSV-tk and endostatin gene therapy eradicates orthotopic human renal cell carcinomas in nude mice Cancer Gene Ther 2002; 9:908–916.

137. Gao JQ, Tsuda Y, Katayama K, Nakayama T, Hatanaka Y, Tani Y, Mizuguchi H, Hayakawa T, Yoshie O, Tsutsumi Y, Mayumi T, Nakagawa S. Antitumor effect by interleukin-11 receptor α-locus chemokine/CCL27, introduced into tumor cells through a recombinant adenovirus vector Cancer Res 2003; 63:4420–4425.

138. Fukuda K, Abei M, Ugai H, Seo E, Wakayama M, Murata T, Todoroki T, Tanaka N, Hamada H, Yokoyama KK. E1A, E1B double-restricted adenovirus for oncolytic gene therapy of gallbladder cancer Cancer Res 2003; 63:4434–4440.

139. Tango Y, Fujiwara T, Itoshima T, Takata Y, Katsuda K, Uno F, Ohtani S, Tani T, Roth JA, Tanaka N. Adenovirus-mediated $p14^{ARF}$ gene transfer cooperates with Ad5CMV-p53 to induce apoptosis in human cancer cells Hum Gene Ther 2002; 13: 1373–1382.

140. Huang Y, Tyler T, Saadatmandi N, Lee C, Borgstrom P, Gjerset RA. Enhanced tumor suppression by a p14ARF/p53 bicistronic adenovirus through increased p53 protein translation and stability Cancer Res 2003; 63:3646–3653.

141. Flynn JR, Ramanitharan A, Moparty K, Davis R, Sikka S, Agrawal KC, Abdel-Mageed AB. Adenovirus-mediated inhibition of NF-κB confers chemo-sensitization and apoptosis in prostate cancer cells Int J Oncol 2003; 23:317–323.

142. Ramachandra M, Rahman A, Zou A, Vaillancourt M, Howe JA, Antelman D, Sugarman B, Demers GW, Engler H, Johnson D, Shabram P. Re-engineering adenovirus regulatory pathways to enhance specificity and efficacy Nature Biotechnol 2001; 19:1035–1041.

143. Wirth T, Zender L, Schulte B, Mundt B, Plentz R, Rudolph KL, Manns M, Kubicka S, Kuhnel F. A telomerase-dependent

conditionally replicating adenovirus for selective treatment of cancer Cancer Res 2003; 63:3181–3188.

144. Kitazono M, Rao VK, Robey R, Aikou T, Bates S, Fojo T, Golsnith ME. Histone decetylase inhibitor FR901228 enhances adenovirus infection of hematopoietic cells Blood 2002; 99: 2248–2251.

145. Yamamoto M, Davydova J, Wang M, Siegal GP, Krasnykh V, Vickers SM, Curiel DT. Infectivity enhanced, cyclooxygenase-2 promoter-based conditionally replicative adenovirus for pancreatic cancer Gastroenterology 2003; 125:1203–1218.

146. Ahmed A, Thompson J, Emiliusen L, Murphy S, Beauchamp RD, Suzuki K, Alemany R, Harrington K, Vile RG. A conditionally replicating adenovirus targeted to tumor cells through activated RAS/P-MAPK-selective mRNA stabilization Nat Biotechnol 2003; 21:771–777.

147. Blanc-Brude OP, Mesri M, Wall NR, Plescia J, Dohi T, Altieri DC. Therapeutic targeting of the survivin pathway in cancer: initiation of mitochondrial apoptosis and suppression of tumor-associated angiogenesis Clin Cancer Res 2003; 9:2683–2692.

148. Mesri M, Wall NR, Li J, Kim RW, Altier DC. Cancer gene therapy using a survivin mutant adenovirus J Clin Invest 2001; 108:981–990.

149. Rieger RR, Kipps TJ. CpG oligodeoxynucleotides enhance the capacity of adenovirus-mediated CD154 gene transfer to generate B-cell lymphoma vaccines Cancer Res 2003; 63: 4128–4135.

150. Johnson L, Shen A, Boyle L, Kunich J, Pandey K, Lemmon M, Hermiston T, Giedlin M, McCormick F, Fattaey A. Selectively replicating adenoviruses targeting deregulated E2F activity are potent, systemic antitumor agents Cancer Cell 2002; 1: 325–337.

151. Perez D, White E. E1A sensitizes cells to tumor necrosis factor alpha by downregulating c-FLIP$_s$ J Virol 2003; 77:2651–2662.

152. Hu Z, Garen A. Intratumoral injection of adenoviral vectors encoding tumor-targeted immunoconjugates for cancer immunotherapy Proc Natl Acad Sci USA 2000; 97:9221–9225.

153. Habib NA, Mitry R, Seth P, Kuppuswamy M, Doronin K, Toth K, Krajcsi P, Tollefson AE, Wold WS. Adenovirus replication-competent vectors (KD1, KD3) complement the cytotoxicity and transgene expression from replication-defective vectors (Ad-GFP, Ad-Luc) Cancer Gene Ther 2002; 9:651–654.

154. Hamel Y, Blake N, Gabrielsson S, Haigh T, Jooss K, Martinache C, Caillat-Zucman S, Rickinson AB, Hacein-Bey S, Fischer A, Cavazzana-Calvo M. Adenovirally transduced dendritic cells induce bispecific cytotoxic T lymphocyte responses against adenovirus and cytomegalovirus pp65 or against adenovirus and Epstein-Barr virus EBNA3C protein: a novel approach for immunotherapy Hum Gene Ther 2002; 13:855–866.

155. Olive M, Eisenlohr L, Flomenberg N, Hsu S, Flomenberg P. The adenovirus capsid protein hexon contains a highly conserved human $CD4^+$ T-cell epitope Hum Gene Ther 2002; 13: 1167–1178.

156. Ferrari S, Griesenbach U, Geddes DM, Alton E. Immunological hurdles to lung gene therapy Clin Exp Immunol 2003; 132: 1–8.

157. Kohl NE, Omer CA, Conner MW, Anthony NJ, Davide JP, deSolms SJ, Giuliani EA, Gomez RP, Graham SL, Hamilton K, et al. Inhibition of farnesyltransferase induces regression of mammary and salivary carcinomas in ras transgenic mice Nat Med 1995; 1:792–797.

158. Khosravi-Far R, Cox AD, Kato K, Der CJ. Protein prenylation: key to ras function and cancer intervention Cell Growth Differ 1992; 3:461–469.

159. Adjei AA. Re: Blocking oncogenic Ras signaling for cancer therapy J Natl Cancer Inst 2002; 94:1031–1032; author reply 1032.

160. Rowinsky EK, Windle JJ, Von Hoff DD. Ras protein farnesyltransferase: A strategic target for anticancer therapeutic development J Clin Oncol 1999; 17:3631–3652.

161. Long SB, Casey PJ, Beese LS. Reaction path of protein farnesyltransferase at atomic resolution Nature 2002; 419:645–50.

162. Purcell WT, Donehower RC. Evolving therapies: farnesyltransferase inhibitors Curr Oncol Rep 2002; 4:29–36.

163. Downward J. Targeting RAS signalling pathways in cancer therapy Nat Rev Cancer 2003; 3:11–22.

164. Chang F, Lee JT, Navolanic PM, Steelman LS, Shelton JG, Blalock WL, Franklin RA, McCubrey JA. Involvement of PI3K/Akt pathway in cell cycle progression, apoptosis, and neoplastic transformation: a target for cancer chemotherapy Leukemia 2003; 17:590–603.

165. Eskens FA, Awada A, Cutler DL, de Jonge MJ, Luyten GP, Faber MN, Statkevich P, Sparreboom A, Verweij J, Hanauske AR, Piccart M. European Organization for Research and Treatment of Cancer Early Clinical Studies Group. Phase I and pharmacokinetic study of the oral farnesyl transferase inhibitor SCH 66336 given twice daily to patients with advanced solid tumors J Clin Oncol 2001; 19:1167–1175.

166. Smalley KS, Eisen TG. Farnesyl transferase inhibitor SCH66336 is cytostatic, pro-apoptotic and enhances chemosensitivity to cisplatin in melanoma cells Int J Cancer 2003; 105:165–175.

167. Nakajima A, Tauchi T, Sumi M, Bishop WR, Ohyashiki K. Efficacy of SCH66336, a farnesyl transferase inhibitor, in conjunction with Imatinib against BCR-ABL-positive cells Mol Cancer Ther 2003; 2:219–224.

168. Caponigro F, Casale M, Bryce J. Farnesyl transferase inhibitors in clinical development Expert Opin Investg Drugs. 2003; 12:943–954.

169. Chun KH, Lee HY, Hassan K, Khuri F, Hong WK, Lotan R. Implication of protein kinase B/Akt and Bcl-2′Bcl-XL suppression by the farnesyl transferase inhibitor SCH66336 in apoptosis induction in squamous carcinoma Cancer Res 2003; 63:4796–4800.

170. Gunning WT, Kramer PM, Lubet RA, Steele VE, End DW, Wouters W, Pereira MA. Chemoprevention of benzo(a)pyrene-induced lung tumors in mice by the farnesyltransferase inhibitor R115777 Clin Cancer Res 2003; 9:1927–1930.

171. Lancet JE, Karp JE. Farnesyl transferase inhibitors in myeloid malignancies Blood Rev 2003; 17:123–129.

172. Karp JE, Lancet JE, Kaufmann SH, End DW, Wright JJ, Bol K, Horak I, Tidwell ML, Liesveld J, Kottke J, Ange D, Bud-

dharaju L, Gojo I, Highsmith WE, Belly RT, Hohl RJ, Rybak ME, Thibault A, Rosenblatt J. Clinicakl and biological activity of the farnesyltransferase inhibitor R115777 in adults with refractory and relapsed acute leukemias: a phase I clinical-laboratory correlative trial Blood 2001; 97:3361–3359.

173. Kurzrock R, Kantarjian HM, Cortes JE, Singhania N, Thomas DA, Wilson EF, Wright JJ, Freireich EJ, Talpaz M, Sebti SM. Farnesyltransferase inhibitor R115777 in myelodysplastic syndrome: clinical and biologic activities in the phase I setting Blood 2003; 102:4527–4534.

174. Cortes J. Farnesyltransferase inhibitors in acute myeloid leukemia and myelodysplastic syndromes Clin Lymphoma 2003; 4S:30–35.

175. Morgan MA, Ganser A, Reuter CW. Therapeutic efficacy of prenylation inhibitors in the treatment of myeloid leukemias Leukemia 2003; 17:1482–1498.

176. Smith V, Rowlands MG, Barrie E, Workman P, Kelland LR. Establisment and characterization of acquired resistance to the farnesyl protein transferase R11577 in a human colon cancer cell line Clin Cancer Res 2002; 8:2002–2009.

177. Bondar VM, Sweeney-Gotsch B, Andreeff M, Mills GB, McConkey DJ. Inhibition of the phosphatidylinositol 3′-kinase-AKT pathway induces apoptosis in pancreatic carcinoma cells in vitro and in vivo Mol Cancer Ther 2002; 1:989–997.

178. Katayose K, Seki T, Ohba N, Funatomi H, Goto N, Mitamura K. Growth-inhibitory effect of phosphatidylinositol 3-kinase inhibitor on human pancreatic cancer cells and expression of Bcl-2 family Anticancer Res 2003; 23:2383–2387.

179. Cohen SJ, Ho L, Ranganathan S, Abbruzzese JL, Alpaugh RK, Beard M, Lewis NL, McLaughlin S, Rogatko A, Perez-Ruixo JJ, Thistle AM, Verhaeghe T, Wang H, Weiner LM, Wright JJ, Hudes GR, Meropol NJ. Phase II and pharmacodynamic study of the farnesyltransferase inhibitor R115777 as initial therapy in patients with metastatic pancreatic adenocarcinoma J Clin Oncol 2003; 21:1301–1306.

180. Adjei AA, Croghan GA, Erlichman C, Marks RS, Reid JM, Sloan JA, Pitot HC, Alberts SR, Goldberg RM, Hanson LJ, Bruzek LM, Atherton P, Thibault A, Palmer PA, Kaufmann

SH. A Phase I trial of the farnesyl protein transferase inhibitor R115777 in combination with gemcitabine and cisplatin in patients with advanced cancer Clin Cancer Res 2003; 9: 2520–2526.

181. Adjei AA, Mauer A, Bruzek L, Marks RS, Hillman S, Geyer S, Hanson LJ, Wright JJ, Erlichman C, Kaufmann SH, Vokes EE. Phase II study of the farnesyl transferase inhibitor R115777 in patients with advanced non-small-cell lung cancer J Clin Oncol 2003; 21:1760–1766.

182. Zujewski J, Horak ID, Bol CJ, Woestenborghs R, Bowden C, End DW, Piotrovsky VK, Chiao J, Belly RT, Todd A, Kopp WC, Kohler DR, Chow C, Noone M, Hakim FT, Larkin G, Gress RE, Nussenblatt RB, Kremer AB, Cowan KH. Phase I and pharmacokinetic study of farnesyl protein transferase inhibitor R115777 in advanced cancer J Clin Oncol 2000; 18: 927–941.

183. Johnston SRD, Hickish T, Ellis P, Houston S, Kellard L, Dowsett M, Salter J, Michelis B, Perez-Ruixo JJ, Palmer P, Howes A. Phase II study of the efficacy and tolerability of two dosing regimens of the farnesyl transferase inhibitor R115777 in advanced breast cancer J Clin Oncol 2003; 21:2492–2499.

184. Dempke WC. Farnesyltransferase inhibitors – novel compounds for the treatment of myeloid malignancies Anticancer Res 2003; 23:813–818.

185. Crul M, de Klerk GJ, Swart M', Veer LJ, de Jong D, Boerrigter L, Palmer PA, Bol CJ, Tan H, de Gast GC, Beijnen JH, Schellens JH. Phase I clinical and pharmacologic study of chronic oral administration of th farnesyl protein transferase inhibitor R115777 in advanced cancer J Clin Oncol 2002; 20:2726–2735.

186. Kelland LR. Farnesyl transferase inhibitors in the treatment of breast cancer Expert Opin Investig Drugs. 2003; 12: 413–421.

187. Haluska P, Dy GK, Adjei AA. Farnesyl transferase inhibitors as anticancer agents Europ J Cancer 2002; 38:1685–1700.

188. Britten CD, Rowinsky EK, Soignet S, Patnaik A, Yao SL, Deutsch P, Lee Y, Lobell RB, Mazina KE, McCreery H, Pezzuli S, Spriggs D. A phase I and pharmacological study of the farne-

syl protein transferase inhibitor L-778,123 in patients with solid malignancies Clin Cancer Res 2001; 7:3894–3903.

189. Lobell RB, Liu D, Buser CA, Davide JP, DePuy E, Hamilton K, Koblan KS, Lee Y, Mosser S, Motzel SL, Abbruzzese JL, Fuchs CS, Rowinsky EK, Rubin EH, Sharma S, Deutsch PJ, Mazina KE, Morrison BW, Wildonger L, Yao SL, Kohl NE. Preclinical and clinical pharmacodynamic assessment of L-778,123, a dual inhibitor of farnesyl:protein transferase and geranylgeranyl:protein transferase type-I Mol Cancer Ther 2002; 1:747–758.

190. Hahn SM, Bernhard EJ, Regine W, Mohiuddin M, Haller DG, Stevenson JP, Smith D, Pramanik B, Tepper J, DeLaney TF, Kiel KD, Morrison B, Deutsch P, Muschel RJ, McKenna WG. A Phase I trial of the farnesyltransferase inhibitor L-778,123 and radiotherapy for locally advanced lung and head and neck cancer Clin Cancer Res 2002; 8:1065–1072.

191. Cortes JE, Kurzrock R, Kantarjian HM. Farnesyltransferase inhibitors: novel compounds for the treatment of myeloid malignancies Semin Hematol 2002; 39:26–30.

192. Lunghi P, Tabilio A, Dall'Aglio PP, Ridolo E, Carlo-Stella C, Pelicci PG, Bonati A. Downmodulation of ERK activity inhibits the proliferation and induces the apoptosis of primary acute myelogenous leukemia blasts Leukemia 2003; 17:1783–1793.

193. van Golen KL, Bao L, DiVito MM, Wu Z, Prendergast GC, Merajver SD. Reversion of RhoC GTPase-induced inflammatory breast cancer phenotype by treatment with a farnesyl transferase inhibitor Mol Cancer Ther 2002; 1:575–583.

194. Jansen B, Schlagbauer-Wadl H, Kahr H, Heere-Ress E, Mayer BX, Eichler H, Pehamberger H, Gana-Weisz M, Ben-David E, Kloog Y, Wolff K. Novel Ras antagonist blocks human melanoma growth Proc Natl Acad Sci U S A 1999; 96:14019–14024.

195. Zhang B, Prendergast GC, Fenton RG. Farnesyltransferase inhibitors reverse Ras-mediated inhibition of Fas gene expression Cancer Res 2002; 62:450–458.

196. Girgert R, Wittrock J, Pfister S, Schweizer P. Farnesyltransferase inhibitor FTI-277 prevents autocrine growth stimulation of neuroblastoma by BDNF J Cancer Res Clin Oncol 2003; 129:227–233.

197. Nagasu T, Yoshimatsu K, Rowell C, Lewis MD, Garcia AM. Inhibition of human tumor xenograft growth by treatment with the farnesyl transferase inhibitor B956 Cancer Res 1995; 55:5310–5314.

198. Fischbach MA, Settleman J. Specific biochemical inactivation of oncogenic Ras proteins by nucleoside diphosphate kinase Cancer Res 2003; 63:4089–4094.

199. Chen J, Freeman A, Liu J, Dai Q, Lee RM. The apoptotic effect of HA14–1, a Bcl-2-interacting small molecular compound, requires Bax translocation and is enhanced by PK11195 Mol Cancer Ther 2002; 1:961–967.

200. Okaro AC, Fennell DA, Corbo M, Davidson BR, Cotter FE. PK11195, a mitochondrial benzodiazepine receptor antagonist, reduces apoptosis threshold in Bcl-X(L) and Mcl-1 expressing human cholangiocarcinoma cells Gut 2002; 51: 556–561.

201. Banker DE, Cooper JJ, Fennell DA, Willman CL, Appelbaum FR, Cotter FE. PK11195, a peripheral benzodiazepine receptor ligand, chemosenzitizes acute myeloid leukemia cells to relevant therapeutic agents by more than one mechanism Leuk Res 2002; 26:91–106.

202. Muller-Tidow C, Kiehl M, Sindermann JR, Probst M, Banger N, Zuhlsdorf M, Chou TC, Berdel WE, Serve H, Koch OM. Synergistic growth inhibitory effects of interferon-alpha and lovastatin on bcr-abl positive leukemic cells Int J Oncol 2003; 23:151–158.

203. Abrams TJ, Murray LJ, Pesenti E, Howay VW, Colombo T, Lee LB, Cherrington LM, Pryer NK. Preclinical evaluation of the tyrosine kinase inhibitor SU11248 as a single agent and in combination with "standard of care" therapeutic agents for the treatment of breast cancer Mol Cancer Res 2003; 2: 1011–1021.

204. Gupta Ak, Cerniglia GJ, Mick R, Ahmed MS, Bakanauskas VJ, Muschel RJ, McKenna WG. Radiation sensitization of human cancer cells in vivo by inhibiting the activity of PI3K using LY294002 Int J Rad Oncol Biol Phys 2003; 56:846–853.

205. Yip-Schneider MT, Wiesenauer CA, Schmidt CM. Inhibition of the phosphatidylinositol 3′-kinase signaling pathway in-

creases the responsiveness of pancreatic carcinoma cells to sulindac Gastrointest Surg 2003; 7:354–363.

206. Gao N, Zhang Z, Jiang BH, Shi X. Role of PI3K/AKT/mTOR signaling in the cell cycle progression of human prostate cancer Biochem Biophys Res Commun 2003; 310:1124–1132.

207. Yang CC, Lin HP, Chen CS, Yang YT, Tseng PH, Rangnekar VM, Chen CS. Bcl-xl mediates a survival mechanism independent of the phosphoinositide 3-kinase Akt pathway in prostate cancer cells J Biol Chem 2003; 278:25872–25878.

208. Toyonaga T, Nakano K, Nagano M, Zhao G, Yamaguchi K, Kuroki S, Eguchi T, Chijiiwa K, Tsuneyoshi M, Tanaka M. Blockade of constitutively activated Janus kinase signal transducer and activator of transcription-3 pathway inhibits growth of human pancreatic cancer Cancer Lett 2003; 291: 107–116.

209. Amin HM, Medeiros LJ, Ma Y, Feretzaki M, Das P, Leventaki V, Rassidakis GZ, O'Connor SL, McDonnell TJ, Lai R. Inhibition of JAK3 induces apoptosis and decreases anaplastic lymphoma kinase activity in anaplastic large cell lymphoma Oncogene 2003; 22:5399–5404.

210. Arbel R, Rojansky N, Klein BY, Levitzky R, Hartzstark Z, Laufer N, Ben-Bassat H. Inhibitors that target protein kinases for the treatment of ovarian carcinoma Am J Obstet Gynecol 2003; 188:1283–1290.

211. Wagner WM, Ouyang Q, Pawelec G. The abl/bcr gene product as a novel leukemia-specific antigen: peptides spanning the fusion region of abl/bcr can be recognized by both CD4+ and CD8+ T lymphocytes Cancer Immunol Immunother 2003; 52: 89–96.

212. Sinkovics JG. The graft-versus-leukemia (GvL) reaction; its early history, value in human bone marrow transplantation and recent developments concerning its mechanism and induction In Dicke KA, Keating A, Eds. Autologous Marrow and Blood Transplantation Proceedings 7th International Symposium, Aug 17–20, 1994. Arlington TX, 1995:305–318.

213. Wisniewski D, Lambek CL, Liu C, Strife A, Veach DR, Nagar B, Young MA, Schindler T, Bornmann WG, Bertino JR, Kuriyan J, Clarkson B. Characterization of potent inhibitors of the

Bcr-Abl and the c-kit receptor tyrosine kinases Cancer Res 2002; 62:4244–4255.

214. Kantarjian HM, Talpaz M, Cortes J, O'Brien S, Faderl S, Thomas D, Giles F, Rios MB, Shan J, Arlinghaus R. Quantitative polymerase chain reaction monitoring of BCR-ABL during therapy with imatinib mesylate (STI571; gleevec) in chronic-phase chronic myelogenous leukemia Clin Cancer Res 2003; 9:160–166.

215. Kantarjian HM, O'Brien S, Cortes JE, Smith TL, Rios MB, Shan J, Yang Y, Giles FJ, Thomas DA, Faderl S, Garcia-Manero G, Jeha S, Wierda W, Issa JP, Kornblau SM, Keating M, Resta D, Capdeville R, Talpaz M. Treatment of philadelphia chromosome-positive, accelerated-phase chronic myelogenous leukemia with imatinib mesylate Clin Cancer Res 2002; 8:2167–2176.

216. Kantarjian HM, Talpaz M, O'Brien S, Smith TL, Giles FJ, Faderl S, Thomas DA, Garcia-Manero G, Issa JP, Andreeff M, Kornblau SM, Koller C, Beran M, Keating M, Rios MB, Shan J, Resta D, Capdeville R, Hayes K, Albitar M, Freireich EJ, Cortes JE. Imatinib mesylate for Philadelphia chromosome-positive, chronic-phase myeloid leukemia after failure of interferon-alpha: follow-up results Clin Cancer Res 2002; 8: 2177–2187.

217. Deininger MW, O'Brien SG, Ford JM, Druker BJ. Practical management of patients with chronic myeloid leukemia receiving imatinib J Clin Oncol 2003; 21:1637–1647.

218. Zonder JA, Pemberton P, Brandt H, Mohamed AN, Schiffer CA. The effect of dose increase of imatinib mesylate in patients with chronic or accelerated phase chronic myelogenous leukemia with inadequate hematologic or cytogenetic response to initial treatment Clin Cancer Res 2003; 9:2092–2097.

219. Hernandez-Boluda JC, Cervantes F. Imatinib mesylate (Gleevec, Glivec): a new therapy for chronic myeloid leukemia and other malignancies Drugs Today (Barc) 2002; 38:601–613.

220. Hahn EA, Glendenning GA, Sorensen MV, Hudgens SA, Druker BJ, Guilhot F, Larson RA, O'Brien SG, Dobrez DG, Hensley ML, Cella D. IRIS Investigators. Quality of life in patients with newly diagnosed chronic phase chronic myeloid

leukemia on imatinib versus interferon alfa plus low-dose cytarabine: results from the IRIS Study J Clin Oncol 2003; 21: 2138–2146.

221. Johnson JR, Bross P, Cohen M, Rothmann M, Chen G, Zajicek A, Gobburu J, Rahman A, Staten A, Pazdur R. Approval Summary: imatinib mesylate capsules for treatment of adult patients with newly diagnosed philadelphia chromosome-positive chronic myelogenous leukemia in chronic phase Clin Cancer Res 2003; 9:1972–1979.

222. Mohamed AN, Pemberton P, Zonder J, Schiffer CA. The effect of imatinib mesylate on patients with Philadelphia chromosome-positive chronic myeloid leukemia with secondary chromosomal aberrations Clin Cancer Res 2003; 9:1333–1337.

223. Marcucci G, Perrotti D, Caligiuri MA. Understanding the molecular basis of imatinib mesylate therapy in chronic myelogenous leukemia and the related mechanisms of resistance. (Commentary re: A. N. Mohamed et al., The effect of imatinib mesylate on patients with Philadelphia chromosome-positive chronic myeloid leukemia with secondary chromosomal aberrations. Clin. Cancer Res 9: 1333–1337, 2003.) Clin Cancer Res 2003; 9:1248–1252.

224. Pardanani A, Reeder T, Porrata LF, Li CY, Tazelaar HD, Baxter EJ, Witzig TE, Cross NC, Tefferi A. Imatinib therapy for hypereosinophilic syndrome and other eosinophilic disorders Blood 2003; 101:3391–3397.

225. Druker BJ. Overcoming resistance to imatinib by combining targeted agents Mol Cancer Ther 2003; 2:225–226.

226. Hoover RR, Mahon FX, Melo JV, Daley GQ. Overcoming STI571 resistance with the farnesyl transferase inhibitor SCH66336 Blood 2002; 100:1068–1071.

227. La Rosee P, Corbin AS, Stoffregen EP, Deininger MW, Druker BJ. Activity of the Bcr-Abl kinase inhibitor PD180970 against clinically relevant Bcr-Abl isoforms that cause resistance to imatinib mesylate (Gleevec, STI571) Cancer Res. 2002; 62: 7149–7153.

228. Sausville EA. Is another bcr-abl inhibitor needed for chronic myelogenous leukemia Clin Cancer Res 2003; 9:1233–1234.

229. Huron DR, Gorre ME, Kraker AJ, Sawyers CL, Rosen N, Moasser MM. A novel pyridopyrimidine inhibitor of Abl kinase is a picomolar inhibitor of Bcr-abl-driven K562 cells and is effective against STI571-resistant Bcr-abl mutants Clin Cancer Res 2003; 9:1267–1273.

230. Golas JM, Arndt K, Etienne C, Lucas J, Nardin D, Gibbons J, Frost P, Ye F, Boschelli DH, Boschelli F. SKI-606, a 4-anilino-3-quinolinecarbonitrile dual inhibitor of Src and Abl kinases, is a potent antiproliferative agent against chronic myelogenous leukemia cells in culture and causes regression of K562 xenografts in nude mice Cancer Res 2003; 63:375–381.

231. Gambacorti-Passerini CB, Gunby RH, Piazza R, Galietta A, Rostagno R, Scapozza L. Molecular mechanisms of resistance to imatinib in Philadelphia-chromosome-positive leukaemias Lancet Oncol 2003; 4:75–85.

232. Liu W-H, Makrigiorgos GM. Sensitive and quantitative detection of mutations associated with clinical resistance to STI-571 Leukemia Res 2003; 27:979–982.

233. Samoszuk M, Corwin M-A. Acceleration of tumor growth and peri-tumoral blood clotting by imatinib mesylate (Gleevec) Int J Cancer 2003; 106:647–652.

234. Akin C, Brockow K, D'Ambrosio C, Kirshenbaum AS, Ma Y, Longley BJ, Metcalfe DD. Effects of tyrosine kinase inhibitor STI571 on human mast cells bearing wild-type or mutated c-kit Exp Hematol 2003; 31:686–692.

235. Paner GP, Silberman S, Hartman G, Micetich KC, Aranha GV, Alkan S. Analysis of signal transducer and activator of transcription 3 (STAT3) in gastrointestinal stromal tumors Anticancer Res 2003; 23:2253–2260.

236. von Mehren M. Recent advances in the management of gastrointestinal stromal tumors Curr Oncol Rep 2003; 5:288–294.

237. Sapi Z, Kovacs RB, Bodo M. Gastrointestinal stromal tumors. Observations on the basis of 29 cases Orv Hetil 2001; 142: 2479–2485.

238. Noguchi T, Sato T, Takeno S, Uchida Y, Kashima K, Yokoyama S, Muller W. Biological analysis of gastrointestinal stromal tumors Oncol Rep 2002; 9:1277–1282.

239. Schneider-Stock R, Boltze C, Lasota J, Miettinen M, Peters B, Pross M, Roessner A, Gunther T. High prognostic value of p16INK4 alterations in gastrointestinal stromal tumors J Clin Oncol 2003; 21:1688–1697.

240. Dei Tos AP. The reappraisal of gstrointestinal stromal tumors: from Stout to the KIT revolution Virchows Arch 2003; 442: 421–428.

241. Joensuu H, Fletcher C, Dimitrijevic S, Silberman S, Roberts P, Demetri G. Management of malignant gastrointestinal stromal tumours Lancet Oncology 2002; 3:655–664.

242. Patel SR. Early results from randomized phase III trial of imatinib mesylate for gastrointestinal stromal tumors Curr Oncol Rep 2003; 5:273.

243. Dagher R, Cohen M, Williams G, Rothmann M, Gobburu J, Robbie G, Rahman A, Chen G, Staten A, Griebel D, Pazdur R. Approval summary: imatinib mesylate in the treatment of metastatic and/or unresectable malignant gastrointestinal stromal tumors Clin Cancer Res 2002; 8:3034–3038.

244. Rossi CR, Mocellin S, Mencarelli R, Foletto M, Pilati P, Nitti D, Lise M. Gastrointestinal stromal tumors: from a surgical to a molecular approach Int J Cancer 2003; 107:171–176.

245. Gill S, Thomas RR, Goldberg RM. New targeted therapies in gastrointestinal cancers Curr Treat Options Oncol 2003; 4: 393–403.

246. Attoub S, Rivat C, Rodrigues S, Van Bocxlaer S, Bedin M, Bruyneel E, Louvet C, Kornprobst M, Andre T, Mareel M, Mester J, Gespach C. The c-kit tyrosine kinase inhibitor STI571 for colorectal cancer therapy Cancer Res 2002; 62: 4879–4883.

247. Zhang P, Gao WY, Turner S, Ducatman BS. Gleevec (STI-571) inhibits lung cancer cell growth (A549) and potentiates the cisplatin effect in vitro Mol Cancer 2003; 2:1.

248. Lonardo F, Pass HI, Lucas DR. Immunohistochemistry frequently detects c-Kit expression in pulmonary small cell carcinoma and may help select clinical subsets for a novel form of chemotherapy Appl Immunohistochem Mol Morphol 2003; 11: 51–55.

249. Kijima T, Maulik G, Ma PC, Tibaldi EV, Turner RE, Rollins B, Sattler M, Johnson BE, Salgia R. Regulation of cellular proliferation, cytoskeletal function, and signal transduction through CXCR4 and c-Kit in small cell lung cancer cells Cancer Res 2002; 62:6304–6311.

250. Hwang JH, Hwang JH, Chung HK, Kim DW, Hwang ES, Suh JM, Kim H, You KH, Kwon OY, Ro HK, Jo DY, Shong M. CXC chemokine receptor 4 expression and function in human anaplastic thyroid cancer cells J Clin Endocrinol Metab 2003; 88:408–416.

251. Mouriaux F, Kherrouche Z, Maurage C-A, Demailly F-X, Labalette P, Saule S. Expression of the c-Kit receptor in choroidal melanomas Melanoma Res 2003; 13:161–166.

252. Smithey BE, Pappo AS, Hill DA. C-kit expression in pediatric solid tumors: a comparative immunohistochemical study Am J Surg Pathol 2002; 26:486–492.

253. Komdeur R, Hoekstra HJ, Molenaar WM, Van Den Berg E, Zwart N, Pras E, Plaza-Menacho I, Hofstra RM, Van Der Graaf WT. Clinicopathologic assessment of postradiation sarcomas: KIT as a potential treatment target Clin Cancer Res 2003; 9:2926–2932.

254. McGary EC, Weber K, Mills L, Doucet M, Lewis V, Lev DC, Fidler IJ, Bar-Eli M. Inhibition of platelet-derived growth factor-mediated proliferation of osteosarcoma cells by the novel tyrosine kinase inhibitor STI571 Clin Cancer Res 2002; 8: 3584–3591.

255. Maki RG, Awan RA, Dixon RH, Jhanwar S, Antonescu CR. Differential sensitivity to imatinib of 2 patients with metastatic sarcoma arising from dermatofibrosarcoma protuberans Int J Cancer 2002; 100:623–626.

256. Sawyers CL. Imatinib GIST keeps finding new indications: successful treatment of dermatofibrosarcoma protuberans by targeted inhibition of the platelet-derived growth factor receptor J Clin Oncol 2002; 20:3568–3569.

257. Demetri GD. Targeting c-kit mutations in solid tumors: scientific rationale and novel therapeutic options Semin Oncol 2001; 28((5 Suppl 17)):19–26.

258. Nimmanapalli R, O'Bryan E, Huang M, Bali P, Burnette PK, Loughran T, Tepperberg J, Jove R, Bhalla K. Molecular characterization and sensitivity of STI-571- (imatinib mesylate-, Gleevec-) resistant, Bcr-Abl-positive, humamn acute leukemia cells to SRC kinase inhibitor PD180970 and 17-allylamino-17-demethoxygeldanamycin Cancer Res 2002; 62:5761–5769.

259. Huang M, Dorsey JF, Epling-Burnette PK, Nimmanapalli R, Landowski TH, Mora LB, Niu G, Sinibaldi D, Bai F, Kraker A, Yu H, Moscinski L, Wei S, Djeu J, Dalton WS, Bhalla K, Loughran TP, Wu J, Jove R. Inhibition of Bcr-Abl kinase activity by PD180970 blocks constitutive activation of Stat5 and growth of CML cells Oncogene 2002; 21:8804–8816.

260. La Rosee P, Corbin AS, Stoffregen EP, Deininger MW, Druker BJ. Activity of the Bcr-Abl kinase inhibitor PD180970 against clinically relevant Bcr-Abl isoforms that cause resistance to imatinib mesylate (Gleevec, STI5711) Cancer Res 2002; 62: 7149–7153.

261. Giles FJ. New drugs in acute myeloid leukemia Curr Oncol Rep. 2002; 4:369–374.

262. Goodsell DS. The molecular perspective: histone deacetylase Oncologist 2003; 8:389–391.

263. Zhao S, Venkatasubbarao K, Li S, Freeman JW. Requirement of a specific Sp1 site for histone deacetylase-mediated repression of transforming growth factor beta Type II receptor expression in human pancreatic cancer cells Cancer Res 2003; 63:2624–2630.

264. Tsurutani J, Soda H, Oka M, Suenaga M, Doi S, Nakamura Y, Nakatomi K, Shiozawa K, Yamada Y, Kamihira S, Kohno S. Antiproliferative effects of the histone deacetylase inhibitor FR901228 on small-cell lung cancer lines and drug-resistant sublines Int J Cancer 2003; 104:238–242.

265. Peart MJ, Tainton KM, Ruefli AA, Dear AE, Sedelies KA, O'Reilly LA, Waterhouse NJ, Trapani JA, Johnstone RW. Novel mechanisms of apoptosis induced by histone deacetylase inhibitors Cancer Res 2003; 63:4460–4471.

266. Rosato RR, Almenara JA, Grant S. The histone deacetylase inhibitor MS-275 promotes differentiation or apoptosis in human leukemia cells through a process regulated by genera-

tion of reactive oxygen species and induction of p21$^{CIP1/WAF1}$ Cancer Res 2003; 63:3637–3645.

267. Yu C, Rahmani M, Almenara J, Subler M, Krystal G, Conrad D, Varticovski L, Dent P, Grant S. Histone deacetylase inhibitors promote STI571-mediated apoptosis in STI571-sensitive and -resistant Bcr/Abl^{+} human myeloid leukemia cells Cancer Res 2003; 63:2118–2126.

268. Nimmanapalli R, Fuino L, Stobaugh C, Richon V, Bhalla K. Cotreatment with the histone deacetylase inhibitor suberoylanilide hydroxamide acid (SAHA) enhances imatinib-induced apoptosis of Bcr-Abl-positive human acute leukemia cells Blood 2003; 101:3236–3239.

269. Shelton JG, Moye PW, Steelman LS, Blalock WL, Lee JT, Franklin RA, McMahon M, McCubrey JA. Differential effects of kinasae cascade inhibitors on neoplastic and cytokine-mediated cell proliferation Leukemia 2003; 17:1765–1782.

270. Imamura F, Mukai M, Ayaki M, Akedo H. Y-27632, an inhibitor of Rho-associated protein kinase, suppresses tumor cell invasion via regulation of focal adhesion and focal adhesion kinase Japan J Cancer Res 2000; 91:811–816.

271. Sawada K, Morishige K, Tahara M, Ikebuchi Y, Kawagishi R, Tasaka K, Murata Y. Lysophosphatic acid induces focal adhesion assembly through Rh9/Rho-associated kinase pathway in human ovarian cancer cells Gynecol Oncol 2002; 87: 252–259.

272. Abrams TJ, Lee LB, Murray LJ, Pryer NK, Cherrington JM. SU11248 inhibits Kit and platelet-derived growth factor receptor beta in preclinical models of human small cell lung cancer Mol Cancer Ther 2003; 2:471–478.

273. O'Farrell A-M, Abrams T, Yu H-A, Ngai TJ, Louie SG, Yee KWH, Wong LM, Hong W, Lee LB, Town A, Smolik BD, Manning WC, Murray LJ, Heinrich MC, Cherrington JM. SU11248 is a novel FLT tyrosine kinase inhibitor with potent activity in vitro and in vivo Blood 2003; 101:3597–3605.

274. Rusnak DW, Lackey K, Affleck K, Wood ER, Alligood KJ, Rhodes N, Keith BR, Murray DM, Knight WB, Mullin RJ, Gilmer TM. The effects of the novel, reversible epidermal growth factor receptor/ErbB-2 tyrosine kinase inhibitor,

GW2016, on the growth of human normal and tumor-derived cell lines in vitro and in vivo Mol Cancer Ther 2001; 1:85–94.

275. Murata K, Kumagai H, Kawashima T, Tamitsu K, Irie M, Nakajima H, Suzu S, Shibuya M, Kamihira S, Nosaka T, Asano S, Kitamutra T. Selective cytotoxic mechanism of GTP-14564, a novel tyrosine kinase inhibitor in leukemia cells expressing a constitutively active Fms-like tyrosine kinase 3 (FLT3) J Biol Chem 2003; 278:32892–32898.

276. Levis M, Small D. FLT3: ITDoes matter in leukemia Leukemia 2003; 17:1738–1752.

277. Stirewalt DL, Radich JP. The role of FLT3 in haematopoietic malignancies Nature Rev Cancer 2003; 3:650–665.

278. Song L, Turkson J, Karras JG, Jove R, Haura EB. Activation of Stat3 by receptor tyrosinase kinases and cytokines regulates survival in human non-small cell carcinoma cells Oncogene 2003; 22:4150–4165.

279. Barbacci EG, Pustilnik LR, Rossi AM, Emerson E, Miller PE, Boscoe BP, Cox ED, Iwata KK, Jani JP, Provoncha K, Kath JC, Liu Z, Moyer JD. The biological and biochemical effects of CP-654577, a selective erbB2 kinase inhibitor, on human breast cancer cells Cancer Res 2003; 63:4450–4459.

280. Agus DB, Akita RW, Fox WD, Lewis GD, Higgins B, Pisacane PI, Lofgren JA, Tindell C, Evans DP, Maiese K, Scher HI, Sliwkowski MX. Targeting ligand-activated ErbB2 signaling inhibits breast and prostate tumor growth Cancer Cell 2002; 2:127–137.

281. Weber KL, Doucet M, Price JE, Baker C, Kim SJ, Fidler IJ. Blockade of epidermal growth factor receptor signaling leads to inhibition of renal cell carcinoma growth in the bone of nude mice Cancer Res 2003; 63:2940–2947.

282. Jakubkova J, Duraj J, Hunakova L, Chorvath B, Sedlak J. PK11195, an isoquinoline carboxamide ligand of the mitochondrial benzodiazepine receptor, increased drug uptake and facilitated drug-induced apoptosis in human multidrug-resistant leukemia cells in vitro Neoplasma 2002; 49:231–236.

283. Nepomuceno RR, Balatoni CE, Natkunam Y, Snow AL, Krams SM, Martinez OM. Rapamycin inhibits the interleukin 10 sig-

nal transduction pathway and the growth of Epstein Barr virus B-cell lymphomas Cancer Res 2003; 63:4472–4480.

284. Huang S, Bjornsti M-A, Houghton PJ. Rapamycins Cancer Biol Ther 2003; 2/3:222–232.

285. Van Der Poel G, Hanrahan C, Zhong H, Simons W. Rapamycin induces Smad activity in prostate cancer cell lines Urol Res 2003; 30:380–386.

286. Peralba JM, DeGraffenried L, Friedrichs W, Fulcher L, Grunwald V, Weiss G, Hidalgo M. Pharmacodynamic evaluation of CCI-779, an inhibitor of mTOR, in cancer patients Clin Cancer Res 2003; 9:2887–2892.

287. Mita MN, Mita A, Rowinsky EK. Mammalian target of rapamycin: a new molecular target for breast cancer Clin Cancer Res 2003; 4:126–137.

288. Elit L. CCI-779 Wyett Curr Opin Investig Drugs 2002; 3: 1249–1253.

289. Huang S, Houghton PJ. Targeting mTOR signaling for cancer therapy Curr Opin Pharmacol 2003; 3:371–377.

290. Punt CJA, Boni J, Bruntsch U, Peters M, Thielert C. Phase I and pharmacokoinetic study of CCI-779, a novel cytostatic cell-cycle inhibitor in combination with 5-fluorouracil and leucovorin in patients with advanced solid tumors Ann Oncol 2003; 14:931–937.

291. Gao N, Zhang Z, Jiang B-H, Shi X. Role of PI3K/AKT/mTOR signaling in the cell cycle progression of human prostate cancer Biochem Biophys Res Commun 2003; 310:1124–1132.

292. Georger B, Kerr K, Tang C-B, Fung K-M, Powell B, Sutton LN, Phillis PC, Janss AJ. Antitumor activity of the rapamycin analog CCI-779 in human primitive neuroectodermal tumor/medulloblastoma models as single agent and in combination chemotherapy Cancer Res 2001; 61:1527–1532.

293. Modi S, Seidman AD. An update on epidermal growth factor receptor inhibitors Curr Oncol Rep 2002; 4:47–55.

294. Kari C, Chan TO, Rocha de Quadros M, Rodeck U. Targeting the epidermal growth factor receptor in cancer: apoptosis takes center stage Cancer Res 2003; 63:1–5.

295. Arteaga CL, Baselga J. Clinical trial design and end points for epidermal growth receptor-targeted therapies: implications for drug development and practice Clin Cancer Res 2003; 9:1579–1589.

296. Skorski T. Oncogenic tyrosine kinases and the DNA-damage response Nat Rev Cancer 2002; 2:351–360.

297. Mendelsohn J, Baselga J. Status of epidermal growth factor receptor antagonists in the biology and treatment of cancer J Clin Oncol 2003; 21:2787–2799.

298. Allen TM. Ligand-targeted therapeutics in anticancer therapy Nat Rev Cancer 2002; 2:750–763.

299. Goel S, Mani S, Perez-Soler R. Tyrosine kinase inhibitors: a clinical perspective Curr Oncol Rep 2002; 4:9–19.

300. Moulder SL, Yakes FM, Muthuswamy SK, Bianco R, Simpson JF, Arteaga CL. Epidermal growth factor receptor (HER1) tyrosine kinase inhibitor ZD1839 (Iressa) inhibits HER2/neu (erbB2)-overexpressing breast cancer cells in vitro and in vivo Cancer Res 2001; 61:8887–8895.

301. Wakeling AE, Guy SP, Woodburn JR, Ashton SE, Curry BJ, Barker AJ, Gibson KH. ZD1839 (Iressa): an orally active inhibitor of epidermal growth factor signaling with potential for cancer therapy Cancer Res 2002; 62:5749–5754.

302. Campiglio M, Locatelli A, Olgiati C, Normanno N, Somenzi G, Vigano L, Fumagilli M, Menard S, Gianni L. Inhibition of proliferation and induction of apoptosis in breast cancer cells by the epidermal growth factor receptor (EGFR) tyrosine kinase inhibitor ZD1839 ('Iressa') is independent of EGFR expression level J Cell Physiol 2004; 198:259–268.

303. Suzuki T, Nakagawa T, Endo H, Mitsudomi T, Masuda A, Yatabe Y, Sugiura T, Takahashi T, Hida T. The sensitivity of lung cancer cell lines to the EGFR-selective tyrosine kinase inhibitor ZD1839 ('Iressa') is not related to the expression of EGFR or HER-2 or to K-ras gene status Lung Cancer 2003; 42:35–41.

304. Fujimura M, Hidaka T, Saito S. Selective inhibition of the epidermal growth factor receptor by ZD1839 decreases the growth and invasion of ovarian clear cell adenocarcinoma cells Clin Cancer Res 2002; 8:2448–2454.

305. Sirotnak FM, She Y, Lee F, Chen J, Scher HI. Studies with CWR22 xenografts in nude mice suggest that ZD1839 may have a role in the treatment of both androgen-dependent and androgen-independent human prostate cancer Clin Cancer Res 2002; 8:3870–3876.

306. Zembutsu H, Ohnishi Y, Daigo Y, Katagiri T, Kikuchi T, Kakiuchi S, Nishime C, Hirata K, Nakamura Y. Gene-expression profiles of human tumor xenografts in nude mice treated orally with the EGFR tyrosine kinase inhibitor ZD1839 Int J Oncol 2003; 23:29–39.

307. Herbst RS, Maddox AM, Rothenberg ML, Small EJ, Rubin EH, Baselga J, Rojo F, Hong WK, Swaisland H, Averbuch SD, Ochs J, LoRusso PM. Selective oral epidermal growth factor receptor tyrosine kinase inhibitor ZD1839 is generally well tolerated and has activity in non-small-cell lung cancer and other solid tumors: results of a phase I trial J Clin Oncol 2002; 20:3815–3825.

308. LoRusso PM, Herbst RS, Rischin D, Ranson M, Calvert H, Raymond E, Kieback D, Kaye S, Gianni L, Harris A, Bjork T, Maddox AM, Rothenberg ML, Small EJ, Rubin EH, Feyereislova A, Heyes A, Averbuch SD, Ochs J, Baselga J. Improvements in quality of life and disease-related symptoms in phase I trials of the selective oral epidermal growth factor receptor tyrosine kinase inhibitor ZD1839 in non-small cell lung cancer and other solid tumors Clin Cancer Res 2003; 9:2040–2048.

309. Kukunoor R, Shah J, Mekhail T. Targeted therapy for lung cancer Curr Oncol Rep 2003; 5:326–333.

310. Dy GK, Adjei AA. Novel targets for lung cancer therapy: parts I & II J Clin Oncol 2002; 20:2881–2894; 3016–3028.

311. Johnson DH, Arteaga CL. Gefitinib in recurrent non-small-cell lung cancer: an IDEAL trial J Clin Oncol 2003; 21: 2227–2229.

312. Miller VA, Johnson DH, Krug LM, Pizzo B, Tyson L, Perez W, Krozely P, Sandler A, Carbone D, Heelan RT, Kris MG, Smith R, Ochs J. Pilot trial of the epidermal growth factor receptor tyrosine kinase inhibitor gefitinib plus carboplatin and paclitaxel in patients with stage IIIB or IV non-small-cell lung cancer J Clin Oncol 2003; 21:2094–100.

313. Janmaat ML, Kruyt FA, Rodriguez JA, Giaccone G. Response to epidermal growth factor receptor inhibitors in non-small cell lung cancer cells: limited antiproliferative effects and absence of apoptosis associated with persistent activity of extracellular signal-regulated kinase or Akt kinase pathways Clin Cancer Res 2003; 9:2316–2326.

314. Gelibter A, Ceribelli A, Milella M, Mottolesa M, Vocaturo A, Cognetti F. Clinically meaningful resonse to the EGFR tyrosine kinase inhibitor gefitinib ('Iressa', ZD1839) in non small cell lung cancer J Exp Clin Cancer Res 2003; 22:481–485.

315. Simon GR, Ruckdeschel JC, Williams C, Cantor A, Chiappori A, Rocha Lima CM, Antonia S, Haura E, Wagner H, Robinson L, Sommers E, Alberts M, Bepler G. Gefitinib (ZD1839) in previously treated advanced non-small-cell lung cancer: experience from a single institution Cancer Control 2003; 10: 388–395.

316. Kris MG, Natale RB, Herbst RS, Lynch TJ, Prager D, Belani CP, Schiller JH, Kelly K, Spiridonidis H, Sandler A, Albain KS, Cella D, Wolf MK, Averbuch SD, Ochs JJ, Kay AC. Efficacy of gefitinib, an inhibitor of the epidermal growth factor receptor tyrosine kinase, in symptomatic patients with non-small lung cancer: a randomized trial JAMA 2003; 290: 2149–2158.

317. Fukuoka M, Yano S, Giaccone G, Tamura T, Nakagawa K, Douillard JY, Nishiwaki Y, Vansteenkiste J, Kudoh S, Rischin D, Eek R, Horai T, Noda K, Takata I, Smit E, Averbuch S, Macleod A, Feyereislova A, Dong RP, Baselga J. Multi-institutional randomized phase II trial of gefitinib for previously treated patients with advanced non-small-cell lung cancer J Clin Oncol 2003; 21:2237–2246.

318. Chang GC, Yang TY, Wang NS, Huang CM, Chiang CD. Successful treatment of multifocal bronchioalveolar cell carcinoma with ZD1839 (Iressa) in two patients J Formos Med Assoc 2003; 102:407–411.

319. Yano S, Kanematsu T, Miki T, Aono Y, Azuma M, Yamamoto A, Uehara H, Soone S. A report of two bronchioalveolar carcinoma cases which were rapidly improved by treatment with the epidermal growth factor receptor tyrosine kinase inhibitor ZD1839 ("Iressa") Cancer Sxci 2003; 94:453–458.

320. Baselga J, Rischin D, Ranson M, Calvert H, Raymond E, Kieback DG, Kaye SB, Gianni L, Harris A, Bjork T, Averbuch SD, Feyereislova A, Swaisland H, Rojo F, Albanell J. Phase I safety, pharmacokinetic, and pharmacodynamic trial of ZD1839, a selective oral epidermal growth factor receptor tyrosine kinase inhibitor, in patients with five selected solid tumor types J Clin Oncol 2002; 20:4292–4302.

321. Bianco C, Tortora G, Bianco R, Caputo R, Veneziani BM, Caputo R, Damiano V, Troiani T, Fontanini G, Raben D, Pepe S, Bianco AR, Ciardiello F. Enhancement of antitumor activity of ionizing radiation by combined treatment with the selective epidermal growth factor receptor-tyrosine kinase inhibitor ZD1839 (Iressa) Clin Cancer Res 2002; 8:3250–3258.

322. Barnes CJ, Bagheri-Yarmand R, Mandal M, Yang Z, Clayman GL, Hong WK, Kumar R. Suppression of epidermal growth factor receptor, mitogen-activated protein kinase, and Pak1 pathways and invasiveness of human cutaneous squamous cancer cells by the tyrosine kinase inhibitor ZD1839 (Iressa) Mol Cancer Ther 2003; 2:345–351.

323. Daneshmand M, Parolin DA, Hirte HW, Major P, Goss G, Stewart D, Batist G, Miller WH, Matthews S, Seymour L, Lorimer IA. A pharmacodynamic study of the epidermal growth factor receptor tyrosine kinase inhibitor ZD1839 in metastatic colorectal cancer patients Clin Cancer Res 2003; 9:2457–2464.

324. Hopfner M, Sutter AP, Gerst B, Zeitz M, Scherubi H. A novel approach in the treatment of neuroendocrine gastrointestinal tumors. Targeting the epidermal growth factor receptor by gefitinib (ZD1839) Bit J Cancer 2003; 89:1766–1775.

325. Sintani S, Li C, Mihara M, Terakado N, Yano J, Nakashiro K, Hamakawa H. Enhancement of tumor radioresponse by combined treatment with gefitinib (Iressa, ZD1839), an epidermal growth factor tyrosine kinase inhibitor, is accompanied by inhibition of DNS damage repair and cell growth in oral cancer Int J Cancer 2003; 107:1030–1037.

326. Magne N, Fischel JL, Dubreuil A, Forment P, Ciccolini J, Forment JL, Tiffon C, Renee N, Marchetti S, Etienne MC, Milano G. ZD1839 (Iressa) modifies the activity of key enzymes linked to fluoropyrimidine activity: rational basis for a new combination therapy with capecitabine Clin Cancer Res 2003; 9:4735–4742.

327. Xu JM, Azzariti A, Severino M, Lu B, Colucci G, Paradiso A. Characterization of sequence-dependent synergy between ZD1839 ("Iressa") and oxaliplatin Biochem Pharmacol 2003; 66:551–563.

328. Cohen EE, Rosen F, Stadler WM, Recant W, Stenson K, Huo D, Vokes EE. Phase II trial of ZD1839 in recurrent or metastatic squamous cell carcinoma of the head and neck J Clin Oncol 2003; 21:1980–1987.

329. Douglass EC. Development of ZD1839 in colorectal cancer Semin Oncol 2003; 30S:17–22.

330. Huang SM, Li J, Armstrong EA, Harari PM. Modulation of radiation response and tumor-induced angiogenesis after epidermal growth factor receptor inhibition by ZD1839 (Iressa) Cancer Res 2002; 62:4300–4306.

331. Burton A. What went wrong with Iressa Lancet Oncol 2002; 3:708.

332. Okamoto I, Fujii K, Matsumoto M, Terasaki Y, Kihara N, Kohogi H, Suga M. Diffuse alveolar damage after ZD1839 therapy in a patient with non-small cell lung cancer Lung Cancer 2003; 40:339–342.

333. Cohen MH, Williams GA, Sridhara R, Chen G, Pazdur R. FDA drug approval summary: gefitinib (ZD1839) (Iressa) tablets Oncologist 2003; 8:303–306.

334. Malik SN, Siu LL, Rowinsky EK, deGraffenried L, Hammond LA, Rizzo J, Bacus S, Brattain MG, Kreisberg JI, Hidalgo M. Pharmacodynamic evaluation of the epidermal growth factor receptor inhibitor OSI-774 in human epidermis of cancer patients Clin Cancer Res 2003; 9:2478–2486.

335. Ng SS, Tsao MS, Nicklee T, Hedley DW. Effects of the epidermal growth factor receptor inhibitor OSI-774, Tarceva, on downstream signaling pathways and apoptosis in human pancreatic adenocarcinoma Mol Cancer Ther 2002; 1:777–783.

336. Grunwald V, Hidalgo M. Development of the epidermal growth factor receptor inhibitor Tarceva (OSI-7744) Adv Exp Med Biol Semin Oncol 2003; 30S:23–31.

337. Adis International Ltd. Erlotinib (CP358774; NSC718781; OSI774; R1415) Drugs Res Developm 2003; 4:243–248.

338. Huang XF, Ren W, Rollins L, Pittman P, Shah M, Shen L, Gu Q, Strube R, Hu F, Chen S-Y. A broadly applicable, personalized heat shock protein-mediated oncolytic tumor vaccine Cancer Res 2003; 63:7321–7329.

339. Wang Y, Hu JK, Krol A, Li Y-P, Li C-Y, Yuan F. Systemic dissemination of viral vectors during intratumoral injection Molec Cancer Therap 2003; 2:1233–1242.

340. Goldsmith ME, Kitazonoi M, Fok P, Aikou T, Bates S, Fojo T. The histone deacetylase inhibitor FK228 preferentially enhances adenovirus transgene expression in malignant cells Clin Cancer Res 2003; 9:5394–5401.

341. Qin M, Chen S, Yu T, Escuardo B, Sharma S, Batra RK. Coxsackievirus adenovirus receptor expression predicts the efficiency of adenoviral gene transfer into non-small cell lung cancer xenografts Clin Cancer Res 2003; 9:4992–4999.

342. Moon C, Oh Y, Roth JA. Current status of gene therapy for lung cancer and head and neck cancer Clin Cancer Res 2003; 9:5055–5067.

343. Sah NK, Munshi A, Nishikawa T, Mukhopadhyay T, Roth JA, Meyn RE. Adenovirus-mediated wild-type p53 radiosensitizes human tumor cells by suppressing DNA repair capacity Molec Cancer Therap 2003; 2:1223–1231.

344. Cao M-Y, Lee Y, Feng N-P, Xiong K, Jin H, Wang M, Vassilakos A, Viau S, Wright JA, Young AH. Adenovirus-mediated ribonucleotide reductase R1 gene therapy of human colon adenocarcinoma Clin Cancer Res 2003; 9:4553–4561.

345. Cascallo M, Capellá G, Mazo A, Alemany R. Ras-dependent oncolysis with an adenovirus VAI mutant Cncer Res 2003; 63: 5544–5550.

346. Jetzt A, Howe JA, Horn MT, Maxwell E, Yin Z, Johnson D, Kumar CC. Adenoviral-mediated expression of a kinase-dead mutant of Akt induces apoptosis selectively in tumor cells and suppresses tumor growth in mice Cancer Res 2003; 63: 6697–6706.

347. Cuevas Y, Hernández-Aloceba R, Aragones J, Naranjo-Suárez S, Castellanos MC, Estaban MA, Martin-Puig S, Landazuri MO, del Peso L. Specific oncolytic effect of a new hypoxia-

inducible factor-dependent replicative adenovirus on von Hippel-Lindau-defective renal cell carcinomas Cancer Res 2003; 63:6877–6884.

348. Freytag SO, Stricker H, Pegg J, Paielli D, Pradham DG, Peabody J, DePeralta-Venturina M, Xia X, Brown S, Lu M, Kim JH. Phase I study of replication-competent adenovirus-mediated double-suicide gene therapy in combination with conventional-dose three-dimensional radiation therapy for the treatment of newly diagnosed, intermediate- to high-risk prostate cancer Cancer Res 2003; 63:7497–7506.

349. Cowen RL, Patterson AV, Telfer BA, Airley RE, Hobbs S, Phillips RM, Jaffar M, Stratford IJ, Willimas KJ. Viral delivery of P450 reductase recapitulates the ability of constitutive overexpression of reductase enzymes to potentiate the activity of mitomycin C in human breast cancer xenografts Molec Cancer Therap 2003; 2:901–909.

350. Lee W-P, Tai D-I, Tsai S-L, Yeh C-T, Chao Y, Lee S-D, Hung M-C. Adenovirus type 5 E1A sensitizes hepatocellular carcinoma cells to gemcitabine Cancer Res 2002; 63:6229–6236.

351. Stubdal H, Perin N, Lemmon M, Holman P, Rauzon M, Potter PM, Danks MK, Fattaey A, Dubensky T, Hohnson L. A prodrug strategy using ONYX-015-based replicating adenovirus to deliver rabbit carboxylesterase to tumor cells for conversion of CPT-11 to SN-38 Cancer Res 2003; 63:6900–6908.

352. Baldwin PJ, van der Berg SH, Boswell CM, Offrings R, Hickling JK, Dobson J, Roberts JStC, Latimer JA, Moseley RP, Coleman N, Stanley MA, Sterling JC. Vaccinia-expressed human papillomavirus 16 and 18 E6 and E7 as a therapeutic vaccination for vulval and vaginal intraepithelial neoplasia Clin Cancer Res 2003; 9:5205–5213.

353. Yang AS, Monken CE, Lattime EC. Intratumoral vaccination with vaccinia-expressed tumor antigen and granulocyte macrophage colony-stimulating factor overcomes immunological ignorance to tumor antigen Cancer Res 2003; 63: 6956–6961.

354. Astsaturov I, Petrella T, Bagriacik EU, de Benedette M, Uger R, Lumber G, Berinstein N, Elias I, Iscoe N, Hammond C, Hamilton P, Spaner DE. Amplification of virus-induced anti-

melanoma T-cell reactivity by high-dose interferon-α2b: implications for cancer vaccines Clin Cancer Res 2003; 9:4347–4355.

PART II: FLAVOPIRIDOLS, PROTEIN KINASE C, METALLOPROTEINASES, AND CYCLOOXYGENASE INHIBITORS

FLAVOPIRIDOLS

Cyclins

Cycline-dependent kinases (CDK) in their complexes with cyclins and their inhibitors (p15, 16, 21, 27, and especially $p16^{INK4}$) regulate the cell cycle [1]. CDK2, 4, and 6, in combination with cyclins E and D, promote progression of the cell cycle from G to S; CDK2 and cyclin A and CDK1 and cyclin B carry the cell through the S phase to conclude mitosis. When the cycline kinase inhibitors (CKI) suffer mutation, the cell cycle progresses through mitoses incessantly—non-stop.

Flavonoids occur naturally and can be isolated from the stem bark of the Indian plant *Dysoxylum binectariferum*. Flavopiridols (FP) (flavopiridol, L868275, Aventis), including olomoucin-roscovitine (R-roscovitine, CYC202, Cyclacel, Dundee, UK), the semisynthetic flavonoids, imitate the activity of, and can serve as replacement for, mutated CKI; they also inhibit EGF-R kinases and PKC. Some leukemic cells, when their mutated CKI is replaced by FP, activate the mitochondrial apoptotic death cascade. Flavopiridols, roscovitine, and the PI3K inhibitor LY294002, together, acted most efficiently against the PI3K/Akt, Bcl-2, and Mcl-1 pathways. PI3K phosphorylates membrane phosphatidyl inositol, thereby creating polyphosphoinositides. The lipid phosphatase gene product protein MMAC/PTEN (mutated in cultiple advanced cancers/phosphatase on chromosome ten; myotubularin; phosphatase and tensin homolog deleted on chromosome ten) [2] dephosphorylates polyphosphoinositides, but MMAC/PTEN is often mutated or deleted in human cancers. Wortmannin, LY294002, and PD98059 could inhibit PI3K-Akt cell survival pathway and thereby induce apoptosis in cancer (pancreatic and ovarian carcinoma) cells. Flavopiridols synergize with TRAIL, the TNF-related apoptosis-inducing ligand, and inhibit NF-κB-dependent gene transcription, so TNFα is not antagonized in FP-treated cells [3]. Flavopiridols are forceful inducers of IL-6 [4]. Substratum-attached cancer cells operate with an upregulated PI3K-Akt pathway but LY294002 could induce apoptosis in these cells [5] (see Part I of this chapter). The genomic sequence v-*akt* was discovered in the directly transforming AKR murine retrovirus AKT8 isolated from thymic lymphomas of the mouse; v-*akt* is the derivative of the cellular genomic sequence c-*akt*, which was transduced by the retrovirus. The human AKT2 protein is a serine-threonine kinase with *src* homology and it is activated by PI3K, PDGF, bFGF, EGF, and IGF. The PI3K inhibitor LY294002 kills acute myeloblastic leukemia cells by inducing apoptosis. Flavopiridol inhibitors (see Part I) activate Rho-B proteins, regulators of cytoskeletal actin organization and cell adhesion; the geranylgeranylated Rho-B-GG proteins mediate arrest of the cell cycle and promote apoptosis, especially when a combination of FT and PI3K-Akt inhibitors is applied. RhoGD12 has been recognized as a metastasis-suppressor protein [6]. In vitro and xenograft assays show that FP induce apoptotic death of SCLC cells [7], potentiate the effect of STI571 in BCR-ABL$^+$ CML cells [8–12] and that of docetaxel in pancreatic and gastric cancer cells [13], enhances EGF-R suppression by trastuzumab in breast cancer cells [14], and increases radiosensitivity of ovarian cancer cells [15,16]. However, in a clinical trial, FPs were ineffective against advanced colorectal cancers [17]. Adenovirus ONYX certainly promises a better performance in this tumor category (see Part I).

Flavopiridols synergize with the histone deacetylase (HDAC) inhibitor Na-butyrate and intensify apoptosis induction in human leukemia cells [18], in which they reduced Rb protein/E2F1 complex formation, downregulated Mcl-1, and induced caspase-mediated cleavage of Bcl-2. Histone deacetylase 5 is an early p53 transcript; in HDAC 5-transfected hosts, osteosarcoma, neuroblastoma, and breast carcinoma cell growth is suppressed through the TNFα-to-its-receptor pathway. Histone deacetylase 1, 2, and 3 deacetylate p53 (an anti-apoptotic event), whereas they stabilize p53 (a pro-apoptotic event). In pancreatic adenocarcinoma cells, HDAC-mediated downregulation of TGFβR occurs; TGFβ is inhibitory to epithelial cell growth and its inhibitory effect on pancreatic carcinoma cells is thus annulled. Histone deacetylase inhibitors (depsipeptide, trichostatin) cause cell cycle arrest by inducing the wild-type p53-activated protein $p21^{WAF1}$ and dephosphorylation of Rb. In imatinib-resistant CML K562 cells, the addition of HDAC inhibitors [suberoylanilide hydroxamide acid (SAHA)] synergistically induced apoptosis. The benzamid derivative MS-275 (Nihon Schering KK, Chiba, Japan) is a HDAC inhibitor and inducer of mitochondrial cytochrome C-mediated apoptosis of human leukemia cells. Of the HDAC inhibitors (aphicidin, Na-butyrate, oxamflatin, SAHA, trichostatin A, and FR901228), a depsipeptide from *Chromobacteriun violaceum* upregulates the coxsackievirus and adenovirus receptors CAR and $\alpha\nu$ integrins on the surface of cancer cells thus rendering these cells susceptible to adenoviral infection; this compound offers itself to cooperate with oncolytic adenoviruses. The E1, E3 gene-depleted replication-defective type 5 Ad5.CMV-LacZ adenovirus (Qbiogene, Carlsbad, CA) could enter FR901228-pretreated cells. FR901228 exerts strong antiproliferative effect on lung cancer cells; in multidrug-resistant (MDR) SCLC, it inhibited telomerase mRNA; cycloheximide antagonized this effect. Depsipeptide failed, but SAHA or oxamflatin could kill P-glycoprotein-positive MDR malignant (T cell leukemia, colon carcinoma) cells (see Part I).

Flavopiridol-Induced Apoptosis

Flavopiridols induce apoptosis in B-CLL and multiple myeloma cells through reducing the expression of anti-apoptotic proteins Bcl-2, Bcl_{XL}, and Mcl-1. Overexpression of Mcl-1 protects multiple myeloma cells against apoptosis; FP target Mcl-1 (*vide infra*). In human breast carcinoma cells, FPs downregulate the apoptosis-inhibitory protein (IAP), Bcl_{XL}, Mcl-1, and survivin.

Flavopiridols synergize with the tubulin-polymerizing agent epothilone [19]. Epothilones are naturally occurring macrolides that polymerize tubulins, cause cell-cycle arrest at G_2-M, and therefore elicit apoptosis even in cancer cells resistant to taxanes; they may be synergizing also with gemcitabine, capecitabine, or with the platinols when tested in patients with pancreatic cancer. A preliminary report indicates that patients with visceral metastases of breast carcinoma responded to epothilone (BMS-247550, Bristol-Myers Squibb) with PR and stabilization of disese (ASCO #30 2003). Cells protect themselves against TNF-induced apoptosis by activation of NFκB. In human tumor cells, FPs synergize with TNFα. Flavopiridols inhibit NFκB and overcome apoptosis resistance [3]. In colon carcinoma xenografts, FPs synergized with CPT-11 (irinotecan) [20]. STI571 (imatinib) and L86-8275 (FP) together were more effective in apoptosis induction in Ph^+ human leukemic cells than any of them alone [10]. Flavopiridols synergize with trastuzumab in breast cancer cells but breast cancer cells overexpressing both EGF-R and HER2/*neu*(*erb*2) are resistant to this combination [14].

Inhibitors

In melanoma cells, CDK1, CDK2, CDK4, and CDK6 act uninhibited; the Rb protein is phosphorylated because the suppressor protein p16 is not produced because of mutations in its gene; FP inhibits CDK in melanoma cells, which undergo apoptotic death [21]. Roscovitine (Alexis Coger Co) is a radiosensitizer as tested against xenografted human breast or ovarian carcinoma cells [15,16]. The bioflavonoid quercetin (Sigma) sensitizes Ewing's sarcoma cells to hyperthermia [22]. Quercetin suppresses proliferation of prostate cancer cells; in this latter case, signal transduction from Erb2, Erb3, and TGFα were mitosis-inducing for the tumor cells and quercetin antagonized this effect. The isoflavonoid genistein of soy products inhibits Akt after its phosphorylation by EGF-R kinases or by PI3K, therefore also blocking NFκB activation. High-grade intra-epithelial prostatic hyperplasia in its peripheral zone may advance to invasive carcinoma, transgresses the capsule, and eventually metastasizes and gains hormone-refractory state. Overexpression of cyclo- and lipooxygenases because of arachidonic enzyme activities carries major responsibility for these changes. In cultured human prostate cancer cells, genistein induced apoptosis but left nontumorigenic prostate epithelial cells intact. The endogenous lipid anandamide down-regulates EGF-R and exerts an antiproliferative effect in prostate cancer cells [23–26].

Survivin is a member of the IAP family; it is highly expressed in embryonic cells but is silent in differentiated adult cells. The anti-apoptotic survivin inhibits caspase 9 and is overexpressed in many human cancers. It can be inhibited by its antisense oligonucleotides (see Part III of this article) or by the nonphosphorylable T34A adenovirally delivered survivin gene (pAdT34A). When it reappears overexpressed in tumor cells, survivin is recognized by the tumor-bearing host, which mobilizes anti-survivin immune T cells (HLA-A24-restricted cytotoxic T cells); these cells expanded by IL-2 could be used for adoptive immunotherapy. Human leukocyte antigen-A24-positive and survivin2B-positive cancer cells are killed by these immune T cells. Soft-tissue sarcomas frequently exhibit elevated survivin levels, predicting shortened life expectancy attributable to death with tumors. Mesothelioma cells thrive on the homozygous deletion of the INK4a/ARF locus: $p14^{ARF}$ is lost and the anti-p53 MDM2 protein is overactive in these tumor cells; in addition, 87% of mesotheliomas overexpress survivin. Anti-survivin oligonucleotides restored the susceptibility of mesothelioma cells to apoptosis. In gastric carcinoma cells, survivin mRNA expression is inversely correlated with the apoptotic index. Flavopiridols suppress survivin and inhibit its phosphorylation, depriving the tumor cell of its apoptosis-exempt status [27–36].

Flavopiridols synergize with taxanes and doxorubicin in inducing apoptosis in xenografted human breast cancer cells. In case of xenografted human colon carcinoma cells, FP synergized with the topoisomerase inhibitor CPT11 camptothecin (irinotecan). In *bcr-abl*–mutated MDR leukemic cells, FP synergized with STI571 (imatinib mesylate) and induced intrinsic (mitochondrial) apoptosis. Of 28 patients with mantle cell lymphoma, i.v. FP induced 3 (11%) PR and stabilized the disease in 20 (71%) patients. Intravenously infused FP in 55 patients with advanced neoplasms induced, in 12 cases, long stabilization of disease for 3 to 11 months (median 6 months).

Flavopiridol-induced diarrhea is preventable with cholestyramine and lymphopenia is reversible [37–39]. If FP were administered to patients with oncolytic viruses, no possible antagonism between them can be foreseen. Would the attenuated measles virus synergize with FP in killing multiple myeloma cells?

PROTEIN KINASE C INHIBITORS

A Bad Kinase

Protein kinase C (PKC) is a member of the serine-threonine kinase family. When induced by the diacylglycerol phorbol myristate ester, PKC contributes to the cell two criteria of the neoplastic phenotype: anti-apoptotic and pro-telomeric states. PKCδ, a member of the PKC family, is activated by IFN-γ [40] and functions as an antagonist of ionizing radiation in breast carcinoma and glioblastoma cells [41,42]. Raf-1 (*vide supra*), which phophorylates Bcl-2, is a substrate of PKC. Protein kinase C activates telomerase reverse transcriptase (TeRT) and telomerase-associated protein (TeP-1) by phosphorylation, rejuvenating the cell. The Raf kinase inhibitor BAY 43-9006 performs well against colon carcinoma xenografts, especially when combined with irinotecan, vinorelbin, or gemcitabin [43].

Thromboxane (THX) A_2 is captured by its receptors as expressed by glioma cells, leading to the activation of phospholipase Cβ and the activation of PKC. Phosphorylation of focal adhesion kinase (FAK) follows, leading to integrin activation and cell migration [44,45]. Thromboxane synthetase inhibitors arrest glioma cell migration and the immobilized cell may succumb to programmed death (see Part V of this chapter). Protein kinase C overexpressed in breast carcinoma and glioblastoma cells functions as a promoter of cell division. Protein kinase C expression in ovarian carcinoma reaches 52% of the tumors examined and represents an adverse prognostic factor [46]: This tumor awaits clinical trials with PKC inhibitors and is targeted for virotherapy (mumps, NDV, measles, and reovirus and genetically engineered HSV; discussed in this volume).

Inhibitors

Protein kinase C inhibitors are bryostatin (from the marine bryozoan *Bugula maritima*); the staurosprin derivatives OH, benzoyl UCN-01, and PKC-412; and ISIS3521, a phosphorothioate antisense oligodeoxynucleotide binding to PKC mRNA. Bryostatin suppresses PKC and upregulates TNFα, IL-2, and 6, but induced only a few partial remissions (PR) in patients with solid tumors. In patients with metastatic kidney carcinoma, bryostatin induced remissions very seldom (6% PR), but durable (≥ 6 months) stabilization of disease occurred in 25% to 47% of patients. In 38 patients with advanced malignancies, prolonged infusions of bryostatin failed to induce remissions [47–50].

The hydroxystaurosporine UCN-01 and PKC412 synergize with 5FU by downregulating thymidylate synthetase in some adenocarcinomas (colon and cholangiocarcinomas). A preliminary report indicates that PKC412 inhibits a transmembrane tyrosine kinase active in AML and significantly reduces blast count in the blood (ASCO #2265 2003). UCN-01 blocks the cell cycle at G_1 and inhibited the growth of xenografted human squamous cell carcinomas of the head and neck, while stagnating tumor cells with suppressed CDK-2 activity died programmed cell death. Staurosporins downregulate XIAP (X-linked inhibitor of apoptosis). UCN-01 inhibits DNA repair in B-CLL and other tumor cells and offers itself as an adjuvant to cytotoxic chemotherapeutic drugs. ISIS3521 induced two CR in cases of non-Hodgkin's lymphoma; it synergized with carboplatin in 53 patients with NSCLC, inducing 42% complete remission (CR), PR, and minor responses (MR), with median survival of 19 months. It suppressed rat gliosarcomas and breast carcinomas but failed to induce remission in the human counterparts of these tumors. Tamoxifen suppresses PKC in cultured human glioblastoma cells and synergizes with the nitrosourea BCNU. Accordingly, clinical trials are in progress with tamoxifen and cisplatin for pediatric brain tumors. The flavonoid quercetin inhibits PKC and suppresses the growth of ovarian carcinoma cells in vitro and in vivo it synergizes with cisplastin in xenografted human lung cancer cells; it inhibits the growth of human prostate cancer cells [51–57].

Neuropeptides (bombesin, bradykinin, vasopressin, neurotensin) are mitogenic for some (small-cell lung) cancer cells and tumor cells may make use of them through paracrine or autocrine growth loops. Neurotensin (NT) is a growth factor of certain K-*ras*– and c-*raf*–driven pancreatic cancer cells and exerts its effect through PKC activation. Protein kinase C inhibitors GF-1 (bisindolylmaleimide), U0126, and Ro 31-8220 (Calbiochem) counteracted NT [58,59].

METALLOPROTEINASE INHIBITORS

Pathophysiology

The endopeptidase matrix metalloproteinases (MMP)—collagenases, gelatinases, stromelysins, and matrilysins—promote tumor cell locomotion and invasion as they cleave laminin, hyaluronan, and cadherin. Some of the MMP, such as MMP-9, the gelatinase that degrades type IV collagen, also cleave IL-2R of immune T cells or activate TGFβ to inhibit expansion of immunoreactive T cell clones; or cleave CCL7, the monocyte chemoattractant protein-3 (MCP-3), thereby exerting tumor-protective functions [60–64]. Indeed, lung adenocarcinoma cells overexpressing MMP-9 are highly malignant [65].

CXCR-4 chemokine receptor captures its ligand CXCL-12, the stromal cell-derived factor-1. This factor protects B-CLL cells against apoptotic death and promotes metastatic spread of breast cancer cells. This factor is cleaved by MMP-1, 3, 9, 13, and 14. When MMPs remove a tumor-protective factor, their blockade could promote tumor growth. Whereas most MMPs are expressed by stromal cells, MMP-7 (Matrilysin) is a product of epithelial cells such as certain adenocarcinoma cells; it promotes the growth of pancreatic carcinoma cells. Some murine mammary carcinoma cells co-express FasL and FasR, yet do not succumb to self-induced apoptosis except when exposed to MMP-7. In animals bearing this tumor, continuous treatment with MMP-7 selected out an apoptosis-resistant cell population [66]. Insulinlike growth factor-l upregulates and IL-10 downregulates MMP-2 production in tumor cells [67]. Matrix metalloproteinase-2 production by lung and gastic cancers in China is associated with a single nucleotide polymorphism in the gene promoter of MMP-2 [68]. Phorbols (myristate acetate) induce MMP-9 but a fungal metabolite, dykellic acid, antagonizes this effect [69]. Bladder carcinoma cells overproduce MMP-9 and titrated high levels of MMP-9 in the urine indicate recurrence with invasive tumors [70]. These tumors are promoted by VEGF, bFBF, MMP-9, and IL-8 and are inhibited by IFNs, endostatin, angiostatin, and thrombospondin. Transfection of bladder cancer cells with the IL-8 gene increases, whereas IL-8 antisense sequences reduce, their malignant potentials; the ABX-IL-8 human anti–IL-8 monoclonal antibody downregulates MMP-9 production by, and inhibits the growth of, transitional carcinoma cells of the urinary bladder [71]. Prostate cancer cells produce membrane-type MMP that cleaves laminin and thus promotes cancer cell migration [72]. Matrix metalloproteinase-19 is expressed in endothelial and vascular smooth muscle cells and is not associated with malignant transformation [73]. Some MPs can cleave TNFα or Fas ligand (L); adenoviral vectors of the FasL gene are available (RAdFasL) and induce tumor cell death, with release of sFasL from the apoptotic cells. The adenoviral vector RAdD4 encodes a noncleavable FasL that could not be antagonized by MMP [74]. FasL-expressing tumor cells kill $FasR^+$ host immune T lymphocytes [75]; however $FasL^+$ tumor cells are immunogenic in their hosts and induce rejection-strength immune reactions toward inocula of $FasL^-$ autologous tumor cells. $FasL^+$ tumor cells interact with the Fas receptor of DCs to elicit this reaction [76]. $FasL^+$ tumors are infiltrated by polymorphonuclear leukocytes and suffer cell death but tumor cells can kill the leukocytes as well [77]. Some human melanoma cells have expropriated the FasL⇒Fas receptor system to function as an autocrine growth loop [78]. This is possible through the formation of a new putative fusion oncoprotein in which the extracellular domain is the Fas receptor and the intracellular domain is the G-CSF receptor: t(10;1)(q23;p32). Capture of the FasL by this fusion oncoprotein would translate into mitogenic signal transduction [75]. Multiple–drug-resistant tumor cells overexpress the extracellular MMP inducer and MMP-1, 2, and 9 and thereby engage in accelerated invasiveness [79].

Reed-Sternberg (RS) cells express MMP-7, which activates MMP-2 in the stroma. Although RS cells are negative for MMP-9, T lineage lymphoma cells express it and the degree of expression correlates with advancement of the disease in the liver [80,81].

Inhibitors

Trophoblasts and tumor cells degrade their extracellular matrix to promote their invasiveness. Of numerous inhibitors of MMPs (antisense oligonucleotides and ribozymes, anthrax toxin, peptidomimetic Batimastat and Marimastat, nonpeptidomimetic BAY12-9566, Prinomastat AG3340, tetracycline derivatives, Æ-941 Neovastat, bisphosphonates, acetylsalicylic acid), Marimastat prolonged life in certain subgroups of patients with pancreatic and gastric carcinomas; matrilysin promotes the growth of pancreatic carcinoma cells. When BAY12-9566 was compared with gemcitabine in patients with pancreatic carcinoma, it proved to be weaker than chemotherapy. Batimastat is effective against metastases of a rat mammary carcinoma only when treatment starts before removal of the primary tumor and is administered continuously thereafter. Its performance in human clinical trials is so far unsatisfactory [82–86]. Tabulated clinical trials of several MMPs show prolongation of life at the best [62]. Tetracycline derivative CMT-3 (Metastat) inhibitor of COX-2 and PGE2 (Hospitals of Joint Diseases, New York, NY) is being tested against Kaposi's sarcoma and high-grade brain tumors. BAY12-9566 and ONO-4817 are called *selective third generation MMP inhibitors* (broad-spectrum first-generation: batimastat; second-generation: marimastat and prinomastat) [87]. The hydroxamate inhibitor prinomastat (AG3340) inhibited vitronectin proteolysis but did not antagonize the increased migration of human breast cancer cells, whereas inhibition of integrin degradation by MTI-MMP halted cancer cell motility [88]. ONO-4817 (third-generation) inhibits the growth of human NSCLC (adeno- and squamous cell carcinoma lines) micrometastases in the lungs of nude mice and synergizes with chemotherapeuticals [89,90]. ONO-4817 inhibited regional lymph node metastases of an orthotopically xenografted squamous cell carcinoma cell line [89]. Testicans are heparin-chondroitin proteoglycans; N-Tes inhibits membrane-type matrix metalloproteinases (MT-MMPs) and testican-2 antagonizes this effect, therefore promoting invasiveness of glioma cells [91,92]. These weak agents do not compete with oncolytic viruses; the medical oncologist probably would favor the selection of an oncolytic virus for a clinical trial rather than one of the MMP inhibitors; it is unpredictable whether or not oncolytic viruses or MMP inhibitors would cooperate.

CYCLOOXYGENASE INHIBITORS

Eicosanoids

Epidermal growth factor (EGF)-R and H-*ras* or K-*ras* overexpressions are essential for the maintenance of the malignant phenotype in colon adenocarcinoma cells. Signals emanating from EGF-R lead to phosphorylation of Extracellular regulated kinases (ERK)1, ERK2, and MAPK. Extracellular regulated kinases 1 and 2 and MAP kinases secure the anti-apoptotic state of the cell. Extracellular regulated kinases 1 and 2 phosphorylate and thereby inactivate the pro-apoptotic protein BAD at serine residue 112. Serine 112-phosphorylated BAD dissociates from the anti-apoptotic proteins Bcl-2 and Bcl_{XL}, liberating them to act [93].

Of COX-1, the constitutive isoform, and COX-2, the inducible isoform of cyclooxygenases, COX-2 is overexpressed in adenomatous polyps and adenocarcinoma cells converting arachidonic acid to prostaglandins (and other eicosanoids), inducers of mitosis, cell locomotion (by adherence junction changes in the relationship of β-catenins to actin cytoskeleton and cadherin transmembrane cell-to-cell adhesion receptors), and neoangiogenesis by VEGF overexpression in the cell's microenvironment. Colorectal adenomas already manufacture excessive amounts of COX-2 [94]. Interleukin-1β is responsible for COX-2 overexpression in colorectal carcinoma cells [95].

Cyclooxygenase-2 is overexpressed or upregulated by IL-1β in Peutz-Jeghers hamartomas [96] and human colorectal and head and neck carcinomas and in murine gastric adenomas and colon carcinomas; COX-2 is suppressed by celecoxib, refecoxib, and diclofenac. Chemoprevention of colon cancer with COX-2 inhibitors [nonsteriodal anti-inflammatory drugs (NSAIDs)] is now widely practiced, while mature results are awaited. Stromal fibroblasts of colonic carcinomas are enticed to produce IL-1β and COX-2 [97]. Cyclooxygenase-2 facilitates VEGF (see Part III) expression by tumor or stromal cells and COX-2 inhibitor NS-398 suppressed this effect [98]. Cyclooxygenase-2 activity in lung cancer correlates with microvessel density, VEGF expression, and tumor size [99]. Insulinlike growth factor-II and angiotensin II induce Akt phosphorylation in colon cancer cells [100]. Indomethacin, celecoxib (*vide infra*), wortmannin, or LY294002 suppress prostaglandin (PG)E-2 production and inhibit tumor cell growth. Cyclooxygenase-2 of tumor cell origin suppresses the antigen-presenting function of DCs [101]. Prostate cancer cells enlist platelet type 12 lipooxygenase (LOX) as their growth factor; these cells express integrins (αvβ3,5) and anti-integrin monoclonal antibodies (Chemicon) induced their apoptotic death [102,103]. Lipooxygenase inhibitors are baicalein (Biomol) and benzylhydroxy-phenyl pentamide (Biomide). The "scull-cap" from Lake Baikal yields the flavonoid baicalin that inhibits LOXs and Raf-1–mediated phosphorylating signal transductions. Lipooxygenase-producing pancreatic carcinoma xenografts suffered apoptotic death through the mitochondrial pathway upon exposure to LOX inhibitors nordihydroguaiaretic acid, Rev-5901, and baicalein [104–106]. Co-expression of COX-2 and MMP-2 in kidney carcinoma cells contributes to neoangiogenesis [107]. Ovarian carcinoma cells overexpressing COX-2 are chemotherapy-resistant and neoangiogenic [108,109]. Endothelial cells within lung cancers overexpress COX-2 but COX-2 expression by lung tumor cells did not predict higher incidence of brain metastases [99–112]. Cyclooxygenase-2 expression directly correlates with the intensity of neovasculature in breast cancers: COX-2, VEGF, and cyclin D1 are often co-expressed. Cyclooxygenase-2–positive ductal carcinomas in situ exhibit higher nuclear grades, and COX inhibitors licensed to treat arthritis may have a role in prevention of breast cancer. Because COX-2 also increases aromatase production, its inhibition by NSAIDs may suppress (chemoprevention) in situ breast cancers [113–118].

Nonsteroidal Anti-inflammatory Agents

Epidermal growth factor-R and ERK/MAPK inhibitors U0126 and COX inhibitors (sulindac, acetylsalicylic acid, celecoxib, rofecoxib, indomethacin, ibuprofen) induce apoptotic death in colonic adenoma and xenografted adenocarcinoma cells [93,119]. Ursodecholic acid is an inhibitor of *ras* gene product proteins and COX-2 in colon carcinoma cells [120]. Diclofenac, a dual inhibitor of COX-1 and COX-2, given in drinking water, inhibited the growth of murine colon carcinoma in vivo [121]. Benzylamide sulindac analogues (CP461) induce apoptotic death of CLL cells [122]. Sulindac sulfide inhibits Se112 phosphorylation of the Bcl-2/Bcl_{XL}-associated death promoter BAD; dephosphorylated BAD enhances apoptosis. BAD phosphorylated at serine 136 is anti-apoptotic; its phosphorylation at serine 128 results in enhanced pro-apoptotic activity [93]. To be fully effective, NSAIDs should disconnect COX-2 expression regulated by the Wnt pathway as well (see Part IV) [123].

The selective COX-2 inhibitor JT-522 (Japan Tobacco Inc, Tokyo) inhibited the growth of NSCLC cells both in tissue cultures and in xenografts; it inhibited endothelial cells and their inducer bFGF [124]. Exisulind (sulindac sulfone, Aptosyn) is pro-apoptotic, exerting its effect without COX 1 or COX 2 inhibition; it induced apoptosis in human NSCLC orthotopic xenografts and synergized with docetaxel. Exisulind analogues CP-461 and CP-255 are broad-spectrum tumor-inhibitory agents acting by suppressing Raf and NFκB. These drugs are active when administered orally and induced long stabilization of metastatic solid tumors in Phase I clinical trials. Tumor suppressor efficacy of exisulinds encompasses adenocarcinomas of the breast, colon, esophagus, lung, and prostate; squamous carcinomas; and gliomas [125–130].

Celecoxib (or rofecoxib) acts additively in patients with paclitaxel, a carboplatin, against NSCLC by reducing PGE_2

synthesis in the microenvironment of the tumor. Cyclooxygenase-2 inhibitors increase tumor cell radiosensitivity. Doses of celecoxib lower than that of sulindac killed human head and nack squamous carcinoma cells and synergized with chemotherapeuticals [131–134]. Sulindac augmented apoptosis induction of TNFα in xenografts of human lung cancer cells and, in orthotopically xenotransplanted human lung cancer cells, it exerted augmenting effects on docetaxel-induced remissions, resulting in prolonged survival [135]. The COX-2 inhibitor SC-236 (Pharmacia-Upjohn; Searle/Monsanto, Skokie, IL) in combination with EGF-R inhibitor ZD1839 (Iressa) and a PKA antisense oligonucleotide (AS-PKAI) synergized with each other against human lung and breast cancer cells grown in soft agar or as xenografts in nude mice [136]. The $p185^{neu}$ (from HER2/neu) TKI emodin synergized with celecoxib in rat cholangiocarcinoma cells [133]. The selective COX-2 inhibitor NS-398 (Sigma Pharmaceuticals, South Croydon, Australia; Cayman Chemicals, Ann Arbor, MI) suppresses the proliferation of oral squamous carcinoma cells and hepatocellular carcinoma cells; in the latter tumor, the malignant cells overexpressed FasL and underwent self-induced apoptosis [137,138]. The exisulind (sulindac) analog CP-461 is a pro-apoptotic inhibitor of PKC (*vide supra*) and NFκB; in an AUC (area under the curve) of 200 to 800 mg/day dosage range, it induced prolonged stable disease in patients with advanced solid tumors (colon cancer) [125,139]. If efficacy of rofecoxib in suppressing murine colon carcinoma metastases can be translated to humans, the mechanisms of action included suppression of PGE_2 and IL-10 and induction of IL-12 [140]. ZD1839 (Iressa, see Part I) and COX-2 inhibitor SC236 caused apoptotic death of $HER2/neu^+$ breast cancer cells [141]. Doxycycline acting as a MMP inhibitor and celecoxib in combination suppressed the growth of osteosarcoma and rhabdomyosarcoma xenografts [142]. In breast cancer cells, COX-inhibitors upregulated ceramide expression, a pro-apoptotic event [143].

In comparison, in clinical trials, intratumorally injected ONYX-015 in combination with gemcitabine given i.v. for the treatment of nonresectable pancreatic carcinoma induced 2 PR, 2 MR, and 6 stabilizations of disease against 11cases with progressions of disease. For adenoviral therapy of human tumors, the tumor cells have to express the coxsackie-adenovirus receptor hCAR (see Part I). Conditionally replication-competent adenoviruses have been constructed (COX-2 CRAds); their E1A expression depends on overexpressed COX-2, which is a feature of *ras*-driven tumor cells: The $E3^-$ Ad-E1A-COX replicates almost exclusively in H-*ras*–driven tumor cells and is oncolytic in tumor cells with an overactive MAPK pathway. Adenoviral vectors delivering HSV *tk* gene into tumor cells (AdCOX-2MTk) render tumor cells susceptible to ganciclovir in the HSV-TK/GCV protocols [144–147]. Cyclooxygenase-2 inhibitors and oncolytic viruses may act additively or synergistically in the tumor-bearing human host.

RECENT DEVELOPMENTS IN MOLECULAR ONCOLOGY: PARTS I AND II

In signet-ring colorectal carcinomas, codon 12 and 13 *ras* gene mutations are infrequent, but a new A:T transversion at the third base of K-*ras* codon 61 (CAA to CAT; Gln to His) emerges [148]; would these tumors support the growth of oncolytic viruses? Ras-CAAX and its interactions with FT inhibitors are depicted and explained; FT inhibitors are tabulated and, in conclusion, FT inhibitors in rational combination with cytotoxic chemotherapy are recommended [149]; would FT inhibitors antagonize or act additively with oncolytic viruses? Farnesyl transferase inhibitor SCH66336 acts as a pro-apoptotic and chemotherapy-sensitizing agent in melanoma cells [150]. Tipifarnib (R115777, Zarnestra, Johnson & Johnson, New Brunswick, NJ) and gemcitabine induce PR in patients with pancreatic and nasopharyngeal carcinomas [151]. Overexpression of β-catenin in colon cancer cells is a nearly ubiquitous event; sulindac induces degradation of this oncoprotein in the proteasomes [152].

Interactions of Bcr-Abl kinases with translation regulator protein S6 and with the mammalian target of rapamycin (mTOR) are explained [153]. Although point mutations of the *bcr-abl* fused gene render CML cells imatinib-resistant, these cells remain susceptible to pyrido-pyrimidine inhibitors (PD173955; PD166326) [154]. Imatinib does not protect Ph^+ ALL patients from CNS relapses: CNS prophylaxis remains obligatory [155]. In GIST, *kit* gene mutations are further analyzed to show that exon 9 mutations predict large and recurrent tumors [156]. The EGF-R is targeted by monoclonal antibodies and immunotoxins [157] and by small molecular inhibitors of its tyrosine kinase signal transduction [158]. Gefitinib-resistant breast adenocarcinoma cells (MDA-468) regained susceptibility to the drug after transfection with lipofectamine 2000 (Invitrogen), reinserting the tumor suppressor gene PTEN, or by suppressing the PI3K/Akt pathway with LY294002 [159]. 5′-deoxy-5-fluorouridine (5′-DFUR) and ZD1839 (gefitinib, Iressa) interact synergistically. Iressa potentiates radiation sensitivity of human tumor xenografts. Internal tandem duplication (a D835Y substitution) in AML cells is treatable with FLT3 tyrosine kinase indolinone inhibitors SU5416 and SU11248, which also target VEGF-R; with the piperazinyl quinazoline CT53518; with the indolocarbazole CEP-701; or with the staurosporine-derivative PKC412. This latter compound also inhibits VEGF-R, PDGF-R, and c-*kit* tyrosine kinase, reduces blast cell count, and induces PR (but not without toxicity) [160,161]. FTL3-mutated leukemic cells are susceptible to herbimycin and geldanamycin (17-AAG) [162,163].

The histone deacetylase inhibitor apicidin induces internal (mitochondrial) apoptosis in Ph chromosome-positive CML cells [164]. The CDK inhibitor flavopiridol interacts with the Rb/E2F-1 system, causing persistent expression of dephosphorylated E2F-1, thereby inducing apoptotic death of Rb-deficient tumor cells. The CDK-1 inhibitors BMI-1026 and BMI-1042 (Laboratory of Metabolism, Center Cancer Research, NCI, Bethesda, Maryland) arrest tumor cells in G_2-M, leading to apoptosis (for Cancer Res 2003; 63:7384 See Erratum in Cancer Res 2004; 64:3725). Breast cancer resistance protein (BCRP), its relationship to P-gp (P-glycoprotein product of the MDR-1 gene), and its overexpression in FP-resistant AL cells are explained. Lung cancer cells (SCLC) exposed to flavopiridol in vitro die apoptotic death [167–168].

The histone deacetylase inhibitor SAHA given i.v. induced MR in patients with chemotherapy-resistant bladder and breast cancers and Hodgkin's disease [169]. Apicidin (Calbiochem) induces mitochondrial apoptosis in Bcr-Abl^{+} CML cells [164]. Suberoylanilide hydroxamide acid or trichostatin (TSA) increased susceptibility of various tumor cells to a long list of chemotherapeutical agents [170]. Epothilone (EPO906) deriving from the myxobacterium *Sorangium cellulosum* causes mitotic arrest and microtubular derangements in cultured breast cancer cells [171]; STI571 (imatinib) suppresses receptors c-Kit and PDGF and enhances the therapeutc index of epothilone [172]. Lamellarin D isolated from the marine mollusk *Lamellaria* emerges as a potent new inhibitor of topoisomerase I [173].

The macrocyclic lactone bryostatin isolated from *Bugula neritina* (*maritima*) stimulates COX-2 transcription, an observation calling for protocols in which bryostatin is combined with COX-2 inhibitors. Cyclooxygenase-2 upregulates VEGF expression and stimulates neoangiogenesis in gastric carcinoma. Celecoxib induces nuclear localization of p53 in colon cancer cells, a pro-apoptotic event. In premalignant lesions of the oral cavity, COX-2 is already overexpressed, an observation calling for its inhibitors to be used to prevent malignant transformation. Rofecoxib effectively reduced the number and size of rectal polyps in patients with familial adenomatous polyposis [174–178]. Exisulind in combination with capecitabine failed to significantly improve response rate in patients with metastatic breast cancer [126]. Uterine cervical cancer cells express peroxisome proliferator-activated receptor-γ (PPARγ), but binding its cognate ligand results in downregulation of COX-2 [179] (Tables 2 and 3).

Reasons for the increased virulence of acquired immunodeficiency syndrome (AIDS)-associated Kaposi's sarcoma (KS) in comparison with the relative indolence of classical KS [180,181] begin to unfold. In addition to immunodeficiency in AIDS occurring because of the depletion of CD4/CD8 T cells, the human immunodeficiency virus (HIV)-1 structural protein Tat induces the overexpression of metalloproteinases in KS cells and, by activating the PI3K/Akt pathway, it secures the

Table 2 Prominent Targets and Molecular Inhibitors

Targets	Inhibitors
Bcr/Abl fusion oncoprotein	STI-571 imatinib mesylate, Gleevec) PD166326 PD173955 PD180940
basic fibroblast growth factor (bFGF)	PD173074 PD166285 baicalein
benzodiazepine receptor	PK11195
Bcl-2	HA-14-1 antisense oligonucleotides
Bradykinin	CU201 (B9870)
BRAF kinase	U0126 CI-1040 PD184352 UCN-01
cycline-dependent kinase I (CDK-I)	NMI-1026 BMI-1042
cycline deacetylase	flavopiridols BMI-1026 BMI-1042
c-Kit CD117 R for stem cell factor	STI-571 ZD1839 OSI-774 SU11248
c-MET R for hepatocyte growth factor/scatter factor (HGF/SF)	PHA-665752 SU11271 SU11606
cyclooxygenase-2 (COX-2) lipooxygenase (LOX)	U0126 sulindac & analogues CP255 CP 248 CP461 Exisulind Aptosyn celecoxib rofecoxib diclofenac ursodecholic acid indomethacin acetyl salicylic acid ibuprofen JT-522 SC-236 NS-398 nordihydroguaiaretic acid Rev-5901
CXCL12/SDF-1 chemokine stroma-derived factor-1	NSC651016 (Suradista) distamycin PNU-145156E
epidermal GF receptor (EGF-R)	ZD1839 gefitinib Iressa OSI-774 erlotinib Tarceva PK11166 CCP-358,774 GW2016 SU11248 AG1478
EGF-R/HER2	GW572016 (Tables 5 & 6)
Erb2 kinase	CP-65477 CI1033
extracellular signal-regulated kinase (ERK)	PD98059
farnesyl transferase (FT)	R115777 tipifarnib Zarnestra SCH66336 lonafarnib Sarasar L778123 L774123 SCH44342 PD083176 PD98059 BMS-214662 BMS-186511 SU11248 SU5416 LB42722 FT-277 B956 B1086

(*Continued*)

Table 2 Continued

Targets	Inhibitors
fm-like tyrosine kinase-3 (FLT-3)	GTP14564 SU11248 SU5416 MLNS18 PKC412 CT53518 PKC412 herbimycins geldanamycins
HER2/neu	PKI166 CI1033 GW572016
histone deacetylase (HDAC)	aphicidin depsipeptide trichostatins suberoylanilide hydroxamide (SAHA) sodium butyrate oxamflatin FR901228 MS-275
h telomerase reverse transcriptase (hTERT)	rapamycins
insulin-like growth factor receptor-I (IGF-R)	tyrphostins AG1024 enalapril perindoprilat candesartan (CV11974)
Janus kinase (JAK)	AG490 cytokine signaling suppressor proteins (SOCS)
neutrophilic factor (NFκB)	PS-341
matrix metalloproteinases (MMP)	BAY12-9566 AG3340 Æ-941 (BMS-275291, Neovastat) ONO-4817 CP255 CP461 testicans
mitogen-activated protein kinase (MAPK)	PD98059 GGTI-298 FTI-277
mTOR (target of rapamycin)	rapamycins CCI-779 RAD001
protein kinase C (PKC)	PKC-412 UCN-01 ISIS 3521 quercetin bryostatin
platelet-derived growth factor receptor (PDGF-R)	STI571 (imatibib mesylate, Gleevec) CT52923 SU11248 AG-013736
phosphatidyl inositol-3 kinase (PI3K)	LY294002 PD98059 wortmannin genistein
proteasome	PS-341 (bortezomib, Velcade)
Raf kinase	L779450 BAY43-9006
Ras kinase	LY294002
Rb protein phosphorylated	tyrphostin AG1024
Rho G-triphosphatase (GTP)	L774123 L774832
Rho kinase	L27632
signal transducer/activator of transcription (STAT) proteins	cucurbitacin antisense oligonucleotides protein tyrosine phosphatases (SHP1,2) phosphotyrosyl peptides SOCS PIAS: proteins inhibitory to activated STAT

(*Continued*)

Table 2 Continued

Targets	Inhibitors
Rho-associated kinase	L27632 STI571 Y27632
Src kinase	PD1666326 PD173955 PD180970 SU6656 SKI-606
vascular endothel GF-receptor (VEGF-R)	CP-547,632 VGA1155 CEP7055 SU11248 SU5416 SU6668 PTK787/ZK222584/ZD6474
VEGF system	ZD6126 (NG453) ZD6474 ZD4190 ZK222584 PNU-145156E AG-013736
Wnt signaling	dickkopf gene product protein DKK1

anti-apoptotic state of these cells. Highly active anti-retroviral therapy (HAART), with inhibitors of viral reverse transcriptase, viral packaging, virus and cell membrane fusion, and viral protease, ameliorates the clinical course of AIDS-KS, because it also allows for restoration of NK and T cell counts [182–184]. Reactivation of latent or newly acquired HHV-8 in both classical and AIDS-KS initiates the process. The HHV-8-encoded protein ORF-74 transforms fibroblasts and endothelial cells. The HHV-8-encoded vFLIP (FLICE-inhibitory protein), also referred to as K13, activates the anti-apoptotic NFκB pathway. Virally induced para- or auto-

Table 3 Activators, Promoters, Replacements (APR)

Targets	APR
benzodiazepine receptor	Ro5-4864 (inhibitor PK11195)
insulin-like growth factor receptor	angiotensin II (inhibitors in Table 2)
neutrophilic factor (NFκB)	FLICE* inhibitory protein FLIP
mutated cyclin-dependent kinases	flavopiridols L868275 roscovitine
peroxisome proliferator-activated nuclear receptor (peroxisome proliferator-activated nuclear receptors)	ciglitazone poglitazone rosiglitazone troglitazone

* Fas-associated death domain-like IL-1β-converting enzyme (caspase 8).

crine growth factors include IFN-γ, IL-6, stem cell factor (c-*kit* ligand), and bFGF [182,185,186]. Classical KS is a chemotherapy-sensitive tumor; AIDS-KS responds to doxorubicin and paclitaxel [187] but also is treatable with IFN-α (not with IFN-γ), metalloproteinase inhibitors, *cis*-retinoic acid, and neo-angiogenesis inhibitors. The recent demonstration of CD117 (cluster of differentiation) known as c-Kit receptor and PDGF-R on KS cells calls for clinical trials with imatinib mesylate [188–190].

REFERENCES: PART II

1. Vermeulen K, Van Bockstaele DR, Borneman ZN. The cell cycle: a review of replication, deregulation and therapeutic targets in cancer Cell Prolif 2003; 36:131–149.
2. Wishart MJ, Dixon JE. PTEN and myotubularin phosphatases: from 3-phosphoinositide dephosphorylation to disease. Phosphatase and tensin homolog deleted on chromosome ten Trends Cell Biol 2002; 12:579–585.
3. Kim DM, Koo SY, Jeon K, Kim MH, Lee J, Hong CY, Jeong S. Rapid induction of apoptosis by combination of flavopiridol and tumor necrosis factor (TNF)-α or TNF-related apoptosis-inducing ligand in human cancer cell lines Cancer Res 2003; 63:621–626.
4. Messmann RA, Ullmann CD, Lahusen T, Kalehua A, Wasfy J, Melillo G, Ding I, Headlee D, Figg WD, Sausville EA, Senderowicz AM. Flavopiridol-related proinflammatory syndrome is associated with induction of interleukin-6 Clin Cancer Res 2003; 9:562–570.
5. Yu C, Rahmani M, Dai Y, Conrad D, Krystal G, Dent P, Grant S. The lethal effects of pharmacological cyclin-dependent kinase inhibitors in human leukemia cells proceed through a phosphatidylinositol 3-kinase/Akt-dependent process Cancer Res 2003; 63:1822–1833.
6. Gildea JJ, Seraj MJ, Oxford G, Harding MA, Hampton GM, Moskaluk CA, Frierson HF, Conaway MR, Theodorescu D. RhoGDI2 is an invasion and metastasis suppressor gene in human cancer Cancer Res 2002; 62:6418–6423.

7. Litz J, Carlson P, Warshamana-Greene GS, Grant S, Krystal GW. Flavopiridol potently induces small cell lung cancer apoptosis during S phase in a manner that involves early mitochondrial dysfunction Clin Cancer Res 2003; 9:4586–4594.

8. Rosato RR, Almenara JA, Cartee L, Betts V, Chellappan SP, Grant S. The cyclin-dependent kinase inhibitor flavopiridol disrupts sodium butyrate-induced $p21^{WAF1/CIP1}$ expression and maturation while reciprocally potentiating apoptosis in human leukemia cells Mol Cancer Ther 2002; 1:253–266.

9. Rosato RR, Almenara JA, Grant S. The histone deacetylase inhibitor MS-275 promotes differentiation or apoptosis in human leukemia cells through a process regulated by generation of reactive oxygen species and induction of $p21^{CIP1/WAF1}$ Cancer Res 2003; 63:3637–3645.

10. Yu C, Krystal G, Dent P, Grant S. Flavopiridol potentiates STI571-induced mitochondrial damage and apoptosis in BCR-ABL-positive human leukemia cells Clin Cancer Res 2002; 8: 2976–2984.

11. O'Dwyer M. Multifaceted approach to the treatment of bcr-abl-positive leukemias Oncologist 2002; 7S:30–38.

12. Fischer PM, Gianella-Borradori A. CDK inhibitors in clinical development for the treatment of cancer Expert Opin Investig Drugs 2003; 121:955–970.

13. Motwani M, Rizzo C, Sirotnak F, She Y, Schwartz GK. Flavopiridol enhances the effect of docetaxel in vitro and in vivo in human gastric cancer cells Mol Cancer Ther 2003; 2:549–555.

14. Nahta R, Trent S, Yang C, Schmidt EV. Epidermal growth factor receptor expression is a candidate target of the synergistic combination of trastuzumab and flavopiridol in breast cancer Cancer Res 2003; 63:3626–3631.

15. Raju U, Nakata E, Mason KA, Ang KK, Milas L. Flavopiridol, a cyclin-dependent kinase inhibitor, enhances radiosensitivity of ovarian carcinoma cells Cancer Res 2003; 63:3263–3267.

16. Maggiorella L, Deutsch E, Frascogna V, Chavaudra N, Jeanson L, Milliat F, Eschwege F, Bourhis J. Enhancement of radiation response by roscovitine in human breast carcinoma in vitro and in vivo Cancer Res 2003; 63:2513–2517.

17. Aklilu M, Kindler HL, Donehower RC, Mani S, Vokes EE. Phase II study of flavopiridol in patients with advanced colorectal cancer Ann Oncol 2003; 14:1270–1273.

18. Kitazono M, Rao VK, Robey R, Aikou T, Bates S, Fojo T, Goldsmith ME. Histone deacetylase inhibitor FR901228 enhances adenovirus infection of hematopoietic cells Blood 2002; 99: 2248–2251.

19. Wittmann S, Bali P, Donapaty S, Nimmanapalli R, Guo F, Yamaguchi H, Huang M, Jove R, Wang HG, Bhalla K. Flavopiridol down-regulates antiapoptotic proteins and sensitizes human breast cancer cells to epothilone B-induced apoptosis Cancer Res 2003; 63:93–99.

20. Motwani M, Jung C, Sirotnak FM, She Y, Shah MA, Gonen M, Schwartz GK. Augmentation of apoptosis and tumor regression by flavopiridol in the presence of CPT-11 in Hct116 colon cancer monolayers and xenografts Clin Cancer Res 2001; 7:4209–4219.

21. Robinson WA, Miller TL, Harrold EA, Bemis LT, Brady BM, Nelson RP. The effect of flavopiridol on the growth of p16+ and p16− melanoma cell lines Melanoma Res 2003; 13: 231–238.

22. Debes A, Oerding M, Willers R, Gobel U, Wessalowski R. Sensitization of human Ewing's tumor cells to chemotherapy and heat treatment by the bioflavonoid quercetin Anticancer Res 2003; 23:3359–3366.

23. Lee SC, Kuan CY, Yang CC, Yang SD. Bioflavonoids commonly and potently induce tyrosine dephosphorylation/inactivation of oncogenic proline-directed protein kinase FA in human prostate carcinoma cells Anticancer Res 1998; 18: 1117–1121.

24. Farhan H, Wahala K, Cross HS. Genistein inhibits vitamin D hydroxylases CYP24 and CYP27B1 expression in prostate cells J Steroid Biochem Mol Biol 2003; 84:423–429.

25. Yu L, Blackburn GL, Zhou JR. Genistein and daidzein down-regulate prostate androgen-regulated transcript-1 (PART-1) gene expression induced by dihydrotestosterone in human prostate LNCaP cancer cells J Nutr 2002; 133:389–392.

26. Mimeault M, Pommery N, Wattez N, Bailly C, Henichart JP. Anti-proliferative and apoptotic effects of anandamide in human prostatic cancer cell lines: implications of epidermal growth factor receptor down-regulation and ceramide production Prostate 2003; 56:1–12.

27. McKay TR, Bell S, Tenev T, Stoll V, Lopes R, Lemoine NR, McNeish IA. Procaspase 3 expression in ovarian carcinoma cells increases survivin transcription which can be countered with a dominant-negative mutant, survivin T34A; a combination gene therapy strategy Oncogene 2003; 22:3539–3547.

28. Yang CT, You L, Lin CL, McCormick F, Jablons DM. A comparison analysis of anti-tumor efficacy of adenoviral gene replacement therapy ($p1^{ARF}$ and $p16^{INK4A}$) in human mesothelioma cells Anticancer Res 2003; 23:33–38.

29. Cao XX, Mohuiddin I, Chada S, Mhashilkar AM, Ozvaran MK, McConkey DJ, Miller SD, Daniel JC, Smythe WR. Adenoviral transfer of mda-7 leads to BAX up-regulation and apoptosis in mesothelioma cells, and is abrogated by overexpression of BCL-XL Mol Med 2002; 8:869–876.

30. Nowak AK, Lake RA, Kindler HL, Robinson BW. New approaches for mesothelioma: biologics, vaccines, gene therapy and other novel agents Semin Oncol 2002; 29:82–96.

31. Mesri M, Wall NR, Li J, Kim RW, Alteri DC. Cancer gene therapy using a survivin mutant adenovirus J Clin Invest 2001; 108:981–990. Comments:. 2001; 108:965–9. DC. 2002; 109:285–286.

32. Kappler M, Kotzsch M, Bartel F, Fussel S, Lautenschlager C, Schmidt U, Wurl P, Bache M, Schmidt H, Taubert H, Meye A. Elevated expression level of survivin protein in soft-tissue sarcomas is a strong independent predictor of survival Clin Cancer Res 2003; 9:1098–1104.

33. Wakana Y, Kasuya K, Katayanagi S, Tsuchida A, Aoki T, Koyanagi Y, Ishii H, Ebihara Y. Effect of survivin on cell proliferation and apoptosis in gastric cancer Oncol Rep 2002; 9: 1213–1218.

34. Casati C, Dalerba P, Rivoltini L, Gallino G, Deho P, Rini F, Belli F, Mezzanzanica D, Costa A, Andreola S, Leo E, Parmi-

ani G, Castelli C. The apoptosis inhibitor protein survivin induces tumor-specific $CD8^+$ and $CD4^+$ T cells in colorectal cancer patients Cancer Res 2003; 63:4507–4515.

35. Wall NR, O'Connor DS, Plescia J, Pommier Y, Altieri DC. Suppression of survivin phosphorylation on Thr^{34} by flavopiridol enhances tumor cell apoptosis Cancer Res 2003; 63: 230–235.

36. Hirohashi Y, Torigoe T, Maeda A, Nabeta Y, Kamiguchi K, Sato T, Yoda J, Ikeda H, Hirata K, Yamanaka N, Sato N. An HLA-A24-restricted cytotoxic T lymphocyte epitope of a tumor-associated protein, survivin Clin Cancer Res 2002; 8: 1731–1739.

37. Senderowicz AM, Headlee D, Stinson SF, Lush RM, Kalil N, Villalba L, Hill K, Steinberg SM, Figg WD, Tompkins A, Arbuck SG, Sausville EA. Phase I trial of continuous infusion flavopiridol, a novel cyclin-dependent kinase inhibitor, in patients with refractory neoplasms J Clin Oncol 1998; 16: 2986–2999.

38. Tan AR, Headlee D, Messmann R, Sausville EA, Arbuck SG, Murgo AJ, Melillo G, Zhai S, Figg WD, Swain SM, Senderowicz AM. Phase I clinical and pharmacokinetic study of flavopiridol administered as a daily 1-hour infusion in patients with advanced neoplasms J Clin Oncol 2002; 20:4074–4082.

39. Senderowicz AM. Novel direct and indirect cyclin-dependent kinase modulators for the prevention and treatment of human neoplasms Cancer Chemother Pharmacol 2003; 52S1:61–73.

40. Deb DK, Sassano A, Lekmine F, Majchrzak B, Verma A, Kambhampati S, Uddin S, Rahman A, Fish EN, Platanias LC. Activation of protein kinase Cδ by IFN-γ J Immunol 2003; 171: 267–273.

41. McCracken MA, Miraglia LJ, McKay RA, Strobl JS. Protein kinase C δ is a prosurvival factor in human breast tumor cell lines Mol Cancer Ther 2003; 2:273–281.

42. Grana TM, Rusyn EV, Zhou H, Sartor CI, Cox AD. Ras mediates radioresistance through both phosphatidylinositol 3-kinase-dependent and Raf-dependent but mitogen-activated protein kinase/extracellular signal-regulated kinase kinase-

independent signaling pathways Cancer Res 2002; 62: 4142–4150.

43. DeGrendele H. Activity of the Ras kinase inhibitor BAY43–9006 in patients with advanced solid tumors Clin Colorect Cancer 2003; May 16–17.
44. Giese A, Hagel C, Kim EL, Zapf S, Djawaheri J, Berens ME, Westphal M. Thromboxane synthase regulates the migratory phenotype of human glioma cells Neurooncology 1999; 1:3–13.
45. Yoshizato K, Zapf S, Westphal M, Berens ME, Giese A. Thromboxane synthase inhibitors induce apoptosis in migration-arrested glioma cells Neurosurgery 2002; 50:343–354.
46. Weichert W, Gekeler V, Denkert C, Dietel M, Hauptmann S. Protein kinase C isoform expression in ovarian carcinoma correlates with indicators of poor prognosis Int J Oncol 2003; 23: 633–639.
47. Marshall JL, Bangalore N, El-Ashry D, Fuxman Y, Johnson M, Norris B, Oberst M, Ness E, Wojtowicz-Praga S, Bhargava P, Rizvi N, Baidas S, Hawkins MJ. Phase I study of prolonged infusion bryostatin-1 in patients with advanced malignancies Cancer Biol Ther 2002; 1:409–416.
48. Madhusudan S, Protheroe A, Propper D, Han C, Corrie P, Earl H, Hancock B, Vasey P, Turner A, Balkwill F, Hoare S, Harris AL. A multicenter phase II trial of bryostatin in patients with advanced renal cancer Br J Cancer 2003; 89:1418–1422.
49. Mackay HJ, Twelves CJ. Protein kinase C: a target for anticancer drugs Endocrine-Related Cancer 2003; 10:389–396.
50. Senderowicz AM. Small-molecule cycline-dependent kinase modulators Oncogene 2003; 22:6609–6620.
51. Swannie HC, Kaye SB. Protein kinase C inhibitors Curr Oncol Rep 2002; 4:37–46.
52. Yamauchi T, Keating MJ, Plunkett W. UCN-01 (7-hydroxystaurosporine) inhibits DNA repair and increases cytotoxicity in normal lymphocytes and chronic lymphocytic leukemia lymphocytes Mol Cancer Ther 2002; 1:287–294.
53. Keague A, Wilson DJ, Nelson J. Staurosporine-induced apoptosis and hydrogen peroxide-induced necrosis in two human breast cell lines Brit J Cancer 2003; 88:125–131.

54. Patel V, Lahusen T, Leethanakul C, Igishi T, Kremer M, Quintanilla-Martinez L, Ensley JF, Sausville EA, Gutkind JS, Senderowicz AM. Antitumor activity of UCN-01 in carcinomas of the head and neck is associated with altered expression of cyclin D3 and $p27^{KIP1}$ Clin Cancer Res 2002; 8:3549–3560.

55. Propper DJ, McDonald AC, Man A, Thavasu P, Balkwill F, Braybrooke JP, Caponigro F, Graf P, Dutreix C, Blackie R, Kaye SB, Ganesan TS, Talbot DC, Harris AL, Twelves C. Phase I and pharmacokinetic study of PKC412, an inhibitor of protein kinase C. J Clin Oncol 2001; 19:1485–1492.

56. Nemunaitis J, Holmlund JT, Kraynak M, Richards D, Bruce J, Ognoskie N, Kwoh TJ, Geary R, Dorr A, Von Hoff D, Eckhardt SG. Phase I evaluation of ISIS 3521, an antisense oligodeoxynucleotide to protein kinase C-alpha, in patients with advanced cancer J Clin Oncol 1999; 17:3586–3595.

57. Haas NB, Smith M, Lewis N, Littman L, Yeslow G, Joshi ID, Murgo A, Bradley J, Gordon R, Wang H, Rogatko A, Hudes GR. Weekly bryostatin-1 in metastatic renal cell carcinoma: a phase II study Clin Cancer Res 2003; 9:109–114.

58. Hu Y, Bally M, Dragowska WH, Mayer L. Inhibition of mitogen-activated protein kinase/extracellular signal-regulated kinase kinase enhances chemotherapeutic effects on H460 human non-small cell lung cancer cells through activation of apoptosis Mol Cancer Ther 2003; 2:641–649.

59. Guha S, Lunn JA, Santiskulvong C, Rozengurt E. Neurotensin stimulates protein kinase C-dependent mitogenic signaling in human pancreatic carcinoma cell line PANC-1 Cancer Res 2003; 63:2379–2387.

60. Wall SJ, Jiang Y, Muschel RJ, DeClerck YA. Meeting report: Proteases, extracellular matrix, and cancer: an AACR Special Conference in Cancer Research Cancer Res 2003; 63: 4750–4755.

61. Hernandez-Barrantes S, Bernardo M, Toth M, Fridman R. Regulation of membrane type-matrix metalloproteinases Semin Cancer Biol 2002; 12:131–138.

62. Smutzer G. Molecular demolition. Matrix metalloproteinases and their inhibitors play key roles in tissue remodelling and

pathogenesis of metastatic and inflammatory diseases The Scientist 2003:34–36.

63. Egeblad M, Werb Z. New functions for the matrix metalloproteinases in cancer progression Nat Rev Cancer 2002; 2: 161–174.

64. Nelson AR, Fingleton B, Rothenberg ML, Matrisian LM. Matrix metalloproteinases: biologic activity and clinical implications J Clin Oncol 2000; 18:1135–1149.

65. Pinto CA, Carvalho PE, Antonangelo L, Garippo A, Da Silva AG, Soares F, Younes R, Beyruti R, Takagaki T, Saldiva P, Vollmer RT, Capelozzi VL. Morphometric evaluation of tumor matrix metalloproteinase 9 predicts survival after surgical resection of adenocarcinoma of the lung Clin Cancer Res 2003; 9:3098–3104.

66. Vargo-Gogola T, Fingleton B, Crawford HC, Matrisian LM. Matrilysin (matrix metalloproteinase-7) selects for apoptosis-resistant mammary cells in vivo Cancer Res 2002; 62: 5559–5563.

67. Stearns ME, Wang M, Hu Y, Garcia FU, Rhim J. Interleukin 10 blocks matrix metalloproteinase-2 and membrane type 1-matrix metalloproteinase synthesis in primary human prostate tumor lines Clin Cancer Res 2003; 9:1191–1199.

68. Miao X, Yu C, Tan W, Xiong P, Liang G, Lu W, Lin D. A functional polymorphism in the matrix metalloproteinase-2 gene promoter (-1306C/T) is associated with risk of development but not metastasis of gastric cardia adenocarcinoma Cancer Res 2003; 63:3987–3990.

69. Woo JH, Park JW, Lee SH, Kim YH, Lee IK, Gabrielson E, Lee SH, Lee HJ, Kho YH, Kwon TK. Dykellic acid inhibits phorbol myristate acetate-induced matrix metalloproteinase-9 expression by inhibiting nuclear factor κB transcriptional activity Cancer Res 2003; 63:3430–3434.

70. Durkan GC, Nutt JE, Marsh C, Rajjayabun PH, Robinson MC, Neal DE, Lunec J, Mellon JK. Alteration in urinary matrix metalloproteinase-9 to tissue inhibitor of metalloproteinase-1 ratio predicts recurrence in nonmuscle-invasive bladder cancer Clin Cancer Res 2003; 9:2576–2582.

71. Mian BM, Dinney CP, Bermejo CE, Sweeney P, Tellez C, Yang XD, Gudas JM, McConkey DJ, Bar-Eli M. Fully human anti-interleukin 8 antibody inhibits tumor growth in orthotopic bladder cancer xenografts via down-regulation of matrix metalloproteases and nuclear factor-κB Clin Cancer Res 2003; 9:3167–3175.

72. Udayakumar TS, Chen ML, Bair EL, Von Bredow DC, Cress AE, Nagle RB, Bowden GT. Membrane type-1-matrix metalloproteinase expressed by prostate carcinoma cells cleaves human laminin-5 β3 chain and induces cell migration Cancer Res 2003; 63:2292–2299.

73. Impola U, Toriseva M, Suomela S, Jeskanen L, Hieta N, Jahkola T, Grenman R, Kahari VM, Saarialho-Kere U. Matrix metalloproteinase-19 is expressed by proliferating epithelium but disappears with neoplastic dedifferentiation Int J Cancer 2003; 103:709–716.

74. Knox PG, Milner AE, Green NK, Eliopoulos AG, Young LS. Inhibition of metalloproteinase cleavage enhances the cytotoxicity of Fas ligand J Immunol 2003; 170:677–685.

75. Sinkovics JG, Horvath JC. Virological and immunological connotations of apoptotic and anti-apoptotic forces in neoplasia Int J Oncol 2001; 19:473–488.

76. Tada Y, Wang JO, Takiguchi Y, Tatsumi K, Kuriyama T, Okada S, Tokuhisha T, Sakiyama S, Tagawa M. Cutting edge: a novel role for Fas ligand in facilitating antigen acquisition by dendritic cells J Immunol 2002; 169:2241–2245.

77. Chen Y-L, Chen S-H, Wang J-L, Yang B-C. Fas ligand on tumor cells mediates inactivation of neutrophils J Immunol 2003; 171:1183–1191.

78. Horvath JC, Horvath E, Sinkovics JG, Horak A, Pendleton S, Mallah J. Human melanoma cells (HMC) eliminate autologous host lymphocytes (Ly^{FasR+}) and escape apoptotic death (A) by utikizing the Fas⇒FasR system as an autocrine growth loop ($HMC^{Fas \Rightarrow FasR}$), 89th Annual Meeting American Association for Cancer Reseaerch, New Orleans, LA, March 28–April 1, 1998 Proceedings 1998; 39:584 #3971.

79. Yang J-M, Xu Z, Wu H, Zhu H, Wu X, Hait WN. Overexpression of extracellular metalloproteinase inducer in multidrug resistant cancer cells Mol Cancer Res 2003; 1:420–427.

80. Thorns C, Bernd HW, Hatton D, Merz H, Feller AC, Lange K. Matrix-metalloproteinases in Hodgkin lymphoma Anticancer Res 2003; 23:1555–1558.

81. Arlt M, Kopitz C, Pennington C, Watson KL, Krell HW, Bode W, Gansbacher B, Khokha R, Edwards DR, Kruger A. Increase in gelatinase-specificity of matrix metalloproteinase inhibitors correlates with antimetastatic efficacy in a T-cell lymphoma model Cancer Res 2002; 62:5543–5550.

82. Bramhall SR, Rosemurgy A, Brown PD, Bowry C, Buckels JA. Marimastat Pancreatic Cancer Study Group. Marimastat as first-line therapy for patients with unresectable pancreatic cancer: a randomized trial J Clin Oncol 2001; 19:3447–3455.

83. Yamamoto H, Itoh F, Iku S, Adachi Y, Fukushima H, Sasaki S, Mukaiya M, Hirata K, Imai K. Expression of matrix metalloproteinases and tissue inhibitors of metalloproteinases in human pancreatic adenocarcinomas: clinicopathologic and prognostic significance of matrilysin expression J Clin Oncol 2001; 19:1118–1127.

84. Hess KR, Abbruzzese JL. Matrix metalloproteinase inhibition of pancreatic cancer: matching mechanism of action to clinical trial design J Clin Oncol 2001; 19:3445–3446.

85. Rowinsky EK, Humphrey R, Hammond LA, Aylesworth C, Smetzer L, Hidalgo M, Morrow M, Smith L, Garner A, Sorensen JM, Von Hoff DD, Eckhardt SG. Phase I and pharmacologic study of the specific matrix metalloproteinase inhibitor BAY 12–9566 on a protracted oral daily dosing schedule in patients with solid malignancies J Clin Oncol 2000; 18: 178–186.

86. Wojtowicz-Praga S, Torri J, Johnson M, Steen V, Marshall J, Ness E, Dickson R, Sale M, Rasmussen HS, Chiodo TA, Hawkins MJ. Phase I trial of Marimastat, a novel matrix metalloproteinase inhibitor, administered orally to patients with advanced lung cancer J Clin Oncol 1998; 16:2150–2156.

87. Moore MJ, Hamm J, Dancey J, Eisenberg PD, Dagenais M, Field A, Hagan K, Greenberg B, Colwell B, Zee B, Tu D, Ottaway J, Hunphrey R, Seymour L. Comparison of gemcitabine versus the matrix metalloproteinase inhibitor BAY12–9566 in patients with advanced or metastatic adenocarcinoma of

the pancreas; a phase III trial of the National Institue of Canada Clinical Trials Group J Clin Oncol 2003; 21:3296–3302.

88. Deryugina EI, Ratnikov BI, Strongin AY. Prinomastat, a hydroxamate inhibitor of matrix metalloproteinases, has a complex effect on migration of breast carcinoma cells Int J Cancer 2003; 104:533–541.

89. Yamashita T, Fujii M, Tomita T, Ishiguro R, Tashiro M, Tokumaru Y, Imanishi Y, Kanke M, Ogawa K, Kameyama K, Otani Y. The inhibitory effect of matrix metalloproteinase inhibitor ONO-4817 on lymph node metastasis in tongue carcinoma Anticancer Res 2003; 23:2297–2302.

90. Yamamoto A, Yano S, Shiraga M, Ogawa H, Goto H, Miki T, Zhang H, Sone S. A third-generation matrix metalloproteinase (MMP) inhibitor (ONO-4817) combined with docetaxel suppresses progression of lung micrometastasis of MMP-expressing tumor cells in nude mice Int J Cancer 2003; 103:822–828.

91. Nakada M, Yamada A, Takino T, Miyamori H, Takahashi T, Yamashita J, Sato H. Suppression of membrane-type 1 matrix metalloproteinase (MMP)-mediated MMP-2 activation and tumor invasion by testican 3 and its splicing variant gene product, N-Tes Cancer Res 2001; 61:8896–8902.

92. Nakada M, Miyamori H, Yamashita J, Sato H. Testican-2 abrogates inhibition of membrane-type matrix metalloproteinases by other testican family proteins Cancer Res 2003; 63: 3364–3369.

93. Rice PL, Washington M, Schleman S, Beard KS, Driggers LJ, Ahnen DJ. Sulindac sulfide inhibits epidermal growth factor-induced phosphorylation of extracellular-regulated kinase 1/2 and Bad in human colon cancer cells Cancer Res 2003; 63: 616–620.

94. Einspahr JG, Krouse RS, Yochim JM, Danenberg PV, Danenberg KD, Bhattacharyya AK, Martinez ME, Alberts DS. Association between cyclooxygenase expression and colorectal adenoma characteristics Cancer Res 2003; 63:3891–3893.

95. Liu W, Reinmuth N, Stoeltzing O, Parikh AA, Tellez C, Williams S, Jung YD, Fan F, Takeda A, Akagi M, Bar-Eli M, Gallick GE, Ellis LM. Cyclooxygenase-2 is up-regulated by in-

terleukin-1β in human colorectal cancer cells via multiple signaling pathways Cancer Res 2003; 63:3632–3636.

96. deLeng WWJ, Westerman AM, Westerman MAJ, deRooij FWM, van Dekken H, deGoeij AFPM, Gruber SB, Wilson JHP, Offerhaus GJA, Giardello FM, Keller JJ. Cyclooxygenase-2 expression and molecular alterations in Peutz-Jeghers hamartomas and carcinomas Clin Cancer Res 2003; 9:3065–3072.

97. Zhu Y, Hua P, Lance P. Cyclooxygenase-2 expression and prostanoid biogenesis reflect clinical phenotype in human colorectal fibroblast strains Cancer Res 2003; 63:522–526.

98. Ma DH, Chen JI, Zhang F, Hwang DG, Chen JK. Inhibition of fibroblast-induced angiogenic phenotype of cultured endothelial cells by the overexpression of tissue inhibitor of metalloproteinase (TIMP)-3 J Biomed Sci 2003; 10:526–534.

99. Yoshida S, Amano H, Hayashi I, Kitasato H, Kamata M, Inukai M, Yoshimura H, Majima M. COX-2/VEGF-dependent facilitation of tumor-associated angiogenesis and tumor growth in vivo Lab Invest 2003; 83:1385–1394.

100. Yasumaru M, Tsuji S, Tsuji M, Irie T, Komori M, Kimura A, Nishida T, Kakiuchi Y, Kawai N, Murata H, Horimoto M, Sasaki Y, Hayashi N, Kawano S, Hori M. Inhibition of angiotensin II activity enhanced the antitumor effect of cyclooxygenase-2 inhibitors via insulin-like growth factor I receptor pathway Cancer Res 2003; 63:6726–6734.

101. Sharma S, Stolina M, Yang SC, Baratelli F, Lin JF, Atianzar K, Luo J, Zhu L, Lin Y, Huang M, Dohadwala M, Batra RK, Dubinett SM. Tumor cyclooxygenase 2-dependent suppression of dendritic cell function Clin Cancer Res 2003; 9:961–968.

102. Pidgeon GP, Tang K, Cai YL, Piasentin E, Honn KV. Overexpression of platelet-type 12-lipooxygenase promotes tumor cell survival by enhancing $\alpha_v\beta_3$ and $\alpha_v\beta_5$ integrin expression Cancer Res 2003; 63:4258–4267.

103. Shappell SB, Olson SJ, Hannah SE, Manning S, Roberts RL, Masumori N, Jisaka M, Boeglin WE, Vader V, Dave DS, Shook MF, Thomas TZ, Funk CD, Brash AR, Matusik RJ. Elevated expression of 12/15-lipooxygenase and cyclooxygenase-2 in a transgenic mouse model of prostate carcinoma Cancer Res 2003; 63:2256–2267.

104. Tong WG, Ding XZ, Witt RC, Adrian TE. Lipooxygenase inhibitors attenuate growth of human pancreatic cancer xenografts and induce apoptosis through the mitochondrial pathway Mol Cancer Ther 2002; 1:929–935.

105. Zhang DY, Wu J, Ye F, Xue L, Jiang S, Yi J, Zhang W, Wei H, Sung M, Wang W, Li X. Inhibition of cancer cell proliferation and prostaglandin E2 synthesis by Scutellaria baicalensis Cancer Res 2003; 63:4037–4043.

106. Chen S, Ruan Q, Bedner E, Deptala A, Wang X, Hsieh TC, Traganos F, Darzynkiewicz Z. Effects of flavonid baicalin and its metabolite baicalein on androgen receptor expression, cell cycle progression and apoptosis of prostate cancer cell lines Cell Prolif 2001; 34:293–304.

107. Miyata Y, Koga S, Kanda S, Nishikido M, Hayashi T, Kanetake H. Expression of cyclooxygenase-2 in renal cell carcinoma: correlation with tumor cell proliferation, apoptosis, angiogenesis, expression of matrix metalloproteinase-2, and survival Clin Cancer Res 2003; 9:1741–1749.

108. Ferrandina G, Lauriola L, Zannoni GF, Fagotti A, Fanfani F, Legge F, Maggiano N, Gessi M, Mancuso S, Ranelletti FO, Scambia G. Increased cyclooxygenase-2 (COX-2) expression is associated with chemotherapy resistance and outcome in ovarian cancer patients Ann Oncol 2002; 13:1205–1211.

109. Gupta RA, Tejada LV, Tong BJ, Das SK, Morrow JD, Dey SK, DuBois RN. Cyclooxygenase-1 is overexpressed and promotes angiogenic growth factor production in ovarian cancer Cancer Res 2003; 63:906–911.

110. Kim HS, Youm HR, Lee JS, Min KW, Chung JH, Park CS. Correlation between cyclooxygenase-2 and angiogenesis in non-small cell lung cancer Lung Cancer 2003; 42:163–170.

111. Milas I, Komaki R, Hachiya T, Bubb RS, Ro JY, Langford L, Sawaya R, Putnam JB, Allen P, Cox JD, McDonnell TJ, Brock W, Hong WK, Roth JA, Milas L. Epidermal growth factor receptor, cyclooxygenase-2, and BAX expression in the primary non-small cell lung cancer and brain metastases Clin Cancer Res 2003; 9:1070–1076.

112. Ermert L, Dierkes C, Ermert M. Immunohistochemical expression of cyclooxygenase isoenzymes and downstream en-

zymes in human lung tumors Clin Cancer Res 2003; 9: 1604–1610.

113. Lim SC. Role of COX-2, VEGF and cyclin D1 in mammary infiltrating duct carcinoma Oncol Rep 2003; 10:1241–1249.

114. Davies G, Martin LA, Sacks N, Dowsett M. Cyclooxygenase-2 (COX-2), aromatase and breast cancer: a possible role for COX-2 inhibitors in breast cancer chemoprevention Ann Oncol 2002; 13:669–678.

115. Davies G, Salter J, Hills M, Martin LA, Sacks N, Dowsett M. Correlation between cyclooxygenase-2 expression and angiogenesis in human breast cancer Clin Cancer Res 2003; 9: 2651–2656.

116. Shim V, Gauthier ML, Sudilovsky D, Mantei K, Chew KL, Moore DH, Cha I, Tlsty TD, Esserman LJ. Cyclooxygenase-2 expression is related to nuclear grade in ductal carcinoma in situ and is increased in its normal adjacent epithelium Cancer Res 2003; 63:2347–2350.

117. Lewis R. COX-2 inhibitors tackle cancer The Scientist 2002: 28–29.

118. Keller JJ, Giardiello FM. Chemoprevention strategies using NSAIDs and COX-2 inhibitors Cancer Biol Ther 2002; 2: S140–149.

119. Whitehead CM, Earle KA, Fetter J, Xu S, Hartman T, Chan DC, Zhao TL, Piazza G, Klein-Szanto AJ, Pamukcu R, Alila H, Bunn PA, Thompson WJ. Exisulind-induced apoptosis in a non-small cell lung cancer orthotopic lung tumor model augments docetaxel treatment and contributes to increased survival Mol Cancer Ther 2003; 2:479–488.

120. Khare S, Cerda S, Wali RK, von Lintig FC, Tretiakova M, Joseph L, Stoiber D, Cohen G, Nimmagadda K, Hart J, Sitrin MD, Boss GR, Bissonnette M. Ursodeoxycholic acid inhibits Ras mutations, wild-type Ras activation, and cyclooxygenase-2 expression in colon cancer Cancer Res 2003; 63:3517–3523.

121. Falkowski M, Skogstad S, Shahzidi S, Smedsrod B, Sveinbjornsson B. The effect of cyclooxygenase inhibitor diclofenac on experimental murine colon carcinoma Anticancer Res 2003; 23:2303–2308.

122. Moon EY, Lerner A. Benzylamide sulindac analogues induce changes in cell shape, loss of microtubules and G_2-M arrest in a chronic lymphocytic leukemia (CLL) cell line and apoptosis in primary CLL cells Cancer Res 2002; 62: 5711–5719.

123. Araki Y, Okamura S, Hussain SP, Nagashima M, He P, Shiseki M, Miura K, Harris CC. Regulation of cyclooxygenase-2 expression by the Wnt and ras pathways Cancer Res 2003; 63:728–734.

124. Hida T, Kozaki K, Ito H, Miyaishi O, Tatematsu Y, Suzuki T, Matsuo K, Sugiura T, Ogawa M, Takahashi T, Takahashi T. Significant growth inhibition of human lung cancer cells both in vitro and in vivo by the combined use of a selective cyclooxygenase 2 inhibitor, JTE-522, and conventional anticancer agents Clin Cancer Res 2002; 8:2443–2447.

125. Liu L, Li H, Underwood T, Lloyd M, David M, Sperl G, Pamukcu R, Thompson WJ. Cyclic GMP-dependent protein kinase activation and induction by exisulind and CP461 in colon tumor cells Pharmacol Exp Ther 2001; 299:583–592.

126. Pusztai L, Zhen JH, Arun B, Rivera E, Whitehead C, Thompson WJ, Nealy KM, Gibbs A, Symmans WF, Esteva FJ, Booser D, Murray JL, Valero V, Smith TL, Hortobagyi GN. Phase I and II study of exisulind in combination with capecitabine in patients with metastatic breast cancer J Clin Oncol 2003; 21: 3454–3461.

127. Joe AK, Liu H, Xiao D, Soh JW, Pinto JT, Beer DG, Piazza GA, Thompson WJ, Weinstein IB. Exisulind and CP248 induce growth inhibition and apoptosis in human esophageal adenocarcinoma and squamous carcinoma cells J Exp Ther Oncol 2003l; 3:83–94.

128. Lim JT, Piazza GA, Pamukcu R, Thompson WJ, Weinstein IB. Exisulind and related compounds inhibit expression and function of the androgen receptor in human prostate cancer cells Clin Cancer Res 2003; 9:4972–4982.

129. Yoon JT, Palazzo AF, Xiao D, Delohery TM, Warburton PE, Bruce JN, Thompson WJ, Sperl G, Whitehead C, Fetter J, Pamukcu R, Gundersen GG, Weinstein IB. CP248, a derivative of exisulind, causes growth inhibition, mitotic arrest, and

abnormalities in microtubule polymerization in glioma cells Mol Cancer Ther 2002; 1:393–404.

130. Bunn PA, Chan DC, Earle K, Zhao TL, Helfrich B, Kelly K, Piazza G, Whitehead CM, Pamukcu R, Thompson W, Alila H. Preclinical and clinical studies of docetaxel and exisulind in the treatment of human lung cancer Semin Oncol 2002; 29: 87–94.

131. Altorki NK, Keresztes RS, Port JL, Libby DM, Korst RJ, Flieder DB, Ferrara CA, Yankelevitz DF, Subbaramaiah K, Pasmantier MW, Dannenberg AJ. Celecoxib, a selective cyclooxygenase-2 inhibitor, enhances the response to preoperative paclitaxel and carboplatin in early-stage non-small-cell lung cancer J Clin Oncol 2003; 21:2645–2650.

132. Hashitani S, Urade M, Nishimura N, Maeda T, Takaoka K, Noguchi K, Sakurai K. Apoptosis induction and enhancement of cytotoxicity of anticancer drugs by celecoxib, a selective cyclooxygenase-2 inhibitor, in human head and neck carcinoma cell lines Int J Oncol 2003; 23:665–672.

133. Lai GH, Zhang Z, Sirica AE. Celecoxib ats in a cyclooxygenase-2-independent manner and in synergy with emodin to suppress rat cholangiocarcinoma growth in vitro through a mechanism involving enhanced akt inactivation and increased activation of caspases-9 and -3 Mol Cancer Ther 2003; 2:265–271.

134. Saha D, Pyo H, Choy H. COX-2 inhibitor as a radiation enhancer: new strategies for the treatment of lung cancer Am J Clin Oncol 2003; 26:S70–74.

135. Yasui H, Adachi M, Imai K. Combination of tumor necrosis factor-α with sulindac augments its apoptotic potential and suppresses tumor growth of human carcinoma cells in nude mice Cancer 2003; 97:1412–1420.

136. Tortora G, Caputo R, Damiano V, Melisi D, Bianco R, Fontanini G, Veneziani BM, De Placido S, Bianco AR, Ciardiello F. Combination of a selective cyclooxygenase-2 inhibitor with epidermal growth factor receptor tyrosine kinase inhibitor ZD1839 and protein kinase A antisense causes cooperative antitumor and antiangiogenic effect Clin Cancer Res 2003; 9: 1566–1572.

137. Minter HA, Eveson JW, Huntley S, Elder DJ, Hague A. The cyclooxygenase 2-selective inhibitor NS398 inhibits prolifera-

tion of oral carcinoma cell lines by mechanisms dependent and independent of reduced prostaglandin E2 synthesis Clin Cancer Res 2003; 9:1885–1897.

138. Cheng AS, Chan HL, Leung WK, Wong N, Johnson PJ, Sung JJ. Specific COX-2 inhibitor, NS-398, suppresses cellular proliferation and induces apoptosis in human hepatocellular carcinoma cells Int J Oncol 2003; 23:113–119.

139. Sun W, Stevenson JP, Gallo JM, Redlinger M, Haller D, Algazy K, Giantonio B, Alila H, O'Dwyer PJ. Phase I and pharmacokinetic trial of the proapototic sulindac analog CP-461 in patients with advanced cancer Clin Cancer Res 2002; 8: 3100–3104.

140. Yao M, Kargman S, Lam EC, Kelly CR, Zheng Y, Luk P, Kwong E, Evans JF, Wolfe MM. Inhibition of cyclooxygenase-2 by rofecoxib attenuates the growth and metastatic potential of colorecteal carcinoma in mice Cancer Res 2003; 63:586–592.

141. Rosato FE, Dicker A, Burd R, Miller S, Lanza-Jacoby S. Combining the epidermal growth factor (EGF) tyrosine kinase inhibitor ZD1839 with a selective cyclooxygenase (COX)-2 inhibitor SC236, causes a cooperative antitumor effect in breast cancer cells derived from HER2/neu mice J Surg Res 2003; 114:274.

142. Dickens DS, Cripe TP. Effect of combined cyclooxygenase-2 and matrix metalloproteinase inhibition on human sarcoma xenografts J Pediatr Hematol Oncol 2003; 25:709–714.

143. Kundu N, Smyth MJ, Samsel L, Fulton AM. Cyclooxygenase inhibitors block cell growth, increase ceramide and inhibit cell cycle Breast Cancer Res Treat 2002; 76:57–64.

144. Mulvihill S, Warren R, Venook A, Adler A, Randlev B, Heise C, Kirn D. Safety and feasibility of injection with an E1B-55 kDa-deleted, replication-selective adenovirus (ONYX-015) into proimary carcinomas of the pancreas: a phase I trial Gene Ther 2001; 8:308–315.

145. Yamamoto M, Davydova J, Wang M, Siegal GP, Krasnykh V, Vickers SM, Curiel DT. Infectivity enhanced, cyclooxygenase-2 promoter-based conditionally replicative adenovirus for pancreatic cancer Gastroenterology 2003; 125:1203–1218.

146. Hecht JR, Bedford R, Abbruzzese JL, Lahorti S, Reid TR, Soetikno RM, Kirn DH, Freeman SM. A phase I/II trial of intratumoral endoscopic ultrasound injection of ONYX-015 with intravenous gemcitabine in unresectable pancreatic carcinoima Clin Cancer Res 2003; 9:555–561.

147. Nagi P, Vickers SM, Davydova J, Adachi Y, Takayama K, Barker S, Krasnykh V, Curiel DT, Yamamoto M. Development of a therapeutic adenoviral vector for cholangiocarcinoma combining tumor-restricted gene expression and infectivity enhancement J Gastrointest Surg 2003; 7:364–371.

148. Wistuba II, Behrens C, Albores-Saavedra J, Delgado R, Lopez F, Gazdar AF. Distinct K-ras mutation pattern characterizes signet ring cell colorectal carcinoma Clin Cancer Res 2003; 9: 3615–3619.

149. Brunner TB, Hahn SM, Gupta AK, Muschel RJ, McKenna WG, Bernhard EJ. Farnesyltransferase inhibitors: an overview of the results of preclinical and clinical investigations Cancer Res 2003; 63:5656–5668.

150. Smalley KSM, Eisen TG. Farnesyl transferase inhibitor Sch66336 is cytostatic, pro-apoptotic and enhances chemosensitivity to cisplatin in melanoma cells Int J Cancer 2003; 105: 165–175.

151. Patnaik A, Eckhardt SG, Izbicka E, Tolcher AA, Hammond LA, Takimoto CH, Schwartz G, McCreery H, Goetz A, Mori M, Terada K, Gentner L, Rybak M-E, Richards H, Zhang S, Rowinsky EK. A phase I, pharmacokinetic, and biological study of the farnesyltransferase inhibitor ipifarnib in combination with gemcitabine in patients with advanced malignancies Clin Cancer Res 20093; 9:4761–4771.

152. Rice PL, Kelloff J, Sullivan H, Driggers LJ, Beard KS, Kuwada S, Piazza G, Ahnen DJ. Sulindac metabolites induce caspase- and proteasome-dependent degradation of β-catenin protein in human colon cancer cells Mol Cancer Ther 2003; 2:885–892.

153. Ly C, Arechiga AF, Melo JV, Walsh CM, Ong ST. Bcr-Abl kinase modulates the translation regulators ribosomal protein S6 and 4E-BPI in chronic myelogenous leukemia cells via the mammalian target of rapamycin Cancer Res 2003; 63: 5716–5722.

154. von Bubnoff N, Veach DR, Todd W, Li W, Sänger J, Peschel C, Bornmann WG, Clarkson B, Duyster J. Inhibition of wild-type and mutant Bcr-Abl by pyrido-pyrimidine-type small molecule kinase inhibitors Cancer Res 2003; 63:6395–6404.

155. Pfeifer H, Wassmann B, Hofmann W-K, Komor M, Scheuring U, Brück P, Binckebanck A, Schleyer E, Gökbuget N, Wolff T, Lübbert M, Leimer L, Gschaidmeier H, Hoelzer D, Ottmann OG. Risk and prognosis of central nervous system leukemia in patients with Philadelphia chromosome-positive acute leukemias treated with imatinib mesylate Clin Cancer Res 2003; 9:4674–4681.

156. Antonescu CR, Sommer G, Sarran L, Tschernyavsky SJ, Riedel E, Woodruff JM, Robson M, Maki R, Brennan MF, Ladanyi M, DeMatteo RP, Besmer P. Association of KIT exon 9 mutations with nongastric primary site and aggressive behavior: KIT mutation analysis and clinical correlates of 120 gastrointestinal stromal tumors Clin Cancer Res 2003; 9:3329–3337.

157. Bruell D, Stöcker M, Huhn M, Redding N, Küpper M, Schumacher P, Paetz A, Bruns CJ, Haisma HJ, Fischer R, Finnern R, Barth S. The recombinant anti-EGF receptor immunotoxin 425(scFv)-ETA suppresses growth of a highly metastatic pancreatic carcinoma cell line Int J Oncol 2003; 23:1179–1186.

158. Hirsch FR, Bunn PA, Rosell R, Giaccone G, Johnson DH, Eds Invited Papers of the First IASLC/ASCO International Conference on Molecular Targeted Therapies in Lung Cancer, Marbella, Spain, January 15–19, 2003 Lung Cancer. Vol. 41S, 2003:1–186.

159. She Q-B, Solit D, Bass A, Moasser MM. Resistance to gefitinib in PTEN-null HER-overexpressing tumor cells can be overcome through restoration of PTEN function or pharmacologic modulation of constitutive phosphatidylinositol 3′-kinase/Akt pathway signaling Clin Cancer Res 2003; 9:4340–4346.

160. She Y, Lee F, Chen J, Haimovitz-Friedman A, Miller VA, Rusch VR, Kris MG, Sirotnak FM. The epidermal growth factor receptor tyrosine kinase inhibitor ZD1839 selectively potentiates radiation response of human tumors in nude mice with a marked improvement in therapeutic index Clin Cancer Res 2003; 9:3773–3778.

161. Magné N, Fischel J-L, Dubreuil A, Forment P, Ciccolini J, Forment J-L, Tiffon C, Reneé N, Marchetti S, Etienne M-C, Milano G. ZD1839 (Iressa) modifies the activity of key enzymes linked to fluoropyrimidine activity: rational basis for a new combination therapy with capecitabine Clin Cancer Res 2003; 9:4735–4742.

162. Stone RM, Wadleigh M, DeAngelo DJ. FLT-3 tyrosine kinase (TK) inhibition: an emerging therapy in acute myeloid leukemia (AML). In: Greenspan EM, Ed.. Chemotherapy Foundation Symposium XXI Innovative Cancer Therapy for Tomorrow. New York. NY, November 12–15, 2003:20–21 (abstracts to be printed in Cancer Investigation vol. 22).

163. Yao Q, Nishiuchi R, Li Q, Kumar AR, Hudson WA, Kersey JH. FLT3-expressing leukemias are selectively sensitive to inhibitors of the molecular chaperone heat shock protein 90 through destabilization of signal transduction-associated kinases Clin Cancer Res 2003; 9:4483–4493.

164. Cheong J-W, Chong SY, Kim JY, Eom JI, Jeung HK, Maeng HY, Lee ST, Min YH. Induction of apoptosis by apicidin, a histone deacetylase inhibitor, via the activation of mitochondria-dependent caspase cascades in human Bcr-Abl-positive leukemia cells Clin Cancer Res 2003; 9:5018–5027.

165. Jiang J, Matranga CB, Cai D, Latham VM, Zhang X, Lowell AM, Martelli F, Shapiro GI. Flavopiridol-induced apoptosis during S phase requires E2F-1 and inhibition of cyclin A-dependent kinase activity Cancer Res 2003; 63:7410–7422.

166. Seong Y-S, Min C, Li L, Yang JY, Kim S-Y, Cao X, Kim K, Yuspa SH, Chung H-H, Lee KS. Characterization of a novel cyclin-dependent kinase 1 inhibitor, BM-1026 Cancer Res 2003; 63:7384–7391.

167. Nakanishi T, Karp JE, Tan M, Doyle LA, Peters T, Yang W, Wei D, Ross DD. Quantitative analysis of breast cancer resistance protein and cellular resistance to flavopiridol in acute leukemia patients Clin Cancer Res 2003; 9:3320–3328.

168. Kitz J, Carlson P, Warshamana-Greene GS, Grant S, Krystal GW. Flavopridol potently induces small cell lung cancer apoptosis during S phase in a manner that involves early mitochondrial dysfunction Clin Cancer Res 2002; 9:4586–4594.

169. Kelly WK, Richon VM, O'Connor O, Curley T, MacGregor-Curtelli B, Tong W, Klang M, Schwartz L, Richardson S, Rosa E, Drobnjak M, Cordon-Cordo C, Chiao JH, Rifkind R, Marks PA, Scher H. Phase I clinical trial of histone deacetylase inhibitor: suberoylanilide hydroxamic acid administered intravenously Clin Cancer Res 2003; 9:3578–3588.

170. Kim MS, Blake M, Baek JH, Kohlhagen G, Pommier Y, Carrier F. Inhibition of histone deacetylase increases cytotoxicity to anticancer drugs targeting DNA Cancer Res 2003; 63: 7291–7300.

171. Kamath K, Jordan MA. Suppression of microtubule dynamics by epothilone B is associated with mitotic arrest Cancer Res 2003; 63:6026–6031.

172. Pietras K, Stumm M, Hubert M, Buchdunger E, Rubin K, Heldin C-H, McSheehy P, Wartmann M, Östman A. STI571 enhances the therapeutic index of epothilone B by a tumor-selective increase of drug uptake Clin Cancer Res 2003; 9: 3779–3787.

173. Facomopré M, Tardy C, Bal-Mahleu C, Colson P, Perez C, Manzanares I, Cuevas C, Bailly C. Lamellarin D: a novel potent inhibitor of topoisomerase I. Cancer Res 2003; 63: 7392–7399.

174. De Lorenzo MS, Yamaguchi K, Subharamaiah K, Dannenberg AJ. Bryostatin-1 stimulates the transcription of cyclooxygenase-2: evidence for an activator protein-1-dependent mechanism Clin Camncer Res 2003; 9:5036–5043.

175. Swamy MY, Herzog CR, Rao CV. Inhibition of COX-2 in colon cancer cell lines by celecoxib increases the nuclear localization of active p53 Cancer Res 2003; 63:5239–5242.

176. Leung WK, To KF, Go MYY, Chan K-K, Chan FKL, Ng EKW, Chung SCS, Sung JJY. Cyclooxygenase-2 upregulates vascular endothelial growth factor expression and angiogenesis in human gastric carcinoma Int J Oncol 2003; 23:1317–1322.

177. Banerjee AG, Gopalakrishnan VK, Bhattacharya I, Vishwanatha JJ. Deregulated cyclooxygenase-2 in oral premalignant tissues Mol Cancer Ther 2002; 1:1265–1271.

178. Higachi T, Iwama T, Yoshinaga K, Toyooka T, Taket MM, Sugihara K. A randomized, double-blind trial of the effects

of rofecoxib, a selective cyclooxygenase-2 inhibitor, on rectal polyps in familial adenomatous polyposis patients Clin Cancer Res 2003; 9:4756–4700.

179. Han S, Inoue H, Flowers LC, Sidell N. Control of COX-2 gene expression through peroxisome proliferator-activated receptor γ in human cervical cancer cells Clin Cancer Res 2003; 9: 4627–4635.

180. Sinkovics JG. Kaposi's sarcoma: its 'oncogenes' and growth factors Crit Rev Oncol/Hematol 1991; 11:87–1076.

181. Sinkovics JG. Contradictory concepts in the etiology and regression of Kaposi's sarcoma. The Ferenc Györkey Memorial lecture Pathol Oncol Res 1996; 2:249–267.

182. Barillari G, Sgadari C, Toschi E, Monini P, Ensoli B. HIV protease inhibitors as new treatment options for Kaposi's sarcoma Drug Resist Update 2003; 6:173–181.

183. Sgodari C, Monini P, Barilari G, Ensoli B. Use of HIV protease inhibitors to block Kaposi's sarcoma and tumour growth Lancet Oncol 2003; 4:537–547.

184. Nasti G, Martellotta F, Berretta M, Mena M, Fasan M, Di Perri M, Talamini R, Pagano G, Montroni M, Cinelli R, Vaccher E', Monforte A, Tirelli U. Impact of highly active antiretroviral therapy on the presenting features and outcome of patients with acquired immunodeficiency syndrome-related Kaposi sarcoma Cancer 2003; 98:2440–2446.

185. Sun Q, Zachariah S, Chaudhary PM. The human herpes virus 8-encoded viral FLICE-inhibitory protein induces cellular transformation via NF-kappaB activation J Biol Chem 2003; 278:52437–52445.

186. Sinkovics JG, Horvath JC. Kaposi's sarcoma: breeding ground of Herpesviridae. A tour de force over viral evolution Int J Oncol 1999; 14:615–646.

187. Stebbing J, Wildfire A, Portsmouth S, Powles T, Thirlwell C, Hewitt P, Nelson M, Patterson S, Mandalia S, Gotch F, Gazzard BG, Bower M. Paclitaxel for anthracycline-resistant AIDS-related Kaposi's sarcoma: clinical and angiogenic correlations Ann Oncol 2003; 14:1660–1666.

188. Moses AV, Jarvis MA, Raggo C, Bell YC, Ruhl R, Luukkonen BG, Griffith DJ, Wait CL, Druker BJ, Heinrich MC, Nelson

JA, Fruh K. Kaposi's sarcoma-associated herpes virus-induced upregulation of the c-kit proto-oncogene, as identified by gene expression profiling, is essential for the transformation of endothelial cells J Virol 2002; 76:8393–8399.

189. Koon HB, Bubley G, Pantanowitz L, Masiello D, Proper J, Weeden W, Tahan S, Dezube B. Imatinib mesylate in AIDS-related Kaposi's sarcoma, Proc Am Soc Clin Oncol 39th Annual Meeting, May 31– June 3, 2003, Chicago IL, 2003; 22:195 abstract #782.

190. Noy A. Update in Kaposi sarcoma Curr Opin Oncol 2003; 15: 379–381.

PART III: ANTISENSE OLIGONUCLEOTIDES AND ANTI-ANGIOGENESIS

ANTISENSE OLIGONUCLEOTIDES

Viral Blockade

Antisense oligonucleotides are synthesized single-stranded DNA sequences of 12 to 25 mononucleotides binding to mRNA via Watson-Crick base-pairing, thereby inhibiting translation in a manner similar to nucleoside analogues (fludarabine, cytarabine, fluorouracil, capecitabine) or nucleobases (thiopurines, deoxyadenosine derivatives, deoxycytidines). The short synthesized DNA sequences hybridize with specific mRNA strands.

First, replication of Rous sarcoma virus was blocked in chicken fibroblasts, followed by the inhibition of many—practically all—viruses, among them HTL-I and HIV-1, against which antisense oligonucleotides were tested. Cytomegalovirus retinitis of patients with AIDS is now treatable with the licensed phosphorothioate oligonucleotide fomivirsen (ISIS-2922; Vitravine) (ISIS Pharmaceuticals Carlsbad, CA; Novartis' CIBA Vision) [1–8]. The oligonucleotide enters the cell through energy-consuming endocytosis and escapes within the cytoplasm from endosomal or lysosomal vesicles. Cleavage of target mRNA from the DNA-RNA heteroduplex in the ribosomes occurs through the action of the endonuclease endogenous RNAase. The antisense oligonucleotide should hybridize with mRNA sequences that do not have homology with sequences of healthy genes at the single-stranded AUG site of the mRNA. In phosphorothioates, a sulphur atom in the PO moiety replaces the oxygen atom; in methylphosphonates, CH_3 group; in phosphoramitades, amines occupy this position [7–11].

From Anti- to Pro-Apoptosis

The Human Genome Project yielded some 100,000 mRNA species and from this large library target sequences can now be selected. The *bcl-2* and *bcl*$_{XL}$ gene→mRNA sequences are primary targets. The phosphorothioate Bcl-2 antisense oligonucleotide G3139 (Oblimersen, Genasense, Genta at *www.genta-.com*) is being tested against melanoma, NSCLC, and leukemias. In 21 patients with non-Hodgkin's lymphoma, one CR, two PR, and nine stabilizations of disease occurred; in nine patients the disease progressed. When given as continuous infusion combined with DTIC (dacarbazine) to patients with melanoma, complete (or almost complete) regression of disease was observed. In combination with rituximab (Rituxan, Idec), it exerted enhanced cytotoxicity toward EBV-induced human lymphoblasts in SCID mice. Bcl-2 is translocated or amplified in B lymphoblasts driven by EBV genomic sequences, especially in lymphoma cells of an immunosuppressed organ transplant recipient, or in patients with AIDS. G3139 exerts antiproliferative and pro-apoptotic effects in xenografted lymphoma cells as it downregulates Bcl-2 expression and cooperates with rituximab. G3139 targets the first six codons of the Bcl-2 mRNA and, while suppressing the lymphoma, may cause thrombocytopenia. G3139 could induce apoptotic death of xenografted imatinib-resistant Bcr-Abl$^+$ CML blasts. G3139 stimulated IL-12 production in the host harboring human melanoma xenografts, which regressed. A programmable fusogenic vesicle-encapsulated G3139 reduced by 80% Bcl-2 expression in human melanoma cells. In PC3 prostatic carcinoma cells, G3139 reduced Bcl-2 expression but failed to induce apoptotic death. A Bcl-2 antisense oligonucleotide targeting hTERT "devitalized" human bladder carcinoma cells in vitro [12–19].

Kidney carcinoma cells, like many other tumor cells, kill defensive host T cells by secreting TNFα-related ligands (FasL; TRAIL) [20]; among lymphocyte-killer tumors (melanoma or sarcoma cells) [21], ovarian carcinoma cells exude FasL-containing vesicles, which kill CD3-χ host T cells [22]. A *bcl-2* transgene protects lymphocytes from apoptotic death induced by kidney carcinoma cells [20]. A Fas-receptor agonist antibody (CH-11) in subtoxic doses increases the sensitivity of Fas receptor-expressor sarcoma cells (but not that of normal fibroblasts) to chemotherapy (doxorubicin; paclitaxel) [23]. Although FasL-expressing tumors may escape death by immune T cells of the host, they seldom metastasize and remain immunogenic [24]. Natural killer cells expressing Fas or FasL and the inhibitory killer immunoglobulin-type receptor (KIR2DL/CD158a) protect kidney carcinoma cells from $CD8^+$ immune T cells of the host by inducing sustained c-FLIP (the cellular Fas-associated death domainlike IL-1β converting enzymelike inhibitory protein) and by downregulating caspase 8. Antibody blocking the NK cell KIR/HCL-Cwd receptor restored the tumor cell-directed cytotoxicity of immune T cells [25]. Cisplatin downregulates c-Flip, restoring tumor cell susceptibility to TRAIL [26].

ISIS3521 (Affinitac, Eli Lilly, *www.isispharma.com*) is an anti-PKCα (see Part II) agent tested against solid tumors. This tetrameric enzyme consists of catalytic (C_2) and regulatory (R_2) subunits, which are overexpressed in breast cancer. ISIS3521 suppressed the growth of glioblastoma xenografts (see Part V). It induced CR in patients with non-Hodgkin's lymphoma. When given in combination with paclitaxel and carboplatin for the treatment of metastatic NSCLC, survival of 8 months with chemotherapy alone was prolonged to 18 months in the combination-treated patients. Genasense and mitoxantrone induced PR in patients with hormone-refractory prostate cancers [27].

Soft-tissue sarcoma cells re-express survivin, the IAP family protein that becomes silent in adult life after extensive activity in the fetus. This is an adverse prognostic factor [28]. An open reading frame (ORF K7) of HHV-8, the Kaposi's sarcoma-associated herpesvirus, encodes a survivinlike protein

[29]. The replication-defective adenovirus pAd-T34A inhibits survivin expression in a variety of human cancer cells and, upon intratumoral injection, matches the efficacy of doxorubicin or paclitaxel in inducing cell death [30,31]. Telomerase, its reverse transcriptase and protein (hTR; hTERT) (see Part IV) are co-expressed with survivin in sarcoma cells [32]. Phosphorothioate antisense oligodeoxynuleotides (As-ODN) (Invitrogen, Karlsruhe, Germany) inhibit hTERT expression in bladder carcinoma [33]. Perhaps Ad-T34A and As-ODN in combination—with or without doxorubicin or ifosfamide—will cure metastatic soft-tissue sarcoma?

Mesothelioma cells are attacked by LAK (lymphokine-activated NK) cells; killed by ganciclovir after insertion of HSV *tk* gene; and respond to IFN$\alpha\gamma$ or GM-CSF, but remain chemotherapy- and radiotherapy-resistant and incurable [34]. Even metastatic sarcomas sometimes are curable, however [35,36]. Ad-IFNγ could induce remissions in mesothelioma [37]. In some 87% of mesotheliomas, the anti-apoptotic survivin is overexpressed. Anti-survivin oligonucleotides restored apoptosis-susceptibility in mesothelioma cells by downregulating survivin protein expression [30,31,38]. The growth of EGF-R^+ mesothelioma cells could also be suppressed by gefitinib (see Part I) [39,40]. SV40 genome-carrier or SV40^- mesotheliomas arise in patients exposed to asbestos. SV40 induces hepatocyte growth factor/scatter factor (HGF/SF) release from mesothelioma cells because of its interaction with proto-oncogenes *met* and *ras*. *Met* is the receptor of this ligand; it was first identified in a human osteosarcoma. The ligand is commonly secreted by mesenchymal cells, whereas the receptor is commonly expressed by epithelial cells. For example, leukocytes infiltrating bronchioalveolar carcinomas secrete HGF/SF, which is captured by *Met* receptors expressed by the tumor cells (see Part IV)—a paracrine growth loop. Tyrosine kinase signal transduction from *Met* can be inhibited by small molecules (PHA-665752; SU11271, SU11606, Sugen, Pharmacia; Pfizer). A replication-defective, *c-met* ribozyme-expressing adenovirus (Ad-v-Met) suppresses the growth of prostate cancer xenografts [41–44]. A c-*met* ribozyme delivered by an adenovirus inhibited the growth and metastatic potential of prostatic carcinoma cells [45]. This vector could suppress mesothelioma cells, if these tumor cells used the HGF/SF⇒*Met* pathway as an autocrine growth loop. Mesothelioma cells (see Part IV) thrive on the homozygous deletion of the cell-cycle regulatory molecules INK40/ARF: $p14^{ARF}$ is lost and the anti-p53 MDM2 (HMD2 in the human genome: 12q14.3-15) enzyme is overactive and eliminates the p53 protein (an anti-apoptotic event). An antisense oligonucleotide directed at MDM2 oncogene of prostate cancer cells reduced tumor cell proliferation, increased apoptotic death rate and potentiated the effects of irinotecan and paclitaxel [46–49]; it should be tested against mesothelioma cells. Bcl-2 and survivin overexpression protect mesothelioma cells from apoptotic death. Anti–Bcl-2/Bcl_{XL} and antisurvivin phosphorothioate oligonucleotides targeting the 232–251 nucleotides of the survivin mRNA suppress survivin expression in mesothelioma cells and restore the cell's susceptibility to apoptosis [30,31,50]. Adenoviruses pAdARF and Adp16INK4A reinsert the homozygously eliminated gene, and adenovirus Ad-mda7 transfects the *mda* tumor suppressor gene into the genome of mesothelioma cells [51,52]. Imatinib suppresses in mesothelioma cells, the autocrine growth loop of PDGF, and its receptor; pAd-T34A releases cytochrome C from the mitochondria of mesothelioma cells, a pro-apoptotic event. A simian sarcoma retrovirus carries the transduced PDGF gene: *c-sis*⇒*v-sis*. Perhaps adenoviruses (AD16INK4A or Ad-mda7 with anti-survivin As-ODN) will cure mesothelioma (Table 4)?

In Ewing's sarcoma cells, IGF-I receptors are overexpressed [53]. These receptors could be downregulated by antibody or by an antisense mRNA expressor plasmid. Ewing's sarcoma cells so treated showed increased sensitivity to doxorubicin and failed to form tumors in nude mice. Another intervention is the use of IGF-receptor inhibitor AG1024, which accelerates the degradation of phosphorylated Rb protein and restores the cell-cycle–arresting function of Rb. A plasmid-expressing IGF-I–receptor antisense RNA inhibited the growth of xenografted human melanoma cells [54–57]. Adenoviral constructs Ad-IGF-Ir482st and, to a lesser degree, Ad-IGF-Ir950st suppressed the growth of pancreatic carcinoma xeno-

Table 4 Mesothelioma

Standard therapy not curative (decortication; XRt; chemotherapy: doxorubicin, platinols; taxanes, pemetrexed: PMTX).
PMTX (LY231514, Alimta) folate metabolism enzyme inhibitor additive to cisplatin in remission induction[1]
Pathogenesis: SV40 genome (when present[2]): p53 p14 Wnt
↑EGF-R Rx: ZD1839 OSI-774 ↑PDGF-R Rx: STI-571 (Tables 2, 5)
↑Survivin Rx: Ad.survivin.As-ODN ↑Bcl-2/Bcl$_{XL}$ Rx[3]: As-ODN *bcl-2*
↑*mdm2* MDM2 ↓p53 Rx: Adenovirus.anti-*mdm*.As-ODN
↓*mda*-7 suppressor gene Rx: Ad-*mda*7 ↑BAX ↑caspase3→apoptosis[3]
del INK40/ARF ↓p14^{ARF} Rx: Adp14ARF; Adp16INk4A[4]
↑c-Met ←HGF/SF Rx: PHA-665752 SU11271 SU11606 (Table 2)
Malfunctioning Wnt: ↑glycogen synthetase kinase 3 β ↑βcatenin ↑c-*myc* ↑*cyclinD* Rx: HSV*tk* ←ganciclovir
↑K(keratin)19 gene Rx: liposome pK19-*tk* ←ganciclovir[5]
pAd-734A to release mitochondrial cytochrome C→ apoptosis
Ad.mIFNγ: ↑LAK cells ↑immune T cells[6] Ad.mIFNβ: ↑NK ↑Mϕ ↑PMN[7]
Pro-apoptotic adenoviral gene Rx[3]: AdBax AdBak Adp53
Persistence of NDV in stabilized intraperitoneal tumor[8]

As-ODN = antisense oligonucleotide Wnt = *wingless/int* fusion gene
MDM = human equivalent of mouse double minute chromosome
ARF = alternative reading frame INK = inhibitor of cyclin DK
EGF-R = epidermal growth factor receptor PDGF-R = platelet-derived growth factor receptor
HGF/SF = hepatocyte GF/scatter factor HSV = Herpes simplex virus
tk = thymidine kinase gene IFN = interferon NK = natural killer MΦ = macrophage
LAK = lymphokine-activated killer PMN = polymorphonuclear (leukocytes)
BAK, BAX = pro-apoptotic members of the Bcl-2 protein family.

1 Thoraxklinik, Heidelberg, Germany; University Western Australia, Australia. Medical University S. Carolina, Charleston, SC. 2 Loyola University Cardinal Bernardin CC, Maywood, IL. 3 Section for Thoracic Molecular Oncology, University Texas M. D. Anderson CC, Houston TX. 4 Chang Gung Memorial Hospital, Taipei, Taiwan. 5 Tokyo Medical University, Bunkyo-ku, Tokyo, Japan. 6 Faculté de Medecine Creteil Cedex, Paris, France; Institut Gustave Roussy, Villejuif, France. 7 Thoracic Oncology, University of Pennsylvania, Philadelphia, PA. 8 Hackensack University Medical Center, Hackensack, NJ.

Table 4 Attack on mesothelioma is dominated by virotherapeutical agents

grafts and enhanced chemo- and radiotherapy sensitivity of these tumors [58].

Tumor growth factor-α drives rhabdomyosarcoma cell growth through ligation to EGF-R, but this signal transduction could be inhibited by an anti–EGF-R antisense oligonucleotide [59]. Tumor growth factor-α, the ligand of EGF-R, induces dimerization of the intracellular domain of the receptor and autophosphorylation of its tyrosine residues. This process is constitutively active in squamous carcinoma cells of the head and neck. An antisense oligonucleotide of the EGF-R mRNA and docetaxel acted additively in reducing autophosphorylation, signal transduction, AKT expression and VEGF secretion, resulting in cessation of growth of tumor xenografts. This combination acted more effectively than TKI or anti–EGF-R monoclonal antibodies [60].

Colony stimulating factor-1 (CSF-1) is produced by tumor-infiltrating macrophages and exerts mitogenic effect in tumor cells overexpressing its receptor, which is encoded by the proto-oncogene c-*fms* (found first in the McDonough feline sarcoma retrovirus, v-*fms*). Antisense oligonucleotides directed at mRNA copies of the CSF-1 gene inhibited CSF-1 production of stromal cells and of MMP in colon carcinoma cells [61]. The FTL3-L (Amgen, Thousand Oaks, CA) (see Parts I and II) stimulates hematopoietic cells in the peritoneal cavity to produce IL-12 (and IL-10) in patients with gynecologic tumors; it stabilized the tumor in the case of an intraperitoneal mesothelioma [62]. Newcastle disease virus, on occasion, can stabilize intraperitoneal mesothelioma [63]. Could these two modalities of treatment be combined?

Raf-1 encoded by *c-raf* (3p25) downstream in the Ras→-MAPK cascade is inhibited by ISIS5132 antisense oligonucleotid, which prolonged the life of patients with chemotherapy-refractory metastatic tumors. ISIS2503 is an antisense inhibitor of H-*ras* mRNA; when given in combination with gemcitabine it prolonged the lives of patients with various advanced tumors. Liposomally encapsulated As-ODN targeting RAF-1 signaling increases radiosensitivity of the tumor. BRAF mutations (3p25) are extremely common (80%) in metastatic melanomas and result in the activation of the Ras-Raf-MAPK pathway (whereas the inherited form of melanoma is consequential to germline mutation of CDKN2A gene at the locus 9p21); BRAF is not mutated in uveal melanoma [64–69]. The MEK/ERK1, 2 (mitogen-activated protein kinase; extracellular signal-regulated kinase) inhibitor U0126 suppresses the growth of BRAF-mutated melanoma cells [65,68] and human gallbladder cancer xenografts [70]. BRAF kinase inhibitors (PD184352, CI-1040, staurosporine derivatives UCN-01) are to be tested in early clinical trials in competition or in combination with As-ODN.

The phosphorothioate liposomally encapsulated antisense *c-myc* (Burnaby, British Columbia, Canada) targets *c-myc* mRNA (INX-6295) (as tested at Regina Elena Cancer Institute, Rome, Italy, in 2001). In human melanoma xenografts, the *c-myc* As-ODN downregulated Bcl-2 and *c-myc* and upregulated p53, leading to extensive apoptotic deaths of the tumor cells [71]. Anti–N-*myc* strategy is used against neuroblastoma cells [72]. When *c-myc* is knocked out of ovarian carcinoma cells, the tumor cells lose resistance to, and undergo growth arrest by, TGFβ [73]. An As-ODN, ISIS2503, targets H-*ras* and acts additively with gemcitabine in patients with metastatic cancers [74]. An anti–*c-myb* As-ODN is in clinical trial for the treatment of neuroblastoma [72]. The proto-oncogenes *c-myb, c-myc,* and *c-raf* were transduced by ancient retroviruses (avian myeloblastosis and myelocytosis and the murine sarcoma MSV3611 retroviruses) [75].

The COX-2 inhibitor SC-236 (Pharmacia-Upjohn) combined with ZD1839 (Iressa, AstraZeneca) and with anti-PKA AS-PKAI hybrid antisense oligonucleotide induced 60% durable CR in nude mice xenografted with human colon carcinoma cells [76]. Cyclooxygenase-17 is the assembly protein of cytochrome C oxydase. It is overexpressed in NSCLC. A synthetic short, double-stranded RNA (siRNA) AS-ODN vector targeting the COX-17 gene and transfecting NSCLC cells (Lipofectam, Invitrogen, Carlsbad, CA) suppressed gene expression [77]. Short, double-stranded RNAs trigger RNA interference, reduce VEGF synthesis in fibrosarcoma cells, and synergize with the tumor-suppressor thrombospondin-1of tumor cell origin [78].

The FLICE (Fas-associated death-domainlike IL-1β converting enzyme) inhibitory protein (FLIP) activates NFκB and exerts a strong anti-apoptotic effect in tumor cells; so does survivin, and when mRNAs of both are targeted by oligonucleotides, cells' susceptibility to apoptosis is restored. The fusion oncoprotein $p210^{Bcr\text{-}Abl}$ (see Part I) can be attacked by monoclonal antibodies, by oligonucleotides directed at the mRNA emanating from the genes of the translocated (t9;22) fusion oncoprotein, and by the inhibitor of the phosphorylated tyrosine kinase, imatinib mesylate (Gleevec, Novartis).

Adenoviral tyrosine kinase has the ligand FGF linked to its knob region; pancreatic carcinoma cells overexpressing FGF-R readily admit this adenoviral vector, which renders them susceptible to ganciclovir-induced death [79]. The antisense oligonucleotide G4ASODN eliminates the expression of the X-linked XIAP in human lung cancer cells, which either undergo spontaneous apoptosis or become increasingly susceptible to chemotherapeuticals (paclitacel, doxorubicin, etoposide, vinorelbine) [80].

Colon cancer xenografts produce CSF-1 to invite macrophages; MMPs produced by macrophages facilitate the growth of the xenografts. Colony stimulating factor-1 antisense oligonucleotides reduced CSF-1 production in the xenografts, depriving them of macrophages and neoangiogenesis: The xenografts remained dormant or regressed [61]. Expression of H-*ras* is inhibited by the antisense oligonucleotide ISIS2503 [74].

The phosphorothioate oliogonucleotides synthesized against VEGF receptors KDR/Flk-1-ASO inhibited the intraperitoneal dissemination of human gastric carcinoma cells in nude mice [81]. The EGF-R antisense plasmid administered in liposomes to mice harboring tumors of a human head and neck squamous carcinoma cell line inhibited tumor growth; endostatin given subcutaneously suppressed neoangiogenesis and synergized with the antisense oligonucleotide in inducing CR [60,82].

The AKR mouse thymoma-inducing retroviral v-*akt* derives from c-*akt*. The gene product protein AKT2 is overexpressed in human gastric, pancreatic, and ovarian carcinoma cells, where it interacts with PI3K, PDGF, bFGF, and IGF

[83,84]. Antisense RNA oligonucleotides abolished AKT2 encoding in pancreatic carcinoma cell lines; cancer cells so treated lost their ability to grow in xenografts [85]. Tumor cells with activated insulin receptor secrete annexin II [86]. The expression of IGF-IR in human melanoma cells was significantly reduced by an antisense RNA or oligodeoxynucleotide directed at the gene of IGF-IR. Melanoma cells so treated could not form xenografted tumors in nude mice [56]. The anti-apoptotic survivin inhibits caspase 9 and is overexpressed in many human cancers, but the tumor-bearing host mobilizes CD8 immune T cells against peptide epitopes of the survivin molecule (see Part II). Antisense cDNA, or adenovirus pAdT34A, downregulate survivin expression in tumor cells [30,31,38]. Oligonucleotide GEM231, targeting PKA (see Part I), is in Phase I and II clinical trials for patients with refractory solid tumors [87]. Antisense strategies can obliterate the PKC pathway (see Part II) [88] or attack telomerase⇒telomere preservation in tumor cells [89]. Tyrphostin (AG1024) promotes the elimination of phosphorylated Rb proteins in melanoma cells and restores the control of the cell cycle [90]. Although histone deacetylase inhibitors activate $p21^{Waf1}$ [91], an As-ODN directed at $p21^{Waf1/Cip1}$ (inhibitors of cyclin-dependent kinases) forces breast cancer cells to undergo the programmed death process [92]. Gene expression of cells infected by HIV-1 was inhibited by an AAV vector encoding antisense RNA directed to the trans-activation response (TAR) sequence and to polyadenylation signals of the virus. A trinucleotide RNA bulge of the viral regulatory sequence TAR positioned within the viral long terminal repeat (LTR) binds the viral regulatory protein Tat, which induces a conformational rearrangement of the TAR RNA and stimulates a 100-fold increase in viral RNA replication. In infected cells treated with the AAV vector, no viral replication occurred and the cells suffered no cytotoxic side effects [93,94].

c-*myc*

The "oncogene from hell" [95], c-*myc* (8q24), was found first in the genome of the acutely transforming avian myelocytomatosis retrovirus MC29 (v-*myc*), in which it hybridizes with the viral *gag* gene. It encodes a p62 DNA-binding protein. In

Burkitt's lymphoma, c-*myc* translocates to chromosome 14q32, next to the heavy-chain immunoglobulin locus, but it may translocate to 2p12 or to 22q11, next to the κ or λ light-chain loci, where it acts liberated from its natural repressors operating at locus 8q24. Not translocated, but amplified, c-*myc* is overactive in many human tumors, prominently in SCLC, breast carcinomas, and some sarcomas. Janus-faced c-*myc* is pro-apoptotic in growth factor-deprived cells, but anti-apoptotic in the presence of abundant growth factors. ConA, EGF, IL-2, and PDGF activate c-*myc*. In embryonal life and ontogenesis, c-*myc* is expressed in cytotrophoblasts of the placenta and in endo- and mesodermal cells, but becomes silent in differentiated cells. Cycline-dependent kinases and their inhibitors (CDKN2), which are encoded by the frequently mutated "multiple tumor suppressor gene" at 9p13-22, interact with the c-Myc gene product protein, allowing uninhibited progression of the cell cycle. Interferon-α downregulates c-*myc* in promyelocytic leukemia cells and malignant melanoma cells, in which chromosome 9p13-22 is frequently mutated. c-Myc suppresses the CDK 4 and 6 inhibitor $p15^{INK4b}$. Tumor growth factor-β antagonizes c-Myc and liberates $p15^{INK4b}$, but overexpressed c-Myc subverts TGFβ and $p15^{INK4b}$ and the cell cycle rolls forward uninhibited [96–99]. In ovarian carcinoma cells, c-Myc repression is lost and the tumor cells become insensitive to TGFβ [71–73]. The c-Myc protein induces reactive oxygen species (ROS), leading to DNA damage and genomic instability, but it counteracts p53-induced growth arrest, so that ineluctable advancement of these cells through S phase ensues. In p53-deficient Saos-2 osteosarcoma cells, c-Myc inhibits NFκB, which then fails to activate the antioxidant MnSOD (or N-acetyl-cysteine) [96–101].

Inhibitors of c-*myc* can be as simple as doxycycline inducing irreversible differentiation of c-*myc*, overexpressing murine osteosarcoma cells into osteocytes. Temporary inactivation of c-*myc* indicates that oncogene-induced malignant transformation may be reversible and, when the oncogene is reactivated, it may exert pro-apoptotic effects [100,101]. The chemokine CCL6 is regulated by the Myc proteins. Overexpression of CCL6 liberates malignant cells of their growth-factor dependence and promotes their metastatic potential [102]. When TGFβ exerts its antiproliferative effect, it down-regulates c-Myc. The Ser/Thr kinase TOR proteins are positioned downstream of PI3K and are targets of rapamycin, an inhibitor of IL-2. Insulin and IGF autophosphorylate IR and IGF-R, which activate the Ras/MAPK/PI3K cascade. The N-Myc and IGF systems are cross-linked: IGF activates N-*myc*. The PI3K inhibitor wortmannin and the TOR (target of rapamycin) inhibitor rapamycin block this reaction; IGF also counteracts the PI3K/AKT inhibitor LY294002. N-Myc is active in neuroblastomas and rapamycin inhibits the proliferation of neuroblastoma cells. Rapamycin-induced apoptosis of rhabdomyosarcoma cells is prevented by IGF. Rapamycin analogues CC1-779 and RAD001 target mTOR and induce apoptotic death of tumor cells (see Part II). In malignant cells (breast, ovarian, pancreatic, and squamous cell carcinomas), TIE (transforming growth factor-β inhibitory element) in the c-*myc* promoter protects c-Myc. Antiestrogens (tamoxifen; ICI182780) arrest c-*myc* overexpressing breast carcinoma cells by downregulating cyclin D1, CDK4 and 6, while free CDK inhibitors accumulate [103–106]. Antisense oligonucleotides targeting c-*myc* mRNA inhibit proliferation of cancer cells, including either estrogen receptor positive (ER^+) or ER^- breast carcinoma cells [107]. Breast cancer, SCLC, hepatoma, Burkitt's and non-Hodgkin's lymphomas, and melanoma are awaiting clinical trials with As-ODN [108,109].

ANTI-NEOANGIOGENESIS

Vasculogenesis; Angiogenesis

The remarkable history indicating that tumor growth depends on its vascular supply dates back to the 1920s and 1930s [110]. Tumors unable to recruit and sustain a supportive vascular network of their own remain dormant. Vasogenic angiogenesis (vasculogenesis) occurs in embryonic life *de novo* from mesodermal stem cells. Angioblasts of endothelial lineage proliferate, form lumina, and organize themselves into a primitive vascular network under the guidance of VEGF) by forming autocrine growth loops, as they express receptors for these mediators (VEGF-R2 or KDR = kinase insert domain-containing receptor in humans). The eight exons of VEGF-A (chromosome 6), by alternative splicing, produce the VEGF family. Vascular endothelial growth factors mediate vascular sprouting; endothelial cell proliferation, migration, and intussusception; and increase vascular permeability. Their expression is induced by hypoxia, IL-1β, TGFβ, epidermal growth factor (EGF) and signals emanating from EGF-R. Vascular endothelial growth factor receptors (KDR), in capturing their ligands, induce endothelial cell proliferation [111–113]. Hypoxia induces high levels of carbonic anhydrase 9 (CA9) and promotes VEGF gene expression [114,115]. Hypoxia-inducible factor-1 (HIF-1) protects tumor cells against apoptosis and induces VEGF production in the tumor bed. Tumor cells overexpress thioreduxin (Trx-1), which upregulates HIF-1 production [116]. Kidney carcinoma cells with deleted VHL gene and its gene product protein (pVHL) constitutively express HIF-1; reinsertion of the VHL gene induces tumor cell death [117]. In these and other (breast and colon) carcinoma cells, HIF-1 overproduction is inhibited by PX-12 (methylpropyl-2-imidazolyl disulfide) and the protein pleurotin [118]. The flavoprotein enzyme Trx-1 is induced by TNFα; it activates NFκB [116], an anti-apoptotic event. Organotellurium steroids inhibit thioreductases [119]. Neurolipin on glia cells also binds $VEGF_{165}$ or PlGF-2. VEGF-C and D (but not A) induce lymphangiogenesis. Tumor growth factor-β1 is stimulatory to VEGF gene transcription in cholangiocarcinoma; the tumor produces VEGF and endothelial cells of the tumor bed express VEGF-R (a paracrine growth loop), whereas the tumor cells use the TGFβ1→R pathway as an autocrine loop [120]. In breast carcinoma cells, estrogens upregulate VEGF production and VEGF-R expression, creating autocrine growth circuits [121]. Activated STAT3 is a major contributor to neoangiogenesis (Figure 3). Platelets store VEGF produced elsewhere and release it where the platelets are consumed [122]. When platelets are transfused in myelo-suppressed patients, is tumor vasculature inadvertently stimulated by exogenous VEGF?

Angiogenesis is initiated from a pre-existing vascular plexus. In the process of neoangiogenesis, extracellular matrix is degraded, and endothelial cells proliferate, invade, and dif-

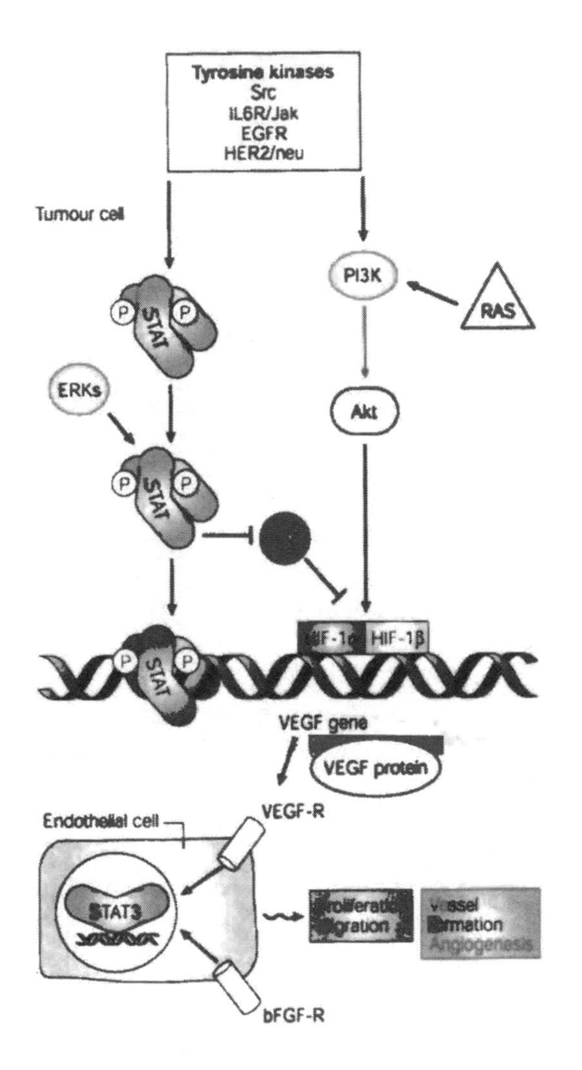

Figure 3 Figures 1 and 2 (see Part I) depicted the STAT system as a major contributor to the inhibition of proinflammatory cytokines and chemokines and to dendritic cell maturation/differentiation. STATs upregulate c-*myc*, the cyclins, and the anti-apoptotic survivin.This figure shows the delicate balance of VEGF and bFGF inducers and inhibitors in the cell and gives a rationale for blocking STAT3 and thus reducing neoangiogenesis a tumor cell would otherwise induce. (The permission of Drs. Hua Yu and Richard Jove to reprint these figures from the February 2004 issue of *Nature Reviews Cancer*, with the approval of the publisher, is hereby gratefully acknowledged.)

ferentiate after their branches form lumina. Nonsprouting angiogenesis results from intussusception of pre-existing blood vessels. Functional tyrosine kinase receptor TIE-2 (*tunica interna* endothelial cell kinase) is overexpressed in tumor cells [123–127]. *Tunica interna* endothelial cell kinase-2 and its agonist angiopoietin-1 (Ang-1) are essential for sprouting neoangiogenesis and for interaction among endothelial cells, smooth muscle cells, and pericytes. Their interaction can be antagonized by the compound muTekdelta FC (μTekδFC) [126]. Soluble sVEGF-A, sVEGF-R, and sTIE-2 are at high levels in cancer patients [122], but these substances are detectable in the blood supply of blood banks [128]. In malignant gliomas, Ang-1, Ang-2, VEGF, and TIE-2 are co-expressed [126]. The TIE-2 antagonist Ang-2 mediates regression of blood vessels. Could its gene be delivered into tumor cells by a viral vector? EphrinB2 in arteries and ephrinB4 in veins promote interaction of endothelial cells with smooth muscle cells [129]. Small primary tumors, including leukemic cells in the bone marrow themselves, release or induce the release from stromal cells in their microenvironment, of angiogenic growth factors in addition to VEGF. These include basic fibroblast growth factor (bFGF), placental growth factor (PlGF), and TGFα. Some tumor cells expropriate the entire VEGF→R system and use it as their own autocrine growth loop, with or without an additional autocrine growth circuit using TGFβ and its receptor [130,131].

In cases in which TGFβ acts as a growth factor, its blockade is inhibitory to tumor growth [132]. When TGFβ exerted inhibitory effects on tumor growth, disruption of signaling from its receptor, as in hepatocellular carcinoma, could lead to enhanced malignant behavior [133]. Some other tumors (uveal melanoma cells) may de-differentiate and redifferentiate toward endothelial cell morphology, phenotype, and function, thereby contributing to blood flow within the tumor bed [134,135]. Tumor-induced neoangiogenesis is usually disorganized and inchoate, and as such, is recognizable as the "tumor blush" in angiograms.

Vascular permeability factor (now known as VEGF-A) and VEGFs belong to a large family of cytokines stimulatory to neoangiogenesis ("serpentine" or "glomeruloid" from preexisting blood vessels, both for blood and lymphatic vessels [136,137]. In endometrial carcinoma cells, both tumor and stromal cells express VEGF-D and its receptor VEGF-R-3; the receptor is active in the adjacent stromal cells as well. Myometrial invasiveness of the tumor correlates directly with the expression of VEGF-D and VEGF-R-3 [138]. The understanding of *Wnt* signaling (ligands to "fizzled" receptors; Fz) in physiological (placenta; wound healing) and pathological (neoplasia; see Parts IV and V) angiogenesis and its relationship to COX-2, MMP-7, and VEGF is taking shape as a new territory to explore [139–142]. The catalytic subunit of PI3K acts as an oncogene in ovarian carcinoma cells, promoting proliferation and inhibiting apoptosis while upregulating VEGF production. The PI3K inhibitor LY294002 suppresses all these effects and restores the susceptibility to apoptosis of the treated tumor cells [143].

Established large primary tumors switch to the production of anti-angiogenic factors—angiostatin, a proteolytic fragment of plasminogen; and endostatin, the COOH-terminal of collagen XVIII—to suppress the growth of their own micrometastases. Rapid growth of these metastases sometimes occurs after the surgical removal of the large primary tumor, inasmuch as the small metastatic tumors are in that stage of their growth in which the production of neoangiogenesis factors, not that of anti-angiogenesis factors, dominates. For example, established endometrial carcinoma switches off VEGF production and switches on angiostatin production [144–146]. In these tumors, there is an operational VEGF-D⇒VEGF-R3 gowth loop. The plasminogen derivative angiostatin (AS), which is now cloned and is expressed in *Pichia pastoris*, inhibits endothelial cell migration, proliferation, and tube formation; it induces both intrinsic (mitochondrial) and extrinsic apoptosis of vascular endothelial cells [147–149].

Inhibitors of Neoangiogenesis

The most important natural angiogenic mediators are VEGF, bFGF, PlGF-2, PDGF, TGFα, IL-3, IL-8, G-CSF, and, in AIDS–Kaposi's sarcoma, HIV-1 *tat* gene product protein Tat. Signaling from VEGF-R runs the PI3K pathway and provides protection against apoptosis to replicating vascular endothelial cells by Bcl-2 activation [150,151]. Blockade of VEGF-R in endothelial cells may disrupt similar signal conduction in coexisting tumor cells.

The most important natural angiogenesis inhibitors are IFNα, IFNβ, angiostatin, endostatin, vasostatin, platelet factor-4, thrombospondin-1, IL-12, IL-18, and prothrombin-1 and -2. Placental growth factor acts as an inhibitor of angiogenesis when it forms inactive PlGF/VEGF heterodimers [152,153]. Results of the Entremed-sponsored endostatin–angiostatin clinical trials remain below the high expectations whipped up by irresponsible and premature daily press releases.

In the early 1980s, when IFNα (but not IFNγ) [154,155] suppressed the growth of AIDS-associated Kaposi's sarcoma, but failed to reduce HIV-1 viral load or to reconstitute CD4 and CD8 cells, this author suggested that IFNα acted as an anti-angiogenic agent by suppressing the *int* gene family encoding bFGF [156,157]. Unfortunately, an assistant editor rejected his first manuscript. Since then the anti-angiogenic effects of IFNα are widely accepted and publicized [158]. Interferon-α is now licensed for the treatment of hairy cell leukemia, CML, certain lymphomas and multiple myeloma, Kaposi's sarcoma, melanoma, and kidney carcinoma and it is used for its anti-neoangiogenic effects in glioblastoma multiforme (see Part V).

For patients with SCLC, IFNα and IFNγ failed to prolong survival, except for one cohort of patients in which 22% of patients treated with chemotherapy and IFNα survived 2 years, whereas 15% of patients treated with chemotherapy only survived 2 years (ASCO #1832 1999). Interferon-γ retarded the growth of mesothelioma [159] and ovarian carcinoma [160]. In Th1-type tumor rejection, IFNγ cooperates with IL-12, DCs, immune T cells, and NK cells. As to IL-12 (Wyeth-Ayerst), one opinion holds that its mechanism of action consists of upregulation of INFγ; another view considers it inhibitory to IFNγ. If IL-12 acts against Kaposi's sarcoma, inhibition of IFNγ is its mechanism of action, given that IFNγ is an auto- and paracrine growth factor of Kaposi's sarcoma cells; IFNγ promotes the growth of endothelial cells in response to the HIV-1 Tat protein [161,162]. Fibroblast-induced angiogenesis could be inhibited by the tissue-inhibitory protein TIMP-3 [163]. Because Kaposi's sarcoma cells receive paracrine growth factors, would TIMP-3 suppress the growth of this tumor?

Newer trials for SCLC include STI-571 (imatinib), ZD1839 (gefitinib) (see Part I); monoclonal antibodies to VEGF, VEGF-R, and gastrin-bombesin (G17DT); VEGF-R and EGF-R inhibitors (ZD6474); the G3139 antisense oligodeoxynucleotide against Bcl-2; the proteasome inhibitor PS-341; and the mTOR inhibitor rapamycin-derivative CI-779 (see Part II). In the majority of NSCLC, COX-2 expression correlated with neovasculogenesis and poor prognosis. Cyclooxygenase-2 inhibitors (see Part II) suppressed tumor cell (sarcoma) growth, not through direct killing of tumor cells, but consequentially to antiproliferative effects on endothelial cells in the tumor bed [165–170].

VEGF-A also transactivates the anti-apoptotic genes of Bcl-2, survivin, and COX-2 [164]. Cyclooxygenase-1 and COX2 inhibitors (see Part II) hinder wound healing on account of neoangiogenesis inhibition. When PlGF-1 forms heterodimers with VEGF-A, it inactivates VEGF-A, whereas PlGF-2 is neoangiogenic and also anti-apoptotic [150,151].

The South African tree *Combretum caffrum* yielded the tubulin-binding combretastatins (CA)-seeking colchicin receptors. The tubular-binding combretastatins (Oxigene) induce endothelial cell apoptosis by selectively destroying tumor vasculature and disorganizing endothelial cell cytoskeletons. The water-soluble CA4P depolymerizes microtubules of vascular endothelial cells, especially those that form tumor neovasculature. Kaposi's sarcoma xenografts are readily eliminated in CA4R-treated hosts. In patients, dose-limiting toxicities (over 50 mg/m^2) are ataxia, vasovagal syncope, and bowel ischemia [171–176]. The fumagillol derivative (*Aspergillus fumigatus*) TPN-470 (AGM-147 Pharmaceuticals) suppresses endothelial cell proliferation. It inhibited the growth in vitro and in vivo of several human cancers (transitional cell carcinoma of the urinary bladder, gastric, colonic, pancreatic, renal, endometrial and hepatocellular carcinomas, and melanoma). In a clinical trial with paclitaxel, TNP-470 induced PR in 25% to 38% of previously treated patients with metastatic solid tumors [177,178]. Paradoxically, FGF-R, upon capturing its cognate ligand FBF, may act as an inhibitor to mitosis in certain breast carcinoma cells [179]. The flavonoid baicalein (see Part II) antagonizes bFGF-induced neoangiogenesis [180]: Could it inadvertently stimulate the growth of FGF-inhibited breast cancer cells? Isoflavones are anti-angiogenic agents [181]. Of other inhibitors of bFGF or its receptor—PD173074 and PD166285—the first is exclusively specific to bFGF-R, whereas the second also blocks PDGF-R [182].

The lipophilic methylated indolinone angiogenesis inhibitor SU5416 (Sugen) performed poorly against kidney carcinoma and soft-tissue sarcomas, inducing stabilization of disease but no CR or PR. SU5416 directed at VEGF-R prevents neoangiogenesis of developing new tumors but is ineffective against the vascular bed of established tumors. It was tested in combination with IFNα against kidney carcinom, but the toxicity of the treatment outweighed its benefit. It did not score well in trials against colon and prostate cancers or against hemangioblastoma of the retina. SU6668 and SU5416 inhibit VEGF-R2, but contributed to severe toxicity (pulmonary embolism, myocardial infarction, cerebrovascular accidents), so these compounds are not in any further clinical trials. Monotherapy with the SU preparations, or with VEGF inhibitors PTK787/ZK222584 or ZD6474, is so far not promising. Oncolytic viruses certainly appear to be much safer than these agents. SU5416, a selective inhibitor of KDR (VEGF-R) tyrosine kinase, also inhibits endothelial cell migration induced by PlGF. The indolinone SU11248 (Sugen) counteracts c-Kit, FLT3, flk-1/KDR VEGF-R, and PDGF-R in preclinical trials for SCLC and other tumors [183–186].

Tyrosine kinase inhibitor SU11248 suppresses growth factor receptors for VEGF, PDGF, Kit, and FLT-3 ligands. In 42 patients with advanced solid tumors, SU11248 induced 4 PR, 3 MR, and 23 stabilizations of disease, while 11 patients continued to progress. Fatigue, asthenia, and neutropenia occurred in 21% to 26 % of the patients [187].

Other molecular inhibitors of the VEGF system in preclinical trials are ZD6126, ZD6474 (aimed at flk-1/KDR VEGF-R2), and ZD4190; PTK787/ZK222584; and PNU-145156E. These agents are extensively reviewed [188–198]. ZD1839 (gefitinib, Iressa) has been approved for the treatment of bronchogenic carcinoma [199] (see Part I). Overexpression of VEGF and Flt-1 receptors in gastric carcinoma [200] would render this tumor to be a reasonable candidate for a clinical trial with SU11248 and SU5416 because these compounds inhibit Flt-1 receptors in gastric carcinoma [201]. Gefitinib, in combination with COX-2 inhibitor SC-236 and the antisense oligonucleotide As-PKAI, all given *per os* to mice bearers of human colon cancer xenografts, induced CR in 60% of these experimental subjects [76]. Joint administration of SU5416 and IFNα2b to patients with kidney carcinoma failed to improve survival and exerted significant toxicity [202]. In contrast, IFNα acted additively with gemcitabine against orthotopically xenografted human pancreatic carcinoma [158]. Blockade of EGF-R could lead to suppressed neoangiogenesis in orthotopically xenografted human kidney cancer [203]. Treatment with PK1166 inhibits autophosphorylation of a tyrosine kinase and suppresses the growth of these human kidney carcinoma xenografts by downregulating IL-8, BCL_{XL}, VEGF, and bFGF. Metastases of this tumor to lungs were inhibited when PK1166 was administered in combination with gemcitabine [203]. The colchicine derivative ZD6126 (NG453) inhibited the growth of xenografted human NSCLC (both adeno- and squamous cell carcinomas) and colorectal, prostatic, ovarian, and breast carcinomas by inducing apoptotic deaths of $CD31^+$ vascular endothelial cells of the tumors' microenvironment; it is in Phase I clinical trials. The VEGF-R signaling through tyrosine kinases is inhibited by ZD6474 and ZD4190; these compounds inhibit the growth of *ras*-mutated breast carcinoma cells and in combination with paclitaxel, they suppressed the growth of human tumor (breast, colon, gastric and ovarian cancer) xenografts. When ZD6474 was given *per os* to human tumor xenograft-bearing mice, it induced regression of the xenografts [188,189]. So far, no major impact on the course of metastatic breast carcinoma occurred: ZD6474 and SU11248, blocking flk-1/

kKDR (VEGF-R) and FGF-R; ZD6126 (AstraZeneca), an inducer of apoptosis in endothelial cells; 2MEL (2-methoxyestradiol, Introgen Therapeutics), and the ribozyme Angiozyme failed to induce any major remissions [204].

Ecteinascidin (ET)-743 combined with plasminogen-related protein B additively suppresses tumor growth through antagonism exerted on tumor-associated microvessel formation. Ecteinascidin is in clinical trials for metastatic sarcomas [205–209]. A dipeptide L-glutamyl-L-tryptophan (IM862) is anti-angiogenic and exerts systemic effects in patients treated intranasally for AIDS-associated Kaposi's sarcoma, exhibiting 36% major responses [210]. The bisphosphonate zoledronic acid is in use for the treatment of patients with bone metastases; it reduced circulating VEGF and PDGF levels [211]. The sulfonated distamycin-A derivative suppresses the growth of human tumor xenografts [212–214]. NSC651016 (suradista) antagonizes stromal cell-derived factor 1α (CXCL12; SDF-1), the ligand for chemokine receptor CXCR4 (which also serves as a co-receptor for HIV-1 entry). Suradista inhibits endothelial cell chemotaxis and proliferation. The sulfonated distamycin derivative is more effective than suramin as a FGF inhibitor. The ligand SDF-1 promotes breast cancer cell migration through induction of tumor cell pseudopodia. Non-Hodgkin's malignant lymphoma cells migrate and proliferate under the effect of SDF-1, but a monoclonal antibody neutralizing CXCR4 inhibits the growth of these lymphoma cells. Some Chinese herbal preparations rich in tannic acid are inhibitory to the SDF-1 (CXCL12→CXCR4 interaction) and suppress the growth of tumor cells that overexpress the receptor [215]. In Auckland, New Zealand, dimethyl xanthenone acetic acid (DMXAA) is in Phase I and II clinical trials; the compound destroys established tumor vasculature [216–218]. The β-sheet-forming peptide anginex exerts anti-angiogenic effects additively or synergistically with carboplatin or angiostatin in ovarian cancer cells [219,220]. Aplidin blocks VEGF⇒R autocrine loops in human leukemic cells (MOLT) grown as established cell lines [221]. Leukemia cell death occurred because of mitochondrial and FasL⇒CD95 pathway-induced apoptosis [222].

Human papilloma virus (HPV)-infected keratinocytes and uterine cervical squamous carcinoma cells utilize endothelin and its receptor as an autocrine growth factor and gain resistance to paclitaxel. The vasoconstrictor endothelin (ET) and its receptor, ET(A)R, are frequently expropriated by malignant tumors (ovarian and prostatic carcinomas) to serve as their autocrine growth loop. Atrasentan (ABT-627) inhibits the growth, counteracts the neoangiogenesis-inducing effects and restores the paclitaxel sensitivity of these tumor cells in xenografts. In hormone-refractory prostate cancer, the ET-1→R pathway functions as an autocrine growth loop. Atrasentan acts as an ET inhibitor and stabilizes the disease. Side effects are rhinitis, peripheral edema, and anemia. In ovarian carcinoma cell lines or xenografts, atrasentan reduced microvessel density, downregulated VEGF expression and MMP-2, and increased the rate of apoptotic tumor cell deaths [223–226].

The aminosterol, squalamine, obtained from the dogfish shark (*Squalus acanthias*), is not without dose-dependent liver toxicity, but it induced transient tumor regression in patients with synovial cell sarcoma and breast carcinoma. Squalamine does not kill tumor cells in vitro, but in vivo, because of its anti-angiogenic activity, it suppresses tumor growth. It is not acting against VEGF or its receptors; it inhibits sodium-proton exchange pumps of endothelial cells. In patients, 500 mg/m^2/day squalamine is tolerated without major toxicity. In patients with NSCLC, squalamine was added to paclitaxel and carboplatin; it was tolerated well and it might have potentiated the effects of chemotherapy [227,228].

The MMP-inhibitor BMS-275291 (Æ-941 Neovastat) is anti-angiogenic and pro-apoptotic to vascular endothelial cells by caspase 3, 8, and 9 activation [229]. MMP inhibitors elicit self-induced apoptosis of Ewing's sarcoma cells by upregulating and retaining transmembrane FasL [230]. MMP inhibitors other than BMS-275291 (such as BB3103; TIMP-2) expose tumor (Ewing's sarcoma) cells to apoptosis. The sarcoma cell expresses FasL for its protection; MMP-7 cleaves and solubilizes FasL but when MMP-7 is blocked, FasL is retained and accumulates in the tumor cell, which then commits the act of programmed cell death [230]. In patients with kidney carcinoma, Neovastat suppressed MMP and VEGF production, induced angiostatin production, and increased the rate of apoptotic tumor cell deaths [231,232]. The sulfonated distamycin-A derivative PNU-145156E, tested in a Phase I clinical trial, induced tumor stabilization in 4 of 29 patients with solid tumors but it did not lower bFGF blood levels, and it caused thrombophlebitis leading to pulmonary embolism [196,214]. However, a tumor as resistant to therapy as chondrosarcoma could be brought to the state of massive cell deaths elicited by a combination of anti-angiogenic and chemotherapeutical agents [233].

Thalidomide (Celgene, Warren, NJ) downregulates VEGF, TNFα, and bFGF production in the tumor's microenvironment and proved to be an effective therapeutic agent for multiple myeloma, myelodysplasia, glioblastoma, KS, kidney carcinoma, and other solid tumors (see Part V). Thalidomide (α-N-phthalylimido-glutarimide) and its tetrafluorinated analogues block bFGF-induced angiogenesis and TNFα release from monocytes; these compounds are effective in the pathological entities listed above and in metastatic melanoma [234–238]. The IMID derivatives are in clinical trials (see Part V).

Whereas PI3K is a promoter of tumor neoangiogenesis, placenta GF (PlGF) is an antagonist of VEGF. Placenta GF forms nonfunctional PlGF/VEGF heterodimers [152,153]. Angiopoietins 1 and 2 exert more inhibitory than promoting effects on angiogenesis, and by removing pericytes from the vascular wall, angiopoietins may destabilize neovasculature [125,137]. They should be tested in the treatment of hemangiopericytomas, malignant tumors of the pericytes of Zimmermann. VEGF-producing breast cancer cells become estrogen independent [121] as they express constitutively active AKT3 [239] and generate TGFβ (inhibitory) and VEGF (stimulatory) autocrine growth loops [121,179].

The antiangiogenic thrombospondin-1 (TSP-1) induces radiosensitivity and inhibits the growth of human tumor (melanoma) xenografts [78]. VGF antagonizes TSP-, but small interfering RNA (siRNA), by suppressing VEGF production, restores the tumor's susceptibility toTSP-1 [78,240]. Tumor growth factor β blockade [132], growth inhibition and apoptosis induction by staurosporine [241] (see Part II), and treatment with TSP-1 together may subdue some breast cancer cell colonies. In HER2/*neu*-positive murine breast cancer, ZD1839 and COX-2 inhibitor SC236 exerted strong antitumor effect [242]. However, TGFβ was inhibitory to the growth of hepatocellular carcinoma cells, which show accelerated malignant behavior upon disruption of TGFβ signals [133]. For breast carcinoma cells, TGFβ could be stimulatory to growth [131], whereas the FGF receptor may transmit growth-inhibitory signals [179]; TGFβ blockade therefore is expected to suppress, but FGF blockade may release from inhibition, the growth potential of these tumor cells.

Some chemotherapeutic agents administered frequently in low dosage (doxorubicin, cyclophosphamide, paclitaxel, docetaxel, vincristine) inhibit blood vessel growth. The DC101 anti–VEGF-R2 monoclonal antibody and dose-dense (frequent) low-dose "metronomic" chemotherapy together suppress tumor neovascularization in an additive to synergistic manner [243].

Monoclonal Antibodies

Anti-VEGF monoclonal antibody (bevacizumab, Avastin; rhuMab, Genentech) and anti–VEGF-R2 monoclonal antibodies are in use and act additively or even synergize with 5FU in the treatment of colon carcinoma (ASCO #3646 2003) [244]. The combination of the humanized anti-VEGF monoclonal antibody (RhuMab anti-VEGF) and chemotherapy (paclitaxel; carboplatin) resulted in increased response rates of NSCLC in Stages III and IV (ASCO #1896 2000): Response rates were: chemotherapy alone, 31%; Mab (15 mg/kg) alone, 22%; and chemotherapy with Mab (7.5 mg/kg), and 40%. Median survival in these three groups of patients, in months, was 15, 12, and 18, respectively. However, a few patients with squamous cell carcinoma developed life-threatening pulmonary hemorrhage; therefore this tumor category is not treated at this time with anti-VGF Mab. These monoclonal antibodies may synergize with TKIs (erlotinib) for the treatment of lung adenocarcinomas as well. Rhu-α-VEGF suppresses the growth of prostatic carcinoma xenografts [245]. Trastuzumab (Herceptin, Genentech) also inhibits neoangiogenesis. Bevacizumab (Avastin) stabilized disease and induced minor responses in 17% of patients with metastatic breast cancers and acted additively in combination with capecitabine [204]. Capecitabine and bevacizumab induced objective responses in 19% of patients with metastatic breast cancer vs. a 9% response rate for capecitabine alone, but relapses occurred so rapidly that progression-free survival was not improved [246]. Tyrosine kinase inhibitors (see Part I) of signal transduction emanating from EGF-R or from HER2/neu/erbB2 suppress VEGF production in targeted tumor cells. The VEGF-Trap system eradicated the vascular bed of established xenografts [247]. Anaplastic thyroid carcinomas resist chemotherapy with doxorubicin, paclitaxel, and cisplatin (the Mayo Clinics protocol) but may respond to combretastatin or monoclonal antibody to VEGF and thalidomide [248–250]. Adenovirus ONYX enhances radiosensitivity of these tumors [251]. Could ONYX and combretastatin be combined for therapy? The monoclonal antibody infliximab (Remicade, Centocor, Malvern, PA) (Physicians Drug Reference, 2004; pp 11145–1148) neutralizes TNFα and thereby pre-empts the induction of those inflammatory and neoangiogenic cytokines that are mobilized by TNFα; it ameliorates the self-destructive inflammatory processes of rheumatoid arthritis and Crohn's disease; it has been insinuated but without proof that it contributes to lymphomagenesis.

Antisense Technology

A 40-bp liposome-encapsulated antisense plasmid aimed at the mRNA of the EGF-R gene administered in combination with endostatin to mice bearing xenografts of human squamous cell carcinomas of the head and neck exerted synergistic antitumor effects [82]. The VEGF-R phosphorothioate antisense oligonucleotide for KDR inhibited the growth of intraperitoneally xenografted human gastric carcinoma cells in nude mice [81]. Another As-ODN targeting VEGF suppresses the growth of mesothelioma xenografts [252]. As microvessel density was reduced, tumor cells succumbed to programmed death. The DNA/RNA mixed backbone As-ODN AsPKAI in combination with ZD1839 (gefitinib, Iressa) and with COX-2 inhibitor SC-238 exerts direct antiproliferative and anti-angiogenic effects on human colon carcinoma xenografts [76].

The *tat*-free retrolentiviral vector PEX (noncatalytic fragment of MMP-2) with HIV-1 *gag, pol,* and *rev* genes within VSV envelope blocks bFGF and MMP-2 production, inhibits angiogenesis in the chicken chorioallantoic membrane, and suppresses the growth of xenografted human melanoma cells [253–255]. Binding of MMP-2 to integrin $\alpha\nu\beta3$ is disrupted and neovascularization is inhibited. Murine leukemia viral vectors fail to transduce nondividing cells, but retrolentiviral vectors could infect both dividing and nondividing cells. Retrolentiviral vectors integrated in the host cell genome exert long-term effects of the transgene. Adenoviral vectors delivering angiostatin or endostatin genes into tumor cells achieve similar effects (see Part I). The HIV-1 Tat protein potentiates bFGF in promoting endothelial cell proliferation mediated by activated MMP-2 in AIDS-associated KS. The recombinant adenovirus (AdTIMP-3) inserting the TIMP-3 gene (tissue inhibitor of metalloproteinase) into endothelial cells suppressed fibroblast-induced neoangiogenesis [256–258] and could kill tumor cells. Of TIMP-1, -2, and -3, TIMP-1 and -2 inhibit vasculogenesis and TIMP-3 is pro-apoptotic for endothelial cells. Spindle cells of AIDS-associated KS express VEGF and form tumors in SCID mice; anti-VEGF Ig suppresses the growth of these tumors [259]. Could AdTIMP-3 and anti-VEGF monoclonal antibody be applied in combination?

When VEGF production is suppressed by ds siRNA interference, TSP-1 secreted by the tumor (fibrosarcoma) inhibits further neovascularization of the tumor bed. Melanoma xenografts suppress the growth of their own micrometastases by producing TSP-1. When the TSP-1–producing parent tumor is surgically removed or eliminated by radiotherapy, the hitherto dormant metastases begin to grow. Thrombospondin-1 administered to the tumor-bearing host inhibits the growth of the micrometastases. Thrombospondin-1 given before and during the elimination of the primary tumor suppresses the growth of both the primary tumor and its micrometastases. Established tumors produce more VEGF than TSP-1, but VEGF production could be suppressed by a ds siRNA preparation [78,240]. Either endogenously induced (by ds siRNA), or exogenously administered TSP-1 should be administered to patients at the time when large, established primary tumors are removed. Retrovirally mediated siRNA silencing of the HER2/neu oncogene in breast and ovarian carcinoma cells induced cell cycle arrest, decreased PI3K and AKT signaling and cyclin D1 levels, and increased and restored levels of levels of p27 and Rb protein, leading to apoptotic death of the cancer cells [260]. A plasmid encoding the VHL tumor-suppressor protein and the As-ODN directed at hypoxia-inducible factor 1α eradicated large lymhomas in mice [117]. The hypoxia-inducible factor-dependent replicative adenovirus Ad9xHRE1A kills VHL gene-defective kidney cancer cells [261]. The ribozyme Angiozyme (Ribozyme Pharmaceuticals, Boulder, CO) cleaves the flt-1 mRNA for VEGF-R1 but failed to induce remissions in a clinical trial for breast cancer [262]. Angiostatin mediates apoptotic death in vascular endothelial cells [263]; adenoviral vectors are being constructed to insert its gene into the vascular network of the tumor bed.

Vaccines

Active angiogenic factor-specific immunizations consist of xenogeneic endothelial cell vaccines, DNA vaccines against VEGF-R, and liposomal FGF-2 peptide vaccine [264–266]. Xenogeneic endothelial cell vaccines protect mice against various transplanted tumors (which were antigenically unrelated to the endothelial cells constituting the vaccine) and the immunoprotective effect could be transferred from immune to nonimmune mice with blood serum (antibodies). Liposome-encapsulated FGF peptide vaccines also induced protection against transplanted tumors [266]. Will it be active vascular endothelial-specific immunization with vaccines, or treatment with antibodies (monoclonal antibodies directed at endothelial cells or at their vasculogenic growth factors) or anti-angiogenic molecular mediators, that will control tumor-induced neovascularity best [267–270]? Tumor necrosis factor-producing cells coated by infliximab (*vide supra*) are lysed by complement or by NK cells in an ADCC reaction. The TNF family is considered to be a new target for cancer therapy [271]. So are vasculogenic growth factor-producer cells coated by antibodies (exogenously administered or produced by the host vaccinated against the growth factor) that surrender these cells to ADCC-mediated reactions. Although actively replicating tumor cells are susceptible to many oncolytic viruses, would dormant tumors deprived of their blood supply be accessible and, if so, found still vulnerable to a viral attack?

REFERENCES, PART III

1. Zamecnik PC, Stephenson ML. Inhibition of Rous sarcoma virus replication and cell transformation by a specific oligodeoxynucleotide Proc Natl Acad Sci U S A 1978; 83:4143–4146.
2. Zamecnik PC, Goodchild J, Taguchi Y, Sarin PS. Inhibition of replication and expression of human T-cell lymphotropic virus type III in cultured cells by exogenous synthetic oligonucleotides complementary to viral RNA Proc Natl Acad Sci U S A 1986; 3:4143–4146.
3. von Ruden T, Gilboa E. Inhibition of human T-cell leukemia virus type I replication in primary human T cells that express antisense RNA J Virol 1989; 63:677–682.
4. Jansen B, Zangemeister-Wittke U. Antisense therapy for cancer - the time of truth Lancet Oncol 2002; 3:672–683.
5. Galmarini CM, Mackey JR, Dumontet C. Nucleoside analogues and nucleobases in cancer treatment Lancet Oncol 2002; 3:415–424.
6. Dias N, Stein CA. Antisense oligonucleotides: basic concepts and mechanisms Mol Cancer Ther 2002; 1:347–355.
7. Uhlmann E, Peyman A. Antisense oligonucleotides, structure and function. In: Meyers RE, Ed.. Molecular Biology and Biotechnology. New York:: Wiley VCH, 1995:38–45.
8. Sazani P, Kole R. Therapeutic potential of antisense oligonucleotides as modulators of alternative splicing J Clin Invest 2003; 112:481–486.

9. Biroccio A, Leonetti C, Zupi G. The future of antisense therapy: combination with anticancer treatments Oncogene 2003; 22:6579–6588.

10. Stephens AC, Rivers RP. Antisense oligonucleotide therapy in cancer Curr Opin Mol Ther 2003; 5:118–122.

11. Stahel RA, Zangemeister-Wittke U. Antisense oligonucleotides for cancer therapy: an overiew Lung Cancer 2003; S1: 81–88.

12. Ciardiello F, Tortora G. Inhibition of Bcl-2 as cancer therapy Ann Oncol. 2002; 13:501–502.

13. Loomis R, Carbone R, Reiss M, Lacy J. Bcl-2 antisense (G3139, Genasense) enhances the in vitro and in vivo response of Epstein-Barr virus-associated lymphoproliferative disease to rituximab Clin Cancer Res 2003; 9:1931–1939.

14. Chi KN, Gleave ME, Klasa R, Murray N, Bryce C, Lopes de Menezes DE, D'Aloisio S, Tolcher AW. A phase I dose-finding study of combined treatment with an antisense Bcl-2 oligonucleotide (Genasense) and mitoxantrone in patients with metastatic hormone-refractory prostate cancer Clin Cancer Res 2001; 7:3920–3927.

15. Waters JS, Webb A, Cunningham D, Clarke PA, Raynaud F, di Stefano F, Cotter FE. Phase I clinical and pharmacokinetic study of Bcl-2 antisense oligonucleotide therapy in patients with non-Hodgkin's lymphoma J Clin Oncol 2000; 18: 1812–1823.

16. Lai JC, Benimetskaya L, Santella RM, Wang Q, Miller PS, Stein CA. G3139 (oblimersen) may inhibit prostate cancer cell growth in a partially bis-CpG-dependent manner Mol Cancer Ther 2003; 2:1031–1043.

17. Bettaieb A, Dubrez-Daloz L, Launay S, Plenchette S, Rebe C, Cathelin S, Solary E. Bcl-2 proteins: targets and tools for chemosensitization of tumor cells Curr Med Chem Anti-Cancer Agents 2003; 3:307–318.

18. Hu Q, Bally MB, Madden TD. Subcellular trafficking of antisense oligonucleotides and down-regulating of Bcl-2 gene expression in human melanoma cells using fusogenic liposome delivery system Nucleic Acids Res 2002; 30:3632–3641.

19. Tauchi T, Sumi M, Nakajima A, Sashida G, Shimamoto T, Ohyashiki K. BCL-2 antisense oligonucleotide Genasense is active against imatinib-resistant BCR-ABL-positive cells Clin Cancer Res 2003; 9:4267–4273.

20. Molto L, Rayman P, Paszkiewicz-Kozik E, Thornton M, Reese L, Thomas JC, Das T, Kudo D, Bukowski R, Finke J, Tannenbaum C. The Bcl-2 transgene protects T cells from renal cell carcinoma-mediated apoptosis Clin Cancer Res 2003; 9: 4060–4068.

21. Sinkovics JG. Malignant lymphoma arising from natural killer cells: report of the first case in 1970 and newer developments of the FasL⇒FasR (CD95) system Acta Microbiol Immunol Hung 1997; 44:295–306.

22. Taylor DD, Gercel-Taylor C, Lyons KS, Stanson J, Whiteside TL. T-cell apoptosis and suppression of T-cell receptor CD3-ζ by Fas ligand-containing membrane vesicles shed from ovarian tumors Clin Cancer Res 2003; 9:5113–5119.

23. Li WW, Bertino JR. Fas-mediated signaling enhances sensitivity of human soft tissue sarcoma cells to anticancer drugs by activation of p38 kinase Mol Cancer Ther 2002; 1:1343–1348.

24. Tada Y, O-Wang J, Wada A, Takiguchi Y, Tatsumi K, Kuriyama T, Sakiyama S, Tagawa M. Fas ligand-expressing tumors induce tumor-specific protective immunity in the inoculated host but vaccination with the apoptotic tumors suppress antitumor immunity Cancer Gene Ther 2003; 10:134–140.

25. Gati A, Guerra N, Gaudin C, Da Rocha S, Escudier B, Lécluse Y, Bettaieb A, Chouaib S, Caignard A. CD158 receptor controls cytotoxic T-lymphocyte susceptibility to tumor-mediated activation-induced cell death by interfering with Fas signaling Cancer Res 2003; 63:7475–7482.

26. Song JH, Song DK, Herlyn M, Petruk K, Hao C. Cisplatin down-regulation of cellular Fas-associated death domain-like interleukin-1β-converting enzyme-like inhibitory proteins to restore tumor necrosis factor-related apoptosis-inducing ligand-induced apoptosis in human melanoma cells Clin Cancer Res 2003; 9:4255–4266.

27. Nemunaitis J, Holmlund JT, Kraynak M, Richards D, Bruce J, Ognoskie N, Kwoh TJ, Geary R, Dorr A, Von Hoff D, Eck-

hardt SG. Phase I evaluation of ISIS 3521, an antisense oligodeoxynucleotide to protein kinase C-alpha, in patients with advanced cancer J Clin Oncol 1999; 17:3586–3595.

28. Kappler M, Kohler T, Kampf C, Diestelkotter P, Wurl P, Schmitz M, Bartel F, Lautenschlager C, Rieber EP, Schmidt M, Taubert H, Meye A. Increased survivin transcript levels: an independent negative predictor of survival in soft tissue sarcoma patients Int J Cancer 2001; 95:360–363.

29. Wang HW, Sharp TV, Koumi A, Koentgas G, Boshoff C. Characterization oif an anti-apoptotic glycoprotein encoded by Kaposi's sarcoma-associated herpesvirus which resembles a spliced variant of human survivin EMBO J 2002; 21: 2602–2615.

30. Mesri M, Wall NR, Li J, Kim RW, Altieri DC. Cancer gene therapy using a survivin mutant adenovirus J Clin Invest 2001; 108:981–990.

31. Blanc-Brude OP, Mesri M, Wall NR, Plescia J, Dohi T, Altieri DC. Therapeutic targeting of the survivin pathway in cancer: initiation of mitochondrial apoptosis and suppression of tumor-associated angiogenesis Clin Cancer Res 2003; 9: 2683–2692.

32. Wurl P, Kappler M, Meye A, Bartel F, Kohler T, Lautenschlager C, Bache M, Schmidt H, Taubert H. Coexpression of survivin and TERT and risk of tumour-related death in patients with soft-tissue sarcomas Lancet 2002; 350: 943–945.

33. Kraemer K, Fuessel S, Schmidt U, Kotzsch M, Schwenzer B, Wirth MP, Meye A. Antisense-mediated hTERT inhibition specifically reduces the growth of human bladder cancer cells Clin Cancer Res 2003; 9:3794–3800.

34. Nowak AK, Lake RA, Kindler HL, Robinson BW. New approaches for mesothelioma: biologics, vaccines, gene therapy, and other novel agents Semin Oncol 2002; 29:82–96.

35. Yap BS, Sinkovics JG, Benjamin RS, Bodey GP. Survival and relapse patterns of complete responders in adults with advanced soft tissue sarcomas Proc 15th Annual Meeting Am Soc Clin Oncol 1979; 20:352 #C250.

36. Sinkovics JG. Complete remissions lasting over three years in adult patients treated for metastatic sarcomas. In: Crispen RG, Ed.. The 6th Chicago Cancer Symposium: Tumor Progression. Chicago Il Oct 3–5, 1979. New York, Amsterdam, London: Elsevier/North Holland, 1980:315–331.

37. Gattacceca F, Pilotte Y, Billard C, La Carrou J, Eliotr M, Jaurand M-C. Ad-IFN-γ induces proliferative and antitumoral responses in malignant mesothelioma Clin Cancer Res 2002; 8: 3298–3304.

38. Xia C, Xu Z, Yuan X, Uematsu K, You L, Li K, Li L, McCormick F, Jablond DM. Induction of apoptosis in mesothelioma cells by antisurvivin oligonucleotides Mol Cancer Ther 2002; 1: 687–684.

39. Jänne PA, Taffaro ML, Salgia R, Johnson BE. Inhibition of epidermal growth factor receptor signaling in malignant pleural mesothelioma Cancer Res 2002; 62:5242–5247.

40. Scappaticci FA, Marina N. New molecular targets and biological therapies in sarcomas Cancer Treat Rev 2001; 27:317–326.

41. Sattler M, Pride YB, Ma P, Gramlich JL, Chu SC, Quinnan LA, Shiazian S, Liang C, Podar K, Christensen JG, Salgia R. A novel small molecule Met inhibitor induces apoptosis in cells transformed by the oncogenic TPR-MET tyrosine kinase Cancer Res 2003; 63:5462–5469.

42. Shinomiya N, Vande Woude GF. Suppression of Met expression: a possible cancer treatment Clin Cancer Res 2003; 9: 5085–5090.

43. Christensen JG, Schreck R, Burrows J, Kuruganti J, Chan E, Le P, Chen J, Wang X, Ruslim L, Blake R, Lipson KE, Ramphal J, Do S, Cui JJ, Cherrington JM, Mendel DB. A selective molecule inhibitor of c-Met kinase inhibits c-met-dependent phenotypes in vitro and exhibits cytoreductive antitumor activity in vivo Cancer Res 2003; 63:7345–7355.

44. Wang X, Le P, Liang C, Chan J, Kiewlich D, Miller T, Harris D, Sun L, Rice A, Vasile S, Blake RA, Howlett AR, Patel N, McMahon G, Lipson KE. Potent and selective inhibitors of the Met [hepatocyte growth factor/scatter factor (HGF/SF) receptor] tyrosine kinase block HGF/SF-induced tumor cell growth and invasion Gene Ther 2003; 2:1085–1095.

45. Kim SJ, Johnson M, Koterba K, Herynk MH, Uehara H, Gallick GE. Reduced c-Met expression by an adenovirus expressing a c-Met ribozyme inhibits tumorigenic growth and lymph node metastases of PC3-LN4 prostate tumor cells in an orthotopic mouse model Clin Cancer Res 2003; 9:5161–5170.

46. Perry ME, Levine AJ. P53 and mdm-2: interactions between tumor suppressor gene and oncogene products Mt Sinai J Med 1994; 61:291–299.

47. Michael D, Oren M. The p53-Mdm2 module and the ubiquitin system Semin Cancer Biol 2003; 13:49–58.

48. Chéne P. Inhibiting the p53-MDM2 interaction: an important target for cancer therapy Nature Rev Cancer 2003; 3:102–109.

49. Zhang Z, Li M, Wang H, Agrawal S, Zhang R. Antisense therapy targeting MDM2 oncogene in prostate cancer: effects on proliferation, apoptosis, multiple gene expression and chemotherapy Proc Natl Acad Sci U S A 2003; 100:11636–11642.

50. Hopkins-Donaldson S, Cathomas R, Simoes-Wust AP, Kurtz S, Belyanskya L, Stahel RA, Zangemeister-Wittke U. Induction of apoptosis and chemosensitization of mesothelioma cells by Bcl-2 and Bcl-xL antisense treatment Int J Cancer 2003; 106:160–166.

51. Yang CT, You L, Lin YC, Lin CL, McCormick F, Jablone DM. A comparison analysis of anti-tumor efficacy of adenoviral gene replacement therapy (p14ARF and p16INK4A) in human mesothelioma cells Anticancer Res 2003; 23:33–38.

52. Cao XX, Mohuiddin I, Chada S, Mhashilkar AM, Ozvaran MK, McConkey DJ, Miller SD, Daniel JC, Smythe WR. Adenoviral transfer of mda-7 leads to BAX up-regulation and apoptosis in mesothelioma cells, and is abrogated by overexpression of BCL-XL Mol Med 2002; 8:869–870.

53. Scotlandi K, Maini C, Manara MC, Benini S, Serra M, Cerisano V, Strammiello R, Baldini N, Lollini PL, Nanni P, Nicoletti G, Picci P. Effectiveness of insulin-like growth factor I receptor antisense strategy against Ewing's sarcoma cells Cancer Gene Ther 2002; 9:296–307.

54. Andrews DW, Resnicoff M, Flanders AE, Kenyon L, Curtis M, Merli G, Baserga R, Iliakis G, Aiken RD. Results of a pilot

study involving the use of an antisense oligodeoxynucleotide directed against the insulin-like growth factor type I receptor in malignant astrocytomas J Clin Oncol 2001; 19:2189–2200.

55. Hailey J, Maxwell E, Koukouras K, Bishop WR, Pachter JA, Wang Y. Neutralizing anti-insulin-like growth factor receptor 1 antibodies inhibit receptor function and induce receptor degradation in tumor cells Mol Cancer Ther 2002; 1:1349–1353.

56. Resnicoff M, Coppola D, Sell C, Rubin R, Ferrone S, Baserga R. Growth inhibition of human melanoma cells in nude mice by antisense strategies to the type I insulin-like growth factor receptor Cancer Res 1994; 54:4848–4850.

57. Surmacz E. Growth factor receptors a therapeutic targets: strategies to inhibit the insulin-like growth factor I receptor Oncogene 2003; 22:6589–6597.

58. Min Y, Adachi Y, Yamamoto H, Ito H, Itoh F, Lee C-T, Nadaf S, Carbone DP, Imai K. Genetic blockade of the insulin-like growth factor-I receptor: a promising strategy for human pancreatic cancer Cancer Res 2003; 63:6432–6441.

59. De Giovanni C, Landuzzi L, Frabetti F, Nicoletti G, Griffoni C, Rossi I, Mazzotti M, Scotto L, Nanni P, Lollini PL. Antisense epidermal growth factor receptor transfection impairs the proliferative ability of human rhabdomyosarcoma cells Cancer Res 1996; 56:3898–38901.

60. Niwa H, Wentzel AL, Li M, Gooding WE, Lui VWY, Grandis JR. Antitumor effects of epidermal growth factor receptor antisense oligonucleotides in combination with docetaxel in squamous cell carcinoma of the head and neck Clin Cancer Res 2003; 99:50287–5035.

61. Aharinejad S, Abraham D, Paulus P, Abri H, Hofmann M, Grossschmidt K, Schäfer R, Stanley ER, Hofbauer R. Colony-stimulating factor-1 antisense treatment suppresses growth of human tumor xenografts in mice Cancer Res 2002; 62: 5317–5324.

62. Freedman RS, Vadhan-Raj S, Butts C, Savary C, Melichar B, Verschragen C, Kavanagh JJ, Hicks ME, Levy LB, Folloder JK, Garcia ME. Pilot study of Flt3 ligand comparing intraperitoneal with subcutaneous routes on hematologic and immuno-

logic responses in patients with peritoneal carcinomatosis and mesotheliomas Clin Cancer Res 2003; 9:5228–5237.

63. Pecora AL, Rizvi N, Cohen GI, Meropol NJ, Sterman D, Marshall JL, Goldberg S, Gross P, O'Neil JD, Groene WS, Roberts MS, Rabin H, Bamat MK, Lorence RM. Phase I trial of intravenous administration of PV701, an oncolytic virus in patients with advanced solid cancers J Clin Oncol 2002; 20:2251–2266.

64. Stevenson JP, Yao K-S, Gallagher EP, Friedland D, Mitchell EP, Cassella A, Monia B, Kevoh TJ, Yu R, Holmlund J, Dorr FA, O'Dwyer PJ. Phae I clinical pharmacokinetic and pharmacodynamic trial of the c-raf-1 antisense oligonucleotide ISIS5132 (CGP69846A) J Clin Oncol 1999; 17:2227–2236.

65. Kumar R, Angelini S, Czene K, Sauroja I, Hahka-Kemppinen M, Pyrhönen S, Hemminki K. BRAF mutations in metastatic melanoma: a possible association with clinical course Clin Cancer Res 2003; 9:3362–3368.

66. Rimoldi D, Salvi S, Liénard D, Lejeune FJ, Speiser D, Zografos L, Cerotini J-C. Lack of BRAF mutations in uveal melanoma Cancer Res 2003; 9:5712–5715.

67. Cruz F, Rubin BP, Wilson D, Town A, Schroeder A, Haley A, Bainbridge T, Heinrich MC, Corless CL. Absence of BRAF and NRAS mutations in uveal melanoma Cancer Res 2003; 9: 5761–5766.

68. Tuveson DA, Weber B, Herlyn M. BRAF as a potential therapeutic target in melanoma and other malignancies Cancer Cell 2003; 4:95–98.

69. Kasid U, Dritschilo A. RAF antisense oligonucleotide as a tumor radiosensitizer Oncogene 2003; 22:5876–5884.

70. Horiuchi H, Kawamata H, Fujimori T, Kuroda Y. A MEK inhibitor (U0126) prolongs survival in mice bearing human gallbladder cancer cells with K-ras mutation: analysis in a novel orthotopic inoculation model Int J Oncol 2003; 23:957–963.

71. Pastorino F, Brignole C, Marimpietri D, Pagnan G, Morando A, Ribatti D, Semple SC, Gambini C, Allen TM, Ponzoni M. Targeted liposomal c-myc antisense oligodeoxynucleotides induce apoptosis and inhibit tumor growth and metastases in human melanoma models Clin Cancer Res 2003; 9:4595–4605.

72. Brignole C, Pagnan G, Marimpietri D, Cosimo E, Allen TM, Ponzoni M, Pastorino F. Targeted delivery system for antisense oligonucleotides: a novel experimental strategy for neuroblastoma treatment Cancer Lett 2003; 197:231–235.

73. Baldwin RL, Tran H, Karlan BY. Loss of c-myc repression coincides with ovarian cancer resistance to transforming growth factor β growth arrest independent of transforming growth factor β/Smad signaling Cancer Res 2003; 63: 1413–1419.

74. Adjei AA, Dy GK, Erlichman C, Reid JM, Sloan JA, Pitot HC, Alberts SR, Goldberg RM, Hanson LJ, Atherton PJ, Watanabe T, Geary RS, Holmlund J, Dorr FA. A phase I trial of ISIS 2503, an antisense inhibitor of H-ras, in combination with gemcitabine in patients with advanced cancer Clin Cancer Res 2003; 9:115–123.

75. Rapp UR, Goldsborough MD, Mark GE, Bonner TI, Groffen J, Reynolds FH, Stephenson JR. Structure and biological activity of v-raf, a unique oncogene transduced by a retrovirus Proc Natl Acad Sci U S A 1983; 80:4218–4222.

76. Tortora G, Caputo R, Damiano V, Melisi D, Bianco R, Fontanini G, Veneziani BM, De Placido S, Bianco AR, Ciardiello F. Combination of a selective cyclooxygenase-2 inhibitor with epidermal growth factor receptor tyrosine kinase inhibitor ZD1839 and protein kinase A antisense causes cooperative antitumor and antiangiogenic effect Clin Cancer Res 2003; 9: 1566–1572.

77. Suzuki C, Daigo Y, Kikuchi T, Katagiri T, Nakamura Y. Identification of COX17 as a therapeutic target for non-small cell lung cancer Cancer Res 2003; 63:7038–7041.

78. Filleur S, Courtin A, Ait-Si-Ali S, Guglielmi J, Merle C, Harel-Bellan A, Clezardin P, Cabon F. SiRNA-mediated inhibition of vascular endotelial growth factor severely limits tumor resistance to antiangiogenic thrombospondin-1 and slows tumor vascularization and growth Cancer Res 2003; 63:3919–3922.

79. Kleef J, Fukahi K, Lopez ME, Friess H, Büchler MW, Sosnowski BA, Korc M. Targeting suicide gene delivery in pancreatic cancer via FGF receptors Cancer Gene Ther 2002; 9:22–532.

80. Hu Y, Cherton-Horvat G, Dragowska V, Baird S, Korneluk RG, Durkin JP, Mayer LD, LaCasse EC. Antisense oligonucle-

otides targeting XIAP induce apoptosis and enhance chemotherapeutic activity against human lung cancer cells in vitro and in vivo Clin Cancer Res 2003; 9:2826–2836.

81. Kamiyama M, Ichikawa Y, Ishikawa T, Chishima T, Hasegawa S, Hamaguchi Y, Nagashima Y, Miyagi Y, Mitsuhashi M, Hyndman D, Hoffman RM, Ohki S, Shimada H. VEGF receptor antisense therapy inhibits angiogenesis and peritoneal dissemination of human gastric cancer in nude mice Cancer Gene Ther 2002; 9:197–201.

82. Li M, Ye C, Feng C, Riedel F, Liu X, Zeng Q, Grandis JR. Enhanced antiangiogenic therapy of squamous cell carcinoma by combined endostatin and epidermal growth factor receptor-antisense therapy Clin Cancer Res 2002; 8:3570–3578.

83. Arboleda MJ, Lyons JF, Kabbinavar FF, Bray MR, Snow BE, Ayala R, Daninoi M, Karlan BY, Slamon DJ. Overexpression of AKT2 protein kinase B beta leads to up-regulation of beta 1 integrin, increased invasion and metastasis of human breast and ovarian cancer cells Cancer Res 2003; 63:196–206.

84. Staal SP. Molecular cloning of the akt oncogene and its human homologues AKT1 and AKT2: amplification of AKT1 in a primary human gastric adenocarcinoma Proc Natl Acad Sci U S A 1987; 84:5034–5037.

85. Cheng JQ, Ruggeri B, Klein WM, Sonoda G, Altomare DA, Watson DK, Testa JR. Amplification of AKT2 in human pancreatic cells and inhibition of AKT2 expression and tumorigenicity by antisense RNA Proc Natl Acad Sci U S A 1996; 93: 3636–3641.

86. Zhao WQ, Chen GH, Chen H, Pascale A, Ravindranath L, Quon MJ, Alkon DL. Secretion of Annexin II via activation of insulin receptor and insulin-like growth factor receptor J Biol Chem 2003; 278:4205–4215.

87. Goel S, Desai K, Bulgaru A, Fields A, Goldberg G, Agrawal S, Martin R, Grindel M, Mani S. A safety study of a mixed-backbone oligonucleotide (GEM231) targeting the type I regulatory subunit of protein kinase A using a continuous infusion schedule in patients with refractory solid tumors Clin Cancer Res 2003; 9:4069–4076.

88. Tortora G, Ciardiello F. Antisense strategies targeting protein kinase C: preclinical and clinical development Semin Oncol 2003; 30:26–31.

89. Mo Y, Gan Y, Song S, Johnston J, Xiao X, Wientjes MG, Au JL. Simultaneous targeting of telomeres and telomerase as a cancer therapeutic approach Cancer Res 2003; 63:579–585.

90. von Willebrand M, Zacksenhaus E, Cheng E, Glazer P, Halaban R. The tyrphostin AG1024 accelerates the degradation of phosphorylated forms of retinoblastoma protein (pRb) and restores pRb tumor suppressive function in melanoma cells Cancer Res 2003; 63:1420–1429.

91. Ju R, Muller MT. Histone deacetylase inhibitors activate $p21^{WAF1}$ expression via ATM Cancer Res 2003; 63:2891–2897.

92. Fan Y, Borowsky AD, Weiss RH. An antisense oligodeoxynucleotide to $p21^{Waf1/Cip1}$ causes apoptosis in human breast cancer cells Mol Cancer Ther 2003; 2:773–782.

93. Chatterjee S, Johnson PR, Wong KK. Dual-target inhibition of HIV-1 in vitro by means of an adeno-associated virus antisense vector Science 1992; 258:1485–1488.

94. Duzubges N, Simoes S, Konopka K, Rosai JJ, de Lima PMC. Delivery of novel macromolecular drugs against HIV-1 Expert Opin Biol Ther 2001; 1:949–970.

95. Soucek L, Evan G. Myc – is the oncogene from Hell Cancer Cell 2002; June:406–408.

96. Greenwood E. The many faces of Myc Nature Rev Cancer 2002; 2:485.

97. Wilkinson E. Myc inactivation stops tumorigenesis Lancet Oncol 2002; 3:451.

98. Alaminos M, Mora J, Cheung NK, Smith A, Qin J, Chen L, Gerald WL. Genome-wide analysis of gene expression associated with MycN in human neuroblastoma Cancer Res 2003; 63:4538–4546.

99. Hermeking H. The MYC oncogene as a cancer drug target Current Cancer Drug Targets 2003; 3:163–175.

100. Jain M, Arvanitis C, Chu K, Dewey W, Leonhardt E, Trinh M, Sundberg CD, Bishop JM, Felsher DW. Sustained loss of

a neoplastic phenotype by brief inactivation of MYC Science 2002; 297:102–104.

101. Felsher DW, Bishop JM. Transient excess of MYC activity can elicit genomic instability and tumorigenesis Proc Natl Acad Sci U S A 1999; 96:3940–3944.
102. Yi F, Jaffe R, Prochownik EV. The CCL6 chemokine is differently regulated by c-Myc and L-Myc and promotes tumorigenesis and metastasis Cancer Res 2003; 63:2923–2932.
103. Misawa A, Hosoi H, Tsuchiya K, Sugimoto T. Rapamycin inhibits proliferation of human neuroblastoma cells without suppression of MycN Int J Cancer 2003; 104:233–237.
104. Harding MW. Immunophilins, mTOR, and pharmacodynamic strategies for a targeted cancer therapy Clin Cancer Res 2003; 9:2882–2886.
105. Oldham S, Hafen E. Insulin/IGF and target of rapamycin signaling: a TOR de force in growth control Trends Cell Biol 2003; 13:79–85.
106. Thimmaiah KN, Easton J, Huang S, Veverka KA, Germain GS, Harwood FC, Houghton PJ. Insulin-like growth factor I-mediated protection from rapamycin-induced apoptosis is independent of Ras-Erk1-Erk2 and phosphatidylinositol 3′-kinase-Akt signaling pathways Cancer Res 2003; 63:364–374.
107. Carroll JS, Swarbrick A, Musgrove EA, Sutherland RL. Mechanisms of growth arrest by c-myc antisense oligonucleotides in MCF-7 breast cancer cells: implications for the antiproliferative effect of antiestrogens Cancer Res 2002; 62:3126–3131.
108. Crooke ST. Antisense Drug Technology: Principles, Strategies and Applications. New York. NY. USA: Marcel Dekker Inc, 2001.
109. Chana JS, Grover R, Tulley P, Lohrer H, Sanders R, Grobbelaar AO, Wilson GD. The c-myc oncogene: use of a biological prognostic marker as a potential target for gene therapy in melanoma Br J Plast Surg 2002; 55:623–627.
110. Ferrara N. VEGF and the quest for tumour angiogenesis factors Nat Rev Cancer 2002; 2:795–803.
111. Ferrara N. Role of vascular endothelial growth factor in physiologic and pathologic angiogenesis: therapeutic implications Semin Oncol 2002; 29(Suppl 16):10–14.

112. Jain RK. Tumor angiogenesis and accessibility: role of vascular endothelial growth factor Semin Oncol 2002; 29(Suppl 16): 3–9.

113. Thorpe PE, Chaplin DJ, Blakey DC. The first international conference on vascular targeting: meeting overview Cancer Res 2003; 63:1144–1147.

114. Turner KJ, Crew JP, Wykoff CC, Watson PH, Poulsom R, Pastorek J, Ratcliffe PJ, Cranston D, Harris Al L. The hypoxia-inducible genes VEGF and CA9 are differentially regulated in superficial vs invasive bladder cancer Br J Cancer 2002; 86:1276–1282.

115. Giatromanolaki A, Koukourakis MI, Sivridis E, Turley H, Wykoff CC, Gatter KC, Harris AL. DEC1 (STRA13) protein expression relates to hypoxia-inducible factor 1-alpha and carbonic anhydrase-9 overexpression in non-small cell lung cancer J Pathol 2003; 200:222–228.

116. Sakurai A, Yuasa K, Shoji Y, Himeno S, Tsujimoto M, Kunimoto M, Imura N, Hara S. Overexpression of thioreduxin reductase 1 regulates NF-κB activation J Cell Physiol 2004; 198: 22–30.

117. Sun X, Kanwar JR, Leung E, Vale M, Krissansen GW. Regression of solid tumors by engineered overexpression of von Hippel-Lindau tumor suppressor protein and antisense hypoxia-inducible factor-1alpha Gene Ther 2003; 10:2081–2089.

118. Welsh SJ, Williams RR, Birmingham A, Newman DJ, Kirkpatrick DL, Powis G. The thioreduxin redox inhibitors 1-methylpropyl 2-imidazolyl disulfide and pleutotin inhibit hypoxia-induced factor 1α and vascular growth factor formation Mol Cancer Therap 2003; 2:235–243.

119. Engman L, Al-Maharik N, McNaughton M, Birmingham A, Powis G. Thioreduxin reductase and cancer cell growth inhibition by organotellurium compounds that could be selectively incorporated into tumor cells Bioorg Med Chem 2003; 11: 5091–5100.

120. Benckert C, Jonas S, Cramer T, Von Marschall Z, Schäfer G, Peters M, Wagner K, Radke C, Wiedenmann B, Neuhaus P, Hocker M, Rosewicz S. Transforming growth factor beta 1 stimulates vascular endothelial growth factor gene transcrip-

tion in human cholangiocellular carcinoma cells Cancer Res 2003; 63:1083–1092.

121. Guo P, Fang Q, Tao HQ, Schafer CA, Fenton BM, Ding I, Hu B, Cheng SY. Overexpression of vascular endothelial growth factor by MCF-7 breast cancer cells promotes estrogen-independent tumor growth in vivo Cancer Res 2003; 63: 4684–4691.

122. Poon RT, Lau CP, Cheung ST, Yu WC, Fan ST. Quantitative correlation of serum levels and tumor expression of vascular endothelial growth factor in patients with hepatocellular carcinoma Cancer Res 2003; 63:3121–3126.

123. Wada M, Ebihara Y, Ma F, Yagasaki H, Ito M, Takahashi T, Mugishima H, Takahashi S, Tsuji K. Tunica interna endothelial cell kinase expression and hematopoietic and angiogenic potentials in cord blood $CD34^+$ cells Int J Hematol 2003; 77: 245–252.

124. Zhang L, Yang N, Park JW, Katsaros D, Fracchioli S, Cao G, O'Brien-Jenkins A, Randall TC, Rubin SC, Coukos G. Tumor-derived vascular endothelial growth factor up-regulates angiopoietin-2 in host endothelium and destabilizes host vasculature, supporting angiogenesis in ovarian cancer Cancer Res 2003; 63:3403–3432.

125. Tse V, Xu L, Yung YC, Santarelli JG, Juan D, Fabel K, Silverberg G, Harsh G. The temporal-spatial expression of VEGF, angiopoietin-1 and 2, and Tie-2 during tumor angiogenesis and their functional correlation with tumor neovascular architecture Neurol Res 2003; 25:729–738.

126. Sussman LK, Upalakalian JN, Roberts MJ, Kocher O, Benjamin LE. Blood markers for vasculogenesis increase with tumor progression in patients with breast carcinoma Cancer Biol Ther 2003; 2:255–256.

127. Chin KF, Greenman J, Reusch P, Gardiner E, Marme D, Monson JR. Vascular endothelial growth factor and soluble Tie-2 receptor in colorectal cancer: associations with disease recurrence Eur J Surg Res 2003; 29:497–505.

128. Larsson A, Sköldenberg E, Ericson H. Serum and plasma levels of FGF-2 and VEGF in healthy blood donors Angiogenesis 2002; 5:107–110.

129. Steinle JJ, Meininger CJ, Forough R, Wu G, Wu MH, Granger HJ. Eph B4 receptor signaling mediates endothelial cell migration and proliferation via the phosphatidylinositol 3-kinase pathway J Biol Chem 2002; 277(46):43830–43835.

130. Nilsson EE, Skinner MK. Role of transforming growth factor beta in ovarian surface epithelium biology and ovarian cancer Reprod Biomed Online 2002; 5:254–258.

131. Lei X, Bandyopadhyay A, Le T, Sun L. Autocrine TGF beta supports growth and survival of human breast cancer MDA-MB-231 cells Oncogene 2002; 21:7514–7523.

132. Muraoka RS, Dumont N, Ritter CA, Dugger TC, Brantley DM, Chen J, Easterly E, Roebuck LR, Ryan S, Gotwals PJ, Koteliansky V, Arteaga CL. Blockade of TGF-β inhibits mammary tumor cell viability, migration, and metastases J Clin Invest 2002; 109:1551–1559.

133. Sugano Y, Matsuzaki K, Tahashi Y, Furukawa F, Mori S, Yamagata H, Yoshida K, Matsushita M, Nishizawa M, Fujisawa J, Inoue K. Distortion of autocrine transforming growth factor β signal accelerates malignant potential by enhancing cell growth as well as PAI-1 and VEGF production in human hepatocellular carcinoma cells Oncogene 2003; 22:2309–2321.

134. Maniotis AJ, Folberg R, Hess A, Seftor EA, Gardner LM, Pe'er J, Trent JM, Meltzer PS, Hendrix MJ. Vascular channel formation by human melanoma cells in vivo and in vitro: vasculogenic mimicry Am J Pathol 1999; 155:739–752.

135. Hendrix MJC, Seftor EA, Seftor REB. Vasculogenic mimicry and tumor-cell plasticity: lessons from melanoma Nat Rev Oncol 2003; 3:411–421.

136. Dvorak HF. Vascular permeability factor/vascular endothelial growth factor: a critical cytokine in tumor angiogenesis and a potential target for diagnosis and therapy J Clin Oncol 2002; 20:4368–4380.

137. Stoelzig O, Ahmad SA, Liu W, McCarty MF, Wey JS, Parikh AA, Fan F, Reinmuth N, Kawaguchi M, Bucana CD, Ellis LM. Angiopoietin-1 inhibits vascular permeability, angiogenesis and growth of hepatic colon cancer tumors Cancer Res 2003; 63:3370–3377.

138. Yokoyama Y, Charnock-Jones DS, Licence D, Yanaihara A, Hastings JM, Holland CM, Emoto M, Sakamoto A, Sakamoto T, Maruyama H, Sato S, Mizunuma H, Smith SK. Expression of vascular endothelial growth factor (VEGF)-D and its receptor, VEGF receptor 3, as a prognostic factor in endometrial carcinoma Clin Cancer Res 2003; 9:1361–1369.

139. Araki Y, Okamura S, Hussain SP, Nagashima M, He P, Shiseki M, Miura K, Harris CC. Regulation of cyclooxygenase-2 expression by the Wnt and ras pathways Cancer Res 2003; 63:728–734.

140. Goodwin AM, D'Amore PA. Wnt signaling in the vasculature Angiogenesis 2002; 5:1–9.

141. Lustig B, Behrens J. The Wnt signaling pathway and its role in tumor development J Cancer Res Clin Oncol 2003; 129: 199–221.

142. Giles RH, van Es JH, Clevers H. Caught up in a Wnt storm: Wnt signaling in cancer Biochim Biophys Acta 2003; 1653: 1–24.

143. Zhang L, Yang N, Katsaros D, Huang W, Park JW, Fracchioli S, Vezzani C, Rigault de la Longrais IA, Yao W, Rubin SC, Coukos G. The oncogene phosphatidylinositol 3′-kinase catalytic subunit α promotes angiogenesis via vascular endothelial growth factor in ovarian carcinoma Cancer Res 2003; 63: 4225–4231.

144. O'Reilly MS, Boehm T, Shing Y, Fukai N, Vasios G, Lane WS, Flynn E, Birkhead JR, Olsen BR, Folkman J. Endostatin: an endogenous inhibitor of angiogenesis and tumor growth Cell 1997; 88:277–285.

145. Berges G, Benjamin LE. Tumorigenesis and the angiogenic switch Nat Rev Cancer 2003; 3:401–410.

146. Yabushita H, Noguchi M, Kinoshita S, Kishida T, Sawaguchi K, Noguchi M. Angiostatin expression in endometrial cancer Oncol Rep 2002; 9:1193–1196.

147. O'Reilly MS, Holmgren L, Shing Y, Chen C, Rosenthal RA, Moses M, Lane WS, Cao Y, Sage EH, Folkman J. Angiostatin: a novel angiogenesis inhibitor that mediates the suppression of metastases by a Lewis lung carcinoma Cell 1994; 79: 315–328.

148. Jung SP, Siegrist B, Hornick CA, Wang YZ, Wade MR, Anthony CT, Woltering EA. Effect of human recombinant endostatin protein on human angiogenesis Angiogenesis 2002; 5: 111–118.

149. Folkman J. Role of angiogenesis in tumor growth and metastasis Semin Oncol 2002; 29(Suppl 16):15–18.

150. Adini A, Kornaga T, Firoozbakht F, Benjamin LE. Placental growth factor is a survival factor for tumor endothelial cells and macrophages Cancer Res 2002; 62:2749–2752.

151. Cai J, Ahmad S, Jiang WG, Huang J, Kontos CD, Boulton M, Ahmed A. Activation of vascular endothelial growth factor receptor-1 sustains angiogenesis and Bcl-2 expression via the phosphatidylinositol 3-kinase pathway in endothelial cells Diabetes 2003; 52:2959–2968.

152. Eriksson A, Cao R, Pawliuk R, Berg SM, Tsang M, Zhou D, Fleet C, Tritsaris K, Dissing S, Leboulch P, Cao Y. Placenta growth factor-1 antagonizes VEGF-induced angiogenesis and tumor growth by the formation of functionally inactive PlGF-1/VEGF heterodimers Cancer Cell 2002; 1:99–108.

153. DiSalvo J, Bayne ML, Conn G, Kwok PW, Trivedi PG, Soderman DD, Palisi TM, Sullivan KA, Thomas KA. Purification and characterization of a naturally occurring vascular endothelial growth factor.placenta growth factor heterodimer J Biol Chem 1995; 270:7717–7723.

154. Krown SE. Therapy of AIDS-associated Kaposi's sarcoma: targeting pathogenetic mechanisms Hematol Oncol Clin North Am 2003; 17:763–783.

155. Sinkovics JG, Campos LT, Gyorkey F, Melnick JL. Kaposi's sarcoma: virology, pathogenesis and interferon α and γ therapy. Abstract Book, International Congress Infectious Diseases (WHO) Cairo, Egypt April 20–24, 1985:133 #472.

156. Sinkovics JG. Oncogenes and growth factors CRC Clin Rev Immunol 1988; 8:217–298.

157. Sinkovics JG. Interferons: antiangiogenic agents Canad J Infect Dis. 1992; 3S:128–132.

158. Solorzano CC, Hwang R, Baker CH, Bucana CD, Pisters PW, Evans DB, Killion JJ, Fidler IJ. Administration of optimal

biological dose and schedule of interferon α combined with gemcitabine induces apoptosis in tumor-associated endothelial cells and reduces growth of human pancreatic carcinoma implanted orthotopically in nude mice Clin Cancer Res 2003; 9:1858–1867.

159. Cordier Kellerman L, Valeyrie L, Fernandez N, Opolon P, Sabourin JC, Maubec E, Roy PL, Kane A, Legrand A, Abina MA, Descamps V, Haddada H. Regression of AK7 malignant mesothelioma established in immunocompetent mice following intratumoral gene transfer of interferon gamma Cancer Gene Ther 2003; 10:481–490.

160. Wall L, Burke F, Barton C, Smyth J, Balkwill F. IFN-gamma induces apoptosis in ovarian cancer cells in vivo and in vitro Clin Cancer Res 2003; 9:2487–2496.

161. Bream JH, Curiel RE, Yu CR, Egwuagu CE, Grusby MJ, Aune TM, Young HA. IL-4 synergistically enhances both IL-2- and IL-12-induced IFN-gamma expression in murine NK cells Blood 2003; 102:207–214.

162. Fiorelli V, Barillari G, Toschi E, Sgadari C, Monini P, Sturzl M, Ensoli B. IFN-gamma induces endothelial cells to proliferate and to invade the extracellular matrix in response to the HIV-1 Tat protein: implications for AIDS-Kaposi's sarcoma pathogenesis J Immunol 1999; 162:1165–1170.

163. Ma DH, Chen JI, Zhang F, Hwang DG, Chen JK. Inhibition of fibroblast-induced angiogenic phenotype of cultured endothelial cells by the overexpression of tissue inhibitor of metalloproteinase (TMP)-3 J Biomed Sci 2003; 10:526–534.

164. Tran J, Rak J, Sheehan C, Saibil SD, LaCasse E, Korneluk RG, Kerbel RS. Marked induction of the IAP family antiapoptotic proteins survivin and XIAP by VEGF in vascular endothelial cells Biochem Biophys Res Commun 1999; 264: 781–788.

165. Yoshida S, Amano H, Hayashi I, Kitasato H, Kamata M, Inukai M, Yoshimura H, Majima M. COX-2/VEGF-dependent facilitation of tumor-associated angiogenesis and tumor growth in vivo Lab Invest 2003; 83:1385–1394.

166. Kim HS, Youm HR, Lee JS, Min KW, Chung JH, Park CS. Correlation between cyclooxygenase-2 and tumor angiogene-

sis in non-small cell lung cancer Lung Cancer 2003; 42: 163–170.

167. Luo X, Belcastro R, Cabacungan J, Hannam V, Negus A, Wen A, Plumb J, Hu J, Steer B, Koehler DR, Downey GP, Tanswell AK. Transfection of lung cells in vitro and in vivo: effect of antioxidants and intraliposomal bFGF Am J Physiol Lung Celk Mol Physiol 2003; 284:817–825.

168. Cappuzzo F, Bartolini S, Crino L. Emerging drugs for non-small cell lung cancer Expert Opin Emerg Drugs 2003; 8: 179–192.

169. Shepherd FA, Sridhar SS. Angiogenesis inhibitors under study for the treatment of lung cancer Lung Cancer 2003; 41: S63–S72.

170. Herbst RS, O'Reilly MS. The rationale and potential of combining novel biologic therapies with radiotherapy: focus on non-small cell lung cancer Semin Oncol 2003; 30:113–123.

171. Iyer S, Chaplin DJ, Rosenthal DS, Boulares AH, Li LY, Smulson ME. Induction of apoptosis in proliferating human endothelial cells by the tumor-specific anti-angiogenesis agent combretastatin A-4 Cancer Res 1998; 58:4510–4514.

172. Ahmed B, Van Eijk LI, Bouma-Ter Steege JC, Van Der Schaft DW, Van Esch AM, Joosten-Achjanie SR, Lambin P, Landuyt W, Griffioen AW. Vascular targeting effect of combretastatin A-4 phosphate dominates the inherent angiogenesis inhibitory activity Int J Cancer 2003; 105:20–25.

173. Rojiani AM, Li L, Rise L, Siemann DW. Activity of the vascular targeting agent combretastatin A-4 disodium phosphate in a xenograft model of AIDS-associated Kaposi's sarcoma Acta Oncol 2002; 41:98–105.

174. Rustin GJ, Galbraith SM, Anderson H, Stratford M, Folkes LK, Sena L, Gumbrell L, Price PM. Phase I clinical trial of weekly combretastatin A4 phosphate: clinical and pharmacokinetic results J Clin Oncol 2003; 21:2815–2822.

175. Anderson HL, Yap JT, Miller MP, Robbins A, Jones T, Price PM. Assessment of pharmacodynamic vascular response in a phase I trial of combretastatin A4 phosphate J Clin Oncol 2003; 21:2823–2830.

176. Galbraith SM, Maxwell RJ, Lodge MA, Tozer GM, Wilson J, Taylor NJ, Stirling JJ, Sena L, Padhani AR, Rustin GJ. Combretastatin A4 phosphate has tumor antivascular activity in rat and man as demonstrated by dynamic magnetic resonance imaging J Clin Oncol 2003; 21:2831–2842.

177. Inoue K, Chikazawa M, Fukata S, Yoshikawa C, Shuin T. Frequent administration of angiogenesis inhibitor TNP-470 (AGM-1470) at an optimal biological dose inhibits tumor growth and metastasis of metastatic human transitional cell carcinoma in the urinary bladder Clin Cancer Res 2002; 8: 2389–2398.

178. Herbst RS, Madden TL, Tran HT, Blumenschein GR, Meyers CA, Seabrooke LF, Khuri FR, Puduvalli VK, Allgood V, Fritsche HA, Hinton L, Newman RA, Crane EA, Fossella FV, Dordal M, Goodin T, Hong WK. Safety and pharmacokinetic effects of TNP-470, an angiogenesis inhibitor, combined with paclitaxel in patients with solid tumors: evidence for activity in non-small-cell lung cancer J Clin Oncol 2002; 20: 4440–4447.

179. McLeskey SW, Ding IY, Lippman ME, Kern FG. MDA-MB-134 breast carcinoma cells overexpress fibroblast growth factor (FGF) receptors and are growth-inhibited by FGF ligands Cancer Res 1994; 54:523–530.

180. J.-J. Liu, T.-S. Huang, W.-F. Cheng, F.-J. Lu. Baicalein and baicalin are potent inhibitors of angiogenesis: inhibition of endothelial cell proliferation, migration and differentiation Int J Cancer 2003; 106:559–565.

181. Whatmore JL, Swann E, Barraja P, Newsome JJ, Bunderson M, Beall HD, Tooke JE, Moody CJ. Comparative study of isoflavone, quinoxaline and oxindole families of anti-angiogenic agents Angiogenesis 2002; 5:45–51.

182. Bansal R, Magge S, Winkler S. Specific inhibitor of FGF receptor signaling: FGF-2-mediated effects on proliferation, differentiation and MAPK activation are inhibited by PD173074 in oligodendrocyte-lineage cells J Neurosci Res 2003; 74: 486–493.

183. DePrimo SE, Wong LM, Khatry DB, Nicholas SL, Manning WC, Smolich BD, O'Farrell AM, Cherrington JM. Expression

profiling of blood samples from SU5416 phase III metastatic colorectal cancer clinical trial: a novel strategy for biomarker identification BMC Cancer 2003; 3:3.

184. Huss WJ, Barrios RJ, Greenberg NM. SU5416 selectively impairs angiogenesis to induce prostate cancer-specific apoptosis Mol Cancer Ther 2003; 2:611–616.

185. Kuenen BC, Tabernero J, Baselga J, Cavalli F, Pfanner E, Conte PF, Seeber S, Madhusudan S, Deplanque G, Huisman H, Scigalla P, Hoekman K, Harris AL. Efficacy and toxicity of the angiogenesis inhibitor SU5416 as a single agent in patients with advanced renal cell carcinoma, melanoma, and soft tissue sarcoma Clin Cancer Res 2003; 9:1648–1655.

186. Mendel DB, Laird AD, Xin X, Louie SG, Christensen JG, Li G, Schreck RE, Abrams TJ, Ngai TJ, Lee LB, Murray LJ, Carver J, Chan E, Moss KG, Haznedar JÖ, Sukbuntherng J, Blake RA, Sun L, Tang C, Miller T, Shirazian S, McMahon G, Cherrington JM. In vivo antitumor activity of SU11248, a novel tyrosine kinase inhibitor targeting vascular endothelial growth factor and platelet-derived growth factor receptors: determination of a pharmaco-kinetic/pharmacodynamic relationship Clin Cancer Res 2003; 9:327–337.

187. Ellis LM. Antiangiogenic therapy at a cross roads: clinical trial results and future directions J Clin Oncol 2003; 21/23S: 281–283.

188. Ciardiello F, Caputo R, Damiano V, Caputo R, Troiani T, Vitagliano D, Carlomagno F, Veneziani BM, Fontanini G, Bianco AR, Tortora G. Antitumor effects of ZD6474, a small molecule vascular endothel growth factor receptor tyrosine kinase inhibitor, with additional activity against epidermal growth factor receptor tyrosine kinase Clin Cancer Res 2003; 9: 1546–1556.

189. Pradel C, Siauve N, Bruneteau G, Clement O, de Bazelaire C, Frouin F, Wedge SR, Tessier JL, Robert PH, Frija G, Cuenod CA. Reduced capillary perfusion and permeability in human tumor xenografts treated with the VEGF signaling inhibitor ZD4190: an in vivo assessment using dynamic MR imaging and macromolecular contrast media Magn Res Imaging 2003; 21:845–851.

190. Robinson SP, McIntyre DJ, Checkley D, Tessier JJ, Howe FA, Griffiths JR, Ashton SE, Ryan AJ, Blakey DC, Waterton JC. Tumour dose response to the antivascular agent ZD6126 assessed by magnetic resonance imaging Br J Cancer 2003; 88: 1592–1597.

191. Davis PD, Dougherty GJ, Blakey DC, Galbraith SM, Tozer GM, Holder AL, Naylor MA, Nolan J, Stratford MR, Chaplin DJ, Hill SA. ZD6126: a novel vascular-targeting agent that causes selective destruction of tumor vasculature Cancer Res 2002; 62:7247–7253.

192. Blakey DC, Westwood FR, Walker M, Hughes GD, Davis PD, Ashton SE, Ryan AJ. Antitumor activity of the novel vascular targeting agent ZD6126 in a panel of tumor models Clin Cancer Res 2002; 8:1974–1983.

193. Goto H, Yano S, Zhang H, Matsumori Y, Ogawa H, Blakey DC, Sone S. Activity of a new vascular targeting agent, ZD6126, in pulmonary metastases by human lung adenocarcinoma in nude mice Cancer Res 2002; 62:3711–3715.

194. Micheletti G, Poli M, Borsotti P, Martinelli M, Imberti B, Taraboletti G, Giavazzi R. Vascular-targeting activity of ZD6126, a novel tubulin-binding agent Cancer Res 2003; 63:1534–1537.

195. Wedge SR, Ogilvie DJ, Dukes M, Kendrew J, Chester R, Jackson JA, Boffey SJ, Valentine PJ, Curwen JO, Musgrove HL, Graham GA, Hughes GD, Thomas AP, Stokes ES, Curry B, Richmond GH, Wadsworth PF, Bigley AL, Hennequin LF. ZD6474 inhibits vascular endothelial growth factor signaling, angiogenesis, and tumor growth following oral administration Cancer Res 2002; 62:4645–4655.

196. Groen HJ, de Vries EG, Wynendaele W, van der Graaf WT, Fokkema E, Lechuga MJ, Poggesi I, Dirix LY, van Oosterom AT. PNU-145156E, a novel angiogenesis inhibitor, in patients with solid tumors: a phase I and pharmacokinetic study Clin Cancer Res 2001; 7:3928–3933.

197. Thomas AL, Morgan B, Drevs J, Unger C, Wiedenmann B, Vanhoefer U, Laurent D, Dugan M, Steward WP. Vascular endothelial growth factor receptor tyrosine kinase inhibitors: PTK787/ZK 222584 Semin Oncol 2003; 30:32–38.

198. Glade-Bender J, Kandel JJ, Yamashiro DJ. VEGF blocking therapy in the treatment of cancer Expert Opin Biol Ther 2003; 3:263–276.

199. Herbst RS, Maddox AM, Rothenberg ML, Small EJ, Rubin EH, Baselga J, Rojo F, Hong WK, Swaisland H, Averbuch SD, Ochs J, LoRusso PM. Selective oral epidermal growth factor receptor tyrosine kinase inhibitor ZD1839 is generally well-tolerated and has activity in non-small-cell lung cancer and other solid tumors: results of a phase I trial J Clin Oncol 2002; 20:3815–3825.

200. Zhang H, Wu J, Meng L, Shou CC. Expression of vascular endothelial growth factor and its receptors KDR and Flt-1 in gastric cancer cells World J Gastroenterol 2002; 8:994–998.

201. Itokawa T, Nokihara H, Nishioka Y, Sone S, Iwamoto Y, Yamada Y, Cherrington J, McMahon G, Shibuya M, Kuwano M, Ono M. Antiangiogenic effect by SU5416 is partly attributable to inhibition of Flt-1 receptor signaling Mol Cancer Ther 2002; 1:295–302.

202. Lara PN, Quinn DI, Margolin K, Meyers FJ, Longmate J, Frankel P, Mack PC, Turrell C, Valk P, Rao J, Buckley P, Wun T, Gosselin R, Galvin I, Gumerlock PH, Lenz HJ, Doroshow JH, Gandara DR. SU5416 plus interferon alpha in advanced renal cell carcinoma: a phase II California Cancer Consortium Study with biological and imaging correlate of angiogenesis inhibition Clin Cancer Res 2003; 9:4772–4781.

203. Kedar D, Baker CH, Killion JJ, Dinney CP, Fidler IJ. Blockade of the epidermal growth factor receptor signaling inhibits angiogenesis leading to regression of human renal cell carcinoma growing orthotopically in nude mice Clin Cancer Res 2002; 8: 3592–3600.

204. Schneider BP, Houck WA, Sledge G. Targeted therapy in breast cancer Diseases of the Breast Updates 2003; 6:1–16.

205. Scotlandi K, Perdichizzi S, Manara MC, Serra M, Benini S, Cerisano V, Strammiello R, Mercuri M, Reverter-Branchat G, Faircloth G, D'Incalci M, Picci P. Effectiveness of Ecteinascidin-743 against drug-sensitive and -resistant bone tumor cells Clin Cancer Res 2002; 8:3893–3903.

206. Kanzaki A, Takebayashi Y, Ren XQ, Miyashita H, Mori S, Akiyama S, Pommier Y. Overcoming multidrug drug resis-

tance in P-glycoprotein/MDR1-overexpressing cell lines by ecteinascidin 743 Mol Cancer Ther 2002; 1:1327–1334.

207. Takahashi N, Li W, Banerjee D, Guan Y, Wada-Takahashi Y, Brennan MF, Chou TC, Scotto KW, Bertino JR. Sequence-dependent synergistic cytotoxicity of ecteinascidin-743 and paclitaxel in human breast cancer cell lines in vitro and in vivo Cancer Res. 2002; 62:6909–6915.

208. Laverdiere C, Kolb EA, Supko JG, Gorlick R, Meyers PA, Maki RG, Wexler L, Demetri GD, Hesley JH, Huvos AG, Gorin AM, Bagatell R, Ruiz-Casado A, Guzman C, Jimeno J, Harmon D. Phase II study of ecteinascidin 743 in heavily pretreated patients with recurrent osteosarcoma Cancer 2003; 98: 832–840.

209. Twelves C, Hoekman K, Bowman A, Vermorken JB, Anthoney A, Smyth J, van Kesteren C, Beijnen JH, Uiters J, Wanders J, Gomez J, Guzman C, Jimenoi J, Hanauske A. Phase I and pharmacokinetic study of Yondelis (ecteinascidin-743; ET-743) administered as an infusion over 1 h or 3 h every 21 days in patients with solid tumours Eur J Cancer 2003; 39: 1842–1851.

210. Tulpule A, Scadden DT, Espina BM, Cabriales S, Howard W, Shea K, Gill PS. Results of a randomized study of IM862 nasal solution in the treatmemnt of AIDS-related Kaposi's sarcoma J Clin Oncol 2000; 18:716–723.

211. Santini D, Vincenzi B, Dicuonzo G, Avvisati G, Massacesi C, Battistoni F, Gavasci M, Rocci L, Tirindelli MC, Altomare V, Tocchini M, Bonsignori M, Tonini G. Zoledronic acid induces significant and long-lasting modifications of circulating angiogenic factors in cancer patients Clin Cancer Res 2003; 9: 2893–2897.

212. Ganju RK, Brubaker SA, Meyer J, Dutt P, Yang Y, Qin S, Newman W, Groopman JE. The alpha-chemokine, stromal cell-derived factor alpha, binds to the transmembrane G-protein-coupled CXCR-4 receptor and activates multiple signal transduction pathways J Biol Chem 1998; 273:23169–23175.

213. Schneider GP, Salcedo R, Dong HF, Kleinman HK, Oppenheim JJ, Howard OM. Suradista NSC 651016 inhibits the angiogenic activity of CXCL12-stromal cell-derived factor 1α Clin Cancer Res 2002; 8:3955–3960.

214. Finch PW, Yee LK, Chu MY, Chen TM, Lipsky MH, Maciag T, Friedman S, Epstein MH, Calabresi P. Inhibition of growth factor mitogenicity and growth of tumor cell xenografts by a sulfonated distamycin A derivative Pharmacology 1997; 55: 269–278.

215. Chen X, Beutler JA, McCloud TG, Loehfelm A, Yang KL, Dong H-F, Chertov OY, Salcedo R, Oppenheim JJ, Howard OMZ. Tannic acid is an inhibitor of CXCL12 (SDF-1alpha)/CXCR4 with antiangiogenic activity Clin Cancer Res 2003; 9: 3115–3123.

216. Zhou S, Kestell P, Baguley BC, Paxton JW. Preclinical factors affecting the individual variability in the clearance of the investigational anti-cancer drug 5.6-dimethylxanthenone-4-acetic acid Biochem Pharmacol 2003; 65:1853–1865.

217. Baguley BC. Antivascular therapy of cancer: DMXAA Lancet Oncol 2003; 4:1410–1148.

218. Jameson MB, Thompson PI, Baguley BC, Evans BD, Harvey VJ, Porter DJ, McCrystal MR, Small M, Bellenger K, Gumbrelkl L, Halbert GW, Kestell P. Clinical aspects of a phase I trial of 5.6-dimethylxanthenone-acetic acid (DMXAA), a novel antivascular agent Br J Cancer 2003; 88:1844–1850.

219. van der Schaft DW, Dings RP, de Lussanet QG, van Eijk LI, Nap AW, Beets-Tan RG-, Steege JC, Wagstaff J, Mayo KH, Griffioen AW. The designer anti-angiogenic peptide anginex targets tumor endothelial cells and inhibits tumor growth in animal models FASEB 2002; 16:1991–1932.

220. Dings RP, Yokoyama Y, Ramakrishnan S, Griffioen AW, Mayo KH. The designed angiostatic peptide anginex synergistically improves chemotherapy and anti-angiogenesis therapy with angiostatin Cancer Res 2003; 63:382–385.

221. Broggini M, Marchini SV, Galliere E, Borsotti P, Taraboletti G, Erba E, Sironi M, Jimeno J, Faircloth GT, Giavazzi R, D'Incalci M. Aplidine, a new anticancer agent of marine origin, inhibits vascular endothelial growth factor (VEGF) secretion and blocks VEGF-VEGF-R-1 (flt-1) autocrine loop in human leukemia cells MOLT-4 Leukemia 2002; 17:52–59.

222. Gajate C, An F, Mollinedo F. Rapid and selective apoptosis in human leukemic cells induced by aplidine through a Fas/

CD95- and mitochondrial-mediated mechanism Clin Cancer Res 2003; 9:1535–1545.

223. Ellis LM. Antiangiogenic therapy at a cross roads: clinical trial results and future directions J Clin Oncol 2003; 21/23S: 281–283.

224. Zonnenberg BA, Groenewegen G, Janus TJ, Leahy TW, Humerickhouse RA, Isaacson JD, Carr RA, Voest E. Phase I dose-escalation study of the safety and pharmacokinetics of atrasentan: an endothelin receptor antagonist for refractory prostate cancer Clin Cancer Res 2003; 9:2965–2972.

225. Nelson JB. Endothelin inhibition: novel therapy for prostate cancer J Urol 2003; 170:S65–S67.

226. Rosano L, Spinella F, Salani D, Di Castro V, Venuti SA, Nicotra MR, Natali PG, Bagnato A. Therapeutic targeting of the endothelin a receptor in human ovarian carcinoma Cancer Res 2003; 63:2447–3453.

227. Hao D, Hammond LA, Eckhardt SG, Patnaik A, Takimoto CH, Schwartz GH, Goetz AD, Tolcher AW, McCreery HA, Mamun K, Williams JI, Holroyd KJ, Rowinsky EK. A Phase I and pharmacokinetic study of squalamine, an aminosterol angiogenesis inhibitor Clin Cancer Res 2003; 9:2465–2471.

228. Herbst RS, Hammond LA, Carbone DP, Tran HT, Holroyd KJ, Deasi A, Willliams JI, Bekele BN, Hait H, Allgood V, Solomon S, Schiller JH. A phase I/IIA trial of continuous five-day infusion of squalamine lactate (MSI-1256F) plus carboplatin and paclitaxel in patients with advanced non-small cell lung cancer Clin Cancer Res 2003; 9:4108–4115.

229. Boivin D, Gendron S, Beaulieu E, Gingras D, Beliveau R. The antiangiogenic agent Neovastat (AE-941) induces endothelial cell apoptosis Mol Cancer Ther 2002; 1:795–802.

230. Mitsiades N, Yu WH, Poulaki V, Tsokos M, Stamenkovic I. Matrix metalloproteinase-7-mediated cleavage of Fas ligand protects tumor cells from chemotherapeutic drug cytotoxicity Cancer Res 2001; 61:577–581.

231. Batist G, Patenaude F, Champagne P, Croteau D, Levinton C, Hariton C, Escudier B, Dupont E. Neovastat (AE-941) in refractory renal cell carcinoma patients: report of a phase II trial with two dose levels Ann Oncol 2002; 13:1259–1263.

232. Bukowski RM. AE0–941, a multifunctional antiangiogenic compound: trial in renal cell carcinoma Expert Opin Investig Drugs 2003; 12:1403–1411.

233. Morioka H, Weissbach L, Vogel T, Nielsen GP, Faircloth GT, Shao L, Hornicek FJ. Anti-angiogenesis treatment combined with chemotherapy produces chondrosarcoma necrosis Clin Cancer Res 2003; 9:1211–1217.

234. Amato RJ. Thalidomide: an antineoplastic agent Curr Oncol Rep 2002; 4:56–62.

235. Eisen T. Thalidomide in solid malignancies J Clin Oncol 2002; 20:2607–2609.

236. Marriott JB, Clarke IA, Czajka A, Dredge K, Childs K, Man HW, Schafer P, Govinda S, Muller GW, Stirling DI, Dalgleish AG. A novel subclass of thalidomide analogue with anti-solid tumor activity in which caspase-dependent apoptosis is associated with altered expression of Bcl-2 family proteins Cancer Res 2003; 63:593–599.

237. Mall JW, Philipp AW, Mall W, Pollmann C. Long-term survival of a patient with small-cell lung cancer (SCLC) following treatment with thalidomide and combination chemotherapy Angiogenesis 2002; 5:11–13.

238. Ng SS, Gütschow M, Weiss M, Hauschildt S, Teubert U, Hecker TK, Luzzio FA, Kruger EA, Eger K, Figg WD. Antiangiogenic activity of N-substituted and tetrafluorinated thalidomide analogues Cancer Res 2003; 63:3189–3194.

239. Faridi J, Wang L, Endemann G, Roth RA. Expression of constitutively active Akt-3 in MCF-7 breast cancer cells reverses the estrogen and tamoxifen responsivity of these cells in vivo Clin Cancer Res 2003; 9:2933–2939.

240. Rofstad EK, Henriksen K, Galappathi K, Mathiesen B. Antiangiogenic treatment with thrombospondin-1 enhances primary tumor radiation response and prevents growth of dormant pulmonary micrometastases after curative radiation therapy in human melanoma xenografts Cancer Res 2003; 63: 4055–4061.

241. McKeague AL, Wilson DJ, Nelson J. Staurosporine-induced apoptosis and hydrogen peroxide-induced necrosis in two human breast cell lines Br J Cancer 2003; 88:125–131.

242. Rosato FE, Dicker A, Burd R, Miller S, Lanza-Jacoby S. Combining the epidermal growth factor (EGF) tyrosine kinase inhibitor ZD1839 with a selective cyclooxygenase (COX)-2 inhibitor SC236, causes a cooperative antitumor effect in breast cancer cells derived from HER2/neu mice J Surg Res 2003; 114:274.

243. Klement G, Huang P, Mayer B, Green SK, Man S, Bohlen P, Hicklin D, Kerbel RS. Differences in therapeutic indexes of combination metronomic chemotherapy and an anti-VEGFR-2 antibody in multidrug-resistant human breast cancer xenografts Clin Cancer Res 2002; 8:221–232.

244. Kabbinavar F, Hurwitz HI, Fehrenbacher L, Meropol NJ, Novotny WF, Lieberman G, Griffing S, Bergsland E. Phase II randomized trial comparing bevacizumab plus fluorouracil (FU)/leucovorin (LV) with FU/LV alone in patients with metastatic colorectal cancer J Clin Oncol 2003; 21:60–65.

245. Fox WD, Higgins B, Maiese KM, Drobnjak M, Cordon-Cardo C, Scher HI, Agus DB. Antibody to vascular endothelial growth factor slows growth of an androgen-independent xenograft model of prostate cancer Clin Cancer Res 2002; 8: 3226–3231.

246. Miller KDet al, cited by Schneider BP, Houck WA, G. Sledge. (204).

247. Huang J, Frischer JS, Serur A, Kadenhe A, Yokoi A, McCrudden KW, New T, O'Toole K, Zabski S, Rudge JS, Holash J, Yancopoulos GD, Yamashiro DJ, Kandel JJ. Regression of established tumors and metastases by potent vascular endothelial growth factor blockade Proc Natl Acad Sci USA 2003; 100: 7785–7790.

248. McIver B, Hay ID, Giuffrieda DF, Dvorak CE, Grant CS, Thompson GB, van Heerden JA, Goellner JR. Anaplastic thyroid carcinoma: a 50-year experience at a single institution Surgery 2001; 130:1028–1034.

249. Dziba JM, Marcinek R, Venkataraman G, Rbinson JA, Ain KB. Combretastatin A4 phosphate has primary antineoplastic activity against human anaplastic carcinoma cell lines and xenograft tumors Thyroid 2002; 12:1063–1070.

250. Bauer AJ, Terrell R, Donipathi NK, Patel A, Tuttle RM, Saji M, Ringel MD, Francis GL. Vascular endothelial growth factor

monoclonal antibody inhibits growth of anaplastic thyroid cancer xenografts in nude mice Thyroid 2002; 12:953–961.

251. Portella G, Pacelli R, Libertini S, Cella L, Vecchio G, Salvatore M, Fusco A. ONYX-015 enhances radiation-induced death of human anaplastic thyroid carcinoma cells J Clin Endocrinol Metab 2003; 88:5027–5032.

252. Massod R, Kundra A, Zhu S, Xia G, Scalia P, Smith DL, Gill PS. Malignant mesothelioma growth inhibition by agents that target the VEGF and VEGF-C autocrine loop Int J Cancer 2003; 104:603–610.

253. Kessler TA, Pfeifer A, Silletti S, Mesters RM, Berdel WE, Verma I, Cheresh D. Matrix metalloproteinase/integrin interactions as target for anti-angiogenic treatment strategies Ann Hematol 2002; 81S2:69–70.

254. Pfeifer A, Kessler T, Silletti S, Cheresh DA, Verma IM. Suppression of angiogenesis by lentiviral delivery of PEX, a noncatalytic fragment of matrix metalloproteinase 2 Proc Natl Acad Sci U S A 2000; 97:12227–12232.

255. Silletti S, Kessler T, Goldberg J, Boger DL, Cheresh DA. Disruption of matrix metalloproteinase 2 binding to integrin$\alpha_v\beta_3$ by an organic molecule inhibits angiogenesis and tumor growth in vivo Proc Natl Acad Sci USA 2001; 98:119–124.

256. Toschi E, Barillari G, Sgadari C, Bacigalupo I, Cereseto A, Carlei D, Palladino C, Zietz C, Leone P, Sturzl M, Butto S, Cafaro A, Monini P, Ensoli B. Activation of matrix-metalloproteinase-2 and membrane-type-1-matrix metalloproteinase in endothelial cells and induction of vascular permeability in vivo by human immunodeficiency virus-1 Tat protein and basic fibroblast growth factor Mol Biol Cell 2001; 12:2934–2946.

257. Sgadari C, Barillari G, Toschi E, Carlei D, Bacigalupo I, Baccarini S, Palladino C, Leone P, Bugarini R, Malavasi L, Cafaro A, Falchi M, Valdembri D, Rezza G, Bussolini F, Monini P, Ensolini B. HIV protease inhibitors are potent anti-angiogenic molecules and promote regression of Kaposi sarcoma Nature Med 2002; 8:225–232.

258. Baker AH, George SJ, Zaltsman AB, Murphy G, Newby AC. Inhibition of invasion and induction of apoptotic cell death of

cancer cell lines by overexpression of TIMP-3 Br J Cancer 1999; 79:1347–1355.

259. Samaniego F, Young D, Grimes C, Prospero V, Christofidou-Solomidou M, DeLisser HM, Prakash O, Sahin AA, Wang S. Vascular endothelial growth factor and Kaposi's sarcoma cells in human skin grafts Cell Growth Differ 2002; 13:387–395.

260. Yang G, Cai KQ, Thompson-Lanza JA, Bast RC, Liu J. Inhibition of breast and ovarian tumor growth through multiple signaling pathways by using retrovirus-mediated siRNA silencing against Her-2/neu gene expression J Biol Chem 2003; November (in print).

261. Cuevas Y, Hernández-Alcoceba R, Aragones J, Naranjo-Suárez S, Castellanos MC, Esteban MA, Martin-Puig S, Landazuri MO, del Peso L. Specific oncolytic effect of a new hypoxia-inducible factor-dependent replicative adenovirus on von Hippel-Lindau-defective renal cell carcinomas Cancer Res 2003; 63:6877–6884.

262. Hortobagyi, G. et al: cited by Schneider BP, Houck WA, G. Sledge. (204).

263. Hanford HA, Wong CA, Kassan H, Cundiff DL, Chandel DL, Chandel N, Underwood S, Mitchell CA, Soff GA. Angiostatin4,5-mediated apoptosis of vascular endothelial cells Cancer Res 2003; 63:4275–4281.

264. Niethammer AG, Xiang R, Becker JC, Wodrich H, Pertl U, Karsten G, Eliceiri BP, Reisfeld RA. A DNA vaccine against VEGF receptor 2 prevents effective angiogenesis and inhibits tumor growth Nat Med 2002; 8:1369–1375.

265. Wei YQ, Wang QR, Zhao X, Yang L, Tian L, Lu Y, Kang B, Lu CJ, Huang MJ, Lou YY, Xiao F, He QM, Shu JM, Xie XJ, Mao YQ, Lei S, Luo F, Zhou LQ, Liu CE, Zhou H, Jiang Y, Peng F, Yuan LP, Li Q, Wu Y, Liu JY. Immunotherapy of tumors with xenogeneic endothelial cells as a vaccine Nat Med 2000; 6:1160–1166.

266. Plum SM, Holaday JW, Ruiz A, Madsen JW, Fogler WE, Fortier AH. Administration of a liposomal FGF-2 peptide vaccine leads to abrogation of FGF-2-mediated angiogenesis and tumor development Vaccine 2000; 19:1294–1303.

267. Ciardiello F. An update of new targets for cancer treatment: receptor-mediated signals Ann Oncol 2002; 13(Suppl 4): 29–38.

268. Kerbel R, Folkman J. Clinical translation of angiogenesis inhibitors Nat Rev Cancer 2002; 2:727–739.

269. Margolin K. Inhibition of vascular endothelial growth factor in the treatment of solid tumors Curr Oncol Rep 2002; 4:20–28.

270. Scappaticci FA. Mechanisms and future direction for angiogenesis-based cancer therapies J Clin Oncol 2002; 20: 3906–3927.

271. Younes A, Kadin ME. Emerging applications of the tumor necrosis factor family of ligands and receptors in cancer therapy J Clin Oncol 2003; 21:3526–3534.

PART IV. THE PREDICTIVE VALUE OF TISSUE CULTURE AND XENOGRAFT ASSAYS

CULTURE SHOCK

Tissue Cultures

When cells are explanted from their microenvironment of their host [1,2], they may replicate first in an accelerated fashion, but eventually activate growth-inhibitory mechanisms and surrender to the limited life span rules of Hayflick. The inhibitory mechanisms ensue by the activation of $p16^{INK4a}$ and $p21^{Cip1}$, inhibitors of CDK, and 19^{ARF}, the inducer of p53. In the population of cell cultures, hypotetraploid individual cells with suppressed ARF and/or MDM2, regulators of p53, are selected out. Some cells with nonfunctional p53 bypass the replication block. If Myc or Ras gene product proteins appear on the scene (with or without actual mutation of the corresponding genes), the cell assumes the malignant phenotype. H-*ras*–transformed cells in suspension and deprived of cell-to-substratum contact (anoikis or "homelessness") are exquisitely susceptible to FT inhibitors (see Part I). Tropomyosins (TM) allow cell detachment and promote detachment-induced apoptosis caused by unligated integrins. Tropomyosins are downregulated in primary breast cancers; consequentially, tumor cells remain adherent to the supportive stroma. Resistance to anoikis is implicated in the progression of osteosarcoma cells that thwart apoptosis, even when the sarcoma cells are separated from their subverted and tumor-supportive stroma. Adrenoceptor-α1 antagonists (doxazosin, terazosin) decrease the adhesion of prostate cancer cells to gelatin and collagen and inhibit tumor cell invasiveness. Doxazosin-induced anoikis was counteracted by Bcl-2 overexpression in the tumor cells [3–5]. All these complex events are consequential to the transfer of cell populations, after the disruption of their genuine cell-to-substratum contacts and interrelationships, from their natural host to plastic surfaces of tissue culture flasks, where they are nourished with fetal calf serum-containing media, are exposed to air, and are transilluminated with fluorescent light under the microscope. Explanted cells undergo "culture shock" [6].

Trypsin

Trypsinogen expression is frequently silenced in gastric and other cancer cells. The demethylating agent 5-aza-dC and a trichostatin preparation (Sigma) reactivated trypsinogen expression in tumor cells [7]. Tumor cells exposed to trypsin lose cell-surface integrin α_v expression. Intraperitoneal dissemination of trypsin-treated gastric carcinoma cells is inhibited, whereas tumor cell susceptibility to apoptosis is increased. Thus, cultured tumor cells trypsinized to detach them from glass or plastic surfaces for transfer as suspensions of rounded-up cells may be altered (downregulated) as to their malignant potentials. Yet in vitro cell line models appeared to be somewhat more acceptable predictors regarding the outcome of subsequent in vivo clinical trials of the agents that were tested first in vitro than were murine allograft (xenograft) models [8].

THE TELOMERES ARE GETTING SHORTER

Telomerase

The enzyme telomerase consists of its reverse transcriptase and protein (hTR; hTERT). It maintains the length of telomeres at chromosomal ends. Immortalized (or malignantly transformed) cells preserve their remaining telomeres through the action of their constitutively expressed telomerase. Cells undergoing natural senescence gradually lose their telomeres. Exempt from telomere loss are germ cells and some stem cells. Physiological telomere loss is 50 bp to 100 bp during each cell division. The TTAGGG telomere repeats are synthesized at the 3′ ends of the chromosomes by the RNA-containing reverse transcriptase (RT) telomerase. Without telomere preservation, unprotected (uncapped) DNA ends may either induce cell senescence through Rb and p53 activation or join with one another, resulting in fused chromosomes (dicentrics). Overexpressed telomerase, the ribonucleoprotein DNA polymerase, preserves chromosomal length in cancer cells, despite the frequent mitoses these cells go through. Antisense oligonucleotides complementary to the template region and directed at the RNA component of telomerase reduce tumor growth and increase paclitaxel sensitivity of cultured human tumor cells, especially when given in combination with the RT inhibitor 3′-azido-3′ deoxythymidine (AZT); however, clinical trial results have not yet been published. Telomerase-specific replication-deficient oncolytic adenovirus containing hTERT replicates preferantially in cells overexpressing telomerase as it adds TTAGGG repeats onto the 3′ ends of chromosomes (see Part III). In the host, but not in tissue cultures or xenografts, the enzymatic RT catalytic subunit hTERT and RNA (hTR) elicit immune T cell response. Telomerase inhibitors are antisense RNAs antagonizing hTR. Phosphorothioate oligonucleotides or ribozymes target hTR or hTERT mRNAs. Short RNA duplexes exist and interfere with gene expression: siRNAs (see Part III). These dsRNAs, homologous to the RT or its template RNA, inhibit telomerase activity in cancer (colon carcinoma, uterine cervical carcinoma or fibrosarcoma) cells [9,10].

Rapamycins, As-ODN, Lymphocytes

The serine-threonine kinase, rapamycin, an inhibitor of mTOR, inactivated mRNA of telomerase in endometrial carcinoma cells. Rapamycin inhibits mTOR kinases that regulate the cell cycle and translation and transcription. mTOR is activated when tumor suppressor gene PTEN is eliminated (a frequent event in cancer cells). Rapamycin in nanomolecular concentrations decreases telomerase hTERT mRNA levels. Rapamycin arrests the cell cycle idependently from PTEN: It is active in both $PTEN^+$ and $PTEN^-$ tumor cells [11]. By synthesizing the telomeric TTAGGG/GTTAGG repeats to cap chromosomal ends, the ribonucleoprotein enzyme telomerase endows the cell with limitless replicative potential, as it occurs in rejuvenated cells, stem cells, and malignantly transformed cells. The catalytic subunit of telomerase (hTERT) is targeted by As-ODN in bladder cancer cells [9,12]. The genetically engineered adenovirus Ad-RB94 is hTERT promoter-driven and kills tumor cells overexpressing telomerase. In bladder carcinoma cells, telomerase erosion leads to apoptosis, but healthy cells suffer no harm. If IFNα protects telomeres by blocking telomere erosion, it may counteract the efficacy of telomere-targeted virotherapy [10,13], either by natural induction or by additional therapeutic administration. The 2′-0-methylethyl oligonucleotide, by binding telomerase RNA template, inhibits the ribonucleoprotein enzyme. Some peptides of hTERT induce MHC class II-restricted recognition by host CD4 lymphocytes [14]; it may be possible to increase this lymphocyte population by administration of a vaccine or IL-2. The oligonucleotide thiophosphoramidate GRN163 arrests the growth of multiple myeloma and lymphoma cells; tumor cells suffer telomerase shortening and undergo senescence. This oligonucleotide hybridizes with human telomerase RNA (hTR); tumor cells treated with GRN163 senesced rapidly and died apoptotic deaths [15–19].

Natural telomere loss may continue in tissue cultures, but a subpopulation of cultured cells may learn to preserve its telomeres. Populations consisting of healthy or reprogrammed stromal cells in admixture with a heterogeneous assembly of tumor cells may perish during the “culture shock” or transform in vitro, whereas a subpopulation of mutated cells endowed with the mechanisms of telomerase reactivation escape senescence and death and proliferate in an immortalized fashion. These are the new cell populations submitted to experimentation. Tumor cells established in culture do not represent the entire cell population of the natural tumor that was residing in the microenvironment of its host. Furthermore, the complex interactions of the tumor cells with immune reactions (suppressive or enhancing) of the host are absent in vitro. Drug or oncolytic virus assays of tumor cell populations established in culture or in xenografts may not reflect the responses elicited when tumor-bearing hosts are treated with the same agents in vivo.

ONE DAY IN THE LIFE OF "IVAN XENOGRAFTOVICH"*

Paracrine Circuits

Tumor cell populations consist of individual tumor cells dependent on paracrine growth factors of stromal cell origin and of tumor cells independent of the reprogrammed stroma and able to drive their own growth by autocrine ligand-to-receptor growth loops. This is most evident in cultures of sarcoma cells [20,21]. Tumor cells dependent on paracrine growth factors of stromal cell origin become extremely vulnerable when they are forcefully extricated from their natural environment and are xenotransplanted; a selection process ensues, resulting in the survival of the tumor cell subpopulation that is able to drive its own growth by autocrine growth loops. The gene of human VEGF-A transferred by an adeno-associated viral vector can induce neoangiogenesis in the mouse, but other growth factors may not transgress species barriers: If produced by the xenotransplanted human cells, they may not induce a response in the mouse or, if produced by the new murine host of the

---

Dedicated to Alexandr Solzhenitsyn on his 85th birthday, after his "One Day in the Life of Ivan Denisovich" in the "GULAG Arkhipelag" of the Soviet Union. In the xenograft, the exiled cancer cell struggles for survival and sustains forceful attacks on its life. Cancer cells are considered to be subversive criminal elements, whereas prisoners of the GULAG were innocent victims of a tyrannical regime that subdued them into the brutality of forced labor camps of immense proportions.

xenotransplanted human tumor cells, they may not find a cognate receptor in the xenograft to act upon.

Let us consider the human bronchioalveolar carcinoma (BAC) of the lung (Table 5) with purported but unproven viral (herpes and retroviral) etiology. Indeed, this multifocal tumor occurring in non-smokers and recurring after excisions, resembles the retrovirally induced jaagsiekte of sheep. Sheep retrovirus RNA or proviral DNA could not be found in the human tumor, but antibodies directed to the capsid proteins of the ovine retrovirus stain the cytoplasms of some human BAC cells. The ovine retroviral provirus DNA transforms rodent fibroblasts [22–26]. A comparable situation exists in the feline host [27]. Human BAC cells express Met receptor, the gene product protein of the c-*met* proto-oncogene. The avian erythroblastosis retrovirus S13 posseses a related oncogene—the v-*sea* (for sarcoma, erythroblastosis, and anemia). The cognate ligand for Met is the hepatocyte growth factor (HGF) produced by platelets and neutrophil leukocytes. Bronchioalveolar carcinoma tumors attract neutrophil leukocytes and receive HGF from them [28]. Xenografted BAC tumors may be deprived of such support and therefore succumb readily to a kinase inhibitor that these tumor cells would have resisted in their natural environment. However, Met receptor and HGF antagonists (K252a; HGF/NK4) may suppress these tumors best in their natural environment [29–32]. Ribozyme-carrier adenovirus suppressed the growth of c-*met*–mutated other (prostate cancer) tumor xenografts [33]. Brochioalveolar carcinoma is considered to be susceptible to gefitinib (Iressa) (see Part I). The c-*met* oncogene is mutated also in SCLCs [34]. This latter tumor frequently uses the bombesin/gastrin-releasing peptide receptor (BN/GRP) and its ligand for an autocrine growth loop [35]. Monocyte Fc receptor-mediated ADCC, in combination with cisplatin, etoposide, and paclitaxel chemotherapy, acted additively against SCLC. Non-SCLCs are driven by the glycoprotein osteonectin (SPARC; secreted protein acidic and rich in cystein) of stromal cell origin. The locus of the osteonectin gene is at 5q31-33; it dictates cell-matrix interactions during ontogenesis. When reactivated during oncogenesis, its activities correlate with worsening prognosis [36]. Lung adenocarcino-

Table 5 Target: Epidermal Growth Factor Receptor

cetuximab (C225)→ E G F-R
↓ ↓ ↓
LY294002→ PI3K STAT Ras ←Gefitinib (AstraZeneca)
↓ ↓ erlotinib (Genentech)
←EKB −589 (W-A)
PKI-166 (CGP59326) (N)
←GW2016 (GSK) CI1033
(PD183805) (Pf)
PTEN→ Akt Raf
↓
U0126→ MEK/MAPK/ERK

jun/fos/myc cyclin D1

anti-apoptosis neo-angiogenesis invasiveness
↑VEGF ↑MMP-9
↑bFGF
↑IL −8

EGF-R 170 kD transmembrane protein; dimerizes with HER; ligates EGF & TGFα. Overexpressed in human neoplasms: H&N squamous cell carcinoma 80%–100%; NSCLC, 40%–80%; prostate cancer, 40%–80%; ovarian cancer 30%–70%; colorectal cancer 20%–70%; pancreatic cancer, 30%–50%; breast cancer, 30%. W-A = Wyeth-Ayerst; N = Novartis; GSK = GlaxoSmithKline; Pf = Pfizer

Summary of clinical trials in advanced NSCLC: standard chemotherapy plus EGF-R inhibitors

Response (%)	erlotinib	gefitinib	cetuximab	best supportive care
PR/CR	12	12–18	22	0
Stabilization	39	31–36	33	no data
Overall survival (mo.)	8.4	6.5–7.9	7.6	4.6
Alive at 1 year (%)	40	24	no data	11

Bronchioalveolar carcinoma (BAC): of 50 patients treated with erlotinib 13 responded: 26%; response rate of BAC to gefitinib 38%; to paclitaxel 14%. Median duration of response >6 months. Grade of skin rash positively correlates with survival. Gefitinib and erlotinib: clinical response does not necessarily correlate with grade of EGF-R expression (responding ↓EGF -R tumors; nonresponding ↑EGF -R tumors). Lack of significant additivity to chemotherapy results. In H&N squamous cell carcinoma, ↑EGF -R dictates adverse prognosis. Response to gefitinib correlated with skin reaction; severe: 11 months, none 5 months median survival.

Table 5. Data presented at the 10th World Conference on Lung Cancer are based on the bulletin issued on January 2004 by the American Academy of Continuing Medical Education www.academycme.org

mas evade host immunity by overexpressing FasL to kill host NK and T cells and downregulate IL-13Rα2 to facilitate metastatic growth in the brain [37]. The phytoalexin resveratrol (present in red wine) inhibits the anti-apoptotic NFκB and the phosphorylation of the Rb protein and suppresses the growth of lung (and other) human cancer cells in xenografts [38,39].

Natural Killer Cells

Athymic and other immunodeficient hosts of xenotransplanted human tumors cannot launch an immune T cell-mediated attack on their newly introduced parasite. If tumor cells defended themselves in their natural host by expressing FasL to kill Fas receptor-positive immune T cells of the host [40], now they may safely downregulate the expression of FasL, but may overexpress survivin instead [41]. If the tumor concealed certain tumor antigens (fusion oncoproteins, point mutated oncoproteins, survivin), now these antigens may reappear without being targeted by host immune T cells. The newly transplanted tumor should protect itself against NK cell- and macrophage-mediated attacks of its new host and modify its cell surface accordingly by concealing those antigens that NK cell receptors (KIRs, NKp30, NKp44, NKp46, NKG2D) recognize to escape ADCC reactions of host NK cells and macrophages.

Insulinlike Growth Factor-R

Tumor cells surviving in their new host may switch from the dominant oncogenic cascade that they used in their natural host to a secondary, substitute cascade serving them better in their new microenvironment [42–46]. For example, tumor cell populations deprived of their natural contact with their matrix (anoikis) undergo apoptotic death unless they switch to EGF-R–mediated signal transduction for survival. These cells are exquisitely susceptible to the blockade of EGF-R signal transduction and succumb to it, whereas in their natural environment integrin-linked kinases and IGF factor-induced kinase cascades could have protected them. For example, some colon cancers use IGF and its receptor for an autocrine growth loop.

Signal transduction from the IGF-R activates the c-*src* oncogene. Insulinlike growth factor and its binding protein promote the growth of glioblastoma cells. High levels of IGF and its binding protein in the blood predispose to malignant transformations (breast cancer and squamous cell carcinoma of the uterine cervix) [47–50]. Angiotensin II activates the IGF-R. In colon carcinoma cells, the COX/PGE_2 pathway activates the IGF-R; these tumor cells produce IGF-II and its ligand for an autocrine growth loop. Angiotensin II inhibitors (enalapril, perindoprilat) downregulate IGF-R expression in colon cancer cells. The growth of prostate cancer xenografts is inhibited by angiotensin II receptor-1 blockade by candesartan (CV11974, Takeda Pharmaceuticals North America, Lincolnshire, IL, subsidiary of Takeda Chemical Industries, Japan) [51,52]. Do xenografted cancer cells continue the production of IGF and express its receptor? If not, these cells become more vulnerable to noxious agents. Differences between EGF-R and IGF-R expression in the tumor's natural environment and in its new host after xenografting may increase the susceptibility of the tumor cell population surviving in the xenograft to the agents that are being assayed for therapy.

Activated Stroma

Tumors "activate the stroma" and fibroblasts of the tumor's native microenvironment express FAP, the fibroblast activation protein, a protease against which a monoclonal antibody (sibrotuzumab) has been developed [53,54]. When we assay for an oncolytic virus, do we check whether the xenograft's fibroblasts are activated and express FAP? The relationship of melanoma cells and stromal fibroblasts entails direct contact and "cross-talk" by means of molecular mediators and their receptors (cadherins, bFGF, PDGF, VEGF, etc.), resembling a "conspiracy" between tumor and stromal cells [55]. Not only tumor cells but also stromal cells, even when only moderately positive for HER2/neu, and even without HER2 amplification, could react to trastuzumab (Herceptin, Genentech): Treated stromal cells ceased to support breast cancer cell adhesion and growth in co-cultures [56]. There is a formidable lack of knowledge regarding the behavior of stromal cells toward the heterologous xenografts.

Most of these events, and more, are depicted in a table and in a cartoon [57,58], which propose that a fusion oncoprotein is generated in melanoma cells. The extracellular domain of this putative oncoprotein is supposed to be the Fas receptor (CD95); its intracellular domain is the G-CSF receptor (G-CSF-R). Chromosome breaks occurring in melanoma cells at the G-CSF-R and Fas loci (1p32-34 and 10q23-26, respectively) would bring about the formation of this fusion oncoprotein: t(1; 10)(p32-34;q23-26). Attachment of FasL to this receptor would induce cell divisions. The FasL may be part of an autocrine growth circuit of the tumor cell or it may derive from the host. Such a fusion oncoprotein in melanoma cells would make a good target for TKI and monoclonal antibodies. Is it the tumor's natural environment or its tissue cultures or xenografts that are most conducive to natural hybridoma formation?

Intercellular Adhesions

It was in the 1940s that Professor T. Huzella published in Hungarian and German his first astute observations on cell-matrix interaction, as he visualized them in his tissue cultures [59]. The Hungarian school continues to make contributions to the understanding of these relationships in a recent volume edited by Eva Klein in Stockholm [60–62]. In elastin-matrix interactions, the tumor cells produce elastin; of proteoglycans, decorin inhibits and perlecan promotes, tumor cell locomotion; and the aging process alters the extracellular matrix, rendering it more favorable for invasion by incipient tumors. Fc receptor expression and metalloproteinase interactions in the tumor bed are major determinants of the host–tumor relationship [63]. Laminin isoforms interact with MMPs [64,65]. Legumain overexpression by the tumor promotes its invasiveness [66]. All these features may be lost or altered in the xenograft. The heterodimeric transmembrane receptors of cell surfaces, the integrins, connect cells and matrix; fibrinogen, fibronectin, vitronectin, and collagens are the major ligands of integrins. Alpha-v integrin inhibitors (the humanized monoclonal antibody Vitaxin directed to $\alpha_v\beta_3$; or cilengitide, mimicking the recognition of the RGD ligand) prevent tumor cell migration [67–70]. The integrin antagonist S247 inhibits the formation of liver metastases by colon carcinoma cells in mice and it is anti-angiogenic [71]. Obtustatin purified from vipera venom is a selective inhibitor of $\alpha1\beta1$ integrins [72]. Integrin-linked kinase expression by melanoma cells correlates with enhanced invasiveness and poor prognosis [73]. Periostin (member of the family of axon-guiding proteins in mammals and fascilin in insects) reappears in ovarian carcinoma cells, serves as a ligand to integrins $\alpha_v\beta_{3,5}$, and mediates tumor cell adhesiveness and migration; anti-integrin monoclonal antibodies counteract its effects [74]. Sugen (SU6668), by antagonizing VEGF-R2, FGF-R1, and PDGF-R, suppresses the intraperitoneal spread of, and ascites formation by, ovarian carcinoma xenografts [75]. Thus, once recognized, factors that promote invasiveness and metastasis formation of tumor cells in their natural microenvironment can be effectively antagonized in vitro and in vivo, as their assays in this case appear to bear clinical validity.

Natural Hybridomas

Tumor cells under attack within their microenvironment may resort to dramatic escape maneuvers (that may or may not take place in xenografts). When tumor cells express the FasL gene encoding the ligand, they not only may kill host lymphocytes, they may utilize the FasL→Fas receptor system as an autocrine growth loop, such as in human melanoma cells [58,76]. Tumors expressing FasL attract neutrophil leukocyte infiltrates; when rejected, these tumors may immunize their host against live tumor cell challenge. However, these $FasL^+$ tumors were able to deflect all host immune reactions when they produced TGFβ and reduced costimulatory CD80 expression [43,77]. Fusion of tumor cells with subverted defensive or other cells of the host is referred to as "natural hybridoma formation" [78]. Cells of a retrovirus-producer murine lymphoma fused with plasma cells of the host; the tetraploid cells remained highly malignant and continued to secrete the specific antiretroviral immunoglobulins originating from the plasma cell component of the fusion product. If it were called "hybridoma" (it was not) at the time when this phenomenon was first observed, reproduced, and reported in 1969–1970, these continuously antibody-producing fused cells would be cited as the very first example of "natural hybridomas" (J. Sinkovics, Chapter 1 in this volume) [79–82]. Committed or noncommitted lymphocytes may fuse with macrophages or DCs (in a case of a murine lymphoma; in AIDS under the effect of HIV-1; in Reed-Sternberg cells under the effect of an as-yet-unidentified retrovirus) [83–85]. In the case of a murine sarcoma, lymphocytes "spontaneously" fused with semiallogeneic host macrophages and thereby gained increased malignancy [86].

THE MICROVASCULAR TUMOR BED

Microvasculature

The microvasculature of the tumor bed, consisting of active vascular endothelial and stromal cells, whether developing or established in the natural host or in xenografts, regulates the influx of nutrients and efflux of metabolites; the degree of hypoxia; and the access or exclusion of lymphocytes, macrophages, antibodies, and chemo- and cytokines to the tumor. Of the VEGF-R family, VEGF-R3/Flt-4 and its ligands VEGF-C and VEGF-D induce lymphangiogenesis and spread of tumor cells through lymphatic vessels [87]. Of the molecular mediators in interplay, endostatin exerts, through its COOH terminal fragment inhibitory and through its fragment III, stimulatory effects on new blood vessel formation. Endostatin preparations from Calbiochem (EMD Biosciences, San Diego, CA) and EntreMed (Prince William, MD) differ: Only the former preparation in a HIF-independent manner inhibits endothelial tube formation [88,89]. Vasculogenic agents may be expropriated by malignant cells to function as growth factors: Acute myeloid leukemia cells produce TGFβ and its receptor, and chronic myelogenous leukemia cells produce VEGF and its receptor for autocrine growth loops [90,91]. The 21 amino acid vasoactive peptides, the ETs, work in breast cancer cells as overexpressed autocrine and paracrine growth loops [92,93]. Angiotensin (Ang) 1 and 2 are ligands for Tie 2 (see Part III), and Ang 1 stimulates the glomeruloid microvessel proliferation in the tumor bed of NSCLC [94]. Thrombospondin-1 is anti-angiogenic. Tumors expressing no TSP-1 at their deep invading margins spread and metastasize [95]. Interleukin-10 is referred to as an immunosuppressive cytokine; it inhibits TGFβ-dependent neoangiogenesis and downregulates VEGF. Yet its low expression correlates with tumor spread and metastasis formation. VEGF overexpression at the deepest invading edge of a tumor forecasts heavy microvessel density and rapid tumor growth, with short survival of the patient. Low TSP-1 and high VEGF expressing tumor cells exhibit the most rapid rate of expansion. The TSP-1 gene is silenced in neuroblastoma cells by methylation; 5-aza-2-deoxycytidine demethylates the gene and restores TSP-1 expression, with regression of tumor xenografts [96]. Angiogenins and tissue factors, the glycoproteins that initiate blood coagulation, are responsible for the hypervascularization of hepatocellular carcinomas [97,98]. Ovarian carcinoma cells under stress (exposed to norepinephrine) increase their VEGF production; the process is reversed by propranolol [99]. The hemangioblasts of VHL disease are VHL gene-defective neoplastic cells that co-express erythropoietin receptors and ligands for an autocrine growth loop, but they may differentiate toward erythrocyte (red blood cell) formation [100]. Vascular mimicry practiced by melanoma cells is modulated by the tissue factor pathway inhibitors-1 and 2 (TFPI-1 and -2) and may be inhibited by antibody to TFPI-1 or TFPI-2 [101]. The tumor microvasculature may form a barrier to antitumor immune reactions of the host. Histamine receptor antagonists famotidine and cimetidine allow transgression of the barrier by tumor-infiltrating immune T cells [102]. Solubilized Flt-1/VEGF-R1 inhibits HER2/neu$^+$ mouse mammary tumor growth. Immune T cells directed at HER2/neu antigens were most efficient in suppressing tumor growth in mice that were overproducing sFlt-1 [103].

Suppressors

Antiangiogenic agents include the COX-2 inhibitors celecoxib [104] and SU5416 (Sugen) [105]. Treated cervical or lung cancer cells fail to produce prostaglandins. The dipeptide boronate proteasome inhibitor PS-341 (Proscript, Millenium Pharmaceuticals) inhibits NFκB activation, depriving pancreatic cancer cells of an anti-apoptotic survival mechanism [106]. The dithiolethione antischistosomal agent oltipraz inhibited the growth of a murine angiosarcoma [107]. CP-547,632 (Pfizer) inhibits autophosphorylation of tyrosine kinases and their signal transmittal from VEGF-R-2 and from bFGF-R. It is strongly anti-angiogenic [108]. VGA1155 (Taisho Pharmaceuticals, Tokyo, Japan) blocks binding of VEGF to receptor KDR/Flk-1 [109]. The synthetic glycine ester CEP-7055 (Cephalon, USA; Sanofi, France) decreased microvessel density and inhibited the growth of glioblastoma, breast, colon, lung, pancreatic, and prostate cancers and an angiosarcoma in xenografts [110]. Orthotopic human pancreatic cancer xenografts are suppressed by a VEGF-neutralizing antibody [111]. The growth of orthotopic human breast cancer xenografts is inhibited by a monoclonal antibody blocking VEGF binding to VEGF-R2 (KDR/Flk-1) [112]. EntreMed's rh-angiostatin given twice daily subcutaneously was nontoxic and stabilized advanced cancers in 6 of 24 patients [113]. Platelet-activating factor (PAF) mediates inflammatory cytokines and vasculogenesis. It is inactivated by the enzyme PAF-acetylhydrolase (AH). Kaposi's sarcoma xenografts transfected to overexpress PAF-AH grew very slowly because of the lack of neovascularization [114]. It appears that most established xenografts elicited neovascularization, so agents that act against formed blood vessels can be reliably tested in this system; suppression of new blood vessel formation during the first days after xenograft implantation is much more difficult to interpret correctly.

CHEMOKINE AND CYTOKINE GROWTH FACTORS

Expropriated Cytokines

Tumor cells expropriate inflammatory cytokines for their auto- or paracrine-driven growth loops. These growth factors may not be readily available to them in their new microenvironment. Examples of cytokine and other common growth factors expropriated by tumors for their own growth advantage have been cited elsewhere (J. Sinkovics, Chapter 1 in this volume) [115,116]. Tumor cells can convert molecular mediators that are physiological inducers of cell death to serve as their growth factors: TNFα for neuroblastoma and melanoma [117]; FasL for melanoma [58,76]; IL-2 for T cell lymphoma-leukemia and for some melanomas; IL-6 for Kaposi's sarcoma, multiple myeloma, kidney carcinoma, and hormone-refractory prostatic carcinoma [118]; IL-7 for Sézary cell lymhoma and cutaneous T cell lymphomas; IL-8 for Kaposi's sarcoma [119]; IFN-γ for B-CLL [120], Kaposi's sarcoma, and some osteosarcomas (see Part III). In contrast, IFNγ exerted cytotoxicity on ovarian carcinoma and mesothelioma cells [121,122]. In ovarian carcinoma patients in clinical trials, INF-γ could induce PR [122]. Interferon-γ cooperates with IL-12 in inducing Th1-type antitumor immunity [123]. Low levels of IFN-γ are generated in the bladder mucosa during BCG treatment of bladder carcinoma. High levels of INF-γ are tumor suppressive, but low levels may serve as growth factors for the tumor [124]. Hormone receptor-negative breast cancer cells overproduce IL-1α and attract metalloproteinases to facilitate tumor cell invasiveness [125]. Some melanoma cells produce constitutively IL-1. The natural IL-1R antagonist IL-1ra suppressed the metastatic potential of these melanoma xenografts [126]. Autocrine production of IL-4 and IL-10 by thyroid carcinoma cells upregulates Bcl-2 and Bcl_{XL}, rendering tumor cells to be chemotherapy-resistant [127]. Interleuken-6 soluble receptor and TGFβ blood levels are high in patients with prostate cancer and correlate with adverse postoperative prognosis [128]. Interleuken-6 may be neutralized by monoclonal antibodies BE-8 or CNTO328 in clinical trials for those tumors in which IL-6 may have a promotional role [129]. Interleukin-8 protects ovarian carcinoma cells against the TNFα-related apoptosis-inducing ligand TRAIL-induced apoptosis [130]. The double agent IL-10 displays a dual role in tumor immunology [131]. While it stimulates NK cells, it suppresses the maturation of DC. Antigens presented by immature DC are often tolerogenic, whereas antigens presented by mature DC are immunogenic.

In some gastric cancers, EBV-induced small RNA mediates a mitotic cascade [132]. In some Burkitt lymphoma cell lines (Akita), EBER-induced IL-10 acts as an autocrine growth factor [133]. Interleukin-15–producer colon carcinoma cells induce neoangiogenesis in the mucosa adjacent to the tumor [134]. The heterodimer IL-23, consisting of p40 (IL-2–like protein) and of the novel p19 protein (distantly related to IL-6, G-CSF, and IL-12 p35) exerted tumor-growth inhibition as expressed in murine colon carcinoma and melanoma cells, and rendered these hosts immune against challenge with tumor cells [135]. Replication-defective adenovirus Ad.*mda*-7 expressing IL-24, in combination with glutathione transferase treatment, synergistically suppressed the growth of kidney carcinoma cells by generating free radicals; N-acetyl cysteine, by neutralizing free radicals, abolished this effect [136]. Human mesothelioma cells transduced by Ad-IFN-γ, the E1E3-deleted replication-defective recombinant adenovirus type 5, delivering the IFN-γ gene inserted into a plasmid and also containing a CMV promoter/enhancer and a SV40 polyadenylation signal, inhibited tumor cell growth in vitro and induced antitumor immune reactions in the nude mouse, consisting of NK cells and macrophages (see Parts I and III) [121,137–139]. VEGF production and survivin overexpression in mesothelioma cells could be suppressed effectively (see Part III). Are cytokines induced in xenografts by oncolytic viruses the same or different from those that arise in patients who are enrolled in a clinical trial with these viruses?

Interleukin-18, the IFN-γ–inducing factor, is produced by osteoblasts and macrophages and acts through GM-CSF in inhibiting the multinucleation of osteoclasts. By upregulating osteoprotegerin (a TNF family member) and obilizing GM-CSF–producing T cells, IL-18 antagonizes osteoclastogenesis. Interleukin-18 inhibited the lodging of multiple myeloma cells (see Part V) in the vertebrae of SCID mice through the activation of NK cells. With IL-12, it inhibited bone resorption by osteoclasts. Physiologically, IL-18 serves as an auto- and paracrine growth factor for osteoblasts, whereas M-CSF and IL-11 are osteoclast growth factors. If effective IL-18 inhibitors could be generated, would they suppress the growth of osteosarcoma cells? In mice bearing human MDA-231 osteolytic breast cancer cell xenografts, IL-18 inhibited osteolytic bone metastasis formation [140–147]. Therefore, if the xenografted host produces IL-18, osteoclasts will be suppressed, IFN-γ will be produced, NK cells will be activated, and growth of the xenograft will be naturally inhibited. In multiple myeloma cells, FasL and Fas receptor co-expression may be upregulated, leading to self-destruction by apoptosis of the xenografted myeloma cells (see Part V). Interleukin-18 production by the xenografted host may invalidate an assay in which an agent is tested against a transplanted tumor.

Matrix Metalloproteinases and Chemokines

Matrix metalloproteinase-2 and MMP-9 (see Part II) stimulate VEGF release from ovarian carcinoma cells and batimastat (BB-94) inhibited this process. Endothelin or VEGF-producer ovarian carcinoma xenografts invaded the abdominal cavity and grew as ascitic tumors [148–150]. Matrix metalloproteinase-2 and MMP-14 promote melanoma cells to assume tubular formations referred to as "vascular mimicry" (see Part III). Antisense oligonucleotides directed at laminin and antibodies directed at MMP-2 and MMP-14 inhibited the process of vascular mimicry [65,151]. Gelatinolytic MMP-2 (gelatinase A) and MMP-9 (gelatinase B) promote the progression of uterine cervical carcinomas and liver metastases of colon carcinomas [152,153]. The production of macrophage migration inhibitory factor (MIF) is increased by lysophosphatidic acid [154]. Macrophage migration inhibitory factor promotes neoangiogenesis and tumor growth. Antisense-MIF plasmids and antibodies neutralizing MIF inhibited tumor (murine colon carcinoma) growth [155]. The Canadian preparation Virulizin derived from bovine bile (Lorus Therapeutics, Toronto) suppresses the growth of human tumor xenografts and stabilizes the disease in patients with advanced cancers. The mechanism of action is claimed to be macrophage activation [156]. Indeed, macrophages could be acting, as designed, against auto- or allogeneic foreign invaders, or yield to subversion and support the growth of autologous or transplanted tumors.

Before modern technology (labeled monoclonal antibodies specific to the gene product protein) it was not easy to detect and titrate cytokines [157]. Macrophage migration inhibitory factor drives the growth of certain colon carcinoma cells and an antisense oligonucleotide directed at the mRNA of MIF inhibited the growth of these tumors. Furthermore, tumor-infiltrating macrophages secrete VEGF and induce neoangiogenesis in favor of the tumor [158–160]. Proinflammatory cytokines (TNFα, IL1β, and IL-6) block growth of breast carcinoma cells by inhibiting IGF-1–induced DNA synthesis [161,162]. Insulin growth factors and their receptors serve many tumors in the role of autocrine growth circuits. Ligand- or receptor-specific monoclonal antibodies, As-ODN (see Part III), or the mTOR-targeting rapamycins could neutalize or switch off the signal cascade emanating from the ligand-bound IGF-R [163–171]. Are these growth factors readily available in the murine host for xenotransplanted human breast cancer cells? Stromal cell cultures deriving from rat lymph nodes could provide such growth factors (IGF-1 and EGF) for human breast cancer cell lines established in culture [46]. Desmoplastic reaction [172] to certain tumors (breast carcinoma, carcinoid, melanoma, pancreatic carcinoma) may be defensive or may be a host reaction promotional to tumor growth, which may be absent in xenotransplanted tumors.

Chemokine receptor CXCR4 is activated by its ligand CXCL12 and the system is operational in metastatic melanoma. Tumor cell growth is promoted by interactions between melanoma cells and β1 integrins of endothelial cells. Other inflammatory chemokines (CCL3, CCl20) attract CD8 immune T cells and NK cells into a colon carcinoma tumor model. Expression of CCR7 chemokine receptor in esophageal carcinoma cells endows the tumor cells with increased invasiveness and metastatic potential. Human pancreatic cancer cells express the chemokine MCP-1/CCL2 and attract macrophages. High CCL2 levels and extensive macrophage invasion correlated with increased IFN-γ and TNF-α production and increased the rate of apoptotic tumor cell deaths [173–176]. If stromal cells and macrophages of the xenograft-bearer host are not immediately subverted, they may not provide cell growth promoting MMPs and chemokines for the tumor. The tumor deprived of this support may experience increased vulnerability from noxious substances assayed against it.

Mesothelioma

Antisense Bcl-2/Bcl_{XL}-directed interventions disable xenografted mesothelioma cells (see Part III). Mesotheliomas often contain SV40 genomic sequences (acquired from the first-generation polio vaccines?) [177], inactivate the wild-type p53 by deletion of p14, and activate the Wnt pathway. Antisense Bxcl-2/Bcl_{XL}-directed therapy disables xenografted mesothelioma cells (see Part III) [178–180]. The term Wnt stands for the derivatives of the *Drosophila* gene *wingless* and the mouse gene *int*; these two genes found fused are called *wnt*; they encode glycoprotein ligands that are cognate to the "fizzled" receptors. The Wnt pathway is essential in embryonic life; when it malfunctions, fruitflies emerge without wings or tadpoles grow two heads. The Wnt-related gene "dickkopf" directs development of the head and eyes (see Parts III and V). The Wnt pathway suppresses the enzyme glycogen synthetase kinase 3β. This enzyme, when functional, eliminates β-catenin, the cadherin-associated protein, by ubiquitination and proteasomic degradation. Catenin mediates the linking of cadherin to the cytoskeleton. In mesothelioma cells, β-catenin is overexpressed and induces transcription of c-*myc* and *cyclin*D genes [181–184]. In addition to mesothelioma, Wnt is malfunctioning in melanoma cells, colonic polyps (APC; FAP), gastric adenomas, hepatocellular carcinoma, medulloblastoma, aggressive fibromatosis, and soft-tissue sarcomas [185,186]. In colonic polyps (APC), loss of APC protein liberates Wnt signal transduction, as β-catenin translocates into the nucleus and the downregulating effect of APC on COX-2 ceases to exist [186]. Cyclooxygenase-2 is active in colon carcinogenesis. The gene of fibroblast growth factor 18 (FGF18) is targeted by β-catenin and is overexpressed in colon carcinoma cells, where it functions as a growth factor. The 23 FGF genes regulate organogenesis in embryonic life. Of EGF-R inhibitors (C225, Erbitux, Imclone, New York, NY; gefitinib, Iressa, AstraZeneca; OSI-774, erlotinib, Tarceva, Genentech) (see Part I), ZD1839 suppressed EGF-mediated signaling in EGF-R^+ mesothelioma cells (see Part III) [187] (Table 4).

Catenins, Caveolins, Connexins

The Sloan Kettering defective retrovirus possesses the oncogene v-*ski*. The human cellular homologue c-*ski* resides at 1p36 [184]. The SKI protein antagonizes TGFβ, inactivates the Rb pathway, and activates [1] Wnt/β-catenin signaling, [2] the tissue-specific coactivator of the androgen receptor FHL2, which binds β-catenin and functions as a transcriptional repressor, and [3] the melanocyte master regulator MITF (microphthalmia-associated transcription factor) in melanoma cells. These activated pathways render an essential contribution, in addition to mutated RAS/MAPK and BRAF pathways (see Part I), to the maligancy of this tumor. In early stages of melanoma, there is cytoplasmic β-catenin but its expression is reduced in the advanced stage of superficial spreading melanoma [185]. The gene-targeting vectors β-cateninKO-neo and –hyg delete catenin exons II-IV from colon carcinoma cells with potential therapeutic valuue [188]. Co-activation of β-catenin, c-*myc,* and PKC (see Parts II and III) downregulates caveolin-1 (Cav-1). Caveolae, the caveolin-coated cell surface invaginations, interact with lipid rafts, participate in endocytosis and molecular trafficking, emit signals, and exercise tumor suppressive functions, as cells in their terminally differentiated stage overexpress caveolins. Caveolin-1 inactivates H-Ras, Src kinases, G proteins, EGF-R, nitric oxide, and PKC. In the thyroid gland, caveolins act as tumor suppressors, and they are downregulated in follicular thyroid carcinoma [189–193]. Caveolin-1 expression is depleted in HPV-transformed squamous cells of the uterine cervix [194]. In the thyroid gland, caveolins act as tumor suppressors and are downregulated in follicular thyroid carcinoma [189]. In contrast, in hormone-resistant prostate cancer cells, Cav-1 is overexpressed and soluble Cav-1 stimulated the growth of prostate cancer cells [195]. The behavior of caveolins in xenografted tumors is not known. Do they retain their tumor-suppressive function, or are they switched to be subverted to function as tumor promoters? The connexin proteins form gap junction channels within the membranes of adjacent cells, allowing molecular trafficking. Connexin proteins are inhibitory to cell divisions and act as tumor suppressors; the connexin 43 gene (*Cx43*) is a recognized tumor-suppresor gene. Connexins are depleted in lung cancer and glioblastoma cells. Cessation of growth and loss of tumorigenicity was displayed by lung cancer cells transfected with *Cx43* cDNA [196,197]. It is not known how connexins of xenografted human tumor cells relate to the heterologous stromal cells of their new host.

Autocrine Circuits

Tumor cell populations derive paracrine growth factors from their stroma (fibroblasts, endothelial cells, monocyte-macrophages, microglia cells) and manufacture their own growth factors in autocrine loops. Examples for this latter event abound. Among many others, these are G-CSF for transitional cell carcinoma of the urinary bladder [198], acetylcholine or bombesin for small-cell carcinoma of the lung [199], and erythropoietin (induced by hypoxia) for breast cancer cells [299,201]. The EBV-encoded small RNA (EBER) induces the production of IGF-I in gastric carcinoma cells, which also express IGF-IR and use the system as an autocrine growth loop [132]. Insulinlike growth factor and its receptor are highly promotional for tumor growth and targeted for therapy [202], but their behavior in xenografts is not well understood. When tumor cell populations are extricated from their natural environment, the subpopulation dependent on paracrine growth support is expected to perish; the subpopulation self-sustaining by autocrine growth loops would prevail. This latter tumor cell subpopulation is the one that is assayed in tissue cultures or in xenografts.

The EGF homologue TGFα promotes tumor cell growth and invasiveness (colon carcinoma, glioma, melanoma). TGFβ inhibits tumor cell growth but paracrine growth factor-independent tumor cells may gain resistance to it and some tumor cells (colon, breast, cholangiocarcinoma) may expropriate

TGFβ and its receptor for an autocrine growth loop [42,203,204]. The growth regulatory role of TGFαβ encompasses many tumor categories, from breast and ovarian carcinomas to brain and neuroectodermal tumors [203–208]. Tumor growth factor-β transactivates the VEGF gene in cholangiocarcinoma cells, whereas vascular endothelial cells in the tumor bed overexpress VEGF-R (see Part III) [209]. Tumor growth factor-β is an inhibitor of hepatocellular carcinoma cells; when it is switched off, the tumor cells produce VEGF and gain enhanced malignant potential [210]. TGFβ-R kinases are inhibited by SB431542; PDGF-R kinases are inhibited by AG1296, and c-Kit kinases are inhibited by imatinib (see Part I) in osteosarcoma cells that replicate in response to TGFβ and PDGF [211–213]. The sensitive TGFβ receptors are Alks; sphingosine-1 phosphate and AKT drive PDGR-R [213]. How do xenotransplanted human tumors set the balance of TGFαβ in their new host?

Stromal cell-derived factor-1 (SDF-1) is produced in the bone marrow and is the cognate ligand of chemokine receptor CXCR4, which is expressed in B-lineage lymphocytes and CD34 hematopoietic progenitor (stem) cells. It stimulates formation of pseudopodia and cell migration, especially in lymphoid cells. It activates the anti-apoptotic NFκB system [214]. The c-*rel* gene encodes the p50 and p65 subunits of the NFκB heterodimer. The v-*rel* gene was discovered in the RevT reticuloendothelial retrovirus, inducer of bursal lymphomas in turkeys. NF-κB is constitutively activated in Reed-Sternberg cells of Hodgkin's disease and is under the promotional control of SDF-1 in multiple myeloma and in B-CLL cells, protecting these cells in collaboration with IL-6 from dexamethasone-induced apoptosis. The activation of NF-κB in muliple myeloma cells can be inhibited by anti-CXCR4 monoclonal antibodies rendering the tumor cells vulnerable to pro-apoptotic stimuli (see Part V). Stromal cell-derived factor-1α (SDF-1; CXCL.12) is the ligand of chemokine receptor CXCR4 expressed on endothelial cells responding to angiogenic stimuli and on monocytes obeying chemotactic stimuli. Stromal cell-derived factor-1 does not limit its stimulatory effect to B-lineage lymphoid cells; under its effect, breast cancer cells emit pseudopodia and migrate [215]. It protects from death explanted B-CLL cells and myeloid progenitor cells of the bone marrow [58,216]. PDGF secreted by tumor cells and endothelial cells stimulates the pericyte to release VEGF, which induces endothelial cells to form neovasculature in the tumor bed (see Part III). The distamycin analogue NSC651016 (Suradista, Pharmacia; Upjohn; Pharmitalia Milan, Italy) inhibits neoangiogenesis induced by CXCL.12. Suradista and its analogues (see Part III) exert an antitumor effect for xenografted tumors driven by bFGF, but without the toxicity of suramin[217,218]. Tannic acid in nontoxic concentrations inhibits CXCL.12 (the ligand) and suppresses the migration of monocytes and breast cancer cells [219].

Virus-Induced Cytokines and Chemokines

Chemokine receptor CXCR4 serves as coreceptor with CD4 for the entry of HIV-1; some of the molecular inhibitors of HIV-1 entry may also downregulate the NFκB anti-apoptotic system in some human malignant tumors. Human immunodeficiency virus-1 protease inhibitors exert anti-angiogenic effects in KS (see Part III). Acquired immunodeficiency virus-associated KS cells are driven by SDF-1 of capillary endothelial cell origin. Stromal cell-derived factor-1, the ligand of receptor CXCR4, inhibits entry of HIV-1 into susceptible cells. Polyphemusin (from *Limulus polyphemus*) also interacts with CXCR4 in an inhibitory fashion regarding HIV-1 entry. The viral macrophage inflammatory protein (vMIP-II) encoded by the KS-associated HHV-8 prevents signal transduction from CXCR4. Viral macrophage inflammatory protein-I is the ligand of chemokine receptor CCR8, which is expressed on KS cells. Whereas IFN-γ is listed as a growth factor for KS cells (see Part III), the IFN-γ–inducible chemokine protein (IP-10), vMIP-II, or SDF-1α inhibit signal transduction from KS cells' G-protein-coupled receptors (GPCR) [220–225]. Human immunodeficiency virus-1–infected macrophages or primary effusion lymphoma cells (with or without HHV-8 co-infection) use NGF for an autocrine growth circuit [226,227]. Is there any connection between the NGF and SDF-1 to CXCR4 circuits and, if so, how is it altered in xenografts of KS cells? Alveolar rhabdomyosarcoma cells driven by the PAX3, 7/FKHR fused-gene product protein growth factor also overexpress CXCR4 and accept SDF-1 as a growth factor [228]; the CXCR4 inhibitor T140 blocks this growth circuit. Stromal cell-derived factor-1 facilitates the formation of bone metastases by prostate cancer cells [229]. When we infect patients or xenografts of their tumors with genetically engineered or naturally oncolytic viruses, are chemokines induced in both situations the same or different? Some chemokines exert tumor survival-enhancing effects, whereas others are detrimental to the tumor.

Viral infections induce proinflammatory cytokines: IFN-α and TNF-α by NDV; and these with IFN-β, IL-1β, IL-6, IL-10, MIP-1αβ, MCP1, MCP3, and chemokines by influenza A virus (especially H5N1) [230,231]. Investigators of the Wellstat Biologics Corporation (Gaithersburg, MD)"desensitize" patients who experience "acute toxicity" upon the first i.v. inoculation of PV701 oncolytic NDV preparation [232]. Certainly, these virally induced proinflammatory cytokines, contrary to their expected antitumor effect, may, on occasion, promote the growth of some tumor cells so programmed, given that the use of defensive cytokines or chemokines in the auto- or paracrine growth circuits of tumors is not uncommon.

HORMONE-RELATED EVENTS IN BREAST AND PROSTATE CANCERS

Complex Interactions

Allo- and xenotransplantations for the transfer of organs or lympho- and hematopoietic cells are extensively studied as potential therapeutic interventions [233,234]. Allo-transplanted lymphoid cells at the price of graft-vs.-host reaction may exert graft-vs.-leukemia or even graft-vs.-tumor immune reactions [235]. For preclinical assays of therapeutic agents, including oncolytic viruses, human tumors are transplanted into athymic hairless ("nude"), SCID (severe combined immunodeficient), or X-irradiated mice. There appears to be a discrepancy between the results of these preclinical tests and those of Phase I and II clinical trials. Although complete eradications of xeno-transplanted human tumors frequently occur, complete clinical responses remain a rarity, the most common clinical response being stabilization of the disease and, less frequently, PR and MR.

Androgen-dependent and independent prostate cancer xenotransplants responded equally well to the EGFR tyrosine kinase inhibitor ZD1839 (Iressa, Astra Zeneca) (see Part I) and exhibited increased susceptibility to bicalutamide, carboplatin, and paclitaxel. The polyphenolic flavonoid antioxidant silibinin/silymarin inhibited the growth of prostate cancer xenografts by neutralizing IGF through the upregulation of IGF-binding proteins. This substance is available as a dietary supplement and its consumption may lower PSA levels in patients with indolent prostatic carcinomas [236]. Quercetin inhibited Erb-2 and Erb-3 expression and antagonized TGFα and EGF-R signal transduction in, and thereby inhibited the proliferation of, prostate cancer cells [237]. Some Chinese herbal formulas contain substances that induce apoptosis in prostate cancer cells [238,239]. The isoflavonoid genistein and grape seed extract inhibit NFκB and the AKT signaling pathway in cultured prostate cancer cells [240–242]. The growth of prostate cancer xenografts was inhibited by combined treatment of the hosts with 1α,25-dihydroxyvitamin D_3 and 9-cis-retinoic acid [243]. Adenoviral insertion of the prolactin gene encoding the 16-kDa prolactin fragment into prostate cancer cells abolished xenograft formation of these tumor cells, presumably because of anti-angiogenic effects [244]. The apoptosis-inducing natural protein U19 is frequently underexpressed in prostate cancer cells because of loss of hetero- or homozygosity or hypermethylation of its gene; gene expression is testosterone-regulated and the gene product protein is pro-apoptotic [245]. The clinical use of the system remains unresolved.

Ansamycin antibiotics, especially 17-demethoxygeldanamycin (17-AAG), are inhibitors of chaperone proteins of the Hsp90 family [246]. HER2-positive and androgen receptor (AR)-mutated prostate cancer xenografts were inhibited by 17-AAG. Nonmutated AR tyrosine kinases can be phosphorylated within the pathways of other growth-factor receptors: HER2, EGF-R, IGF-R, and keratinocyte GF-R, and by IL-6. 17-Demethoxygeldanamycin interferes with these phosphorylation pathways (see Part III).

Monoclonal antibody rhuαVEGF inhibits the growth of several xenografted human tumors, among them the androgen-independent CWR22R prostate cancer cell line [247]. Certain prostate cancer cells express VEGF receptors for auto- or paracrine growth stimulation. Further, the VEGF gene is transactivated by androgens; VEGF-producing cells upon abrupt withdrawal of androgen support may succumb to apoptosis. In contrast, both VEGF and PSA levels rise in patients harboring androgen-independent tumors.

Human breast cancer xenografts did not grow or metastasize and are rejected consequentially to peritumoral or intraperitoneal injections of betaglycan, the soluble TGFβIII receptor that binds TGFβ with high affinity [248]. In certain tumor cells, including those of pancreatic carcinoma, TGFβ functions as an auto- or paracrine growth factor [249]. Some breast cancer cell lines developed at M. D. Anderson Hospital in the 1970s overexpress the *fgfr* genes encoding bFGF receptors; the MDA-MB-134 cell line is estrogen receptor positive and expresses elevated levels of *fgfr*-1 and *fgfr*-4 mRNAs because of an amplified *fgfr*-1 gene; however the ligands FGF-1 and FGF-2 inhibited the growth of these tumor cells even though the receptor FGF-R was functional [250]. Fibroblast growth factor-negative breast cancer cells transfected with *fgf*-1 or *fgf*-2 genes induced neovascularization and gained a rapid growth rate [251]. The histone deacetylase inhibitor PXD101 suppresses the growth of human colon and ovarian carcinoma xenografts; it was also tested against breast cancer cells [252,253] because another histone deacetylase inhibitor suberoylanilide hydroxamide acid (HDAC SAHA) (see Part II) induced differentiation of human breast cancer cells. Breast cancer cells with *ras* mutation are resistant to radiotherapy because of constitutively overactive PI3K and AKT signaling. PI3K inhibitor LY294002 reverted radioresistance and restored radiosensitivity of this tumor [254]. This molecular mediator also is strongly pro-apoptotic in pancreatic carcinoma cells [255]. Constitutively active AKT pathway deprives breast cancer cells of their estrogen and tamoxifen sensitivity [256]. The new antiestrogen ICI182780 (fulvestrant, AstraZeneca) downregulates cyclin D, inhibits c-*myc*, and transfers the CDK inhibitory protein $p21^{WAF/CIP}$ (wild-type p53-activated fragment/cycline-dependent kinase-interacting protein) from cyclin D/cdk4 to the cyclin E/cdk2 complex, resulting in inceased levels of CDK inhibitor $p27^{KIP1}$, which brings the cell to quiescence ($Q = G_0$) [257–259]. Stromal cell-derived factor-1 (*vide supra*) interacts with estrogen receptors and with estradiol in promoting breast cancer cell mitoses [259]. The PKC inhibitor staurosporine (see Part II) causes apoptotic death of human breast cancer cells [260]. The lectin galectin-3 is overexpressed on fibrosarcoma and on MDA-MB-435 breast carcinoma cells and facilitates metastatic spread; treatment of the xenograft-bearing host with a genetically engineered truncated form of galectin inhibited metastasis formation [261]. These estrogen-independent breast carcinoma cells overexpress Erb2/HER2 receptors and the c-*jun* and c-*fos* oncogenes, but respond to bombesin antagonists with cessation of growth [262,263].

Astonishing Observations

Two recent articles review stromal cell and tumor cell interactions in prostate and breast cancers [264,265]. Carcinoma-associated fibroblasts promote malignant transformation of benign hypertrophic prostatic epithelial cells. Reactive fibromuscular stroma predicts recurrence-free survival of patients with prostatic carcinoma [266]. Reduced desmin and smooth muscle α-actin in the stroma were adverse, whereas high labeling indices for these markers in the stroma were very favorable prognostic predictors. Fibroblast growth factor-R1 activation promoted and FGF-R2 activation inhibited the proliferation of prostatic carcinoma cells. Fibroblast growth factor-R1 signaling upregulated osteopontin, thereby enhancing anchorage-independent tumor cell invasiveness and VEGF overexpression; it downregulated the tumor-suppressive thrombospondin [267] (see Part III). Proteasomes prevailed throughout evolution from archaea (archebacteria) to the mammalian cell, where these structures degrade proteins, among them products of oncogenes (Myc) and those of tumor-suppressor genes (p53; PTN). The proteasome inhibitor bortezomib (PS-341, Velcade) (see Part V) induces apoptosis of androgen-dependent and -independent prostate cancer cell lines and xenografts thereof and synergizes with TNFα and TRAIL [268,269]. ER^+ mammary gland fibroblasts expressing HGF, collagens, fibronectin, EGF, and IGF-1 promote the proliferation of mammary epithelial cells, but laminin exerts antiproliferative activity through "loss of estrogen responsiveness" [64,270,271]. Transgenes encoding ribozymes that specifically target the HGF gene or the Met (the receptor) gene, transduced by retroviral vector into HGF-producer stromal fibroblasts or into MET-expressing breast cancer cells, silenced this paracrine growth loop [272]. HER2/neu could be suppressed by R11577 (tipifarnib, Zarnestra, Janssen) (see Part I) and trastuzumab or by the monoclonal antibody 2C4 [262]. The rapamycin analogue CCI-779 (Wyett-Ayerst) [273], the inhibitor of mTOR and the AKT cascade, suppresses the growth of breast cancer xenografts and, in a clinical trial, it showed activity (PR inductions) against hormone- and chemotherapy-refractory breast cancer, serving as one of the examples of a xenograft assay correctly predicting subsequent clinical results. Heat shock proteins (Hsp 90) chaperone the estrogen receptor (and also promote AKT and Raf-1 proteins) [274]. The geldanamycin derivative 17AAG is an inhibitor of Hsp 90 by occupying the ATP-binding pocket of the chaperone. When ER loses its chaperone, it is condemned to degradation by proteasomes. The same fate awaits the Akt, Raf-1, IGF-R1, and erbB2 proteins after the loss of their Hsp 90 protection. Tamoxifen opposes 17AAG and preserves ER. Tamoxifen-resistant breast cancer cells still may be inhibited by other estrogen-receptor antagonists (ICI 164,384; ICI 182,780, fulvestrant), stimulating degradation of ER (Table 6). In addition, aromatase inhibitors suppress estrogen-2 synthesis. The luteinizing hormone-releasing hormone analogues (SB-75, Cetrorelix) imitate surgical oophorectomy. Not with tamoxifen, but in combination with fulvestrant, 17AAG may act additively in the treatment of estrogen-sensitive and refractory breast cancers [275–277]. In other systems cited, normal fibroblasts could induce differentiation of basal cell carcinomas, urogenital sinus mesenchymal cells induced differentiation of transitional carcinoma cells of the urinary bladder, and human colon carcinoma cells underwent differentiation in the presence of fetal rat mesenchymal cells [264,265].

In the male fetus, the Müllerian ducts regress under the effect of the Müllerian inhibiting substance (MIS). In the fe-

Table 6 Pathobiology of the Breast Cancer Cell and Its Repair

HER2 (ErbB2/c-*erb*B2/HER2/*neu*): Its genes are mapped at 17q12–21. Its genes were discovered transduced by avian *erythroblastosis* retroviruses (c$\rightarrow$ v-*erb*). HER2 is a member of the ErbB family: ErbB1, 2, 3, 4. Responds to ligands epidermal growth factor (EGF), transforming growth factor-alpha (TGFα), amphiregulin, β-cellulin, epiregulins, heregulin (HRGβ)/ neuregulins. ?HER2 expression is equated with adverse prognosis and chemotherapy resistance. Trastuzumab (Herceptin, Genentech) downregulates HER2 expression and upregulates cyclin kinase inhibitor $p27^{Kip1}$ and retinoblastoma (Rb) protein p130.
The $\uparrow$HER2 $\uparrow$EGF-R expressing BCC converts the estrogen receptor (ER) competitive antagonist tamoxifen (TXF, Nolvadex) into agonist-stimulating breast cancer cell (BCC) mitosis. The ligands to EGF-R/HER2, EGF and HRGβ, signal cross-phosphorylation of the TXF-ligated ER, resulting in BCC mitosis-induction. Fulvestrant (Faslodex) retains its ER antagonism in HER2 EGF-R BCC. ZD1839 (gefitinib, Iressa) synergizes with fulvestrant in suppressing growth of HER2 EGF -R BCC stimulated by TXF[1,2]. In BCC with ER or downregulated insulinlike growth factor-I receptor (IGF -IR), fulvestrant resistance develops and BCC mitoses are resumed. HRGβ-stimulated HER2/EGF-R signaling through the anti-apoptotic AKT kinase pathway (inhibitory to pro-apoptotic BAD) counteracts gefitinib-mediated growth inhibition. Phosphatidylinositol-3 kinase (PI3K) inhibitor LY294002 synergized with gefitinib in suppressing mitoses of TXF-resistant, HRGβ1-stimulated BCC. Insulinlike growth factor-IR phosphorylation by ligation of IGF-II cross-phosphorylates EGF-R signaling pathway, but EGF-R pathway phosphorylation does not cross-phosphorylate the IGF-IR pathway. AG1024 (Table 2) inhibits IGF-IIR, but leaves EGF-R signaling unaffected. TXF-resistant BCCs overexpress IGF-II but not IGF-I[3,4,5]. $\uparrow$HER2-specific immune T cells are cytotoxic to HER2 autologous BCC[6]. HER2/*neu* oncoprotein is antigenic and induces immune reactions to BCC when used as a vaccine[7].

(*Continued*)

Table 6 Continued

Letrozole (Femara, Novartis) extends relapse-free survival of TXF-naïve and TXF-pretreated patients[8]. Dual inhibition of EGF-R/HER2 by GW572016 suppresses growth of even trastuzumab-resistant BCC in vitro[9]. Farnesyltransferase inhibitor R115777 synergized with TXF; flavopiridol synergized with epothilone in suppressing growth and inducing apoptosis of BCC[10] (Table 2). Although erythropoietin receptor (EpoR) is overexpressed in some BCC (Part IV) and it activates signal transducers and activators of transcription (STAT) (Figures 1 and 2) and nuclear factor kappa B (NFκB), thus exerting cell-survival and anti-apoptotic effects; it does not enable BCC to resist chemotherapy.[11]

Receptors for adeno-associated viral (AAV) vectors are expressed on BCC and AAV vectors are very good candidates for viro- and gene therapy for these (and other—glioblastoma) tumors.[12] AdEHT2 and Ad-CMV-CD are used for virotherapy (Table 1).

1 Baylor College of Medicine, Houston, TX; AstraZeneca, Maclesfield; Imperial College of Science, Technology and Medicine, London, United Kingdom. 2 Kawasaki Medical School, Kurshiki, Okayama, Japan. 3,4,5 Tenovus Centre for Cancer Research, Cardiff; City Hospital Nottingham; AstraZeneca, Alderley Edge; United Kingdom. 6 Technical University of Munich, Bavaria, Germany. 7 GlaxoSmithKline Biologicals, Rixensart, Belgium; Ringshospitalet, Herlev Hospital, Aalborg Hospital, Odense University Hospital, Pharmexa A/S, Denmark; Leiden University Medical Center, The Netherlands; Hammersmith Hospital, Leicester Royal Infirmary, London, United Kingdom. 8 Ringshospitalet, Denmark; Chinese Academy of Medical Sciences, Beijing, China; Petrov Research Institute of Oncology, St. Petersburg, Russia; Hospital Universitario de la Princesa, Madrid, Spain; CHG Andre Boulloche, Mondbeliard, France; Novartis Pharma AG, Basel, Switzerland. 9 University of California, Los Angeles, CA; GlaxoSmithKline, Research Triangle Park, NC. 10 Institute for Cancer Research; Royal Marsden Hospital, London, United Kingdom; Johnson & Johnson, Titusville, NJ; H. L. Moffitt Cancer Center, University of South Florida, Tampa, FL. 11 Molecular Radiation Biology Lab, University Clinic Freiburg, Germany. 12 Haukeland Hospital, Bergen, Norway; National Institutes of Health, Bethesda, MD; University of Washington, Seattle, WA; University of Pennsylvania, Philadelphia, PA.

Table 6. Condensed and selected summaries of lectures and abstracts of the San Antonio Breast Cancer Symposia, 25th (2002) and 26th (2003), Lippman, Marc E., editor; printed in Breast Cancer Research and Treatment 2002; Supplement, volume 76 and 2003; Supplement, volume 82.

male fetus, the coelomic epithelium invaginates to form anlagen from which the uterus and the upper vagina develop. Highly purified recombinant MIS inhibited the growth of human ovarian carcinoma cell lines xenografted into immunosuppressed mice [278].

Intratumoral injection of the liposome-encapsulated proapoptotic virally encoded protein E1A of adenovirus type 5 (see Part I) into xenografted HER2/neu-positive breast adeno- and head and neck squamous cell carcinomas resulted in tumor regressions. Phase I clinical trials confirm apoptosis induction of injected breast, ovarian, and squamous cell carcinomas, but without curative effect. Small molecular inhibitors and monoclonal antibodies compete for targeting the EGF-R in breast (and other) cancer cells [279–284]. These modalities of treatment will eventually be given in combined protocols (Table 5).

MISCELLANEOUS AGENTS TESTED IN XENOGRAFTS

Xenografts of Human Tumors Are Very Sensitive to Noxious and Natural Substances

The mushroom toxin illudin contains the DNA alkylating agent irofulven (MGI-114), which completely destroys human cancer xenografts in nude mice. In clinical trials, only occasional PR occurred (one in a patient with pancreatic carcinoma) but stabilization of disease was more common, but renal and hematologic toxicities (thrombocytopenia) disallowed dose escalation. Subsequent French studies indicated unique cytotoxicity of irofulven to cancers of epithelial cell origin (squamous cell carcinoma of the head and neck; hormone-refractory prostatic carcinoma and others) (285, 286). K-ras mutation in pancreatic cancer cells activates the anti-apoptotic transcription factor NKκB. The endogenous polypeptide NFκB inhibitor is degraded by proteasomes. Proteasome inhibitors sustain the downregulation of NFκB, allowing apoptosis to ensue when induced. The proteasome inhibitor PS-341 (see Part V) suppressed the growth of orthotopically xenografted pancreatic carcinomas because of caspase activation and DNA fragmentation; further extensive necrosis of the microvessel network occurred [287,288]. This compound is active against xenografted colon, head and neck, lung and prostatic carcinomas. The proteasome inhibitor PS-341 induces Bcl-2 phosphorylation and cleavage, depriving NSCLC cells of their anti-apoptotic machinery. Another NFκB suppressor is the vitamin E analogue α-tocopheryl succinate; it sensitizes lymhoma cells to TRAIL-induced apoptosis [289]. Leukotriene B4 stimulates, but its receptor antagonist LY293111 inhibits, the growth of pancreatic cancer xenografts by abrogating ERK1 and ERK2 phosphorylation [255]. The benzimidazole derivative mebendazole (Vermox, McNeil, Ft. Washington, PA) induced apoptotic death of xenotransplanted lung cancer cells and suppressed neoangiogenesis within the xerotransplants [290,291].

The somatostatin analogue RC-160 and the luteinizing hormone-releasing hormone agonist SB-75 ([D-Trp^6]-LH-RH) inhibited the growth of human pancreatic cancer xenografts in nude mice [292]. Interferonα and gemcitabine acted additively against orthotopically implanted human pancreatic carcinoma xenografts in nude mice; endothelial cells of the tumor bed died with apoptosis [293]. Both NCLS and, more frequently, SCLC express neuropeptide receptors and utilize bradykinin as their para- or autocrine growth factor. The bradykinin antagonist CU201 suppressed the growth and induced apoptosis in cultures or in xenografts of these cancer cells, and acted additively or synergistically with chemotherapeutical or biological agents (paclitaxel; ZD1839). Bradykinin under physiological conditions induces IL-12 production by DCs with subsequent Th1 immune reaction to follow. It was previously known that the nonapeptide bradykinin contributed to the development of endotoxin shock and that bradykinin antagonists (angiotensin-II; B2 receptor antagonists) ameliorated the outcome of endotoxin shock [294–297]. Now the bradykinin antagonist CU201 (B9870) emerges as the strongest inhibitor of lung cancer cells that exhibit neuroendocrine features. This compound acts additively or synergistically with chemotherapeuticals (vinorelbine, etoposide, cisplatin, paclitaxel) and with the EGF-R TKI ZD1839 (gefitinib; Iressa) [294]. The diterpene glycoside cotylenin A of plant origin induces differentiation of human leukemic cells (AML, acute promyelocytic leukemia). In combination with IFN-α, cotylenin induced the DR5 gene encoding the receptor for TRAIL in lung cancer cells. In xenografts, lung cancer cells underwent programmed cell death, whereas normal lung epithelial cells were spared [298].

In the endoplasmic reticulum, the Ca^{2+}-ATPase pathway is inhibited by thapsigargin (THG); from mitochondria, THG liberates pro-apoptotic proteins that activate caspases, leading to cell death through DNA fragmentation. In the machinery of apoptosis, pro-apoptotic proteins, the cytoplasmic-to-mitochondria BAX, and the mitochondrial Bak are antagonized by anti-apoptotic proteins Bcl-2 and Bcl_{XL}.When THG acts, Bid is cleaved and Bax initiates the cascade, beginning with the release of cytochrome C from mitochondria and leading through caspase activation to programmed cell death. A chemically modified form of THG (L12ADT) is coupled with a peptide that is a substrate of the protease PSA, which hydrolyses the prodrug to release THG in prostatic tissues; there, THG/L12ADT induces apoptosis and when it was tested against human prostate cancer xenografts, THG/L12ADT retarded the growth of these tumors [299,300]. This mechanism of THG-induced death may result in a clinical trial for prostate and colon carcinomas at the H. L. Moffitt Cancer Center in Tampa, FL.

Flavonoids (see Part II), such as favopiridol, genistein, or quercetin, are pro-apoptotic in cancer cells; the benzopyron-hydroxyphenyl phenoxodiol shows intensive activity in human breast, prostate, and ovarian cancer cells. It induces apoptosis in chemotherapy-resistant ovarian carcinoma cells by inhibiting the X-linked inhibitor of apoptosis (XIAP), liberating FLICE (caspase 9) by eliminating its inhibitory protein FLIP, and thereby activating the intrinsic pathway to apoptosis. The PI3K/Akt pathway is anti-apoptotic because it activates NFκB and inactivates by phosphorylation BAD, the Bcl-2 antagonist of cell death, and suppresses caspase 9 and FasL [301]. Phenoxodiol, by antagonizing the translocation of Akt into the nucleus, reduces the high levels of FLIP and liberates the extrinsic Fas-mediated apoptotic pathway through caspase 8. Ovarian carcinoma xenografts regress in the hosts treated with phenoxodiol [302]; the Yale University School of Medicine is enlisting patients with chemotherapy-resistant ovarian carcinoma for a clinical trial with phenoxodiol.

The combination of HA14-1 and PK11195 (see Parts I and II) inhibits Bcl-2, induces translocation of Bax to mitochondria and antagonizes the peripheral benzodiazepine receptor (PBR) of mitochondria, resulting in cytochrome C release for the initiation of the apoptotic cascade [303]. The PBR agonist Ro5-4864 antagonizes PK11195 but pan-caspase inhibitors (z-VAD-fmk) fail to counteract HA14-1. This combination of apoptosis inducers awaits clinical trial at the Hunstman Cancer Institue in Salt Lake City, UT. The pentacyclic triterpenoid of the cyclosqualenoid family, ursolic acid, inactivates the anti-apoptotic NFκB and consequentially suppresses the expression of cyclin D, COX-2, and MMP-9, rendering tumor cells chemotherapy-sensitive [304]. It may be tested for such efficacy in a clinical trial at the M. D. Anderson Cancer Center in Houston, TX. The anti-angiogenic agent squalamine passed its preclinical assys and is being tested in Phase I clinical trials [305]. The COX-2 inhibitor NS398 acts as an antiproliferative and pro-apoptotic agent in hepatocellular and head and neck squamous cell carcinomas [306,307].

Murine NK cell-depleted SCID mice accept i.v. injected human melanoma xenografts that grow as lung metastases. Purified human NK cells from healthy donors given i.v. with IL-2 and IL-12 eradicated these metastases [308]. Immune T cells (as tumor-infiltrating lymphocytes) and lymphokine-activated killer cells (LAK) are in clinical use at the National Cancer Institute and elsewhere [309,310]. Although the xenograft may be subdued by the injected cytotoxic lymphocytes, in the tumor-bearing natural host, suppresssor T cell clones may arise for the protection of the tumor [311,312]. Helper T cells may overcome the suppressor T cell clone [313].). Soluble viral peptides may induce immune tolerance [314]. If a non-pathogenic oncolytic virus induces immune tolerance toward itself, the oncolytic process could advance unimpeded, with lack of virus-neutralizing antibodies. These events may or may not be reproducible in a xenograft.

Endless List of Anti-Tumor Agents and Interventions

Interleukin-4 receptors are overexpressed in pancreatic and breast carcinoma cells. Pseudomonas exotoxin fused with IL-4 forms a chimeric protein attached to and endocytosed by the receptor. By intraperitoneal and intratumoral injections, the IL-4 cytotoxin eradicated 60% of xenografted human pancreatic carcinomas and reduced the growth of breast carcinoma xenografts. It induced macrophage and neutrophil leukocyte infiltrations into the regressing tumors. Its deleterious actions against normal IL-4 receptor-expressing cells is unknown. Because malignant glioma cells constitutively overexpress the receptor for IL-13, these cells readily accept attachment of the cytotoxin consisting of the fused proteins of IL-13 and Pseudomonas exotoxin; because of apoptotic death of the tumor cells, even established tumors regress [315–317] (see Part V).

Xenografted arginine succinate synthetase-deficient arginine auxotrophic human melanoma cells do not grow in pegylated arginine deaminase-treated animals [318]. The dithiocarbamate disulfiram (DSF) left melanocytes intact but activated the apoptotic pathway in cultured melanoma cells. N-acetyl-cysteine (NAC) reversed the inhibitor of γ-glutamylcysteine synthetase, and buthionine sulfoximide (BSO) augmented this effect. Dithiocarbamate disulfiram combined with cisplatin or nitrosoureas awaits clinical trials [319]. The tryphostin AG1024 inhibits IGF-R; in human melanoma cells it inhibits the MAPK/ERK pathway and restores tumor-suppressive Rb function [320,321]. Accordingly, the IGF-R system of colon carcinoma cells is similarly targeted [202]. The chloroindolyl sulfonamide cell-cycle inhibitor E7070 arrests the cell at $G_1 \rightarrow S$, upregulates p53, and releases apoptotic pathways in colon carcinoma cells. In a Phase I clinical trial it induced long stabilization of disease in patients with advanced cancers [322].

Ecteinascidin-743 (ET-743, Yondelis; PharmaMar, Madrid, Spain) (see Part III) from the Caribbean tunicate *Ecteinascidia turbinata* inhibits nucleotide excision repair and downregulates the P-gp MDR-1 gene product protein by forming covalent guanine adducts as it attaches to the minor groove of DNA. It can overcome multidrug resistance. In xenografted breast carcinoma cells, it synergized with paclitaxel; in Ewing's sarcoma cells it induced massive apoptosis [323–325]. Dexamethasone protected rats against ET hepatotoxicity [326]. Naamidine, an alkaloid from the sponge *Leucetta,* targets the extracellular signal-regulated kinases ERK1 and ERK2; it is antimitogenic in human squamous carcinoma cells [327]. The tunicate *Aplidium albicans* from the Mediterranean sea yields the depsipeptide didemnin, aplidine; it cleaves the protein Bid, the connector of extrinsic (CD95 Fas-mediated) and intrinsic (mitochondrial) apoptosis in human leukemic cells (see Part III) [328]. Epigallocatechin-3-gallate induces, in human head and neck squamous carcinoma cells, a drop of cyclin D1 and rise of $p21^{Cip1}$ and $p27^{Kip1}$; a reduction of hyperphosphorylated Rb, resulting in G_1 arrest of the cell cycle [320,329]; a decrease of Bcl-2 and Bcl_{XL} and increase in Bax proteins; inhibition of EGF-R, STAT, and ERK phosphorylation; and inhibition of TGFα-induced stimulation of the c-*fos* proto-oncogene. It synergized with 5-fluorouracil.

Statins are 3-hydroxy-3-methylglutaryl CoA reductase inhibitors synthesized and licensed for the treatment of hypercholesterolemia. In malignant tumors, these compounds show anti-invasive, antiproliferative, pro-apoptotic, and radiosensitizing efficacy [330]. The Princess Margaret Hospital in Toronto, Canada, tests statins in combination with chemo-biotherapeuticals for antineoplastic efficacy. The cardiac glycoside bufadienolide (hellebrin) induces mitochondrial apoptosis preferentially in malignant T (Jurkat) cells, and overexpression of Bcl-2 or the administration of the pan-caspase inhibitor z-Vad-fmk antagonized this effect [331,332].

Many new TKIs (see Parts I-II) have passed through preclinical assays and are in Phase I or Phase II clinical trials. The major targets of these TKI are the EGF-R from sarcomas [187] to lung cancer [333–335] (see Part I); CI-1033 for Erb2 [336]; PTK787/ZK222584 (PTK/ZK) for VEGF-R [337]; PKI166; CI1033 for HER2/neu [338–340]; and wortmannin for PI3K [341]. The pan-erbB-R inhibitor CI-1033 irreversibly blocks signal transduction from the receptor in chemotherapy-resistant breast cancer cells [336]. The water-soluble quinazoline CI-1033 (Pfizer) binds irreversibly to EGF-R tyrosine kinase and is a pan-erbB2-4 inhibitor. It inhibited the growth of human carcinoma cell lines in vitro and eradicated human cancer xenografts [342].

In clinical trials, it induced only one PR but many cases of stabilizations of metastatic disease (once in a case of osteosarcoma). The pyrollo-pyrimidine derivative PKT-166 (Novartis) [343], a reversible inhibitor of EGF-R, is performing well against xenografted human cancers; in nude mice it inhibited the growth of orthotopically xenografted human pancreatic carcinoma cells and reduced VEGF and bFGF production in the xenograft. No remissions were induced in a Phase I clinical trial involving 44 patients with metastatic cancers. Against xenografted kidney carcinoma, the drug performed best when it was given in combination with gemcitabine [344]. However, the combination of somatostatin analogue RC160, luteinizing hormone-releasing hormone antagonist Cetrorelix, and bombesin antagonist RC-3940-II acts on xenografted kidney carcinoma cells by reducing EGF-R expression in the tumor cells and reducing growth hormone and IGF levels in the host, resulting in tumor regression [345].

The quinazolamine derivative GW2016 (Glaxo SmithKline, Brentford, UK) inhibits signal transduction from EGF-R and HER2/Erb/B2. In nude mice, the growth of squamous cell carcinomas was inhibited. Tumor cells pretreated with the drug failed to grow in xenografts [346]. OSI-774 (Tarceva, OSI Pharmaceuticals) (see Part I) in combination with a PI3K inhibitor (wortmannin) (see Part II) inhibited the growth of, and induced apoptosis in, xenografted human pancreatic carcinoma cells [340]. The drug that appears to recognize in vitro a difference between adeno- and squamous cell carcinomas in the heterogeneous group of NSCLC is ubenimex (Bestatin). It is more effective against squamous than adenocarcinoma cells [347–349]. In large clinical trials, it significantly prolonged 5 years disease-free and overall survival of patients whose primary squamous cell carcinomas of the lung were completely resected: 81% vs. 74% and 71.6% vs. 62% in treated vs. control patients, respectively [348,349].

How can oncolytic viruses compete (or cooperate) with such a formidable display of new biologicals that are being tested in tissue cultures and xenografts and are entering Phase I or Phase II clinical trials by the hundreds? Xenografted osteosarcoma cells were infected with adenovirus Ad-sCE2 constructed to deliver the cDNA of the human enzyme carboxylesterase-2. The semisynthetic camptothecin derivative CPT-11 (irinotecan) is converted into its active form by carboxylesterase-2. Osteosarcoma cells expressing the enzyme become highly susceptible to CPT-11 and exhibited much retarded growth rate in xenografts [350]. A similar adenovirally delivered gene enhances susceptibility of Ewing's sarcoma cells to VP-16 (etoposide) [351]. Will this technology work in clinical trials (Table 1)?

REFERENCES, PART IV

1. Sherr CJ, DePinho RA. Cellular senescence: mitotic clock or culture shock? Cell 2000; 102:407–410.
2. Sinkovics JG, Gyorkey F, Kusyk C, Siciliano M. Growth of human tumor cells in established cultures. In: Busch H, Ed.. Methods in Cancer Research. Vol. 14. New York: Academic Press, 1978:243–323.
3. Diaz-Montero CM, McIntyre BW. Acquisition of anoikis resistance in human osteosarcoma cells Eur J Cancer 2003; 39: 2395–2402.
4. Raval GN, Bharadwaj S, Levine EA, Willingham MC, Geary RL, Kute T, Prasad GL. Loss of expression of tropomyosin-1, a novel class I tumor suppressor that induces anoikis, in primary breast tumors Oncogene 2003; 22:6194–6203.
5. Sage H. Culture shock J Biol Chem 1986; 261:7082–7092.
6. Keledjian K, Kyprianu N. Anoikis induction by quinazoline based alpha 1-adrenoceptor antagonist in prostate cancer cells: antagonistic effect of bcl-2 J Urol 2003; 169:1150–1154.
7. Yamashita K, Mimori K, Inoue H, Mori M, Sidransky D. A tumor-suppressive role for trypsin in human cancer progression Cancer Res 2003; 63:6575–6578.
8. Voskoglou-Nomikos T, Pater JL, Seymour L. Clinical predictive value of the *in vitro* cell line, human xenograft, and mouse

allograft preclinical cancer models Clin Cancer Res 2003; 9: 4227–4239.

9. Kraemer K, Fussel S, Schmidt U, Kotzsch M, Schwezener B, Wirth MP, Meye A. Antisense-mediated hTERT inhibition specifically reduces the growth of human bladder cancer cells Clin Cancer Res 2003; 9:3794–3800.

10. Zhang X, Multani AS, Zhou J-H, Shay JW, McConkey DM, Dong L, Kim C-S, Rosser CJ, Pathak S, Benedict WF. Ad-RB94 produces rapid telomere erosion, chromosomal crisis and caspase dependent apoptosis in bladder cancer and immortalized human urothelial cells, but not in normal urothelial cells Cancer Res 2003; 63:760–765.

11. Zhou C, Gehring PA, Whang YE, Boggess JF. Rapamycin inhibits telomerase activity by decreasing the hTERT mRNA level in endometrial cancer cells Mol Cancer Ther 2003; 2: 789–795.

12. Shay JW. Telomerase therapeutics: telomeres recognized as a DNA damage signal Clin Cancer Res 2003; 9:3521–3525.

13. Igarashi M, Suda T, Hara H, Takimoto M, Nomoto M, Takahashi T, Okoshi S, Kawai H, Mita Y, Waguri N, Aoyagi Y. Interferon can block telomere erosion and in rare cases result in hepatocellular carcinoma development with telomeric repeat binding factor 1 overexpresasion in chronic hepatitis C. Clin Cancer Res 2003; 9:5264–5270.

14. Schroers R, Shen L, Rollins L, Rooney CM, Slawin K, Sonderstrup G, Huang XF, Chen S-Y. Human telomerase reverse transcriptase-specific T-helper responses induced by promiscuous major histocompatibility complex class II-restricted epitopes Clin Cancer Res 2003; 9:4743–4755.

15. Asai A, Oshima Y, Yamamoto Y, Uochi TA, Kusaka H, Akinaga S, Yamashita Y, Pongracz K, Pruzan R, Wunder E, Piatyszek M, Li S, Chin AC, Harley CB, Gryaznov S. A novel telomerase template antagonist (GRN163) as a potential anticancer agent Cancer Res 2003; 63:3931–3939.

16. Chen Z, Koenemann KS, Corey DR. Consequences of telomerase inhibition and combination treatment for the proliferation of cancer cells Cancer Res 2003; 63:5917–5925.

17. Wang ES, Wu K, Chin AC, Chen-Kiang S, Pongracz K, Gryaznov S, Moore MA. Telomerase inhibition with an oligonucleotide telomerase template antagonist: in vitro and in vivo studies in multiple myeloma and lymphoma Blood 2004; 103: 258–266.

18. Kosciolek BA, Kalantidis K, Tabler M, Rowley PT. Inhibition of telomerase activity in human cancer cells by RNA interference Mol Cancer Ther 2003; 2:209–216.

19. Seimiya H, Oh-hara T, Suzuki T, Naasani I, Shimazaki T, Tsuchiya K, Tsuruo T. Telomere shortening and growth inhibition of human cancer cells by novel synthetic telomerase inhibitors MST-312, MST-295, and MST-1991 Mol Cancer Ther 2002; 1:657–665.

20. Hu M, Nicolson GL, Trent JC II, Yu D, Zhang L, Lang A, Killary A, Ellis LM, Bucana CD, Pollock RE. Characterization of 11 human sarcoma cell strains Cancer 2002; 95:1569–1576.

21. Sinkovics JG. Comments on cultured sarcoma cells Sarcoma 2003; 7:75–77.

22. Hiatt KM, Highsmith WE. Lack of DNA evidence for jaagziekte sheep retrovirus in human bronchioalveolar carcinoma Hum Pathol 2002; 33:680.

23. Yousem SA, Finkelstein SD, Swalsky PA, Bakker A, Ohori NP. Absence of jaagziekte sheep retrovirus DNA and RNA in bronchioalveolar and conventional human pulmonary adenocarcinoma by PCR and RT-PCR analysis Hum Pathol 2002; 33:680.

24. York DF, Querat G. A history of ovine pulmonary adenocarcinoma (jaagziekte) and experiments leading to the deduction of the JSRV nucleotide sequence Curr Top Microbiol Immunol 2003; 275:1–23.

25. Maeda N, Palmarini M, Murgia C, Fan H. Direct transformation of rodent fibroblats by jaagziekte sheep retrovirus DNA Proc Nat Acad Sci U S A 2001; 98:4449–4454.

26. De las Heras M, Barsky SH, Hasleton P, Wagner M, Larson E, Egan J, Ortin A, Gimenez-Mas JA, Palmarini M, Sharp JM. Evidence for a protein related immunologically to the jaagziekte sheep retrovirus in some human lung tumors Eur Respir J 2000; 16:330–332.

27. Grossman DA, Hiti AL, McNiel EA, Ye Y, Alpaugh ML, Barsky SH. Comparative oncological studies of feline bronchioalveolar lung carcinoma, its derived cell line and xenograft Cancer Res 2002; 62:3826–3833.

28. Wislez M, Rabbe N, Marchal J, Milleron B, Crestani B, Mayaud C, Antoine M, Soler P, Cadranel J. Hepatocyte growth factor production by neutrophils infiltrating bronchioalveolar subtype pulmonary adenocarcinoma: role in tumor progression and death Cancer Res 2003; 63:1405–1412.

29. Sattler M, Pride YB, Ma P, Gramlich JL, Chu SC, Quinnan LA, Shiazian S, Liang C, Podar K, Christensen JG, Salgia R. A novel small molecule Met inhibitor induces apoptosis in cells transformed by the oncogenic TPR-MET tyrosine kinase Cancer Res 2003; 63:5462–5469.

30. Christensen JG, Schreck R, Burrows J, Kuruganti J, Chan E, Le P, Chen J, Wang X, Ruslim L, Blake R, Lipson KE, Ramphal J, Do S, Cui JJ, Cherrington JM, Mendel DB. A selective molecule inhibitor of c-Met kinase inhibits c-met-dependent phenotypes in vitro and exhibits cytoreductive antitumor activity in vivo Cancer Res 2003; 63:7345–7355.

31. Wang X, Le P, Liang C, Chan J, Kiewlich D, Miller T, Harris D, Sun L, Rice A, Vasile S, Blake RA, Howlett AR, Patel N, McMahon G, Lipson KE. Potent and selective inhibitors of the Met [hepatocyte growth factor/scatter factor (HGF/SF) receptor] tyrosine kinase block HGF/SF-induced tumor cell growth and invasion Mol Gene Ther 2003; 2:1085–1095.

32. Shinomiya N, Vande Woude GF. Suppression of Met expression: a possible cancer treatment Clin Cancer Res 2003; 9: 5085–5090.

33. Kim SJ, Johnson M, Koterba K, Herynk MH, Uehara H, Gallick GE. Reduced c-Met expression by an adenovirus expressing a c-Met ribozyme inhibits tumorigenic growth and lymph node metastases of PC3-LN4 prostate tumor cells in an orthotopic mouse model Clin Cancer Res 2003; 9:5161–5170.

34. Ma PC, Kijama T, Maulik G, Fox EA, Sattler M, Griffin JD, Johnson BE, Salgia R. c-Met mutational analysis in small cell lung cancer: novel juxtamembrane domain mutations regulating cytoskeletal functions Cancer Res 2003; 63:6272–6281.

35. Zhou J, Chen J, Mokotoff M, Zhong R, Shultz LD, Ball ED. Bombesin/gastrin-releasing peptide receptor: a potential target for antibody-mediated therapy of small cell lung cancer Clin Cancer Res 2003; 9:4953–4960.

36. Koukourakis MI, Giatromanolaki A, Brekken RA, Sivridis E, Gatter KC, Harris AL, Sage EH. Enhanced expression of SPARC/osteonectin in the tumor-associated stroma of non-small cell lung cancer is correlated with markers of hypoxia/acidity and with poor prognosis of patients Cancer Res 2003; 63:5376–5380.

37. Oshima K, Hamasaki M, Makimoto Y, Yoneda S, Fujii A, Takamatsu Y, Nakashima M, Watanabe T, Kawahara K, Kikuchi M, Shirakusa T. Different chemokine, chemokine receptor, cytokine and cytokine receptor expression in pulmonary adenocarcinoma: diffuse down-regulation is associated with immune evasion and brain metastasis Int J Oncol 2003; 23: 965–973.

38. Kim Y-A, Lee WH, Choi TH, Rhee S-H, Park K-Y, Choi YH. Involvement of p21WAF1/CIP1, pRB, Bax and NF-κB in induction of growth arrest and apoptosis by resveratrol in human lung carcinoma A549 cells Int J Oncol 2003; 23: 1143–1149.

39. Kaneuchi M, Sasaki M, Tanaka Y, Yamamoto R, Sakuragi N, Dahiya R. Resveratrol suppresses growth of Ishikawa cells through down-regulation of EGF Int J Oncol 2003; 23: 1167–1172.

40. Li WW, Bertino JR. Fas-mediated signaling enhances sensitivity of human soft tissue sarcoma cells to anticancer drugs by activation of p38 kinase Mol Cancer Ther 2002; 1:1343–1348.

41. Altieri DC. Validating survivin as a cancer therapeutic target Nat Rev Cancer 2003; 3:46–54.

42. Brattain MG, Howell G, Sun LZ, Willson JK. Growth factor balance and tumor progression Curr Opin Oncol 1994; 6: 77–81.

43. Tada Y, O-Wang J, Wada A, Takiguchi Y, Tatsumi K, Kuriyama T, Sakiyama S, Tagawa M. Fas ligand-expressing tumors induce tumor-specific protective immunity in the inoculated

hosts but vaccination with the apoptotic tumors suppresses antitumor immunity Cancer Gene Ther 2003; 10:134–140.

44. Ahmad SA, Jung YD, Liu W, Reinmuth N, Parikh A, Stoeltzing O, Fan F, Ellis LM. The role of the microenvironment and intercellular cross-talk in tumor angiogenesis Semin Cancer Biol 2002; 12:105–112.

45. Roskelley CD, Bissell MJ. The dominance of the microenvironment in breast and ovarian cancer Semin Cancer Biol 2002; 12:97–104.

46. LeBedis C, Chen K, Fallavollita L, Boutros T, Brodt P. Peripheral lymph node stromal cells can promote growth and tumorigenicity of breast carcinoma cells through the release of IGF-I and EGF Int J Cancer 2002; 100:2–8.

47. Wang H, Wang H, Shen W, Huang H, Hu L, Ramdas L, Zhou YH, Liao WS, Fuller GN, Zhang W. Insulinlike growth factor binding protein 2 enhances glioblastoma invasion by activating invasion-enhancing genes Cancer Res 2003; 63: 4315–4321.

48. Decensi A, Veronesi U, Miceli R, Johansson H, Mariani L, Camerini T, Di Mauro MG, Cavadini E, De Palo G, Costa A, Perloff M, Malone WF, Fornelli F. Relationship between plasma insulin-like growth factor-I and insulin-like growth factor binding protein-3 and second breast cancer risk in a prevention trial of fenretidine Clin Cancer Res 2003; 9: 4722–4720.

49. Wu X, Tortolero-Luna G, Zhao H, Phatak D, Spitz R, Follen M. Serum levels of insulin-like growth factor I and risk of squamous intraepithelial lesions of the cervix Clin Cancer Res 2003; 9:3356–3361.

50. Sekharan M, Nasir A, Kaiser HE, Coppola D. Insulinlike growth factor I receptor activates c-SRC and modifies transformation and motility of colon cancer in vitro Anticancer Res 2003; 23:1517–1524.

51. Yasumaru M, Tsuhi S, Tsuji M, Irie T, Komori M, Kimura A, Nishida T, Kakiuchi Y, Kaewai N, Murata H, Horimoto M, Sasaki Y, Hayashi N, Kawano S, Hori M. Inhibition of angiotensin II activity enhanced the antitumor effect of cyclooxy-

genase-2 inhibitors via insulin-like growth factor I receptor pathway Cancer Res 2003; 63:6726–6734.

52. Uemura H, Ishiguro H, Nakaigawa N, Nagashima Y, Miyoshi Y, Fujinami K, Sakaguchi A, Kubota Y. Angiotensin II receptor blocker shows antiproliferative activity in prostate cancer cells: a possibility of tyrosine kinase inhibitor of growth factor Mol Cancer Ther 2003; 3:1139–1147.

53. Scott AM, Wiseman G, Welt S, Adjei A, Lee FT, Hopkins W, Divgi CR, Hanson LH, Mitchell P, Gansen DN, Larson SM, Ingle JN, Hoffman EW, Tanswell P, Ritter G, Cohen LS, Bette P, Arvay L, Amelsberg A, Vlock D, Rettig WJ, Old LJ. A Phase I dose-escalation study of sibrotuzumab in patients with advanced or metastatic fibroblast activation protein-positive cancer Clin Cancer Res 2003; 9:1639–1647.

54. Cheng JD, Weiner LM. Tumors and their microenvironments: tilling the soil: Commentary re: A. M. Scott et al., A phase I dose-escalation study of sibrotuzumab in patients with advanced or metastatic fibroblast activation protein-positive cancer Clin Cancer Res 2003; 9:1590–1595, 1639–1647.

55. Li G, Satyamoorthy K, Meier F, Berking C, Bogenrieder T, Herlyn M. Function and regulation of melanoma-stromal fibroblast interactions: when seeds meet soil Oncogene 2003; 22:3162–3171.

56. Corsini C, Mancuso P, Paul S, Burlini A, Martinelli G, Pruneri G, Bertolini F. Stroma cells: a novel target of herceptin activity Clin Cancer Res 2003; 9:1820–1825.

57. Sinkovics JG. Malignant lymphoma arising from natural killer cells: report of the first case in 1970 and newer developments of the FasL⇒FasR system Acta Microbiol Immunol Hung 1997; 44:295–306.

58. Sinkovics JG, Horvath JC. Virological and immunological connotations of apoptotic and anti-apoptotic forces in neoplasia Int J Oncol 2001; 19:473–488.

59. Huzella T, Sinkovics JG. On the threshold of the door of "no admittance". In: Friedman H, Szentivanyi A, Eds. The Immunologic Revolution: Facts and Witnesses. Boca Raton. FL: CRC Press, 1994:241–286.

60. Lapis K, Timár J. Role of elastin-matrix interactions in tumor progression Semin Cancer Biol 2002; 12:209–217.

61. Timár J, Lapis K, Dudás J, Sebestyén A, Kopper L, Kovalszky I. Proteoglycans and tumor progression: Janus-faced molecules with contradictory functions in cancer Semin Cancer Biol 2002; 12:173–186.

62. Robert L. Cell-matrix interactions in cancer spreading—effect of aging Semin Cancer Biol 2002; 12:157–163.

63. Eshel R, Neumark E, Sagi-Assif O, Witz IP. Receptors involved in microenvironment-driven molecular evolution of cancer cells Semin Cancer Biol 2002; 12:139–147.

64. Patarroyo M, Tryggvason K, Virtanen I. Laminin isoforms in tumor invasion, angiogenesis and metastasis Semin Cancer Biol 2002; 12:197–207.

65. Seftor RE, Seftor EA, Koshikawa N, Meltzer PS, Gardner LM, Bilban M, Stetler-Stevenson WG, Quaranta V, Hendrix MJ. Cooperative interactions of laminin 5 $\gamma 2$ chain, matrix metalloproteinase-2, and membrane type-1-matrix/metalloproteinase are required for mimicry of embryonic vasculogenesis by aggressive melanoma Cancer Res 2001; 61:6322–6327.

66. Liu C, Sun C, Huang H, Janda K, Edgington T. Overexpression of legumain in tumors is significant for invasion/metastasis and a candidate enzymatic target for prodrug therapy Cancer Res 2003; 63:2957–2964.

67. Tucker GC. Alpha v integrin inhibitors and cancer therapy Curr Opin Invest Drugs 2003; 4:722–731.

68. Posey JA, Khazaeli MB, DelGrosso A, Saleh MN, Lin CY, Huse W, LoBuglio AF. A pilot trial of Vitaxin, a humanized anti-vitronectin receptor (anti alpha v beta 3) antibody in patients with metastatic cancer Cancer Biother Radiopharm 2001; 16: 125–132.

69. Patel SR, Jenkins J, Papadopoulos N, Burgess MA, Plager C, Gutterman J, Benjamin RS. Pilot study of vitaxin, an angiogenesis inhibitor, in patients with advanced leiomyosarcomas Cancer 2001; 92:1347–1348.

70. Gutheil JC, Campbell TN, Pierce PR, Watkins JD, Huse WD, Bodkin DJ, Cheresh DA. Targeted antiangiogenic therapy for

cancer using Vitaxin: a humanized monoclonal antibody to the integrin alphavbeta3 Clin Cancer Res 2000; 6:3056–3061.

71. Reinmuth N, Liu W, Ahmad SA, Fan F, Stoeltzing O, Parikh AA, Bucana CD, Gallick GE, Nickols MA, Westlin WF, Ellis LM. $\alpha_v\beta_3$ integrin antagonist S247 decreases colon cancer metastasis and angiogenesis and improves survival in mice Cancer Res 2003; 63:2079–2087.

72. Marcinkiewicz C, Weinreb PH, Calvete JJ, Kisiel DG, Mousa SA, Tuszynski GP, Lobb RR. Obtustatin: a potent selective inhibitor of $\alpha 1\beta 1$ integrin in vitro and angiogenesis in vivo Cancer Res 2003; 63:2020–2023.

73. Dai DL, Makretsov N, Campos EI, Huang C, Zhou Y, Huntsman D, Martinka M, Li G. Increased expression of integrin-linked kinase is correlated with melanoma progression and poor patient survival Clin Cancer Res 2003; 9:4409–4414.

74. Gillan L, Matei D, Fishman DA, Gerbin CS, Karlan BY, Chang DD. Periostin secreted by epithelial ovarian carcinoma is a ligand for alpha(V)beta(3) and alpha(V)beta(5) integrins and promotes cell motility Cancer Res 2002; 62:5358–5264.

75. Garofalo A, Naumova E, Manenti L, Ghilardi C, Ghisleni G, Caniatti M, Colombo T, Cherrington JM, Scanziani E, Nicoletti MI, Giavazzi R. The combination of the tyrosine kinase receptor inhibitor SU6668 with paclitaxel affects ascites fornmation and tumor spread in ovarian carcinoma xenografts growing orthotopically Clin Cancer Res 2003; 9:3476–3485.

76. Horvath JC, Horvath E, Sinkovics JG, Horak A, Pendleton S, Mallah J. Human melanoma cells (HMC) eliminate autologous host lymphocytes (Ly^{FasR+}) and escape apoptotic death (A) by utilizing the Fas⇒FasR system as an autocrine growth loop ($HMC^{Fas \Rightarrow FasR}$), 89th Annual Meeting, American Association Cancer Reasearch, March 28-April 1, 1998, New Orleans, LA Proceedings 1998; 39:584 #3971.

77. Chen Y-L, Chen S-H, Wang J-L, Yang B-C. Fas ligand on tumor cells mediates inactivation of neutrophils J Immunol 2003; 171:1183–1191.

78. Wainwright M. The Sinkovics hybridoma The discovery of the first "natural hybridoma." Perspect Biol Med 1992; 35: 372–379.

79. Sinkovics JG, Drewinko B, Thornell E. Immunoresistant tetraploid lymphoma cells Lancet 1970; i:139–140.

80. Sinkovics JG. Early history of specific antibody-producing lymphocyte hybridomas Cancer Res 1981; 41:1246–1247.

81. Sinkovics JG. The earliest concept of the 'hybridoma principle' recognized in 1967–1968 Front Rad Ther Oncol 1990; 24: 18–31.

82. Sinkovics JG. Discovery of the hybridoma pinciple in 1968–69: immortalization of the specific antibody-producing cell by fusion with a lymphoma cell J Med 1985; 16:509–524.

83. Larizza L, Schirrmacher V. Somatic cell fusion as a source of genetic rearrangement leading to metastatic variants Cancer Metastasis Rev 1984; 3:193–201.

84. Sinkovics JG. Hodgkin's disease revisited: Reed-Sternbereg cells as natural hybridomas Crit Rev Immunol 1991; 11: 33–63.

85. Sinkovics JG, Dreesman G. Monoclonal antibodies of hybridomas Rev Infect Dis 1983; 5:9–34.

86. Busund LT, Killie MK, Bartnes K, Seljelid R. Spontaneously formed tumorigenic hybrids of Meth A sarcoma cells and macrophages in vivo Int J Cancer 2003; 106:153–159.

87. Nakamura Y, Yasuoka H, Tsujimoto M, Yang Q, Imabun S, Nakahara M, Nakao K, Nakamura M, Mori I, Kakudo K. Flt-4-positive vessel density correlates with vascular endothelial growth factor-D expression, nodal status, and prognosis in breast cancer Clin Cancer Res 2003; 9:5313–5317.

88. Morbidelli L, Donnini S, Chilemi F, Giachetti A, Ziche M. Angiosuppressive and angiostimulatory effects exerted by synthetic partial sequences of endostatin Clin Cancer Res 2003; 9:5358–5369.

89. Macpherson GR, Ng SSW, Forbes SL, Melillo G, Karpova T, McNally J, Conrads TP, Veenstra TD, Martinez A, Cuttita TD, Price DK, Figg WD. Anti-angiogenic activity of human endostatin is HIV-1-independent *in vitro* and sensitive to timing of treatment in a human saphenous vein assay Mol Cancer Ther 2003; 2:845–854.

90. Bieker R, Padró T, Steins M, Kessler T; Retzlaff S, Herrera F, Kienast J, Berdel WE, Mesters RM. Overexpression of basic fibroblast growth factor and autocrine stmulation in acute myeloid leukemia Cancer Res 2003; 63:7241–7246.

91. Ebos JML, Tran J, Master Z, Dumond D, Melo JV, Buchdunger E, Kerbel RS. Imatinib mesylate (STI-571) reduces Bcr-Abl-mediated vascular endothel growth factor secretion in chronic myelogenous leukemia Mol Cancer Res 2003; 1:89–95.

92. Grimshaw MJ, Naylor S, Balkwill FR. Endothelin-2 is a hypoxia-induced autocrine survival factor for breast cancer cells Mol Cancer Ther 2002; 1:1273–1281.

93. Wülfing Pia, Diallo R, Kersting C, Wülfing C, Porembe C, Rody A, Greb RR, Böcker W, Kiesel L. Expression of endothelin-1, endothelin-A, and endothelin-B recptor in human breast cancer and correlation with long-term follow-up Clin Cancer Res 2003; 9:4125–4131.

94. Tanaka F, Oyanagi H, Takenaka K, Ishikawa S, Yanagihara K, Miyahara R, Kawano Y, Li M, Otake Y, Wada H. Glomeruloid microvascular proliferation is superior to intratumoral microvessel density as a prognostic marker in non-small cell lung cancer Cancer Res 2003; 63:6791–6794.

95. Kaio E, Tanaka S, Oka S, Hiyama T, Kitadai Y, Haruma K, Chayama K. Clinical significance of thrombospondin-1 expression in relation to vascular endothelial growth factor and interleukein-10 expression at the deepest invasive tumor site of advanced colorectal carcinoma Int J Oncol 2003; 23:901–911.

96. Yang Q-W, Liu S, Tian Y, Salwen HR, Chlenski A, Weinstein J, Cohn SL. Methylation-associated silencing of the *thrombospondin-1* gene in human neuroblastoma Cancer Res 2003; 63: 6299–6310.

97. Hisai H, Kato J, Kobune M, Murakami T, Miyanashi K, Takahashi M, Yoshizaki N, Takimoto R, Terui T, Niitsu Y. Increased expression of angiogenin in hepatocellular carcinoma in correlation with tumor vascularity Clin Cancer Res 2003; 9:4852–4859.

98. Poon RT-P, Lau CP-Y, Ho JW-Y, Yu W-C, Fan S-T, Wong J. Tissue factor expression correlates with tumor angiogenesis

and invasiveness in human hepatocellular carcinoma Clin Cancer Res 2003; 9:5339–5345.

99. Lutgendorf SK, Cole S, Costanzo E, Bradley S, Coffin J, Jabbari S, Rainwater K, Ritchie JM, Yang M, Sood AK. Stress-related mediators stimulate vascular endothelial growth factor secretion in two ovarian cancer cell lines Clin Cancer Res 2003; 9:4514–4521.

100. Vortmeyer AO, Frank S, Jeong S-Y, Yuan K, Ikejiri B, Lee Y-S, Bhowmick D, Lonser RR, Smith R, Rodgers G, Oldfield EH, Zhuang Z. Developmental arrest of angioblastic lineage initiates tumorigenesis in von Hippel-Lindau disease Cancer Res 2003; 63:7051–7055.

101. Ruf W, Seftor EA, Petrovan RJ, Weiss RM, Gruman LM, Margaryan NV, Seftor REB, Miyagi Y, Hendrix MJC. Differential role of tissue factor pathway inhibitors 1 and 2 in melanoma vasculogenic mimicry Cancer Res 2003; 63:5381–5389.

102. Chen Q, Wang W-C, Evans SS. Tumor microvasculature as a barrier to antitumor immunity Cancer Immunol Immunother 2003; 52:670–679.

103. Cuadros C, Dominguez AL, Frost GI, Borgstrom P, Lustgarten J. Cooperative effect between immunotherapy and antiangiogenic therapy leads to effective tumor rejection in tolerant Her-2/neu mice Cancer Res 2003; 63:5895–5901.

104. Ferrandina G, Ranelletti FO, Legge F, Lauriola L, Salutari V, Gessi M, Testa AC, Werner U, Navarra P, Tringali G, Battaglia A, Scambia G. Celecoxib modulates the expression of cyclooygenase-2, KI67, apoptosis-related marker, and microvessel density in human cervical cancer: a pilot study Clin Cancer Res 2003; 9:4324–4331.

105. Saha D, Sekhar KR, Cao C, Morrow JD, Choy H, Freeman ML. The antiangiogenic agent SU5416 down-regulates phorbol ester-mediated induction of cyclooxygenase 2 expression in inhibiting nicotinamide adenine dinucleotide phosphate oxidasae activity Cancer Res 2002; 63:6920–6927.

106. Nawrocki ST, Bruns CJ, Harbison MT, Bold RJ, Gotsch BS, Abbruzzese JL, Elliott P, Adams J, McConkey DJ. Effects of the proteasome inhibitor PS-341 on apoptosis and angiogene-

sis in orthotopic human pancreatic tumor xenografts Mol Cancer Ther 2002; 1:1243–1253.

107. Ruggeri B, Robinson C, Angeles T, Wilkinson IV J, Clapper MI. The chemopreventive agent oltipraz possesses potent antiangiogenic activity *in vitro*, *ex vivo* and *in vivo* and inhibits tumor xenograft growth Clin Cancer Res 2002; 8:267–274.

108. Beebe JS, Jani JP, Knauth E, Goodwin P, Higdon C, Rossi AM, Emerson E, Finkelstein M, Floyd E, Harriman S, Atherton J, Hillerman S, Soderstrom C, Kou K, Gant T, Noe MC, Foster B, Rastinjad F, Marx MA, Schaeffer T, Whalen PM, Roberts WG. Pharmacological characterization of CP-547,632, a novel vascular endothelial growth factor receptor-2 tyrosine kinase inhibitor for cancer therapy Cancer Res 2003; 63:7301–7309.

109. Ueda Y, Yamagishi T, Samata K, Ikeya H, Hirayama N, Takashima H, Nakaike S, Tanaka M, Saiki I. A novel low molecular weight antagonist of vascular endothelial growth factor receptor binding: VGA155 Mol Cancer Ther 2003; 2:105–111.

110. Ruggeri B, Singh J, Gingrich D, Angeles T, Albom M, Chang H, Robinson C, Hunter K, Dobrzanski P, Jones-Bolin S, Aimone L, Klein-Szanto A, Herbert J-M, Bono F, Schaeffer P, Casellas P, Bourie B, Pili R, Isaacs J, Ator M, Hudkins R, Vaught J, Mallamo J, Dionne C. CEP-7055: a novel, orally active pan inhibitor of vascular endothelial growth factor receptor tyrosine kinase with potent antiangiogenic activity and antitumor efficacy in preclinical models Cancer Res 2003; 63: 5978–5991.

111. Bockhorn M, Tsuzuki Y, Xu L, Frilling A, Broelsch CE, Fukumura D. Differential vascular transcriptional responses to anti-vascular endothelial growth factor antibody in orthotopic human pancreatic cancer xenografts Clin Cancer Res 2003; 9: 4221–4226.

112. Zhang WM, Ran S, Sambade M, Huang X, Thorpe PE. A monoclonal antibody that blocks VEGF binding to VEGFR2 (KDR/Flk-1) inhibits vascular expression of Flk-1 and tumor growth in an orthotopic human breast cancer model Angiogenesis 2002; 5:35–44.

113. Beerepoot L, Witteven EO, Groenewegen G, Fogler WE, Sim BKL, Sidor C, Zonnenberg BA, Schramel F, Gebbink MFBG,

Voest EE. Recombinant human angiostatin by twice-daily subcutaneous injection in advanced cancer: a pharmacokinetic and long-term safety study Clin Cancer Res 2003; 9: 4025–4033.

114. Biancone L, Cantaluppi V, Del Sorbo L, Russo S, Tjoelker LW, Camussi G. Platelet-activating factor inactivation by local expression oof platelet-activating factor acetyl-hydrolase modifies tumor vascularization and growth Clin Cancer Res 2003; 9:4214–4220.

115. Sinkovics JG, Horvath JC. Vaccination against human cancers Int J Oncol 2000; 16:81–96.

116. Sinkovics JG, Horvath JC. Virus therapy of human cancers Melanoma Research 2003; 13:431–432.

117. Katerinski E, Evans GS, Lorigan PC, MacNeil S. TNF-alpha increases human melanoma cell invasion and migratiom in vitro: the role of proteolytic enzymes Br J Cancer 2003; 89: 1123–1129.

118. Lee SO, Lou W, Hou M, de Miguel F, Gerber L, Gao AC. Interleukin-6 promotes androgen-independent growth in LNCaP human prostate cancer cells Clin Cancer Res 2003; 9: 370–376.

119. Masood R, Cai J, Tulpule A, Zheng T, Hamilton A, Sharma S, Espina BM, Smith DL, Gill PS. Interleukin 8 is an autocrine growth factor and a surrogate marker for Kaposi's sarcoma Clin Cancer Res 2001; 7:2693–2702.

120. Ghia P, Granziero L, Chilosi M, Caligaris-Cappio F. Chronic B cell malignancies and bone marrow microenvironment Semin Cancer Biol 2002; 12:149–155.

121. Gattacceca F, Pilotte Y, Billard C, La Carrou J, Eliot M, Jaurand M-C. Ad-IFN-γ induces proliferative and antitumor al responses in malignant mesothelioma Clin Cancer Res 2002; 8: 3298–3304.

122. Wall L, Burke F, Barton C, Smyth J, Balkwill F. Interferon-gamma induces apoptosis in ovarian cancer cells in vivo and in vitro Clin Cancer Res 2003; 9:2487–2496.

123. Segal JG, Lee NC, Tsung YL, Norton JA, Tsung K. The role of IFN-γ in rejection of established tumors by IL-12 : source of production and target Cancer Res 2002; 62:4696–4703.

124. Champelovier P, Simon A, Garrel C, Levacher G, Praloran V, Seigneurin D. Is interferon γ one key potential increase in human bladder carcinoma? Clin Cancer Res 2003; 9: 4562–4569.

125. Singer CF, Kronsteiner N, Hudelist G, Marton E, Walter I, Kubista M, Czerwenka K, Schreiber M, Seifert M, Kubista E. Interleukin 1 system and sex steroid receptor expression in human breast cancer: intrerleukin 1α protein secretion is correlated with malignant phenotype Clin Cancer Res 2003; 9: 4877–4883.

126. Weinreich DM, Elaraj DM, Puhlman M, Hewitt SM, Carroll NM, Feldman ED, Turner EM, Spies PJ, Alexander HR. Effect of interleukin 1 receptor antagonist gene transduction on human melanoma xenograft in nude mice Cancer Res 2003; 63:5957–5961.

127. Stassi G, Todaro M, Zerilli M, Ricci-Vitiani L, Di Libert D, Patti M, Florena A, Di Gaudino F, Di Gesú G, De Maria R. Thyroid cancer resistance to chemotherapeutic drugs via autocrine production of inerleukin-4 and interleukin-10 Cancer Res 2003; 63:6784–6790.

128. Kattan MW, Shariat SF, Andrews B, Zhu K, Canto E, Matsumoto K, Muramoto M, Scardino PT, Ohori M, Wheeler TM, Slavin LM. The addition of interleukin-6 soluble receptor and transforming growth factor $beta_1$ improves a preoperative nomogram for predicting biochemical progressioin in patients with clinically localized prostate cancer J Clin Oncol 2003; 21: 3573–3579.

129. Trikha M, Corringham R, Klein B, Rossi J-F. Targeted anti-interleukin-6 monoclonal antibody therapy for cancer: a review of the rationale and clinical evidence Clin Cancer Res 2003; 9:4653–4665.

130. Aldred MA, Ginn-Pease ME, Morrison CD, Popkie AP, Gimm O, Hoang-Vu C, Krause U, Dralle H, Jhiang SM, Plass C, Eng C. Caveolin-1 and caveolin-2, together with three bone morphogenetic protein-related genes, may encode novel tumor suppressors down-regulated in sporadic follicular thyroid carcinogenesis Cancer Res 2003; 63:2864–2871.

131. Mocellin S, Panelli MC, Wang E, Nagorsen D, Marincola FM. The dual role of IL-10 Trends Immunol 2003; 24:36–43.

132. Iwakiri D, Eizuru Y, Tokunaga M, Takada K. Autocrine growth of Epstein-Barr virus-positive gastric carcinoma cells mediated by an Epstein-Barr virus encoded small RNA Cancer Res 2003; 63:7062–7067.

133. Komano J, Maruo S, Koizumi K, Takada K. Oncogenic role of Epstein-Barr virus-encoded RNA in Burkitt's lymphoma cell line Akata J Virol 1999; 73:9827–9831.

134. Kuniyasu H, Ohmori H, Sasaki T, Sasahira T, Yoshida K, Kitadai Y, Fidler IJ. Production of interleukin-15 by human colon cancer cells is associated with induction of mucosal hyperplasia, angiogenesis and metastasis Clin Cancer Res 2003; 9:4802–4810.

135. Lo CH, Lee SC, Wu PY, Pan WY, Su J, Cheng CW, Roffler SR, Chiang BL, Lee CN, Wu CW, Tao MH. Antitumor and antimetastatic activity of IL-23 J Immunol 2003; 171: 600–607.

136. Yacoub A, Mitchell C, Brannon J, Rosenberg E, Qiao L, McKinstry R, Linehan WM, Su ZS, Sarkar D, Lebedeva IV, Valerie K, Gopalkrishnan RV, Grant S, Fisher PB, Dent P. MDA-7 (interleukin-24) inhibits the proliferation of renal carcinoma cells and interacts with free radicals to promote cell death and loss of reproductive capacity Mol Cancer Ther 2003; 2:623–632.

137. Nowak AK, Lake RA, Kindler HL, Robinson BW. New approaches for mesothelioma: biologics, vaccines, gene therapy and other novel agents Semin Oncol 2002; 29:82–96.

138. Massod R, Kundra A, Zhu S, Xia G, Scalia P, Smith DL, Gill PS. Malignant mesothelioma growth inhibition by agents that target the VEGF and VEGF-C autocrine loop Int J Cancer 2003; 104:603–610.

139. Xia C, Xu Z, Yuan X, Uematsu K, You L, Li K, Li L, McCormick F, Jablond DM. Induction of apoptosis in mesothelioma cells by antisurvivin oligonucleotides Mol Cancer Ther 2002; 1: 687–684.

140. Horwood NJ, Elliott J, Martin TJ, Gillespie MT. IL-12 alone and in synergy with IL-18 inhibits osteoclast formation in vitro J Immunol 2001; 166:4915–4921.

141. Martin TJ, Romas E, Gillespie MT. Interleukins in the control of osteoclast differentiation Cit Rev Eukaryot Gene Expr 1998; 8:107–123.

142. Nakata A, Tsujimura T, Sugihara A, Okamura H, Iwasaki T, Shinkai K, Iwata N, Kakishita E, Akedo H, Terada N. Inhibition by interleukin 18 of osteolytic bone metastasis by human breast cancer cells Anticancer Res 1999; 19:4131–4138.

143. Nakiishi-Shimobayashi C, Tsujimura T, Iwasaki T, Yamada N, Sugihara A, Okamura H, Hayashi S, Terada N. Interleukin-18 up-regulates osteoprotegerin expression in stromal/osteoblastic cells Biochem Biophys Res Commun 2001; 281: 361–366.

144. Yamada N, Niwa S, Tsujimura T, Iwasaki T, Sugihara A, Futani H, Hayashi S, Okamura H, Akedo H, Terada N. Interleukin-18 and interleukin-12 synergistically inhibit osteoclastic bone-resorbing avtivity Bone 2002; 30:901–908.

145. Cornish J, Gillespie MT, Callon KE, Horwood NJ, Moseley JM, Reid IR. Interleukin-18 is a novel mitogen of osteogenic and chondrogenic cells Endocrinology 2003; 144:1194–1201.

146. Yamashita K, Iwasaki T, Tsujimura T, Sugihara A, Yamada N, Ueda H, Okamura H, Futani H, Maruoi S, Terada N. Interleukin-18 inhibits lodging and subsequent growth of human multiple myeloma cells in the bone marrow Oncol Rep 2002; 9:1237–1244.

147. Udagawa N, Horwood NJ, Elliott J, Mackay A, Owens J, Okamura H, Kurimoto M, Chambers TJ, Martin TJ, Gillespie MT. Interleukin-18 (interferon gamma-inducing factor) is produced by osteoblasts and acts via granulocyte/macrophage colony-stimulating factor and not via interferon gamma to inhibit osteoclast formation J Exp Med 1997; 185:1005–1012.

148. Belotti D, Paganoni P, Manenti L, Garofalo A, Marchini S, Taraboletti G, Giavazzi R. Matrix metalloproteinases (MMP9 and MMP2) induce the release of vascular endothelial growth factor (VEGF) by ovarian carcinoma cells: implications for ascites formation Cancer Res 2003; 63:5224–5229.

149. Glade-Bender J, Kandel JJ, Yamashiro DJ. VEGF blocking therapy in the treatment of cancer Expert Opin Biol Ther 2003; 3:263–276.

150. Rosano L, Spinella F, Salani D, Di Castro V, Venuti SA, Nicotra MR, Natali PG, Bagnato A. Therapeutic targeting of the endothelin a receptor in human ovarian carcinoma Cancer Res 2003; 63:2447–3453.

151. Hendrix MJC, Seftor EA, Hess EA, Seftor REB. Vasculogenic mimicry and tumor-cell plasticity: lessons from melanoma Nat Rev Oncol 2003; 3:411–421.

152. Tien Y-W, Lee P-H, Hu R-H, Hsu S-M, Chang K-J. The role of gelatinase in hepatic metastasis of colorectal cancer Clin Cancer Res 2003; 9:4891–4896.

153. Sheu B-C, Lien H-C, Ho H-N, Lin H-H, Chow S-N, Huang S-C, Hsu S-M. Increased expression and activation of gelanolytic matrix metalloproteinases is associted with the progression and recurence of human cervical cancer Cancer Res 2003; 63:6537–6542.

154. Sun B, Nishihara J, Suzuki M, Fukushima N, Ishibashi T, Kondo M, Sato Y, Todo S. Induction of macrophage migration inhibitory factor by lysophosphatidic acid : relevance to tumor growth and angiogenesis Int J Mol Med 2003; 12:633–641.

155. Sasaki Y, Kasuya K, Nishihira J, Magami Y, Tsuchida A, Aoki T, Koyanagi Y. Suppression of tumor growth through introduction of an antisense plasmid of macrophage migration inhibitory factor Int J Mol Med 2002; 10:579–583.

156. Du C, Feng N, Jin H, Lee V, Wang M, Wright JA, Young AH. Macrophages play a critical role in the anti-tumor activity of Virulizin[R] Int J Oncol 2002; 23:1341–1346.

157. Sinkovics JG. Monitoring in vitro of cell-mediated immune reactions to tumors Methods Cancer Research 1973; 8: 107–175.

158. Takahashi N, Nishihira J, Sato Y, Kondo M, Ogawa H, Oshihima H, Une Y, Todo S. Involvement of macrophage migration inhibitory factor (MIP) in the mechanism of tumor cell growth Mol Med 1998; 4:707–714.

159. Ogawa H, Nishihara J, Sato Y, Kondo M, Takahashi N, Oshima T, Todo S. An antibody for macrophage migration inhibitory factor suppresses tumour growth and inhibits tumour-associated angiogenesis Cytokines 2000; 12:309–314.

160. Nishihara J, Ishibashi T, Fukusima T, Sun B, Sato Y, Todo S. Macrophage migration inhibitory factor (MIF): its potential role in tumor growth and tumor-associated angiogenesis Ann N Y Acad Sci 2003; 995:171–182.

161. Mitchell RA, Bucala R. Tumor growth-promoting properties of macrophage migration inhibitory factor (MIF) Semin Cancer Biol 2000; 10:359–366.

162. Shen WH, Zhou JH, Broussard SR, Freund GG, Dantzer R, Kelley KW. Proinflammatory cytokines block growth of breast cancer cells by impairing signals from a growth factor receptor Cancer Res 2002; 62:4746–4756.

163. Hailey J, Maxwell E, Koukouras K, Bishop WR, Pachter JA, Wang Y. Neutralizing anti-insulin-like growth factor receptor 1 antibodies inhibit receptor function and induce receptor degradation in tumor cells Mol Cancer Ther 2002; 1:1349–1353.

164. Resnicoff M, Coppola D, Sell C, Rubin R, Ferrone S, Baserga R. Growth inhibition of human melanoma cells in nude mice by antisense strategies to the type I insulin-like growth factor receptor Cancer Res 1994; 54:4848–4850.

165. Surmacz E. Growth factor receptors as therapeutic targets: strategies to inhibit the insulin-like growth factor I receptor Oncogene 2003; 22:6589–6597.

166. Min Y, Adachi Y, Yamamoto H, Ito H, Itoh F, Lee C-T, Nadaf S, Carbone DP, Imai K. Genetic blockade of the insulin-like growth factor-I receptor: a promising strategy for human pancreatic cancer Cancer Res 2003; 63:6432–6441.

167. Harding MW. Immunophilins, mTOR, and pharmacodynamic strategies for a targeted cancer therapy Clin Cancer Res 2003; 9:2882–2886.

168. Oldham S, Hafen E. Insulin/IGF and target of rapamycin signaling: a TOR de force in growth control Trends Cell Biol 2003; 13:79–85.

169. Thimmaiah KN, Easton J, Huang S, Veverka KA, Germain GS, Harwood FC, Houghton PJ. Insulinlike growth factor I-mediated protection from rapamycin-induced apoptosis is independent of Ras-Erk1-Erk2 and phosphatidylinositol 3′-kinase-Akt signaling pathways Cancer Res 2003; 63: 364–374.

170. Scotlandi K, Maini C, Manara MC, Benini S, Serra M, Cerisano V, Strammiello R, Baldini N, Lollini PL, Nanni P, Nicoletti G, Picci P. Effectiveness of insulin-like growth factor I receptor antisense strategy against Ewing's sarcoma cells Cancer Gene Ther 2002; 9:296–307.

171. Andrews DW, Resnicoff M, Flanders AE, Kenyon L, Curtis M, Merli G, Baserga R, Iliakis G, Aiken RD. Results of a pilot study involving the use of an antisense oligodeoxynucleotide directed against the insulin-like growth factor type I receptor in malignant astrocytomas J Clin Oncol 2001; 19:2189–2200.

172. Iacobuzio-Donahue CA, Argani P, Hempen PM, Jones J, Kern SE. The desmoplastic response to infiltrating breast carcinoma: gene expression at the site of primary invasion and implications for comparisons between tumor types Cancer Res. 2002; 62:5351–5357.

173. Cardones AAr, Murakami T, Hwang ST. CXCR4 enhances adhesion of B16 tumor cells to endothelial cells *in vitro* and *in vivo* via β_1 integrin Cancer Res 2003; 63:6751–6757.

174. Crittenden M, Gough M, Harrington K, Olivier K, Thompson J, Vile RG. Expression of inflammatory chemokines combined with local tumor destruction enhances tumor regression and long term immunity Cancer Res 2003; 63:5505–5512.

175. Ding Y, Shimada Y, Maeda M, Kaeabe A, Kaganoi J, Komoto I, Hashimoto Y, Miyake M, Hashida H, Imamura M. Association of CC chemokine receptor 7 with lymph node metastasis of esophageal squamous cell carcinoma Clin Cancer Res 2003; 9:3406–3412.

176. Monti P, Leone BE, Marchesi F, Balzano G, Zerbi A, Scaltrini F, Pasquali C, Calori G, Pessi F, Sperti C, Di Carlo V, Allavena P, Piemonti L. The CC chemokine MCP-1/CCL2 in pancreatic cancer progression: regulation of expression and potential mechanisms of antimalignant activity Cancer Res 2003; 63: 7451–7461.

177. Powers A, Carbone M. The role of environmental carcinogens, viruses and genetic predisposition in the pathogenesis of mesothelioma Cancer Biol Ther 2002; 1:348–353.

178. Hopkins-Donaldson S, Cathomas R, Simoes-Wust AP, Kurtz S, Belyanskya L, Stahel RA, Zangemeister-Wittke U. Induc-

tion of apoptosis and chemosensitization of mesothelioma cells by Bcl-2 and Bcl-xL antisense treatment Int J Cancer 2003; 106:160–166.

179. Cao XX, Mohuiddin I, Chada S, Mhashilkar AM, Ozvaran MK, McConkey DJ, Miller SD, Daniel JC, Smythe WR. Adenoviral transfer of mda-7 leads to BAX up-regulation and apoptosis in mesothelioma cells, and is abrogated by overexpression of BCL-XL Mol Med 2002; 8:869–8701.

180. Xia C, Xu Z, Yuan X, Uematsu K, You L, Li K, Li L, McCormick F, Jablons DM. Induction of apoptosis in mesothelioma cells by antisurvivin oligonucleotides Mol Cancer Ther 2002; 1: 687–694.

181. Lustig B, Behrens J. The Wnt signaling pathway and its role in tumor development J Cancer Res Clin Oncol 2003; 129: 199–221.

182. Giles RH, van Es JH, Clevers H. Caught up in a Wnt storm: Wnt signaling in cancer Biochim Biophys Acta 2003; 1653: 1–24.

183. Uematsu K, Kanazawa S, You L, He B, Xu Z, Li K, Peterlin BM, McCormick F, Jablons DM. Wnt pathway activation in mesothelioma: evidence of dishevelled overexpression and transcriptional activity of beta-catenin Cancer Res 2003; 63: 4547–4551.

184. Chen D, Xu W, Bales E, Colmenares C, Conacci-Sorrell M, Ishii S, Stavnezer E, Campisi J, Fisher DE, Ben-Ze'ev A, Medrano EE. SKI activates Wnt/β-catenin signaling in human melanoma Cancer Res 2003; 63:6626–6634.

185. Mælandsmo GM, Holm R, Nesland JM, Fodstad Ø, Flørenes VA. Reduced β-catenin expression in the cytoplasm of advanced-stage superficial spreading melanoma Clin Cancer Res 2003; 9:3383–3388.

186. Shimokawa T, Furukawa Y, Sakai M, Li M, Miwa N, Lin Y-M, Nakamura Y. Involvement of the FGF18 gene in colorectal carcinogenesis, as a novel downstream target of the β-catenin/T-cell factor complex Cancer Res 2003; 63:6116--6120.

187. Jänne PA, Taffaro ML, Salgia R, Johnson BE. Inhibition of epidermal growth factor receptor signaling in malignant pleural mesothelioma Cancer Res 2002; 62:5242–5247.

188. Kim J-S, Crooks H, Foxworth A, Waldman T. Proof-of-principle: oncogenic β-catenin is a valid molecular target for the development of pharmacological inhibitors Mol Cancer Ther 2002; 1:1355–1359.

189. Aldred MA, Ginn-Pease ME, Morrison CD, Popkie AP, Gimm O, Hoang-Vu C, Krause U, Dralle H, Jhiang SM, PLass C, Eng C. *Caveolin*-1 and caveolin-2, together with three bone morphogenetic protein-related genes, may encode novel tumor suppressors down-regulated in sporadic follicular thyroid carcinogenesis Cancer Res 2003; 63:2864–2871.

190. Xie Z, Zeng X, Waldman T, Glazer RI. Transformation of mammary epithelial cells by 3-phosphoinositide-dependent protein kinase-1 activates β-catenin and c-Myc, and down-regulates caveolin-1 Cancer Res 2003; 63:5370–5375.

191. Carver LA, Schnitzer JE, Anderson RGW, Mohla S. Role of caveolae and lipid rafts in cancer: workshop summary and future needs Cancer Res 2003; 63:6571–6574.

192. Carver LA, Scnitzer JE. Caveolae: mininig little caves for new cancer targets Nat Rev Cancer 2003; 3:571–581.

193. van Deurs B, Roepstorff K, Hommelgaard AM, Sandvig K. Caveolae: anchored, multifunctional platforms in the lipid ocean Trends Cell Biol 2003; 13:92–93.

194. Chan TF, Su TH, Yeh KT, Chang JY, Lin TH, Chen JC, Yuang SS, Chang JG. Mutational, epigenetic and expressional analyses of caveolin-1 gene in cervical cancers Int J Oncol 2003; 23: 599–604.

195. Tahir SA, Ren C, Timme TL, Gdor Y, Hoogeveen R, Morrisett JD, Frolov A, Ayala G, Wheeler TM, Thompson TC. Development of an immunoasssat for serum caveolin-1: a novel biomarker for prostate cancer Clin Cancer Res 2003; 9: 3653–3659.

196. Chen J-T, Cheng Y-W, Chou M-C, Sen-Lin T, Lai W-W, Ho WL, Lee H. The correlation between aberrant connexin 43 mRNA expression induced by promoter methylation and nodal micrometastasis in non-small cell lung cancer Clin Cancer Res 2003; 9:4200–4204.

197. Zhang YW, Nakayama K, Nakayama K, Morita I. A novel route for connexin 43 to inhibit cell proliferation: negative reg-

ulation of S-phase kinase-associated protein (Skp 2) Cancer Res 2003; 63:1623–1630.

198. Tachibana M, Murai M. G-CSF production in human bladder cancer and its ability to promote autocrine growth: a review Cytokines Cell Mol Ther 1998; 4:113–120.

199. Song P, Sekhon HS, Jia Y, Keller JA, Blusztajn JK, Mark GP, Spindel ER. Acetylcholine is synthesized by and acts as an autocrine growth factor for small cell lung carcinoma Cancer Res 2003; 63:214–221.

200. Acs G, Zhang PJ, Rebbeck TR, Acs P, Verma A. Immunohistochemical expression of erythropoietin and erythropoietin receptor in breast carcinoma Cancer 2002; 95:969–981.

201. Acs G, Acs P, Beckwith SM, Pitts RL, Clements E, Wong K, Verma A. Erythropoietin and erythropoietin receptor expression in human cancer Cancer Res 2001; 61:3561–3565.

202. Hassan AB, Macaulay VM. The insulin-like growth factor system as a therapeutic target in colorectal cancer Ann Oncol. 2002; 13:349–356.

203. Nilsson EE, Skinner MK. Role of transforming growth factor beta in ovarian surface epithelium biology and ovarian cancer Reprod Biomed Online 2002; 5:254–258.

204. Lei X, Bandyopadhyay A, Le T, Sun L. Autocrine TGF beta supports growth and survival of human breast cancer MDA-MB-231 cells Oncogene 2002; 21:7514–7523.

205. Awwad RA, Sergina N, Yang H, Ziober B, Willson JK, Zborowska E, Humphrey LE, Fan R, Ko TC, Brattain MG, Howell GM. The role of transforming growth factor α in determining growth factor independence Cancer Res 2003; 63:4731–4738.

206. Muraoka RS, Dumont N, Ritter CA, Dugger TC, Brantley DM, Chen J, Easterly E, Roebuck LR, Ryan S, Gotwals PJ, Koteliansky V, Arteaga CL. Blockade of TGF-β inhibits mammary tumor cell viability, migration, and metastases J Clin Invest 2002; 109:1551–1559.

207. Jennings MT, Kaariainen IT, Gold L, Maciunas RJ, Commers PA. TGFβ1 and TGF-β2 are potential growth regulators for medulloblastomas, primitive neuroectodermal tumors, and

ependymomas: evidence in support of an autocrine hypothesis Hum Pathol 1994; 25:464–475.

208. Kurimoto M, Endo S, Arai K, Horie Y, Nogami K, Takaku A. TM-1 cells from an established human malignant glioma cell line produce PDGF, TGF-α, and TGF-β which cooperatively play a stimulatory role for an autocrine growth promotion J Neurooncol 1994; 22:33–44.

209. Benckert C, Jonas S, Cramer T, Von Marschall Z, Schafer G, Peters M, Wagner K, Radke C, Wiedenmann B, Neuhaus P, Hocker M, Rosewicz S. Transforming growth factor beta 1 stimulates vascular endothelial growth factor gene transcription in human cholangiocellular carcinoma cells Cancer Res 2003; 63:1083–1092.

210. Sugano Y, Matsuzaki K, Tahashi Y, Furukawa F, Mori S, Yamagata H, Yoshida K, Matsushita M, Nishizawa M, Fujisawa J, Inoue K. Distortion of autocrine transforming growth factor beta signal accelerates malignant potential by enhancing cell growth as well as PAI-1 and VEGF production in human hepatocellular carcinoma cells Oncogene 2003; 22:2309–2321.

211. Inman GJ, Nicolas FJ, Callahan JF, Harling JD, Gaster LM, Reith AD, Laping NJ, Hill CS. SB-431542 is a potent and specific inhibitor of transforming growth factor-beta superfamily type I activin receptor-like kinase (ALK) receptors ALK4, ALK5, and ALk7 Mol Pharmacol 2002; 62:65–74.

212. Baudhuin LM, Jiang Y, Zaslavsky A, Ishii I, Chun J, Xu Y. S1P3-mediated Akt activation and crosstalk with platelet-derived growth factor receptor (PDGFR) FASEB J 2004; 18: 341–343.

213. Matsuyama S, Iwadate M, Kondo M, Saitch M, Hanyu A, Shimiu K, Aburatani H, Mishima HK, Imamura T, Miyazone K, Miyazawa K. SB-431542 and Gleevec inhibit transforming growth factor-beta-induced proliferation of human osteosarcoma cells Cancer Res 2003; 63:7791–7798.

214. Ganju RK, Brubaker SA, Meyer J, Dutt P, Yang Y, Qin S, Newman W, Groopman JE. The α-chemokine, stromal cell-derived factor-1α, binds to the transmembrane G-protein-coupled CXCR-4 receptor and activates multiple signal transduction pathways J Biol Chem 1998; 273:23169–23175.

215. Muller A, Homey B, Soto H, Ge N, Catron D, Buchanan ME, McClanahan T, Murphy E, Yuan W, Wagner SN, Barrera JL, Mohar A, Verastegui E, Zlotnik A. Involvement of chemokine receptors in breast cancer metastasis Nature 2001; 410:24–25.

216. Broxmeyer HE, Cooper S, Kohli L, Hangoc G, Lee Y, Mantel C, Clapp DW, Kim CH. Transgenic expression of stromal cell-derived factor-1/CXC chemokiine ligand 12 enhances myeloid progenitor cell survival/antiapoptosis in vitro in response to growth factor withdrawal and enhances myelopoiesis in vivo J Immunol 2003; 170:421–429.

217. Finch PW, Yee LK, Chu MY, Chen TM, Lipsky MH, Maciag T, Friedman S, Epstein MH, Calabresi P. Inhibition of growth factor mitogenicity and growth of tumor cell xenografts by a sulfonated distamycin A derivative Pharmacology 1997; 55: 269–278.

218. Schneider GP, Salcedo R, Dong HF, Kleinman HK, Oppenheim JJ, Howard OM. Suradista NSC 651016 inhibits the angiogenic activity of CXCL12-stromal cell-derived factor 1α Clin Cancer Res 2002; 8:3955–3960.

219. Chen X, Beutler JA, McCloud TG, Loehfelm A, Yang KL, Dong H-F, Chertov OY, Salcedo R, Oppenheim JJ, Howard OMZ. Tannic acid is an inhibitor of CXCL12 (SDF-1alpha)/CXCR4 with antiangiogenic activity Clin Cancer Res 2003; 9: 3115–3123.

220. Yao L, Salvucci O, Cardones AR, Hwang ST, Aoki Y, De La Luz Sierra M, Sajewicz A, Pittaluga S, Yarchoan R, Tosato G. Selective expression of stromal-derived factor-1 in the capillary vascular endothelium plays a role in Kaposi sarcoma pathogenesis Blood 2003; 102:3900–3905.

221. Signoret N, Oldridge J, Pelchen-Matthews A, Klasse PJ, Tran T, Brass LF, Rosenkilde MM, Schwartz TN, Holmes W, Dallas W, Luther MA, Wells TN, Hoxie JA, Marsh M. Phorbol esters and SDF-1 induce rapid endocytosis of the chemokine receptor CXCR4 J Cell Biol 1997; 139:651–664.

222. Murakami T, Nakajima T, Koyanagi Y, Tachibana K, Fuji N, Tamamura H, Yoshida N, Waki M, Matsumoto A, Yoshie O, Kishimoto T, Yamamoto N, Nagasawa T. A small molecule CXCR4 inhibitor that blocks T cell line-tropic HIV-1 infection J Exp Med 1997; 186:1389–1393.

223. Zhou N, Luo Z, Luo J, Hall JW, Huang Z. A novel peptide antagonist of CXCR4 derived from the N-terminus of viral chemokine vMIP-II Biochemistry 2000; 39:3782–3787.

224. Haque NS, Fallon JT, Taubman MB, Harpel PC. The chemokine receptor CCR8 mediates human endothelial cell chemotaxis induced by I-309 and Kaposi sarcoma herpesvirus-encoded vMIP-I and by lipoprotein(s) stimulated endothelial cell conditioned medium Blood 2001; 97:39–45.

225. Geras-Raaka E, Varma A, Clark-Lewis I, Gershengorn MC. Kaposi's sarcoma-associated herpesvirus (KSHV) chemokine vMIP-II and human SDF-1alpha inhibit signaling by KSHV G protein-coupled receptor Biochem Biophys Res Commun 1998; 253:725–727.

226. Garaci E, Caroleo MC, Aloe L, Aquaro S, Piacentini M, Costa N, Amendola A, Micera A, Calio R, Perno CF, Levi-Montalcini R. Nerve growth factor is an autocrine factor essential for the survival of macrophages infected with HIV Proc Natl Acad Sci U S A 1999; 96:14013–14018.

227. Pica F, Volpi A, Serafino A, Fraschetti M, Franzese O, Garaci E. Autocrine nerve growth factor is essential for cell saurvval and viral mutation in HIV-infected primary effusion lymphoma cells Blood 2000; 95:2905–2912.

228. Libura J, Drukala J, Majka M, Tomescu O, Navenot JM, Kucia M, Marquez L, Peiper SC, Barr FG, Janowska-Wieczorek A, Ratajczak MZ. CXCR4-SDF-1 signaling is active in rhabdomyosarcoma cells and regulates locomotion, chamotaxis and adhesion Blood 2002; 100:2597–2606.

229. Taichman RS, Cooper C, Keller ET, Pienta KJ, Taichman NS, McCauley LK. Use of stromal cell-derived factor-1/CXCR4 pathway in prostate cancer metastasis to bone Cancer Res 2002; 62:1832–1837.

230. Cheung CY, Poon LL, Lau AS, Luk W, Lau YL, Shortridge KF, Gordon S, Guan Y, Peiris JS. Induction of proinflammatory cytokines in human macrophages by influenza A (H5N1) viruses: a mechanism for the unusual severity of human disease? Lancet 2002; 360:1831–1837.

231. Sprenger H, Bacher M, Rischkowsky E, Bender A, Nain M, Gemsa D. Characterization of a high molecular weight tumor

necrosis factor-α mRNA in influenza A virus-infected macrophages J Immunol 1994; 152:280–289.

232. Pecora AL, Rizvi N, Cohen GI, Meropol NJ, Sterman D, Marshall JL, Goldberg S, Gross P, O'Neil JD, Groene WS, Roberts MS, Rabin H, Bamat MK, Lorence RM. Phase I trial of intravenous administration of PV701, an oncolytic virus in patients with advanced solid cancers J Clin Oncol 2002; 20:2251–2266.

233. Auchincloss H, Sachs DH. Xenogeneic transplantation Annu Rev Immunol 1998; 16:433–470.

234. Platt JL. Xenotransplantation. Totowa. NJ: Humana, 2002: 1–264.

235. Slavin S. New strategies for bone marrow transplantatiom Curr Opin Immunol 2000; 121:542–551.

236. Singh RP, Dhanalakshmi S, Tyagi AK, Chan DC, Agarwal C, Agarwal R. Dietary feeding of silibinin inhibits advance human prostate carcinoma growth in athymic nude mice and increases plasma insulin-like growth factor-binding protein-3 levels Cancer Res 2002; 62:3063–3069.

237. Huynh H, Nguyen TT, Chan E, Tran E. Inhibition of ErbB-2 and ErbB-3 expression by quercetin prevents transforming growth factor alpha (TGF-α)- and epidermal growth factor (EGF)-induced human PC-3 prostate cancer cell proliferation Int J Oncol 2003; 23:821–829.

238. Dhanalakshmi S, Agarwal R, Agarwal C. Inhibition of NF-κB pathway in grape seed extract-induced apoptotic death of human prostate carcinoma DU145 cells Int J Oncol 2003; 23: 721–727.

239. Zhu YS, Huang Y, Cai LQ, Zhu J, Duan Q, Duan Y, Imperato-McGinley J. The Chinese medicinal herbal formula ZYD88 inhibits cell growth and promotes cell apoptosis in prostatic tumor cells Oncol Rep 2003; 10:1633–1639.

240. Li Y, Sarkar FH. Inhibition of nuclear factor kappaB activation in PC3 cells by genistein is mediated via Akt signaling pathway Clin Cancer Res 2002; 8:2369–2377.

241. Farhan H, Wahala K, Cross HS. Genistein inhibits vitamin D hydroxylases CYP24 and CYP27B1 expression in prostate cells J Steroid Biochem Mol Biol 2003; 84:423–429.

242. Yu L, Blackburn GL, Zhou JR. Genistein and daidzein downregulate prostate androgen-regulated transcript-1 (PART-1) gene expression induced by dihydrotestosterone in human prostate LNCaP cancer cells J Nutr 2002; 133:389–392.

243. Ikeda N, Uemura H, Ishiguro H, Hori M, Hosaka M, Kyo S, Miyamoto K, Takeda E, Kubota Y. Combination treatment with 1α,25-dihydroxyvitamin D_3 and 9-cis-retinoic acid directly inhibits human telomerase reverse transcriptase transcription in prostate cancer cells Mol Cancer Ther 2003; 2: 739–746.

244. Kim J, Luo W, Chen DT, Earley K, Tunstead J, Yu-Lee LY, Lin SH. Antitumor activity of the 16-kDa prolactin fragment in prostate cancer Cancer Res 2003; 63:386–393.

245. Xiao W, Zhang Q, Jiang F, Pins M, Kozlowski JM, Wang Z. Suppression of prostate tumor growth by U19, a novel testosterone-regulated apoptosis inducer Cancer Res 2003; 63: 4698–4704.

246. Solit DB, Zheng FF, Drobnjak M, Munster PN, Higgins B, Verbel D, Heller G, Tong W, Cordon-Cardo C, Agus DB, Scher HI, Rosen N. 17-Allylamino-17-demethoxygeldanamycin induces the degradation of androgen receptor and HER-2/neu and inhibits the growth of prostate cancer xenografts Clin Cancer Res 2002; 8:986–993.

247. Fox WD, Higgins B, Maiese KM, Drobnjak M, Cordon-Cardo C, Scher HI, Agus DB. Antibody to vascular endothelial growth factor slows growth of an androgen-independent xenograft model of prostate cancer Clin Cancer Res 2002; 8: 3226–3231.

248. Bandyopadhyay A, Lopez-Casillas F, Malik SN, Montiel JL, Mendoza V, Yang J, Sun LZ. Antitumor activity of a recombinant soluble betaglycan in human breast cancer xenograft Cancer Res 2002; 62:4690–4695.

249. Ellenrieder V, Hendler SF, Ruhland C, Boeck W, Adler G, Gress TM. TGF-beta-induced invasiveness of pancreatic cancer cells is mediated by matrix metalloproteinase-2 and the urokinase plasminogen activator system Int J Cancer 2001; 93:204–211.

250. McLeskey SW, Ding IY, Lippman ME, Kern FG. MDA-MB-134 breast carcinoma cells overexpress fibroblast growth fac-

tor (FGF) receptors and are growth-inhibited by FGF ligands Cancer Res 1994; 54:523–530.

251. Okunieff P, Fenton BM, Zhang L, Kern FG, Wu T, Greg JR, Ding I. Fibroblast growth factors (FGFs) increase breast tumor growth rate, metastases, blood flow, and oxygenation without significant change in vascular density Adv Exp Med Biol 2003; 530:593–601.

252. Plumb JA, Finn PW, Williams RJ, Bandara MJ, Romero MR, Watkins CJ, La Thangue NB, Brown R. Pharmacodynamic response and inhibition of growth of human tumor xenografts by the novel histone deacetylase inhibitor PXD101 Mol Cancer Ther 2003; 2:721–728.

253. Ju R, Muller MT. Histone deacetylase inhibitors activate $p21^{WAF1}$ expression via ATMN Cancer Res 2003; 63: 2891–2897.

254. Liang K, Jin W, Knuefermann C, Schmidt M, Mills GB, Ang KK, Milas L, Fan Z. Targeting the phosphatidylinositol 3-kinase/Akt pathway for enhancing breast cancer cells to radiotherapy Mol Cancer Ther 2003; 2:353–360.

255. Tong WG, Ding XZ, Witt RC, Adrian TE. Lipooxygenase inhibitors attenuate growth of human pancreatic cancer xenografts and induce apoptosis through the mitochondrial pathway Mol Cancer Ther 2002; 1:929–935.

256. Faridi J, Wang L, Endemann G, Roth RA. Expression of constitutively active Akt-3 in MCF-7 breast cancer cells reverses the estrogen and tamoxifen responsivity of these cells in vivo Clin Cancer Res 2003; 9:2933–2939.

257. Bross PF, Baird A, Chen G, Jee JM, Lostritto RT, Morse DE, Rosario LA, Williams GM, Yang P, Rahman A, Williams G, Pazdur R. Fulvestrant in postmenopausal women with advanced breast cancer Clin Cancer Res 2003; 9:4309–4317.

258. Jones SE. Fulvestrant: an estrogen receptor antagonist that downregulates the estrogen receptor Semin Oncol 2003; 30S16:14–20.

259. Hall JM, Korach KS. Stromal cell-derived factor-1, a novel target of estrogen receptor action, mediates the mitogenic effects of estradiol in ovarian and breast cancer cells Mol Endocrinol 2003; 17:792–803.

260. McKeague AL, Wilson DJ, Nelson J. Staurosporine-induced apoptosis and hydrogen peroxide-induced necrosis in two human breast cell lines Br J Cancer 2003; 88:125–131.

261. John CM, Leffler H, Kahl-Knutsson B, Svensson I, Jarvis GA. Truncated galectin-3 inhibits tumor growth and metastasis in orthotopic nude mouse model of human breast cancer Clin Cancer Res 2003; 9:2374–2383.

262. Nahta R, Esteva FJ. HER-2-targeted therapy: lessons learned and future directions Clin Cancer Res 2003; 9:5078–5084.

263. Bajo AM, Schally AV, Krupa M, Hebert F, Groot K, Szepeshazi K. Bombesin antagonists inhibit growth of MDA-MB-435 estrogen-independent breast cancers and decrease the expression of the ErbB-2/HER-2 oncoprotein and c-jun and c-fos oncogenes Proc Natl Acad Sci U S A. 2002; 99:3836–3841.

264. Cunha GR, Hayward SW, Wang YZ, Ricke WA. Role of the stromal microenvironment in carcinogenesis of the prostate Int J Cancer 2003; 107:1–10.

265. Haslam SZ, Woodward TL. Epithelial-cell-stromal cell interactions and steroid hormone action in normal and cancerous mammary gland Breast Cancer Res 2003; 5:208–215.

266. Ayala G, Tuxhorn JA, Wheeler TM, Frolov A, Scardino PT, Ohori M, Wheeler M, Spitler J, Rowley DR. Reactive stroma as a predictor of biochemical-free recurrence in prostate cancer Clin Cancer Res 2003; 9:4792–4801.

267. Freeman KW, Gangula RD, Welm BE, Ozen M, Foster BA, Rosen JM, Ittman M, Greenberg NM, Spencer DM. Conditional activation of fibroblast growth factor receptor (FGFR) 1, but not FGFR2, in prostate cancer cells leads to increased osteopontin induction, extracellular signal-regulated kinase actrivation, and *in vivo* prolifetation Cancer Res 2003; 63: 6237–6243.

268. An J, Sun Y-P, Adams J, Fisher M, Belldegrun A, Rettig MB. Drug interactions between the proteasome inhibitor bortezomib and cytotoxic chemotherapy, tumor necrosis factor (TNF) α, and TNF-related apoptosis-inducing ligand in prostate cancer Clin Cancer Res 2003; 9:4537–4545.

269. Williams S, Pettaway C, Song R, Papandreou C, Logothetis C, McConkey DJ. Differential effects of the proteasome inhibitor

bortezomib on apoptosis and angiogenesis in human prostate tumor xenografts Mol Cancer Ther 2003; 2:835–843.

270. Arihiro K, Oda H, Kaneko M, Inai K. Cytokines facilitate chemotactic motility of breast carcinoma cells Breast Cancer 2000; 7:221–230.

271. Carroll JS, Lynch DK, Swarbrick A, Renoir J-M, Sarcevic B, Daly RJ, Musgrove EA, Sutherland RL. $p27^{kip1}$ induces quiescence and growth factor insensitivity in tamoxifen-treated breast cancer cells Cancer Res 2003; 63:4322–4326.

272. Jiang WG, Grimshaw D, Martin TA, Davies G, Parr C, Watkins G, Lane J, Abounder R, Laterra J, Mansel RE. Reduction of stromal fibroblast-induced mammary tumor growth, by retroviral ribozyme transgenes to hepatocyte growth factor/scatter factor and its receptor, c-Met Clin Cancer Res 2003; 9: 4274–4281.

273. Elit L. CCI-779 Wyett Curr Opin Invest Drugs 2002; 3: 1249–1253.

274. Beliakoff J, Bagatell R, Paine-Murrieta G, Taylor CW, Lykkesfeldt AE, Whitesell L. Hormone-refractory breast cancer remains sensitive to the antitumor activity of heat shock protein 90 inhibitors Clin Cancer Res 2003; 9:4961–4971.

275. Nimmanapalli R, O'Bryan E, Huang M, Bali P, Burnette PK, Loughran T, Tepperberg J, Jove R, Bhalla K. Molecular characterization and sensitivity of STI-571 (imatinib mesylate, Gleevec)-resistant, Bcr-Abl-positive, human acute leukemia cells to SRC kinase inhibitor PD180970 and 17-allylamino-17-demethoxygeldanamycin Cancer Res 2002; 62:5761–5769.

276. Munster PN, Marchion DC, Basso AD, Rosen N. Degradation of HER2 by ansamycins induces growth arrest and apoptosis in cells with HER2 overexpression via a HER3, phosphatidylinositol 3′-kinase-AKT-dependent pathway Cancer Res 2002; 62:3132–3137.

277. Basso AD, Solit DB, Munster PN, Rosen N. Ansamycin antibiotics inhibit Akt activation and cyclin D expression in breast cancer cells that overexpress HER2 Oncogene 2002; 21: 1159–1166.

278. Stephen AE, Pearsall LA, Christian BP, Donahoe PK, Vacanti JP, MacLaughlin DT. Highly purified müllerian inhibiting

substance inhibits human ovarian cancer in vivo Clin Cancer Res 2002; 8:2640–2646.

279. Ueno NT, Bartholomeusz C, Xia W, Anklesaria P, Bruckheimer EM, Mebel E, Paul R, Li S, Yo GH, Huang L, Hung MC. Systemic gene therapy in human xenograft tumor models by liposomal delivery of the E1A gene Cancer Res 2002; 62: 6712–6716.

280. Hortobagyi G, Ueno NT, Xia W, Zhang S, Wolf JK, Putnam JB, Widen PL, Willey JS, Carey M, Branham DL, Payne JY, Tucker SD, Bartholomeusz C, Kilbourn RG, De Jager RL, Sneige R, Katz RL, Anklesaria P, Ibrahim NK, Murray JL, Theriault RL, Valero V, Greshenson DM, Bevers MW, Huang L, Lopez-Berestein G, Hung M-C. Cationic liposome-mediated E1A gene transfer to human breast and ovarian cancer cells and its biological effects: a phase I clinical trial J Clin Oncol 2001; 19:3422–3433.

281. Nemunaitis J, Khuri F, Ganly I, Arseneau J, Posner M, Vokes E, Kuhn J, McCarty T, Landers S, Blockburn A, Romel L, Randlev B, Kaye S, Kim D. Phase II triakl of intratumoral administration of ONYX-015, a replication-selective adenovirus, in patients with refractory head and neck cancers J Clin Oncol 2001; 19:289–298.

282. Pels H, Schulz H, Manzke O, Hom E, Thall A, Engert A. Intraventricular and intravenous treatment of a patient with refractory primary CNS lymphoma using rituximab J Neurooncol 2002; 59:213–216.

283. Mendelsohn J. Targeting the epidermal growth factor receptor for cancer therapy J Clin Oncol 2002; 20(Suppl):1S–13S.

284. Solbach C, Roller M, Ahr A, Loibl S, Nicoletti M, Stegmueller M, Kreysch HG, Knecht R, Kaufmann M. Anti-epidermal growth factor receptor-antibody therapy for treatment of breast cancer Int J Cancer 2002; 101:390–394.

285. Leggas M, Stewart CF, Woo MH, Fouladi M, Cheshire PJ, Peterson JK, Friedman HS, Billups C, Houghton PJ. Relation between Irofulven (MGI-114) systemic exposure and tumor response in human solid tumor xenografts Clin Cancer Res 2002; 8:3000–3007.

286. Poindessous V, Koeppel F, Raymond E, Comisso M, Waters SJ, Larsen AK. Marked activity of irofulven toward human

carcinoma cells: comparison with cisplatin and ecteinascidin Clin Cancer Res 2003; 9:2817–2825.

287. Ling YH, Liebes L, Jiang JD, Holland JF, Elliott PJ, Adams J, Muggia FM, Perez-Soler R. Mechanisms of proteasome inhibitor PS-341-induced G_2-M-phase arrest and apoptosis in human non-small cell lung cancer cell lines Clin Cancer Res 2003; 9:1145–1154.

288. Ling YH, Liebes L, Ng B, Buckley M, Elliott PJ, Adams J, Jiang JD, Muggia FM, Perez-Soler R. PS-341, a novel proteasome inhibitor, induces Bcl-2 phosphorylation and cleavage in association with G_2-M phase arrest and apoptosis Mol Cancer Ther 2002; 1:841–849.

289. Dalen H, Neuzil J. α-Tocopheryl succinate sensitises a T lymphoma cell line to TRAIL-induced apoptosis by suppressing NF-kappaB activation Br J Cancer 2003; 88:153–158.

290. Mukhopadhyay T, Sasaki J, Ramesh R, Roth JA. Mebendazole elicits a potent antitumor effect on human cancer cell lines both in vitro and in vivo Clin Cancer Res 2002; 8:2963–2969.

291. Sasaki J, Ramesh R, Chada S, Gomyo Y, Roth JA, Mukhopadhyay T. The anthelmintic drug mebendazole induces mitotic arrest and apoptosis by depolymerizing tubulin in non-small cell lung cancer cells Mol Cancer Ther. 2002; 1:1201–1209.

292. Radulovic S, Comaru-Schally AM, Milovanovic S, Schally AV. Somatostatin analogue RC-160 and LH-RH antagonist SB-75 inhibit growth of MIA PaCa-2 human pancreatic cancer xenografts in nude mice Pancreas 1993; 8:88–97.

293. Solorzano CC, Hwang R, Baker CH, Bucana CD, Pisters PW, Evans DB, Killion JJ, Fidler IJ. Administration of optimal biological dose and schedule of interferon α combined with gemcitabine induces apoptosis in tumor-associated endothelial cells and reduces growth of human pancreatic carcinoma implanted orthotopically in nude mice Clin Cancer Res 2003; 9:1858–1867.

294. Chan DC, Gera L, Stewart JM, Helfrich B, Zhao TL, Feng WY, Chan KK, Covey JM, Bunn PA. Bradykinin antagonist dimer, CU201, inhibits the growth of human lung cancer cell lines in vitro and in vivo and produces synergistic growth inhibition

in combination with other antitumor agents Clin Cancer Res 2002; 8:1280–1287.

295. Chan D, Gera L, Stewart J, Helfrich B, Verella-Garcia M, Johnson G, Baron A, Yang J, Puck T, Bunn P. Bradykinin antagonist dimer, CU201, inhibits the growth ogf human lung cancer cell lines by a "biased agonist" mechanism Proc Nat Acad Sci U S A 2002; 99:4608–4613.

296. Aliberti J, Viola JP, Vieira-de-Abreu A, Bozza PT, Sher A, Scharfstein J. Cutting edge: Bradykinin induces IL-12 production by DCs: a danger signal that drives Th1 polarization J Immunol 2003; 170:5349–5353.

297. Sinkovics JG. Monoclonal antibodies in the treatment of endotoxin shock Acta Microbiol Hung 1990; 37:247–261.

298. Honma Y, Ishii Y, Yamamoto-Yamaguchi Y, Sassa T, Asahi K, Cotylenin A. a differentiation-inducing agent, and IFN-α cooperatively induce apoptosis and have an antitumor effect on human non-small cell lung carcinoma cells in nude mice Cancer Res 2003; 63:3659–3566.

299. Yamaguchi H, Bhalls K, Wang HG. Bax plays a pivotal role in thapsigargin-induced apoptosis of human colon cancer HCT116 cells by controlling Smac/Diablo and Omi/HtrA2 release from mitochondria Cancer Res 2003; 63:1483–1489.

300. Denmeade SR, Jakobsen CM, Janssen S, Khan SR, Garrett ES, Lilja H, Christensen SB, Isaacs JT. Prostate-specific antigen-activated thapsigargin prodrug as targeted therapy for prostate cancer J Natl Cancer Inst 2003; 95:990–1000.

301. Vivanco I, Sawyers CL. The phosphatidylinositol 3-kinase-Akt pathway in human cancer Nat Rev Cancer 2002; 2:489–501.

302. Kamsteeg M, Rutherford T, Sapi E, Hanczaruk B, Shahabi S, Flick M, Brown D, Mor G. Phenoxodiol--an isoflavone analog--induces apoptosis in chemoresistant ovarian cancer cells Oncogene 2003; 22:2611–2620.

303. Chen J, Freeman A, Liu J, Dai Q, Lee RM. The apoptotic effect of HA14–1, a Bcl-2-interacting small molecular compound, requires Bax translocation and is enhanced by PK11195 Mol Cancer Ther 2002; 1:961–967.

304. Shishodia S, Majumdar S, Banerjee S, Aggarwal BB. Ursolic acid inhibits nuclear factor-κB activation induced by carcino-

genic agents through suppression of IκBα kinase and p65 phosphorylation: correlation with down-regulation of cyclooxygenase 2, matrix metalloproteinase 9, and cyclin D1 Cancer Res 2003; 63:4375–4383.

305. Bhargava P, Marshall JL, Dahut W, Rizvi N, Trocky N, Williams JI, Hait H, Song S, Holroyd KJ, Hawkins MJ. A phase I and pharmacokinetic study of squalamine, a novel antiangiogenic agent, in patients with advanced cancers Clin Cancer Res 2001; 7:3912–3919.

306. Minter HA, Eveson JW, Huntley S, Elder DJ, Hague A. The cyclooxygenase 2-selective inhibitor NS398 inhibits proliferation of oral carcinoma cell lines by mechanisms dependent and independent of reduced prostaglandin E2 synthesis Clin Cancer Res 2003; 9:1885–1897.

307. Cheng AS, Chan HL, Leung WK, Wong N, Johnson PJ, Sung JJ. Specific COX-2 inhibitor, NS-398, suppresses cellular proliferation and induces apoptosis in human hepatocellular carcinoma cells Int J Oncol 2003; 23:113–119.

308. Hill LL, Perussia B, McCue PA, Korngold R. Effect of human natural killer cells on the metastatic growth of human melanoma xenografts in mice with severe combined immunodeficiency Cancer Res 1994; 54:763–770.

309. Horvath J, Szabo-Szabari M, Sinkovics JG. Autologous activated lymphocyte therapy in a community hospital Acta Microbiol Hung 1994; 41:205–214.

310. Horvath J, Sinkovics JG. Adoptive immunotherapy with activated peripheral blood lymphocytes Leukemia 1994; 8S1: S121–S126.

311. Javia LR, Rosenberg SA. $CD4^+CD25^+$ suppressor lymphocytes in the circulation of patients immunized against melanoma antigens J Immunother 2003; 26:85–93.

312. D'Ambrosio D, Sinigaglia F, Adorini L. Special attractions for suppressor T cells Trends Immunol 2003; 24:122–126.

313. Mattes J, Hulett M, Xie W, Hogan S, Rothenberg ME, Foster P, Parish C. Immunotherapy of cytotoxic T cell-resistant tumors by T helper 2 cells: an eotaxin and STAT6-dependent process J Exp Med 2003; 197:387–393.

314. Kyburz D, Aichele P, Speiser DE, Hengartner H, Zinkernagel RM, Pircher H. T cell immunity after a viral infection versus T cell tolerance induced by soluble viral peptides Eur J Immunol 1993; 23:1956–2962.

315. Kawakami M, Kawakami K, Puri RK. Intratumor administration of interleukin 13 receptor-targeted cytotoxin induces apoptotic cell death in human malignant glioma tumor xenografts Mol Cancer Ther 2002; 1:999–1007.

316. Kawakami K, Kawakami M, Husain SR, Puri RK. Targeting interleukin-4 receptors for effective pancreatic cancer therapy Cancer Res 2002; 62:3575–3580.

317. Kawakami K, Kawakami M, Husain SR, Puri RK. Effect of interleukin (IL)-4 cytotoxin on breast tumor growth after in vivo gene transfer of IL-4 receptor α chain Clin Cancer Res 2003; 9:1826–1836.

318. Ensor CM, Holtsberg FW, Bomalaski JS, Clark MA. Pegylated arginine deiminase (ADI-SS PEG20,000 mw) inhibits human melanomas and hepatocellular carcinomas in vitro and in vivo Cancer Res 2002; 62:5443–5450.

319. Cen D, Gonzalez RI, Buckmeier JA, Kahlon RS, Tohidian NB, Meyskens FL. Disulfiram induces apoptosis in human melanoma cells: a redox-related process Mol Cancer Ther. 2002; 1: 197–204.

320. Chau BN, Wang JY. Coordinated regulation of life and death by RB Nat Rev Cancer 2003; 3:130–138.

321. von Willebrand M, Zacksenhaus E, Cheng E, Glazer P, Halaban R. The tyrphostin AG1024 accelerates the degradation of phosphorylated forms of retinoblastoma protein (pRb) and restores pRb tumor suppressive function in melanoma cells Cancer Res 2003; 63:1420–1429.

322. Raymond E, ten Bokkel Huinink WW, Taieb J, Beijnen JH, Faivre S, Wanders J, Ravic M, Fumoleau P, Armand JP, Schellens JH. European Organization for the Research and Treatment of Cancer Early Clinical Study Group. Phase I and pharmacokinetic study of E7070, a novel chloroindolyl sulfonamide cell-cycle inhibitor, administered as a one-hour infusion every three weeks in patients with advanced cancer J Clin Oncol 2002; 20:3508–3521.

323. Scotlandi K, Perdichizzi S, Manara MC, Serra M, Benini S, Cerisano V, Strammiello R, Mercuri M, Reverter-Branchat G, Faircloth G, D'Incalci M, Picci P. Effectiveness of ecteinascidin-743 against drug-sensitive and -resistant bone tumor cells Clin Cancer Res 2002; 8:3893–3903.

324. Kanzaki A, Takebayashi Y, Ren XQ, Miyashita H, Mori S, Akiyama S, Pommier Y. Overcoming multidrug drug resistance in P-glycoprotein/MDR1-overexpressing cell lines by ecteinascidin 743 Mol Cancer Ther 2002; 1:1327–1334.

325. Takahashi N, Li W, Banerjee D, Guan Y, Wada-Takahashi Y, Brennan MF, Chou TC, Scotto KW, Bertino JR. Sequence-dependent synergistic cytotoxicity of ecteinascidin-743 and paclitaxel in human breast cancer cell lines in vitro and in vivo Cancer Res. 2002; 62:6909–6915.

326. Donald S, Verschoyle RD, Graeves P, Gant TW, Colombo T, Zaffaromnui M, Frapoli R, Zucchetti M, D'Inclci M, Meco D, Riccardi R, Lopez-Azaro L, Jimeno J, Gescher AJ. Complete protection by high-dose dexamethasone against the hepatotoxicity of the novel antitumor drug Yondelis (ET-743) in the rat Cancer Res 2003; 63:5902–5908.

327. James RD, Jones DA, Aalbersberg W, Ireland CM. Naamidine A intensifies the phosphotransferase activity of extracellular signal-regulated kinases causing A-431 cells to arrest in G1 Mol Cancer Ther 2003; 2:747–751.

328. Gajate C, An F, Mollinedo F. Rapid and selective apoptosis in human leukemic cells induced by aplidine through a Fas/CD95- and mitochondrial-mediated mechanism Clin Cancer Res 2003; 9:1535–1545.

329. Masuda M, Suzui M, Lim JTE, Weinstein IB. Epigallocatechin-3-gallate inhibits activation of HER-2/neu and downstream signaling pathways in human head and neck and breast carcinoma cells Clin Cancer Res 2003; 9:3486–3491.

330. Chan KK, Oza AM, Siu LL. The statins as anticancer agents Clin Cancer Res 2003; 9:10–19.

331. Daniel D, Susal C, Kopp B, Opelz G, Terness P. Apoptosis-mediated selective killing of malignant cells by cardiac steroids: maintenance of cytotoxcity and loss of cardiac activity

of chemically modified derivatives Int Immunopharmacol 2003; 3:1791–1801.

332. Segura-Pacheco B, Trejo-Becerril C, Perez-Cardenas E, Taja-Chayeb L, Mariscal I, Chavez A, Acuna C, Salazar AM, Lizano M, Duenas-Gonzalez A. Reactivation of tumor suppressor genes by the cardiovascular drugs hydralazine and procainamide and their potential use in cancer therapy Clin Cancer Res 2003; 9:1596–1603.

333. Lorusso PM. Phase I studies of ZD1839 in patients with common solid tumors Semin Oncol 2003; 30:21–29.

334. Khalil MY, Grandis JR, Shin DM. Targeting epidermal growth factor receptor: novel therapeutics in the treatment of cancer Expert Rev Anticancer Ther 2003; 3:367–380.

335. Bonomi P. Clinical studies with non-Iressa EGFR tyrosine kinase inhibitors Lung Cancer 2003; 41:S43–S48.

336. Allen LF, Eiseman IA, Fry DW, Lenehan PF. CI-1033, an irreversible pan-erbB receptor inhibitor and its potential application for the treatment of breast cancer Semin Oncol 2003; 30: 65–78.

337. Morgan B, Thomas AL, Drevs J, Hennig J, Buchert M, Jivan A, Horsfield MA, Mross K, Ball HA, Lee L, Mietlowski W, Fuxuis S, Unger C, O'Byrne K, Henry A, Cherryman GR, Laurent D, Dugan M, Marme D, Stewart WP. Dynamic contrast-enhanced magnetic resonance imaging as a biomarker for the pharmacological response of PTK787/ZK222584, an inhibitor of the vascular endothelial growth factor receptor tyrosine kinases, in patients with advanced colorectal cancer and liver metastasaes: results from two phase I studies J Clin Oncol 2003; 21:3955–3964.

338. Baselga J, Hammond LA. Her-targeted tyrosine-kinase inhibitors Oncology 2002; 63S:6–16.

339. Nahta R, Hortobagyi GN, Esteva FJ. Signal transduction inhibitors in the treatment of breast cancer Curr Med Chem Anti-Cancer Agents 2003; 3:201–216.

340. Ng SS, Tsao MS, Nicklee T, Hedley DW. Effects of the epidermal growth factor receptor inhibitor OSI-774, Tarceva, on downstream signaling pathways and apoptosis in human pancreatic adenocarcinoma Mol Cancer Ther 2002; 1:777–783.

341. Fabbro D, Ruetz S, Buchdunger E, Cowan-Jacob SW, Fendrich G, Liebetanz J, Mestan J, O'Reilly T, Traxler P, Chaudhuri B, Fretz H, Zimmerman J, Meyer T, Caravatti G, Furet P, Manley PW. Protein kinases as targets for anticancer agents: from inhibitors to useful drugs Pharmacol Ther 2002; 93: 79–98.

342. Ehrlichman C, Boerner SA, Hallgren CG, Spieker R, Wang XY, James CD, Scheffer GI, Maliepaard M, Ross DD, Bible KC, Kaufmann SH. The HER tyrosine kinase inhibitor CI-1033 enhances cytotoxicity of 7-ethyl-10-hydroxycamptothecin and topotecan by inhibiting breast cancer resistance prorein-mediated drug efflux Cancer Res 2001; 61:739–748.

343. Drevs J. PTK/ZK (Novartis) Drugs 2003; 6:787–794.

344. Kedar D, Baker CH, Killion JJ, Dinney CP, Fidler IJ. Blockade of the epidermal growth factor receptor signaling inhibits angiogenesis leading to regression of human renal cell carcinoma growing orthotopically in nude mice Clin Cancer Res 2002; 8: 3592–3600.

345. Jungwirth A, Schally AV, Halmos G, Groot K, Szepeshazi K, Pinski J, Armatis P. Inhibition of the growth of Caki-I human renal adenocarcinoma in vivo by luteinizing hormone-releasing hormone antagonist Cetrorelix, somatostatin analog RC-160, and bombesin antagonist RC-3940-II Cancer 1998; 82: 909–917.

346. Rusnak DW, Lackey K, Afflick K, Wood ER, Alligood KJ, Rhodes N, Keith BR, Murray DM, Knight WB, Mullin RJ, Gilmer TM. The effects of the novel, reversible epidermal growth factor receptor/Erb-2 tyrosine kinase inhibitor, GW2016, on the growth of human normal and tumor-derived cell lines in vitro and in vivo Mol Cancer Ther 2001; 1:85–94.

347. Ezawa K, Minato K, Dobashi K. Induction of apoptosis by ubenimex (Bestatin) in human non-small-cell lung cancer cell lines Biomed Pharmacother 1996; 50:283–289.

348. Ichinose Y, Genka K, Koike T, Kato H, Watanabe Y, Mori T, Iioka S, Sakuma A, Ohta M. Randomized double-blind placebo-controlled trial of bestatin in patients with resected stage I squamous-cell lung carcinoma J Natl Cancer Inst 2003; 95: 605–610.

349. DeGrendele H. Highlights from the 27th Congress of the European Society of Medical Oncology. Nice, France October 18–22, 2002 Clin Lung Cancer 2003; Jan:205–210.

350. Oosterhoff D, Witlox MA, van Beusechen VW, Haisma HJ, Schaap GR, Bras J, Kruyt FA, Molenaar B, Boven E, Wuisman PI, Pinedo HM, Gerristen WR. Gene-directed enzyme prodrug therapy for osteosarcoma: sensitization to CPT-11 in vitro and in vivo by adenoviral delivery of a gene encoding secreted carboxylesterase Mol Cancer Ther 2003; 2:765–771.

351. Zhou RR, Jia SF, Zhou Z, Wang Y, Bucana CD, Kleinerman ES. Adenovirus-E1A gene therapy enhances the in vivo sensitivity of Ewing's sarcoma to VP-16 Cancer Gene Ther 2002; 9:407–413.

PART V: CLINICAL EXAMPLES: MULTIPLE MYELOMA AND GLIOBLASTOMA MULTIFORME

MULTIPLE MYELOMA

Pathogenesis

Despite high-dose chemotherapy (melphalan 140 mg/m^2 with, or 200 mg/m^2 without total body radiotherapy given after four courses of the VAD regimen) consisting of vincristine adriamycin and dexamethasone and hematopoietic stem-cell rescue, either autologous or allogeneic, this disease remains incurable, although survival with subclinical disease is prolonged [1–5]. Even double high-dose chemotherapy with stem-cell rescue, at the risk of death consequentially to treatment morbidity, fails to promise cure, but definitely prolongs survival up to 42% at the seventh year. Patients who failed to achieve a CR at the first treatment benefited most from the second hematopoietic stem-cell rescue. Allogeneic lymphoid cells (NK cells), at the risk of morbidity and mortality, given after non-myeloablative pretreatment, could induce PR and CR ("graft-vs.-myeloma" reaction) at a high rate. However, multiple myeloma (MM) is responding to biological interventions and is targeted for many modalities of biological therapy, including viral (genetically engineered measles virus; naturally oncolytic reovirus) agents [6]. Will viral agents match the efficacy of these biological interventions and is there a good rationale to combine virotherapy with some of the biological treatment modalities?

Monoclonal gammopathy of uncertain significance (MGUS) with karyotypic instability may progress to MM. There are seemingly chaotic and promiscuous translocations of chromosomal segments in MM cells—less in the natural tumor than what is found in its established human myeloma cell lines (HMCL). The most common translocations involve cyclin D1 (11q13) and cyclin D3 (6p21) genes; the MM nuclear SET domain gene (4p16) and the FGF receptor 3 (FGF-R3) gene (der 14); and the c-*maf* genes (16q23; 20q11) (*vide infra*). The t(11;14)(q13;q32) translocation dysregulates cyclin D1, albeit at different break points, both in MM and in mantle cell lymphoma. The translocation t(4;14)(p16.3;q32.3) dysregulates the SET nuclear protein and the FGF-R3 genes. Translocation t(11;14) associated with ectopic expression of the FGF-R3 gene occurs in 20% of MGUS plasma cells, in 50% in MM cells, and in 70% in HMCL. Next to the FGF-R3 gene is the locus of the transforming acidic coiled coil-containing gene (TACC3), which is normally active during embryonic life, falls silent, but becomes re-amplified in malignantly transformed cells [7,8]. The t(14;16)(q32;q23) translocation liberates the c-*maf* proto-oncogene encoding a transcription factor for IL-6. This oncogene was discovered in the avian retrovirus AS42. This retrovirus induces musculoaponeurotic fibrosarcomas in chicken and carries v-*maf* fused with the viral *gag* gene, but originally derives from c-*maf*. Next to the c-*maf* gene lies the tumor suppressor gene WWOX/FOR (16q23), encoding an oxidoreductase, but suffering a monoallelic inactivation in the translocation process. Translocations t(6;14)(p21;q32) dysregulates cyclin D3. [9,10].

Although no single mutation characterizes myelomagenesis, N-*ras* or K-*ras* mutations occur in about 50% of the cases. Patients with constitutively active N-*ras* mutation are resistant to doxorubicin and dexamethasone. Physiological activation of the Ras pathway by IL-6 is transient and the cell remains sensitive to doxorubicin. Microarray technology could distinguish four to six groups of MM patients stratified according to gene mutations and stromal cell and IL-6 dependence, given that some mutations render the MM cell independent from paracrine growth factors, including IL-6 [11,12].

Translocations of c-*myc:* t(8;14), such as in Burkitt's lymphoma, or *Bcl*-2: t(14:18), such as in follicular lymphoma, seldom occur in MM. Deletions occur in $13q14^-$, resulting in monoallelic loss of the Rb gene (13q14). Its nuclear phosphoprotein suppresses the cell cycle at $G_1 \rightarrow S$; P^- pRB is active and arrests the cell cycle; P^+ pRb is inactive and lets the cell cycle continue. Multiple myeloma cells express P^+ pRb or deletion (monoallelic loss) of the Rb gene. Cycline D kinase inhibitors (p15, 16, 18^{INK4A}) are lost through deletions. Mutations of K-*ras* and N-*ras* (codons 12, 13, 61) occur in fewer than or up to 5% in MGUS and in 40% in MM cells (especially after culture in vitro). Regarding p53 and its sequence-specific DNA-binding protein, it is mutated in HMCL frequently, but seldom in native MM cells. The PTEN gene (*vide infra*) encoding a pro-apoptotic phosphatase counteracting the anti-apoptotic Akt system, is lost in MM cells [13,14]. Although telomerase is active in MM cells, adding TTAGGG repeats to chromosomal ends, the telomeres are unusually short (perhaps because plasma cells are terminally differentiated). The importance of telomerase activity in oncogenesis was documented when SV40 and H-*ras* genomic sequences and those of the ribonucleoprotein telomerase were combined to create malignant cells in vitro [15]. Telomerase inhibitors; the G-quadruplex intercalating tetramethyl-4-pyridyl porphyrin chloride; and the thiophosphoramidate oligonucleotide, which raises levels of p21 and phosphorylated p53 for apoptosis induction, are in clinical trials (*vide infra*) [16–18].

Interleukin-6 protects MM cells against dexamethasone-induced apoptosis. Wild-type Rb suppresses IL-6 production, whereas IL-6 shifts Rb from P^- to P^+ form. Hyaluronic acid produced by stromal cells promotes IL-6 binding to its receptors IL-6Rα and gp130. The subunit gp130 accepts as its ligand IL-6, IL-11, ciliary neurotrophic factor, cardiolipin, oncostatin M and leukemia inhibitory factor. It activates through the Janus family tyrosine kinases STAT1 and STAT3 (Figure 4). In normal B-lineage lymphocytes, IL-6R can be downregu-

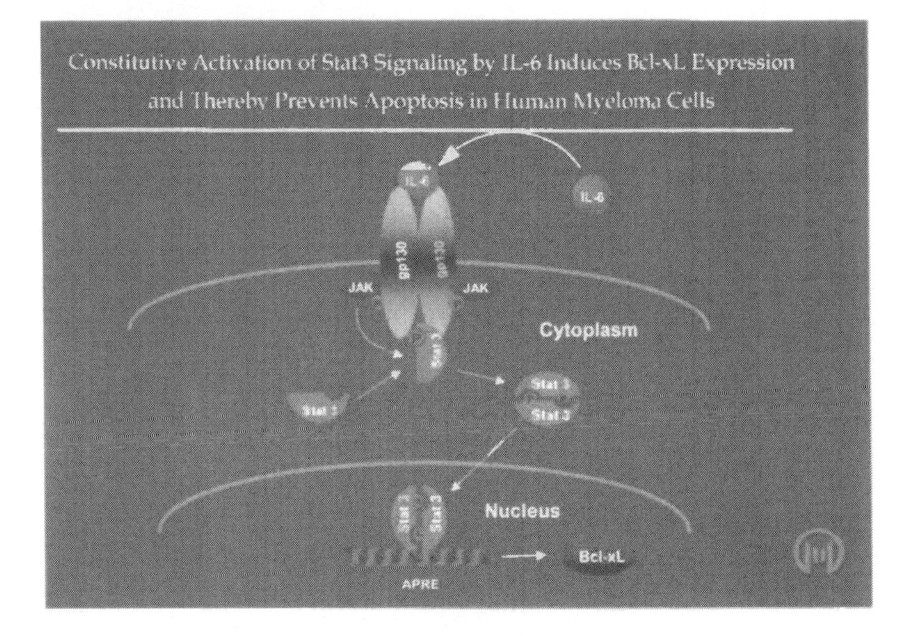

Figure 4 Interleukin-6 activates STAT3, which in turn transactivates the *BCL2* gene for overproduction of the anti-apoptotic Bcl-2 protein. The myeloma cell thus gains resistance to programmed cell death (apoptosis). (Courtesy of Professor Richard Jove Ph.D. of the University of South Florida, H.L. Moffitt Cancer Center, Tampa, FL.)

lated, but in MM cells, IL-6R remains permanently overexpressed. Although IL-6 can function as an autocrine growth factor in MM cells, it frequently is provided to MM cells as a paracrine growth factor by stromal cells, together with VEGF, IL-1β, IL-10, TNFα, TGFβ, MMP-1, osteoprotegerin OPG-R-activator of NFκB (RANKL), macrophage (MΦ) inflammatory protein 1α (MIF), FGF, IGF, and HGF. Statins (coenzyme A inhibitor cholesterol-lowering agents lovastatin, pravastatin) inhibit the growth of human myeloma cells, but IL-6 antagonizes this effect [19–24]. Human IL-6R does not accept mouse IL-6, so xenografted MM cells are deprived of the growth-promoting effect of paracrine IL-6 and become more vulnerable to pro-apoptotic insults that MM cells existing in their natural microenvironment could withstand. There is no evidence that HHV-8–infected stromal cells provide paracrine IL-6 to MM cells.

The multinucleated giant osteoclast differentiates from monocyte-Mϕ stem cells. T_H2 helper CD4 lymphocytes produce the pleiotropic immunomodulatory cytokine IL-4, which primarily stimulates the proliferation of B-lineage lymphocytes. Other sources of IL-4 production are mast cells and eosinophil leukocytes (as in the stroma of Reed-Sternberg cells). In order to act, IL-4 binds either one of its receptors—IL-4Rαγ or IL-4Rα/IL-13Rα1. Signaling from these receptors runs through the Janus kinases and involves both insulin receptor substance (IRS) and STAT activation. STAT6 suppresses mRNA for RANK (the RANKL receptor). In response, B cells produce immunoglobulins (first IgE). The effects of IL-4 on osteoclast precursors are suppressive. RANKL, the TNF realeted ligand, and IL-4 are antagonists. Whereas RANK emits signals to activate NFκB, IL-4 suppresses this pathway. In osteoclasts, the IRS pathway and cell differentiation are inhibited by IL-4 because IL-4 suppresses the expression of RANK, the receptor, in immature osteoclasts. Under the effects of IL-4, even fully mature osteoclasts stop resorbing bone. Thus, IL-4 deprives osteoclasts of stimulation by RANKL to mature and differentiate. Even the M-CSF pathway, another stimulator of osteoclastogenesis, is suppressed by IL-4 [25]. Here, well-known proto-oncogenes interact. Proto-oncogene c-*fms* (locus at 5q21-32) was transduced by the McDonough strain of the feline sarcoma retrovirus (v-*fms*), encoding a transmembrane glycoprotein receptor for the ligand M-CSF. Proto-oncogene c-*fos* (locus at 14q21-22) was transduced by the Finkel-Biskis-Really (FBR) mouse osteosarcoma retrovirus (v-*fos*). The gene product protein c-Fos transactivates nuclear genes, which dictate terminal differentiation of monocytes-Mϕ. The c-Kit^+, c-Fms^+, c-Fos^+, $RANK^-$ stem cell gives rise to DCs under the effects of GM-CSF, which acts as a suppressor to osteoclastogenesis, whereas TNFα; M-CSF; and, later, RANKL promote osteoclastogenesis from the same stem cell. Receptor activation of NFκB (RANK) is expressed by osteoclasts; the cognate receptor activator of NKκB ligand (mRANKL⇒sRANKL) derives from osteoblasts, stromal cells, and malignant (MM) plasma cells. Osteoprotegerin (OPG), a member of the TNF-R family, acts as a soluble decoy receptor for binding RANKL and, by depleting the supply of the ligand, it protects against osteolysis. Apoptosis of myeloma cells induced by TRAIL/Apo21 (TNF-related apoptosis-inducing ligand) was averted by OPG, therefore exerting a tumor-protective effect. Bone resorption markers are the tartrate-resistant acid phosphatase and bone collagen degradation products; bone alkaline phosphatase and osteocalcin are markers for new bone formation. Additional osteoclast-activating ligands are IL1, IL-6, and TNFα deriving from the myeloma cell or its stroma. The osteoclastogenesis inhibitory factor (OCIF), a member of the TNF family, is identical to OPG: Both bind RANKL. Osteoprotegerin plasma or serum levels are decreased in patients with clinically active MM, but the technology (enzyme-linked immunosorbent assay; ELISA) for titrating the levels of this substance is very difficult; for example, in patients with colorectal or pancreatic carcinomas, higher than normal OPG levels were measured. TRAIL is also depleted by its binding to the decoy receptor OCIF/OPG. Removal of RANKL favors the host by suppressing osteolysis by the osteoclast; removal of TRAIL favors the myeloma cell by eliminating the events leading to its programmed cell death. Receptors for TRAIL are complex: DR4 and DR5 transmit signals leading to apoptosis; DcR1 and DcR2 are decoy receptors that capture TRAIL, but remain silent. The two-faced OPG, the recombinant human rhOCIF (Sankyo, Tokyo, Japan), and a RANK-HuTgG fusion protein (RANK-Fc) are in clinical trials [26–33].

The Wnt-related (see Part III & IV) dickkopf ("fathead") gene product protein DKK1 is an inhibitor of osteoblast differentiation and activity. The gene is reactivated in myeloma cells; DKK1 levels are high in the bone marrow and in the blood. High DKK1 levels correlate with the development of lytic bone lesions caused by the loss of osteoblast activitites that are inhibitory to osteaclasts and reparatory to osteolysis [34].

In summary, myeloma cells and stromal cells produce RANKL and thus promote osteoclastogenesis. Production of the RANKL antagonist OPG is downregulated. The Wnt antagonist DKK1 blocks osteoblast differentiation and is overproduced by myeloma cells. Undifferentiated osteoblasts also secrete large amounts of RANKL. Plasma cells in MGUS syndrome do not (as yet) produce DKK1. Patients with far-advanced MM and bone lesions show decreased DKK1 blood levels, inasmuch as overactivated osteoclasts can downregulate DKK1 output. However, the molecular mediator of the therapeutic value with which osteoclasts suppress DKK1 production has not as yet been isolated and identified.

Interleukin-18, the IFN-γ–inducing factor, is of osteoblast or Mϕ origin (see Part IV). It suppresses osteoclast maturation not by IFN-γ, but by GM-CSF induction from subclones of T lymphocytes or from osteoblasts. Interleukin-18 is related to IL-1 and IL-12 and synergizes with IL-12 in inducing IFN-γ secretion from T cell subclones. Mature osteoclasts express the IL-18 receptor MyD88; IL-18 inhibits the bone-resorbing activity of mature osteoclasts. When IL-18 induces IFN-γ, it will suppress osteoblasts [35–41]. In a human myelo-monocytic cell line (KG-1 cells) that expresses Fas, the receptor, IL-18 induced Fas ligand co-expression and upregulated the p53 protein. These events drive KG-1 cells to die apoptotoic deaths [39]. It is presumed that such events could also occur in myeloma cells exposed to IL-18. Proteasome antagonists of NFκB activation (P341; *vide infra*) protect against bone resorption. The bisphosphonates are licensed products for the same purpose [42]. Bisphosphonates are in use to treat lytic-destructive bone metastases of patients with breast cancer or blastic bone metastases of patients with prostate cancer. In the latter entity, Osteoprotegerin/osteoclastogenesis inhibitory factor (OPG/OCIF) emerges as a potential therapeutic agent [27].

The CD95 Fas receptor gene may undergo mutations in MM cells. Multiple myeloma cells attached to fibronectin in their native environment mobilize high FLIP levels and thereby gain protection against nonmutated Fas receptor-mediated apoptotic death induced by doxorubicin and alkylator chemotherapeutic agents, whereas suspended MM cells taken out of their microenvironment exhibit low FLIP levels and become defenseless against apoptosis induction. In their microenvironment, protection of MM cells against apoptotic death is mediated by β1 integrins; antagonists of this substance exist and are readied for clinical trials [43]. Human MM cells injected intravenously into SCID mice cannot lodge in the bone marrow in animals treated with IL-18 (IFN-γ–inducing agent) because of NK cell activation [44].

In evaluating the results of clinical trials, the natural history of MM should be considered. Patients without chromosome 13 deletion, low β2 microglobulin and high $p27^{Kip1}$ cyclin/CDK complex-inhibitor levels without anemia or hypercalcemia and with normal kidney function at presentation respond better to treatment and live longer; their life expectancy is from 48 to 80 months. Patients with monoallelic loss of 13q, t(4;14) or t(14;16) involving the IgH gene (14q32) and low $p27^{Kip1}$ levels usually live only from 16 to 25 months [45].

Biotherapy

The catalytic proteinase enzyme system referred to as *proteasomes* is being targeted for cancer therapy [46]. The ubiquitin-proteasome proteolytic system removes mutant, damaged, and misfolded proteins from the cytoplasm. When the proteasome system is inhibited in a tumor cell, however, the cell activates an apoptotic pathway. In MM cells, the proteasome inhibitor PS-341 overcomes the anti-apoptotic effect of IL-6; phosphorylates the p53 protein on Ser15 (a process inhibited by caffeine); degrades MDM2, the p53 antagonist; inhibits TNFα-induced NFκB activation; switches off DNA repair by cleaving certain phosphatidyl inositol kinases; activates caspases 3 and 8; and inhibits paracrine growth factor release from bone marrow stromal cells. Multiple myeloma cells so afflicted die apoptotic deaths. Further, PS-341 attacks and phosphorylates the gene product protein Bcl-2, especially when *bcl-2* is translocated: t(14;18), as in certain lymphoma cells. The phosphorylated Bcl-2 protein is condemned to proteolytic cleavage, resulting in the arrest of the cell cycle in G_2-M and apoptotic cell death (for illustration see cited references). Other agents that phosphorylate and thereby inactivate Bcl-2 are the phosphatase inhibitor okadaic acid, the PKC-inhibitor bryostatin (see Part II),IL-3, and *Vinca* alkaloids). The dipeptide Phe-Leu pyrazylcarbonyl boronate 20S proteasome-inhibitor PS-341 (bortezomib, Velcade, Millennium Pharmaceuticals), in its first trial, induced one CR in nine heavily pretreated MM patients and reduced the paraprotein levels in others, while restoring other immunoglobulin levels, but also inducing maculopapular skin rash. One heavily pretreated patient with mantle cell lymphoma experienced PR [47]. In a later trial of 54 MM patients, the response rate, including significant reduction of paraprotein levels, was 85% (ASH, Orlando, FL, 2003).

Bortezomib was tested in hundreds of MM patients (heavily preatreated patients in the SUMMIT trial, with 20% response rate; and minimally pretreated patients in the CREST trial, with 30% response rate). Its approval was based on the APEX trial, in which it induced 3% CR and 25% PR; with stabilization of disease, response rate rose to over 90%. Malaise with nausea and diarrhea, peripheral neuropathy, anemia, and thrombocytopenia are adverse side effects, involving 50% of the patients [48–52]. Bortezomib-resistant MM cells remain sensitive to CDDO-imidazolide treatment. This compound also eliminates IL-6–producing stromal cells and synergizes with bortezomib. Interleukin-6 or dexamethasone could not block CDDO-Im–induced apoptosis of MM cells. The pan caspase inhibitor z-VAD-fmk protected MM cells from bortezomib- and CDDO-Im–induced apoptosis. The licensed Velcade (Millenium) was included in the *Physicians' Desk Reference* in 2004 (pp 2136–2139). The next targets of bortezomib are CLL [53] and some solid tumors (including prostatic and NSCLC) because it exerts pro-apoptotic effects on these tumor cells in vitro or in xenografts [54,55].

Bone marrow endothelial cells proliferate in patients with MM [56,57]. Thalidomide (see Part III), the inhibitor of neoangiogenesis and TNFα-induced NFκB activation, elicited 32% clinical response in MM patients who were refractory to conventional therapy. Patients with elevated bFGF blood levels respond to thalidomide [58–61]. The DTPACE chemotherapy regimen works additively with thalidomide [62]. Thalidomide analogues, the immunomodulatory drugs (IMiDs), are more effective because of stimulation of IFN-γ, IL-2, and MM cell-specific immune T cell clonal expansion and NK cell proliferation [63,64]. CC-5013 (Revimid, Celgene) induces a significant drop of paraprotein levels in 63% of relapsed MM patients; if leukopenia occurs, G-CSF (filgrastim, Neupogen, Amgen) restores white blood cell count.

Neovastat (Æ-941, AEterna Laboratories, Quebec, Canada) inhibits MMPs and neoangiogenesis by suppressing bFGF and VEGF, and its as-yet-uncompleted Phase II trial shows paraprotein level reductions in MM patients [65]. S-3APG (S-3 amino-phthalimido-glutarimide) inhibits both MM and Burkitt's lymphoma cells and is awaiting Phase I and Phase II clinical trials. Multiple myeloma cells express VEGF-R-1 and VEGF stimulates their growth. Multiple myeloma cells also secrete VEGF for autocrine growth loop and for IL-6 induction in the bone marrow stromal cells [66]. The VEGF-R tyrosine kinase inhibitor PTK787/ ZK222584 antagonized the growth of MM cells even in co-cultures with bone marrow stromal cells [67]. Combretastatin-A4 (CA4P) (see Part III) arrests the cell cycle in G_1-M and inhibits microtubule assembly in B-CLL cells, which die in a "mitotic catastrophe" [68]. It is awaiting clinical trials for B-lineage malignancies. The cycline-dependent kinase inhibitor flavopiridol (see Part II) suppressed Mcl-1 overexpression and induced apoptosis in MM and B-CLL cells; it is in clinical trials [69]. The GAG glycosaminoglycans (syndecan-1 heparan; neoglycan) may inhibit cell growth (hyaluronic acid for melanoma cells) or stimulate cell growth (hyaluronic acid for MM cells). The carbodimide-modified GAG neoglycan inhibited MM cell growth.

Rapamycin analogue CCI-779 suppressed the growth of PTEN-deficient myeloma cells. Phosphatase on chromosome ten (PTEN) functions in a tumor-suppressor system by inactivating the PI3K/Akt pathway, which is liberated and is constitutively operational in myeloma cells [70]. However, PTEN-deficient myeloma cells are sensitive to mTOR (mammalian target of rapamycin) inhibition. Rapamycin synergized with dexamethasone in apoptosis induction in myeloma cells [71]. Inhibitors of constitutively active STAT pathway exist and are piceatannol, a stilbene from *Euphorbia*, and the synthetic tyrphostin AG490. They abolish the exemption from apoptosis and restore chemotherapy sensitivity of MM cells [72].

Apoptosis induction in MM cells by arsenic trioxide (As_2O_3) is intensified by coadminstration of ascorbic acid and buthionine sulfoximide, which reduces glutathione levels [73,74]. Alemtuzumab (Campath-1H) reacting with CD52 could attack myeloma cells and could induce C'- or ADCC-mediated tumor cell lysis [75]. Some myeloma cells express low levels of HER2 [76]. $CD20^+$ MM cells usually are of the small mature plasma cell morphology, but suffered translocations t(11;14), have $13q^-$, and breaks at 14q32. These patients may respond to rituximab (Rituxan, Biogen Idec, San Diego, CA) or trastuzumab (Herceptin, Genentech), respectively [77].

The chaperone molecule, heat shock protein Hsp90, blocks apoptotic protease activity of factor-1 (APAF-1) and promotes cell survival. Geldanamycin (GA) attaches to the ATP-binding pocket of Hsp90 and inactivates it. Defenseless MM cells succumb to apoptotic death. The 17-demethoxygeldanamycin (17-AAG) is in Phase I trials [78]. The histone deacetylase inhibitors (see Part II) FR901228 depsipeptide and SAHA are apoptosis inducers in MM cells and are in preclinical trials [79,80]. 2-Methoxyestradiol ($2ME_2$, Panzem, Tetrionics, Madison, WI) is anti-angiogenic, causes G_2-M arrest of the cell cycle in dexamethasone-resistant MM cells, and counteracts the anti-apoptotic effects of IL-6. It also upregulates the extrinsic pathway of apoptosis induction. In SCID mouse xenografts, it inhibited MM cell growth by 40%, including that of primary plasma cells taken directly from patients. Inhibited MM cells died apoptotic deaths. The 14-dehydro-2-ME2 analogue is 10 times more potent than the parental compound [81,83]. The TNF-related apoptosis-inducing TRAIL/Apo21 ligand elicits apoptotic cascade in MM cells; the protein-synthesis inhibitor cycloheximide and the PKC-inhibitor bisindolylmaleinimide restore MM cell sensitivity to TRAIL/Apo21 ligand [28].

Some proteins attach to membranes to function, but membrane attachment requires prenylation. Such lipid modification of these proteins—that is, addition of isoprenoid moieties to cystein-containing motifs at the C-terminus of these proteins—is carried out by the enzymes FT and geranylgeranyl transferase (GGT). The Rho family of Ras-related GTP-binding proteins and the Ras family proteins (see Part I) need geranylgeranylation and farnesylation, respectively, for their operational activities within the cell [84,85]. Geranylgeranyl transferase and FT carry out the tasks of prenylation. In myeloma cells, the pro-apoptotic molecular mediators BAX and BAD are neutralized by the formation of their heterodimers with the anti-apoptotic molecular mediators Bcl-2, Bcl_{XL}, and Mcl-1 [86–89]. The latter emerges as the most important anti-apoptotic molecular mediator mobilized for survival in MM cells. The mevalonate enzyme inhibitor lovastatin, the product of *Aspergillus terreus*, which is licensed to lower cholesterol, is pro-apoptotic for myeloma cells because it downregulates Mcl-1 expression and leaves Bcl-2 expression unaffected [19,20]. The GGT inhibitor GGTI-298 and the FT inhibitor FTI-277 (Calbiochem, Schwallbach, Germany) further inhibit prenylation and reduce Mcl-1 levels, driving myeloma cells to apoptotic deaths. Expression of Mcl-1 in myeloma cells also could be antagonized with As-ODN (see Part III). Lovastatin is pro-apoptotic in some other (AML, lung adenocarcinoma) cells. Farnyl transferase inhibitors SCH66335 and R115777 are in clinical trials in many tumor categories (see Part I), including MM. The latter, R115777 (tipifarnib, Zarnestra, Johnson & Johnson, Raritan, NJ), frequently (60%) stabilizes the disease, with reduction of the monoclonal protein levels. When tipifarnib inhibits phosphorylation of Akt, the PI3K/Akt pathway is interrupted, pro-caspase 3 cleavage occurs, and apoptotic death of the myeloma cell ensues. Farnyl transferase inhibitors show early favorable clinical results in the treatment of acute and chronic myelogenous leukemias and the myelodysplasia syndromes [89].

Rising to the ranks of effective inducers of apoptotic death in myeloma cells are the histone deacetylasae inhibitors (SAHA, depsipeptide) [79,80] (see Part II) and the peroxisome proliferator-activated γ receptor agonists (PPARγ) (see Part I) ciglitazone and rosiglitazone. These agents are in clinical trials for MM and for Waldenstrom's macroglobulinemia [90].

Chemotherapy resistance of tumor cells, including myeloma cells, consists of the efflux or extrusion of the drugs from the cell mediated by drug-transport glycoproteins (P-gp) or multidrug-resistance protein pumps (MDR-1). Another mechanism of chemotherapy resistance entails the plugging of pores in the nuclear membrane, blocking access of the drugs to chromosomal DNA. Breast cancer resistance protein (BCRP) and lung cancer resistance protein (LCRP) accomplish these tasks [91–93]. All of these mechanisms are functional in myeloma cells. Levels of P-gp rise after chemotherapy with the VAD regimen, but MDR-1 expression remains low. Expression of BCRP is much lower than that of LCRP. The IL-2 antagonist cyclosporin, the product of the fungus *Beauveria nivea*, and its derivatives modulate P-gp and MDR gene expression. PSC833 (valspodar) inhibited P-gp in healthy liver and kidney cells as well, exposing these cells to excessive drug toxicity; it also caused cerebellar ataxia. Although the efforts to antagonize the drug-resistance pathways remain unattainable, apoptosis induction in myeloma cells with or without chemotherapy has been achieved. The anti-NFκB PS-341 and As_2O_3, the JAK/STAT inhibitor AG490, and the Bcl-2 inhibitor As-ODN (see Part III) appear to be the most promising interventions. The de novo (not chemotherapy-induced) phenomenon of cell adhesion-mediated drug resistance is mediated by the contact between α4β1 integrin of myeloma cells and fibronectin of matrix cells. This contact activates NFκB in the myeloma cell and induces IL-6 secretion from the stromal cell. PS-341 (bortezomib, Velcade) exerts anti-adhesive effect between myeloma and stromal cells and suppresses IL-6 production by stromal cells. SDF-1α is neoangiogenic and growth-supportive for myeloma cells. It can be inhibited by NSC65016 (Suradista) and by specific antibodies [94,95] (see Part IV). By the way, integrin α5β1 serves as the entry site for parvovirus B19 [96]. The South American tree *Tabebuia* yields β-lapachone, a DNA repair inhibitor and inducer of cell death in many human tumor cells, including those of MM [63,64].

Of antisense oligonucleotides [97], the anti–*Bcl*-2 ODN G3139 (Genasense, Genta, Berkeley Heights, NJ) and the telomerase hTERT inhibitors are in Phase I through Phase III clinical trials (see Part III). The oligonucleotide thiophosphoramidate GRN163 targets the RNA of the human telomerase (see Part IV) functioning as a template antagonist; myeloma cells suffering telomere loss enter "crisis" and die [17,18].

These As-ODN (see Part III) enhance the antilymphoma efficacy of rituximab, especially in EBV^+ lymphoproliferative diseases [98] (Medical Education Resources Inc., *www.uemeded.com/event* 727669677424).

Polymorphic epithelial mucin MUC1-expressing MM cells elicit the rise of HLA-I–restricted autologous immune T cell clones [99]; these T cells expanded in vitro may be used for adoptive immunotherapy. The autologous myeloma-follicular lymphoma idiotype globulin proteins are tumor-specific neoantigens eliciting antibody- and immune T cell-mediated reactions against tumor cells expressing these antigens. Multiple myleoma and lymphoma vaccines, including idiotype antigen-pulsed DCs, coadministered with immunological adjuvants such as keyhole limpet hemocyanin (KLH) as a carrier for GM-CSF (sargramostim, Leukine, Immunex, Kenilworth, NJ) induce molecular CR or maintain PR→CR induced by chemotherapy (MyVax®, Genitope, Redwood City, CA; *http://www.genitope.com/myvax.html*). CD6 T cell-depleted hematopoietic allografts from histocompatible donors elicit graft-vs.-myeloma reactions in patients who received high-dose chemotherapy. The incidence of grade ≥2 graft-vs.-host disease is 17% and transplant-related mortality is 5%, but the CR rate is 28% and PR rate, 57%. The reduced-intensity conditioning regimens are administered before allotransplantation and aimed at facilitating acceptance of the allotransplant and its graft-vs.-tumor reaction at the price of a graft-vs.-host disease of varying severity, but with the reward of 65% CR or PR. Chimerism with donor cells occurs, yet 35% of patients relapsed (ASCO #2345 2003). Graft-vs.-myeloma–reacting lymphocytes are present before donor lymphocyte infusions and are $CD3^+$, $CD8^+$, and $CD4^-$ immune T cells. They expand after donor lymphocyte infusion and differ from the T cell clone responsible for graft-vs.-host disease [100]. In patients with MGUS advancing to MM, the invariant $V\alpha24^+V\beta11^+$ NKT cells lose their ability to produce IFN-γ under the effect of glycolipids released by the plasma cells, or by their subverted stroma; these NKT cells normally respond to glycolipid ligands (CD1d) or can be activated by a synthetic ligand, α-galactosyl ceramide (α-GalCer) [101]. Would α-GalCer prevent the MGUS→MM transition?

GLIOBLASTOMA MULTIFORME

Pathogenesis

In the human brain, there are 10 times more glia cells than neurons. Although glia cells are electrically inexcitable, they are far from being silent: There is active communication among glia cells and between glia cells and neurons. There are myelinating glia cells (oligodendrocytes in the brain and Schwann cells in peripheral nerves) and protoplasmic and fibrillary astrocytic glia cells, providers of neurotrophic molecular mediators in the brain.Microglia is considered to be a defensive, inflammatory, and phagocytic faculty [102].

Glioblastoma multiforme (GBM) is a heavily vascularized, highly malignant tumor with more than 99% mortality despite surgery, radiotherapy, and chemotherapy (carmustine and cisplatin; temozolomide and 13-cis-retinoic acid; irinotecan, etc.) [103–105]. Tumor cells show limited susceptibility to irradiation, nitrosoureas, vincristine, procarbazine, and temozolomide (especially with $6q^-$ and $10q^-$ chromosomal deletions). However, arsenic trioxide (As_2O_3) induces acidic vesicular organelle (AVO) formation in GBM cells, resulting in autodigestion (autophagy or programmed cell death type II). The autophagy inhibitor bafilomycin A1 inhibits AVO formation and contributes to the type I caspase-mediated apoptosis [106]. Clinical trials with bafilomycin should begin at Mount Sinai School of Medicine, New York, NY, and elsewhere.

Glioblastoma multiforme tumor cells express amplified or mutated EGF-R and PDGF-R, and a truncated EGF-R is constitutively autophosphorylated in some GBM cells. The PI3K/Akt cascade is overactivated. Overexpressed EGF-RvIII predicts poor survival. In *de novo*-arising primary GBM, EGF-R is overexpressed and genes for INK4a and $p14^{ARF}$ proteins are deleted. In secondary GBM, evolving from another preexisting brain tumor, p53 is mutated or deleted and PDGF and its receptor are overexpressed. Genes coding for proteins of the p53 pathway are TP53, $p14^{ARF}$, and MDM2. Genes coding for proteins of the Rb pathway are RB1, CDKN2A/B, and CDK4. Both lines may suffer mutations and deletions in GBM. Tumor suppressor genes for $p14^{ARF}$ (p53 pathway) or $p15^{INK4ab}$ (9p21) (Rb pathway) are silenced by hypermethylation in some brain tumors. Gene losses characterize the arising tumors: 1p36 and 19q13 leads to oligodendroglioma; 17p13, astrocytoma; 10^{-}, high-grade glioma. Signal transduction from EGF-R in GBM cells could not be completely suppressed by ZD1839 (Iressa) or the anilinoquinazole PD153035 (see Part I): PI3K/Akt and MAPK/Erk remained active, whereas the same dosage of these inhibitors suppressed EGF-R of squamous carcinoma cells [107–117].

Hypoxic tumors overproduce the DNA-binding heterodimers HIF-1α and HIF-1β. Hypoxia-induced factor-1α recognizes the genomic sequence of the hypoxia-responsive element 5′-RCGTG-3′. Hypoxia-inducible factor -1α is ubiquitinated and surrendered to proteasomal degradation by the VHL protein, except when the gene encoding this protein is mutated or deleted. The nuclear coactivator CBP/p^{300} interacts with HIF-1α and an asparagine hydroxylase mediates the HIF-1α and CBP/p^{300} relationship. Phosphatidyl inositol 3′ kinase and Akt promote HIF-1α, whereas PTEN gene product protein would suppress it, were it not that this tumor-suppressor gene is often mutated in GBM cells. The PTEN gene (locus at 10q23.3) suppresses hyaluronic acid-induced MMP-9 expression in GBM cells. Loss of heterozygosity at chromosome 10q11-23 results in the deficiency of the gene(s) that upregulates the gene at 11p13 encoding the PAX6 protein, which is a DNA-binding transcription factor very active in embryonic life (directing development of the eye). Matrix metalloproteinase-9 is more active in primary than in secondary GBM. The notorious PI3K/Akt pathway can be inhibited by quercetin and its derivative LY294002; the fungal metabolite wortmannin; the Akt inhibitors (Calbiochem); and rapamycins, the macrolide antibiotics from *Streptomyces hygroscopicus*. In colon and prostatic carcinoma cells, PGE_2 induces HIF-1α→VEGF production. Wortmannin, rapamycin (Sigma), and the c-*src* inhibitor PP2 (Calbiochem) inhibit this reaction. Hypoxia also induces the HIF1α→VEGF reaction in these tumors [118–126] (Figure 5).

Glioblastoma multiforme cells suffer early p53 gene mutation or the balance of proteins MDM2 and $p14^{ARF}$ favors MDM2 because the gene for the p53 protector $p14^{ARF}$ is frequently deleted [123,124]. Patients do not produce autoantibodies to the p53 protein [127]. Mutations of p53 correlate with EGF-R amplification in GBM cells, but without explicit growth advantage. In contrast, IGF-binding protein-2 overexpression enhances glioblastoma cell invasivesness caused by MMP-2 gene transcription and MMP-2 overproduction [128]. The aggregate response culminates in the massive overproduction of VEGF and other neoangiogenic growth factors (bFGF), hepatocytes, and scatter growth factors (HGF/SF), when c-*met* encoding the HGF/SF receptor is co-expressed for an autocrine growth circuit. A decoy HGF/SF molecule, NK4, can antagonize this growth loop and inhibits growth rate of GBM cells [129]. Glioblastoma multiforme cells downregulate the expression of the melanoma differentiation-associated 7 protein, a relative of IL-24, which is encoded by the gene *mda*-7. Such deficiency occurs in many other malignant tumors (breast, prostate, and gastrointestinal tract carcinomas). In *mda*-7–deficient cells, the antiapoptotic Bcl-2 and Bcl_{XL} dominate over pro-apoptotic factors (BAX; BAK). When the genetically engineered adenovirus Ad.*mda*-7 reinserts the gene, however, the pro-apoptotic elements gain the upper hand and the tumor cell succumbs to programmed death [130]. The STAT3 inhibitor AG490 downregulates Bcl-2, Bcl_{XL}, and Mcl-1 in GBM cells [131], offering itself to a combined protocol with Ad.*mda*-7 (not yet initiated).

Peculiarly, GBM cells also express the neoangiogenesis inhibitor endostatin, the COOH terminal globular domain of collagene XVIII [132]; perhaps this is one reason why metatstases of GBM do not grow. Endostatin given intramuscularly suppressed the growth of some brain tumors [133]. Glioblastoma multiforme is the prototype of neoangiogenesis-inducing tumors: It transactivates the hTERT mRNA in the vascular endothelial cells of the tumor's microenvironment (see Part

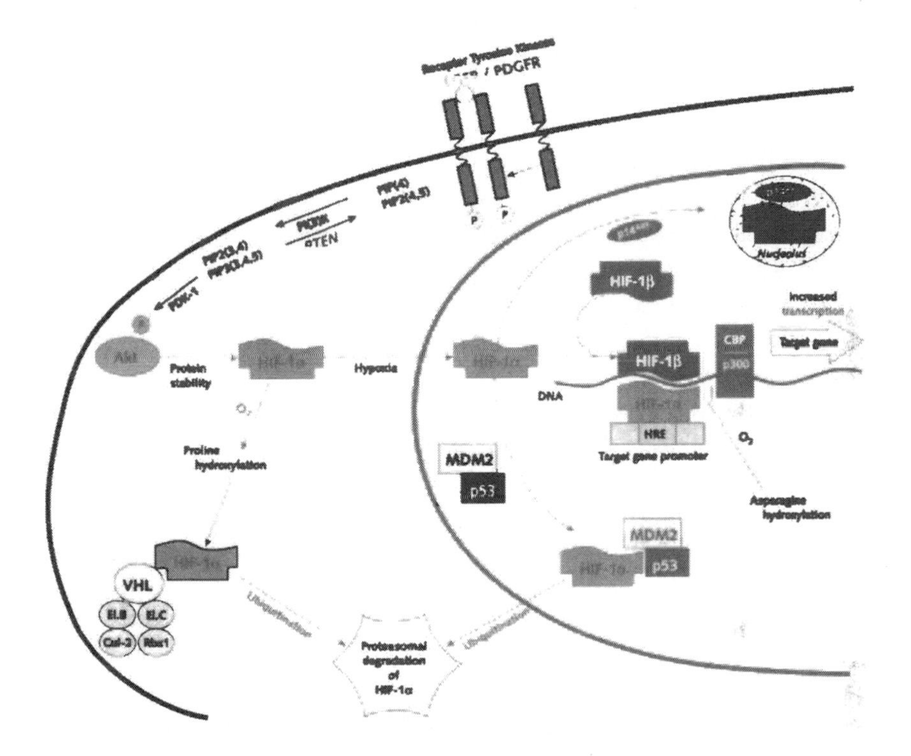

Figure 5 Promoters and inhibitors of cell growth interact in a glioblastoma cell. Hypoxia-inducible factor 1β (HIF-1β) acts prominently, unless it is degraded by ubiquitination (a favorable event). MDM2 and p53 compete; preserved p53 would remain pro-apoptotic, unless it is surrendered to the proteasome (an unfavorable event). If MDM2 is digested in the proteasome, the tumor cell may remain vulnerable to apoptotic death. PTEN is disabled by mutation and the PI3K/Akt pathway remains open unopposed. Two major overexpressed receptors (EGF-R; PDGF-R) constitutively emit promitotic signals. (Courtesy of Drs. Daniel J. Brat and Timothy B. Mapstone, Emory University Hospital, Atlanta, GA. Reprinted with the publisher's permission from the *Annals of Internal Medicine* 2003; 138: 659–668.)

IV). $VEGF_{121, 165}$ isoforms contribute to vascularization and oxygenation of H-*ras*–mutated anaplastic astrocytic tumors, but do not promote their transformation into GBM, whereas tumors transfected by retrovirus expressing VEGF isoform genes grew faster, to a large size, and became hypervascular [134]. Tumor necrosis factor-α fulfills a proneo-angiogenic role in GBM [135]. Angiopoietin-2 is overexpressed by GBM cells. Angiopoietin-1 binds endothelial cell receptor Tie-2 and stabilizes established vasculature. Angiopoietin-2 inhibits Ang-1 binding by occupying the same receptor and thereby activating MMP-2 production by the tumor cell (see Part III). Matrix metalloproteinase-2–producing GBM cells gain enhanced invasiveness [136].

Further, GBM cells suck in excess sodium through inward sodium ion conductance channels because intracellular syntaxin 1A supplies are downregulated; GBM cells therefore change their shape and plasticity, enabling them to invade. Of those protein kinases of the C class that activate or inhibit sodium conductance through the ion channels, only the activator kinases are operational in GBM cells [120].

The INK4α-ARF tumor suppressor locus encodes two proteins– $p16^{INK4a}$ and $p14^{ARF}$ –that modulate Rb/p53; these genomic sequences are often deleted in GBM cells. Exons 19-22 of Rb2/130 may remain intact, yet diabled Rb1 pathway and loss of PTEN predict brief survival [137,138]. Cyclin-dependent kinase inhibitor p27 is often downregulated by the overactive PI3K/Akt cascade in GBM cells, whereas genes for CDK4 and CDK6 are amplified. Overactive Ki-67 provides for accelerated $G_1 \rightarrow S$ progression. The PTEN (phosphatase and tensin homologue deleted from chromosome ten) gene is often mutated in, or is deleted from, GBM cells. Wild PTEN/MMAC is antineo-angiogenic, whereas constitutively overactive EGF-R is proneo-angiogenic. Mutated or deleted PTEN on 10q23 and amplified or mutated EGF-R cooperate in transactivating the VEGF promoters via the PI3K/Akt pathway: Glioblastoma multiforme cells express increased levels of VEGF mRNA; LY294002 counteracts this deficiency [139–142]. Cyclooxygenase-2 and TNFα function as additional neo-angiogenic factors in GBM [135]. The chicken retrovirus CT10 harbors the oncogene v-*crk*, which derives from its cellular counterpart c-*crk*. The gene product protein CrkI is overexpressed in GBM cells, where it induces the PI3K/Akt cascade and promotes invasiveness by rendering GBM cells free of cell-to-cell contact [143]. The avian sarcoma retrovirus UR2 yielded the v-*ros* proto-oncogene [144]; c-*ros* encodes a 260-kDa glycoprotein transmembrane receptor for an as-yet-unknown ligand. In some glioblastoma cells, consequentially to a microdeletion on 6q21, Ros is fused with the FIG protein, which is functional in the Golgi apparatus and forms the FIG-ROS fusion oncoprotein [145]. The signaling pathway of the constitutively active fusion oncoprotein is unknown, but Ros signaling normally is aimed toward phospholipase Cγ and the insulin-receptor substrates (IRS).

In GBM cells arising *de novo*, the EGF-R gene is mutated, the receptor is truncated, and the result is EGF-RνIII. It is without its extracellular domain, so it is ligand-independent and is constitutively activated [107]. In secondarily progressing GBM arising from lower grade astrocytomas, the EGF-R is not mutated, but is amplified. In these tumors, p53 is deleted or mutated. Matrix metalloproteinase-9 expression is high (69% to 83%) in *de novo* tumors and low (14%) in secondary tumors [146,147]. The cascade EGF-R→Ras→MAPK/PI3K/Akt leads to resistance to irradiation and chemotherapy. Mitogen-activated protein kinase overactivity and overexpression of MAPK-related proteins PKC (see Part II) and phospholipase C correlated with rapid tumor growth. Peculiarly, the nitrosourea BCNU may inhibit XRt-induced apoptosis and vice versa. XRt may inhibit BCNU-induced apoptosis [115]. The EGF-R inhibitor AG1478, the PI3K inhibitor LY29400, and the MAPK-1 inhibitor PD98059 (from Cell Signaling Technologies, Beverly, MA), may restore chemo-radiotherapy susceptibility of GBM cells in vitro. The FTI R115777 or SCH66336, blocking the Ras pathway (see Part I), reverse resistance of GBM cells to ionizing radiation [116,117,148–150]. Glioblastoma multiform cells lacking EGF-RνIII cells are less, whereas EGF-RνIII$^+$ GBM cells are more, sensitive to chemo-radiotherapy. The mutant EGF-RνIII is common in native de novo-arising GBM cells but is lost in established cultures of these cells [115,116]. Experiments with transgenic mice indicate that fibrillary astrocytomas grow in H-*ras* transgenic mice; oligodendrogliomas grow in H-*ras* and EGF-RvIII double-transgenic mice, whereas EGF-RvIII alone did not induce gliomagenesis [151].

Platelet-derived factors A, B, C, and D attach to PDGF-R; GBM cells expressing both receptors and their cognate ligands expropriated the system for an autocrine growth loop. The piperazinyl quinazoline PDGF-R inhibitor CT52923 abolishes autophosphorylation of the receptor kinases and may retard growth of GBM cells in vitro [110,113]. Tissue factor pathway inhibitor-2 (TFPI-2) is encoded by its gene on 7q22; phorbols (PMA), diacyl glycerols (DAG), IFN-α, and IFN-γ activate the gene. The serine protease inhibitor TFPI-2 maintains the integrity of the extracellular matrix; it is underexpressed in GBM, contributing to the invasiveness of the tumor, but when the TFPI-2 gene is reactivated, it suppresses tumor cell invasiveness [152].

Nerve growth factor suppresses the growth of rat gliomas and induces differentiation of neuroblastoma, pheochromocytoma, and some glioma cells [153]. But its role in human GBM is not known. Human immunodeficiency virus-1–infected macrophages and AIDS-associated Kaposi's sarcoma cells overexpress its receptor, $p140^{trkA}$, and overproduce the ligand, but die apoptotic death when antibodies neutralize the ligand NGF [154]. Viral etiology for GBM is not proven; polyoma viral genomic sequences (SV40 from the first batches of the original live polio vaccines?) are found only occasionally in GBM cells and there is an absence of activated endogenous or exogenous retroviral agents in GBM tumors. However, refined genetic analysis of GBM cells (like those applied for Kaposi's sarcoma cells) should continue. The *ret*→Ret proto-oncogene receptor captures glial cell-derived neurotrophic factor, active on neuroblastoma cells, and is inhibited by the pyrazolo-pyrimidine PP1; its activity in GBM cells is suspected. Germline mutations of *ret* are associated with multiple endocrine neoplasia (see cartoon in reference) [155,156].

Of the chemokines overexpressed by GBM cells, LARC (liver and activation regulated chemokine) stands out, whereas in normal brain cells there is no LARC mRNA activity. This chemokine attracts host CD4, CD8, and CD45RO lymphocytes into the tumor; it is however, unknown whether these lymphocytes attack and kill GBM cells or vice versa Glioblastoma multiforme cells may be armed with FasL and kill the lymphocytes [157]. The chemokine receptor CXCR4 is activated by its ligand CXCL12 or SDF-1 [158]. Glioblastoma multiforme cells produce and respond to SDF-1 through activation of ERK and PI3K/Akt pathways. Inhibitors of the system (antisense oligonucleotide against the receptor, antibodies against the ligand or receptor, or monoclonal antibody 12G5 against the receptor and LY294002 against PI3K) abolished GBM cell growth [159]. Glioblastoma multiforme cells produce SDF-1 and MCP-1, but it is unknown whether this is a host defense reaction or whether monocytes subverted by the GBM cells provide paracrine growth factors for the tumor cells. Monocyte chemoattractant protein -1 is one of the angiogenic chemokines, the receptors [regulated activation normal T cell expressed and secreted (RANTES), CXCR2, CXCR4) of which are expressed on endothelial cells and on many parenchymal and tumor cells. Matrix metalloproteinase-9 is also overexpressed in GBM tumors. The gene product protein PTEN would suppress its production if it were available in GBM cells (it is deleted from GBM and many other tumor cells) [160]. Matrix metalloproteinases-2 and MMP-9 are overexpressed in GBM cells and in the matrix of the tumor [161,162]. The serine protease urokinase-type plasminogen activator (uPA) and its receptor (uPAR) are overexpressd in GBM cells and are inhibited by phenylbutyrate or the peptide A6 or by an AS-ODN (*vide infra*) [139].

In the natural history of GBM, it is important to recognize the rare long survivors, who exhibit $19q^-$; the function of the deleted chromosomal sequence is not known. Although it is not tested for, this chromosomal change may explain 10 years' survival in subtotally resected Grade III and IV supratentorial astrocytomas. Short survivors have chromosomal deletions—$6q^-$, $10p^-$, and $10q^-$—and chromosomal gains—$19p^+/q^+$ and $20q^+$. Microarray analysis encompassing 12,000 genes relates better to subtype and prognosis than standard histological classification (163—166). In many cases, it was the genetics of the tumor and not the treatment the patient received that allowed for prolonged survival.

Biotherapy

The sulindac sulfone exisulind derivatives CP248 and 461 in cultured GBM cells arrested the cell cycle in the M phase, activated the pro-apoptotic protein kinase G, perturbed microtubule polymerization, and caused cell death [167]. The inhibitory pseudosubstrate sequence within the enzyme PKC (see Part II) is antagonistic to the enzyme and to the growth of GBM cells [168]. The geldanamycin derivative 17-AAG is being tested against many xenografted human tumors (see Part IV), including GBM cells, and it appears to be growth inhibitory in most of the assays.

Glioblastoma multiforme cells overexpress the Th2 cytokine IL-13 receptor. An immunotoxin has been constructed—IL-13-PE38QQR—which consists of IL-13 and Pseudomonas exotoxin fusion protein. It kills xenografted human GBM cells [169–172]. Another immunotoxin targets IL-4R. Stereotactically assisted local injection of the IL-4 (38-37)-PE38KDEL immunotoxin induced extensive necrosis and occasional CR of highly malignant brain tumors. Transferrin linked to diphtheria toxin attaches to its receptor on GBM cells and the complex is endocytosed, with cell death resulting [173–175]. Human leukocyte antigen-A0201–restricted immune T lymphocyte clones cytotoxic to GBM cells expressing IL-13Rα2 could be generated in vitro [176]. Epidermal growth factor-R expression is minimal, if any, in the normal brain, while there is cross-reactivity of IL-13 with IL-3R. Targeting EGF-R with fusion proteins (immunotoxins) therefore offers the highest safety profile [172]. Epidermal growth factor and diphtheria toxin fusion protein $DAB_{389}EGF$ preferentially kills tumor cells with dense expression of EGF-R, but it certainly could induce liver and kidney toxicity; however, its clinical efficacy was inhibited by pre-existing antidiphtheria toxin antibodies [177,178]. Neural stem cells (NSC) exhibit tropism toward GBM cells. Transfected with the IL-12 gene or with the ligand TNFα and injected intracranially into murine gliomas, the genetically engineered NSC cells killed the glioma cells [179,180].

Gap junction communications regulate intercellular relationships through expression of connexins (see Part IV). Connexin 43 (cx43) is underexpressed in GBM cells. Transfection of cx43 gene into GBM cells (also into leiomyo- and rhabdomyosarcoma cells) reversed the malignant phenotype and reduced MCP-1 overexpression by the tumor cells. Connexin 43 downregulates monocyte chemotactic protein in GBM cells, depriving the tumor of one of its prominent growth factors [181,182]. The anti-apoptotic chaperone protein Hsp70 is highly overexpressed in GBM cells. Adenoviral (Ad.asHsp70) introduction of antisense Hsp70 cDNA into GBM cell orthotopic xenografts (also into breast and colon carcinoma xenografts) eradicated the xenografts [183]; GBM cells were wiped out through phagocytosis by macrophages. Neuroblastoma cells are protected from chemotherapy-induced apoptosis by brain-derived neutrophilic factor (BDNF) activation mediated by tyrosine kinase receptors. Inhibitors K252 or LY294002 (aimed at PI3K) reversed the protective effect; MAPK inhibitor PD98059 or peritoneal lymphocyte γ inhibitor U73122 were inefficient [184]. A naturally occurring fragment of human MMP-2 is anti-angiogenic and inhibits cell proliferation and invasiveness. The hemopexin fragment PEX binds to integrin ανβ3 and suppresses growth of GBM xenografts without toxicity; platelet factor-4 (PF-4) acts additively in this system [161,185,186].

The Hungarian team successfully uses adenoviral vectors to deliver genes for specific enzymes (v-*tk*), cytokines (Il-2, IL-4, IL-12, TNFα), or hematopoietic growth factors (GM-CSF) into murine and human glioblastoma cells for the sensitization of these cells to chemotherapeuticals (5 FU; ganciclovir) or to render them more immunogenic [187,188]. Herpes simplex virus-1 *tk* driven by cytomegalovirus (CMV) promoter and delivered by intact E1AB-containing adenoviral vector (IG.Ad5E1+.E3TK) into GBM cells rendered the tumor cells highly susceptible to ganciclovir-induced apoptotic death [189]. Clinical results of *tk* gene transfer into malignant gliomas with replication-defective retroviral vectors were so far unsuccessful because of very low transduction efficiency. Transduction efficiency for p53 adenoviral gene transfer was also low [190]. Modified replication-competent retroviral vectors (MuLV unable to infect quiescent normal cells, but replicating selectively in dividing tumor cells) performed with greatly increased transduction efficiency without infecting normal brain cells and without systemic dissemination in the body (negative PCR for viral genomic segments) [191,192]. Clinical trials are planned in Los Angeles, CA. These retroviral vectors can also insert the yeast cytosine deaminase gene into tumor cells, sensitizing the cells to 5-fluorocytosine.

Adenoviral vectors have been constructed for the restoration of PTEN, the anti-neoangiogenic gene product proteins–(Ad)-wt-PTEN and Ad-G129E-PTEN for the treatment of $PTEN^{-}$ prostate cancer (PC3) and GBM cells (see Part I). In these cells, COX-1 and COX-2 were not inhibited, but VEGF production was decreased. Glioblastoma multiforme cells were arrested in G_1 and lost their metastatic potential, but did not die apoptotic death. Adenoviral vectors of p53 gene performed poorly, even after repeated attempts at transduction of glioma cells, because of underexpression of CAR resulting in low transduction efficiency [190]. Glioblastoma multiforme cells activate MMP-9 secretion after binding uPA to its receptor uPAR. The replication-defective serotype 5 adenovirus Ad-uPAR-MMP-9 contains antisense oligonucleotides disabling transcription of both targeted genes [193]. Matrix metalloproteinase-9 degrades extracellular matrix and facilitates the invasiveness of the tumor cell. Expression of both uPAR and MMP-9 were suppressed in transfected GBM cells and the transfected tumor cell spheroids failed to invade normal neural tissues in co-cultures. As-ODN treatment directed at laminin-8 of glioma cells suppressed their locomotion [194]. Liposome-encapsulated genetically engineered Semliki forest arbovirus delivered the gene of IL-2 into GBM cells and augmented the immune responses these tumor cells evoked [195]. The phosphorothioate antisense oligodeoxynucleotide IGF-IR/AS ODN (Lynx Therapeutics, Des Moines, IA) targets the IGF-R gene. Surgically removed astrocytomas were treated with

As-ODN and reimplanted into the patient's rectus sheath, where the tumor underwent apoptotic death and induced lymphocytic infiltrations [196]; this method of gene therapy may work as a glioma vaccine (*vide infra*). The technology of gene therapy is improving by an enhanced delivery system [197].

No malignant entity mobilized more intensive efforts for viral therapy than GBM and other brain tumors. Adeno-, adeno-associated, herpes-, reo-, and retroviral vectors have been extensively tested. A genetically engineered adenovirus becomes the vector of a transgene, when a selected gene is cloned into a plasmid, which is recombined with another plasmid possessing adenoviral packaging signal; forming virions become vectors of the transgene. Adenovirus with E1B deleted (ONYX) is restricted to replicate only in tumor cells with defective p53 gene (see Part I). Adenoviral genomes occupy extrachromosomal positions in the cell and adenoviral capsids are antigenic in the host; CAR viral entry sites are underexpressed on many tumor cells that remain uninfected. In this way, adenovirally infected cells may get rid of the virus and extracellular virus is often neutralized by antibodies. Adenoviruses delivered the herpesviral *tk* gene, the wild-type p53 gene, and various cytokine genes (IL-2, IL-12, IFN-γ) and hematopoietic growth factor genes (GM-CSF) into brain (and other) tumor cells. Adenoviruses have 10-kb transgene capacity, but grow to high titers. Adeno-associated viruses seek out the 19q chromosomal locus for the integration of their genome in human cells. They cannot accommodate a transgene larger than 4.7 kb and grow in low titers.

The multimutated HSV-1 G207 or the γ34.5 mutant HSV-1 1716 act as oncolytic agents without a transgene incorporated into their genomes. Reoviruses are expected to perform similarly. Retroviruses recognize long terminal repeats (LTR) as packaging signals and incorporate transgenes inserted between two LTR. The NCI murine fibroblast line 3T3 serves as vector producer cells transfected with the retroviral genome (Moloney mouse leukemia virus) and the transgene. Retroviral vectors grow in low titers, accept transgenes not larger than 7.5 kb, but enter the nuclei of dividing (not resting) cells and do not pose a high antigenic impact on the host. The vector-producer cells have to be engrafted into the targeted tumors (requiring many needle punctures); the unstable retroviral particles seldom infect tumor cells otherwise. When tumor cells incorporating the herpes virus *tk* gene are exposed to acyclovir or ganciclovir, they die. The "bystander effect" consists of cell death other than that of transfected tumor cells and that of unintentionally *tk*-transfected vascular endothelial cells. There is host immunity to the herpes viral enzyme protein Tk and there is cell-to-cell transfer of ganciclovir triphosphate through gap junctions. In all tumor categories treated with herpes viral *tk*-transfection, including GBM cells, objective and clinically beneficial response rate barely exceeds 10%. In more than 30 brain tumor clinical trials with various oncolytic viruses and viral vectors, no major adverse effects (no viral encephalitis, not even with the herpes viral mutants) occurred. Decrease in tumor size, stagnation (stabilization), and delayed progression of disease is commonly observed; it is more often a trend, rather than a statistically significant delay in tumor progression that is recorded. Rare cases are on the record with tumor regression and no relapse within 2 years [198]. In 2001, at the 33rd Annual Scientific Meeting of the Hungarian Medical Association of America in Sarasota, FL, the United Cancer Research Institute (Alexandria, VA), and the Hadassah University (Jerusalem, Israel) announced remission inductions in patients with malignant gliomas treated with the attenuated veterinary NDV vaccine, which is used in Hungary to vaccinate poultry (without any further development, this virus was renamed MTH-68H for the purpose of human use). However, no published concise report followed in the ensuing 2 years. Major improvements in viral therapy of GBM (and other tumors) are required for better and confirmed clinical results.

Epidermal growth factor-RvIII yields immunogenic peptides; when it was used with KLH adjuvant for active glioma-specific preventive immunization of mice, it enabled the animals to reject live glioma cell inocula [199,200]. Glioma cell-associated antigens, perhaps those expressed by fusion oncoproteins such as the ROS-FIG oncoprotein [201], when received by mature DC either as downloaded peptides or as tumor genomic mRNA /cDNA to be processed internally in the DCs also co-expressing IL-12 or as fusion partners forming chimeras of autologous irradiated glioma cells with live DC, induced T cell immunity because DCs presented glioma cell antigens to T lymphocytes for IL-2–mediated expansion of the immune T cell clone [202,203]. Clones of CD8 specifically cytotoxic to autologous glioma cells arose in the immunized patients. Glioma cells transfected with the IFN-β gene carried by a retroviral vector plasmid (pSV2-IFN-β/pDONAI, Takara Bio, Otsu, Shiga, Japan) stimulated DC best and induced T cell clones highly cytotoxic to murine glioma cells [204].

Tumor cells, among them GBM cells, are able to express FasL to kill Fas receptor-positive host lymphocytes, but GBM cells are not exempt from FasL-mediated programmed cell death. Taurolidine enhances FasL-mediated apoptosis of glioma cells [205,206].

Glioblastoma multiforme cells obliterate an immune response from the host. These cells do not express the costimulatory molecules B7.1/2; instead they display B7-H1, which is an inhibitor of CD4/CD8 lymphocytes in that it suppresses IL-2 production and the expression of T cell activation marker CD96 [207]. Some malignant gliomas elicit a specific T cell response, which could be intensified by active tumor-specific immunotherapy with DC vaccines [176]. Autotumor-specific T lymphocytes were generated by co-culturing lymphocytes with autologus tumor cells. Topical administration of such lymphocyte populations to five patients with anaplastic astrocytomas resulted in one CR and four PR. In a summary of lymphocyte therapy for GBM and anaplastic astrocytomas (occasionally including melanoma metastatic to the brain), autologous lymphocyte preparations were infused through Ommaya reservoirs or burr holes into the cavities left behind after incomplete surgical removal of the primary tumor or at the time of relapse after postoperative radiotherapy. Interleukin-2 was given systemically to expand the infused lymphocyte population. The University of North Carolina at Chapel Hill used tumor-infiltrating lymphocytes (TIL) in six patients and recorded one CR, two PR and two stabilizations of the disease, with three deaths attributable to eventual progression of the disease. In five clinical trials with LAK cells and reported from Bethesda, MD; New York, NY; India; and Japan, 48 patients were treated; 8 CR, 10 PR, and many stabilizations of disease occurred. Treated groups experienced median survival over 53 to 55 weeks vs. median survival of 25 weeks in the conventionally treated control group. In one trial (Kaplan CCC, New York) 8 of 15 treated and 1 of 18 control patients were alive at the time of the report. Occasionally, brain edema (attributable to the restricted use of corticosteroids) or intratumoral hemorrhage occurred. A young man in Japan with anaplastic glioma survived 9 years after conventional therapy, but relapsed; LAK cell therapy induced another CR [208]. Alloreactive cytotoxic T cells sensitized with the patient's MHC proteins were instilled into the cavitary lesions left behind after surgical removal of malignant gliomas; of five patients, two survived over 28 and 30 months without tumor recurrence [209–218].

Of monoclonal antibodies directed at glioma cells [219], antitenascin immunotoxin has been tested. Tenascin enabled melanoma cells to detach from their matrix and invade; this mediator is also active in GBM. The I^{131}-labeled antitenascin monoclonal antibody (McAb 81C6) was infused into cavities left behind after incomplete surgical removal of highly malignant brain tumors. Duke University (Durham, NC) reports prolongation of life up to 69 to 79 weeks with minimal toxicity if the 120-mCi total dose was not exceeded [220,221].

Standard therapy of GBM is not curative, but is undergoing steady improvements: Irinotecan; nimustine, teniposide, and cytarabine are added to radiotherapy in Germany (Neuro-Oncology Working Group of the German Cancer Society); temozolomide is used at reduced dosage continuously and its efficacy is increased by the co-adminstration of the poly(ADP-ribose) polymerase-1 inhibitor GP 15427 or CEP-6800 [222,223]. Of neoangiogenesis inhibitors, thalidomide and IFN-α are in clinical trials. In its first trials, thalidomide (see Part IV) induced 2 PR, 2 MR, and 12 stabilizations of the disease [224]. SU112448 (Sugen) [225], ZD1839 (gefitinib, Iressa), OSI-774 (erlotinib, Tarceva), anti-ErbB (ABX-EGF) TKIs (see Part I), and anti-VEGF (bevacizumab, Avastin) monoclonal antibody (see Part III) are in advanced clinical trials showing additive, or better— sometimes synergistic—efficacy. Glioma cells overexpress COX-1 and COX-2, and COX or thromboxane synthetase inhibitors suppress their growth and render the tumor cells more susceptible to chemotherapy (camptothecins) [226]. Even nonconventional treatment modalities, such as the alkaloid mixture ruta-6, the extract of *Ruta graveolens,* may score clinical responses (reportedly CR!) [227]. The most accurate way to measure tumor responses in the brain is by ^{201}T-SPECT (single-photon emission computed tomography) scanning [228].

REFERENCES, PART V

1. Kyle RA. Diagnosis of multiple myeloma Semin Oncol 2002; 29(6 Suppl 17):2–4.
2. Barlogie B, Shaughnessy J, Zangari M, Tricot G. High-dose therapy and immunomodulatory drugs in multiple myeloma Semin Oncol 2002; 29(6 Suppl 17):26–33.
3. Rajkumar SV, Kyle RA, Gertz MA. Myeloma and the newly diagnosed patient: a focus on treatment and management Semin Oncol 2002; 29(6 Suppl 17):5–10.
4. Kumar A, Loughran T, Alsina M, Durie BG, Djulbegovic B. Management of multiple myeloma: a systematic review and critical appraisal of published studies Lancet Oncol 2003; 4: 293–304.
5. Attal M, Harousseau JL, Facon T, Guilhot F, Doyen C, Fuzibet JG, Monconduit M, Hulin C, Caillot D, Bouabdallah R, Voillat L, Sotto JJ, Grosbois B, Bataille R. Single versus double autologous stem-cell transplantation for multiple myeloma N Engl J Med 2003; 349:2495–2502.
6. Alain T, Thirukkumaran C, Morris DG, Urbanski SJ, Janowska-Wieczorek A, Lee PWK, Kossakowska AE. Lymphomas and oncolytic virus therapy Clinical Lymphoma 2003; 4: 104–111.
7. Kuehl WM, Bergsagel PL. Multiple myeloma: evolving genetic events and host interactions Nat Rev Cancer 2002; 2:175–187.

8. Pratt G. Molecular aspects of multiple myeloma Mol Pathol 2002; 55:273–283.

9. Kataoka K, Noda M, Nishizawa M. Maf nuclear oncoprotein recognizes sequences related to an AP-1 site and forms heterodimers with both Fos and Jun Mol Cell Biol 1994; 14:700–712.

10. Kataoka K, Shioda S, Yoshitomo-Nakagawa K, Handa H, Nishizawa M. Maf and Jun nuclear oncoproteins share downstream target genes for inducing cell transformation J Biol Chem 2001; 276:36849–36856.

11. Hu L, Shi Y, Hsu JH, Gera J, Van Ness B, Lichtenstein A. Downstream effectors of oncogenic ras in multiple myeloma cells Blood 2003; 101:3126–3135.

12. Croonquist PA, Linden MA, Zhao F, Van Ness BG. Gene profiling of a myeloma cell line reveals similarities and unique signatures among IL-6 response, N-*ras*-activating mutations, and coculture with bone marrow stromal cells Blood 2003; 102: 2581–2592.

13. Filipits M, Pohl G, Stranzl T, Kaufmann H, Ackermann J, Gisslinger H, Greinix H, Chott A, Drach J. Low $p27^{Kip1}$ expression is an independent adverse prognostic factor in patients with multiple myeloma Clin Cancer Res 2003; 9:820–826.

14. Shi Y, Gera J, Hu L, Hsu JH, Bookstein R, Li W, Lichtenstein A. Enhanced sensitivity of multiple myeloma cells containing PTEN mutations to CCI-779 Cancer Res 2002; 62:5027–5034.

15. Hahn WC, Counter CM, Lundberg AS, Beijersbergen RL, Brooks MW, Weinberg RA. Creation of human tumor cells with defined genetic elements Nature 1999; 400:464–468.

16. Shammas MA, Shmookler Reis RJ, Akiyama M, Koley H, Chauhan M, Hideshima T, Goyal RK, Hurley LH, Anderson KC, Munshi NC. Telomerase inhibition and cell growth arrest by G-quadruplex interactive agent in multiple myeloma Mol Cancer Ther 2003; 2:825–833.

17. Akiyama M, Hideshima T, Shammas MA, Hayashi T, Hamasaki M, Tai Y-T, Richardson P, Gryzanov S, Munshi NC, Anderson KC. Effects of oligonucleotide N2′⇒P5′ thio-phosphoramidate (GRN163) targeting telomerase RNA in human multiple myeloma cells Cancer Res 2003; 63:6187–6194.

18. Wang ES, Wu K, Chin AC, Chen-Kiang S, Pongracz K, Gryzanov S, Moore MA. Telomerase inhibition with an oligonucleotide telomerase template antagonist: in vitro and in vivo studies in multiple myeloma and lymphona Blood 2004; 103: 258–266.

19. Otsuki T, Sakaguchi H, Eto M, Fuji T, Hatamaya T, Takata T, Sugihara T, Hyodoh F. IL-6 is a key factor in growth inhibition of human myeloma cells induced by pravastatin, an HMG-CoA reductase inhibitor Int J Oncol 2003; 23:763–768.

20. Hideshima T, Chauhan D, Hayashi T, Podar K, Akiyama M, Gupta D, Richardson P, Munshi N, Anderson KC. The biological sequelae of stromal cell-derived factor-1α in multiple myeloma Mol Cancer Ther 2002; 1:539–544.

21. Anderson KC, Dalton WS. Synopsis of a research roundtable presented on cell signaling in myeloma: regulation of growth and apoptosis—opportunities for new drug discovery Mol Cancer Ther 2002; 1:1361–1365.

22. Anderson KC. Moving disease biology from the lab to the clinic Cancer 2003; 97(3 Suppl):796–801.

23. Koretzky G, Ed. Plasma Cells and Multiple Myeloma Immunol Rev 2003; 194:5–206.

24. Seidl S, Kaufmann H, Drach J. New insights into the pathophysiology of multiple myeloma Lancet Oncol 2003; 4: 557–564.

25. Moreno JL, Kaczmarek M, Keegan AD, Tondravi M. IL-4 suppresses osteoclast dvelopment and mature osteoclast function by a STAT6-dependent mechanism: irreversible inhibition of the differentiation program activated by RANKL Blood 2003; 102:1076–1086.

26. Terpos E, Szydlo R, Apperley JF, Hatjiharissi E, Politou M, Meleteis J, Viniou N, Yataganas X, Goldman JM, Rahemtulla A. Soluble receptor activator of nuclear factor κB ligand-osteoprotegerin ratio predicts survival in multiple myeloma: proposal for a novel prognostic index Blood 2003; 102:1064–1069.

27. Yonou H, Kanomata N, Goya M, Kamijo T, Yokose T, Hasebe T, Nagai K, Hatano T, Ogawa Y, Ochiai A. Osteoprotegerin/osteoclastogenesis inhibitory factor decreases human prostate

cancer burden in human adult bone implanted into nonobese diabetic/severe combined immunodeficient mice Cancer Res 2003; 63:2096–2102.

28. Shipman CM, Croucher PI. Osteoprotegerin is a soluble decoy receptor for tumor necrosis factor-related apoptosis-inducing ligand/Apo2 ligand and can function as a paracrine survival factor for human myeloma cells Cancer Res 2003; 63:912–916.

29. Heider U, Langelotz C, Jakob C, Zavrski I, Fleissner C, Eucker J, Possinger K, Hofbauer LC, Sezer O. Expression of receptor activator of nuclear factor κB ligand on bone marrow plasma cells correlates with osteolytic bone disease in patients with multiple myeloma Clin Cancer Res 2003; 9:1436–1440.

30. Farrugia AN, Atkins GJ, To B, Pan B, Horvath N, Kostakis P, Findlay DM, Bardy P, Zannettino ACW. Receptor activator of nuclear factor-κB ligand expression by human myeloma cells mediates osteoclast formation *in vitro* and correlates with bone destruction *in vivo* Cancer Res 2003; 63:5438–5445.

31. Mitsiades CS, Mitsiades NS, Bronson RT, Chauhan D, Munshi N, Treon SP, Maxwell CA, Pilarski L, Hideshima T, Hoffman RM, Anderson KC. Fluorescence imaging of multiple myeloma cells in a clinically relevant SCID/NOD *in vivo* model: biologic and clinical implications Cancer Res 2003; 63:6689–6696.

32. Tai Y-T, Podar K, Catley L, Tseng Y-H, Akiyama M, Shringarpure R, Burger R, Hideshima T, Chauhan D, Mitsiades N, Richardson P, Munshi NC, Kahn CR, Mitsiades C, Anderson KC. Insulin-like growth factor-1 induces adhesion and migratioin in human multiple myeloma cells via activation of β1-integrin and phosphatidyl 3′-kinase/AKT signaling Cancer Res 2003; 63:5850–5858.

33. Hideshima T, Richardson P, Anderson KC. Novel therapeutic approaches for multiple myeloma Immunol Rev 2003; 194: 164–176.

34. Tian E, Zhan F, Walker R, Rasmussen E, Ma Y, Barlogie B, Shaughnessy JD. The role of the Wnt-signaling antagonist DKK1 in the development of osteolytic lesions in multiple myeloma N Engl J Med 2003; 349:2483–2494.

35. Udagawa N, Horwood NJ, Elliott J, Mackay A, Owens J, Okamura H, Kurimoto M, Chambers TJ, Martin TJ, Gillespie MT.

Interleukin-18 (interferon-gamma-inducing factor) is produced by osteoblasts and acts via granulocyte/macrophage colony-stimulating factor and not via interferon-gamma to inhibit osteoclast formation J Exp Med 1997; 185:1005–1012.

36. Cornish J, Gillespie MT, Callon KE, Horwood NJ, Moseley JM, Reid IR. Interleukin-18 is a novel mitogen of osteogenic and chondrogenic cells Endocrinology 2003; 144:1194–1204.

37. Yamada N, Niwa S, Tsujimura T, Iwasaki T, Sugihara A, Futani H, Hayashi S, Okamura H, Akedo H, Terada N. Interleukin-18 and interleukin-12 synergistically inhibit osteoblastic bone-resorbing activity Bone 2002; 30:901–908.

38. Horwood NJ, Udagawa N, Elliott J, Grail D, Okamura H, Kurimoto M, Dunn AR, Martin T, Gillespie MT. Interleukin-18 inhibits osteoclast formation via T cell production of granulocyte macrophage colony-stimulating factor J Clin Invest 1998; 101:595–603.

39. Ohtsuki T, Micallef MJ, Kohno K, Tanimoto T, Ikeda M, Kurimoto M. Interleukin 18 enhances Fas ligand expression and induces apoptosis in Fas-expressing humaan myelomonocytic KG-1 cells Anticancer Res 1997; 17:3253–3258.

40. Sezer O, Heider U, Zavrski I, Kühne CA, Hofbauer LC. RANK ligand and osteoprotegerin in myeloma bone disease Blood 2003; 101:2094–2098.

41. Dovio A, Sartori ML, Angeli A. Correspondence Re: A. Lipton et al., Serum osteoprotegerin levels in healthy controls and cancer patients Clin Cancer Res 2002; 8:2306–2310 Clin Cancer Res 2003; 9:2384–2385.

42. Ashcroft AJ, Davies FE, Morgan GJ. Aetiology of bone disease and the role of bisphosphonates in multiple myeloma Lancet Oncol 2003; 4:284–292.

43. Dalton WS. The tumor microenvironment: focus on myeloma Cancer Treat Rev 2003; 29(Suppl 1):11–19.

44. Yamashita K, Iwasaki T, Tsujimura T, Sugihara A, Yamada N, Ueda H, Okamura H, Futani H, Maruo S, Terada N. Interleukin-18 inhibits lodging and subsequent growth of human multiple myeloma cells in the bone marrow Oncol Rep 2002; 9:1237–1244.

45. Bergsagel PL. Prognostic factors in multiple myeloma: it is in the genes Clin Cancer Res 2003; 9:533–534.

46. Voorhees PM, Dees EC, O'Neil B, Orlowski RZ. The proteasome as a target for cancer therapy Clin Cancer Res 2003; 9: 6316–6325.

47. Trial Information. A phase 2 study of Velcade in relapsed or refractory mantle cell lymphoma. *http:/www.centerwatch.-com/patient/studies/stu51546.html.*

48. Voorhees PM, Dees EC, O'Neil B, Orlowski RZ. The proteasome as a target for cancer therapy Clin Cancer Res 2003; 9: 6316–6325.

49. Ma MH, Yang HH, Parker K, Manyak S, Friedman JM, Altamirano C, Wu ZQ, Borad MJ, Frantzen M, Roussos E, Neeser J, Mikail A, Adams J, Sjak-Shie N, Vescio RA, Berenson JR. The proteasome inhibitor PS-341 markedly enhances sensitivity of multiple myeloma tumor cells to chemotherapeutic agents Clin Cancer Res 2003; 9:1136–1144.

50. Orlowski RZ, Stinchcombe TE, Mitchell BS, Shea TC, Baldwin AS, Stahl S, Adams J, Esseltine DL, Elliott PJ, Pien CS, Guerciolini R, Anderson JK, Depcik-Smith ND, Bhagat R, Lehman MJ, Novick SC, O'Connor OA, Soignet SL. Phase I trial of the proteasome inhibitor PS-341 in patients with refractory hematologic malignancies J Clin Oncol 2002; 20:4420–4427.

51. LeBlanc R, Catley LP, Hideshima T, Lentzsch S, Mitsiades CS, Mitsiades N, Neuberg D, Goloubeva O, Pien CS, Adams J, Gupta D, Richardson PG, Munshi NC, Anderson KC. Proteasome inhibitor PS-341 inhibits human myeloma cell growth in vivo and prolongs survival in a murine model Cancer Res 2002; 62:4996–5000.

52. Chauhan D, Li G, Podar K, Hideshima T, Shringarpure R, Catley L, Mitsiades C, Munshi N, Tai YT, Suh N, Gribble GW, Honda T, Schlossman R, Richardson P, Sporn MB, Anderson KC. Bortezomib/proteasome inhibitor PS-341 and triterpenoid CDDO-Im induce synergistic anti-multiple myeloma (MM) activity and overcome bortezomib resistance Blood 2004; 103: 3158–3166.

53. Pahler JC, Ruiz S, Niemer I, Calvert LR, Andreeff M, Keating M, Faderl S, McConkey DJ. Effects of the proteasome inhibi-

tor, bortezomib, on apoptosis in isolated lymphocytes obtained from patients with chronic lymphocytic leukemia Clin Cancer Res 2003; 9:4570–4577.

54. Williams SA, McConkey DJ. The proteasome inhibitor bortezomib stabilizes a novel active form of p53 in the human LNCaP-Pro5 prostate cancer cells Cancer Res 2003; 63: 7338–7344.

55. Ling Y-H, Liebes L, Jiang J-D, Holland JF, Elliott PJ, Adams J, Muggia FM, Perez-Soler R. Mechanisms of proteasome inhibitor PS-341-induced G_2-M-phase arrest and apoptosis in human non-small cell lung cancer Clin Cancer Res 2003; 9: 1145–1154.

56. Vacca A, Ria R, Semeraro F, Merchionne F, Coluccia M, Boccarelli A, Scavelli C, Nico B, Gemone A, Battelli F, Tabilio A, Guidolin D, Petrucci MT, Ribatti D, Dammacco F. Endothelial cells in the bone marrow of patients with multiple myeloma Blood 2003; 102:3340–3348.

Heider U, Langelotz C, Jakob C, Zavrski I, Fleissner C, Eucker J, Possinger K, Hofbauer LC, Sezer O. Expression of receptor activator of nuclear factor κB ligand on bone marrow plasma cells correlates with osteolytic bone disease in patients with multiple myeloma Clin Cancer Res 2003; 9:1436–1440.

57. Niemöller K, Jakob C, Heider U, Zavrski I, Eucker J, Kaufmann O, Possinger K, Sezer O. Bone marrow angiogenesis and its correlation with other disease characteristics in multiple myeloma in stage I versus stage II-III J Cancer Res Clin Oncol 2003; 129:234–238.

58. Durie BG. Low-dose thalidomide in myeloma: efficacy and biologic significance Semin Oncol 2002; 29(6 Suppl 17):34–38.

59. Neben K, Moehler T, Egerer G, Kraemer A, Hillengass J, Benner A, Ho AD, Goldschmidt H. High plasma basic fibroblast growth factor concentration is associated with response to thalidomide in progressive multiple myeloma Clin Cancer Res 2001; 7:2675–2681.

60. Neben K, Moehler T, Benner A, Kraemer A, Egerer G, Ho AD, Goldschmidt H. Dose-dependent effect of thalidomide on overall survival in relapsed multiple myeloma Clin Cancer Res 2002; 8:3377–3382.

61. Rajkumar SV, Kyle RA. Thalidomide in the treatment of plasma cell malignancies J Clin Oncol 2001; 19:3593–3595.

62. Lee CK, Barlogie B, Munshi N, Zangari M, Fassas A, Jacobson J, van Rhee F, Cottler-Fox M, Muwalla F, Tricot G. DTPACE: an effective, novel combination chemotherapy with thalidomide for previously treated patients with myeloma J Clin Oncol 2003; 21:2732–2739.

63. Gupta D, Hideshima T, Anderson KC. Novel biologically based therapeutic strategies in myeloma Rev Clin Exp Hematol 2002; 6:301–324.

64. Ryoo JJ, Cole CE, Anderson KC. Novel therapies for multiple myeloma Blood Rev 2002; 16:167–174.

65. Pumphrey CY, Theus AM, Li S, Parrish RS, Sanderson RD. Neoglycans, carbodiimide-modified glysosaminoglycans: a new class of anticancer agents that inhibit cancer cell proliferation and induce apoptosis Cancer Res 2003; 62: 3722–3728.

66. Choi YS, Anderson K, Mufson RA. A workshop on the marrow microenvironment and hematologic malignancy Cancer Res 2003; 63:7539–7542.

67. Lin B, Podar K, Gupta D, Tai YT, Li S, Weller E, Hideshima T, Lentzsch S, Davies F, Li C, Weisberg E, Schlossman RL, Richardson PG, Griffin JD, Wood J, Munshi NC, Anderson KC. The vascular endothelial growth factor receptor tyrosine kinase inhibitor PTK787/ZK222584 inhibits growth and migration of multiple myeloma cells in the bone marrow microenvironment Cancer Res 2002; 62:5019–5026.

68. Nabha SM, Mohammad RM, Dandashi MH, Coupaye-Gerard B, Aboukameel A, Pettit GR, Al-Katib AM. Combretastatin-A4 prodrug induces mitotic catastrophe in chronic lymphocytic leukemia cell line independent of caspase activation and poly(-ADP-ribose) polymerase cleavage Clin Cancer Res 2002; 8: 2735–2741.

69. Gojo I, Zhang B, Fenton RG. The cyclin-dependent kinase inhibitor flavopiridol induces apoptosis in multiple myeloma cells through transcriptional repression and down-regulation of Mcl-1 Clin Cancer Res 2002; 8:3527–3538.

70. Mayo LD, Donner DB. The PTEN, Mdm2, p53 tumor suppressor-oncoprotein network Trends Biochem Sci 2002; 27: 462–466.

71. Stromberg T, Dimberg A, Hammarberg A, Carlson K, Osterborg A, Nilsson K, Jernberg-Wiklund H. Rapamycin sensitizes multiple myeloma cells to apoptosis induced by dexamethasone. 2004; 103:3138–3147.

72. Alas S, Bonavida B. Inhibition of constitutive STAT3 activity sensitizes resistant non-Hodgkin's lymphoma and multiple myeloma to chemotherapeutic drug-mediated apoptosis Clin Cancer Res 2003; 9:316–326.

73. Hayashi T, Hideshima T, Akiyama M, Richardson P, Schlossman RL, Chauhan D, Munshi NC, Waxman S, Anderson KC. Arsenic trioxide inhibits growth of human multiple myeloma cells in the bone marrow microenvironment Mol Cancer Ther 2002; 1:851–860.

74. Dalton WS. Targeting the mitochondria: an exciting new approach to myeloma therapy. Commentary re: N. J. Bahlis et al., Feasibility and correlates of arsenic trioxide combined with ascorbic acid-mediated depletion of intracellular glutathione for the treatment of relapsed/refractory multiple myeloma Clin Cancer Res 2002; 8:3643–3645, 3658–3668.

75. Kumar S, Kimlinger TK, Lust JA, Donovan K, Witzig TE. Expression of CD52 on plasma cells in plasma cell proliferative disorders Blood 2003; 102:1075–1077.

76. Otsuki T, Kurebayashi J, Ohkubo S, Uno M, Fujii T, Sakaguchi H, Hatayama T, Takata A, Tsujioka T, Sugihara T, Hyodoh F. Expression of HER family receptors and effects of anti-HER2-antibody on human myeloma cell lines Int J Oncol 2003; 23:1135–1141.

77. Robillard N, Avet-Loiseau H, Garand R, Moreau P, Pineau D, Rapp M-J, Harousseau J-L, Bataille R. CD20 is associated with a small mature plasma cell morphology and t(11;14) in multiple myeloma Blood 2003; 102:1070–1071.

78. Nimmanapalli R, O'Bryan E, Kuhn D, Yamaguchi H, Wang HG, Bhalla KN. Regulation of 17-AAG-induced apoptosis: role of Bcl-2, Bcl-XL, and Bax downstream of 17-AAG-mediated

down-regulation of Akt, Raf-1, and Src kinases Blood 2003; 102:269–275.

79. Aron JL, Parthun MR, Marcucci G, Kitada S, Mone AP, Davis ME, Shen T, Murphy T, Wickham J, Kanakry C, Lucas DM, Reed JC, Grever MR, Byrd JC. Depsipeptide (FR901228) induces histone acetylation and inhibition of histone deacetylase in chronic lymphocytic leukemia cells concurrent with activation of caspase 8-mediated apoptosis and down-regulation of c-FLIP protein Blood 2003; 102:652–658.

80. Sandor V, Bakke S, Robey RW, Kang MH, Blagosklonny MV, Bender J, Brooks R, Piekarz RL, Tucker E, Figg WD, Chan KK, Goldspiel B, Fojo AT, Balcerzak SP, Bates SE. Phase I trial of the histone deacetylase inhibitor, depsipeptide (FR901228, NSC 630176), in patients with refractory neoplasms Clin Cancer Res 2002; 8:718–728.

81. LaVallee TM, Zhan XH, Johnson MS, Herbstritt CJ, Swartz G, Williams MS, Hembrough WA, Green SJ, Pribluda VS. 2-methoxyestradiol up-regulates death receptor 5 and induces apoptosis through activation of the extrinsic pathway Cancer Res 2003; 63:468–475.

82. Dingli D, Timm M, Russell SJ, Witzig TE, Rajkumar SV. Promising preclinical activity of 2-methoxyestradiol in multiple myeloma Clin Cancer Res 2002; 8:3948–3954.

83. Tinley TL, Leal RM, Randall-Hlubek DA, Cessac JW, Wilkens LR, Rao PN, Mooberry SL. Novel 2-methoxyestradiol analogues with antitumor activity Cancer Res 2003; 63: 1538–1549.

84. Sahai E, Marshall CJ. Rho-GTPases and cancer Nat Rev Cancer 2002; 2:133–135.

85. Moon SY, Zheng Y. Rho GTPase-activating proteins in cell regulation.. 2003; 13:13–20.

86. van de Donk NWC, Kamphuis MMJ, van Kessel B, Kokhorst HM, Bloem AC. Inhibition of protein geranylgeranylation induces apoptosis in myeloma plasma cells by reducing Mcl-1 protein levels Blood 2003; 103:3354–3362.

87. Ochiai N, Uchida R, Fuchida S-i, Okano A, Okamoto M, Ashihara E, Inaba T, Fujita N, Matsubara H, Shimazaki C. Effect

of farnesyl transferase inhibitor R115777 on the growth of fresh and cloned myeloma cells in vitro Blood 2003; 102: 3349–3353.

88. Santucci R, Mackley PA, Sebti S, Alsina M. Farnesyltransferase inhibitors and their role in the treatment of mutliple myeloma Cancer Control 2003; 10:384–387.

89. Lancet JE, Karp JE. Farnesyltransferase inhibitors in hematologic maligancies: new horizons in therapy Blood 2003; 102: 3880–3889.

90. Mitsiades CS, Mitsiades N, Richardson PG, Treon SP, Anderson KC. Novel biologically based therapies for Waldenstrom's macroglobulinemia Semin Oncol 2003; 30:309–312.

91. Yang HH, Ma MH, Vescio RA, Berenson JR. Overcoming drug resistance in multiple myeloma: the emergence of therapeutic approaches to induce apoptosis J Clin Oncol 2003; 21: 4239–4247.

92. Dvorakova K, Payne CM, Tome ME, Briehl MM, Vasquez MA, Waltmire CN, Coon A, Dorr RT. Molecular and cellular characterization of imexon-resistant RPMI8226/I myeloma cells Mol Cancer Ther 2002; 1:185–195.

93. Yamada A, Kawano K, Koga M, Matsumoto T, Itoh K. Multidrug resistance-associated protein 3 is a tumor rejection antigen recognized by HLA-A2402-restricted cytotoxic T lymphocytes Cancer Res 2001; 61:6459–6466.

94. Schneider GP, Salcedo R, Dong HF, Kleinman HK, Oppenheim JJ, Howard OMZ. Suradista NSC 65016 inhibits the angiogenic activity of CXCL12-stromal cell-derived factor 1α Clin Cancer Res 2002; 8:3953–3960.

95. Bertolini F, Dell'Agnola C, Mancuso P, Rabascio C, Burlini A, Monestiroli S, Gobbi A, Pruneri G, Martinelli G. CXCR4 neutralization, a novel therapeutic approach for non-Hodgkin's lymphoma Cancer Res 2002; 62:3106–3112.

96. Weigel-Kelley KA, Yoder MC, Srivastava A. α5β1 integrin as a cellular coreceptor for human parvovirus B19L: requirement of functional activation of β integrin for viral entry Blood 2003; 102:3927–3933.

97. Biroccio A, Leonetti C, Zupi G. The future of antisense therapy: combination with anticancer treatment Oncogene 2003; 22:6579–6588.

98. Loomis R, Carbone R, Reiss M, Lacy J. Bcl-2 antisense (G3139, Genasense) enhances the *in vitro* and *in vivo* response of Epstein-Barr virus-associated lymphoproliferative disease to rituximab Clin Cancer Res 2003; 9:1931–1939.

99. Pellat-Deceunynck C, Jego G, Harousseau JL, Vié H, Bataille R. Isolation of human lymphocyte antigens class I-restricted cytotoxic T lymphocytes against autologous myeloma cells Clin Cancer Res 1999; 5:705–709.

100. Orsini E, Bellucci R, Alyea EP, Schlossman R, Canning C, McLaughlin S, Ghia P, Anderson KC, Ritz J. Expansion of tumor-specific CD8+ T cell clones in patients with relapsed myeloma after donor lymphocyte infusion Cancer Res 2003; 63:2561–2568.

101. Dhodapkar MV, Geller MD, Chang DH, Shimizu K, Fujii S, Dhodapkar KM, Krasovsky J. A reversible defect in natural killer T cell function characterizes the progression of premalignant to malignant multiple myeloma J Exp Med 2003; 197: 1667–1676.

102. Stevens B. Glia: much more than the neuron's side-kick Curr Biol 2003; 13:R469–R472.

103. Jaeckle KA, Hess KR, Yung WK, Greenberg H, Fine H, Schiff D, Pollack IF, Kuhn J, Fink K, Mehta M, Cloughesy T, Nicholas MK, Chang S, Prados M. North American Brain Tumor Consortium. Phase II evaluation of temozolomide and 13-cis-retinoic acid for the treatment of recurrent and progressive malignant glioma: a North American Brain Tumor Consortium study J Clin Oncol 2003; 21:2305–2311.

104. Grossman SA, O'Neill A, Grunnet M, Mehta M, Pearlman JL, Wagner H, Gilbert M, Newton HB, Hellman R. Eastern Cooperative Oncology Group. Phase III study comparing three cycles of infusional carmustine and cisplatin followed by radiation therapy with radiation therapy and concurrent carmustine in patients with newly diagnosed supratentorial glioblastoma multiforme: Eastern Cooperative Oncology Group Trial 2394 J Clin Oncol 2003; 21:1485–1491.

105. Gilbert MR, Supko JG, Batchelor T, Lesser G, Fisher JD, Piantadosi S, Grossman S. Phase I clinical and pharmacokinetic study of irinotecan in adults with recurrent malignant glioma Clin Cancer Res 2003; 9:2940–2949.

106. Kanzawa T, Kondo Y, Ito H, Kondo S, Germano I. Induction of autophagic cell death in malignant glioma cells by arsenic trioxide Cancer Res 2003; 63:2103–2108.

107. Sinojima N, Tada K, Shiraishi S, Kamiryo T, Kochi M, Nakamura H, Makino K, Saya H, Hiranoi H, Kuratsu J-i, Oka K, Ishimaru Y, Ushio Y. Prognostic value of epidermal growth factor receptor in patients with glioblastoma multiforme Cancer Res 2003; 63:6962–6970.

108. Okada Y, Hurwitz EE, Esposito JM, Brower MA, Nutt CL, Louis DN. Selection pressures of TP53 mutation and microenvironmental location influence epidermal growth factor receptor gene amplification in human glioblastomas Cancer Res 2003; 63:413–416.

109. Choe G, Horvath S, Cloughesy TF, Crosby K, Seligson D, Palotie A, Inge L, Smith BL, Sawyers CL, Mischel PS. Analysis of the phosphatidylinositol 3′-kinase signaling pathway in glioblastoma patients in vivo Cancer Res 2003; 63:2742–2746.

110. Halatsch ME, Gehrke E, Borhani FA, Efferth T, Werner C, Nomikos P, Schmidt U, Buchfelder M. EGFR but not PDGFR-β expression correlates to the antiproliferative effect of growth factor withdrawal in glioblastoma multiforme cell lines Anticancer Res 2003; 23:2315–2320.

111. Mawrin C, Diete S, Treuheit T, Kropf S, Vorwerk CK, Boltze C, Kirches E, Firsching R, Dietzmann K. Prognostic relevance of MAPK expression in glioblastoma multiforme Int J Oncol 2003; 23:641–648.

112. Heimberger AB, Learn CA, Archer GE, McLendon RE, Chewning TA, Tuck FL, Pracyk JB, Friedman AH, Friedman HS, Bigner DD, Sampson JH. Brain tumors in mice are susceptible to blockade of epidermal growth factor receptor (EGFR) with the oral, specific, EGFR-tyrosine kinase inhibitor ZD1839 (Iressa) Clin Cancer Res 2002; 8:3496–3502.

113. Lokker NA, Sullivan CM, Hollenbach SJ, Israel MA, Giese NA. Platelet-derived growth factor (PDGF) autocrine signal-

ing regulates survival and mitogenic pathways in glioblastoma cells: evidence that the novel PDGF-C and PDGF-D ligands may play a role in the development of brain tumors Cancer Res 2002; 62:3729–3735.

114. Uhrbom L, Dai C, Celestino JC, Rosenblum MK, Fuller GN, Holland EC. Ink4a-Arf loss cooperates with KRas activation in astrocytes and neural progenitors to generate glioblastomas of various morphologies depending on activated Akt Cancer Res 2002; 62:5551–5558.

115. Chakravarti A, Chakladar A, Delaney MA, Latham DE, Loeffler JS. The epidermal growth factor receptor pathway mediates resistance to sequential administration of radiation and chemotherapy in primary human glioblastoma cells in a RAS-dependent manner Cancer Res 2002; 62:4307–4315.

116. Narita Y, Nagane M, Mishima K, Huang HJ, Furnari FB, Cavenee WK. Mutant epidermal growth factor receptor signaling down-regulates p27 through activation of the phosphatidylinositol 3-kinase/Akt pathway in glioblastomas Cancer Res 2002; 62:6764–6769.

117. Li B, Chang C-M, Yuan MM, McKenna WG, Shu H-KG. Resistance to small molecule inhibitors of epidermal growth factor receptor in malignant gliomas Cancer Res 2003; 63: 7443–7450.

118. Zhou Y-H, Tan F, Hess KR, Yung WKA. The expression of PAX6, PTEN, vascular endothelial growth factor, and epidermal growth factor receptor in gliomas: relationship to tumor grade and survival Cancer Res 2003; 9:3369–3375.

119. Abe T, Terada K, Wakimoto H, Inoue R, Tyminski E, Bookstein R, Basilion JP, Chiocca EA. PTEN decreases in vivo vascularization of experimental gliomas in spite of proangiogenic stimuli Cancer Res 2003; 63:2300–2305.

120. Brat DJ, Mapstone TB. Malignant glioma physiology: cellular response to hypoxia and its role in tumor progression Ann Intern Med 2003; 138:659–668.

121. Park MJ, Kim MS, Park IC, Kang HS, Yoo H, Park SH, Rhee CH, Hong SI, Lee SH. PTEN suppresses hyaluronic acid-induced matrix metalloproteinase-9 expression in U87MG gli-

oblastoma cells through focal adhesion kinase dephosphorylation Cancer Res 2002; 62:6318–6322.

122. Pore N, Liu S, Haas-Kogan DA, O'Rourke DM, Maity A. PTEN mutation and epidermal growth factor receptor activation regulate vascular endothelial growth factor (VEGF) mRNA expression in human glioblastoma cells by transactivating the proximal VEGF promoter Cancer Res 2003; 63:236–241.

123. Mayo LD, Donner DB. The PTEN, Mdm2, p53 tumor suppressor-oncoprotein network Trends Biochem Sci 2002; 27: 462–467.

124. Zhou M, Gu L, Findley HW, Jiang R, Woods WG. PTEN reverses MDM2-mediated chemotherapy resistance by intercalating with p53 in acute lymphoblastic leukemia cells Cancer Res 2003; 63:6357–6362.

125. Rapisarda A, Uranchimeg B, Scudiero DA, Selby M, Sausville EA, Shoemaker RH, Melillo G. Identification of small molecule inhibitors of hypoxia-inducible factor 1 transcriptional activation pathway Cancer Res 2002; 62:4316–4324.

126. Peters KB, Brown JM. Tirapazamine: a hypoxia-activated topoisomerase II poison Cancer Res 2002; 62:5248–5253.

127. Rainov NG, Dobberstein KU, Fittkau M, Bahn H, Holzhausen HJ, Gantchev L, Burkert W. Absence of p53 autoantibodies in sera from glioma patients Clin Cancer Res 1995; 1:775–781.

128. Wang H, Wang H, Shen W, Huang H, Hu L, Ramdas L, Zhou YH, Liao WS, Fuller GN, Zhang W. Insulin-like growth factor binding protein 2 enhances glioblastoma invasion by activating invasion-enhancing genes Cancer Res 2003; 63: 4315–4321.

129. Brockman MA, Papadimitrou A, Brandt M, Fillbrandt R, Westphal M, Lamszus K. Inhibition of intracerebral glioblastoma growth by local treatment with the scatter factor/hepatocyte growth factor-antagonist NK4 Clin Cancer Res 2003; 9: 4578–4585.

130. Yacoub A, Mitchell C, Lister A, Lebedeva IV, Sarkar D, Su Z-Z, Sigmon C, McKinstry R, Ramakrishnan V, Qiao L, Broaddus WC, Gopalkrishnan RV, Grant S, Fisher PB, Dent P. Melanoma differentiation associated 7 (interleukin 24) inhibits

geowth and enhances radiosensitivity of glioma cells *in vitro* and *in vivo* Clin Cancer Res 2003; 9:3272–3281.

131. Rahaman SO, Harbor PC, Chernova O, Barnett GH, Vogelbaum MA, Haque SJ. Inhibition of constitutively active Stat3 suppresses proliferation and induces apoptosis in glioblastoma multifoirme cells Oncogene 2003; 21:8404–8413.

132. Morimoto T, Aoyagi M, Tamaki M, Yoshino Y, Hori H, Duan L, Yano T, Shibata M, Ohno K, Hirakawa K, Yamaguchi N. Increased levels of tissue endostatin in human malignant gliomas Clin Cancer Res 2002; 8:2933–2938.

133. Oga M, Takenaga K, Sato Y, Nakajima H, Koshikawa N, Osato K, Sakiyama S. Inhibition of metastatic brain tumor growth by intramuscular administration of the endostatin gene Int J Oncol 2003; 23:73–79.

134. Sonoda Y, Kanamori M, Deen DF, Cheng SY, Berger MS, Pieper RO. Overexpression of vascular endothelial growth factor isoforms drives oxygenation and growth but not progression to glioblastoma multiforme in a human model of gliomagenesis Cancer Res 2003; 63:1962–1968.

135. Nabors LB, Suswam E, Huang Y, Yang X, Johnson MJ, King PH. Tumor necrosis factor α induces angiogenic factor up-regulation in malignant glioma cells: a role for RNA stabilization and HuR Cancer Res 2003; 63:4181–4187.

136. Hu B, Guo P, Fang Q, Tao H-Q, Wang D, Nagane M, Huang H-JS, Gunji Y, Nishikawa R, Alitalo K, Cavenee WK, Cheng S-Y. Angiopoietin-2 induces human glioma invasion through the activation of matrix metalloproteinase-2 Proc Natl Acad Sci U S A 2003; 100:8904–8909.

137. Gonzalez-Gomez P, Bello MJ, Arjona D, Alonso ME, Lomas J, De Campos JM, Vaquero J, Isla A, Gutierrez M, Rey JA. Retinoblastoma-related gene RB2/p130 exons 19–22 are rarely mutated in glioblastomas Oncol Rep 2002; 9:951–954.

138. Bäcklund LM, Nilsson BR, Goike HM, Schmidt EE, Liu L, Ichimura K, Collins VP. Short postoperative survival for glioblastoma patients with a dysfunctional Rb1 pathway in combination with no wild-type PTEN Clin Cancer Res 2003; 9: 4151–4158.

139. Rao JS. Molecular mechanisms of glioma invasiveness: the role of proteases Nat Rev Cancer 2003; 3:489–501.

140. Godard S, Getz G, Delorenzi M, Farmer P, Kobayashi H, Desbaillets I, Nozaki M, Diserens A-C, Hamou M-F, Dietrich P-Y, Regli L, Janzer RC, Bucher P, Stupp R, de Tribolet N, Domany E, Hegl ME. Classification of human astrocytic gliomas on the basis of gene expression: a correlated group of genes with angiogenic activity emerges as a strong predictor of subtypes Cancer Res 2003; 63:6613–6625.

141. Falchetti ML, Pierconti F, Casalbore P, Maggiano N, Levi A, Larocca LM, Pallini R. Glioblastoma induces vascular endothelial cells to express telomerase in vitro Cancer Res 2003; 63:3750–3754.

142. Lamszus R, Ulbricht U, Matschke J, Brockman MA, Fillbrandt R, Westphal M. Levels of soluble vascular endothelial growth factor (VEGF) receptor 1 in astrocytic tumors and its realtion to malignancy, vascularity and VEGF-A Clin Cancer Res 2003; 9:1399–1405.

143. Takino T, Nakada M, Miyamori H, Yamashita J, Yamada KM, Sato H. CrkI adapter protein modulates cell migration and invasion in glioblastoma Cancer Res 2003; 63:2335–2337.

144. Neckameyer WSW, Wang LH. Molecular cloning and characterization of avian sarcoma virus UR2 and comparison of its transforming sequence with those of other avian sarcoma viruses J Virol 1984; 50:914–921.

145. Charest A, Kheifets V, Park J, Lane K, McMahon K, Nutt CL, Housman D. Oncogenic targeting of an activated tyrosine kinase to the Golgi apparatus in a glioblastoma Proc Natl Acad Sci U S A 2003; 100:916–921.

146. Lakka SS, Gondi CS, Yanamandra N, Dinh DH, Olivero WC, Gujrati M, Rao JS. Synergistic down-regulation of urokinase plasminogen activator receptor and matrix metalloproteinase-9 in SNB19 glioblastoma cells efficiently inhibits glioma cell invasion, angiogenesis, and tumor growth Cancer Res 2003; 63:2454–-2461.

147. Choe G, Park JK, Jouben-Steele L, Kremen TJ, Liau LM, Vinters HV, Cloughesy TF, Mischel PS. Active matrix metallopro-

teinase 9 expression is associated with primary glioblastoma subtype Clin Cancer Res 2002; 8:2894–2901.

148. Shingu T, Yamada K, Hara N, Moritake K, Osago H, Terashima M, Uemura T, Yamasaki T, Tsuchiya M. Synergistic augmentation of antimicrotubule agent-induced cytotoxicity by a phosphoinositide 3-kinase inhibitor in human malignant glioma cells Cancer Res 2003; 63:4044–4047.

149. Delmas C, Heliez C, Cohen-Jonathan E, End D, Bonnet J, Favre G, Toulas C. Farnesyltransferase inhibitor, R115777, reverses the resistance of human glioma cell lines to ionizing radiation Int J Cancer 2002; 100:43–48.

150. Feldkamp MM, Lau N, Roncari L, Guha A. Isotype-specific Ras.GTP-levels predict the efficacy of farnesyl transferase inhibitors against human astrocytomas regardless of Ras mutational status Cancer Res 2001; 61:44250–4431.

151. Ding H, Shannon P, Lau N, Wu X, Roncari L, Baldwin RL, Takebayashi H, Nagy A, Gutmann DH, Guha A. Oligodendrogliomas result from the expression of an activated mutant epidermal growth factor receptor in a RAS transgenic mouse astrocytoma model Cancer Res 2003; 63:1106–1113.

152. Konduri SD, Yanamandra N, Dinh DH, Olivero WC, Gujrati M, Foster DC, Kisiel W, Rao JS. Physiological and chemical inducers of tissue factor pathway inhibitor-2 in human glioma cells Int J Oncol 2003; 22:1277–1283.

153. Marushige Y, Marushige K, Koestner A. Growth inhibition of anaplastic glioma cells by nerve growth factor Anticancer Res 1992; 12:2069–2073.

154. Garaci E, Aquaro S, Lapenta C, Amendola A, Spada M, Covaceuszach S, Perno C-F, Belardelli F. Anti-nerve growth factor Ab abrogates macrophage-mediated HIV-1 infectuion and depletion of CD4 $\pm$ T lymphocytes in hu-SCID mice Proc Natl Acad Sci USA 2003; 100:8927–8932.

155. Santoro M, Melillo RM, Carlomagno F, Visconti R, De Vita G, Salvatore G, Fusco A, Vecchio G. Different mutations of the RET gene cause different human tumoral disesases Biochimie 1999; 81:397–402.

156. Phay JE, Moley JF, Lairmore TC. Multiple endocrine neoplasias Semin Surg Oncol 2000; 18:324–332.

157. Kimura T, Takeshima H, Nomiyama N, Nishi T, Kino T, Kochi M, Kuratsu JI, Ushio Y. Expression of lymphocyte-specific chemokines in human malignant glioma: Essential role of LARC in cellular immunity of malignant glioma Int J Oncol 2002; 21:707–715.

158. Barbero S, Bonavia R, Bajetto A, Porcile C, Pirani P, Ravetti JL, Zona GL, Spaziante R, Florio T, Schettini G. Stromal cell-derived factor 1α stimulates human glioblastoma cell growth through the activation of both extracellular signal-regulated kinases 1/2 and Akt Cancer Res 2003; 63:1969–1974.

159. Bertolini F, Dell'Agnola C, Mancuso P, Rabascio C, Burlini A, Monestiroli S, Gobbi A, Pruneri G, Martinelli G. CXCR4 neutralization, a novel therapeutic approach for non-Hodgkin's lymphoma Cancer Res 2002; 62:3106–3112.

160. Fernandez M, Eng C. The expanding role of PTEN in neoplasia: a molecule for all seasons? Commentary re: M. A. Davies, et al., Adenoviral-mediated expression of MMAC/PTEN inhibits proliferation and metastasis of human prostate cancer cells Clin Cancer Res 2002; 8:1695–1698, 1904–1914.

161. Bello L, Lucini V, Carrabba G, Giussani C, Machluf M, Pluderi M, Nikas D, Zhang J, Tomei G, Villani RM, Carroll RS, Bikfalvi A, Black PM. Simultaneous inhibition of glioma angiogenesis, cell proliferation, and invasion by a naturally occurring fragment of human metalloproteinase-2 Cancer Res 2001; 61:8730–8736.

162. Nie J, Pei D. Direct activation of pro-matrix metalloproteinase-2 by leukolysin/ membrane-type 6 matrix metalloproteinase/matrix metalloproteinae 25 at the Asn^{109}-tyr bond Cancer Res 2003; 63:6758–6762.

163. Walker C, du Plessis DG, Joyce KA, Machell Y, Thomson-Hehir J, Al Haddad SA, Broome JC, Warnke PC. Phenotype *versus* genotype in gliomas displaying inter- or intratumoral histological heterogeneity Clin Cancer Res 2003; 9: 4841–4851.

164. Nutt CL, Mani DR, Betensky RA, Tamayo P, Cairncross JG, Ladd C, Pohl U, Hartmann C, McLaughlin ME, Batchelor TT, Black PM, von Deimling A, Pomeroy SL, Golub TR, Louis DN. Gene expression-based classification of malignant gliomas

correlates better with survival than histological classification Cancer Res 2003; 63:1602–1607.

165. Burton EC, Lamborn KR, Feuerstein BG, Prados M, Scott J, Forsyth P, Passe S, Jenkins RB, Aldape KD. Genetic aberrations defined by comparative genomic hybridization distinguish long-term from typical survivors of glioblastoma Cancer Res 2002; 62:6205–6210.

166. Kyritsis AP, Saya H. Epidemiology, cytogenetics, and molecular biology of brain tumors Curr Opin Oncol 1993; 5:474–480.

167. Yoon JT, Palazzo AF, Xiao D, Delohery TM, Warburton PE, Bruce JN, Thompson WJ, Sperl G, Whitehead C, Fetter J, Pamukcu R, Gundersen GG, Weinstein IB. CP248, a derivative of exisulind, causes growth inhibition, mitotic arrest, and abnormalities in microtubule polymerization in glioma cells Mol Cancer Ther 2002; 1:393–404.

168. Lorimer IA, Parolin DA, Lavictoire SJ. Induction of apoptosis in glioblastoma cells by an atypical protein kinase C pseudosubstrate peptide Anticancer Res 2002; 22:623–631.

169. Debinski W, Obiri NI, Powers SK, Pastan I, Puri RK. Human glioma cells overexpress receptors for interleukin 13 and are extremely sensitive to a novel chimeric protein composed of interleukin 13 and pseudomonas exotoxin Clin Cancer Res 1995; 1:1253–1258.

170. Mintz A, Gibo DM, Madhankumar AB, Debinski W. Molecular targeting with recombinant cytototoxins of interleukin-13 receptor alpha2-expressing glioma J Neurooncol 2002; 64: 117–123.

171. Husain SR, Puri RK. Interleukin-13 receptor-directed cytotoxin for malignant glioma therapy: from bench to bedside J Neurooncol 2003; 65:37–48.

172. Liu TF, Cohen KA, Willingham MC, Tatter SB, Puri RK, Frankel AE. Combination fusion protein therapy of refractory brain tumos J Neurooncol 2003; 65:77–85.

173. Weber F, Asher A, Bucholz R, Berger M, Prados M, Chang S, Bruce J, Hall W, Rainov NG, Westphal M, Warnick RE, Rand RW, Floeth F, Rommel F, Pan H, Hingorani VN, Puri RK. Safety, tolerability, and tumor response of IL4-Pseudomonas

exotoxin (NBI-3001) in patients with recurrent malignant glioma. 2003; 64:125–137.

174. Rand RW, Kreitman RJ, Patronas N, Varricchio F, Pastan I, Puri RK. Intratumoral administration of recombinant circularly permuted interleukin-4-Pseudomonas exotoxin in patients with high-grade glioma Clin Cancer Res 2000; 6: 2157–2165.

175. Puri RK, Hoon DS, Leland P, Snoy P, Rand RW, Pastan I, Kreitman RJ. Preclinical development of a recombinant toxin containing circularly permuted interleukin 4 and truncated Pseudomonas exotoxin for therapy of malignant astrocytoma Cancer Res 1996; 56:56310–5637.

176. Okano F, Storkus WJ, Chambers WH, Pollack IF, Okada H. Identification of a novel HLA-A*0201-restricted, cytotoxic T lymphocyte epitope in a human glioma-associated antigen, interleukin 13 receptor $\alpha 2$ chain Clin Cancer Res 2002; 8: 2851–2855.

177. Liu TF, Cohen KA, Ramage JG, Willingham MC, Thorburn AM, Frankel AE. A diphtheria toxin-epidermal growth factor fusion protein is cytotoxic to human glioblastoma multiforme cells Cancer Res 2003; 63:1834–1837.

178. Liu TF, Willingham MC, Tatter SB, Cohen KA, Lowe AC, Thorburn A, Frankel AE. Diphtheria toxin-epidermal growth factor fusion protein and Pseudomonas exotoxin-interleukin-13 fusion protein exert synergistic toxicity against human glioblastoma multiforme cells Bioconjug Chem 2003; 14: 1107–1114.

179. Ehtesham M, Kabos P, Kabosova A, Neuman T, Black KL, Yu JS. The use of interleukin 12-secreting neural stem cells for the treatment of intracranial glioma Cancer Res 2002; 62: 5657–5663.

180. Ehtesham M, Kabos P, Gutierrez MA, Chung NH, Griffith TS, Black KL, Yu JS. Induction of glioblastoma apoptosis using neural stem cell-mediated delivery of tumor necrosis factor-related apoptosis-inducing ligand Cancer Res 2002; 62: 7170–7174.

181. Sanson M, Marcaud V, Robin E, Valéry C, Sturtz F, Zalc B. Connexin 43-mediated bystander effect in two rat glioma cell models Cancer Gene Ther 2002; 9:149–155.

182. Huang R, Lin Y, Wang CC, Gano J, Lin B, Shi Q, Boynton A, Burke J, Huang RP. Connexin 43 suppresses human glioblastoma cell growth by down-regulation of monocyte chemotactic protein 1, as discovered using protein array technology Cancer Res 2002; 62:2806–2812.

183. Nylandsted J, Wick W, Hirt UA, Brand K, Rohde M, Leist M, Weller M, Jaattela M. Eradication of glioblastoma, and breast and colon carcinoma xenografts by Hsp70 depletion Cancer Res 2002; 62:7139–7142.

184. Jaboin J, Kim CJ, Kaplan DR, Thiele CJ. Brain-derived neurotrophic factor activation of TrkB protects neuroblastoma cells from chemotherapy-induced apoptosis via phosphatidylinositol 3′-kinase pathway Cancer Res 2002; 62:6756–6763.

185. Bello L, Giussani C, Carrabba G, Pluderi M, Lucini V, Pannacci M, Caronzolo D, Tomei G, Villani R, Scaglione F, Carroll RS, Bikfalvi A. Suppression of malignant glioma recurrence in a newly developed animal model by endogenous inhibitors Clin Cancer Res 2002; 8:3539–3548.

186. Breier G, Heidenreich R, Gaumann A, Groot M, Licht A, Nicolaus A, Schmitz J, Reichman E, Plate KH, Vajkoczy P. Regulators of angiogenesis as targets for anti-angiogenic tumor therapy. Abstract. Max-Planck Institut für physiologische und klinische Forschung, Bad Nauheim, Germany.

187. Lumniczky K, Desaknai S, Mangel L, Szende B, Hamada H, Hidvegi EJ, Safrany G. Local tumor irradiation augments the antitumor effect of cytokine-producing autologous cancer cell vaccines in a murine glioma model Cancer Gene Ther 2002; 9:44–52.

188. Desaknai S, Lumniczky K, Esik O, Hamada H, Safrany G. Local tumour irradiation enhances the anti-tumour effect of a double-suicide gene therapy system in a murine glioma model J Gene Med 2003; 5:377–385.

189. Nanda D, Vogels R, Havenga M, Avezant CJ, Bout A, Smitt SS. Treatment of malignant gliomas with a replicating adenoviral vector expressing herpes simplex virus-thymidine kinase Cancer Res 2001; 61:8743–8750.

190. Yamamoto S, Yoshida Y, Aoyagi M, Ohno K, Hirakawa K, Hamada H. Reduced transduction efficiency of adenoviral vec-

tors expressing human p53 gene by repeated transduction into glioma cells Clin Cancer Res 2002; 8:913–921.

191. Wang WJ, Tai CK, Kasahara N, Chen TC. Highly efficient and tumor-restricted gene transfer to malignant gliomas by replication-competent retroviral vectors Hum Gene Ther 2003; 14:117–127.

192. Solly SK, Trajcevski S, Frisén C, Holzer GW, Nelson E, Clerc B, Abordo-Adesida E, Castro M, Lowenstein P, Klatzmann D. Replicative retroviral vectors for cancer gene therapy Cancer Gene Ther 2003; 10:30–39.

193. Adachi Y, Chandrasekar N, Kin Y, Lakka SS, Mohanam S, Yanamandra N, Mohan PM, Fuller GN, Fang B, Fueyo J, Dinh DH, Olivero WC, Tamiya T, Ohmoto T, Kyritsis AP, Rao JS. Suppression of glioma invasion and growth by adenovirus-mediated delivery of a bicistronic construct containing antisense uPAR and sense p16 gene sequences Oncogene 2002; 21: 87–95.

194. Khazenzon NM, Ljubimov AV, Lakhter AJ, Fujita M, Fujiwara H, Sekiguchi K, Sorokin LM, Petajaniemi N, Virtanen I, Black KL, Ljubinova JY. Antisense inhibition of laminin-8 expression reduces invasion of human gliomas in vitro Mol Cancer Ther 2003; 2:985–994.

195. Ren H, Boulikas T, Soling A, Warnke PC, Rainov NG. Immunogene therapy of recurrent glioblastoma multiforme witha liposomally encapsulated replication-competent Semliki forest virus vector carrying the human interleukin-12 gene: phase I/II clinical protocol J Neurooncol 2003; 64:147–154.

196. Andrews DW, Resnicoff M, Flanders AE, Kenyon L, Curtis M, Merli G, Basergo R, Iliakis G, Aiken RD. Results of a pilot study involving the use of an antisense oligodeoxynucleotide directed against the insulin-like growth factor type I receptor in malignant astrocytomas J Clin Oncol 2001; 19:2189–2200.

197. Voges J, Reszka R, Gossman A, Dittmar C, Richter R, Garlip G, Kracht L, Coenen HH, Sturm V, Wienhard K, Heiss WD, Jacobs AH. Imaging-guided convection-enhanced delivery and gene therapy of glioblastoma Ann Neurol 2003; 54:479–487.

198. Chiocca EA, Aghi M, Fulci G. Viral therapy for glioblastoma Cancer J 2003; 9:167–178.

199. Kuan C-T, Wikstrand CJ, Bigner DD. EGF mutant receptor vVIII as a molecular target in cancer therapy. Abstract. Duke University Medical Center, Durham NC 27710.

200. Heimberger AB, Crotty LE, Archer GE, Hess KR, Wikstrand CJ, Friedman AH, Friedman HS, Bigner DD, Sampson JH. Epidermal growth factor receptor VIII peptide vaccination is efficacious against established intracerebral tumors Clin Cancer Res 2003; 9:4247–4254.

201. Charest A, Lane K, McMahon K, Park J, Preisinger E, Conroy H, Housman D. Fusion of FIG to the receptor tyrosine kinase ROS in a glioblastoma with an interstitial del [6] (q21q21) Genes Chromosomes Cancer 2003; 37:58–71.

202. Yamanaka R, Yajima N, Abe T, Tsuchiya N, Homma J, Narita M, Takahashi M, Tanaka R. Dendritic cell-based glioma immunotherapy (Review) Int J Oncol 2003; 23:5–15.

203. Insug O, Ku G, Ertl HC, Blaszczyk-Thurin M. A dendritic cell vaccine induces protective immunity to intracranial growth of glioma Anticancer Res 2002; 22:613–621.

204. Nakahara N, Pollack IF, Storkus WJ, Wakabayashi T, Yoshida J, Okada H. Effective induction of antiglioma cytotoxic T cells by coadministration of interferon-β gene vector and dendritic cells Cancer Gene Ther 2003; 10:549–558.

205. Glaser T, Wagenknecht B, Groscurth P, Kramer PH, Weller M. Death ligand/receptor-independent caspase activation mediates drug-induced cytotoxic cell death in human malignant glioma cell Oncogene 1999; 18:5044–5053.

206. Stendel R, Scheurer L, Stoltenburg-Didinger G, Brock M, Mohler H. Enhancement of Fas-ligand-mediated programmed cell death by taurolidine Anticancer Res 2003; 23:2309–2314.

207. Wintterle S, Schreiner B, Mitsdoerffer M, Schneider D, Chen L, Meyermann R, Weller M, Wiendt H. Expression of the B7-related molecule B7H1 by glioma cells: a potential mechanism of immune paralysis Cancer Res 2003; 63:7462–7467.

208. Naganuma H, Sasaki A, Satoh E, Nagasaka M, Isoe S, Nakano S, Nukui H. Long-term survival in a young patient with anaplastic glioma Brain Tumor Pathol 1997; 14:71–74.

209. Quattrocchi KB, Miller CH, Cush S, Bernard SA, Dull ST, Smith M, Gudeman S, Varia MA. Pilot study of local autolo-

gous tumor infiltrating lymphocytes for the treatment of recurrent malignant gliomas J Neurooncol 1999; 45:141–157.

210. Sankhla SK, Nadkarni JS, Bhagwati SN. Adoptive immunotherapy using lymphokine-activated killer (LAK) cells and interleukin-2 for recurrent malignant primary brain tumors J Neurooncol 1996; 27:133–140.

211. Hayes RL, Koslow M, Hiesiger EM, Hymes KB, Hochster HS, Moore EJ, Pierz DM, Chen DK, Budzilovich GN, Ransohoff J. Improved long term survival after intracavitary interleukin-2 and lymphokine-activated killer cells for adults with recurrent malignant glioma Cancer 1995; 76:840–852.

212. Nakagawa K, Kamezaki T, Shibata Y, Tsunoda T, Meguro K, Nose T. Effect of lymphokine-activated killer cells with or without radiation therapy against malignant brain tumors Neurol Med Chir (Tokyo) 1995; 35:22–27.

213. Thomas C, Schober R, Lenard HG, Lumenta CB, Jacques DB, Wechsler W. Immunotherapy with stimulated autologous lymphocytes in a case of a juvenile anaplastic glioma Neuropediatrics 1992; 23:123–125.

214. Hayes RL. The cellular immunotherapy of primary brain tumors Rev Neurol (Paris) 1992; 148:454–466.

215. Barba D, Saris SC, Holder C, Rosenberg SA, Oldfield EH. Intratumoral LAK cell and interleukin-2 therapy of human gliomas J Neurosurg 1989; 70:175–182.

216. Kruse CA, Cepeda L, Owens B, Johnson SD, Stears J, Lillehei KO. Treatment of recorrent glioma with intracavitary alloreactive cytotoxic T lymphocytes and interleukin-2 Cancer Immunol Immunother 1997; 45:77–87.

217. Tsuboi K, Saijo K, Ishikawa E, Tsurushima H, Takanoi S, Morishita Y, Ohno T. Effects of local injection of *ex vivo* expanded autologous tumor-specific T lymphocytes in cases with recurrent malignant gliomas Clin Cancer Res 2003; 9: 3294–3302.

218. Nagane M, Oyama H, Shibui S, Nomura K. Recurrence with tumor bleeding in a patient with malignant astrocytoma during the treatment with intracranial injection of lymphokine-activated killer cells—a case report No To Shinkei 1993; 45: 547–551.

219. Papanastassiou V, Pizer BL, Chandler CL, Zananiri TF, Kemshead JT, Hopkins KI. Pharmacokinetics and dose estimates following intrathecal administration of 131I-monoclonal antibodies for the treatment of central nervous system malignancies Int J Radiat Oncol Biol Phys 1995; 31:541–552.

220. Bigner DD, Brown MT, Friedman AH, Coleman RE, Akabani G, Friedman HS, Thorstad WL, McLendon RE, Bigner SH, Zhao XG, Pegram CN, Wikstrand CJ, Herndon JE, Vick NA, Paleologos N, Cokgor I, Provenzale JM, Zalutsky MR. Iodine-131-labeled antitenascin monoclonal antibody 81C6 treatment of patients with recurrent malignant gliomas: phase I trial results J Clin Oncol 1998; 16:2202–2212.

221. Cokgor I, Akabani G, Kuan CT, Friedman HS, Friedman AH, Coleman RE, McLendon RE, Bigner SH, Zhao XG, Garcia-Turner AM, Pegram CN, Wikstrand CJ, Shafman TD, Herndon JE, Provenzale JM, Zalutsky MR, Bigner DD. Phase I trial results of iodine-131-labeled antitenascin monoclonal antibody 81C6 treatment of patients with newly diagnosed malignant gliomas J Clin Oncol 2000; 18:3862–3872.

222. Virag L, Szabo C. The therapeutic potential of poly(ADP-ribose) polymerase inhibitors Pharmacol Rev 2002; 54:375–429.

223. Tentori L, Leonetti C, Scarsella M, d'Amati G, Vergati M, Portarena I, Xu W, Kalish V, Zupi G, Zhang J, Graziani G. Systemic administration of GPI 15427, a novel poly(ADP-ribose) polymerase-1 inhibitor, increases the antitumor activity of temozolomide against intracranial melamoma, glioma, lymphoma Clin Cancer Res 2003; 9:5370–5379.

224. Fine HA, Wen PY, Maher EA, Viscosi E, Batchelor T, Lakhani N, Figg WD, Purow BW, Borkowf CB. Phase II trial of thalidomide and carmustine for patients with recurrent high-grade gliomas J Clin Oncol 2003; 21:2299–2304.

225. Schueneman AJ, Himmelfarb E, Geng L, Tan J, Donnelly E, Mendel D, McMahon G, Hallahan DE. SU11248 maintenance therapy prevents tumor regrowth after fractionated irradiation of murine tumor model Cancer Res 2003; 63:4009–4016.

226. Kurzel F, Hagel C, Zapf S, Meissner H, Westphal M, Giese A. Cyclo-oxygenase inhibitors and thromboxane inhibitors differentially regulate migration arrest, growth inhibition and

apoptosis in human glioma cells Acta Neurochir 2002; 144: 71–87.

227. Pathak S, Multani AS, Banerji P, Banerji P. Ruta 6 selectively induces cell death in brain cancer cells but proliferation in normal peripheral blood lymphocytes: a novel treatment for human brain cancer Int J Oncol 2003; 9:975–982.

228. Perry JR, Cairncross JG, Busam KJ. Glioma therapies: how to tell which work? J Clin Oncol 2003; 21:3547–3551.

DISCUSSION

This chapter omitted review of the extensive use of preventive and therapeutic tumor vaccines (however, viral oncolysate cancer vaccines are described) and lymphocyte therapy, extending from melanoma and kidney carcinoma (the National Cancer Institute protocols) to other tumor categories, receives only some brief remarks. Only a few examples of the rapidly enlarging field of monoclonal antibodies and immunotoxins are mentioned because many of these agents are not in Phase I or Phase II trials any longer, but receive now full credit and have become respected and licensed therapeutic modalities. These include rituximab (Rituxan, Genentech) for $CD20^+$ lymphomas and trastuzumab (Herceptin, Genentech) for HER2/ neu^+ breast carcinomas. $Iodine^{131}$-labeled tostitumomab (Bexxar, Aventis Pasteur) is a licensed monoclonal antibody inducing remissions in patients with chemo-radiotherapy–resistant malignant lymphomas. The latest addition is ibritumomab tiuxetan, the Y^{90}-labeled monoclonal antibody (Zevalin, Idec) directed at the CD20 antigen of B-lineage lymphoma cells. Not only cognate ligands of targeted receptors (Tac for IL-2, IL-4, IL-13) linked with cytotoxins, but antibodies directed at cell surface receptors, could also be linked to such a toxin (Pseudomonas or diphtheria exotoxin). These are the LMB immunotoxins, one of which, the anti-CD22 BL22, is awaiting approval for the treatment of hairy cell leukemia. Another, the calicheamicin-containing gemtuzumab ozogamicin (Mylotarg, Wyeth/Genetics), aimed at CD33 of AML cells, has completed all its pre-licensure clinical trials. Denileukin diftitox (Ontak, Ligand Pharmaceuticals, San Diego, CA) attacks IL-2R-overexpressing T lymphoma cells. The anti-CD22 epratuzumab and the anti-CD52 alemtuzumab (Campath-1H, Berlex Laboratories, Seattle, WA (PDR p. 932 2004)) induce remissions in cases of lymphomas and B-CLL refractory to chemotherapy. The anti–CA-125 oregovomab delays postoperative relapses in patients with ovarian carcinoma. This tumor is targeted for mumps, measles, NDV, and reoviral therapy. Is there a rationale to combine viral therapy with the antibody?

Of IMIDs, CC-5013 (Revimid, Celgene) is already in advanced clinical trials for MM and GBM. Of inhibitors of VEGF and its receptor, the indolinones SU5416 (semaxanib) and SU6668; the anilinoquinazoline ZD6474; the anti-Flt-1 ribozyme; and the cyclodepsipeptide aplidine (from *Aplidium albicans*) reached the Phase I and Phase II clinical trial stage. Although the idea to combine TKIs and monoclonal antibodies shines as a most brilliant one (Iressa and Herceptin for breast cancer; Tarceva and Avastin for NSCLC, except squamous cell carcinomas, or for GBM), the preliminary clinical results are somewhat below expectations given that CR remain rare achievements. The question arises whether naturally oncolytic or genetically engineered viruses would fit into any of these combinations.

Omitted for review are the retinoic acid derivatives that induce differentiation of malignant cells. Some of these are licensed therapeutic agents, others are widely used to suppress precancerous lesions, such as oral leukoplakia, and to prevent head and neck squamous cell and many other carcinomas. These trials are dealt with in detail in textbooks of medical oncology. Patients with acute promyelocytic/progranulocytic leukemia (APL) suffer the characteristic translocation t(15;17)(q22;q12), where the breakpoint on chromosome 17 is at the intron of the retinoic acid receptor gene α (RARα). All-transretinoic acid (ATRA) induces differentiation of these leukemia cells (but for durable CR, combination chemotherapy is required). Retinoids selective for Tac-101 RARα occasionally induce a remission in patients with metastatic cancers. Otherwise, 13-cis RA, fenretidine, vitamin D analogue ILX23-7533, and tazarotene failed to elicit any tumor response. However, bexarotene (Targretin) emerges as the retinoid-receptor agonist that induces apoptotic death of squamous carcinoma cells and prolongs the survival of patients with NSCLC who receive it in combination with chemotherapy. Perhaps another protocol (recall the one initiated by M. D. Anderson Cancer Center in Mexico for squamous cell carcinoma of the uterine cervix) combining RA derivatives with IFN-α would score; this combination is being now tested Eastern Cooperative Oncology Group (ECOG) in patients with squamous cell carcinoma of the penis. Could some of the oncolytic viruses be combined with RA derivatives in a clinical trial? If these tumors are caused by certain subtypes of HPV, viral interference and antitumor vaccination applied together may explain the susceptibilty to virally induced regression of these tumors.

This chapter was written at the time when the 2003 American Association for Cancer Research Congress in Toronto was canceled and the ASCO Conference in Chicago was to be held 1 month later. Although the Proceedings of both of these conferences (AACR vol. 44, 2003; ASCO vol. 22, 2003) are replete with abstracts of TKIs and anti-neoangiogenesis agents and other new biotherapeuticals, oncolytic viruses remain underrepresented in them; only adenoviral constructs for gene therapy receive some attention. A new biochemotherapy protocol for per os and out-patient administration emerges for the treatment of metastatic malignant melanoma (ASCO #2881, 2003): temozolomide (Temodar, Schering), GM-CSF (sargramostim, Leukine, Immunex), IFN-α2b (Intron, Schering, Seattle, WA), and IL-2 (aldesleukine, Proleukine, Chiron, Emeryville, CA). This regimen induced 16% CR and 16% PR rates and stabilization of disease in 28% of patients with "acceptable" toxicities consisting of "flu-like" symptoms. Could any oncolytic virus match this efficacy of a treatment modality? If not, either only those patients who fail (progressive disease in 40% of patients) would be acceptable for viral therapy, or viral therapy (in the form of a viral oncolysate or a DC vaccine) should be incorporated in the chemobiotherapy protocol. Bulletins from the Institute of Drug Development issued by the Cancer Therapy and Research Center in San Antonio, TX, publish serial reports on new cancer therapeuticals entering clinical trials. In the last issues of *Oncology Times* (volume XXV/18–19, September 25, October 10, and October 25, 2003) there is a much-abbreviated listing of newly approved clinical trials. It is an overwhelming list in which viral therapy of human cancers is conspicuously underrepresented.

In 1986, Marcel Dekker published the single-authored textbook "*Medical Oncology, an Advanced Course*", second edition, in two volumes comprising 2062 pages and based on some 25,000 references. In the Preface, the author expressed his hope that the immediate and long-term toxicities of chemotherapy will somehow be ameliorated and that biological means of treating cancer will gain ground and that we shall be able to communicate with our cells, even malignantly transformed cells, using the cells' own vocabulary and language. In the past few years, the biological treatment of cancer has become a reality at many levels. Tyrosine kinase inhibitors are suppressing signal transduction from oncogenes: The malignant cell follows our orders to the extent that it surrenders to its apoptotic death. We learned which cytokines or chemokines favor the host and oppose the tumor. The antigenicity of some point-mutated (Ras; Her2/neu), truncated (EGF-RvIII), or fusion (Bcr/Abl) oncoproteins opens up avenues for the use of preventive and therapeutic tumor-specific vaccinations. Autologous immune T cell clones rise against amplified or mutated survivins or telomerases (Inter J Oncol 2004; 25:211–7; POR Sept. 2004) and gefitinib works best against mutated EGF–R (Nature Med. 2004; 10:577–8). Monoclonal antibodies deriving from immune plasma cells fused with immortalized lymphoid cells (B. H. Tom and J. P. Allison, The University of Texas Medical School, Houston, TX, eds; "Hybridomas and Cellular Immortality", Plenum Press, New York and London, 1981; pp 1–302) evolved into tumor-specific immunotoxins and agents of antitumor ADCC reactions. Rituximab-coated lymphoma cells are phagocytized by dendritic cells (DC). These DC then induce immune T cell clones that kill the lymphoma cells in the patient (Th53. 319 Immunology 2004, Montreal, Canada). Lymphocytes (NK cells and immune T cells) found to be cytotoxic to tumor cells in the 1970s (In: J. G. Cory and A. Szentivanyi, The University of South Florida College of Medicine, Tampa, FL, eds.; "Cancer Biology and Therapeutics", Plenum Press, New York and London, 1986; pp 225–253) have become autologous LAK cells and TIL launching successful attacks on metastases. Adoptive immunotherapy with allogeneic NK lymphoid cells eradicates remnants of malignant cell populations at the price of tolerable graft-vs.-host disease. Means are now available to stop the subversion of the stroma by the tumor and to deprive the tumor of its nutrients and paracrine growth factors. Viral vectors deliver healthy genes into our disabled or mutated cells (with modest success rate). Genetically engineered or naturally oncolytic viruses pick and replicate in mutated and malignantly transformed cells and spare healthy cells of the tumor-bearing host. In the past 25 years, since the publication of the first monograph on cancer immunology-immunotherapy (J. E. Harris and J. G. Sinkovics, authors; The University of Texas M. D. Anderson Hospital, Houston, TX, "The Immunology of Malignant Disease" second edition, C. V. Mosby, St Louis, 1976; pp 1-606), our understanding of the cause and treatment of cancer has undergone fundamental changes. Biological therapy of cancer closes ranks with surgery, radio-, and chemotherapy. It will make the decisive contribution to the cure of cancer.

SELECTED REFERENCES

AACR-NCI-EORTC International Conference Molecular Targets and Cancer Therapeutics. Program & Proceedings Clin Cancer Research 2003; 9/16:6069s–6306s.

Bast RC, Kufe DW, Pollock RE, Weichselbaum RR, Holland JF, Frei E. Cancer Medicine. 5th edition. Hamilton. Ontario. Canada: B.C. Decker, 2000:2546.

Bronchud MH, Foote MA, Olopade O, Workman P. Principles of Molecular Oncology. Totowa. N.J.: Human Press, 2003:736.

Figg WD, McLeod HL. Handbook of Anticancer Pharmacokinetics and Pharmacodynamics. Totowa. N.J.: Humana Press, 2004:650.

Giordano A, Soprano KJ. Cell Cycle Inhibitors in Cancer Therapy. Totowa. N.J.: Humana Press, 2003:326.

Greene JN. Infections in Cancer Patients. New York. N. Y.: Marcel Dekker, 2004:500.

Harris RE. COX-2 Blockade in Cancer Prevention and Therapy. Totowa. N.J.: Humana Press, 2003:384.

Levine MM, Kaper JB, Rappuoli R, Liu M, Good MF. New Generation Vaccines. New York. N.Y.: Marcel Dekker, 2004:1000.

Lo BKC. Antibody Engineering. Totowa. N.J.: Humana Press, 2003: 576.

Machida CSA. Viral Vectors for Gene Therapy. Totowa. N.J.: Humana Press, 2002:460.

Murata H. Tumor-Suppressing Viruses, Genes and Drugs. San Diego. CA & New York N.Y.: Academic Press, 2003:325.

Pagé M. Tumor Targeting in Cancer Therapy. Totowa. N.J.: Humana Press, 2002:466.

Phillips MI. Antisense Therapeutics. Totowa. N.J.: Humana Press, 2004:300.

Rak JW. Oncogene-Directed Therapies. Totowa. N.J.: Humana Press, 2003:487.

Pitot HC. Fundamentals of Oncology. New York. N.Y.: Marcel Dekker, 2002:984.

Sinkovics JG. Medical Oncology an Advanced Course. 2nd edition. Vol. I & II. New York. N.Y.: Marcel Dekker, 1986:2062.

Syrigos KN, Harrington KJ. Targeted Therapy of Cancer. London. UK: Oxford University Press, 2003:343.

Templeton N Smyth, Ed. (Baylor College of Medicine, Houston, TX). Gene and Cell Therapy. Marcel Dekker, New York, Basel, 2003-4.

SUMMARY

Medical oncologists are close to being overwhelmed by the complexity of the selection process and by the abundance of the options when they are confronted by the patient who failed conventional therapy and is to be enrolled in Phase I or II protocols; or who, under the influence of the Internet or other sources of sensation-seeking publicity, refuses conventional therapy in favor of a biotherapeutical agent that is in preclinical or Phase I or II trials. Viral therapy remains one of the choices; when applied as monotherapy, it should match or surpass the results of numerous other agents without increased risks, or else it will have to be integrated into the framework of the new premises of biotherapeuticals and with good rationale. There is no doubt now that single-treatment modalities will seldom kill out an entire cancer cell population because the individual cancer cell switches from one growth cascade to another and so reproduces itself. However, simultaneous or dose-dense sequential neutralization of more than two or three of the tumor cell's proliferative pathways, coupled with efforts to modify the tumor's subverted microenvironment so that it ceases supporting the growth of the tumor, are expected to obliterate the entire multiplicity of tumor cell clones.

3

Measles Virus: Improving Natural Oncolytic Properties by Genetic Engineering

CHRISTOPH SPRINGFELD, ADELE FIELDING, KAH-WHYE PENG, EVA GALANIS, STEPHEN J. RUSSELL, and ROBERTO CATTANEO

Molecular Medicine Program,
Mayo Clinic Rochester
Rochester, MN

A PATHOGEN WITH ONCOLYTIC PROPERTIES

Measles is a highly contagious disease with a characteristic erythematous, nonpruritic, maculopapular rash that is preceded by an unspecific prodromal illness of malaise, cough, coryza, and fever [1,2]. Although an effective live vaccine has been available for more than 30 years, measles is still a significant cause of morbidity and mortality in countries with poor or no vaccination coverage. Thirty to forty million cases of measles, resulting in 777,000 deaths, have been estimated for the year 2000 [3]. Mortality is associated with extreme ages, underlying illness, low socioeconomic status, lack of access to medical care, and malnutrition. Fatalities can be caused by secondary infections such as bacterial pneumonia, which are facilitated by an immunosuppression typical for measles. About one in 1000 children develops postinfectious encephalitis that is fatal in about 10% of the patients and frequently leads to sequelae in the surviving patients. An extremely rare (1 in 10^5 infections) fatal complication of measles, subacute sclerosing panencephalitis, can occur several years after wild-type measles virus (MV) infection and is caused by a persisting brain infection with defective virus mutants [4].

Measles vaccination is one of the most cost effective and successful public health interventions. Measles virus was first isolated by J. F. Enders and T. C. Peebles in 1954 [5]. Subsequently, different attenuated life vaccines were developed by passaging the virus on chick cells [6]. Vaccination with the currently used vaccine strains is safe and effective in preventing infection with wild-type MV. Adverse effects more serious than a low-grade fever are very rare and consist mainly of anaphylactic reactions, probably against the gelatin that is present in the vaccine as a stabilizer [7]. Vaccination against MV is even recommended for HIV-infected children unless they are severely immunocompromised [8].

Circumstantial evidence suggests that infection with wild-type MV can lead to tumor regression. Similarly, occasional regressions of human cancer after natural infections with other viruses have been reported [9]. Several cases of regression of hematological malignancies after natural MV infection have been noted at different places around the world. S. A. Hernandez [10] describes a 10-year-old child from Cuba with histologically proven Hodgkin's disease who contracted measles before radiotherapy could be started. The child had a complete remission; however, the Hodgkin's lymphoma relapsed after 4 months. A. Z. Bluming and J. L. Ziegler describe a case of an 8-year-old boy in Uganda with Burkitt's lymphoma that completely resolved without any antineoplastic therapy after accidental MV infection [11]. Z. Zygiert reports the natural measles infection of 3 out of 98 children with Hodgkin's lymphoma in a hospital in Poland; all three children showed a striking remission of the lymphomas [12]. Two children had "one slight relapse" that could be successfully treated by conventional therapy; the third child remained in remission for at least 6 years after diagnosis. H. C. Mota adds a case from Portugal, where a 23-month-old boy with Hodgkin's lymphoma showed a complete remission after natural MV infection; this child also relapsed after 6 months and needed further treatment [13].

Remissions associated with natural MV infection have also been observed in patients with lymphoblastic leukemia. G. Pasquianucci describes two children in Italy who had long-term remissions after measles; however, these children were also treated with chemotherapy, and one of them also contracted rubella [14]. A girl in Cleveland who was no longer responsive to chemotherapy developed measles and had a complete remission. However, she also relapsed 3 weeks later and died shortly after that [15].

Although some of the authors of the case reports discussed the possibility of an experimental infection of cancer patients with MV as a means of cancer therapeutics, to our knowledge this approach has not been implemented. On the other hand, the occasional case reports have triggered in vitro studies and experiments in mouse cancer models that are described in this review. The encouraging results of these studies led to the submission of clinical trial protocols that are open or will open in the near future. Furthermore, the MV reverse genetics system that is available since 1995 allows genetic modification of the MV-vaccine strain to improve its oncolytic properties.

MOLECULAR BIOLOGY AND REVERSE GENETICS

Measles virus is an enveloped RNA virus that is a member of the genus *Morbillivirus* within the family *Paramyxoviridae* and the order *Mononegavirales* [2]. The negative-stranded, nonsegmented viral genome comprises 15,894 bases and encodes six structural and two nonstructural proteins. It is organized into six contiguous, nonoverlapping transcription units separated by three untranscribed nucleotides and codes for the six structural viral proteins in the following order (antigenome): 5′–N–P–M–F–H–L–3′ (Fig. 1A, center). The P cistron additionally encodes two nonstructural proteins: The C protein is translated from an overlapping open reading frame by ribosomal choice [16], whereas the V protein is translated from an RNA that is cotranscriptionally edited by insertion of one nucleotide [17]. The viral genome is tightly encapsidated by the nucleoprotein (N) and further associated with the viral RNA polymerase (L) and the phosphoprotein (P), which serves as a cofactor for the viral polymerase (Fig. 1B). The viral RNA and the three proteins N, P, and L form the ribonucleoprotein complex. The MV envelope contains two glycoproteins, the receptor-binding hemagglutinin (H) and the fusion protein (F). The matrix protein (M) is the assembly organizer, associated with the inner layer of the envelope, and interacts with the cytoplasmatic tails of the glycoproteins [18]. Recently, it has been demonstrated genetically that MV virions are polyploid [19], confirming earlier conclusions drawn for other Paramyxoviruses and based on microscopical and physical examination of virus particles [20]. Another important characteristic is that the levels of transcription of the genes reflect their position in the genome [21]. This fact is being exploited to produce foreign proteins at different levels by inserting their genes in different locations of the MV genome (see below).

Figure 1 (A) Schematic representation of the measles virus (MV) antigenome (center). Selected unique restriction sites are shown. The replacement of the H protein by hybrid proteins displaying specificity domains is depicted at the top, and the insertion of foreign genes in additional transcription units (ATU) at three different locations is illustrated at the bottom. The intergenic nontranscribed trinucleotide (tri) is symbolized by a black bar. (B) Diagram of a polyploid MV particle containing three genomes. The symbols for the nucleocapsid protein (N), phosphoprotein (P), matrix protein (M), fusion protein (F), hemagglutinin (H), and the RNA polymerase (L) are indicated.

A

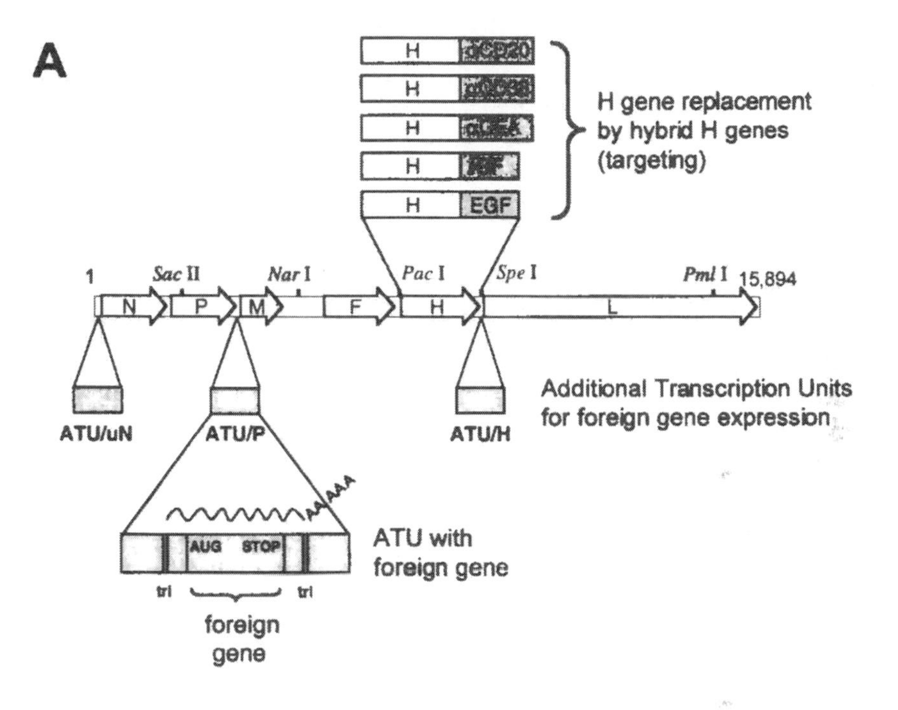
H
H
H
H
H
EGF
H gene replacement
by hybrid H genes
(targeting)
1
Sac II
Nar I
Pac I
Spe I
Pml I
15,894
N
P
M
F
H
L
ATU/uN
ATU/P
ATU/H
Additional Transcription Units
for foreign gene expression
AAAAA
AUG
STOP
ATU with
foreign gene
tri
tri
foreign
gene

B

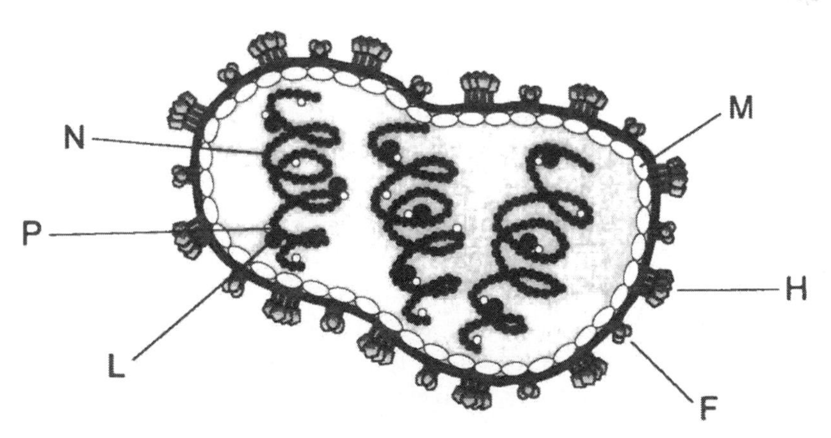
N
P
L
M
H
F

Measles virus enters the cell by binding to different receptors and subsequently fusing its envelope with the cellular membrane, a process executed by the F protein. Two cellular receptors have been identified, the ubiquitous CD46 [22,23] and the immune cell-specific signaling lymphocyte activating molecule (SLAM) [24]. Measles virus wild-type and vaccine strains differ in their receptor usage: Wild-type strains enter the cells preferentially by SLAM, whereas vaccine strains use CD46 more efficiently [25]. However, other MV receptors may exist [26].

Recovery of infectious MV from cloned cDNA was first achieved by the Billeter laboratory in 1995 [27]. The recovery system is based on a stable cell line that expresses the MV proteins N and P as well as T7-polymerase. When this cell line is transfected with a plasmid encoding MV RNA polymerase (L) and a plasmid encoding the full-length MV-antigenome, both driven by a T7-promoter, infectious virus can be recovered. An alternative system that involves modified vaccinia Ankara virus expressing T7 polymerase and three plasmids encoding the N, P, and L proteins under the control of a T7-promoter has also been reported [28]. The first infectious cDNA clone was mainly based on an attenuated laboratory strain of the Edmonston lineage. In the meantime, viruses have also been recovered from cDNA clones encoding wild-type MV [29] and the Moraten/Schwartz vaccine strain (V. von Messling and R. Cattaneo, unpublished data).

The MV reverse genetics systems have been extensively used to characterize mutant viruses with deleted genes [30–33], genes exchanged between different strains [25,34–36], single amino-acid substitutions [37,38], and additional genes (Fig. 1A, bottom). Proteins expressed by recombinant MV include the reporters green fluorescent protein [39], chloramphenicol acetyl transferase [40], beta-galactosidase [41,42], and luciferase (K.-W. Peng, unpublished); the marker peptides soluble human carcinoembryonic antigen (shCEA) and human chorionic gonadotropin (βhCG) [43]; proteins of other viruses—e.g., the hepatitis B surface antigen [44] and human immunodeficiency virus proteins [45]; and immuno-

Table 1 Proteins Expressed from Additional Transcription Units Inserted at Different Positions in the Measles Virus Genome.

Genomic position	Upstream of N	After P	After H	References
Reporters/trackers				
CAT		X		Mrkic et al., 1998
GFP	X	X	X	Duprex et al., 1999; Hangartner and Billeter, personal communication
β-gal		X		Neumeister et al., 2001
CEA	X			Peng et al., 2002
βhCG	X			Peng et al., 2002
luciferase	X			Peng, unpublished
Effectors				
human IL12			X	Singh and Billeter, 1999
mouse IL12			X	Devaux and Cattaneo, unpublished
human GM-CSF			X	Fielding, unpublished
mouse GM-CSF	X	X	X	Grote et al., 2003
HSV-TK	X			Springfeld and Cattaneo, unpublished
E. coli CD	X			Springfeld and Cattaneo, unpublished
NIS		X	X	Dingli et al., 2004

CAT = chloramphenicol acetyl transferase; GFP = green fluorescent protein; β-gal = beta-galactosidase; CEA = carcinoembryonic antigen; βhCG = human chorionic gonadotropin; IL = interleukin; GM-CSF = granulocyte-macrophage colony-stimulatory factor; HSV-TK = herpes simplex virus thymidine kinase; CD = cytosine deaminase; NIS = natrium-iodine symporter.

modulatory proteins—e.g., interleukin-12 [46] and granulocyte-macrophage colony-stimulatory factor (GM-CSF) [47] (Table 1). Until now, up to about 5 kb of foreign genetic material has been added to the MV genome without a profound negative effect on viral replication (L. Hangartner and M. A. Billeter, personal communication). Because of the pleomorphism of MV virions, there are no theoretical size limitations for the viral genome. The maximal amount of additional sequences that is tolerated by MV has not yet been determined.

VIROTHERAPY OF HEMATOLOGICAL MALIGNANCIES

Recently, the possible effect of MV on lymphoma cells suggested by the case reports mentioned above has been further investigated in vitro and in animal models [42]. After confirming that MV can infect and lyse the human lymphoma cell lines Raji and DoHH2 in vitro, MV was injected in tumors of these cell lines established in severe combined immunodeficiency (SCID) mice. In this study, the attenuated MV strain of the Edmonston (MV-Edm) lineage used caused regression of the established tumor xenografts in vivo. It was further shown that passively transferred antibodies against MV did not eliminate the antitumor effect of the intratumorally injected virus. Measles virus proteins and mRNA were detected in the tumors, and infectious virus was recovered from explanted tumors.

In the light of these encouraging results, a Phase I clinical study of intratumoral injections in patients with non-Hodgkin's lymphoma (NHL) has been approved. Patients will be injected with one or two doses of the commercially available Moraten-vaccine strain. The inclusion criteria for this study being relatively strict, patient recruitment has initially been difficult.

A significant antineoplastic activity was also observed recently when human myeloma xenografts were treated with MV [48]. The MV-Edm strain efficiently replicated in a variety of myeloma cell lines and in primary human myeloma cells from six different patients. The in vitro infection of the tumor cells resulted in a strong cythopathic effect with extensive syncytia formation that finally led to apoptotic death. When tumors were established in SCID mice using the myeloma cell line ARH-77 and injected with MV-Edm, they were completely eradicated. Even i.v. inoculation of MV-Edm led to a significant growth inhibition of RPMI 8226 tumors and to a complete regression of ARH-77 tumors in athymic mice.

TREATMENT OF OVARIAN CANCER

Another type of human cancer considered for treatment with MV is ovarian cancer. This disease is commonly diagnosed at an advanced stage [49]. The patients are initially treated with surgery and chemotherapy, but most patients ultimately die from recurrent disease. Recurrent ovarian cancer is a promising target for intraperitoneal virotherapy because, in many patients, the disease does not spread beyond the peritoneal cavity. Viruses injected intraperitoneally will not be eliminated by circulating antibodies in the serum; however, antibodies in ascites may still interfere with infection. The effect of MV on ovarian cancer in vitro and in mouse models has been studied in detail using a trackable MV genetically engineered to produce carcinoembryonic antigen (CEA; MV-CEA) that is described in the following section [50]. In this study, MV was shown to be selectively oncolytic for ovarian cancer cells with minimal effects on normal cells. Treatment with MV enhanced the survival of athymic mice with s.c. and i.p. tumors of the ovarian cancer cell line SKOV3ip.1. Tumor infection in vivo was followed by monitoring the serum CEA levels of the mice. Based on the promising results achieved, a Phase I dose-escalation trial of intraperitoneal administration of MV-CEA in advanced stage or recurrent ovarian cancer patients was planned. Only a minority of ovarian cancer patients (<10%–15%) express CEA, and these will be excluded as per the trial design. Detection of CEA in the serum therefore could serve as a good correlate of viral gene expression. This is the first clinical trial of therapeutic administration of a recombinant MV in humans. Extensive biodistribution [51] and toxicology testing has now been performed, and recombinant virus is being produced under good manufacturing practice conditions. An investigational new drug application has been approved by the Food and Drug Administration (FDA), and the clinical trial has begun in July 2004.

TRACKABLE MEASLES VIRUS

A common problem in the evaluation and development of replicating viral vectors for gene therapy is the lack of tools to monitor viral gene expression and replication in vivo. In the case of MV, this problem has recently been overcome by the development of recombinants that express soluble marker peptides [43]. These viruses possess an additional transcription unit upstream of the first gene that encodes either shCEA (MV-CEA) or βhCG. These two peptides, cloned in the MV genomic position sustaining highest expression, were selected because they are nonimmunogenic, without relevant biological function, and have a relatively short half-life. Moreover, laboratory tests to measure their serum levels are available in nearly every hospital. Human carcinoembryonic antigen is used as a tumor marker for several types of cancer, whereas βhCG is mainly used for the diagnosis of pregnancy and as a tumor marker for certain rare malignancies.

When MV encoding the marker genes was inoculated into transgenic mice susceptible to MV-infection because they express human CD46 and have a defective interferon system [40], viral replication could be monitored by measuring the serum or urine levels of the marker peptides. Furthermore, the trackable viruses elucidated the reasons for different MV antitumor effects in murine tumor models. A rapid rise and fall of serum CEA levels after inoculation of MV was typically associated with a good tumor response, whereas low or absent CEA levels indicated a failure of the virus to infect or replicate in the tumor cells. Some tumor cell types were chronically infected and produced progressively rising CEA levels, but the infection did not lead to tumor regression. Measles virus CEA also recently showed significant oncolytic effects in subcutaneous and intracranial orthotopic mouse models of human glioblastoma [52]. Trackable MV hopefully will allow monitoring of viral replication in clinical trials: It should be possible to correlate marker peptides levels with different doses and routes, with antitumor effects, and with toxicity levels. This will facilitate the development of MV-based virotherapy.

MEASLES VIRUS WITH RETARGETED CELL ENTRY

A major challenge in the development of vectors for oncolytic virotherapy is the production of viruses that are specific for tumor cells. This specificity can theoretically be conferred to different stages in the viral life cycle: replication, transcription, or cell entry. Certain viruses replicate preferentially in tumor cells naturally, others have been genetically engineered to do so [53,54]. A second way to achieve specificity for a given tissue is the use of tissue-specific promoters active only in the targeted cells [55]. The third approach is in principle the most straightforward: to target viral entry, the first step of the viral cycle.

Measles virus entry is mediated by two envelope glycoproteins—the attachment protein H and the F protein that executes fusion. Because the attachment and the fusion function are distributed on two entities, retargeting strategies are based on H protein modification only (Fig. 1A, top). Indeed, we showed that the addition of specificity domains to the viral H protein led not only to binding to cells that express the appropriate ligand but also to subsequent entry [56]. The first specificity domains that were shown to promote attachment and entry of MV were epidermal growth factor (EGF) and insulinlike growth factor 1 (IGF-1) [57]. These domains were connected to the extracellular terminus of the viral H via a flexible linker. The hybrid proteins (Fig. 1, top) were stable and were efficiently incorporated in the viral envelope. Measles virus displaying EGF or IGF-1 on their surface efficiently entered rodent cells that expressed the human EGF or IGF-1 receptors and that are normally not permissive for MV infection.

Single-chain antibodies (scFv) with any desired specificity can be selected, and therefore the next step was to verify whether a single-chain antibody displayed on H was compatible with virus assembly, and elicited targeted entry [58]. The first scFv displayed on MV H was directed against hCEA, a glycoprotein expressed during embryogenesis and in certain tumors of epithelian origin, including colon carcinoma, pancreatic carcinoma, and some types of lung cancer [59]. Indeed, the anti-hCEA scFv displayed on MV H mediated entry in rodent cells that express human CEA. Subsequently, recombinant viruses with H proteins displaying other scFv in place of standard H were produced. One of these was directed against CD38, a myeloma marker [60], and one against CD20, a common lymphoma antigen [61]. These viruses can also enter cells via the corresponding antigen. These results suggest that MV can probably be retargeted to many of the cell-surface antigens against which scFv are available.

The viruses described above are still able to infect cells via the two known natural MV receptors, namely CD46 and SLAM. However, several residues in the H protein that are necessary for efficient fusion mediated by one or the other receptor have been identified [62,63]. Moreover, when the genes for these proteins were exchanged for the standard gene in the MV-infectious cDNA, viruses were recovered that enter the cells selectively via SLAM or CD46. Currently, mutations restricting entry through the natural receptors are being combined with specificity domains to obtain fully retargeted viruses that are no longer able to enter cells via their natural receptors.

MEASLES VIRUSES WITH ENHANCED CYTOTOXICITY

In the last years, several systems to improve the cytolytic effects of viral vectors in gene therapy of cancer have been developed [64]. The most common approach is the introduction of "suicide genes" into the infected cell [65]. The proteins encoded by these genes transform a normally nontoxic prodrug into a toxic metabolite and thus kill the cell that expresses it. The most widely used suicide gene/prodrug combinations are herpes simplex virus thymidine kinase (HSV-TK)/ganciclovir and *Escherichia coli* cytosine deaminase/5-fluorocytosine. Both prodrugs are already used in the clinic to treat human cytomegalovirus or fungal infections, respectively. An important advantage of suicide gene therapy is that the toxic metabolite may not only kill the cell in which the suicide gene is expressed, but also the neighboring cells. This phenomenon is called *bystander effect*. Recently, HSV-TK has been successfully introduced into the genomes of simian virus 5, another member of the *Paramyxoviridae* [66], and vesicular stomatitis virus, a rhabdovirus [67]. Recombinant MV expressing different suicide genes have also been rescued, and the therapeutic effect of these genes in mouse cancer models is currently under investigation (Springfeld and Cattaneo, unpublished data).

An alternative to the "classical" suicide genes mentioned above is the sodium-iodine symporter (NIS) gene, a gene that is normally expressed in the thyroid [68]. The gene product, a large transmembrane glycoprotein in the basolateral membrane of the thyroid follicular cells, is responsible for the transport of iodine into the thyroid, where it is subsequently incorporated in the thyroid hormones. The concentration of iodine in the thyroid tissue by NIS is also the basis for diagnostic thyroid scintigraphy and, more importantly, the treatment of malignant and benign thyroid disorders with radioactive iodine. The treatment option with iodine-131 results in remarkably good prognosis of patients even with metastasized follicular and papillary thyroid cancer [69]. Different groups introduced the NIS gene into several tumor cell lines and showed that these cell lines could be sensitized to the treatment with iodine-131 [70–72]. A recombinant MV expressing NIS (MV-NIS) has been rescued, and intratumoral spread of this virus in myeloma xenografts could be demonstrated by serial gamma-camera imaging of iodine-123 uptake. Furthermore, MM1 myeloma xenografts resistant to MV-NIS treatment alone completely regressed after administration of iodine-131 nine days after a single intravenous injection of MV-NIS [73].

PERSPECTIVES

Cancer is the second-leading cause of death in industrialized countries, and the available therapies–mainly surgery, radiotherapy, and chemotherapy–can cure only a minority of patients. Treatment with replicating oncolytic viruses is a promising new approach to add to the armamentarium of cancer therapies [74]. Among the different viruses that are being developed for oncolytic therapy, MV has several distinctive properties. First, the vaccine strain has an excellent safety record and has been administered without major complications to millions of people. Second, because MV is a human virus with known pathogenicity, there is no risk of transforming an animal virus into a human pathogen, a possible concern with animal viruses. Third, MV has an inherent oncolytic potential, as suggested by several case reports of regression of human tumors after natural MV infections and recently confirmed in animal models. Fourth, the well-established MV reverse genetics system, in combination with the minimal constraints on MV particle structure, have favored the development of several avenues of experimentation, leading to the enhancement of the natural oncolytic properties. Experimental approaches include the introduction of suicide genes to enhance virus cytotoxicity, or of immunomodulatory genes to enhance the response against tumor cells. Also fascinating is the unique possibility to retarget MV by scFv to specific tumor antigens. Up to now, scFv have been shown to efficiently mediate viral entry through a targeted receptor only when displayed on MV; similar experiments with adenoviruses have been confronted with the constraints on viral assembly imposed by the icosahedral symmetry of the capsid. In retroviruses, the concentration of attachment and fusion function in a single protein complicates retargeting.

A major obstacle to the usage of MV and many other viruses for virotherapy of cancer is the presence of neutralizing antibodies in the majority of patients. However, there is evidence that a subclinical measles infection is possible even in seropositive individuals [75–77]. The oncolytic effects of reovirus [78], adenoviruses [79], and herpes viruses [80] were not eliminated by pre-existing antibodies when the viruses were inoculated intratumorally. Several approaches to diminish the patient immune response against viral vectors have been tested recently. A study with replicating reovirus showed that the oncolytic effect of reovirus therapy in a murine model was abolished in the presence of neutralizing antibodies. However, it was restored when the animals where treated with cyclosporin or anti-CD4/CD8 antibodies [81].

It is also possible to exchange the MV envelope proteins with the homologous proteins of related animal viruses, an approach already taken with canine distemper virus, another Morbillivirus (von Messling and Cattaneo, unpublished). However, CDV induces crossreactive antibodies to MV. Tupaia paramyxovirus is a nonpathogenic virus that is not neutralized by antibodies against the common human pathogens in the family *Paramyxoviridae* [82]. Preliminary studies indicate that the TPMV H protein can be retargeted using scFvs (Springfeld and Cattaneo, unpublished results). In contrast to the targeted MV produced up to now, a retargeted chimeric MV-TPMV would be specific for tumor cells because TPMV does not naturally infect human cells.

In conclusion, several preclinical studies have demonstrated that MV is a promising agent for the virotherapy of cancer. Clinical trials in the near future will give indications about the most promising strategies to obtain benefits for cancer patients, the ultimate and most important goal.

REFERENCES

1. Katz M. Clinical spectrum of measles. Curr Top Microbiol Immunol 1995; 191:1–12.
2. Griffin DE. Measles Virus. In: David M. Knipe PMH, Eds Fields Virology. Philadelphia: Lippincott Williams & Wilkins, 2001: 1401–1442 .
3. WHO. Global measles mortality reduction and regional elimination, 2000–2001. Part I. Wkly Epidemiol Rec 2002; 77:49–56.
4. Billeter MA, Cattaneo R, Spielhofer P, Kaelin K, Huber M, Schmid A, Baczko K, ter Meulen V. Generation and properties of measles virus mutations typically associated with subacute sclerosing panencephalitis. Ann N Y Acad Sci 1994; 724: 367–377.
5. Enders JF, Peebles TC. Propagation in tissue culture of cytopathogenic agents from patients with measles. Proc Soc Exp Biol Med 1954; 86:277–286.
6. Hilleman MR, Buynak EB, Weibel RE, Stokes J, Jr., Whitman JE, Jr., Leagus MB. Development and evaluation of the Moraten measles virus vaccine. Jama 1968; 206:587–590.
7. Duclos P, Ward BJ. Measles vaccines: a review of adverse events. Drug Saf 1998; 19:435–454.
8. Moss WJ, Cutts F, Griffin DE. Implications of the human immunodeficiency virus epidemic for control and eradication of measles. Clin Infect Dis 1999; 29:106–112.

9. Sinkovics J, Horvath J. New developments in the virus therapy of cancer: a historical review. Intervirology 1993; 36:193–214.
10. Hernández SA. Observación de un caso de enfermedad de Hodgkin, con regresión de los síntomas e infartos ganglionares, postsarampión. Archivos Cubanos de Cancerologia 1949; 8:26–31.
11. Bluming AZ, Ziegler JL. Regression of Burkitt's lymphoma in association with measles infection. Lancet 1971; 2:105–106.
12. Zygiert Z. Hodgkin's disease: remissions after measles. Lancet 1971; 1:593.
13. Mota HC. Infantile Hodgkin's disease: remission after measles. Br Med J 1973; 2:421.
14. Pasquinucci G. Possible effect of measles on leukaemia. Lancet 1971; 1:136.
15. Gross S. Measles and leukaemia Lancet 1971; 1:397–398.
16. Bellini WJ, Englund G, Rozenblatt S, Arnheiter H, Richardson CD. Measles virus P gene codes for two proteins. J Virol 1985; 53:908–919.
17. Cattaneo R, Kaelin K, Baczko K, Billeter MA. Measles virus editing provides an additional cysteine-rich protein. Cell 1989; 56:759–764.
18. Cathomen T, Naim HY, Cattaneo R. Measles viruses with altered envelope protein cytoplasmic tails gain cell fusion competence. J Virol 1998; 72:1224–1234.
19. Rager M, Vongpunsawad S, Duprex WP, Cattaneo R. Polyploid measles virus with hexameric genome length. EMBO J 2002; 21:2364–2372.
20. Simon EH. The distribution and significance of multiploid virus particles. Prog Med Virol 1972; 14:36–67.
21. Cattaneo R, Rebmann G, Schmid A, Baczko K, ter Meulen V, Billeter MA. Altered transcription of a defective measles virus genome derived from a diseased human brain. EMBO J 1987; 6:681–688.
22. Naniche D, Varior-Krishnan G, Cervoni F, Wild TF, Rossi B, Rabourdin-Combe C, Gerlier D. Human membrane cofactor protein (CD46) acts as a cellular receptor for measles virus. J Virol 1993; 67:6025–6032.

23. Dorig RE, Marcil A, Chopra A, Richardson CD. The human CD46 molecule is a receptor for measles virus (Edmonston strain). Cell 1993; 75:295–305.
24. Tatsuo H, Ono N, Tanaka K, Yanagi Y. SLAM (CDw150) is a cellular receptor for measles virus. Nature 2000; 406:893–897.
25. Schneider U, von Messling V, Devaux P, Cattaneo R. Efficiency of measles virus entry and dissemination through different receptors. J Virol 2002; 76:7460–7467.
26. Oldstone MB, Homann D, Lewicki H, Stevenson D. One, two, or three step: measles virus receptor dance. Virology 2002; 299: 162–163.
27. Radecke F, Spielhofer P, Schneider H, Kaelin K, Huber M, Dotsch C, Christiansen G, Billeter MA. Rescue of measles viruses from cloned DNA. EMBO J 1995; 14:5773–5784.
28. Schneider H, Spielhofer P, Kaelin K, Dotsch C, Radecke F, Sutter G, Billeter MA. Rescue of measles virus using a replication-deficient vaccinia-T7 vector. J Virol Methods 1997; 64:57–64.
29. Takeda M, Takeuchi K, Miyajima N, Kobune F, Ami Y, Nagata N, Suzaki Y, Nagai Y, Tashiro M. Recovery of pathogenic measles virus from cloned cDNA. J Virol 2000; 74:6643–6647.
30. Cathomen T, Mrkic B, Spehner D, Drillien R, Naef R, Pavlovic J, Aguzzi A, Billeter MA, Cattaneo R. A matrix-less measles virus is infectious and elicits extensive cell fusion: consequences for propagation in the brain. EMBO J 1998; 17:3899–3908.
31. Escoffier C, Manie S, Vincent S, Muller CP, Billeter M, Gerlier D. Nonstructural C protein is required for efficient measles virus replication in human peripheral blood cells. J Virol 1999; 73:1695–1698.
32. Tober C, Seufert M, Schneider H, Billeter MA, Johnston IC, Niewiesk S, ter Meulen V, Schneider-Schaulies S. Expression of measles virus V protein is associated with pathogenicity and control of viral RNA synthesis. J Virol 1998; 72:8124–8132.
33. Mrkic B, Odermatt B, Klein MA, Billeter MA, Pavlovic J, Cattaneo R. Lymphatic dissemination and comparative pathology of recombinant measles viruses in genetically modified mice. J Virol 2000; 74:1364–1372.
34. Duprex WP, Duffy I, McQuaid S, Hamill L, Cosby SL, Billeter MA, Schneider-Schaulies J, ter Meulen V, Rima BK. The H gene of rodent brain-adapted measles virus confers neuroviru-

lence to the Edmonston vaccine strain. J Virol 1999; 73: 6916–6922.

35. Johnston IC, ter Meulen V, Schneider-Schaulies J, Schneider-Schaulies S. A recombinant measles vaccine virus expressing wild-type glycoproteins: consequences for viral spread and cell tropism. J Virol 1999; 73:6903–6915.
36. Takeuchi K, Takeda M, Miyajima N, Kobune F, Tanabayashi K, Tashiro M. Recombinant wild-type and edmonston strain measles viruses bearing heterologous H proteins: role of H protein in cell fusion and host cell specificity. J Virol 2002; 76: 4891–4900.
37. Maisner A, Mrkic B, Herrler G, Moll M, Billeter MA, Cattaneo R, Klenk HD. Recombinant measles virus requiring an exogenous protease for activation of infectivity. J Gen Virol 2000; 81: 441–449.
38. Moeller K, Duffy I, Duprex P, Rima B, Beschorner R, Fauser S, Meyermann R, Niewiesk S, ter Meulen V, Schneider-Schaulies J. Recombinant measles viruses expressing altered hemagglutinin (H) genes: functional separation of mutations determining H antibody escape from neurovirulence. J Virol 2001; 75:7612–7620.
39. Duprex WP, McQuaid S, Hangartner L, Billeter MA, Rima BK. Observation of measles virus cell-to-cell spread in astrocytoma cells by using a green fluorescent protein-expressing recombinant virus. J Virol 1999; 73:9568–9575.
40. Mrkic B, Pavlovic J, Rulicke T, Volpe P, Buchholz CJ, Hourcade D, Atkinson JP, Aguzzi A, Cattaneo R. Measles virus spread and pathogenesis in genetically modified mice. J Virol 1998; 72:7420–7427.
41. Neumeister C, Nanan R, Cornu T, Luder C, ter Muelen V, Naim H, Niewiesk S. Measles virus and canine distemper virus target proteins into a TAP- independent MHC class I-restricted antigen-processing pathway. J Gen Virol 2001; 82:441–447.
42. Grote D, Russell SJ, Cornu TI, Cattaneo R, Poland GA, Fielding AK. Live attenuated measles virus induces regression of human lymphoma xenografts in immunodeficient mice. Blood 2001; 97: 3746–3754.
43. Peng KW, Facteau S, Wegman T, O'Kane D, Russell SJ. Non-invasive in vivo monitoring of trackable viruses expressing soluble marker peptides. Nat Med 2002; 8:527–531.

44. Singh M, Cattaneo R, Billeter MA. A recombinant measles virus expressing hepatitis B virus surface antigen induces humoral immune responses in genetically modified mice. J Virol 1999; 73:4823–4828.

45. Wang Z, Hangartner L, Cornu TI, Martin LR, Zuniga A, Billeter MA, Nain HY. Recombinant measles viruses expressing heterologous antigens of mumps and simian immunodeficiency viruses. Vaccine 2001; 19:2329–2336.

46. Singh M, Billeter MA. A recombinant measles virus expressing biologically active human interleukin-12. J Gen Virol 1999; 80: 101–106.

47. Grote D, Cattaneo R, Fielding AK. Neutrophils contribute to the measles virus-induced antitumor effect: enhancement by granulocyte macrophage colony-stimulating factor expression. Cancer Res 2003; 63:6463–6468.

48. Peng KW, Ahmann GJ, Pham L, Greipp PR, Cattaneo R, Russell SJ. Systemic therapy of myeloma xenografts by an attenuated measles virus. Blood 2001; 98:2002–2007.

49. Greenlee RT, Hill-Harmon MB, Murray T, Thun M. Cancer statistics, 2001. CA Cancer J Clin 2001; 51:15–36.

50. Peng KW, TenEyck CJ, Galanis E, Kalli KR, Hartmann LC, Russell SJ. Intraperitoneal therapy of ovarian cancer using an engineered measles virus. Cancer Res 2002; 62:4656–4662.

51. Peng KW, Frenzke M, Myers R, Soeffker D, Harvey M, Greiner S, Galanis E, Cattaneo R, Federspiel MJ, Russell SJ. Biodistribution of oncolytic measles virus after intraperitoneal administration into Ifnar-CD46Ge transgenic mice. Hum Gene Ther 2003; 14:1565–1577.

52. Phuong LK, Allen C, Peng KW, Giannini C, Greiner S, TenEyck CJ, Mishra PK, Macura SI, Russell SJ, Galanis EC. Use of a vaccine strain of measles virus genetically engineered to produce carcinoembryonic antigen as a novel therapeutic agent against glioblastoma multiforme. Cancer Res 2003; 63: 2462–2469.

53. Kirn D, Martuza RL, Zwiebel J. Replication-selective virotherapy for cancer: Biological principles, risk management and future directions. Nat Med 2001; 7:781–787.

54. Ring CJ. Cytolytic viruses as potential anti-cancer agents. J Gen Virol 2002; 83:491–502.

55. Rodriguez R, Schuur ER, Lim HY, Henderson GA, Simons JW, Henderson DR. Prostate attenuated replication competent adenovirus (ARCA) CN706: a selective cytotoxic for prostate-specific antigen-positive prostate cancer cells. Cancer Res 1997; 57:2559–2563.
56. Hammond AL, Plemper RK, Cattaneo R. Targeting Measles Virus Entry. In: Curiel DT, Douglas JT, Eds. Vector Targeting for Therapeutic Gene Delivery. Hoboken: Wiley-Liss Inc., 2002: 321–336.
57. Schneider U, Bullough F, Vongpunsawad S, Russell SJ, Cattaneo R. Recombinant measles viruses efficiently entering cells through targeted receptors. J Virol 2000; 74:9928–9936.
58. Hammond AL, Plemper RK, Zhang J, Schneider U, Russell SJ, Cattaneo R. Single-chain antibody displayed on a recombinant measles virus confers entry through the tumor-associated carcinoembryonic antigen. J Virol 2001; 75:2087–2096.
59. Hammarstrom S. The carcinoembryonic antigen (CEA) family: structures, suggested functions and expression in normal and malignant tissues. Semin Cancer Biol 1999; 9:67–81.
60. Peng KW, Donovan KA, Schneider U, Cattaneo R, Lust JA, Russell SJ. Oncolytic measles viruses displaying a single chain antibody against CD38, a myeloma cell marker. Blood 2003: 101.
61. Bucheit AD, Kumar S, Grote DM, Lin Y, von Messling V, Cattaneo RB, Fielding AK. An oncolytic measles virus engineered to enter cells through the CD20 antigen. Mol Ther 2003; 7:62–72.
62. Lecouturier V, Fayolle J, Caballero M, Carabana J, Celma ML, Fernandez-Munoz R, Wild TF, Buckland R. Identification of two amino acids in the hemagglutinin glycoprotein of measles virus (MV) that govern hemadsorption, HeLa cell fusion, and CD46 downregulation: phenotypic markers that differentiate vaccine and wild-type MV strains. J Virol 1996; 70:4200–4204.
63. Vongpunsawad S, Oezgun N, Braun W, Cattaneo R. Selectively receptor-blind measles viruses: Identification of residues necessary for SLAM- or CD46-induced fusion and their localization on a new hemagglutinin structural model. J Virol 2004; 78: 302–313.
64. Hermiston TW, Kuhn I. Armed therapeutic viruses: Strategies and challenges to arming oncolytic viruses with therapeutic genes. Cancer Gene Ther 2002; 9:1022–1035.

65. Springer CJ, Niculescu-Duvaz I. Prodrug-activating systems in suicide gene therapy. J Clin Invest 2000; 105:116–1167.

66. Parks GD, Young VA, Koumenis C, Wansley EK, Layer JL, Cooke KM. Controlled cell killing by a recombinant nonsegmented negative-strand RNA virus. Virology 2002; 293: 192–203.

67. Fernandez M, Porosnicu M, Markovic D, Barber GN. Genetically engineered vesicular stomatitis virus in gene therapy: application for treatment of malignant disease. J Virol 2002; 76: 895–904.

68. Spitzweg C, Morris JC. The sodium iodide symporter: its pathophysiological and therapeutic implications. Clin Endocrinol (Oxf) 2002; 57:559–574.

69. Schlumberger MJ. Papillary and follicular thyroid carcinoma. N Engl J Med 1998; 338:297–306.

70. Mandell RB, Mandell LZ, Link CJ, Jr. Radioisotope concentrator gene therapy using the sodium/iodide symporter gene. Cancer Res 1999; 59:661–668.

71. Cho JY, Shen DH, Yang W, Williams B, Buckwalter TL, La Perle KM, Hinkle G, Pozderac R, Kloos R, Nagaraja HN, Barth RF, Jhiang SM. In vivo imaging and radioiodine therapy following sodium iodide symporter gene transfer in animal model of intracerebral gliomas. Gene Ther 2002; 9:1139–1145.

72. Boland A, Ricard M, Opolon P, Bidart JM, Yeh P, Filetti S, Schlumberger M, Perricaudet M. Adenovirus-mediated transfer of the thyroid sodium/iodide symporter gene into tumors for a targeted radiotherapy. Cancer Res 2000; 60:3484–3492.

73. Dingli D, Peng KW, Harvey ME, Greipp PR, O'Connor MK, Cattaneo R, Morris JC, Russell SJ. Image-guided radiovirotherapy for multiple myeloma using a recombinant measles virus expressing the thyroidal sodium iodide symporter. Blood 2004; 103:1641–1646.

74. Russell SJ. RNA viruses as virotherapy agents. Cancer Gene Ther 2002; 9:961–966.

75. Pedersen IR, Mordhorst CH, Glikmann G, von Magnus H. Subclinical measles infection in vaccinated seropositive individuals in arctic Greenland. Vaccine 1989; 7:345–348.

76. Muller CP, Huiss S, Schneider F. Secondary immune responses in parents of children with recent measles. Lancet 1996; 348: 1379–1380.

77. Damien B, Huiss S, Schneider F, Muller CP. Estimated susceptibility to asymptomatic secondary immune response against measles in late convalescent and vaccinated persons. J Med Virol 1998; 56:85–90.

78. Coffey MC, Strong JE, Forsyth PA, Lee PW. Reovirus therapy of tumors with activated Ras pathway. Science 1998; 282: 1332–1334.

79. Heise C, Kirn DH. Replication-selective adenoviruses as oncolytic agents. J Clin Invest 2000; 105:847–851.

80. Todo T, Rabkin SD, Sundaresan P, Wu A, Meehan KR, Herscowitz HB, Martuza RL. Systemic antitumor immunity in experimental brain tumor therapy using a multimutated, replication-competent herpes simplex virus. Hum Gene Ther 1999; 10: 2741–2755.

81. Hirasawa K, Nishikawa SG, Norman KL, Coffey MC, Thompson BG, Yoon CS, Waisman DM, Lee PW. Systemic reovirus therapy of metastatic cancer in immune-competent mice. Cancer Res 2003; 63:348–353.

82. Tidona CA, Kurz HW, Gelderblom HR, Darai G. Isolation and molecular characterization of a novel cytopathogenic paramyxovirus from tree shrews. Virology 1999; 258:425–434.

4

Antitumor Immune Memory and its Activation for Control of Residual Tumor Cells and Improvement of Patient Survival

A NEW CONCEPT DERIVED FROM TRANSLATIONAL RESEARCH WITH THE VIRUS-MODIFIED TUMOR VACCINE ATV-NDV

VOLKER SCHIRRMACHER
Division of Cellular Immunology,
Tumor Immunology Program,
German Cancer Research Center (DKFZ)
Heidelberg, Germany

SUMMARY

A new concept is presented that proposes that a certain threshold of antitumor immune memory plays an important role [1] in the control of residual tumor cells that remain after most therapies and [2] for long-term survival of treated cancer patients. This immune memory is T cell based and most likely maintained by persisting tumor-associated antigen (TAA) from residual dormant tumor cells. Such immune memory was prominent in the bone marrow in animal tumor models as well as in cancer patients. Pre-existing antitumor memory T cells from cancer patients could be activated by stimulation with TAA-pulsed autologous dendritic cells (DC) or with virus-infected TAA expressing autologous tumor cell vaccine (ATV-NDV). Antitumor vaccination with ATV-NDV caused augmentation of antitumor memory Delayed Hyper-sensitivity (DTH) responses. In a variety of Phase II vaccination studies, an optimal formulation of this vaccine could improve long-term survival beyond what is seen in conventional standard therapies. Possible reasons for differences in long-term survival of patients treated by immunotherapy vs. chemotherapy are being discussed.

INTRODUCTION

The host immune response to foreign challenge requires the coordinated action of both the innate and acquired arms of the immune system. The innate immune response not only provides the first line of defense against microorganisms but also the biological context–the "danger signal"–that instructs the adaptive immune system to mount a response [1]. The adaptive response is mediated by T and B lymphocytes that have undergone germline gene rearrangements of their antigen-specific receptors. This second line of defense is characterized by exquisite specificity and long-lasting memory.

In innate immune responses, Toll-like receptors (TLR) [2] function as pattern-recognition receptors and allow the recognition of microbial components for activation of the immune system. Activation of TLR leads to the release of several inflammatory mediators, including chemokines from resident tissue-macrophages and DCs and modulates the expression of chemokine receptors on DCs. These TLR-mediated events are essential for both the recruitment of immature DCs to sites of pathogen entry and their ultimate journey back to lymph nodes to activate naïve T cells. In addition, chemokines released by resident tissue cells after TLR activation guide these activated T cells into the site of pathogen entry or replication. In this way, chemokines link innate immune cell activation in the tissue to the recruitment of antigen-specific T cells generated in secondary lymphoid organs [3].

In this Chapter, I review our work and concepts related to the immunotherapy of cancer. We exploit the use of virus infection to introduce danger signals into tumor cells to activate, via antitumor vaccination, innate immune responses and adaptive antitumor immune responses in connection with long-term immunological memory. The review includes a summary of results from many clinical studies. The survival data are interpreted to result from activation and maintenance of long-term immune memory.

DEVELOPMENT OF THE NEW CONCEPT IN ANIMAL TUMOR MODELS

Postoperative Active-Specific Immunotherapy in the ESb Tumor Model

About 20 years ago, we started this work in the murine ESb lymphoma animal tumor model. The ESb lymphoma is one of the most aggressive animal tumors. It metastasizes to visceral organs, in particular, the liver, and kills syngeneic hosts within about 12 days. We used it as a challenge to design new antimetastatic therapy strategies. Treatment with cytostatic drugs that were claimed to have antimetastatic activity were not effective and even reduced the overall survival time. In contrast, postoperative vaccination with a virus-modified—but not with unmodified—ESb cells was able to cause protection from metastases in about 50% of syngeneic mice [4]. The surviving animals developed long-lasting protective immunity specific for the ESb tumor line and did not cross-react with other syngeneic or allogeneic tumor lines. It was based on tumor-specific immune T cell memory.

The effectiveness of this approach of postoperative active specific immunotherapy (ASI) with virus-modified autologous live cell tumor vaccine was thereafter confirmed in other metastasising animal tumors such as murine B16 melanoma [5], 3LL Lewis lung carcinoma [6], and guinea pig L10 hepatocarcinoma [7,8].

The Trick of Virus Infection Using Newcastle Disease Virus

In those studies, we had selected for the purpose of virus infection an avian RNA paramyxovirus, Newcastle disease virus (NDV). Because this was a good choice, we continued to use NDV for clinical application. Newcastle disease virus is an enveloped virus of 150 nm to 300 nm size containing a nonsegmented negative-stranded RNA of 15 kb size (Figure 1A). Virulent strains of NDV are important pathogens for poultry and are widely distributed in naturally occurring bird populations. The NDV genome contains six genes encoding for the following six gene products listed in order from the 3′- end: nucleocapsid protein (NP, 55 kD); phosphorprotein (P, 53 kD); matrix protein (M, 40 kD); fusion protein (F, 67 kD); hemagglutinin-neuraminidase (HN, 74 kD); and large protein (L, 200 kD). The F glycoprotein is synthesized as an inactive precursor (Fo, 67 kD), which undergoes proteolytic cleavage to yield the biologically active protein consisting of the disulfide-linked chains F_1 (55 kD) and F_2 (12.5 kD) [9].

Upon accidental exposure, NDV can also infect humans. Human infections are mild and can cause symptoms of conjunctivitis or laryngitis. There is an extensive safety database for NDV, primarily from low-dose human tumor vaccine trials. Newcastle disease virus is well tolerated in humans in doses of at least 3×10^9 infectious units by the i.v. route and at least 4×10^{12} infectious units by the intratumoral route. Complementing these clinical findings, animal safety data provide evidence of the low pathogenicity of NDV in mammals [10].

An Effective Method of Introducing Danger Signals into Tumor Cells

Figure 1 illustrates the structure of NDV, its surface molecules HN and F (A), and the two steps of infection, namely [1] cell surface binding and [2] replication in the cells' cytoplasm (B).

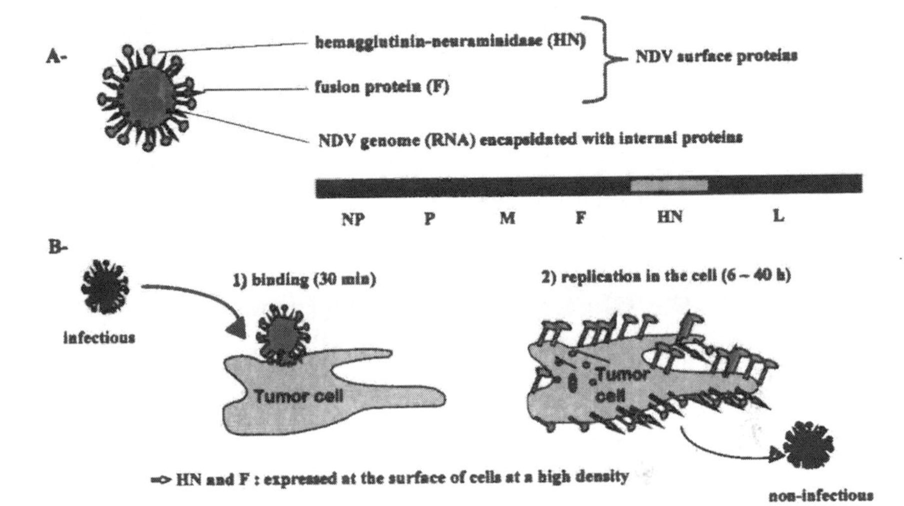

Figure 1 (A) Structure: Newcastle disease virus is an avian paramyxovirus of 150 nm to 300 nm diameter with an envelop containing two surface proteins (HN and F). Its 15-Kb nonsegmented genome contains six genes, two of which code for the surface proteins HN and F. (B) Replication of NDV in tumor cells. [1] Initial binding to a host cell takes place through interaction of HN in the virus coat with sialic acid expressing cellular HN receptors (gangliosides). Infectious virus from the allantoic fluid of infected embryonated eggs express an active fusion protein F, which, upon interaction with HN and HN receptors will cause virus–host cell membrane fusion. This allows the viral genome to enter the cytoplasm of the host cell. Normal cells resist virus replication but tumor cells do not [11]. Virus replication in tumor cells leads to expression of a high density of surface HN and F molecules. The avirulent strain Ulster performs an abortive monocyclic replication cycle [36] within 6 to 40 hours, upon which noninfectious virus progeny is produced before the tumor cells die via apoptosis [12,16]

Newcastle disease virus can replicate up to 10,000 times better in human cancer cells than in most normal human cells. This finding has prompted much interest in this virus as a potential anticancer agent. The resistance of normal cells to infection by Newcastle Disease Virus (NDV) may have to do with their strong interferon (IFN) α response. Whereas normal peripheral blood mononuclear cells (PBMC) produce IFNα immediately after surface contact with inactivated NDV, tumor cells produce IFNα only after true virus infection. Interferon-α is an antiviral factor that can quickly induce a state of virus resistance in other cells. Newcastle disease virus has been labelled as a complementary and alternative medicine (CAM) and much detailed information can be found at a related homepage of the National Cancer Institute USA (*http://www.nci.nih.gov/cancerinfo/pdq/cam/NDV*).

The first hints of the potential anticancer benefit of this virus were noted more than 30 years ago. Three different conceptual uses of NDV in cancer treatment can be distinguished: [1] Use for tumor-selective cytolysis (oncolysis) [10], [2] use of NDV as an adjuvant and danger signal in a tumor vaccine for stimulation of cytotoxic T lymphocyte (CTL) and DTH responses after tumor vaccination [11], [3] use of NDV for non-specific immune stimulation and induction of cytokines and IFNs [8,10].

As a consequence of infection of tumor cells by NDV, tumor cells can be perceived by the immune system as "dangerous" because of the following danger-signalling molecules that we have identified: [1] viral HN-molecules [12,13], [2] double-strand RNA [14], [3] IFN-α[??]β, [13,15], [4] chemokines (RANTES and IP-10) [16]. (Figure 2.)

The release of large amounts of IFN-α by natural IFN-producing cells (NIPC) and plasmacytoid DCs indicates that these cells of the innate immune system have sensed "danger." It has only recently become clear that IFN-α has an important adjuvant function in the immune response [17]. It activates DCs [18] and induces TRAIL in NK cells [19] and the interleukin (IL)-12– receptor β chain in T cells [20]. Together with IL-12, IFN-α polarizes the T cell toward a cell-mediated Th1 response characterized by DTH and CTL activity. In addition, IFN-α induces the upregulation of molecules important for antigen recognition (e.g., human leukocite antigen)(16), cell–cell interaction (e.g., cell adhesion molecules)[16], and cytotoxicity (e.g., TRAIL)[13].

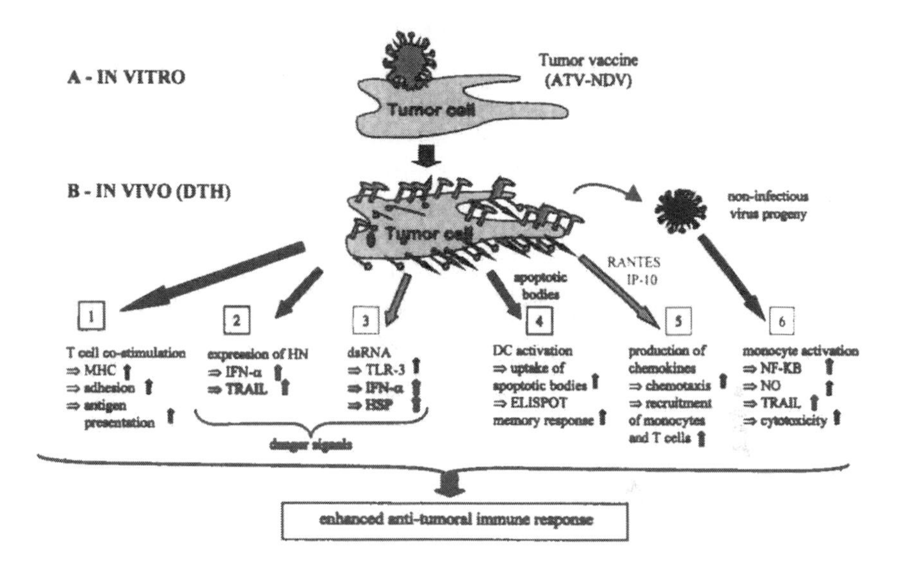

Figure 2 Mechanisms involved in the enhanced antitumoral immune response when tumor cells are infected with Newcastle disease virus (NDV): (**1**) Upregulation of major histocompatibility complex molecules, ICAM-1 (CD54) and LFA-3 (CD58), and an increase in co-stimulation [16]. (**2,3**) Delivery of danger signals through surface protein HN [13] and double-strand RNA [14]. Transfectants expressing HN induce high amounts of interferon (IFN)-α in human peripheral blood mononuclear cells [13]. Responsible for this response is a subset of CD16$^+$/CD64$^+$ cells belonging to the innate immunity system [13,15]. Similar IFN-α responses can be induced by dsRNA via Toll-like receptor 3 in plasmacytoid dendritic cells [14]. Interferon-α and β activate a variety of additional molecules, including heat-shock proteins (HSP) and tumor necrosis factor-related apoptosis-inducing ligand [13,15,41]. [**4**] Apoptotic bodies and lysate from NDV-infected tumor cells are taken up by immature dendritic cells and cause their activation for augmented stimulation of antitumor memory T cell responses in Enzyme-Linked-Immuno-Spot (ELISPOT) assays [52]. (**5**) Infected tumor cells release the chemotactic factors regulated upon activation normal T cell expressed and secreted (RANTES) and IFN-γ–inducible protein-10 (IP-10), which contribute to augmented recruitment of monocytes and T cells [16]. (**6**) Released noninfectious NDV progeny can directly activate monocytes and macrophages to cytotoxic antitumoral activities [41]

Lessons from Tumor Dormancy Studies

As already mentioned, postoperative vaccination in the murine ESb tumor model with ESb-NDV vaccine resulted in about 50 % long-term survivors who had established systemic and tumor-specific immune memory [21]. A detailed analysis later revealed the existence of residual tumor cells in such mice in the bone marrow. These cells were never completely eradicated and were kept under immune control at a low level. Therapeutic antitumor vaccination in these animals had thereby caused the establishment not only of immunological memory but also of tumor dormancy. This was confirmed in prophylactic immunization experiments. In immunocompetent animals, we saw a correlation between the persistence of dormant tumor cells at low levels in the bone marrow and long-term protective immune memory [22]. We could break the status of tumor dormancy in situ by CD8 T cell depletion [23]. In immunocompetent tumor-dormant mice, 21% of the bone marrow-derived tumor cells were positive for the proliferation marker Ki67. The fraction of Ki67-positive tumor cells in diseased bone marrow of immune compromised mice was 40% and, in absence of immune CD8 T cells, the frequency of bone marrow-residing tumor cells increased by two orders of magnitude. These findings suggested that tumor dormancy in this model was attributable to active immune control of tumor cells by CD8 T cells in the bone marrow, keeping them at a low level without elimination [24].

Tumor dormancy in general and in clinical cancer in particular can be caused by different growth constrain mechanisms [24]. One involves macrometastatic lesions that fail to induce angiogenic activity necessary for their expansion. Local hypoxia and undernourishment are insufficient to eradicate the cells but limit their expansion by inducing apoptotic death. Angiogenesis inhibitors such as angiostatin and endostatin can keep small tumor foci dormant over extended periods of time. Another form of tumor dormancy consists of solitary cancerous cells that reside after operation of the primary tumor at distinct sites such as bone marrow or other tissues.

An Interactive Balance Between Residual Tumor Cells and Immune Memory

In our animal tumor studies, we noticed an interactive balance between residual tumor cells and immune memory [24]. Persistence of tumor cells and, therefore, persistence of TAA derived from them correlated with long-term protective immune memory [22–24]. Following injection of live or irradiated tumor cells into the external ear pinna in exemplum (i.e.) of mice—a site where the tumor cells could not grow—comparable numbers of X-Gal–stained cells were detectable in the bone marrow of host animals one week after inoculation. Live tumor cells persisted at a similar level (about 30 cells/10^6 bone marrow cells) in the bone marrow for follow-up periods of up to 2 months, while the number of bone marrow-derived irradiated cells declined within 3 weeks. The persistence or nonpersistence of tumor cells in bone marrow seen with injection of either live or irradiated tumor cells correlated with the presence or absence of long-term memory in these respective groups. Thus, when the tumor challenge was made later than 4 weeks after vaccination, only the live tumor vaccine could effectively protect the mice. The tumor-dormant mice were fully protected even when challenged with parental tumor cells 6 months after primary i.e. tumor cell injection.

Characteristics of Memory T Cells

Whereas naïve T cells can only be primed by antigens that are cross-presented by professional host antigen-presenting cells such as DCs, memory T cells can also be activated directly by a tumor vaccine presenting TAA together with costimulatory signals. Secondary immune responses by memory T cells are faster and stronger than primary responses. Their requirements for activation are less strict (lower dependency on co-stimulation) and they release a broader spectrum of cytokines and are multifunctional after re-activation [25]. Thus, memory T cells are superior to naïve T cells [26] for protective immunity. There is a programmed development of effector and memory CD8 T cells, and different subsets of memory T cells, namely "central" and "effector" memory cells, have recently been distinguished. Of special interest are also stem cell-like

Table 1 Stem Cell-like Properties of Memory T Cells

1. Self-renewal capacity Response to homeostatic signals; causing self-renewal
2. Programmed differentiation Response to antigen-presenting cells; stimulating naive T cells to differentiate to effector cytotoxic T lymphocytes and, finally, to memory T cells
3. Pluripotentiality Response to antigen; activating perforin, granzymes, IFN-γ, IL-2, etc.
4. Longevity/immortality unlimited proliferation; via telomerase upregulation (?)

(From Ref. 25,26.)

properties of memory T cells: Upon response to homeostatic signals they have a self-renewal capacity [26]. Longevity or immortality is another important aspect of memory T cells, which may involve telomerase upregulation (Table 1).

We recently established a novel tumor model system for the study of long-term protective immunity and immune T cell memory [27]. In this adoptive memory T cell transfer system involving as recipients nude (nu/nu) mice, we were able to study [1] the role of persisting antigen for long-term maintenance of peptide epitope-specific CD8 memory T cells and [2] the longevity of the cells.

FROM MOUSE TO MAN

Long-term persistence of tumor cells in a dormant state is suggested also from clinical observations in cancer patients, notably breast cancer [28]. Some patients develop secondary tumors at distant sites many years after successful therapy of the primary tumors; others do not develop recurrencies in spite of disseminated tumor cells at the time of diagnosis. The longest duration of breast cancer dormancy—i.e., the longest interval between primary treatment and tumor recurrence—was calculated in a retrospective study involving 1547 patients, to be between 20 and 25 years [28]. In breast cancer, 25% to 43 % of primary-operated patients exhibit micrometastatic tumor cells in their bone marrow. Detection levels are in the range of 1 to 10 cancer cells per 10^6 to 10^7 bone marrow-derived mononuclear cells. Successful enrichment, reliable identification, and molecular profiling are key issues of ongoing and future studies [29].

Antitumor Memory T Cells in Cancer Patients

To test for the presence of memory T cells in cancer patients, we investigated bone marrow of breast cancer patients with respect to tumor cell content, immune activation status, and memory T cell content [30]. Bone marrow-derived cells from primary-operated breast cancer patients ($n = 90$) were compared with those from healthy donors ($n = 10$) and cells from respective blood samples. Cytokeratine 19-positive tumor cells were detected by nested polymerase chain reaction. Three-color flow cytometry was used to identify numbers and activation state of T cells, natural killer (NK) cells, monocytes/macrophages, and subsets by a panel of monoclonal antibodies. The proportion of memory T cells among the CD4 and CD8 T cells was found to be much higher in bone marrow of cancer patients than in healthy donors ($p < 0.001$). The extent of memory T cell increase was related to the size of the primary tumor. The highest relative memory content was detected in patients with T2 tumor stage and these were significantly different from patients both with T1 and T3/4 stages. Thus, with tumor progression, the memory content in T cell populations first steadily increased and then decreased again. Patients with disseminated tumor cells in their bone marrow had more memory CD4 T cells than patients with tumor cell negative bone marrow [30]. Our proposition from animal studies that bone marrow is a special compartment for immunological memory and tumor dormancy [22] had therefore been supported by our clinical findings.

Therapy of Human Tumors in NOD/SCID Mice with Autologous Memory T cells

Further functional studies with patients' bone marrow-derived memory T cells revealed tumor specificity and functional competence [31]. We could demonstrate the existence of TAA-specific CD8 T cells that specifically bound to tetramers consisting of human leukocyte antigen (HLA)-A2 molecules and peptides derived from two important markers of breast cancer, namely MUC-1, a mucin-type TAA, and HER2/neu, an overexpressed growth-factor receptor. Bone marrow memory T cells from patients could be specifically restimulated in short-term culture (20 to 40 hours) to IFN-γ–producing cells by autologous DCs pulsed with respective tumor lysate or with the aforementoned TAA. The calculated frequency of memory T cells in Enzyme-Linked Immuno-Spot (ELISPOT) among total T lymphocytes from bone marrow of patients responding to autologous TAA was rather high and varied from 1/200 to 1/11000. We could also induce antitumor CTL activity in re-activated bone marrow-derived memory T cells. Their immunotherapeutic competence in vivo was finally demonstrated by showing regression of human breast cancer in non-obese diabetes mellitus/severe combined immuno-deficiency (NOD/SCID) mice after transfer of autologous reactivated bonemarrow-derived memory T cells [31]. Following such cell transfer, human T cells were found to infiltrate the human breast cancer tissue (Fig. 3). Most infiltrating T cells had the memory marker CD45 R0. CD8 T cells were found in close contact with CD4 T cells and within the tumor close to necrotic areas, where, in serial tumor sections, apoptotic cells could be detected [31].

Tolerance to Tumor-Associated Antigens

In some of the cancer patients investigated, it was possible to perform a paired analysis of samples from peripheral blood (PB) and bone marrow. In all these cases, the T cells from the blood were negative and did not produce a significant ELISPOT response, whereas the bone marrow samples were positive. Because tetramer analysis had shown the presence of TAA-specific T cells in PB of breast cancer patients with a frequency similar to that in bone marrow samples, these data suggested that the PB-derived T cells were tolerant (anergic) and not functional [31].

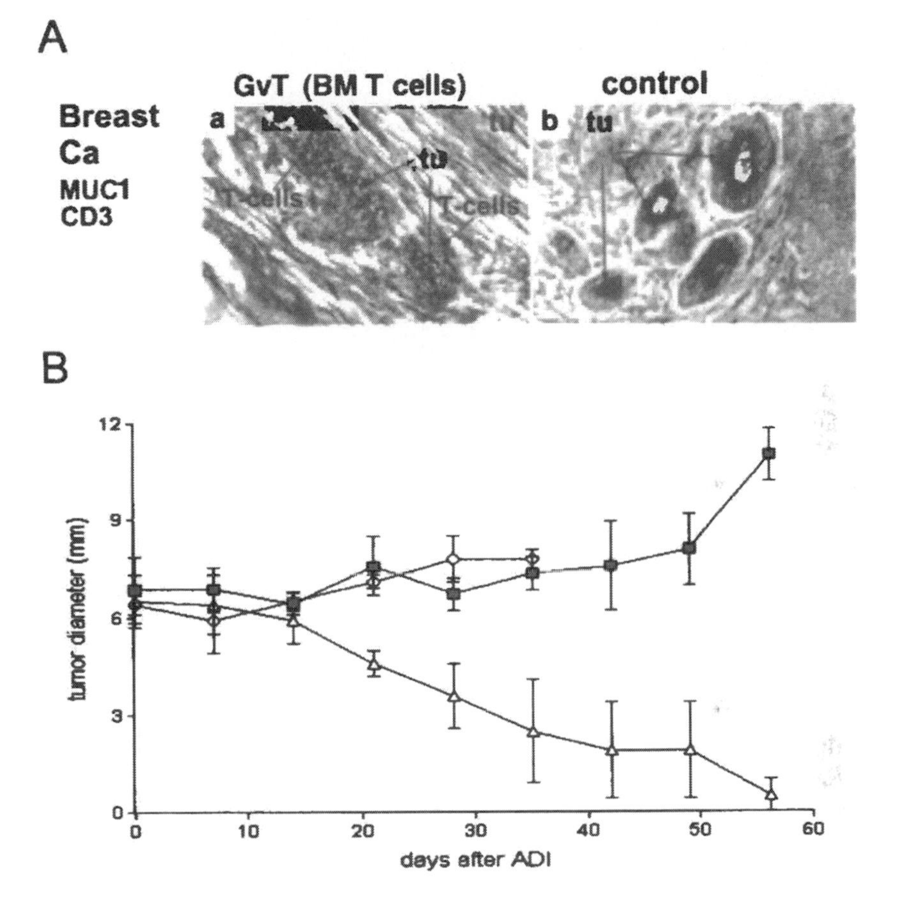

Figure 3 Tumor infiltration (**A**) and therapy of human breast carcinoma (**B**) in non obese diabetes mellitus/severe combined immunodeficiency (NOD/SCID) mice by ex vivo re-activated patient derived memory T cells from bone marrow [31]. (**A**) Tumor cells are stained by mAb to Mucin-1 (MUC) and T cells by mAb anti CD3. Notice that there are no tumor infiltrating T cells in control mice (b) which were either untreated (■) or treated with T cells from peripheral blood mononuclear cells (Γ), while blood marrow-derived T cells infiltrated the tumor (a) and caused graft-vs.-tumor reactivity, leading to complete tumor remission (group Δ in **B**)

Newcastle Disease Virus Infection Breaking Tolerance to Tumor-Associated Antigens

Tumor immunologists have developed a number of strategies to overcome states of anergy or tolerance to TAA. Some used TAA expressed in xenogeneic cells; others tried to xenogenize the tumor cells or to use xenogeneic DNA immunization or recombinant viruses expressing TAA. In cases of non-identified TAA, it is also possible to use whole-cell tumor vaccines, typically transfected with cytokines, costimulatory molecules, or infected with viruses, and to use them in combination with exogenous cytokines.

We could demonstrate in an autologous human anti-melanoma CD4 T helper clone that infection by NDV of autologous melanoma cells induces a T cell costimulatory activity and thereby prevents, in the clone, the induction of anergy by the tumor cell [32]. Without NDV infection, contact with autologous melanoma cells rendered the clone nonreactive and unresponsive even to subsequent stimulation by IL-2. Newcastle disease virus-infection of the melanoma cell line not only completely restored the proliferative response of the tumor-specific T cell clone but also inhibited the induction of anergy. Electrophoretic mobility shift assays of cell lysates from the clone revealed the induction of the CD28-responsive complex by co-incubation with NDV-infected melanoma cells [32].

Conditioning the Immune System to Associate Recognition of Tumor-Associated Antigen with Danger

Based on results of conditioning experiments with immune responses in animal models [33,34] we propose that it might be possible to condition or train the immune system to associate recognition of TAA with danger signals. Training could be performed by repeated vaccination with vaccine expressing TAA and danger signals. We also conclude that NDV is an appropriate reagent to introduce danger signals into tumor cells. Finally, we hypothesize that once the immune system's memory T cells are trained to associate TAA recognition with danger [1], they may not require the danger signal any longer and can react to TAA without being tolerized.

CLINICAL IMMUNOTHERAPY STUDIES

Since 1986, when we first reported on the "prevention of metastatic spread by postoperative immunotherapy with virally modified autologous tumor cells' [4], it took about 10 years of translational research to develop the autologous human tumor vaccine ATV-NDV and to perform "proof of principle" clinical studies at the evidence level of Phase II [35]. This pioneering translational research was everything but easy. It meant entering new, "forbidden" territory. In addition, the scientific community was not prepared for constructive discussion of these new thoughts and approaches. It took a lot of courage, self-confidence, and energy on the part of myself and cooperating clinicians to circumvent the various road blocks that were thrown on our way and to try to continue, mostly without financial or other support.

Development of an Autologous Virus Modified Tumor Vaccine for clinical studies

The specific component of each vaccine for human application that we developed in analogy to the effective animal tumor vaccine are patient-derived (autologous) live tumor cells (ATV). These are infected by the avirulent strain Ulster of NDV and inactivated by 200 Gy γ-irradiation to produce the vaccine ATV-NDV. Tumor cells are isolated from freshly operated tumor specimens by mechanical dissection and enzymatic dissociation and can be stored, after controlled freezing in liquid nitrogen. Although the first clinical studies employed total dissociated cells from a tumor, we later introduced a further tumor cell purification procedure to remove tumor infiltrating lymphocytes by immunomagnetic beads [35]. In some other studies, cell culture-adapted autologous tumor cells were used instead. All studies were approved in advance by local ethical commissions.

We selected the non-lytic strain Ulster for reasons of safety during application in cancer patients and because we intended to develop a virus-infected whole-cell cancer vaccine consisting of intact viable irradiated cancer cells. Newcastle disease virus Ulster has a monocyclic abortive replication cycle in tumor cells [36]. This replication involves the formation of double-strand RNA that functions as a danger signal and activates Toll-like receptors [14]. In this way, we tried to link the expression of TAA with danger signals [1,37].

We decided to use autologous tumor cells because we had experienced that protective antitumor immunity in animal tumor models was highly specific for the autologous tumor and also because autologous tumor cells represent the closest possible match to a patient's tumor. Autologous tumor cells might express not only common TAA but also individually unique TAA derived from mutations or other genetic alterations. The use of whole tumor cells eliminates the need to first identify the respective TAA.

It has been the experience of many tumor immunologists that whole-cell vaccines can stimulate the immune system more efficiently than oncolysates [38]. Tumor cell membrane integrity was found to be important for CTL activation by cancer vaccines [39]. Newcastle disease virus Ulster is first adsorbed to the tumor cells in vitro (1-hour binding) (Fig. 1). Then the virus-modified tumor vaccine is injected intradermally, allowing for virus replication in vivo at the site of vaccine application. Vaccine consisting of tumor cells infected with NDV Ulster should remain in the body long enough to generate effective immune responses that are mainly based on T cell-mediated immunity. Viral replication in the tumor cells takes about 6 to 40 hours [10–2], a time sufficient to generate delayed-type hypersensitivity (DTH) skin responses which are dependent on antigen-specific memory T cells.

During this time period of 6 to 40 hours after vaccine application, the infected tumor cells can produce progeny NDV virus particles. Although these are noninfectious because of uncleaved Fo molecules, they are able to stimulate innate immune responses in NIPC [15,13] and monocytes that are recruited by IP-10 (Fig. 2). The release of interferon-α and other cytokines causes activation of various antitumor effector cells such as NK cells and cytotoxic monocytes/macrophages [40,41]. In this way, we may achieve a link between innate and adaptive immune responses.

Figure 2 illustrates the consequences of human tumor cell infection by NDV Ulster for the immune response. We found that modification with live, but not inactive, NDV induced, in all human tumor cells tested, the production of IFN-β, IFN-α, and the chemokines RANTES and IFN-γ–inducible protein 10 (IP-10) [16]. In addition, infection by live NDV induced upregulation of HLA-ABC molecules on all tumor lines tested and HLA-DR molecules on breast carcinoma lines. Two cell adhesion molecules, ICAM-1 (CD54) and LFA-3 (CD58), were also upregulated on human tumor cells after infection with live NDV. Forty-eight to seventy-two hours after infection of the irradiated tumor cells with live NDV, many tumor cells were dead or in early or late stages of apoptosis [16]. Figure 2 illustrates six different mechanisms through which NDV infection of tumor cells leads to greater immunogenicity and enhanced immune responses.

Phase I Clinical Studies: Optimization of Vaccine Composition

Much of this pioneering work of establishing a human NDV modified tumor vaccine was performed by my clinical colleague T. Ahlert, who now has more than 10 years of experience with the vaccine preparation and application in cancer patients (*http://www.biotech-praxislab.de*).

After having optimized a technical procedure for isolating live tumor cells from fresh operation specimens and having calculated average yields and stability parameters [42], we started to perform Phase I clinical studies [43,44]. In the ESb animal tumor model, we had described before that an optimal vaccine composition that yielded 50% survival benefit after a single inoculation was composed of 10^7 irradiated tumor cells infected by 32 hemagglutinating units (HU) of NDV Ulster. The first systematic optimization studies for the vaccine ATV-NDV were performed in breast carcinoma [44], colorectal carcinoma [45–48], and renal cell carcinoma [49,50] (Table 2). Two to three weeks after tumor operation, we applied, intradermally at the upper thigh at different sites, vaccine composed of different numbers of tumor cells and different doses of virus, as indicated in Table 1. Twenty-four to forty-eight hours later, the local skin reactions to the vaccine were measured by determining the mean diameter of induration. As can be seen from the results of Table 2, optimal skin reactions were observed

Table 2 Phase I Optimization Studies for the Vaccine ATV-NDV (Dose-Response Relationship)

Tumor type	Vaccine composition			Mean skin reactivity[1] (mm diameter)	% responder	
	Cell number	Virus	Dose (HU)		Renal carcinoma (n=20)	Colorectal carcinoma (n=20)
renal cell carcinoma						
"	1×10⁷	–	–	4.5		
'	"	NDV	8	5.8		
"	"	"	16	7.5*		
"	"	"	32	11.0*	90	
colorectal carcinoma						
"	1×10⁷	—	—	3.0		
"	"	NDV	4	4.0		67
"	"	"	32	7.0*		85
"	"	"	64	6.5		60
"	2×10⁷	"	32	6.5		50
"	2×10⁶	"	4	3.0		50
"	"	"	32	4.5		67
"	"	"	64	4.5		67

[1]* $p < 0.05$; optimal skin responses to the vaccine were observed in both studies and tumor types (45–50) with 1×10^7 tumor cells infected with 32 HU NDV Ulster.
n = number of patients.

in both studies with $1X10^7$ tumor cells infected with 32 HU NDV Ulster. With this vaccine formulation, 85% of colorectal carcinoma patients and 90% of renal carcinoma patients showed about 7 mm to 11 mm skin indurations at the vaccination site.

The fact that the optimal vaccine composition was similar in the ESb mouse tumor model and in cancer patients can perhaps best be explained on the basis of a local memory immune response. Such responses, which are similar to skin responses of allergic people, require only low amounts of antigen.

Specificity of Responses

Next we evaluated skin responsiveness in patients, not only to the vaccine ATV-NDV but also to autologous tumor cells without virus (ATV) and to various controls. The results in Table 3 show skin reactivities after the first inoculation. It can

Table 3 Skin Responsiveness to the Vaccine ATV-NDV, to Tumor Cells Alone, and to Various Controls

Cell type	Cell number	Virus	Dose (HU)	Mean induration (mm diameter)	n	% responder	Ref.
breast carcinoma					9		42–44
" ATV-NDV	5×10^6	NDV	32	2.6		89	
" ATV	5×10^6	–	–	0.75		22	
–	–	NDV	32	0.94		33	
ovarian carcinoma					7		42–44
" ATV-NDV	5×10^6	NDV	32	4.5		100	
" ATV	5×10^6	–	–	3.2		86	
–	–	NDV	32	0.0		0	
colorectal carcinoma					20		45–48
" ATV-NDV	1×10^7	NDV	32	7.0		85	
" ATV	1×10^7	–	–	4.0		65	
–	–	NDV	32	2.0		15	
normal colon mucosa	1×10^7	–	–	3.5	10	40	
CEA 30–60 μg				0.0		0	
renal carcinoma					20		49,50
" ATV-NDV	1×10^7	NDV	32	10.8		100	
" ATV	1×10^7	–	–	4.5		90	
normal kidney epithelium	1×10^7	–	–	0.0	0		

ATV = tumor cells alone; n = number of patients; CEA = carcinoembryonic antigen; HU = hemagglutinating unit.

be seen that the reactivity to ATV is much smaller, both in terms of mean induration and of percent responder patients. Reactions to the virus alone were either absent or very small. For specificity control of the DTH test, we performed either the Mérieux test, consisting of several recall antigens, or we used normal cells from either kidney epithelium (in renal carcinoma patients) or from colon-mucosa (in colorectal carcinoma patients), which had to be separated under the same conditions as the tumor cells. In 40% of colorectal carcinoma patients, we saw some reactivity to normal colon mucosa [45,48], whereas in none of the 20 renal carcinoma patients did we detect reactivity to normal kidney epithelium cells [50]. Whether the weak reactivity to normal colon mucosa is attributable to cross-reactions or represents true autoimmune responses remains to be investigated. We thus conclude that DTH responses to TAA can be distinguished from responses to recall antigens and from autoimmune responses. Reactivity to NDV alone was either not seen [42,44] or seen as weak induration in a minority of patients [42–48].

In a separate study, we demonstrated that the recall antigen responses of patients did not correlate with or allow prediction of their responsiveness to autologous tumor cells [48]. It therefore was not a better general immunocompetence that distinguished immunological responder from nonresponder patients. This was recently confirmed with ELISPOT results demonstrating distinct respective memory responses to recall antigens and to TAA in individual patients [51]. The similarity to memory ELISPOT responses suggests a strong involvement of TAA-specific memory cells in the DTH test. With the ELISPOT assay, we also found out that DCs, when pulsed with viral oncolysates from ATV-NDV vaccine, stimulate antitumor memory T cell responses from cancer patients more strongly than when pulsed with tumor lysate [52].

Effects of Repeated Vaccinations

We next tested the effect of repeated vaccinations on local skin reactivity to the vaccine ATV-NDV (Table 4). A positive skin response was defined as being more than 2-mm diameter induration. Although in tests of various tumor types [43] we had

Table 4 Effect of Repeated Vaccinations on Local Skin Reactivity to the Vaccine ATV-NDV

Carcinoma	% of patients			
	Various tumor types* (n = 27)		Colorectal PT (n=31)	LM (n = 20)
	1°	3°	3°	3°
% Nonresponder (< 2 mm)	22	3		
% Responder	77	93	100**	100†
Induration intensity:				
3–5 mm	37	26		
5–10 mm	33	19		
10–20 mm	7	48		

n = number of patients
PT = primary tumor; LM = liver metastasis
* Data from Ref. 43; 1 ° = first vaccination 3° = third vaccination.
** Mean diameter of induration and erythema at first vaccination = 18 mm and at last vaccination 26 mm; 68% of patients showed increased responsiveness. (From Ref. 53.)
† Mean diameter induration at first vaccination 8 mm and at last vaccination 9.5 mm. (From Ref. 45,46.)

seen 22% of nonresponders at the first ATV-NDV inoculation, virtually all patients became responders at the third vaccination. This was also true for studies in colorectal carcinoma patients who were vaccinated with vaccine derived either from the primary tumor [53] or from liver metastases [45–48]. There was not only an increase in the percentage of responder patients but also in the induration intensity. Whereas at the first vaccination, patients showed intensities of 3 mm to 5 mm or 5 mm to 10 mm, at the third vaccination, many showed an induration intensity of 10 to 20 mm [53].

Turning Immunological Nonresponders into Responders

We were able to induce antitumor DTH reactivity in immunological nonresponders (Table 5). Among 90 cancer patients

Table 5 Change of Immunological Nonresponders to Responders Following Vaccination with ATV-NDV

	ATV-NDV Induced DTH Reactivity to ATV in Patients Who Were Negative to ATV at First Test		
	Breast carcinoma*	Colorectal carcinoma with liver metastasis**	Various tumor types†
% of patients	43	9	30
n‡	44	23	23

DTH = delayed-type hypersensitivity.
* (From Ref. 54.)
** In two patients, the reaction increased from 0 to 10 mm and from 0 to 15 mm. (From Ref. 46.)
† Melanoma, colon carcinoma, rectum carcinoma, stomach carcinoma, ovarian carcinoma, breast carcinoma, renal cell carcinoma, and acute myeloblastic leukemia. (From Ref. 43.)
‡ n = number of patients; in total, among 90 patients tested, 31% showed induction of new DTH reactivity (change from nonresponder to responder).

tested who did not react to autologous tumor cells before vaccination, 31% changed from a nonresponder to a responder phenotype after vaccination with ATV-NDV (at least three times) (Table 4). Thus, in these 31% of patients, we were able to induce by vaccination a specific antitumor DTH memory response.

Potentiation of Antitumor Memory Responses

Next, we investigated the effect of antitumor vaccination with ATV-NDV in original responder patients who already reacted to ATV challenge before vaccination. Table 6 summarizes the results obtained from studies in breast carcinoma [54], ovarian carcinoma [55–56], colorectal carcinoma[45,46,48,53], renal carcinoma [49,50], and other tumor types [43]. Among 264 patients tested, 44% showed significantly increased DTH reactivity to ATV (without NDV) after the course of vaccination. A significant increase in DTH reactivity to ATV was defined by an increase of diameter greater than 3 mm. Thus, in these 44% of patients, we had increased the level of systemic antitumor immune memory reactivity.

Table 6 Potentiation of Antitumor Reactivity in Immunological Responders Following Vaccination with ATV-NDV

Potentiation of DTH Responses to ATV (Comparison of Last to First Reaction) % of patients with increases ≥ 3 mm						
	Breast carcinoma	Ovarian carcinoma	Colorectal carcinoma* PT	LM	Renal carcinoma**	Other tumor types†
% of patients	46	39	57	42	35	45
n‡	67	18	14	24	20	24

DTH = delayed-type hypersensitivity.
* PT = primary tumor; LM = liver metastasis. (From Ref. 45–48.)
** In this tumor type, increased DTH reactions were observed only when the tumor vaccine ATV-NDV was supplementedwith interleukin-2. (From Ref. 49,50.)
† See legend to Table 4.
‡ n = number of patients; in total, among 264 patients tested, 44% showed increased DTH reactivity to ATV after vaccination.

Nonresponder Patients

The great majority of cancer patients did not show a detrimental response after vaccination [46,54]. This is reassuring and supporting the present strategy of vaccination. The fraction of patients with a decrease was less than 10%. Table 7 shows the fraction of patients who, after vaccination, showed either no change or a decrease of DTH reactivity to autologous tumor cells.

Correlation Between Memory Delayed-Type Hypersensitivity Responses and Prognosis

A positive correlation between increase in DTH responsiveness to ATV after vaccination and prognosis was seen by us [50,57] and others [59]. A strong increase (>5 mm) in antitumor DTH reactivity to tumor challenge after vaccination (in comparison to the first reactivity) predicted a survival advantage (35 vs. 14 months), a correlation that was significant by Wilcoxon test [50]. It has been assumed that such a correlation reflects a better general immunocompetence and therefore better prognosis in this subgroup of patients [60]. This, however,

Table 7 Fraction of Immunological Nonresponder Patients: No Change or Decrease of DTH Reactivity Following Vaccination with ATV-NDV

	DTH response*	Colorectal carcinoma** PT (n=14)	Colorectal carcinoma** LM (n=24)	Breast carcinoma† (n=67)	Ovarian carcinoma† (n=18)
% of	– decrease	7	8	4	6
patients	– no change	35	50	21	22
with	– increase	57	42	71	61

* Decrease or increase defined as ≥ 3 mm.
** PT = primary tumor, LM = liver metastasis. (From Ref. 46.)
† (From Ref. 54.)
n = number of patients.

is unlikely because we and others did not see a correlation between response to general recall antigens (Mérieux test) or response to autologous tumor cells [57,59]. Because there was no DTH reactivity to normal kidney epithelial cells, the correlation seen in renal carcinoma patients [50] is likely attributable to activation of specific antitumor immune memory.

SIDE EFFECTS

The Phase I clinical studies were also performed to evaluate possible side effects. The intradermal vaccinations were well tolerated and could be repeated many times without causing serious problems. A few patients developed mild fever or mild headache for 1 to 2 days. There was no evidence for autoimmune phenomena such as vasculitis, rheumatoid arthritis, or lymphatic disorders [16,45,48,50,61].

Phase II Clinical Vaccination Studies: Evaluation of Efficacy

Having seen that antitumor vaccination of cancer patients with ATV-NDV could lead to a significant increase of systemic memory DTH responses to ATV, we next engaged in Phase II clinical vaccination studies to evaluate clinical efficacy of this approach. Table 8 through 11 summarize the most important results from a variety of Phase II studies performed over the past 10 years. Because the majority of studies have been published, I focus here only on the most important aspect—the survival data.

Two-Year Survival

Table 8 lists the 2-year survival benefits observed in four clinical studies. The survival rate in the group that received postoperative vaccinations with ATV-NDVwas compared to that of nonvaccinated control groups specified in the legend. Two studies were performed in colorectal carcinoma with either locally advanced tumor [53] or with operated solitary liver metastasis [48]. There was about 25% increase of the 2-year survival rate in both studies. Results from a recent study in glioblastoma multiforme [61] revealed a 32% increase of the 2-year survival rate. In a fourth study of patients with recurrent resectable malignant melanoma [8], there was a 20% increase of survival rate in a group of 21 patients. In this study, the comparison was made between a group treated with a vaccine containing at least 3×10^6 tumor cells and having greater 33% viable cells (b) and another group (a) treated with a vaccine in which both vaccine parameters were below that threshold. This internally controlled study suggests that the observed vaccination effect is not attributable to a placebo effect because both groups were vaccinated. Instead, the better survival was dependent on the dose ($>3 \times 10^6$ tumor cells) and quality (>33% cell viability) of the product.

Five-Year Survival

Table 9 shows the 5-year survival benefits after postoperative vaccination with ATV-NDV in two studies with [1] locally advanced primary breast cancer [62] and [2] locally advanced colorectal carcinoma [8]. In the latter study, we saw a 25% increase of the 5-year survival rate in a group of 20 patients who received a high-quality ATV-NDV vaccine [8]. In the breast carcinoma study, we saw a 36% increase in the 5-year

Table 8 Two-Year Survival Benefits After Postoperative Vaccination with ATV-NDV

Tumor type	Stage	2-Year survival rate		% Increase of survival rate	Reference
Colorectal carcinoma		control* (n=661)	vaccinated (n=57)	24	53
(median follow-up 22 months)	(a) locally advanced	73.8%	97.9%		
		control** (n=23)	vaccinated (n=23)		
(median follow-up 18 months)	(b) solitary liver metastases, RO resectable	61%	87%	26	48
Glioblastoma multiforme	grade IV	control** (n=25)	vaccinated (n=25)		
(median follow up 36 months)		4%	36%	32	61
		vaccinated† a) ATV-NDV < 3X10^{6}	vaccinated b) ATV-NDV > 3X10^{6}		
Malignant melanoma	recurrent resectable				
(median follow-up 18 months)		(n=20) 45%	(n=21) 65%	20	8

n = number of patients;
* Historic control from the same clinic.
** Pair-matched controls.
† This control group was vaccinated with a vaccine containing less than the indicated number of tumor cells and less than 33% cell viability. It is not historic but an actual internal control.

Table 9 Five-Year Surival Benefit After Postoperative Vaccination with ATV-NDV

Tumor type	Stage	5-Year survival rate a	5-Year survival rate b		% Increase of survival rate (b–a)	Reference
Breast carcinoma	locally advanced	ATV-NDV $<1.5\times10^6$ (n = 31) 48%	ATV-NDV $<1.5\times10^6$ (n = 32) 84%		36	35, 62
Colorectal carcinoma	locally advanced	ATV-NDV $>3\times10^6$ (n = 18) 60%	ATV-NDV $>3\times10^6$ (n = 20) 85%	histor. control (n = 48) 50%	25	8

n = number of patients; a = group of patients vaccinated with a vaccine containing less than the indicated number of tumor cells and less than 33% cell viability; as in Table 8, this is an actual internal control with an insufficient formulation of the same vaccine; a historic control is included in addition in the colorectal carcinoma study; b = verum group which received a high-quality formulation of the vaccine.

survival rate in a group of 33 vaccinated patients in whom the vaccine contained a least 1.5×10^6 tumor cells and greater than 33% viability [33,62].

Overall Median Survival

There was not only a benefit in overall survival as expressed by the percentage of patients surviving for 2 years or 5 years but also when calculating the overall median survival (OS) of the respective whole groups. Table 10 shows the results from two phase II clinical studies. Whereas in the control group of the malignant melanoma study, the OS was 63.6 weeks, in the group vaccinated with a high-quality vaccine, the survival was about twice as long–126 weeks. The median survival prolongation was 62.4 weeks. Similar results can be reported from the recent glioblastoma study. Here the OS in a group of 25 patients vaccinated with a high-quality vaccine was more than twice as long as in a pair-matched control group. Also, the progression free survival was about twice as long as in the control group [61].

Long-Term Survival

In four different tumor types for which antitumor vaccinations were performed with ATV-NDV vaccine from autologous

Table 10 Benefit in Median Survival

Tumor type	Stage	Median survival (weeks)		Median survival prolongation (weeks)	Reference
Malignant melanoma	recurrent, resectable	ATV-NDV* $<3\times10^6$ (n = 20)	ATV-NDV $>3\times10^6$ (n = 21)		
OS		63.6	126	62.4	8
glioblastoma multiforme	IV	control (n = 25)	ATV-NDV 1×10^7 (n = 25)		
PFS ($p = 0.00138$)		24	42	18	
OS ($p < 0.00001$)		44	92	48	61

* n = number of patients; OS = overall survival; PFS = progression-free survival; see legend to Table 8.

Table 11 Exceptionally Long-term Surivors Among Patients Treated with ATV-NDV Vaccine from Autologous Cell Cultures

Tumor type	Median survival time expected	Median survival observed*	Reference
Ovarian carcinoma	~ 12 months	30 months DFS in 15/24 (73%)	56
Glioblastoma multi-forme	~ 12 months	> 36 months in 3/25 (12%)	61
Pancreatic carcinoma	~ 12 months	> 36 months in 5/9 (55%)	8
Stomach carcinoma	~ 12 months	> 36 months in 3/7 (42%)	8

* Number of long-term survivors per number of patients vaccinated is indicated for the respective time period.

tumor cell cultures, we observed exceptionally long-term survivors [8,56,60]. Table 11 shows the median survival time that was expected from standard treatment and the median survival observed. It can be seen that a significant fraction of treated patients, varying from 12% to 73%, consisted of exceptionally long-term survivors. In a few long-term survivor patients, we recently succeeded in demonstrating, in peripheral blood, the presence of antitumor immune memory T cell reactivity [61]. This was observed even several years after the last vaccination, a finding consistent with a conditioning effect, as suggested earlier.

Conclusions from Survival Data

We like to critically review the obtained results and ask three relevant questions [60] that should be addressed to any cancer therapy:

1. Has the therapy the potential to cause antitumor effects? This question can be clearly answered positively. We were able to demonstrate in various tumor models as well as in studies with human cells that the vaccine can increase the activation of a variety of antitumor killer cells, including NK cells, macrophages [40,41], and cytotoxic T lymphocytes [63–65].

The HN molecule of the virus was proven to play an important role in the activation of killer activity by induction of a strong IFN-α response[13,15,16], by upregulation of TRAIL on monocytes [41] and T cells [13], and by increasing costimulatory signals in CTL precursors [12] and T helper cells [32].

2. Has this therapy a promising antitumoral activity in patients? Because most of our vaccination studies were performed in the postoperative adjuvant situation without residual tumor mass, no direct tumor responses could be determined. In these studies, an antitumor activity can only be deduced from the survival data. There were, however, singular cases of tumor responses in some studies. From 40 evaluable renal carcinoma patients treated with ATV-NDV and systemic IFN-α, 5 exhibited a complete response (CR); 6, partial remission (PR); and 12, stable disease (SD, median 25 months). Twenty-three of forty (57.5 %) patients (CR, PR, and SD) appeared to have a significant survival advantage compared with patients with progressive disease during the treatment period and with a historic reference group [49]. In two glioblastoma-patients [61] with residual disease after operation, we saw several months after vaccination with ATV-NDV, complete remissions of the residual tumor masses. In addition, a subgroup of seven glioblastoma patients with residual tumor mass showed, after vaccination with ATV-NDV, a median survival twice that of a nonvaccinated comparable group [61]. The second question therefore can also be answered positively.
3. Does this therapy improve survival of treated patients in comparison to standard treatment or to nontreated patients, and what is the effect on quality of life? Because the side effects of this treatment are negligible, we conclude that the quality of life of treated patients is not affected in a negative way. The evidence level for survival benefit reached so far is that of Phase II studies. In some studies, we used pair-matched con-

trols from patients treated at the same clinic during the same time period without vaccination; in others we used historic controls; and in several studies we used actual internal controls with a different formulation of the same vaccine. Prospective randomized controlled studies are needed to validate the findings from the Phase II studies. The observed improvements in survival rates, between 20% and 36%, the observed median survival prolongation by about 100%, together with significant fractions of exceptionally long-term survivors, however, speak in favor of clinical efficacy of this type of therapy.

A CRITICAL EVALUATION OF STANDARD THERAPIES

Detrimental Effects of Chemotherapy on Memory T Cells

Several of our Phase II clinical studies in which we introduced antitumor vaccination had to be combined with standard therapy such as chemotherapy (CT), radiotherapy (RT), or hormone therapy (HT) in the case of breast cancer. We therefore had to find out how to optimally combine immunotherapy with the other treatment procedures. Because CT and RT are known to be immunosuppressive, we tested the effect of these treatments on vaccination-induced DTH responses against autologous tumor cells in breast cancer patients. The results are shown in Table 12. When CT or RT was administered during the course of six vaccinations, there was a clear-cut detrimental effect on antitumor DTH responsiveness [42]. In contrast to the vaccinated control group (I), in which 46% showed significant increase, in the CT or RT group (II), only 8% showed significant increases and another 8% showed significant decreases of DTH reactivity. When testing previous antitumor DTH-responder patients under CT or RT therapy, only 20% responded to challenge by significant increases. Another 10% showed significant decreases. Fifty percent of such patients (II) showed decreased responses.

Table 12 Effect of Chemotherapy or Radiotherapy on DTH Responsiveness in Breast Cancer Patients

	% of patients*				
	I No CT or RT n = 67	II CT or RT during vaccination n=8	n=20**	III CT or RT 6 m before vaccination n = 15	IV CT or RT > 6 m before vaccination n = 9
Increase (> 2 ×)	46	8	20	33	55
Increase (< 2 ×)	22	40	20	14	23
No change	21	16	20	7	22
Decrease (> 2 ×)	4	8	10	–	–
Decrease (< 2 ×)†	3	28	50	47	–

CT = chemotherapy; RT = radiotherapy; DTH = delayed-type hypersensitivity; n = number of patients; m = months; aus. (From Ref. 44,54.)

* Adjuvant CT or RT was either not given (group I), or given during vaccination (between the third and fourth of six vaccinations, group II), or 6 months before the vaccinations (group III), or more than 6 months before the vaccinations (group IV).

** Previous DTH responder patients.

† Notice the increase in percent of patients in group II and III who, after vaccination and chemo- or radiotherapy, show a decrease of DTH reactivity.

When CT or RT was performed in a period of maximally 6 months before vaccination (group III), still 47% showed decreased responsiveness. In contrast, when vaccination was performed after a time period greater than 6 months (group IV) after CT or RT, a recovery of the original DTH responsiveness was noticed [42,54]. Based on these findings, we decided in all Phase II clinical studies in which chemotherapy or radiotherapy was involved to have a minimal distance of 6 months between these treatments and vaccination.

In a separate study, we investigated the effect of adjuvant therapy in breast cancer on the content of memory T cells in bone marrow. Thirty-four patients who received adjuvant CT or HT were tested 24 months after primary therapy and compared with a group of primary operated patients who had not yet received these treatments and with a group of normal healthy donors. The results revealed that the proportion of T cells was significantly lower after adjuvant systemic therapy than in the other two groups. Adjuvant therapy apparently had a significant detrimental effect not only on total T cell number but also on the content of memory T cells [66]. There were also significant differences between the CT and HT groups with respect to CD4 and CD8 T cells [66]. In contrast to chemotherapy, which preferentially affected CD4 T cells, tamoxifen treatment mainly affected CD8 T cells. The mechanisms behind these effects remain to be elucidated. The findings suggest that the two standard therapies in breast cancer, CT and HT, might have long-term detrimental effects on the immune system and affect negatively the memory content in bone marrow.

A Need for Rational Re-evaluation of Standard Therapies that Cannot Provide Evidence for Survival Benefit

The results from our clinical studies suggest that the immune system—and, in particular, antitumor immune memory T cells—can play an important role in long-term survival of treated cancer patients. Because many present-day adjuvant standard treatments such as CT, RT, and even HT are apparently immunosuppressive, the question arises whether these treatments nevertheless have a positive effect on long-term survival. An analysis of clinical trials performed with such treatment modalities reveals that a great majority of randomized trials shows only very little or no significant difference in overall survival between groups of patients treated by different protocols [67–69]. In spite of innovative techniques in tumor destructive therapies such as surgery, chemo- and radiotherapy, and some advances in the treatment of certain tumor types, overall cancer mortality in the United States could not be decreased in the past 20 years [67]. On the contrary, there was an age-corrected increase by about 6% [67]. With regard to chemotherapy of advanced carcinoma, independent meta-analyses revealed hardly any evidence for relevant improvements of overall survival [68,69]. For instance, a systematic review of published randomized trials involving 31,510 women with metastatic breast cancer who received cytotoxic and hormonal treatment revealed no evidence for survival improvement [70]. A systematic review and meta-analysis of individual patient data from 12 randomized trials of adult high-grade glioblastoma revealed very little benefit of chemotherapy [71].

Clinicians are often satisfied when a drug elicits a "tumor response" or relieves clinical symptoms even if there is no effect on overall survival. The assumption that a "tumor response" is linked with a survival benefit has been criticized as a "grand illusion" [68]. Because clinical therapy trials are very expensive, this field is strongly dominated by the pharmaceutical industry and therefore is profit oriented. Companies select trends and topics and also clinics and cooperation partners. Questions that are not product-related often are not financially supported. It does not come as a surprise that negative effects of drug treatments are rarely reported. Genentech Inc. alone reported in 2000, income of $750 million from the sale of its anti-cancer drugs. Companies selling so-called CAM can also make a good profit because patients long for such drugs.They also should provide evidence of effectiveness.

In recent years, the protocols for good manufactural procedures (GMP) and good clinical practise (GCP) have become more stringent. It is generally accepted that only evidence-based medicine should be paid for by society's funds. Because of these requirements for GMP and GCP and because Phase III clinical trials are very expensive, it is difficult for new biological treatment procedures to enter the market.

On the other hand, one realizes that standard therapies that were introduced long ago, before these stringent requirements were established, are not being re-evaluated. It is argued that it would be unethical to compare a treated verum group with a nontreated control. One should oppose, however, that the small benefits reported from old studies that did not fulfill modern standards may not be acceptable anymore. A re-evaluation of survival benefits of adjuvant systemic therapies in comparison to relevant and proper controls is therefore warranted.

A re-evaluation is also necessary because one cannot *a priori* exclude that a certain therapy could even have a negative effect on survival. A hopefully educational example is provided by postoperative radiotherapy of non–small-cell lung cancer. Although this treatment can improve a surrogate parameter—the time interval to local progression—the overall survival of such treated patients is reduced [72]. Radiotherapy can cause severe side effects such as pneumonitis and cannot prevent the development of distant metastases, which are likely more important with regard to overall survival than local recurrencies.

Economic Evaluations

Evaluations of treatments should include costs and consequences, namely health effects of the interventions. In cancer treatments, important health effects are disease-free survival, overall survival, toxicity, and the quality of life of the patients. Breast cancer is a major public health problem and therefore leads to considerable costs. Costs for antitumor vaccination by ATV-NDV are less than $10,000 per patient. In comparison, the health care costs of standard breast cancer treatment in the United States were estimated to be much higher. For instance, the incremental costs per life year gained were calculated to be $100,000 in a comparison of high-dose vs. standard chemotherapy and $300,000 in a comparison of chemotherapy plus tamoxifen for 5 years vs. 2 years [73]. The total health care costs of breast cancer in the Netherlands were estimated at 253 million Dutch guilders in 1988. This was about 13 % of the total health care costs of cancer. The highest costs were found for lymph node-negative and estrogen receptor-negative breast cancer patients, a fact which is linked with the low effectiveness of the treatment [73]. The Institute of Medicine of the U.S. Academy of Sciences provides further information on prospective randomized controlled studies and on routine patient care costs [74]. In recent years, companies have developed additional drugs that should alleviate side effects from the cytostatic drugs. This increases further the total health care costs without improving survival.

In conclusion, treatments of cancer as a chronic disease are very expensive and a big burden to the public health budget. A re-evaluation of the effectiveness, based on survival data, may help to reduce the costs.

GENERAL CONCLUSIONS

As an independent research scientist with more than 30 years of experience in tumor immunology and with great interest in the improvement of cancer therapy, I am used to the interpretation of data from experiments and from clinical trials. I therefore like to make some general conclusions from the situation as it appears to me at present:

1. There is a consensus that evidence-based medicine has to be the rule for any cancer treatment. Improvement of OS should always be the primary goal. This should therefore also be of importance for financial support through our health insurance systems.
2. In nine independent Phase II clinical studies, we observed considerable and significant improvements in OS when employing an immunotherapy protocol with a high-quality cancer vaccine. These studies were performed by different clinical partners at different clinics. The control groups used for comparison were either historic, pair-matched, or internal controls. Furthermore, the Phase II clinical studies were performed on different tumor types, including the most aggressive ones, such as pancreatic carcinoma and glioblastoma. They also included advanced metastasized stages of disease. Had I performed nine different animal experiments with nine different tumor lines and obtained each time significant survival improvements, I would be prepared to conclude that the treatment results are reproducible and that the treatment is broadly applicable.
3. The general health insurance systems in Germany are presently not prepared to pay for such treatments.
4. A comparison with the data basis for survival of the same types of cancer treated by conventional therapy

reveals lower survival rates. If one acknowledges this fact, the question arises as to why this may be.

5. In the following, I try to offer an explanation. The situation may have to do with the different effects that the respective treatments have on the patients immune system: While immunotherapies, performed with an optimized tumor vaccine, are boosting the patient's immune system, in particular their antitumor immune memory T cell responses, conventional systemic standard therapies have detrimental effects on the immune system, including its memory part. Although the immunosuppressive and other side effects of such treatments are known for many years, no serious efforts have been undertaken by pharmaceutical companies or by clinicians to care about reconstituting the immune system after such treatments and of activating it specifically against the tumor by antitumor vaccination.

Time for a Change!

I believe that it is time for a change of paradigm in oncology. Dogma has been prevalent in oncology for a long time. Prominent surgeons dominated for decades the scenery with a theory that cancer can be treated like a localized disease if only the operation procedure is radical enough.

Radical mastectomy, introduced by Halsted [75], was the treatment of choice for breast cancer of any size or type, regardless of the patient's age, for 80 years. Recent randomized studies now reveal that less extensive breast-conserving surgery can achieve similar 20-year survival rates [76,77]. Although the rate of local recurrence was higher in the group that received breast-conserving therapy, the overall survival rate was the same in both groups. The prognosis of breast cancer therefore depends less on the extent of local surgery than on the presence or absence of occult distant foci of metastatic cells and their control. As a result of these trials, about 300,000 women with early breast cancer worldwide each year can now undergo breast-conserving surgery rather than radical mastectomy.

Hopefully, in the future, a similar development will lead to a better application of chemotherapy, avoiding overdosing and avoiding application in patients who do not profit from it. Sensitive in vitro assays allowing pretesting of the chemosensitivity of a tumor [78] may help to more effectively apply chemotherapy to cancer patients.

Conservative treatment modalities such as surgery, chemotherapy, or radiotherapy should be evaluated in the same way as all new treatment modalities, namely, on the basis of evidence for survival benefit in randomized controlled studies. Claims for financial support should be subject to the same critical evaluation procedures. The survival benefit from any cancer treatment must become absolute priority. All other surrogate parameters are secondary.

Let us now go back to the drawing board for an evaluation of the effect of different strategies for cancer treatment on immune memory and long-term survival.

Change of Paradigm

I am proposing that in many situations, in which tumor cells have already disseminated in the body at the time of operation, it is very difficult for any type of additional therapy to eradicate the last disseminated tumor cell in the body. I am convinced that this is not necessary because the host's immune system is basically capable of controlling and preventing outgrowth of tumors from such disseminated cells as long as the tumor burden is low and below a clinically detectable level. Such an immune control of residual tumor burden, however, requires an intact immune system and, in particular, a certain threshold of antitumor immune memory T cells. It has been our experience that the existence of such immune memory can be the basis for a very efficient antitumor immune control being capable of resisting tumor challenge by a lethal tumor dose. It is of great importance to study the many factors that are important for the maintenance of long-term immune memory because this is very relevant for long-term survival. Therefore, the tumor- bearing host and, in particular, the patient's immune system should play an active part and be integrated in any cancer therapy strategy.

The Tumor-Bearing Host Contributes with Immune Memory a Component of Importance for Survival

Figures 4 through 6 are designed to illustrate what is meant by this proposition. Antigen-specific T cell-mediated immune responses are characterized by a first expansion phase, a second retraction phase, and a third memory phase. Figure 4 illustrates these three phases for a primary immune response (left) and a secondary immune response (right). Secondary immune T cell responses are initiated by antigen-experienced memory T cells and are faster and superior to primary T cell

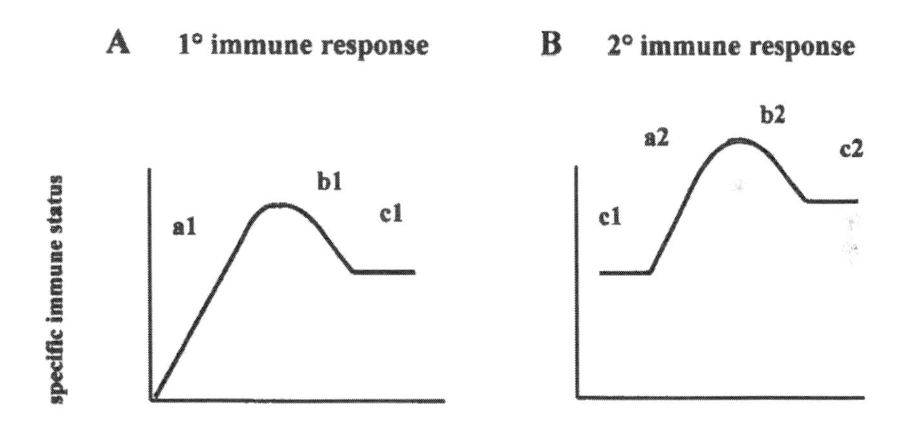

Figure 4 The three phases of immune T cell responses. (**A**) primary T cell responses are initiated by naïve antigen-specific T cells [the frequency of which is usually very low (< 1 in 100,000 T cells)] upon interaction with antigen-presenting dendritic cells in secondary lymphoid organs. (**B**) Secondary T cell responses are initiated by antigen-experienced memory T cells. These responses are faster than those of naïve T cells and have less stringent activation requirements. They can be stimulated by high-quality tumor cell vaccines. Primary and secondary immune responses can be each divided into three phases (**a–c**): (a1,2) expansion phase; (b1,2): retraction phase; (c1,2): memory phase

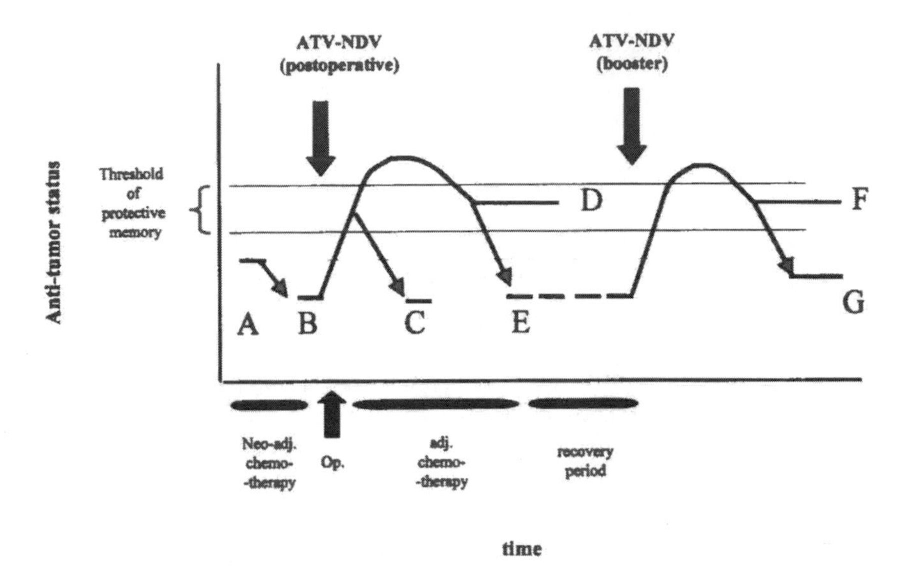

Figure 5 Principle effects of immunotherapy (D, F) or chemotherapy (B,C,E,G) on immune memory in different situations. **A** level of pre-existing immune memory in cancer patient before any therapy **B** detrimental effect of neo-adjuvant chemotherapy on pre-existing memory **C** detrimental effect of adjuvant chemotherapy on pre-existing memory **D** positive effect of postoperative vaccination: increase of memory to relevant threshold **E** detrimental effect of adjuvant chemotherapy during vaccination on maintenance of long- term memory **F** positive effect of booster vaccination if performed more than 6 months after chemotherapy; increase of memory to relevant threshold **G** unsuccessful effect of booster vaccination if performed less than 6 months after chemotherapy; memory decreases below relevant threshold

responses initiated by naïve T cells. Also, the activation requirements for immune memory T cell responses are less stringent than those for naïve T cells (also discussed in "Characteristics of Memory T Cells," in the third section, "Development of the New Concept in Animal Tumor Models," above). Thus, tumor vaccines can be specifically designed to stimulate memory T cell responses and we believe that this is the case with the tumor vaccine ATV-NDV.

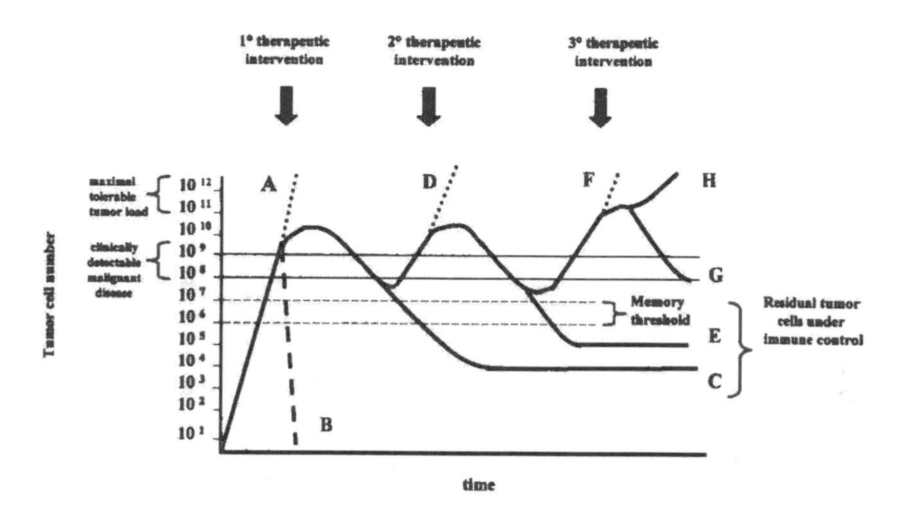

Figure 6 Different effects of therapeutic interventions for cancer treatment. **A** Tumor progression (if untreated or if therapy is ineffective) **B** complete cure (usually after surgery, when tumor has not spread) **C, E, G** stable disease; clinically tumor free; residual tumor cells under control of immune memory **D, F, H** tumor recurrence after 1°, 2° or 3° intervention

Figure 5 illustrates the effects of immunotherapy or chemotherapy on immune memory in different situations of cancer patients. The situations illustrate what we have examined in a variety of clinical situations and summarized in this review.

Figure 6 illustrates, in general, different effects of therapeutic interventions for cancer treatment. Until a tumor becomes clinically detectable, a transformed somatic cell must have undergone approximately 30 cell divisions to reach a tumor cell number of approximately 10^9, making up a tumor mass of about 1 g. If the tumor would be dividing further for 10 more cell divisions (situation A) it would reach the maximal tolerable tumor load of 10^{12} cells, representing a mass of about 1 kg. Clinical therapeutic intervention by surgery usually starts at a tumor size of 1 g to 10 g. If, at the time of surgery, the tumor has not disseminated, a complete cure by surgery can be achieved (situation B). If therapeutic intervention by surgery or any other means achieves a tumor reduction above 99%, there may still be millions of tumor cells left in the body that cannot be reached by these interventions. C illustrates a situation that is stable because immune memory is able to control a remaining tumor load after therapy. Such active immune control could cause long-term survival, as long as a certain memory threshold remains and the tumor is not developing immune escape variants. Situation D illustrates an effect where a tumor response to primary treatment may be greater 90%. Nevertheless, following some time delay, the tumor will grow again and kill the patient unless there is a second therapeutic intervention. E illustrates a situation in which a secondary intervention could reduce the remaining tumor load below the memory threshold so that the cells could become subject to immune control. H and F show situations in which the tumor eventually has become resistant to the type of treatments. G shows a situation in which a third therapeutic intervention could reduce the tumor load below a detectable level.

Figure 6 is shown basically to demonstrate the new concept that therapeutic interventions might be sufficient if they reduce the tumor burden below a certain threshold. This can often be further controlled by the immune system provided that it is not damaged by either the tumor or by the intervention itself. Situations C and E illustrate an important change of paradigm. In these situations, the tumor is controlled by a combination of therapeutic intervention and by the patient's own immune system. This could result in a long-term clinically stable situation.

In support of this new concept, McCoy et al [79] recently demonstrated that cell-mediated immunity to TAA is a better predictor of survival than stage, grade, and lymph node status in early stage breast cancer.

ACKNOWLEDGMENT

Our concepts and procedures of preparation of individual vaccines are admittedly unusual but a conflict of interest cannot be claimed. None of the clinical studies presented has been supported by a pharmaceutical company nor by a public grant. The patents I hold on the ATV-NDV technology (EP 03331 102, US 5273745), on NDV HN DNA and protein (EP/0689596), and on NDV-binding bispecific antibodies (US 5911987; 4407 538) remain unused for many years.

We are particularly grateful for private support from the Mhlschlegel Stiftung and the Dietmar Hopp Stiftung. I would like to thank the following clinicians for their engagement and motivation, without whom none of these studies would have been possible. Their names will be associated with the respective cancer type: Colorectal (primary advanced): D. Ockert, N. Beck, E. Stoelben, J. Flechtenmacher, E. Hagmller, M. Nagel, H.-D. Saeger; Colorectal (liver metastasis): P. Schlag, W. Liebrich, B. Lehner, M. Manasterski; Renal carcinoma: S. Pomer, R. Thiele, D. Brkovic, G. Staehler; Ovarian carcinoma: V. Möbus, S. Horn, M. Stöck; Breast carcinoma: T. Ahlert (http://www.biotech-praxislab.de/), G. Bastert, W. Sauerbrei, S. Ruhland, B. Bartik, N. Simiantonaki, J. Schumacher, B. Hcker, M. Schumacher, I. J. Diel, E.F. Solomayer; Metastatic melanoma: T. Pröbstle, Kamanabrou; Glioblastoma: C. Herold-Mende, H. H. Steiner; Pancreatic and gastric carcinoma: R. Gerhards. For excellent contributions and help in this translational research I would like to thank the following scientists from my group: V. Umansky, P. Beckhove, M. Feuerer, and P. Fournier.

REFERENCES

1. Gallucci S, Matzinger P. Danger signals: SOS to the immune system. Curr Opin Immunol 2001; 13:114–119.
2. Akira S, Takeda K, Kaisho T. Toll-like receptors: critical proteins linking innate and acquired immunity. Nat Immunol 2001; 2:675–680.
3. Luster AD. The role of chemokines in linking innate and adaptive immunity. Curr Opin Immunol 2002; 14:129–135.
4. Heicappell R, Schirrmacher V, von Hoegen P, Ahlen T, Aplehans. Prevention of metastatic spread by postoperative im-

munotherapy virally modified autologous tumor cells. I. Parameters for optimal therapeutic effects. Int J Cancer 1986; 37: 569–577.

5. Plaksin D, Progador A, Vadai E, Feldman M, Shirrmacher V, Eisenbach L. Effective anti metastatic melanoma vaccination with tumor cells transfected with MHC genes and/or infected with Newcastle Disease Virus (NDV). Int J Cancer 1994; 59: 796–801.
6. Shoham J, Hirsch R, Zakay-Rones Z, Osband ME, Brenneth HJ. Augmentation of tumor cell immunogenicity by viruses—an approach to specific immunotherapy of cancer. Nat Immun 1990; 9:165–172.
7. Bier H, Armonat G, Bier J, Shirrmacher V, Ganzer U. Postoperative active-specific immunotherapy of lymph node micrometastasis in a Guinea pig tumor model. Otorhinopharyngology 1989; 51:197–205.
8. Schirrmacher V, Ahlert T, Pröbstle T, Steiner HH, Herold-Mende C, Gerhards R, Hagmüller E. Immunization with virus modified tumor cells. Semin Oncol 1998; 25:677–696.
9. Phillips RJ, Samson AC, Emmerson PT. Nucleotide sequence of the 5′-terminus of Newcastle Disease virus and assembly of the complete genomic sequence: agreement with the "rule of six". Arch Virol 1998; 143:1993–2002.
10. Lorence RM, Roberts MS, Groene WS, Rabin H. Replication-competent, oncolytic Newcastle Disease virus for cancer therapy. In: Hernáiz Driever P, Rabkin SD, Eds. Replication-Competent Viruses for Cancer Therapy. Vol. 22. Basel. Karger: Monographs in Virology, 2001:160–182.
11. Schirrmacher V, Haas C, Bonifer R, Ahlert T, Gerhards R, Ertel C. Human tumor cell modification by virus infection: an efficient and safe way to produce cancer vaccine with pleiotropic immune stimulatory properties when using Newcastle Disease virus. Gene Ther 1999; 6:63–73.
12. Ertel C, Millar NS, Emmerson PT, Schirrmacher V, von Hoegen P. Viral hemagglutinin augments peptide specific cytotoxic T-cell responses. Eur J Immunol 1993; 23:2592–2596.
13. Zeng J, Fournier P, Schirrmacher V. Induction of interferon α and TRAIL in human blood mononuclear cells by HN but not F protein of Newcastle Disease virus. Virology 2002; 297:19–30.

14. Alexopoulou L, Holt AC, Medzhitov R, Flavell RA. Recognition of double-stranded RNA and activation of NF-kappa B by Toll-like receptor. Nature 2001; 413:732–738.

15. Zeng J, Fournier P, Schirrmacher V. Stimulation of human natural interferon-α response via paramyxo-virus hemagglutinin lectin-cell interaction. J Mol Med 2002; 80:443–451.

16. Washburn B, Schirrmacher V. Human tumor cell infection by Newcastle Disease virus leads to upregulation of HLA and cell adhesion molecules and to induction of interferons, chemokines and finally apoptosis. Int J Oncol 2002; 21:85–93.

17. Pulendran B, Banchereau J, Maraskovsky E, Maliszewski C. Modulating the immune response with dendritic cells and their growth factors. Trends Immunol 2001; 22:41–47.

18. Le Bon A, Schiavoni G, D'Agostino G, Gresser I, Belardelli F, Tough DF. Type I interferons potently enhance humoral immunity and can promote isotype switching by stimulating dendritic cells in vivo. Immunity 2001; 14:461–470.

19. Sato K, Hida S, Takayanagi H, Yokochi T, Kayagaki N, Takeda K, Yagita H, Okumura K, Tanaka N, Taniguchi T, Ogasawara K. Antiviral response by natural killer cells through TRAIL gene induction by IFN-alpha/beta. Eur J Immunol 2001; 31: 3138–3146.

20. Rogge L, Barberis-Maino L, Biffi M, Passini N, Presky DH, Gubler DH, Sinigaglia E. Selective expression of an interleukin-12 receptor component by human T helper 1 cells. J Exp Med 1997; 185:825–831.

21. Schirrmacher V, Heicappell R. Prevention of metastatic spread by postperative immunotherapy with virally modified autologous tumor cells. II. Establishment of specific systemic anti tumor immunity. Clin Exp Metastasis 1987; 5([2]):147–156.

22. Khazaie K, Prifti S, Beckhove P, Griesbach A, Russell S, Collins M, Schirrmacher V. Persistence of dormant tumor cells in the bone marrow of tumor-cell-vaccinated mice correlates with long term immunological protection. Proc Natl Acad Sci U S A 1994; 91:7430–7434.

23. Müller M, Gounari F, Prifti S, Hacker HJ, Schirrmacher V, Khazaie K. Eb-lacZ tumor dormancy in bone marrow and lymph nodes: active control of proliferating tumor cells by CD8+ immune T cells. Cancer Res 1998; 58:5439–5446.

24. Schirrmacher V. T-cell immunity in the induction and maintenance of a tumor dormant state. Semin Cancer Biol 2001; 11: 285–295.

25. Veiga-Fernandes WU, Bourgeois C, McLean A, Rocha B. Response of naive and memory CD8+ T cells to antigen stimulation in vivo. Nat Immun 2000; 1:47–53.

26. Kaech SM, Wherry EJ, Ahmed R. Effector and memory T cell differentiation: implications for vaccine development. Nature Rev 2002; 2:251–262.

27. Mahnke Y, Schirrmacher V. A novel tumor model system for the study of long-term protective immunity and immune T cell memory. Cell Immunol 2003; Feb 221(2):89–99.

28. Karrison TG, Ferguson DJ, Meier P. Dormancy of mammary carcinoma after mastectomy. J Natl Cancer Inst 1999; 91: 80–85.

29. Pantel K, Otte M. Occult micrometastasis: enrichment, identification and characterization of single disseminated tumor cells. Cancer Biol 2001; 11:327–337.

30. Feuerer M, Rocha M, Bai L, Umansky V, Solomayer EF, Bastert G, Diel IJ, Schirrmacher V. Enrichment of memory T cells and other profound immunological changes in the bone marrow from untreated breast cancer patients Int J Cancer 2001; 92([4]):1–1.

31. Feuerer M, Beckhove P, Bai L, Solomayer EF, Bastert G, Diel IJ, Heep J, Oberniedermayr M, Schirrmacher V, Umansky V. Therapy of human tumors in NOD/SCID mice with patient derived re-activated memory T cells from bone marrow. Nat Med 2001; 7([4]):452–458.

32. Termeer CC, Schirrmacher V, Bröcker EB, Becker JC. Newcastle-Disease-Virus infection induces a B7-1/ B7-2 independent T-cell-costimulatory activity in human melanoma cells. Cancer Gene Ther 2000; 7([2]):316–323.

33. Ghanta VK, Hiramoto NS, Soong SJ, Hiramoto RN. Conditioning of the secondary cytotoxic T-lymphocyte response to YC8 tumor. Pharmacol Biochem Behav 1995; 50([3]):399–403.

34. Exton MS, von Auer AK, Buske-Kirschbaum A, Stockhorst U, Gobl U, Schedlowski M. Pavlovian conditioning of immune

function: animal investigation and the challenge of human application. Behav Brain Res 2000; 110((1–2)):129–141.

35. Ahlert T, Sauerbrei W, Bastert G, Ruhland S, Bartik B, Simiantonaki N, Schumacher J, Häcker B, Schumacher M, Schirrmacher V. Tumor cell number and viability as quality and efficacy parameters of autologous virus modified cancer vaccines. J Clin Oncol 1997; 15:1354–1366.
36. Ahlert T, Schirrmacher V. Isolation of a human melanoma adapted Newcastle Disease Virus mutant with highly selective replication patterns. Cancer Res 1990; 50:5962–5968.
37. Matzinger P. The danger model: a renewed sense of self. Science 2002; 296:301–305.
38. Kobayashi H, Sendo F, Shirai T, Kaji H, Kodama T, Saito H. Modification in growth of transplantable rat tumors exposed to Friend virus. J Natl Cancer Inst 1969; 42:413–419.
39. Schirrmacher V, von Hoegen P. Importance of tumor cell membrane integrity and viability for CTL activation by cancer vaccines. Vaccine Res 1993; 2:183–196.
40. Schirrmacher V, Bai L, Umansky V, Yu L, Xing Y, Qian Z. Newcastle Disease virus activates macrophages for anti-tumor activity. Int J Oncol 2000; 16:363–373.
41. Washburn B, Wiegand MA, Grosse-Wilde A, Janke M, Stahl H, Rieser E, Sprick MR, Shirrmacher V, Walczak H. TNF-related apoptosis-inducing ligand mediates tumoricidal activity of human monocytes stimulated by Newcastle Disease Virus. J Immunol 2003; Feb. 15; 170(4):1814–21.
42. Ahlert T, Gremm B, Kohler S, Rexin M, Hoffmann R, Terinde R, Rethfeld E, Schirrmacher V, Kaufmann M, Heinrich H, Meisenbacher G, Bastert G. Aktiv spezifische Immuntherapie (ASI). 9. Arbeitsgespräch der klinischen Tumorimmunologie in der Gynäkologie. Aus: Aktuelle Onkologie, Band 79 (Hrsg. U. Koldovski, R. Kreienberg) 1994; 61:236–244.
43. Schirrmacher V, von Hoegen P, Schlag P, Liebrich W, Lehner B, Schumacher K, Ahlert T, Bastert G. Active specific immunotherapy with autologous tumor cell vaccines modified by Newcastle Disease virus: experimental and clinical studies. In: Schirrmacher V, Schwartz-Albiez R, Eds. Cancer Metastasis. Berlin-Heidelberg-New York: Springer Verlag, 1989:157–170.

44. Ahlert T, Bastert G, Schirrmacher V. Mamma- und Ovarialkarzinom mit autologen virusmodifizierten Tumorzellen, aktivspezifische Immuntherapie (ASI). Theorie, Praxis Perspektiven. T. W. Gynäkologie 1989; 2:359–367.

45. Lehner B, Schlag P, Liebrich W, Schirrmacher V. Postoperative active specific immunization in curatively resected colorectal cancer patients with virus-modified autologous tumor cell vaccine. Cancer Immunol Immunther 1990; 32:173–178.

46. Manasterski M, Liebrich W, Möller P, Schirrmacher V, Schlag P. Active specific immunotherapy in colorectal cancer and melanoma. In: Klapdor R, Ed. Recent Results in Tumor Diagnosis and Therapy. Zuckschwerdt Verlag: München:, 1990:499–504.

47. Lehner B, Liebrich W, Mechtersheimer G, Schirrmacher V, Schlag P. Charakterisierung und erste Ergebnisse einer aktiven spezifischen Immuntherapie bei Patienten mit colorectalem Carcinom. Langenbeck's Arch. Chir (Suppl. Chir. Forum) 1989:513–517.

48. Schlag P, Manasterski M, Gerneth TH, Hohenberger P, Dueck M, Herfarth CH, Liebrich W. SchirrmacherV. Active specific Immunotherapy with NDV modified autologous tumor cells following liver metastases resection in colorectal cancer: first evaluation of clinical response of a Phase II trial. Cancer Immunol Immunother 1992; 35:325–330.

49. Pomer S, Thiele R, Staehler G, Löhrke H, Schirrmacher V. Tumor vaccination in renal cell carcinoma with and without interleukin-2 (IL-2) as adjuvant. A clinical contribution to the development of effective active specific immunization. Urologe A 1995; 34:215–220.

50. Pomer S, Schirrmacher V, Thiele R, Löhrke H, Staehler G. Tumor response and 4 year survival data of patients with advanced renal cell carcinoma treated with autologous tumor vaccine and subcutaneous r-IL-2 and IFN-Alpha $_{2b}$. Int J Oncol 1995; 6:947–954.

51. Bai L, Beckhove P, Feuerer M, Umansky V, Solomayer EF, Diel IJ, Schirrmacher V. Cognate interactions between memory T cells and tumor antigen presenting dendritic cells from bone marrow of breast cancer patients: bi-directional cell stimulation survival and anti-tumor activity in vivo. Int J Cancer 2003; 103: 73–83.

52. Bai L, Koopmann J, Fiola C, Fournier P, Schirrmacher V. Dendritic cells pulsed with viral oncolysates potently stimulate autologous T cells from cancer patients. Int J Oncol 2002; 21: 685–694.

53. Ockert D, Schirrmacher V, Beck N, Stoelben E, Ahlert T, Flechtenmacher J, Hagmüller E, Nagel M, Saeger HD. Newcastle Disease Virus infected intact autologous tumor cell vaccine for adjuvant active specific immunotherapy of resected colorectal carcinoma. Clin Cancer Res 1996; 2:21–28.

54. Ahlert T, Striffler H, Bastert G, Kaufmann M, Schirrmacher V. Aktueller Stand gynäkologischer Studien zur aktiv-spezifischen Immuntherapie mit virusmodifizierten autologen Tumorzellen. Aus: Aktuelle Onkologie, 60. "Klinische Tumorimmunologie in der Gynäkologie". Hrsg. Melchert, Neuses, Wischink, 8. Arbeitsgespräch Mannheim 20.–21. Okt., 1989.

55. Möbus V, Kreienberg R, Schirrmacher V. Erfahrungen mit der aktiv-spezifischen Immuntherapie bei gynäkologischen Malignomen. Aktuelle Onkologie 60:196-205.

56. Möbus V, Horn S, Stöck M, Schirrmacher V. Tumor cell vaccination for gynecological tumors. Hybridoma 1993; 12([5]): 543–547.

57. Liebrich W, Schlag P, Manasterski M, Lehner B, Stöhr M, Möller P, Schirrmacher V. In vitro and clinical characterization of a Newcastle Disease virus-modified autologous tumor cell vaccine for treatment of colorectal cancer patients. Eur J Cancer 1991; 27:703–710.

58. Bohle W, Schlag P, Liebrich W, Hohenberger P, Manasterski M, Möller P, Schirrmacher V. Postoperative active specific immunization in colorectal cancer patients with virus-modified autologous tumour cell vaccine: first clinical results with tumour cell vaccines modified with live but avirulent Newcastle Disease Virus. Cancer 1990; 66:1517–1523.

59. McCune CS, O'Donnell RW, Marquis DM, Saharrabudhe DM. Renal cell carcinoma treated by vaccines for active specific immunotherapy: correlation of survival with skin testing by autologous tumor cells. Cancer Immun Immunother 1990; 32:62–66.

60. Abel U. Grundlagen der Biometrie. In: Beuth J, Ed. Hrsg. Grundlagen der Komplementäronkologie, Stuttgart: Hippokrates Verlag. 2002:51.

61. Steiner HH, Bonsanto MM, Beckhove P, Brysch M, Schuele-Freyer R, Geletneky K, Kremer P, Golamrheza R, Bauer H, Kunze S, Schirrmacher V, Herold-Mende C. Anti-tumor vaccination of patients with glioblastoma multiforme in a case-control study: feasibility safety and clinical benefit. J Clin Oncol, in press.

62. Schirrmacher V, Feuerer M, Beckhove P, Ahlert T, Umansky V. T cell memory, anergy and immunotherapy in breast cancer. J Mammary Gland Biol Neoplasia 2002; 7([2]):201–208.

63. Hoegen P, Weber E, Schirrmacher V. Modification of tumor cells by a low dose of Newcastle Disease Virus: augmentation of the tumor-specific T cell response in the absence of an anti-viral response. Eur J Immunol 1988; 18:1159–1166.

64. Schild HJ, von Hoegen P, Schirrmacher V. Modification of tumor cells by a low dose of Newcastle Disease Virus: II. Augmented tumor specific T cell response as a result of CD4+ and CD8+ immune T cell cooperation. Cancer Immunol Immunother 1988; 28:22–28.

65. Von Hoegen P, Zawatzky R, Schirrmacher V. Modification of tumor cells by a low dose of Newcastle Disease Virus: III. Potentiation of tumor specific cytolytic T cell activity via induction of interferon, alfa, beta. Cell Immunol 1990; 126:80–90.

66. Solomayer EF, Feuerer M, Bai L, Umansky V, Meyberg GC, Bastert G, Schirrmacher V, Diel IJ. Influence of adjuvant hormone therapy and chemotherapy on the immune system analysed in the bone marrow of patients with breast cancer. Clin Cancer Res 2003; 9:174–180.

67. Beuth J, Ed Grundlagen der Komplementär-Onkologie. Theorie und Praxis: Stuttgart: Hippokrates Verlag, 2002.

68. Moss RW. The grand illusion of chemotherapy. Deutsche Zeitschrift für Onkologie 2001; 33:15–18.

69. Abel U. Die zytostatische Chemotherapie fortgeschrittener Karzinome: 2. Aufl.. Stuttgart: Hippokrates-Verlag, 1955.

70. Fossati R, Confalonieri C, Torri V, Ghislandi E, Penna A, Pistotti V, Tinazzi A, Liberati A. Cytotoxic and hormonal treatment for metastatic breast cancer: a systematic review of published randomized trials involving 31 510 women. J Clin Oncol 1998; 16:3439–3460.

71. Stewart LA. Chemotherapy in adult high-grade glioma: a systematic review and meta-analysis of individual patient data from 12 randomized trials. Lancet 2002; 359:1001–1018.

72. Port Meta-analysis Trialists Group: Postoperative radiotherapy in non-small-cell lung cancer: systematic review and meta-analysis of individual patient data from nine randomized controlled trials. Lancet 1998; 359:257–263.

73. van Enckevort PJ, TenVergert EM, Schrantee S, Rutten FFH, de Vries EGE. Economic evaluations of systemic adjuvant breast cancer treatments: methodological issues and a critical review. Crit Rev Oncol Hematol 1999; 32:113–124.

74. Committee on routine patient care costs. In: Aaron HJ, Gelband H, Eds Clinical Trials for Medicare Beneficiaries, Extending Medicare Reimbursement in Clinical Trials. Institute of Medicine, National Academy of Sciences, 2000. Erhältlich unter (available at: www.nap.edu/openbook/0309068886/html/15.html).

75. Halsted WS. The results of radical operation for the cure of cancer of the breast. Ann Surg 1907; 46:1.

76. Veronesi U, Cascinelli N, Mariani L, Greco M, Saccozzi R, Luini A, Aguilar M, Marubini E. Twenty-year follow-up of a randomized study comparing breast-conserving surgery with radical mastectomy for early breast cancer. N Engl J Med 2002; 347: 1227–1232.

77. Fisher B, Anderson S, Bryant J, Margolese RG, Deutsch M, Fisher ER, Jeong JH, Wolmark N. Twenty-year follow-up of a randomized trial comparing total mastectomy, and lumpectomy plus irradiation for the treatment of invasive breast cancer. N Engl J Med 2002; 347:1233–1241.

78. Konecny G, Crohns C, Pegram M, Felber M, Lude S, Kurbacher C, Cree IA, Hepp H, Untch M. Correlation of drug response with the ATP tumorchemosensitivity assay in primary FIGO stage III Ovarian cancer. Gynecol Oncol 2000; 77([2]):258–263.

79. McCoy JL, Rucker R, Petros JA. Cell-mediated immunity to tumor-associated antigens is a better predictor of survival in early stage breast cancer than stage, grade, or lymph node status. Breast Cancer Res 2000; 60:227–234.

5

Newcastle Disease Virus: Its Oncolytic Properties

JOSEPH C. HORVATH

St.Joseph's Hospital Cancer Institute Biotherapy
Research Laboratory, University of South Florida
College of Medicine, Department of Medical
Microbiology & Immunology
Tampa, FL

INTRODUCTION

Newcastle disease afflicts at least 236 species from 27 of the 50 orders of birds and is the most serious infectious disease problem in the poultry industry as well as backyard chicken farming worldwide [1]. The causative agent is Newcastle disease virus (NDV), a paramyxovirus. Highly pathogenic (velogenic) strains of NDV are responsible for epidemics, whereas moderate or avirulent (mesogenic and lentogenic) strains are frequently used as vaccines. The molecular characterization of pathogenic and less pathogenic strains is relatively easy and straightforward [2,3]. In spite of high contagiousness and severity of the disease in birds, NDV causes only very mild transient infection in humans, with no difference between pathogenic and less pathogenic strains [4–6]. Medical interest in NDV soared since it was discovered that the virus is oncolytic—i.e., it preferentially kills human tumor cells, while normal cells are spared [7–9].

THE VIRION

Newcastle disease virus is a member of the *Paramyxoviridae* family; it is an enveloped virus with a nonsegmented (−) strand RNA genome of 15186 nucleotides (nt). The order of viral proteins encoded by the genome is nucleoprotein (NP)-phosphoprotein (P)-matrix (M)- fusion (F)- hemagglutinin-neuraminidase (HN)- "large", RNA polymerase (L). The coding region is flanked by regulatory sequences named *leader* and *tailer* at the 3′ and 5′ ends. Further proteins are translated from the P protein gene by mRNA editing (*vide infra*). At the beginning and the end of each gene, there are transcriptional control sequences that are transcribed to the polycistronic mRNA. The core N as well as the P and L proteins are attached to the RNA genome and form the helical nucleocapside structure, which is the template for mRNA and antigenome synthesis. The HN and F proteins embedded in the viral envelope surrounding the nucleocapsid play a role in the attachment of the virus to and fusion with the cell membrane. The M protein fills the space between the nucleocapsid and the envelope.

Until recently, NDV was listed as a member of the *Rubulavirus* genus; however, it was different enough from the rest of the members of that genus to form the *Avulavirus* genus [6].

THE VIRAL REPLICATION CYCLE

The NDV replication cycle starts with binding of the virion to protein and lipid sialoglycoconjugate cell membrane receptors through HN and F proteins in concert. The hydrophobic fusion peptide domains of the F protein homotrimers are embedded in the cell membrane and induce its fusion with the virion and with other surrounding cells, causing syncytium formation (for review see Ref. 10–12). Following fusion, the helical viral nucleocapsid is released into the cytoplasm.

Viral replication takes place in the cytoplasm. It is remarkable, however, that the M protein contains a dipartite nuclear localization signal and a large portion of it is transported to the nucleus and nucleolus [13–15]. The significance of the nuclear localization of the M protein is not known but besides being a structural viral component, M proteins of several paramyxoviruses also inhibit host-cell and (during packaging and budding) viral transcription [16,17]. The viral HN and F proteins are synthesized as precursors and further processing is required for their biological activity. A 90 aa (amino acid) peptide is removed from the C-terminal of HN (a type II membrane protein) by cellular enzymes [18]. The F protein (type I membrane protein) is also cleaved close to the N-terminal, distal to the extensively hydrophobic fusion peptide, resulting in two fragments (F_1 and F_2) held together by a –S–S– bridge. The enzyme(s) responsible for the cleavage is probably furin and/or PC6 [19]. Cleavage of F_o protein renders the virus infective as well as determines host specificity and pathogenicity (*vide infra*). The mRNA transcript from the P gene is either translated unmodified to the P protein or edited at the V ORF internal editing site by insertion of one or two G residues. The edited mRNA with the frameshifts enables the translation of the V and W proteins with an N-terminal identical to that of the P protein and different C-terminal amino acid sequences [20–22]. The primary transcripts and the antigenome (after translation and accumulation of viral proteins) are synthesized by the viral RNA polymerase, a complex of the P and L proteins using the viral RNA genome, covered and protected with the nucleoprotein (N), as template. The genomic RNA–nucleoprotein complex together with the RNA-dependent RNA polymerase (P+L) is encapsidated and migrates to the cell membrane, where it becomes associated with the matrix (M) protein and the viral envelope and leaves the cell by budding and detachment when the HN clips the sialoglycoconjugate cellular receptor (reviewed in Ref. 11). Production of host macromolecules is shut off by the late phase of the viral replication [23].

VIRULENCE, HOST SPECIFICITY

Newcastle disease virus is an avian pathogen and seldom causes disease in mammals; conjunctivitis with preauricular lymph node involvement is a rare and self-resolving consequence of NDV exposure in humans. Sialylglucoconjugate receptors are quite ubiquitous on mammalian cells and shortage in receptor numbers does not seem to be the reason for cell specificity. Some viral precursor proteins need further processing to be functional and changes in the cleavage site influence processing efficiency. Virulent NDV strains with F_0 protein having two or more basic amino acids at the cleavage site are processed by the ubiquitous endoproteases furin and PC6 in avian cells and multiple cycles of viral infection affect different organs [24]. F_0 proteins of less pathogenic NDV strains contain only one basic amino acid residue and are not processed by the same enzyme. Cleavage does not happen in tissue culture; the produced virions remain noninfectious and only one viral cycle occurs. When avirulent NDV replicates in chicken embryos, factor Xa (the necessary enzyme for cleavage of F_0 with one basic amino acid residue) is present in the allantois fluid and surrounding tissues of the embryonated eggs, assuring production of infectious virus particles. In animals, infection with such NDV is restricted to the respiratory and gastrointestinal tracts [25–28]. A recombinant F protein (noncleavable in nature) derived from the avirulent NDV Queensland strain has been expressed in mammalian cells and in vitro processed to mature form with trypsin treatment [29]. Subtilisinlike proprotein convertase enzymes furin, PACE4, PC6, etc. are able to process the F protein of virulent NDV strains in mammalian cells, but PACE4 shows low activity [19,30]. Changing the cleavage site of the F protein from GRQKR (lentigenic phenotype) to RRQKR (mesogenic phenotype) of the LaSota NDV strain increased its neurovirulence but not to the level of the mesogenic Beaudette strain. That clearly shows that the F protein is not the only viral factor that determines virulence [31].

Newcastle disease virus is a strong inducer of interferon (IFN) in mammalian cells, but no IFN response is observed in the natural host. The viral V protein is responsible for shutting off IFN synthesis in avian cells; it has no effect on mammalian cell IFN production [32–34]. Recombinant NDV with a mutated P gene mRNA editing site or a truncated V protein (of which the C-terminal was deleted) shows a heavily compromised replication in chicken embryo fibroblasts, whereas it grows well in IFN^- VERO cells [35]. Interferon induction is not restricted to the dsRNA because the viral proteins HN, M, and the uncleaved F precursor are also IFN inducers [36].

NEWCASTLE DISEASE VIRUS AS AN ONCOLYTIC AGENT

Spontaneous tumor regression observed after different naturally occurring virus infections or vaccination with attenuated lyssa virus of a tumor-bearing person suggested that certain viruses replicate in and kill human tumors (37, also Sinkovics in this volume).

Newcastle disease virus emerged early as an oncolytic agent when it was discovered that the virus was able to kill established Ehrlich ascites and other tumors while sparing normal tissues[8,38,39]. The reason for this tumor-cell specificity is not clear. Clinical trials with several NDV strains as direct oncolytic agents are published (*vide infra*); our clinical experience is with the 73-T strain developed by W. Cassel.

The best-characterized and most-often-used oncolytic NDV 73-T strain was developed and described by W. Cassel (retired, Emory University, Atlanta, GA) from a field isolate of R. Love [8]. The 73-T virus is neurovirulent, killing 1-day-old chicks by paralysis after intracerebral inoculation; after intramuscular inoculation, viral replication occurs in the infected animals but no clinical disease is observed. Nucleotide sequence analysis fragments of the F gene representing the cleavage site, amplified by Reverse Transcription-Polymerase Chain Reaction (RT-PCR), revealed similarity of 73-T to pathogenic strains (J.C. Horvath, unpublished). See comparison of the cleavage site (printed bold) and flanking amino acid sequence of the 73-T strain compared to the consensus sequence between amino acid 97 and amino acid 146 of the F protein, published by D. King and B. Seal. [3]. The two basic amino acids characteristic of the velogenic strains are labeled with *. (This work has been supported, in part, by the Molecular Biology Core Facility at the H. Lee Moffitt Cancer Center & Research Institute, Tampa, FL.)

```
Consensus 97 ESIPRIQGSA TTSGRRQKR FVGAIIGSVA LGVATAAQIT AAAALIQANQ 146
             ::::::: :  :::::::::: : ::::: :: :::::::::: :::::::: :
73-T         ESIPRIQESV TTSGGRRQKR FIGAIIGGVA LGVATAAQIT AAAALIQAKQ
```

Treatment of Ehrlich ascites tumor-bearing mice with 73-T was curative up to 6-day-old tumors and no ascites developed during a 2-month observation period thereafter. The life of mice with tumors older than 6 days when receiving 73-T treatment was prolonged compared with untreated tumor-bearing controls. The first application of 73-T as an oncolytic agent was done by W. Cassel in a woman with a previously untreated, inoperable cervix carcinoma; a single intratumoral NDV injection resulted in extensive sloughing of tumor tissue and regression of lymph node metastases, with considerable subjective improvement; however, the patient expired 7 months later [8]. Genetically immunosuppressed SCID (Severe Combined Immunodeficiency) mice carrying xenografted human neuroblastoma or fibrosarcoma responded well to a single intratumoral injection of NDV 73-T and showed complete or partial remissions [40–42]. The oncolytic property of 73-T was confirmed after intratumoral inoculation into different human cancer xenografts and was observed with systemic application of the virus as well [43]. Sensitivity of lymphoma cells to NDV oncolysis exceeds that of lymphoblastoid cell lines. Resting lymphocytes are nonpermissive to NDV infection. Lymphocytes stimulated with phytohemagglutinin, however, moderately support virus replication (44; J. C. Horvath's own observations).

Newcastle disease virus induces extensive apoptosis (A) and necrosis in infected chicken lymphocytes within 3 hours after infection. The cause of death in infected chicken embryos is A in the brain, heart, thymus, and other organs [45,46]. Newcastle disease virus induces lytic infection in human tumor cell lines, whereas normal fibroblasts are resistant to viral infection. Although the exact mechanism of selective killing of tumor cells is not known, direct effects of the virus on the tumor cells and A are involved. Newcastle disease virus causes A in PC12 rat phaeochromocytoma cells [47,48]. The way NDV induces A is not certain; the authors conclude that major mitogen-activated protein kinase pathways (including the stress-inducible c-Jun N-terminal kinase pathway and p38 pathway) or mechanisms regulated by reactive oxygen species do not appear to have a role in virus-induced cell death.

Cell burst occurs when progeny viruses leave the host cell. The viral replication cycle in natural host chicken embryo fibroblasts is about 6 hours. Some cells start to leak macromolecules 2 hours after NDV infection [49]. Melanoma cells infected with NDV 73-T show intensive syncytium formation and ballooning degeneration, followed by the appearance of apoptotic bodies and infectious virus in the cell-free medium within a 24-hour replication cycle.

In a syngeneic system of transplantable animal tumors, virus-induced innate and adaptive immune mechanisms (NK cells and macrophages, antibody or cell-mediated immune response, cytokine production, etc.) are responsible for oncolysis as well [50].

MECHANISM OF ACTION OF ONCOLYTIC NEWCASTLE DISEASE VIRUS

Interaction of Viral and Host Proteins and Interference with the Cell Cycle

The V protein of Simian virus 5 (member of the *Rubulavirus* genus, closely resembling NDV) interacts with and inactivates the damage-specific DNA binding protein (DDB), a host protein involved in recognition of damaged DNA for nucleotide excision repair [51–53]. In uninfected cells with no DNA damage, DDB interacts with the transcription factor E2F1 and overrides its inhibition by Rb, allowing the cell to cycle [54]. When DDB binds to damaged DNA, the E2F1-Rb complex is unchallenged and the cell cycle slows down during DNA repair. The advantage to slowing down the cell cycle during viral replication is not clear, but viral macromolecule synthesis is probably more efficient when the host cell is not in division [55].

The Interferon Pathway

Definitions

Interferon is part of the innate immune system and is considered to be a tumor suppressor gene; it has cell-cycle regulatory, antiproliferative, and anti-angiogenic effects. Interferons regulate the expression of pro-apoptotic and anti-apoptotic genes and the IFN pathway leads to apoptosis. Interferons are secreted by virus-infected cells and act upon surrounding and distant cells. Viral HN by itself (without viral replication) triggers the production of IFNα and the tumor necrosis factor (TNF)α-related apoptosis-inducing ligand (TRAIL) by mononuclear cells [56].

Interferon-γ induction by NDV in peripheral blood mononuclear cells (PBMC) is weak and IFNγ production by PBMC of cancer patients is diminished compared with healthy controls [57,58]. For that reason, our discussion is focused on type I IFNs. Interleukin (IL)-2 treatment, however, increases IFNγ secretion by activated PBMC [59] and should be tried in combination with virotherapy.

Interferon-α and IFNβ exercise their antiviral, antiproliferative, and immunomodulatory actions by binding to IFN receptors IFNAR1 and IFNAR2, transmembrane proteins associated with protein tyrosine kinases Jak1 and Tyk2. The activated kinases then phosphorylate the tyrosine residue Tyr^{466} on IFNAR1 and the activated receptor–kinase complex attracts the signal transducer and activator of transcription 2 (STAT2). Phosphorylated STAT2 (Tyr^{690}) forms homo- and heterodimers with STAT1 and interferon regulatory factor

(IRF)-9 to form the IFN-stimulated gene factor 3 (ISGF3). Interferon-α–activated factor (homodimer of phosphorylated STAT1) and IGSF3 are transported to the nucleus, associated with other IRFs, and bind to consensus DNA motives TTNCNNNAA found in the promoter region of IFN-regulated protein genes (IFN-stimulated genes; ISGs) to activate their transcription, including IFNα itself (for a review see Ref. 11).

Pro-apoptotic Genes Regulated by Interferons

Interferon-β induces A in melanoma, ovarian carcinoma, and multiple myeloma cell lines in vitro and in xenotransplanted tumors (see review in Ref. 61). Interferon-γ suppresses the growth of both platinum-sensitive and resistant ovarian carcinoma cell lines [62]. Whereas short (20 hours) exposure to IFNγ was not associated with antiproliferative effect in ovarian carcinoma cells and xenotransplanted tumors in mice, continuous exposure over several days resulted in growth suppression and A [63]. Apoptosis induced by IFNγ is mediated by IRF-1 [64]. Encouraging results in small clinical trials warrant further evaluation of the therapeutic value of IFNγ in ovarian carcinoma. In general, IFNs induce the expression of numerous A-related proteins. Apoptosis plays a major role in the antitumor effect of IFN; it is executed independently of p53, making cells with mutated p53 sensitive to IFN-induced A as well. Apoptosis induced by IFNs does not correlate with the cellular level of Bax and Bcl-2 either [65]. Oligonucleotide microarray analysis of IFN-treated tumor cells revealed a set of genes playing a role in A and helped render understanding of the role of IFNs in A [61]. The role of Fas/APO-1/CD95 and its ligand (FasL) in physiological A is well known: [1] peripheral deletion of activated T cells in absence of antigens (but keeping memory T cells intact); [2] elimination of virus-infected or tumor cells; and [3] guarding immunologically privileged sites (testicles, eye, etc.). Expression of Fas is upregulated upon IFNα treatment in hematological malignancies [66,67]. Tumor necrosis factor-related apoptosis-inducing ligand is also induced by IFNs in a tissue or tumor-specific manner. Interferon-β is a better inducer of TRAIL (and

A) in melanoma and ovarian carcinoma, whereas multiple myeloma responds better with upregulated TRAIL to IFNα2 [68]. X-linked inhibitor of apoptosis (XIAP)- associated factor-1 (XAP-1) is also an IFN-induced pro-apoptotic and tumor suppressor gene.). X-linked inhibitor of apoptosis-associated factor-1 neutralizes the inhibition of caspases by XIAP and they co-localize from the cytoplasm to the nucleus. Overexpression of XAP-1 sensitizes the tumor cells to TRAIL-mediated A [69]. Expression of XAP-1 is suppressed in various cancers by promoter hypermethylation [70]. Caspases (cysteine proteases), initiators and effector enzymes of A, are also induced by IFNs. Caspase-4 is induced by IFNα and IFNβ in melanoma, human umbilical vein endothelial cells (mechanism of antiangiogenesis by IFN), and fibrosarcoma cell lines, whereas IFNγ initiates expression of caspase-4 and 8 in breast cancer cell lines [71]. Other proapoptotic genes induced by IFNs include IRFs, double-stranded RNA-dependent protein kinase (PKR, *vide infra*), and 2-5A oligoadenylate synthetases. Oligoadenylate synthetases are activators of DNA and RNA-degrading endonucleases, acting through formation of 2'-5'–linked oligoadenylates from adenosine triphosphate (ATP). Further induced are RNase L (mutated in aggressive prostate carcinoma), galectin-9 (downregulated in melanoma), death-associated protein kinase (a tumor suppressor gene lost in 50% of human malignancies, such as bladder and renal cell carcinomas, B-cell lymphoma, breast cancer, and pituitary tumors) (for review see Ref. 61).

Newcastle disease virus might kill tumor cells with an intact, functioning IFN system through A induction. Such infection would result in no progeny virus production in tumor cells and would explain why mega-dose treatment is needed to achieve clinical response (*vide infra*). Why are normal cells infected with NDV not victims of A?

In search for an explanation as to how NDV preferentially kills tumor cells, we shall hypothesize that in the host, tumor cells escape surveillance of the innate immune system, partly because they are in a nonresponsive state to the antiproliferative action of IFN. The defensive activity has to include blockade both of *endogenous* IFN production and response to *exoge-*

nous IFN. Tumor cells with such a nonfunctional IFN system would fall victim to NDV infection.

Type I Interferon Receptors

Tumor cells are continuously exposed to low amounts of IFNα and IFNβ from surrounding stromal or virus-infected cells [60]. In IFNAR knockout mice, when inoculated with different types of syngeneic transplantable tumor cell lines, tumor development and mortality are increased compared with IFNAR-expressing controls, showing that IFN is a natural mediator of innate immunity against tumor development [72]. Expression of IFNAR in neuroblastoma correlates with grade and appears to be a prognostic factor [73]. S. Navarro et al studied IFNAR-receptor expression in human fetal and adult tissues as well as in different tumors. All normal adult and fetal tissues and most, but not all, tumors (melanomas, adenocarcinomas of bladder, kidney, small intestine, lung, and breast) expressed the IFNAR; however, lymphomas, sarcomas, and endocrine tumors were found to be negative for type I IFN-receptor expression [74]. Exposure of tumor cells to IFN might turn down IFN-receptor expression, as observed with Daudi lymphoblastoid cell lines [75]. Expression of IFNAR is normal in renal cell carcinoma, even in metastatic disease [76].

Tumor Cell Interference with Interferon and the Interferon Signaling Pathway

Interferon induction requires the formation of the ISGF3 complex, which then activates the IRF-7 gene encoding for a transcription factor, inducing the expression of IFN [77]. In a fibrosarcoma cell line, IFNα induction is shut down by hypermethylation of the IRF-7 gene promoter [78]. Our standing concept is that impaired IFN expression renders cancer cells vulnerable to NDV infection.

The antiproliferative effect of IFN on melanoma cells is well documented, but several cell lines become resistant. Different pathways must exist that confer, to the cells, resistance to the growth-inhibitory effect of IFN [79–81]. Signal transducer and activator of transcription-1 is anti-oncogenic; its expression is often downregulated in tumor cells [82]. In uterine cervical cells infected with high-risk oncogenic human papillomavirus serotypes, IFN production is diminished or shut off [83]. Melanoma cell lines with lost STAT2, truncated and nonfunctional Tyk2, or impaired IFNAR are known. Papillomavirus E6 and E7 inhibit IFN production in transformed cells or E7/E7 transgenic mice as well as in transfected mononuclear and NK cells [84,85].

Cells with lost STAT1 might be resistant to both the antiproliferative and antiviral effects of IFN but contradictory observations exist as well about a direct connection between resistance and STAT1. A probable explanation lays in the expression of specific ISGs [80,86]. Interferon regulatory factor-1 (inducer of IFNα and IFNβ) is a tumor suppressor gene; IRF-1 expression reverts the oncogenic phenotype of NIH3T3 cells transformed by c-Ha-ras and c-myc [87]. No oncogene cooperation is needed in IRF-1($^{-/-}$) cells because full transformation may be achieved with the single Ha-ras oncogene [88]. Expression of IRF-1 diminishes with the progression of endometrioid carcinoma from Grade 1 to 3 [89]. In myelodysplastic syndromes and acute myelogenous leukemia, IRF-1 is truncated by skipping exons 2 or 2-3 [90]. Matrix metalloprotease-9 (MMP-9) expression by tumor cells promotes invasion and metastasis as well as angiogenesis by degrading collagens and laminin. Tumor necrosis factor-α induces MMP-9 expression through NFκB, which is a transactivator of the MMP-9 promoter. Interferon inhibits the expression of MMP-9 through competitive binding of IRF-1 to NFκB binding promoter sequences [91]. INF regulatory factor IRF-1 also regulates the expression of the IFN-inducer PKR (double-strand RNA activated protein kinase) through the activation of the ISRE element in its promoter region [92]. Thus, a disabled IFN system in tumor cells is essential for viral oncolysis.

The Ras Connection

Expression of mutated ras seemed to confer sensitivity to NDV oncolysis given that N-*ras*–transformed tumorigenic human fibroblasts become permissive to NDV infection, whereas normal human fibroblasts were 1000-fold less sensitive to the virus [42]. Because *ras*-specific antisense oligonucleotide treatment of melanoma cells did not abolish the oncolytic effect of NDV, obviously there are multiple pathways operating in tumor-cell killing (J. C. Horvath, unpublished data). The point mutations leading to *ras* oncogene activation are seen frequently in cancer. Reovirus takes advantage of the overexpressed Ras oncogene in killing cancer cells. In melanoma, the *ras* gene is frequently mutated [93]. A *ras*-effector homologue, the Ras association domain family 1 (RASSF1) protein is considered to be a tumor suppressor gene. The promoter region and a second region in the first exon of the RASSF1 gene is hypermethylated and its expression is suppressed or abolished in 55% of melanomas [94].

That multiple pathways intertangle in tumor cells, conferring sensitivity to NDV-mediated oncolysis, is supported by L. Klampfer's work showing that, in colon cancer cells, the mutated Ki-*ras* inhibits the expression of STAT1 and 2. Consequently, the expression of all IFNγ-inducible genes, including the tumor suppressor gene IRF-1, is reduced and the tumor cells are nonresponsive to IFN [95]. In colon carcinoma cells with *wt*Ki-*ras*, the STAT1 and 2 genes are silenced by hypermethylation, and these cells would not respond to IFNα treatment [96].

The phenotypic and genotypic changes just mentioned occur in tumor-derived cell lines but not universally, suggesting that either pathway can interchangeably function in viral oncolysis. There may be third or fourth, more universal, causes of sensitivity of tumor cells to NDV infection.

Role of the dsRNA-Dependent Protein Kinase

Double-stranded RNA is a product of viral replication, whereas it is seldom produced in normal cells. Double-stranded RNA-dependent protein kinase is a serin/theronin kinase activated by dsRNA or PACT, a stress-activated kinase [97]. Binding of dsRNA to the two dsRNA binding domains of PKR causes conformational changes and leads to dimerization and autophosphorylation of PKR. The activated PKR phosphorylates its target, the eukaryotic IF2α. The phosphorylated eIF2α in turn makes a stable complex with and inactivates the eIF2 complex and blocks initiation of protein synthesis. Double-stranded RNA-dependent protein kinase has a role in cell differentiation; it is also antiproliferative and pro-apoptotic [98]. The activated PKR phosphorylates and marks IκB for ubiquitination and destruction, unmasking the nuclear localization domain of NFκB. Nuclear factor-κB is transferred to the nucleus, where it binds to the extended promoter sites of target genes [99]. The stress-activated kinase PACT is expressed in all tissues at low level, except in the apical part of the colonic villi, where epithelial cells stop cycling and are prone to A upon stress effect [100]. These cells also show high levels of phosphorylated (active) PKR and eIF2α (inactive) showing inhibition of protein synthesis. The stress-activated kinase PACT activates PKR by heterodimer formation even without dsRNA binding. Because of this heterodimer formation and consequent conformational changes, the PKR ATP binding site opens and PKR becomes phosphorylated [101]. Infection with NDV infection induces PACT and the protein becomes associated and co-localized with the viral N protein, and probably is involved as a host factor in the viral replication process. The stress-activated kinase PACT also activates target genes (like IFNβ) by activating NFκB, IRF-3, and IRF-7 [102–104].

Apoptosis induction by activated PKR is executed through the activation of p38 mitogen-activated protein kinases (MAPKs), stress-activated protein kinases/c-Jun amino-terminal protein kinases. These kinases are all activated through TNFα signaling as well as by lipopolysaccharide (LPS) and dsRNA [105]. Double-stranded RNA-dependent protein kinase is co-localized and interacts with the A signal-regulating kinase-1, which interaction does not require the kinase activity of PKR [106,107]. In PKR $^{-/-}$ fibroblasts, however, poly(I)-poly(C) or LPS could not induce MAPK activation, showing that MAPK activation and A induction by these agents depend on PKR [105].

Influenza virus (and probably NDV) dsRNA-activated PKR induces Fas expression, making cells susceptible to FasL expressed by cytotoxic lymphocytes [108–110].

Role of the Immune System

Macrophages (MΦ) may exert cytotoxic effects or augment tumor growth [111–113]. Newcastle disease virus-infected MΦ express nitric oxide and TNFα and demonstrate cytostatic and cytotoxic activity against a broad range of tumor cell lines. Intravenously transferred NDV-infected MΦ suppressed lung metastases in lung tumor and breast cancer models, whereas noninfected MΦ had no effect or even augmented tumor growth [114]. In NDV-infected mouse cells, the chemokine RANTES (regulated on activation, normal T cell- expressed and secreted) is induced [115–117]. The induction of RANTES does not require viral nucleic acid or protein synthesis and can be achieved by using ultraviolet-inactivated virus as well. Apparently virus binding is sufficient to induce signaling. Interaction of the NDV Ulster strain with the immune system is extensively discussed by V. Schirrmacher in Chapter 4 of this volume.

NEWCASTLE DISEASE VIRUS THERAPY OF HUMAN CANCER

The efficacy of NDV in killing xenografted human tumors in the mouse is very impressive [42]. Results of clinical trials to treat metastatic cancer with NDV, however, are much less encouraging. Furthermore, the virus strains used, dosage, and the method and frequency of application are different and the results are difficult to compare.

The Lorence group studied extensively the oncolytic properties of the 73-T strain (which is a lytic NDV strain, *vide supra*) in mouse tumors and xenotransplanted human tumor tissues [43]. In their Phase I clinical trial, however, the isolate PV701 was used, derived from the "naturally attenuated MK107 vaccine strain" with plaque purification; no further reference on the parental strain was given [118,119]. Intravenous application of single or multiple doses from 1.2×10^{10} to 1.2×10^{11} PFU were tolerated well, although grade 1 and 2 toxicity occurred often, with fatigue, flu-like, and gastrointestinal symptoms being the most frequent side effects. Tumor site-specific toxicity also occurred and was treated with medication. Because toxicity occurred less often after the repeated virus application, a first injection of 1.2×10^{10} PFU PV701 was used for "desensitization." Higher doses were associated with clinical responses: Of 79 patients treated (23 colorectal, 9 pancreatic, 9 renal, 8 breast, 8 non–small-cell lung, several other cancers, and sarcomas), 1 complete remission (CR; a tonsillar squamous cell carcinoma) and 1 partial remission (PR, a colon cancer) were observed and 14 patients showed no progression (4 to 30+ months' duration). Histological examinations of tumor tissues showed inflammatory infiltrations.

The question arises whether multiple viral cycles occur in infected tumor cells or whether virus-induced A is responsible for the cell killing. Why is it necessary to give such a megadose of virus if it is replicating in tumor cells? An even more important but unanswered question is whether or not the repeated large inocula of the virus induce tolerance or "immune paralysis," allowing persistence of the virus [120].

A Hungarian group used an attenuated NDV veterinary commercial vaccine strain (re-named MTH-68/H) to treat various tumors in the forms of inhalation. Unfortunately, the claimed high response rate is supported only with anecdotal case presentations [121]. Repeated application of MTH-68/H induced MRI-documented PR (95% shrinking) in a 14-year-old boy with glioblastoma but was reported without histological proof [122].

Our clinical trial with the 73-T strain, approved by the institutional review board of the hospital, also gave mixed results. Let our patient populations be represented here by two cases: one a dramatic but temporary response to direct intratumoral NDV inoculation; the other, characterized by relentless growth of melanoma, despite various forms of viral inoculations. A single intratumoral injection of 10^7 PFU NDV into the primary melanoma invading the right nasal cavity and the maxillary sinus of a female patient was followed by days of sloughing of necrotic dark brown tissue and resulted in more than a 90% reduction of the tumor 2 months after NDV inoculation. The "thickened tissues" remained in the apex of the nasal cavity, but were negative for melanoma cells by biopsy.

Three months later, the patient relapsed and repeated injections of the virus did not change the course of advancing disease. Would have a higher dose or repeated applications of the virus resulted in a better outcome? Or could surgical removal of the residual tumor, if any, after PR was achieved, have prevented relapse?

A primary malignant melanoma on the left forearm of a 53-year-old man was surgically removed and his axillary lymph nodes with metastases were dissected. The patient refused the IFNα2b protocol but received allogeneic NDV (73-T) oncolysate vaccine instead. He relapsed and local recurrences were removed and he accepted and completed the 1-year IFN protocol. He relapsed again in the epitrochlear lymph node and in-transit metastases appeared on his left upper arm, extending to the left chest wall. Some nodules were injected with dacarbazine or vinblastine and granulocyte-macrophage colony stimulatory factor (GM-CSF) with the hope of generating antigen-presenting dendritic cells (DCs) and in situ vaccine effect by the apoptotic tumor cells. Low-dose IL-2 was given subcutaneously to help raise tumor-specific lymphocytes. None of his nodules regressed. A melanoma cell line established earlier from one of his surgically removed melanoma lesions was killed by NDV inoculation in vitro. He and his family decided then to have some of the nodules injected with purified live 73-T NDV. The injected nodules became red and indurated but the rapid growth of both injected and surrounding nodules continued unchecked. No apoptotic tumor cells were observed in the fine-needle aspirates. Biochemotherapy (vinblastine, *cis*-platin, and dacarbazine; IFNα2b; IL-2) and adoptive lymphocyte therapy with tumor-infiltrating lymphocytes, and lymphokine-activated killer cells were unsuccessful because his lung and liver metastases advanced. He developed brain metastases and died [123].

One can again speculate why the tumor cells failed to respond to virus injection. Because of the NDV oncolysate vaccination, he had virus-specific T cells, NK cells, and antibodies. Why did virus-specific antibodies not induce an antibody dependent cell mediated cytotoxicity (ADCC) type reaction against the virus-infected tumor cells? Why did viral oncolysate (VO) vaccines fail to reactivate his pre-existing mela-

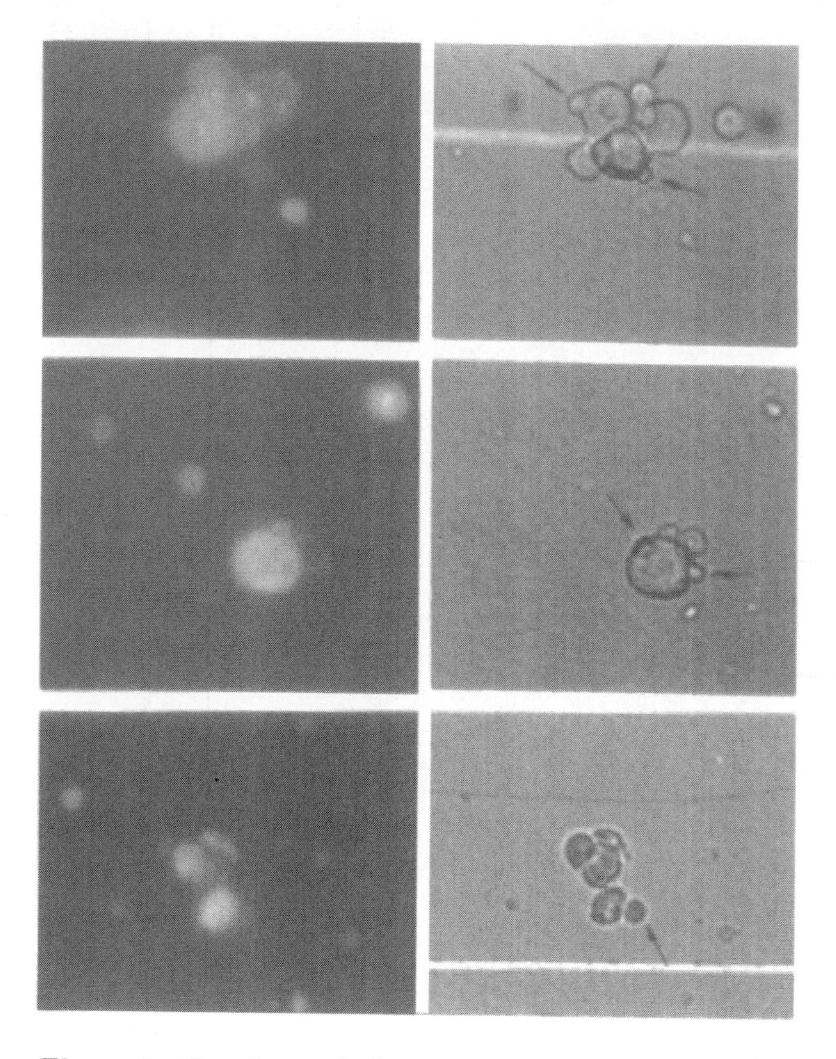

Figure 1 Co-culture of a human melanoma cell line with autologous tumor-infiltrating lymphocytes. Left side of the panel shows lymphocytes, free or attached to large tumor cells, stained with Annexin V and propidium iodide (PI) to detect apoptosis (A). Most tumor cells and some lymphocytes show the sign of late A with PI staining. Healthy lymphocytes, not killed by tumor cell counterattack, are marked with arrows on the right side of the panel (same cells illuminated by visible light).

noma-specific immune memory cells? Could the lymphokines and growth factors released locally in the inflammatory reaction support tumor cell growth? It is known that melanoma and other tumor cells can turn the IL-2→IL-2 receptor or even the apoptotic FasL→Fas system to their own advantage as an autocrine loop, while avoiding apoptotic death [124,129]. Tumor cells producing FasL can kill attacking Fas$^+$ cytotoxic T lymphocytes and escape immune surveillance (Figure 1). Are we the only ones who observe such an unabated tumor growth coincidentally with VO application? Are the patients with progressive disease in clinical trials analyzed for unexpected rapid growth rate after intervention or just excluded from further treatment?

In China, colorectal cancer patients were treated with autologous NDV oncolysate vaccine using the lentigenic La Sota strain. Patients without available surgical sample received NDV injection intradermally [130]. The median survival period for 310 vaccinated patients after resection of colorectal cancer was 5.13 years, vs. 4.15 years for 257 patients with resected tumor but no vaccination. The 5-year survival rate was 80% for patients with positive skin test after vaccination. Virus inoculation to tumor resulted in five stable disease (SD) in six patients with stomach cancer. One CR, three PR, and eight SD occurred in 13 patients with colorectal cancer. Toxic effects were not serious, represented by flu-like symptoms and fatigue.

POSTONCOLYTIC TUMOR IMMUNITY

After virus-induced oncolysis, mice acquire tumor-specific immunity (postoncolytic tumor immunity) and develop resistance to challenge with the same but virally not infected tumor [131,132]. The phenomenon of postoncolytic immunity is not clearly understood. Newcastle disease virus infection of tumor cells changes expression of cellular proteins. The expression of MART-1/Melan A melanoma-specific antigen is augmented four times upon NDV infection (Figure 2). Heat shock proteins (hsp) are stress-induced chaperones capable of binding and presenting antigen-specific peptides to antigen-presenting

37°C + - + -
43°C - + - +
NDV - - + +

Figure 2 Expression of MART-1/MELAN protein in mock- or NDV-infected human melanoma cell line at 37°C or 43°C Western blot with MART-1 specific antibody, chemiluminescence.

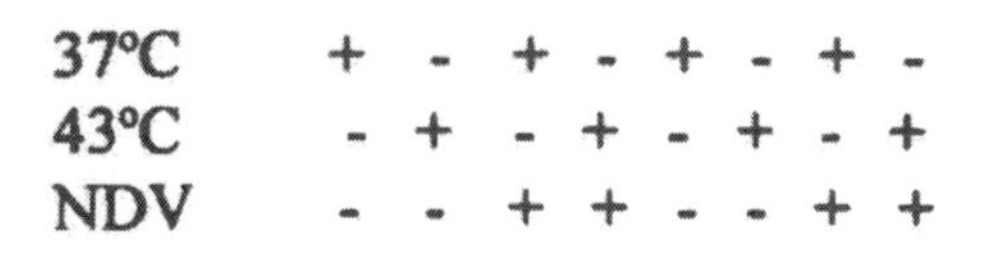

Figure 3 Expression of hsp70 protein in mock- or NDV-infected human neuroblastoma and melanoma cell lines at 37°C or 43°C. Western blot with hsp70-specific antibody, chemiluminescence.

cells (APC) and tumor-specific peptides are among them. Newcastle disease virus infection induces expression of hsp in natural host chicken cells and in infected tumor cells (phosphorylated hsp27) [133]. We observed induction of hsp70 and hsp90 in NDV-infected melanoma and neuroblastoma cell lines (Figure 3). By binding to their ligand CD40, hsp induce DC maturation [134,135]. Viral oncolysate vaccines contain live NDV, which, in turn, induces the production of IFNα, IL-15, and TNFα in the infected tissues upon vaccination [116].

VIRAL ONCOLYSATE VACCINES

W. Cassel started his clinical trial with NDV 73-T oncolysate adjuvant vaccine in patients with surgically resected melanoma with lymph node metastases in the mid-1970s [136,137]. To have most possible melanoma-specific antigens represented, several melanoma cell lines were rotated in the vaccine consisting of three cell lines in a given injection with no chemical or biological adjuvants. One allogeneic cell line was substituted with autologous melanoma cells when available. The vaccine contained live NDV. All patients alive still continue receiving vaccination. The efficacy of the vaccine was compared with historical controls; the patients were not prospectively randomized. The trial gave an unprecedented over-60% survival rate after 10 years and 55% after a 15-year follow-up, whereas only 27% of nonvaccinated patients are alive relapse-free [138,139]. The 20-year evaluation of VO-vaccinated patients of W. Cassel is published in this volume.

Our VO adjuvant melanoma vaccine program with NDV 73-T uses allogeneic melanoma cells developed in our laboratory from autologous or surgical resections. Each vaccine contained oncolysates from three cell lines. The patients received 26 vaccinations, the first eight in weekly intervals, followed by biweekly injections. One allogeneic cell line was substituted with autologous tumor cells, when available. Since 1999, patients are randomized to receive low-dose IL-2 or IFNα2b subcutaneously and the vaccine is admixed with GM-CSF (250 μg). We observed early relapses during vaccination with both

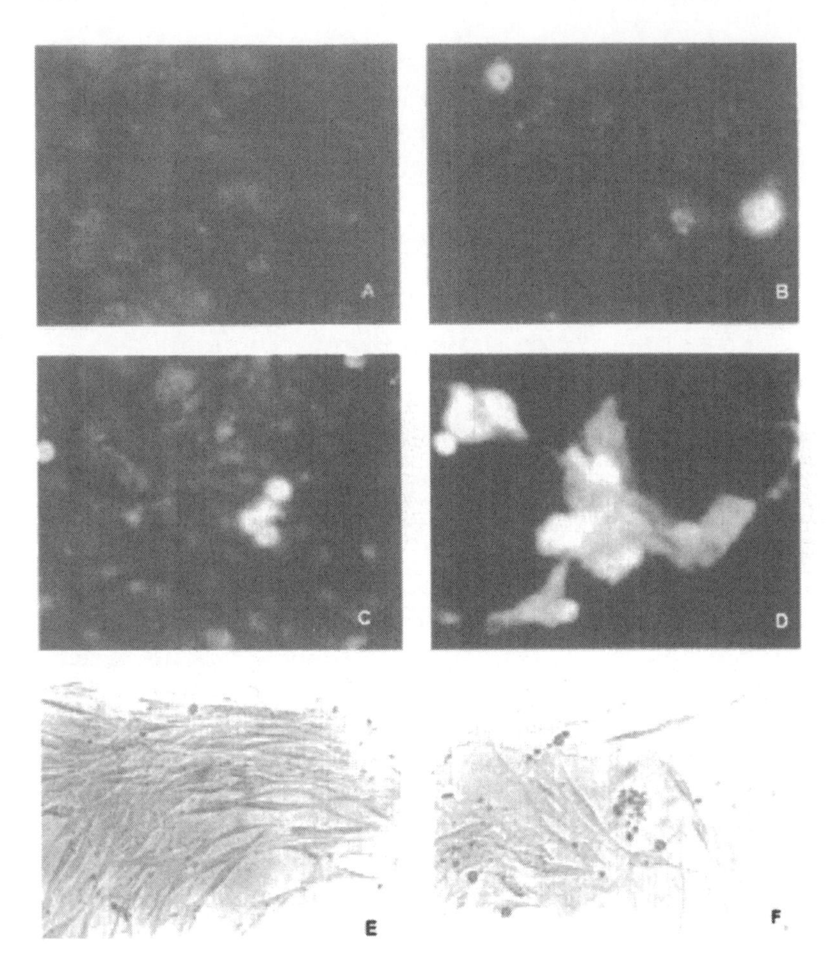

Figure 4 Expression of recombinant hemagglutinine-neuraminidase protein (HN) in human bronchioalveolar carcinoma (*B*) and melanoma (*C,D,E* and *F*) cell lines (established in our laboratory). The cells were stained 24 hours after transfection with rabbit polyclonal antibody against Newcastle disease virus and FITC (fluorescein isothiocyana) conjugated anti-rabbit IgG (*A-D*) or photographed after hemagglutination with chicken red blood cells (*E,F*). The bronchioalveolar carcinoma cells in *A* are transfected with the control plasmid without HN; cells in B–F were transfected with an eukaryotic expression vector carrying the HN gene. Cells on *D* were selected by hemagglutination on a chicken red blood cell monolayer chemically attached to a tissue culture flask (every cell tests positive for HN by immune fluorescence).

vaccines (with or without lymphokines). With the first NDV oncolysate vaccine without lymphokine adjuvants, 63% of the patients (13/15 patients with Stage IIB, 11/21 with Stage III and 1/4 with Stage IV disease) finished the whole course; 34% discontinued vaccine treatment because of early relapse. Of patients who finished the vaccination, 64% are free of disease (4 to 7 years postvaccination). Patients with Stage IV disease did not benefit from the vaccine. Why are our results inferior to the published exceptional results of the W. Cassel trial? Early clinical relapses occurred in patients who were accepted on the protocol on the patient's demand; or because of leniency, despite the fact that some patients were already in the progress of subclinical relapse when entering the program. Therefore, we recognize that [1] Viral oncolysate-induced antitumor immunity develops slowly; and [2] the VO is not a therapeutic vaccine. It may be effective against dormant micrometastases, but it is ineffective against growing and progressive disease. We cannot explain why VO fails to mobilize promptly pre-existing antitumor memory cells. For the valid evaluation of new treatment protocols, the comparison of a prospectively randomized group of patients treated differently is considered to be the most reliable. For the virotherapy of cancer, such protocols seldom have been applied. It is part of medical history that matched historical controls were used for the evaluation of new treatment protocols. For example, in the BCG (Bacillus Calmette-Guérin) of cancer immunotherapy, such control groups were widely relied upon. Clinical investigators who used "matched historical" controls remained as respected heads of departments in leading cancer institutions worldwide. The practice has not been abandoned or condemned. However, a biased selection of matched historical controls weakens the control group and renders the treatment group appear artificially improved. We refrain from such a comparison. However, our VO-vaccinated patients were enrolled with adverse prognostic signs and by the natural history of melanoma they were expected to relapse in 70% to 80% of the cases. We did not selectively enroll patients with favorable prognostic signs to artificially improve the results of the VO vaccination program.

In Germany, the Atzpodien group used irradiated autologous or allogeneic tumor cell oncolysates made with the 73-T strain of Cassel and used it together with low-dose subcutaneous IL-2 and IFNα in an adjuvant setting to protect patients with renal cell carcinoma from recurrence [140]. The ongoing Phase II clinical trial involved 208 patients at the time of the report. Of these, 15% relapsed locally or with lymph node and distant organ metastases; the median survival was 21 months (range 2–64 months), but no further follow-up is available [141]. Testing the tumor-specific cellular and humoral immune responses of vaccinated patients, no tumor-specific antibody or increase in cytotoxicity was found. As an effect of the vaccine, the level of TNFα and IFNγ, but not of IFNα, increased in the serum [142].

The Schirrmacher group in Germany used the non-lytic Ulster strain of NDV to produce adjuvant vaccines for clinical trials of various malignancies. Detailed evaluation of the clinical trials of V. Schirrmacher's group can be found elsewhere in this volume. There are basic differences between their adjuvant vaccine and the lysate vaccines. The lysate vaccine contains soluble components and fragments of tumor cell membranes and live NDV (free and membrane bound), whereas the Schirrmacher vaccine contains intact tumor cells with viral particles fused to the cell membrane. The more intact and viable the tumor cells were, the more efficacious was the vaccine [143]. Virus infection of the tumor cells leads to upregulation of HLA-DR molecules, IFNβ, and the chemokines RANTES and IFNγ-inducible protein-10. The infected tumor cells then die of A. Apoptotic bodies of tumor cells are phagocytosed by APC to present tumor antigens and thus generate immune T cell clones [116].

Newcastle disease virus oncolysate adjuvant vaccination of patients with glioblastoma multiforme increased delayed-type immune reactions and caused no serious side effects, but no improvement in survival was observed [144].

SAFETY

Genomic RNA synthesis of NDV is error-prone because of the lack of 3′-5′ exonuclease "proofreading." However, new virus strains evolve in nature slowly because most of the mutations in nucleic acid sequence cause no changes in amino acid sequences or might result in lethal virus mutants that are never detected [145,146]. At Emory University, Atlanta, GA, more than 12,000 injections with 73-T oncolysate vaccines containing live virus did not result in the appearance of any new or more pathogenic virus variant and did not induce any disease in the patients, in livestock, or among wild birds. Attempts to recover live virus from the blood, saliva, and urine of vaccinated patients by W. Cassel or us were unsuccessful. The mega-doses applied by the Lorence group for oncolytic treatment resulted in transient excretion of infectious virus and require further considerations [119].

FUTURE DIRECTIONS

It is clear from past and present clinical trials with NDV as an oncolytic virus that complete remissions of metastatic disease seldom occur and the natural course of malignant diseases stabilizes, but without cure. Tumor cells could become resistant to repeated applications and these selected clones grow out in relapses or in progressive disease. Producing a better, more oncolytic yet safe agent is on the wish-list of everyone working on the fast-growing field of virotherapy. Selection of more efficient oncolytic variants on specific target cells by mutagenesis can give such results. A melanoma-specific NDV variant was developed from the Ulster strain, however, no further trials are known to show its usefulness [147]. Modifying viral proteins to increase specificity with targeted mutations might be more fruitful when we know the mechanism of action more accurately.

The NDV hemagglutinin by itself is an IFN inducer and strong antigen [148,149]. It can be expressed in tumor cells from encoding eukaryotic expression vectors (Figure 4) [149–152]. In our experiments, the rHN (cloned cDNA was kind gift of Dr. Éva Nagy, University of Guelph, Ontario, Canada, further subcloned into an eukaryotic expression plasmid vector) was found mostly in the cytoplasma but was also expressed on the cell membrane, making it possible to select HN-transfected cells with hemagglutination (Figure 4D) [153]. The vector could also encode a cytokine adjuvant gene such as GM-CSF. Can vector-expressed NDV HN substitute for live NDV oncolysis to make virus (antigen)-modified tumor vaccine?

Several NDV strains are fully sequenced and the viral proteins and their function are well characterized. The relative genomic stability of NDV makes it a good candidate as a virus vector to carry genes encoding antigens of other pathogens or tumor [154,155]. Because of the modular structure of the viral genome, foreign genes can be inserted between viral protein coding regions without a major negative effect on the virus replication [156,157]. Would an oncolytic, lymphokine-armed NDV be more efficacious in generating protective immune response, or attract more NK cells or cytotoxic T lymphocytes to amplify the tumor cell killing and protect from later relapses? Can we make tumor cell-targeted NDV by changes introduced in the HN protein [158]? Until we achieve this goal, we are to use the naturally oncolytic NDV, not necessarily as a therapeutic, but as an adjuvant antitumor vaccine.

Single-agent therapy to treat malignant diseases is rarely successful, whereas multi-agent treatments show additive or even synergistic effects. Would viral oncolysis be more efficacious if different oncolytic viruses were applied in an alternating fashion? Recent drugs coming from extensive translational research and as spin-offs of the human genome project cannot be ignored. They are often more targeted, cause fewer side effects than presently used chemotherapeutic agents, do not induce host immune reactions, and do not replicate in the host. These agents could and should be combined or alternated with VO therapy. This subject is extensively discussed by J. G. Sinkovics elsewhere in this volume.

REFERENCES

1. Kaleta EF, Baldhauf C. Newcastle disease in free-living and pet birds. In Alexander DJ, Ed. Newcastle Disease. Boston: Kluwer Academic Publisher, 1988:197–246.

2. Seal BS. Analysis of matrix protein gene nucleotide sequence diversity among Newcastle disease virus isolates demonstrates that recent disease outbreaks are caused by viruses of psittacine origin. Virus Genes 1995; 11:217–224.
3. King DJ, Seal BS. Biological and molecular characterization of Newcastle disease virus isolates from surveillance of live bird markets in the northeastern United States. Avian Dis 1997; 41:683–689.
4. Burnet FM. Human infection with the virus of Newcastle disease of fowl. Med J Aust 1943; 2:313–314.
5. Radnót M. Maladie oculo-glandulaire jusqu'á présent inconnue. Ophthalmologica 1949; 113:106–108.
6. Swayne DE, King DJ. Avian influenza and Newcastle disease. J Am Vet Med Assoc 2003; 222:1534–1540.
7. Southam CM, Moore AE. Clinical studies of viruses as antineoplastic agents, with particular reference to Egypt 101 virus. Cancer 1952; 5:1025–1034.
8. Cassel WA, Garrett RE. Newcastle disease virus as an antineoplastic agent. Cancer 1965; 18:863–868.
9. Sinkovics JG, Horvath JC. Newcastle disease virus (NDV): brief history of its oncolytic strains. J Clin Virol 2000; 16:1–15.
10. Murakami Y, Kagino T, Niikura M, Mikami T, Ishii K, Matsuura Y. Characterization of Newcastle disease virus envelope glycoproteins expressed in insect cells. Virus Res 1994; 33: 123–137.
11. Lamb NA, Kolakofsky D. Paramyxoviridae: The viruses and their replication. In: Knipe DM, Howley PM, Eds Fields Virology. Fourth edition. 2001. Lippincott Williams & Wilkins, Philadelphia, Baltimore, New York, Louden, Buenos Aires, Hong Kong, Sydney. 1205–1340.
12. Chen L, Gorman JJ, McKimm-Breschkin J, Lawrence LJ, Tulloch PA, Smith BJ, Colman PM, Lawrence MC. The structure of the fusion glycoprotein of Newcastle disease virus suggests a novel paradigm for the molecular mechanism of membrane fusion. Structure (Camb) 2001; 9:255–266.
13. Peeples ME. Differential detergent treatment allows immunofluorescent localization of the Newcastle disease virus matrix

protein within the nucleus of infected cells. Virology 1988; 162: 255–259.

14. Coleman NA, Peeples ME. The matrix protein of Newcastle disease virus localizes to the nucleus via a bipartite nuclear localization signal. Virology 1993; 195:596–607.

15. Peeples ME, Wang C, Gupta KC, Coleman N. Nuclear entry and nucleolar localization of the Newcastle disease virus (NDV) matrix protein occur early in infection and do not require other NDV proteins. J Virol 1992; 66:3263–3269.

16. Peeples ME. Paramyxovirus M proteins: Pulling it all together and taking it on the road. In: Kingsbury D, Ed. The Paramyxoviruses., Plenum Press. 1991:427–456.

17. Ghildyal R, Baulch-Brown C, Mills J, Meanger J. The matrix protein of Human respiratory syncytial virus localizes to the nucleus of infected cells and inhibits transcription. Arch Virol 2003; 148:1419–1429.

18. Romer-Oberdorfer A, Werner O, Veits J, Mebatsion T, Mettenleiter TC. Contribution of the length of the HN protein and the sequence of the F protein cleavage site to Newcastle disease virus pathogenicity. J Gen Virol 2003; 84:3121–3129.

19. Fujii Y, Sakaguchi T, Kiyotani K, Yoshida T. Comparison of substrate specificities against the fusion glycoprotein of virulent Newcastle disease virus between a chick embryo fibroblast processing protease and mammalian subtilisin-like proteases. Microbiol Immunol 1999; 43:133–140.

20. Kondo K, Bando H, Tsurudome M, Kawano M, Nishio M, Ito Y. Sequence analysis of the phosphoprotein (P) genes of human parainfluenza type 4A and 4B viruses and RNA editing at transcript of the P genes: the number of G residues added is imprecise. Virology 1990; 178:321–326.

21. Steward M, Vipond IB, Millar NS, Emmerson PT. RNA editing in Newcastle disease virus. J Gen Virol 1993; 74:2539–2547.

22. Locke DP, Sellers HS, Crawford JM, Schultz-Cherry S, King DJ, Meinersmann RJ, Seal BS. Newcastle disease virus phosphoprotein gene analysis and transcriptional editing in avian cells. Virus Res 2000; 69:55–68.

23. Moore NF, Lomniczi B, Burke DC. The effect of infection with different strains of Newcastle disease virus on cellular RNA and protein synthesis. J Gen Virol 1972; 14:99–101.

24. Nagai Y, Klenk HD. Activation of precursors to both glycoporteins of Newcastle disease virus by proteolytic cleavage. Virology 1977; 77:125–134.

25. Gotoh B, Ogasawara T, Toyoda T, Inocencio NM, Hamaguchi M, Nagai Y. An endoprotease homologous to the blood clotting factor X as a determinant of viral tropism in chick embryo. EMBO J 1990; 9:4189–4195.

26. Kido H, Yokogoshi Y, Sakai K, Tashiro M, Kishino Y, Fukutomi A, Katunuma N. Isolation and characterization of a novel trypsin-like protease found in rat bronchiolar epithelial Clara cells. A possible activator of the viral fusion glycoprotein. J Biol Chem 1992; 267:13573–13579.

27. Ogasawara T, Gotoh B, Suzuki H, Asaka J, Shimokata K, Rott R, Nagai Y. Expression of factor X and its significance for the determination of paramyxovirus tropism in the chick embryo. EMBO J 1992; 11:467–72.

28. Chen L, Colman PM, Cosgrove LJ, Lawrence MC, Lawrence LJ, Tulloch PA, Gorman JJ. Cloning, expression, and crystallization of the fusion protein of Newcastle disease virus. Virology 2001; 290:290–299.

29. Li Z, Sergel T, Razvi E, Morrison T. Effect of cleavage mutants on syncytium formation directed by the wild-type fusion protein of Newcastle disease virus. J Virol 1998; 72:3789–3795.

30. Sakaguchi T, Fujii Y, Kiyotani K, Yoshida T. Correlation of proteolytic cleavage of F protein precursors in paramyxoviruses with expression of the fur, PACE4 and PC6 genes in mammalian cells. J Gen Virol. 1994; 75:2821–2827.

31. Panda A, Huang Z, Elankumaran S, Rockemann DD, Samal SK. Role of fusion protein cleavage site in the virulence of Newcastle disease virus. Microb Pathog 2004; 26:1–10.

32. Didcock L, Young DF, Goodbourn S, Randall RE. The V protein of simian virus 5 inhibits interferon signalling by targeting STAT1 for proteasome-mediated degradation. J Virol 1999; 73:9928–9933.

33. Gotoh B, Komatsu T, Takeuchi K, Yokoo J. Paramyxovirus strategies for evading the interferon response. Rev Med Virol 2002; 12:337–357.

34. Park MS, Shaw ML, Munoz-Jordan J, Cros JF, Nakaya T, Bouvier N, Palese P, Garcia-Sastre A, Basler CF. Newcastle disease virus (NDV)-based assay demonstrates interferon-antagonist activity for the NDV V protein and the Nipah virus V. W, and C proteins. J Virol 2003; 77:1501–1511.

35. Park MS, Garcia-Sastre A, Cros JF, Basler CF, Palese P. Newcastle disease virus V protein is a determinant of host range restriction. J Virol 2003; 77:9522–9532.

36. Wertz K, Buttner M, Mayr A, Kaaden OR. More than one component of the Newcastle disease virus particle is capable of interferon induction. Vet Microbiol 1994; 39:299–311.

37. Dock G. Rabies virus vaccination in a patient with cervical carcinoma. Am J Med Sci 1904; 127:63–92.

38. Flanagan AD, Love R, Tesar W. Propagation of Newcastle disease virus in Ehrlich ascites tumor cells in vitro and in vivo. Proc Soc Exp Biol Med 1955; 90:82–87.

39. Jacotot H. Oncolytic power in vivo of the Newcastle virus with regard to sarcoma ascitic Yoshida. C R Acad Sci Hebd Seances Acad Sci D 1967; 264:2602–2603.

40. Reichard KW, Lorence RM, Cascino CJ, Peeples ME, Walter RJ, Fernando MB, Reyes HM, Greager JA. Newcastle disease virus selectively kills human tumor cells. J Surg Res 1992; 52:448–453.

41. Lorence RM, Reichard KW, Katubig BB, Reyes HM, Phuangsab A, Mitchell BR, Cascino CJ, Walter RJ, Peeples ME. Complete regression of human neuroblastoma xenografts in athymic mice after local Newcastle disease virus therapy. J Natl Cancer Inst 1994; 86:1228–1233.

42. Lorence RM, Katubig BB, Reichard KW, Reyes HM, Phuangsab A, Sassetti MD, Walter RJ, Peeples ME. Complete regression of human fibrosarcoma xenografts after local Newcastle disease virus therapy. Cancer Res 1994; 54:6017–6021.

43. Phuangsab A, Lorence RM, Reichard KW, Peeples ME, Walter RJ. Newcastle disease virus therapy of human tumor xeno-

grafts: antitumor effects of local or systemic administration. Cancer Lett 2001; 172:27–36.

44. Bar-Eli N, Giloh H, Schlesinger M, Zakay-Rones Z. Preferential cytotoxic effect of Newcastle disease virus on lymphoma cells. J Cancer Res Clin Oncol 1996; 122:409–415.
45. Lam KM, Vasconcelos AC. Newcastle disease virus-induced apoptosis in chicken peripheral blood lymphocytes. Vet Immunol Immunopathol 1994; 44:45–56.
46. Lam KM, Vasconcelos AC, Bickford AA. Apoptosis as a cause of death in chicken embryos inoculated with Newcastle disease virus. Microbial Pathogenesis 1995; 19:169–174.
47. Fábián Z, Töröcsik B, Kiss K, Csatary LK, Bodey B, Tigyi J, Csatary C, Szeberényi J. Induction of apoptosis by a Newcastle disease virus vaccine (MTH-68/H) in PC12 rat phaeochromocytoma cells. Anticancer Res 2001; 21:125–135.
48. Szeberényi J, Fábián Z, Töröcsik B, Kiss K, Csatary LK. Newcastle disease virus-induced apoptosis in PC12 pheochromocytoma cells. Am J Ther 2003; 10:282–288.
49. Polos PG, Gallaher WR. A quantitative assay for cytolysis induced by Newcastle disease virus. J Gen Virol 1981; 52: 259–265.
50. Bonina L, Merendino RA, Berlinghieri MC, Arena A. Human mononuclear phagocytic cell interaction with some Paramyxoviridae. G Batteriol Virol Immunol 1985; 78:254–261.
51. Andrejeva J, Poole E, Young DF, Goodbourn S, Randall RE. The p127 subunit (DDB1) of the UV-DNA damage repair binding protein is essential for the targeted degradation of STAT1 by the V protein of the paramyxovirus simian virus 5. J Virol 2002; 76:11379–11386.
52. Lin GY, Paterson RG, Richardson CD, Lamb RA. The V protein of the paramyxovirus SV5 interacts with damage-specific DNA binding protein. Virology 1998; 249:189–200.
53. Keeney S, Eker AP, Brody T, Vermeulen W, Bootsma D, Hoeijmakers JH, Linn S. Correction of the DNA repair defect in xeroderma pigmentosum group E by injection of a DNA damage-binding protein. Proc Natl Acad Sci U S A 1994; 91: 4053–4056.

54. Hayes S, Shiyanov P, Chen X, Raychaudhuri P. DDB, a putative DNA repair protein, can function as a transcriptional partner of E2F1. Mol Cell Biol 1998; 18:240–249.

55. Lin GY, Lamb RA. The paramyxovirus simian virus 5 V protein slows progression of the cell cycle. J Virol 2000; 74: 9152–9166.

56. Zeng J, Fournier P, Schirrmacher V. Induction of interferon-alpha and tumor necrosis factor-related apoptosis-inducing ligand in human blood mononuclear cells by hemagglutinin-neuraminidase but not F protein of Newcastle disease virus. Virology 2002; 297:19–30.

57. Zorn U, Dallmann I, Grosse J, Kirchner H, Poliwoda H, Atzpodien J. Induction of cytokines and cytotoxicity against tumor cells by Newcastle disease virus. Cancer Biother 1994; 9: 225–235.

58. Elsasser-Beile U, von Kleist S, Martin M. Comparison of mitogen- and virus-induced interferon production in whole blood cell cultures of patients with various solid carcinomas and controls. Tumour Biol 1992; 13:358–363.

59. Klimpel GR, Infante AJ, Patterson J, Hess CB, Asuncion M. Virus-induced interferon α/β (IFN-α/β) production by T cells and by Th1 and Th2 helper T cell clones: a study of the immunoregulatory actions of IFN-γ versus IFN-α/β on functions of different T cell populations. Cell Immunol 1990; 128:603–618.

60. Takaoka A, Taniguchi T. New aspects of IFNα/β signalling in immunity, oncogenesis and bone metabolism. Cancer Sci 2003; 94:405–411.

61. Chawla-Sancar M, Lindner DJ, Liu Y-F, Williams BR, Sen GC, Silverman RH, Borden EC. Apoptosis and interferons: role of interferon-stimulated genes as mediators of apoptosis. Apoptosis 2003; 8:237–249.

62. Melichar B, Hu W, Patenia R, Melicharova K, Gallardo ST, Freedman R. rIFN-γ-mediated growth suppression of platinum-sensitive and –resistant ovarian tumor cell lines not dependent upon arginase inhibition. J Transl Med 2003; 1:5–14.

63. Wall L, Burke F, Smyth JF, Balkwill F. The anti-proliferative activity of Interferon-γ on ovarian cancer: in vitro and in vivo. Gynecol Oncol 2003; 88:S149–S151.

64. Kim EJ, Lee JM, Namkoong SE, Um SJ, Park JS. Interferon regulatory factor-1 mediates interferon-γ-induced apoptosis in ovarian carcinoma cells. J Cell Biochem 2002; 85:369–380.

65. Sangfelt O, Erickson S, Castro J, Heiden T, Einhorn S, Grander D. Induction of apoptosis and inhibition of cell growth are independent responses to interferon-α in hematopoietic cell lines. Cell Growth Differ 1997; 8:343–352.

66. Selleri C, Sato T, Del Vecchio L, Luciano L, Barrett AJ, Rotoli B, Young NS, Maciejewski JP. Involvement of Fas-mediated apoptosis in the inhibitory effects of interferon-α in chronic myelogenous leukemia. Blood 1997; 89:957–964.

67. Jedema I, Barge RM, Willemze R, Falkenburg JH. High susceptibility of human leukemic cells to Fas-induced apoptosis is restricted to G1 phase of the cell cycle and can be increased by interferon treatment. Leukemia 2003; 17:576–584.

68. Chawla-Sarkar M, Leaman DW, Borden EC. Preferential induction of apoptosis by interferon (IFN)-β compared with IFN-α2: correlation with TRAIL/Apo2L induction in melanoma cell lines. Clin Cancer Res 2001; 7:1821–1831.

69. Leaman DW, Chawla-Sarkar M, Vyas K, Reheman M, Tamai K, Toji S, Borden EC. Identification of X-linked inhibitor of apoptosis-associated factor-1 as an interferon-stimulated gene that augments TRAIL Apo2L-induced apoptosis. J Biol Chem 2002; 277:28504–28511.

70. Byun DS, Cho K, Ryu BK, Lee MG, Kang MJ, Kim HR, Chi SG. Hypermethylation of XIAP-associated factor 1, a putative tumor suppressor gene from the 17p13.2 locus, in human gastric adenocarcinomas. Cancer Res 2003; 63:7068–7075.

71. de Veer MJ, Holko M, Frevel M, Walker E, Der S, Paranjape JM, Silverman RH, Williams BR. Functional classification of interferon-stimulated genes identified using microarrays. J Leukoc Biol 2001; 69:912–920.

72. Picaud S, Bardot B, De Maeyer E, Seif I. Enhanced tumor development in mice lacking a functional type I interferon receptor. J Interferon Cytokine Res 2002; 22:457–462.

73. Mejia C, Navarro S, Colamonici OR, Pellin A, Castel V, Llombart-Bosch A. Expression of type I interferon receptor and its

relation with other prognostic factors in human neuroblastoma. Oncol Rep 1999; 6:149–153.

74. Navarro S, Colamonici OR, Llombart-Bosch A. Immunohistochemical detection of the type I interferon receptor in human fetal, adult, and neoplastic tissues. Mod Pathol 1996; 9: 150–156.

75. Branca AA, Baglioni C. Down-regulation of the interferon receptor. J Biol Chem 1982; 257:13197–13200.

76. Morell-Quadreny L, Fenollosa-Entrena B, Clar-Blanch F, Navarro-Fos S, Llombart-Bosch A. Expression of type I interferon receptor in renal cell carcinoma. Oncol Rep 1999; 6:639–642.

77. Yoneyama M, Suhara W, Fukuhara Y, Sato M, Ozato K, Fujita T. Autocrine amplification of type I interferon gene expression mediated by interferon stimulated gene factor 3 (ISGF3). J Biochem (Tokyo) 1996; 120:160–169.

78. Lu R, Au WC, Yeow WS, Hageman N, Pitha PM. Regulation of the promoter activity of interferon regulatory factor-7 gene. Activation by interferon and silencing by hypermethylation. J Biol Chem 2000; 275:31805–31812.

79. Wong LH, Krauer KG, Hatzinisiriou I, Estcourt MJ, Hersey P, Tam ND, Edmondson S, Devenish RJ, Ralph SJ. Interferon-resistant human melanoma cells are deficient in ISGF3 components, STAT1, STAT2, and p48-ISGF3gamma. J Biol Chem 1997; 272:28779–28785.

80. Pansky A, Hildebrand P, Fasler-Kan E, Baselgia L, Ketterer S, Beglinger C, Heim MH. Defective Jak-STAT signal transduction pathway in melanoma cells resistant to growth inhibition by interferon-α. Int J Cancer 2000; 85:720–725.

81. Jackson DP, Watling D, Rogers NC, Banks RE, Kerr IM, Selby PJ, Patel PM. The JAK/STAT pathway is not sufficient to sustain the antiproliferative response in an interferon-resistant human melanoma cell line. Melanoma Res 2003; 13:219–229.

82. Huang S, Bucana CD, Van Arsdall M, Fidler IJ. Stat1 negatively regulates angiogenesis, tumorigenicity and metastasis of tumor cells. Oncogene 2002; 21:2504–2512.

83. Cintorino M, Tripodi SA, Romagnoli R, Ietta F, Ricci MG, Paulesu L. Interferons and their receptors in human papillo-

mavirus lesions of the uterine cervix. Eur J Gynaecol Oncol 2002; 23:145–150.

84. Lee SJ, Cho YS, Cho MC, Shim JH, Lee KA, Ko KK, Choe YK, Park SN, Hoshino T, Kim S, Dinarello CA, Yoon DY. Both E6 and E7 oncoproteins of human papillomavirus 16 inhibit IL-18-induced IFN-γ production in human peripheral blood mononuclear and NK cells. J Immunol 2001; 167:497–504.

85. Um SJ, Rhyu JW, Kim EJ, Jeon KC, Hwang ES, Park JS. Abrogation of IRF-1 response by high-risk HPV E7 protein in vivo. Cancer Lett 2002; 179:205–212.

86. Chawla-Sarkar M, Leaman DW, Jacobs BS, Tuthill RJ, Chatterjee-Kishore M, Stark GR, Borden EC. Resistance to interferons in melanoma cells does not correlate with the expression or activation of signal transducer and activator of transcription 1 (Stat1). J Interferon Cytokine Res 2002; 22: 603–613.

87. Kroger A, Dallugge A, Kirchhoff S, Hauser H. IRF-1 reverts the transformed phenotype of oncogenically transformed cells in vitro and in vivo. Oncogene 2003; 22:1045–1056.

88. Tanaka N, Ishihara M, Kitagawa M, Harada H, Kimura T, Matsuyama T, Lamphier MS, Aizawa S, Mak TW, Taniguchi T. Cellular commitment to oncogene-induced transformation or apoptosis is dependent on the transcription factor IRF-1. Cell 1994; 77:829–839.

89. Kuroboshi H, Okubo T, Kitaya K, Nakayama T, Daikoku N, Fushiki S, Honjo H. Interferon regulatory factor-1 expression in human uterine endometrial carcinoma. Gynecol Oncol 200; 91:354–358.

90. Tzoanopoulos D, Speletas M, Arvanitidis K, Veiopoulou C, Kyriaki S, Thyphronitis G, Sideras P, Kartalis G, Ritis K. Low expression of interferon regulatory factor-1 and identification of novel exons skipping in patients with chronic myeloid leukaemia. Br J Haematol 2002; 119:46–53.

91. Sanceau J, Boyd DD, Seiki M, Bauvois B. Interferons inhibit tumor necrosis factor-α-mediated matrix metalloproteinase-9 activation via interferon regulatory factor-1 binding competition with NFκB. J Biol Chem 2002; 277:35766–35775.

92. Beretta L, Gabbay M, Berger R, Hanash SM, Sonenberg N. Expression of PKR is modulated by IRF-1 and is reduced in 5Q-associated leukemias. Oncogene 1996; 12:1593–1596.

93. Eskandarpour M, Hashemi J, Kanter L, Ringborg U, Platz A, Hansson J. Frequency of UV-inducible NRAS mutations in melanomas of patients with germline CDKN2A mutations. J Natl Cancer Inst 2003; 95:790–798.

94. Spugnardi M, Tommasi S, Dammann R, Pfeifer GP, Hoon DS. Epigenetic inactivation of RAS association domain family protein 1 (RASSF1A) in malignant cutaneous melanoma. Cancer Res 2003; 63:1639–1643.

95. Klampfer L, Huang J, Corner G, Mariadason J, Arango D, Sasazuki T, Shirasawa S, Augenlicht L. Oncogenic Ki-ras inhibits the expression of interferon-responsive genes through inhibition of STAT1 and STAT2 expression. J Biol Chem 2003; 278:46278–46287.

96. Karpf AR, Peterson PW, Rawlins JT, Dalley BK, Yang Q, Albertsen H, Jones DA. Inhibition of DNA methyltransferase stimulates the expression of signal transducer and activator of transcription 1, 2, and 3 genes in colon tumor cells. Proc Natl Acad Sci U S A 1999; 96:14007–14012.

97. Gale M, Katze MG. Molecular mechanisms of interferon resistance mediated by viral-directed inhibition of P. K. R. the interferon-induced protein kinase. Pharmacol Ther 1998; 78: 29–46.

98. Jagus R, Joshi B, Barber GN. PKR, apoptosis and cancer. Int J Biochem Cell Biol 1999; 31:123–138.

99. D'Acquisto F, Ghosh S. PACT and PKR: turning on NFκB in the absence of virus. Sci STKE 2001; 2001:RE1.

100. Gupta V, Patel RC. Proapoptotic protein PACT is expressed at high levels in colonic epithelial cells in mice. Am J Physiol Gastrointest Liver Physiol 2002; 283:G801–G808.

101. Patel RC, Ganes CS. PACT, a protein activator of the interferon-induced protein kinase, PKR. EMBO J 1998; 17: 4379–4390.

102. Iwamura T, Yoneyama M, Koizumi N, Okabe Y, Namiki H, Samuel CE, Fujita T. PACT, a double-stranded RNA binding

protein acts as a positive regulator for type I interferon gene induced by Newcastle disease virus. Biochem Biophys Res Commun 2001; 282:515–523.

103. Tian B, Brasier AR. Identification of a nuclear factor kappa B-dependent gene network. Recent Prog Horm Res 2003; 58: 95–130.

104. Tian B, Zhang Y, Luxon BA, Garofalo RP, Casola A, Sinha M, Brasier AR. Identification of NF-kappaB-dependent gene networks in respiratory syncytial virus-infected cells. J Virol 2002; 76:6800–6814.

105. Goh KC, deVeer MJ, Williams BR. The protein kinase PKR is required for p38 MAPK activation and the innate immune response to bacterial endotoxin. EMBO J 2000; 19:4292–4297.

106. Takizawa T, Tatematsu C, Ohashi K, Nakanishi Y. Recruitment of apoptotic cysteine proteases (caspases) in influenza virus-induced cell death. Microbiol Immunol 1999; 43: 245–252.

107. Takizawa T, Tatematsu C, Nakanishi Y. Double-stranded RNA-activated protein kinase interacts with apoptosis signal-regulating kinase 1. Implications for apoptosis signaling pathways. Eur J Biochem 2002; 269:6126–6132.

108. Lee SB, Esteban M. The interferon-induced double-stranded RNA-activated protein kinase induces apoptosis. Virology 1994; 199:491–496.

109. Takizawa T, Fukuda R, Miyawaki T, Ohashi K, Nakanishi T. Activation of apoptotic Fas antigen-encoding gene upon influenza virus infection involving spontaneously produced beta-interferon. Virology 1995; 209:288–296.

110. Wada N, Matsumura M, Ohba Y, Kobayashi N, Takizawa T, Nakanishi Y. Transcription stimulation of the Fas-encoding gene by nuclear factor for interleukin-6 expression upon influenza virus infection. J Biol Chem 1995; 270:18007–18012.

111. Welander CE, Natale RB, Lewis JL. In vitro growth stimulation of human ovarian cancer cells by xenogeneic peritoneal macrophages. J Natl Cancer Inst 1982; 69:1039–1047.

112. Hamburger AW, White CP. Interaction between macrophages and human tumor clonogenic cells. Stem Cells 1982; 1: 209–223.

113. Baetselier P, Kapon A, Katzav S, Tzehoval E, Dekegel D, Segal S, Feldman M. Selecting, accelerating and suppressing interactions between macrophages and tumor cells. Invasion Metastasis 1985; 5([2]):106–124.

114. Schirrmacher V, Bai L, Umansky V, Yu L, Xing Y, Qian Z. Newcastle disease virus activates macrophages for anti-tumor activity. Int J Oncol 2000; 16:363–373.

115. Fisher SN, Vanguri P, Shin HS, Shin ML. Regulatory mechanisms of MuRantes and CRG-2 chemokine gene induction in central nervous system glial cells by virus. Brain Behav Immun 1995; 9:331–344.

116. Washburn B, Schirrmacher V. Human tumor cell infection by Newcastle disease virus leads to upregulation of HLA and cell adhesion molecules and to induction of interferons, chemokines and finally apoptosis. Int J Oncol 2002; 21:85–93.

117. Lokuta MA, Maher J, Noe KH, Pitha PM, Shin ML, Shin HS. Mechanisms of murine RANTES chemokine gene induction by Newcastle disease virus. J Biol Chem 1996; 271:13731–1378.

118. Pecora AL, Rizvi N, Cohen GI, Meropol NJ, Sterman D, Marschall J, Rorence RM. An intravenous phase I trial of a replication-competent virus, PV701 in the treatment of patients with advanced solid cancers. Proceedings ASCO 2001; 20:253a.

119. Pecora AL, Rizvi N, Cohen GI, Meropol NJ, Sterman D, Marshall JL, Goldberg S, Gross P, O'Neil JD, Groene WS, Roberts MS, Rabin H, Bamat MK, Lorence RM. Phase I trial of intravenous administration of PV701, an oncolytic virus, in patients with advanced solid cancers. J Clin Oncol 2002; 20:2251–2266.

120. Fenton RG, Longo DL. Danger versus tolerance: paradigms for future studies of tumor-specific cytotoxic T lymphocytes. J Natl Cancer Inst 1997; 89:272–275.

121. Csatary LK, Eckhardt S, Bukosza I, Czegledi F, Fenyvesi C, Gergely P, Bodey B, Csatary CM. Attenuated veterinary virus vaccine for the treatment of cancer. Cancer Detect Prev 1993; 17:619–627.

122. Csatary LK, Bakacs T. Use of Newcastle disease virus vaccine (MTH-68/H) in a patient with high-grade glioblastoma. JAMA

1999; 281:1588–1589. (Erratum in: JAMA. Vol. 283, 2000: 2107.

123. Sinkovics JG, Horvath JC. Virus therapy of human cancers. Melanoma Res 2003; 13:431–432.

124. Alileche A, Plaisance S, Han DS, Rubinstein E, Mingari C, Bellomo R, Jasmin C, Azzarone B. Human melanoma cell line M14 secretes a functional interleukin 2. Oncogene 1993; 8: 1791–1796.

125. Han D, Pottin-Clemenceau C, Imro MA, Scudeletti M, Doucet C, Puppo F, Brouty-Boye D, Vedrenne J, Sahraoui Y, Brailly H, Poggi A, Jasmin C, Azzarone B, Indiveri F. IL-2 triggers a tumor progression process in a melanoma cell line MELP derived from a patient whose metastasis increased in size during IL-2/IFNα biotherapy. Oncogene 1996; 12:1015–1023.

126. Horvath JC, Horvath E, Sinkovics JG, Horak AI, Pendleton S, Mallah J. Human melanoma cells (HMC) eliminate autologous host lymphocytes (Ly^{FasR+}) and escape apoptotic death (A) by using FasL→FasR system as an autocrin growth loop ($HMC^{FasL \rightarrow FasR}$)., 89th Annual Meeting of the American Association for Cancer Research, April 12–16, 1998. New Orleans, Louisiana. Proceedings 39. Abstract 3971, p.584. 1998.

127. Sinkovics JG, Horvath JC. Virological and immunological connotations of apoptotic and anti-apoptotic forces in neoplasia. Int J Oncol 2001; 19:473–488.

128. Goillot E, Combaret V, Ladenstein R, Baubet D, Blay JY, Philip T, Favrot MC. Tumor necrosis factor as an autocrine growth factor for neuroblastoma. Cancer Res 1992; 52: 3194–3200.

129. Tong LJ, Yamaguchi N, Kita M, Imanishi J. Enhancement of the growth of human osteosarcoma cells by human interferon-γ. Cell Struct Funct 1992; 17:257–261.

130. Liang W, Wang H, Sun TM, Yao WQ, Chen LL, Jin Y, Li CL, Meng FJ. Application of autologous tumor cell vaccine and NDV vaccine in treatment of tumors of digestive tract. World J Gastroenterol 2003; 9:495–498.

131. Lindenmann J, Klein PA. Immunological aspects of viral oncolysis.. Recent Results in Cancer Research.. New York:: Springer Verlag, 1967:1–84.

132. Koprowski H, Love R, Koprowska I. Enhancement of susceptibility to viruses in neoplastic tissues. Eleventh Annual Symposium on Fundamental Cancer Research, University of Texas M.D. Anderson Hospital and Tumor Institute. Viruses and Tumor Growth. 1957:111–128.

133. Bai L, Koopmann J, Fiola C, Fournier P, Schirrmacher V. Dendritic cells pulsed with viral oncolysates potently stimulate autologous T cells from cancer patients. Int J Oncol 2002; 21: 685–694.

134. Zheng H, Dai J, Stoilova D, Li Z. Cell surface targeting of heat shock protein gp96 induces dendritic cell maturation and antitumor immunity. J Immunol 2001; 167:6731–6735.

135. Feng H, Zeng Y, Graner MW, Katsanis E. Stressed apoptotic tumor cells stimulate dendritic cells and induce specific cytotoxic T cells. Blood 2002; 100:4108–4115.

136. Cassel WA, Murray DR, Torbin AH, Olkowski ZL, Moore ME. Viral oncolysate in the management of malignant melanoma. I. Preparation of the oncolysate and measurement of immunologic responses. Cancer 1977; 40:672–679.

137. Murray DR, Cassel WA, Torbin AH, Olkowski ZL, Moore ME. Viral oncolysate in the management of malignant melanoma. II. Clinical studies. Cancer 1977; 40:680–686.

138. Cassel WA, Murray DR. Treatment of stage II malignant melanoma patients with a Newcastle disease virus oncolysate. Nat Immun Cell Growth Regul 1988; 7:351–352.

139. Batliwalla FM, Bateman BA, Serrano D, Murray D, Macphail S, Maino VC, Ansel JC, Gregersen PK, Armstrong CA. A 15-year follow-up of AJCC stage III malignant melanoma patients treated postsurgically with Newcastle disease virus (NDV) oncolysate and determination of alterations in the CD8 T cell repertoire. Mol Med 1998; 4:783–794.

140. Kirchner HH, Anton P, Atzpodien J. Adjuvant treatment of locally advanced renal cancer with autologous virus-modified tumor vaccines. World J Urol 1995; 13:171–173.

141. Anton P, Kirchner H, Jonas U, Atzpodien J. Cytokines and tumor vaccination. Cancer Biother Radiopharm 1996; 11: 315–318.

142. Zorn U, Duensing S, Langkopf F, Anastassiou G, Kirchner H, Hadam M, Knuver-Hopf J, Atzpodien J. Active specific immunotherapy of renal cell carcinoma: cellular and humoral immune responses. Cancer Biother Radiopharm 1997; 12: 157–165.

143. Ahlert T, Sauerbrei W, Bastert G, Ruhland S, Bartik B, Simiantonaki N, Schumacher J, Hacker B, Schumacher M, Schirrmacher V. Tumor-cell number and viability as quality and efficacy parameters of autologous virus-modified cancer vaccines in patients with breast or ovarian cancer. J Clin Oncol 1997; 15:1354–1366 (Erratum in: J Clin Oncol 1997; 15:2763).

144. Schneider T, Gerhards R, Kirches E, Firsching R. Preliminary results of active specific immunization with modified tumor cell vaccine in glioblastoma multiforme. J Neurooncol 2001; 53:39–46.

145. Toyoda T, Sakaguchi T, Hirota H, Gotoh B, Kuma K, Miyata T, Nagai Y. Newcastle disease virus evolution. II. Lack of gene recombination in generating virulent and avirulent strains. Virology 1989; 169:273–282.

146. Sakaguchi T, Toyoda T, Gotoh B, Inocencio NM, Kuma K, Miyata T, Nagai Y. Newcastle disease virus evolution. I. Multiple lineages defined by sequence variability of the hemagglutinin-neuraminidase gene. Virology 1989; 169:260–272.

147. Ahlert T, Schirrmacher V. Isolation of a human melanoma adapted Newcastle disease virus mutant with highly selective replication patterns. Cancer Res 1990; 50:5962–5968.

148. Schirrmacher V, Haas C, Bonifer R, Ertel C. Virus potentiation of tumor vaccine T-cell stimulatory capacity requires cell surface binding but not infection. Clin Cancer Res 1997; 3: 1135–1148.

149. Fournier P, Zeng J, Schirrmacher V. Two ways to induce innate immune responses in human PBMCs: paracrine stimulation of IFN-alpha responses by viral protein or dsRNA. Int J Oncol 2003; 23:673–80.

150. Horvath J, Sinkovics JG. Expression of Newcastle disease virus hemagglutinin in human cells.. Abstract.. Florida: Clinical Virology Symposium, Clearwater, April 25–28, 1993:54.

151. Horvath J, Horak A, Sinkovics JG. Comparison of oncolytic Newcastle Disease Virus strains., 86th Annual Meeting of the American Association for Cancer Research, March 18–22, 1995, Toronto, Ontario, Canada. Proceedings: 439:1995.

152. Risinskaya NV, Vasilenko OV, Fegeding KV, Sudarikov AB. Transfection of the Newcastle disease virus hemagglutinin-neuraminidase gene into murine myeloma cells for induction of host-versus-tumor immune response. Dokl Biochem Biophys 2001; 378:217–220.

153. Nagy E, Derbyshire JB, Dobos P, Krell PJ. Cloning and expression of NDV hemagglutinin-neuraminidase cDNA in a baculovirus expression vector system. Virology 1990; 176:426–438.

154. Huang Z, Krishnamurthy S, Panda A, Samal SK. High-level expression of a foreign gene from the most 3′-proximal locus of a recombinant Newcastle disease virus. J Gen Virol 2001; 82:1729–1736.

155. Huang Z, Elankumaran S, Panda A, Samal SK. Recombinant Newcastle disease virus as a vaccine vector. Poult Sci 2003; 82:899–906.

156. Nakaya T, Cros J, Park MS, Nakaya Y, Zheng H, Sagrera A, Villar E, Garcia-Sastre A, Palese P. Recombinant Newcastle disease virus as a vaccine vector. J Virol 2001; 75: 11868–11873.

157. Zhao H, Peeters BP. Recombinant Newcastle disease virus as a viral vector: effect of genomic location of foreign gene on gene expression and virus replication. J Gen Virol 2003; 84: 781–788.

158. Iorio RM, Field GM, Sauvron JM, Mirza AM, Deng R, Mahon PJ, Langedijk JP. Structural and functional relationship between the receptor recognition and neuraminidase activities of the Newcastle disease virus hemagglutinin-neuraminidase protein: receptor recognition is dependent on neuraminidase activity. J Virol 2001; 75:1918–1927.

6

Influenza A Viruses with Deletions in the NS1 Gene—a Rational Approach to Develop Oncolytic Viruses

MICHAEL BERGMANN
Department of Surgery, University of Vienna Medical School
Vienna, Austria

THOMAS MUSTER
Department of Dermatology, University of Vienna Medical School
Vienna, Austria

THE PRINCIPLE OF INFLUENZA A VIRUS MEDIATED ONCOLYSIS

The major feature typical of oncolytic viruses is their conditionally replicating phenotype, which permits them to grow in malignant cells but not in normal tissue. The recent elucidation of the function of the influenza virus NS1 protein [1] gave us the possibility for a rational development of oncolytic influenza A viruses by virtue of genetic engineering. The antagonism of the cellular antiviral type I interferon (IFN) response turned out to be one of the main functions of NS1. This finding is based on the observation that influenza A viruses, which lack a functional NS1 gene, do not grow in IFN-competent systems but replicate effectively in systems that lack expression of functional IFN [2], contain defects in the Jak/STAT (signal transducer and activator of transcription) pathway [3], or lack expression or activation of the IFN-induced, dsRNA activated protein kinase (PKR) [4]. The link between influenza A virus NS1-deletion viruses and oncolysis derives from the observation that tumor cells frequently contain defects in the IFN pathway [5] as well as in PKR activation [6]. As a consequence, NS1-deletion mutants are capable of selectively destroying such tumor cells. This renders genetically engineered NS1-deletion viruses prototypes for oncolytic influenza A virus strains. Developments that led to genetically modified oncolytic influenza virus strains, the general biology of the virus, its interactions with the IFN system, and the required characteristics of the target tumor are described in this chapter.

CHARACTERIZATION OF THE VIRUS

Influenza A virus is a segmented negative-strand RNA virus. It contains a lipid envelope anchoring the hemagglutinin (HA), neuraminidase (NA), and an iron channel, the M2 protein. The matrix protein M1 forms a protein layer beneath the envelope. The core of the virus is inside the M1 protein layer. It consists of ribonucleoprotein complexes (RNPs), which contain genomic RNA molecules associated with the nucleoprotein (NP) and the viral polymerase subunits PB1, PB2, and PA. In addition to these eight viral structural proteins that form the virus, two more proteins—the nonstructural protein (NS1) and the nuclear export protein (NEP)—are encoded by the genome [1]. According to some reports, the NEP is also present in virions

[7,8]. Recently, influenza A virus was found to encode one more protein, PB1-F2 [9]. Thus, the influenza A virus is a fairly small virus. Its limited number of proteins are well characterized. This is certainly desirable for a virus to be used for virotherapy. The fact that influenza A virus is an RNA virus excludes the possibility of the virus being incorporated in the host genome, which would be an unwanted side effect.

CHARACTERISTICS OF THE GENOME

The genome of influenza A virus consists of eight different RNA molecules. Each RNA segment encodes one viral protein except for the PB1, M, and NS RNA segments, which encode two proteins, PB1 and PB1-F2, M1 and M2, and NS1 and NEP, respectively. Because the genomic viral RNAs (vRNAs) are of negative polarity, they have to be transcribed into mRNA before the viral proteins can be synthesized. This transcription is performed in the nucleus of the infected cell by the incoming viral RNA-dependent RNA polymerase, a trimeric complex formed by the PB1, PB2, and PA proteins. This viral polymerase is also responsible for the replication of the vRNAs. During replication, the viral polymerase makes complementary copies (cRNAs) of the vRNAs. These cRNAs, in turn, are used as templates by the viral polymerase to generate new vRNAs. Both vRNAs and cRNAs are found associated with the viral NP protein. During transcription, mRNAs instead of cRNAs are generated from the vRNA templates. All eight vRNA segments contain short 3′ and 5′ non-coding regions flanking their internal coding sequences. These non-coding regions contain *cis*-acting signals required for the replication, transcription, and packaging of the vRNAs into virus particles [10,11]. The single segments can be exchanged by any two viruses if they infect the same cell. This property of the virus is important if genetically manipulated viruses should be used in humans. A genetically modified influenza A virus segment can be incorporated by reassortment into the circulating wild-type strain. For this reason, the genetic change preferably should be linked to an attenuation marker within the same segment.

INTRODUCING MUTATIONS INTO THE GENOME OF INFLUENZA VIRUS

Reverse genetics was a prerequisite to obtain conditionally replicating oncolytic influenza viruses. These techniques, which permit site-specific genetic manipulation of influenza virus, were first established by P. Palese's group [12]. In the initial experiment, in vitro reconstituted NA-specific RNPs were transfected into cells that previously had been infected with a helper influenza A virus [13]. This method was further developed by intracellular reconstitution of RNP complexes derived from plasmid-based expression vectors. In this system, influenza virus RNA transcripts are derived from plasmids that are transfected into cells. When cells infected with helper virus are cotransfected with conventional eukaryotic protein-expression plasmids expressing the viral polymerase proteins PB1, PB2, PA, and NP, the intracellularly reconstituted RNPs are taken up into new infectious viruses.

Contrary to the RNP-transfection system, the plasmid-based transfection system eliminates the need to purify the viral NP and polymerase proteins required for in vitro reconstitution of RNP complexes [14]. A breakthrough in the field came with the establishment of entirely plasmid-based techniques to rescue influenza viruses in the absence of helper viruses [15,16]. These techniques are based on the method of intracellular reconstitution of RNP complexes just described. Cells are cotransfected with two subsets of plasmids. The first includes eight polymerase I RNA-expression plasmids, each encoding one individual influenza virus vRNA. The second subset includes at least four polymerase II protein-expression plasmids encoding the PB1, PB2, PA, and NP proteins. These viral proteins encapsidate and initiate transcription and replication of the eight vRNAs, mimicking intracellular replication during naturally occurring influenza virus infections. Consequently, cells transfected with the complete set of 12 plasmids produce infectious influenza viruses subsequently infect cells and generate more viruses. As a result, high titers of recombinant influenza viruses are obtained. The generation of genetically manipulated virus entirely from plasmids might be of major importance for the application of such a virus to humans because it facilitates generating the virus under good laboratory practice conditions.

Establishing reverse genetics techniques to genetically manipulate influenza virus genomes has revolutionized the research field of influenza viruses. These techniques were used to analyze the functions of influenza virus proteins and RNAs. Analysis of the function of the influenza virus NS1 protein was the basis for generating conditionally replicating oncolytic influenza virus strains subsequently described.

THE DELNS1 VIRUS

The prototype of an oncolytic influenza A virus is a genetically engineered virus that lacks the NS1 protein. The first NS1-deletion viruses were generated by A. Egorov and colleagues. Surprisingly, these deletion mutants grew to wild-type titers in Vero cells, but were attenuated in Madin Darby Canine Kidney (MDCK) cells [2]. A NS1-knockout influenza A virus (delNS1 virus) was then obtained by replacing the wild-type NS gene with a genetically engineered intron-less NS gene, which can only direct the expression of the NEP protein from its unspliced mRNA. Analysis of the replication properties of this virus allowed defining new functions of the NS1 protein [3], a nonstructural viral protein expressed in high levels in infected cells but not in the virus. Previously, the NS1 protein has been associated with several functions during viral replication—e.g., the inhibition of host mRNA polyadenylation [17], splicing [18,19,20] and nucleocytoplasmic export [18,9], the inhibition of the IFN-inducible antiviral enzyme PKR [21,22,23], and the enhancement of mRNA translation [24], with a higher specificity for viral mRNA [25,26,27].

Despite all the multiple functions that have been ascribed to the NS1 protein, the most obvious characteristic of the delNS1 virus was its conditionally replicating phenotype. The virus was highly impaired in the ability to grow in MDCK cells, 10-day-old embryonated eggs or wild-type mice, but it replicated to titers close to wild-type virus in Vero cells and

6-day-old embryonated eggs, as well as STAT1-knockout cells and STAT-knockout mice. The facts that Vero cells have lost functional expression of their type I IFN genes; 6-day-old embryonated eggs, but not 10-day-old, have an immature-type IFN system and STAT1 is a key protein in the IFN pathway, all indicated that the NS1—in addition to the previously observed functions—is an IFN antagonist.

THE INTERFERON PATHWAY AND THE NS1 PROTEIN

The cellular signal cascade to type I IFNs is initiated by the binding of IFN to its receptor (IFN-AR) [28]. Despite the fact that IFN induces more than 300 different genes in the cell, the IFN signal from the cell surface to the nucleus is mediated by only a few proteins involved in the Jak/STAT pathway. Associated with the intracellular domain of the IFN-receptor are two protein kinases, Jak1 and Tyk2, which are activated upon receptor binding. The signal transducers and activators (STATs) are then recruited to the receptor complex and activated through posphorylation. Signal transducers and activators of transcription STAT1, STAT2, and p48 lead to the formation of the transcription complex ISGF-3, which translocates to the nucleus and promotes the transcription of IFN-responsive genes that have the IFN-responsive signal elements within their promoter regions. The antagonistic effect of NS1 in the IFN pathway was pinpointed by the fact that the delNS1 virus grew to titers similar to wild-type virus in STAT1 knockout mice [3]. Among numerous IFN-responsive genes, some of the essential proteins for the IFN-induced antiviral response are the dsRNA-activated protein kinase (PKR), MxA, and 2′-5′ oligoadenylate synthetase (OAS). Activation of both PKR and OAS results in blockage of translation. Thus, one of the endpoints of the IFN-mediated signal cascade is the inhibition of translation. Further characterization of the NS1 knockout virus revealed that the NS1 protein of influenza A virus is clearly involved in prevention of the activation of PKR, therefore inhibiting PKR-induced translational blockage in infected cells [4].

In addition, the NS1 protein prevents the activation of other dsRNA-activated cellular pathways, such as the activation of transcription factors involved in the induction of type I IFN expression [29,30]. Because the NS1 protein is a dsRNA-binding protein, it is tempting to speculate that inhibition of the IFN system by the NS1 is mediated by its ability to bind dsRNA, a potent activator of the IFN antiviral response. Indeed, the NS1 dsRNA-binding domain of the NS1 protein appears to be the main domain responsible for its type I IFN-antagonistic properties [29,30].

Because its IFN antagonist is eliminated, the NS1 knockout influenza A virus grows selectively in cells with deficiencies in the type I IFN system. Because a large number of tumor cells have an impaired type I IFN response, this mutant influenza A virus replicates in the majority of cell lines derived from tumors. Given the cytopathic nature of influenza virus, replication of the NS1 knockout virus in tumor cells results in selective killing of these cells [31]. Selective oncolytic properties have also been demonstrated with other viruses that are restricted in growth because of their inability to inhibit the type I IFN system or the IFN-inducible PKR-mediated antiviral response [32,33], (for review see 34).

TUMOR-ASSOCIATED DEFECTS OF THE INTERFERON PATHWAY

Given the elimination of its IFN antagonist, the NS1 knockout influenza A virus selectively grows in cells with deficiencies in the type I IFN system. Because a large number of tumor cells have an impaired type I IFN response, the delNS1 virus fails to replicate in normal cells but replicates in the majority of tumor-derived cell lines. Defects in the IFN pathway involve a number of different proteins.

Defects in the Jak/Signal Transducer and Activator of Transcription Pathway

Interferon resistance appears to be a general phenomenon in malignancies such as malignant melanoma, hematological malignancies, and carcinoids. The most frequently observed defects that lead to IFN resistance are alterations in proteins of the Jak/STAT pathway because defects in this pathway cannot be bypassed by an alternative signal cascade. The central transcription factor ISGF-3, which consists of STAT1, STAT2, or p48, is most often affected.

For example, in melanoma cell lines, alterations of ISGF-3 components correlated with IFN non-responsiveness, including a lack of constitutive expression and a lack of activation of these factors [5,35]. The most frequent defect observed in the resistant cell lines was lack of STAT1 protein expression. This observation was extended to cell cultures established from melanoma patient biopsies. Other defects such as a reduced phosphorylation of Tyk2, a tyrosine kinase that binds to the intracellular part of the IFN receptor, have also been reported [35]. However, resistance to IFN can also result from defects of IFN-stimulated genes (ISGs). Similar to melanomas, chronic myeloid leukemia cells resistant to IFN-α also lack STAT1 expression [36].

Because IFN is used as a therapeutic regimen for carcinoids, protein expression of STATs was also investigated in this type of cancer [37]. Tissue immunostaining of STAT1 and STAT2 demonstrated that STATs were significantly increased during IFN-α treatment. The same was observed for leukaemia. Induction of STAT1 by IFN treatment was only observed in patients with objective response as well as in patients with stable disease, but not in those with progressive disease. This supports the clinically relevant role of the Jak/STAT pathway and, at the same time, indicates the need of an alternative therapy for IFN-resistant cancers.

Defects in Protein Kinase

The other group of IFN pathway-associated defects relevant for virus-induced oncolysis is the inhibition of PKR activation by an oncogene. L. J. Mundschau and D. V. Faller demonstrated that dsRNA-mediated activation of PKR is blocked in v-*ras*–containing cells in a manner specific to ras [6]. This inhibitory effect of ras was not attributable to the transformed phenotype. A heat and phenol chloroform-sensitive PKR inhibitory activity could be demonstrated in these cells, that functions in *trans* when mixed with untransformed cell extract prior to stimulation with dsRNA. Thus, even though other functions of the IFN-pathway remain unaltered, the antiviral effect of IFN-induced PKR, which we had proven to be most relevant in the IFN pathway, is inhibited by oncogenic ras [4]. The ras-mediated inhibition of PKR activation is of major importance for a broad range of malignancies, because oncogenic ras has been found in approximately 30% of human tumors. In fact, in pancreatic carcinoma, one of the carcinomas with a very poor overall prognosis attributable to the lack of adequate chemotherapy, ras mutations occur in more than 90% of the tumors. In other frequent cancers, such as lung or colon cancers, oncogenic ras is found in 40% and 50%, respectively. In malignant melanoma, this oncogene is activated in approximately 30% of the cases.

A reduced PKR activity might not be limited to cancer cells expressing oncogenic ras. For example, although expressed at high levels, PKR activity appears to be severely attenuated in several breast cancer cell lines, as judged by its ability to phosphorylate eIF-2a (Savinova, 1999).

INFLUENZA A VIRUS MEDIATED ONCOLYSIS IN IFN RESISTANT TUMORS

Because the delNS1 virus grew in IFN-defective systems as such STAT1-knockout mice, we hypothesized that it might also replicate in IFN-defective tumor cell lines. Different tumor cell lines derived from melanomas, hepatomas, and teratocarcinomas were analyzed. Growth of the delNS1 knockout virus in these cell lines correlated with IFN resistance (current authors, unpublished). In the more IFN-sensitive cell lines, growth of the delNS1 virus was severely attenuated compared with the resistant cell lines. This further indicates that viral growth is mainly dependent on the IFN phenotype and is not restricted by the histological phenotype.

Infection of IFN-sensitive cell lines at high Multiplicity of Infection (MOI) with the NS1 deletion mutant still resulted in lysis of the cell. Importantly, adding IFN completely inhibited the lytic effect of the delNS1 virus in these cells. In contrast, exogenous IFN was not capable to block viral infection in IFN-resistant cells. We believe that the growth of the NS1-deletion mutants in the presence of IFN is relevant for the in vivo situation, given that the delNS1 virus is a potent inductor of IFN in normal cells. To further investigate the influence of factors secreted by stroma cells exposed to delNS1 virus, we infected an endothelial cell line with the delNS1 virus. This cell line does not permit productive replication of the delNS1 virus. Supernatants from the infected endothelial cells inhibited oncolysis in IFN-sensitive but not in IFN-resistant cells. This supports the dominant role of IFN on the inhibition of influenza A virus growth and indicates that the tissue cytokine response against the virus might be circumvented, if IFN-resistant cancers are taken as targets for influenza A-mediated oncolysis.

We also tested the growth of a virus with a truncated NS gene coding for the NH2 terminal 99 aa (NS1-99 virus). This virus was also shown to be highly attenuated in mice, although the level of attenuation was lower than that of the delNS1 virus [38]. The NS1-99 virus still contains the RNA binding site located in a helix and comprising the 71 NH2-terminal aa. However, for efficient RNA binding, a dimerization domain, present in the carboxy terminal half of the protein, is also necessary [39]. Therefore, viruses such as the NS1-99, which only contain the NH2 terminal helix but lack a functional dimerization domain, should be attenuated in their RNA-binding capacity. We hypothesized that this phenotype should have retained partial antagonism of the IFN response. Indeed, the growth of the NS1-99 virus was less dependent on IFN-mediated growth inhibition than the delNS1 virus. This correlates with the fact that the NS1-99 virus is less attenuated in mice than is the delNS1 virus.

Analysis of the oncolytic potential of influenza A deletion viruses on subcutaneously established IFN-resistant tumors in the severe combined immune-deficient (SCID) mouse model revealed that the less attenuated NS1-99 virus was superior to the delNS1 virus. This indicates that efficient influenza A virus-mediated oncolysis can be hampered by overattenuation.

However, the fact that the length of the NS1 deletion correlates with the level of attenuation provides the possibility to tailor influenza deletion mutants with the desired level of attenuation. At present, experiments are being performed to define the optimal-length NS1 protein, which mediates effective oncolysis and is sufficiently attenuated.

INFLUENZA A VIRUS-MEDIATED ONCOLYSIS IN TUMOR EXPRESSION OF ONCOGENIC RAS

Another defect in the IFN-associated signal cascade relevant for delNS1 virus growth results from ras-mediated inhibition of PKR activation. Patrick Lee's group was the first who exploited this defect for virotherapy using reovirus. We reasoned that the delNS1 virus should also grow specifically on ras-expressing tumor cells given that we had previously demonstrated that the NS1 has an antagonistic function toward PKR activation and replicates in PKR-deficient systems. Indeed, N-*ras* transfected cells supported delNS1 virus growth, whereas the parental cell line did not [31]. Abortive growth in the nonpermissive cell line was associated with activated PKR. Apparently, other IFN-induced proteins such as OAS or MxA did not prevent viral growth in these cells. These data show that nonpermissive cells become permissive upon transfection of oncogenic ras. Expression of oncogenic ras correlated with inhibition of PKR activation, which confirms the concept that oncogenic ras is an inhibitor of PKR activation. These data also confirm the hypothesis that PKR is one of the major anti-influenza antagonists within the IFN pathway. Importantly, the delNS1virus induced a complete cytopatic effect (CPE) in the ras-positive cells even when infection was done at a low m.o.i., indicating multi-cycle replication.

Even though no CPE was observed during infection of the nonpermissive wild-type cell line with the delNS1 virus at the low moi, at high moi, the virus still induced CPE. Obviously, infection of nonpermissive cells with the delNS1 virus leads to death of the infected cell, although no detectable virus is produced. The delNS1 virus-induced cell death might be mediated by induction and activation of PKR, given that activated PKR has been shown to be associated with apoptotic cell death.

The conditionally replicating phenotype of the delNS1 virus in ras-postive tumor cells translated to a therapeutic effect in a severe combined immunodeficiency (SCID) mouse model. In this approach, virus was administered by intratumoral injection. Virus replication specific for the tumor was confirmed by immunohistochemisty. We also analyzed whether the ras-dependent growth of the delNS1 virus was restricted to a certain tumor cell line or tumor type or whether this phenomenon was of general interest for other tumor entities. This virus also grew on other malignancies that contained ras mutations, such as lung carcinoma or hepatoma, again suggesting that influenza A virus growth is not restricted by histology, but by the molecular phenotype of the tumor.

In contrast, in nonmalignant human cell lines, such as endothelial cell lines, amelanocytic cell lines, or primary keratinocytes, that are susceptible to wild-type virus infection, the delNS1 virus did not replicate, which underlines the apathogenic phenotype of the delNS1 virus.

PROPERTIES OF INFLUENZA A VIRUS FOR VIROTHERAPY

Influenza A virus derived preferentially from oncolytic strains have only been described very recently and still lack proof of their efficient and safe application in humans. However, we believe that influenza delNS1-deletion mutants have a number of properties that make this virus suitable as an oncolytic agent: (1) DelNS1 virus mutants are apathogenic (delNS1 virus) or highly attenuated (NS1-99) virus in normal mice [38]. Moreover, intratumoral virotherpy was associated with no side effects in SCID mice. Thus, these viruses should be well tolerated even in immunocompromised cancer patients. (2) The genetic change that promotes the preferential replicating phenotype in tumors is an attenuation marker. Thus, it is not possible to introduce a genetically engineered segment into a wild-type virus by reassortment because this segment would attenuate the virus and convert it in a live virus vaccine. (3) The apathogenic phenotype is attributable to a large deletion in its genome, which is unlikely to revert. It should be noted that attenuated influenza A viruses have been administered to humans for vaccine purposes with few adverse effects [40,41]. (4) Given that there are multiple serologically defined subtypes of influenza A viruses, different subtypes of the delNS1 virus can be constructed by exchanging the antigenic surface glycoproteins of the virus, the hemagglutinin, and the neuraminidase. The availability of modified delNS1 viruses may allow repeated administration of these oncolytic strains. Similarily, an existing immunity of the host can be circumvented by choosing strains with appropriate hemagglutinin and neuraminidase.(5) It is possible to construct influenza A viruses with shorter deletions in their NS1 protein. The length of the NS1 protein inversely correlates with the level of attenuation. This feature of the delNS1 virus permits us to choose the optimal length of the NS1 protein that is associated with efficient tumor destruction but is still sufficiently attenuated in the host to allow a safe application of the virus. (6) The delNS1 virus was shown to induce high amounts of NF-κB [30], a key protein of the proinflammatory immune response. Thus, NS1-deletion virus-based virotherapy might be associated with an effective immune stimulation at the site of the tumor could lead to tumor destruction at a distant site by the stimulation of a specific antitumor immune response.

A concern for the use of influenza A viruses as therapeutic agents is the possibility of causing a new epidemic by introducing a new viral strain to mankind. The antigenic epitopes that determine the serologic phenotype of the virus are within the HA and NA proteins. Clearly, the HA protein is immunologically more important. Major changes in HA and NA B-cell epitopes lead to antigenic shift; minor changes cause an antigenic drift. There exist 14 different serological subtypes of the HA protein, of which H1, H2, H3, and H5 are associated with human infections. Concerning the NA protein, two subtypes (N1 and N2) have occurred in humans. At present, two strains, H1N1 and the H3N2, are circulating. In Europe and the United States, influenza epidemics only occur during winter for reasons that are not elucidated yet. Each year, the virus slightly changes because of the antigenic drift. For this reason, the current vaccine formula is modified annually. Given that the new virus strains usually appear in Southeast Asia, a worldwide net is used to predict the phenotype to the annual viral strain and the vaccine formula. To prevent the possibility of introducing new strains in the circulation, we will incorporate the annual HA and NA proteins in the virus, used for oncolytic purposes.

ONCOLYTIC REOVIRUSES

Whereas the influenza NS1-deletion mutants were engineered to selectively replicate in IFN-deficient cells and cells expressing oncogenic Ras, reovirus a nonenveloped, double-stranded RNA virus—possesses inherent selectivity for tumor cells expressing ras. In fact, data on the connection among ras, PKR, and reovirus [42], and our own results showing that the NS1 protein of influenza virus counteracts the PKR-mediated antiviral response [4], led to the hypothesis that an influenza virus lacking the NS1 gene might selectively grow in cells with an activated ras-signalling pathway. For several decades, it was known that reovirus exhibited preferential cytotoxicity for transformed cells over normal cells. However, the mechanism for this selectivity was just discovered recently. Specifically, it was shown that NIH-3T3 cells, which are resistant to reovirus infection, became susceptible when transformed with activated sos or ras. Restriction of reovirus proliferation in untransformed NIH-3T3 cells was at the level of viral protein synthesis. Protein kinase, the phosphorylation of which inhibits translation, was phosphorylated in untransformed NIH-3T3 cells after infection with reovirus, but it was not phosphorylated in infected or uninfected transformed cells. Inhibition of PKR phosphorylation by 2-aminopurine, or deletion of PKR, led to enhancement of reovirus protein synthesis in untransformed cells. These results suggested that, unlike the case in untransformed cells, phosphorylation of PKR is blocked by elements of an activated ras pathway in the transformed cells, allowing viral protein synthesis to ensue and the lytic cycle to proceed. An activated ras-signalling pathway therefore appeared to be the basis of reovirus oncolysis [42].

To investigate whether this property can be exploited for cancer therapy, SCID mice bearing tumors established from v-erbB–transformed murine NIH 3T3 cells or human glioblastoma cells were treated with the virus. A single intratumoral injection of virus resulted in regression of tumors in the majority of mice. Multiple treatments of immune-competent mice bearing tumors established from ras-transformed cells also resulted in tumor regression [33].

Subsequently, it was shown that reovirus also is capable of destroying different types of neoplastic cells in tissue culture and animal models. The result of these studies are summarized below: M.E. Wilcox and colleagues [43] showed that reovirus has potent activity against human malignant gliomas in vitro, in vivo, and ex vivo. Specifically, reovirus was capable of killing more than 80% of malignant glioma cell lines tested. It caused a dramatic and, often, complete tumor regression in vivo in subcutaneous and intracerebral human malignant glioma mouse models. Moreover, survival could also be prolonged. Reovirus was also able to infect and kill primary cultures of brain tumors removed from patients. It killed all glioma specimens but none of the cultured meningiomas tested.

The feasibility of using reovirus as an agent against human colon and ovarian cancer was also evaluated. It was observed that reovirus efficiently infected all human colon cancer cell lines and human ovarian cancer cell lines tested, but did not infect normal colon or ovarian cell lines. Ras activity in the human colon and ovarian cancer cell lines was elevated compared with that in normal colon and ovarian cell lines. In animal models, intratumoral as well as i.v. inoculation of reovirus resulted in significant regression of established human colon and ovarian tumors implanted at the hind flank. Histological studies revealed that reovirus infection was restricted to the tumor. Moreover, in an intraperitoneal human ovarian cancer xenograft model, inhibition of ascites tumor formation and the survival of animals treated with live reovirus were significantly pronounced. Reovirus infection in ex vivo primary human ovarian tumor surgical samples was also confirmed, further demonstrating the potential of reovirus therapy. These results suggest that reovirus holds promise as a novel agent for human colon and ovarian cancer therapy [44].

P. Lee's group also analyzed the potential of reovirus-mediated oncolysis for breast cancer. Direct activating mutations in the ras proto-oncogene in breast cancer are rare [45]. Breast tumor-derived cell lines were efficiently lysed by the virus, whereas normal breast cells resisted infection in vitro. Viral administration could cause tumor regression in a mammary fat pad model in SCID mice. Reovirus could also effect regression of tumors remote from the injection site, raising the possibility of systemic therapy of breast cancer by the oncolytic agent. Finally, reovirus could replicate in ex vivo surgical specimens. Overall, reovirus shows promise as a potential breast cancer therapeutic. One possible explanation for reovirus-induced oncolysis in the absence of oncogenic ras might be a ras-independent inhibition of PKR activation, which has been described for this type of cancer. Alternatively, the ras pathway might be activated by other means, such as an aberrant function of the receptor tyrosine kinase.

To test reovirus as a potential therapeutic agent against lymphoid malignancies, a number of lymphoid cell lines and primary human lymphoid cell cultures, as well as normal lymphocytes and hematopoietic stem/progenitor cells, were characterized with regard to their susceptibility to reovirus infection. There was evidence of efficient reovirus infection and cell lysis in the diffuse large B-cell lymphoma cell lines and in the Burkitt lymphoma cell lines Raji and CA46, but not in Daudi, Ramos, or ST486. Moreover, when Raji and Daudi cell lines were grown subcutaneously in SCID/nonobese diabetic (NOD) mice and subsequently injected with reovirus intratumorally or i.v., significant regression was observed in the Raji-induced, but not the Daudi-induced, tumors. These data are consistent with the in vitro results. Susceptibility to reovirus infection was also detected in the majority of primary lymphoid neoplasias tested but not in the normal lymphocytes or hematopoietic stem/progenitor cells [46]

Finally, the potential to be an antimetastatic cancer agent through remote-site delivery was demonstrated [47]. In particular, the ability of reovirus to replicate in murine cells to test the efficacy of this virus in eliminating distal or metastatic tumors in immune-competent mice was exploited. Intravenous therapy with reovirus not only inhibited metastatic tumor growth, it also led to a significant improvement in animal survival. Moreover combining i.v. reovirus treatment with immune suppression resulted in further reduction of tumor size and a considerable prolongation of survival. Combined therapy was also effective in overcoming an existing immunity to reovirus for inducing metastatic tumor regression. These results show the feasibility of i.v. reovirus therapy as a novel alternative in the treatment of metastatic cancer in humans.

In summary, reoviruses have been shown to effectively destroy many different types of neoplastic cells, including those derived from brain, breast, colon, ovaries, lymphoid malignancies, and prostate. Although it was well tolerated in immune-competent and athymic mice, toxicity was demonstrated in SCID mice. In humans, reovirus either results in an asymptomatic infection or mild gastroenteritis. It can normally be isolated from the respiratory and digestive tracts, however, it is not associated with any disease state. The benign nature of this virus was also confirmed upon experimental intranasal inoculation of reovirus in adult human volunteers, which produced only mild symptoms. Therefore, the safety profile of reovirus in immune-competent humans is promising. However, its safety profile in immunosuppressed humans still has to be demonstrated.

Given its selectivity for the activated Ras pathway, its effectiveness against a variety of human cancer cells in vitro and in vivo as xenografts, as well as its safety profile in humans, reovirus is an attractive candidate for oncolytic treatment. Clinical trials, which are presently being performed, will show whether the pre-clinical results can be translated to a benefit for cancer patients.

REFERENCES

1. Garcia-Sastre A. Inhibition of interferon-mediated antiviral responses by influenza A viruses and other negative-strand RNA viruses. Virology 2001; 279:375–384.

2. Egorov A, Brandt S, Sereinig S, Romanova J, Ferko B, Katinger D, Grassauer A, Alexandrova G, Katinger H, Muster T. Transfectant influenza A viruses with long deletions in the NS1 protein grow efficiently in Vero cells. J Virol 1998; 72: 6437–6441.

3. Garcia-Sastre A, Egorov A, Matassov D, Brandt S, Levy DE, Durbin JE, Palese P, Muster T. Influenza A virus lacking the NS1 gene replicates in interferon- deficient systems. Virology 1998; 252:324–330.

4. Bergmann M, Garcia-Sastre A, Carnero E, Pehamberger H, Wolff K, Palese P, Muster T. Influenza virus NS1 protein counteracts PKR-mediated inhibition of replication. J Virol 2000; 74:6203–6206.

5. Wong LH, Krauer KG, Hatzinisiriou I, Estcourt MJ, Hersey P, Tam ND, Edmondson S, Devenish RJ, Ralph SJ. Interferon-resistant human melanoma cells are deficient in ISGF3 components, STAT1, STAT2, and p48-ISGF3gamma. J Biol Chem 1997; 272:28779–28785.

6. Mundschau LJ, Faller DV. Oncogenic ras induces an inhibitor of double-stranded RNA-dependent eukaryotic initiation factor 2 alpha-kinase activation. J Biol Chem 1992; 267:23092–23098.

7. Richardson JC, Akkina RK. NS2 protein of influenza virus is found in purified virus and phosphorylated in infected cells. Arch Virol 1991; 116:69–80.

8. Yasuda J, Nakada S, Kato A, Toyoda T, Ishihama A. Molecular assembly of influenza virus: association of the NS2 protein with virion matrix. Virology 1993; 196:249–255.

9. Chen W, Calvo PA, Malide D, Gibbs J, Schubert U, Bacik I, Basta S, O'Neill R, Schickli J, Palese P, Henklein P, Bennink JR, Yewdell JW. A novel influenza A virus mitochondrial protein that induces cell death. Nat Med 2001; 7:1306–1312.

10. Luytjes W, Krystal M, Enami M, Pavin JD, Palese P. Amplification, expression, and packaging of foreign gene by influenza virus. Cell 1989; 59:1107–1113.

11. Parvin JD, Palese P, Honda A, Ishihama A, Krystal M. Promoter analysis of influenza virus RNA polymerase. J Virol 1989; 63:5142–5152.

12. Garcia-Sastre A, Palese P. Genetic manipulation of negative-strand RNA virus genomes. Annu Rev Microbiol 1993; 47: 765–790.

13. Enami M, Luytjes W, Krystal M, Palese P. Introduction of site-specific mutations into the genome of influenza virus. Proc Natl Acad Sci U S A 1990; 87:3802–3805.

14. Pleschka S, Jaskunas R, Engelhardt OG, Zurcher T, Palese P, Garcia-Sastre A. A plasmid-based reverse genetics system for influenza A virus. J Virol 1996; 70:4188–4192.

15. Neumann G, Watanabe T, Ito H, Watanabe S, Goto H, Gao P, Hughes M, Perez DR, Donis R, Hoffman E, Hobom G, Kawaoka Y. Generation of influenza A viruses entirely from cloned cDNAs. Proc Natl Acad Sci U S A 1999; 96:9345–9350.

16. Fodor E, Devenish L, Engelhardt OG, Palese P, Brownlee GG, Garcia-Sastre A. Rescue of influenza A virus from recombinant DNA. J Virol 1999; 73:9679–9682.

17. Nemeroff ME, Barabino SM, Li Y, Keller W, Krug RM. Influenza virus NS1 protein interacts with the cellular 30 kDa subunit of CPSF and inhibits 3′end formation of cellular pre-mRNAs. Mol Cell 1998; 1:991–1000.

18. Fortes P, Beloso A, Ortin J. Influenza virus NS1 protein inhibits pre-mRNA splicing and blocks mRNA nucleocytoplasmic transport. EMBO J 1994; 13:704–712.

19. Lu Y, Qian XY, Krug RM. The influenza virus NS1 protein: a novel inhibitor of pre-mRNA splicing. Genes Dev 1994; 8: 1817–1828.

20. Li Y, Chen ZY, Wang W, Baker CC, Krug RM. The 3′-end-processing factor CPSF is required for the splicing of single-intron pre-mRNAs in vivo. RNA 2001; 7:920–931.

21. Hatada E, Saito S, Fukuda R. Mutant influenza viruses with a defective NS1 protein cannot block the activation of PKR in infected cells. J Virol 1999; 73:2425–2433.

22. Tan SL, Katze MG. Biochemical and genetic evidence for complex formation between the influenza A virus NS1 protein and the interferon-induced PKR protein kinase. J Interferon Cytokine Res 1998; 18:757–766.

23. Lu Y, Wambach M, Katze MG, Krug RM. Binding of the influenza virus NS1 protein to double-stranded RNA inhibits the activation of the protein kinase that phosphorylates the elF-2 translation initiation factor. Virology 1995; 214:222–228.

24. Salvatore M, Basler CF, Parisien JP, Horvath CM, Bourmakina S, Zheng H, Muster T, Palese P, Garcia-Sastre A. Effects of influenza A virus NS1 protein on protein expression: the NS1 protein enhances translation and is not required for shutoff of host protein synthesis. J Virol 2002; 76:1206–1212.

25. Enami K, Sato TA, Nakada S, Enami M. Influenza virus NS1 protein stimulates translation of the M1 protein. J Virol 1994; 68:1432–1437.

26. de la Luna S, Fortes P, Beloso A, Ortin J. Influenza virus NS1 protein enhances the rate of translation initiation of viral mRNAs. J Virol 1995; 69:2427–2433.

27. Aragon T, de la Luna S, Novoa I, Carrasco L, Ortin J, Nieto A. Eukaryotic translation initiation factor 4GI is a cellular target for NS1 protein, a translational activator of influenza virus. Mol Cell Biol 2000; 20:6259–6268.

28. Goodbourn S, Didcock L, Randall RE. Interferons: cell signalling, immune modulation, antiviral response and virus countermeasures. J Gen Virol 2000; 81:2341–2364.

29. Talon J, Horvath CM, Polley R, Basler CF, Muster T, Palese P, Garcia-Sastre A. Activation of interferon regulatory factor 3 is inhibited by the influenza A virus NS1 protein. J Virol 2000; 74:7989–7996.

30. Wang X, Li M, Zheng H, Muster T, Palese P, Beg AA, Garcia-Sastre A. Influenza A virus NS1 protein prevents activation of NF-kappaB and induction of alpha/beta interferon. J Virol 2000; 74:11566–11573.

31. Bergmann M, Romirer I, Sachet M, Fleischhacker R, Garcia-Sastre A, Palese P, Wolff K, Pehamberger H, Jakesz R, Muster T. A genetically engineered influenza A virus with ras-dependent oncolytic properties. Cancer Res 2001; 61:8188–8193.

32. Stojdl DF, Lichty B, Knowles S, Marius R, Atkins H, Sonenberg N, Bell JC. Exploiting tumor-specific defects in the interferon pathway with a previously unknown oncolytic virus. Nat Med 2000; 6:821–825.

33. Coffey MC, Strong JE, Forsyth PA, Lee PW. Reovirus therapy of tumors with activated Ras pathway. Science 1998; 282: 1332–1334.

34. Bell JC, Garson KA, Lichty BD, Stojdl DF. Oncolytic viruses: programmable tumour hunters. Curr Gene Ther 2002; 2: 243–254.

35. Pansky A, Hildebrand P, Fasler-Kan E, Baselgia L, Ketterer S, Beglinger C, Heim MH. Defective Jak-STAT signal transduction pathway in melanoma cells resistant to growth inhibition by interferon-alpha. Int J Cancer 2000; 85:720–725.

36. Landolfo S, Guarini A, Riera L, Gariglio M, Gribaudo G, Cignetti A, Cordone I, Montefusco E, Mandelli F, Foa R. Chronic myeloid leukemia cells resistant to interferon-alpha lack STAT1 expression. Hematol J 2000; 1:7–14.

37. Zhou Y, Wang S, Gobl A, Oberg K. Interferon alpha induction of Stat1 and Stat2 and their prognostic significance in carcinoid tumors. Oncology 2001; 60:330–338.

38. Talon J, Salvatore M, O'Neill RE, Nakaya Y, Zheng H, Muster T, Garcia-Sastre A, Palese P. Influenza A and B viruses expressing altered NS1 proteins: A vaccine approach. Proc Natl Acad Sci U S A 2000; 97:4309–4314.

39. Wang X, Basler CF, Williams BR, Silverman RH, Palese P, Garcia-Sastre A. Functional replacement of the carboxy-terminal two-thirds of the influenza A virus NS1 protein with short heterologous dimerization domains. J Virol 2002; 76: 12951–12962.

40. Karron RA, Steinhoff MC, Subbarao EK, Wilson MH, Macleod K, Clements ML, Fries LF, Murphy BR. Safety and immunogenicity of a cold-adapted influenza A (H1N1) reassortant virus vaccine administered to infants less than six months of age. Pediatr Infect Dis J 1995; 14:10–16.

41. Redding G, Walker RE, Hessel C, Virant FS, Ayars GH, Bensch G, Cordova J, Holmes SJ, Mendelman PM. Safety and tolerability of cold-adapted influenza virus vaccine in children and adolescents with asthma. Pediatr Infect Dis J 2002; 21:44–48.

42. Strong JE, Coffey MC, Tang D, Sabinin P, Lee PW. The molecular basis of viral oncolysis: usurpation of the Ras signaling pathway by reovirus. Embo J 1998; 17:3351–3362.

43. Wilcox ME, Yang W, Senger D, Rewcastle NB, Morris DG, Brasher PM, Shi ZQ, Johnston RN, Nishikawa S, Lee PW, Forsyth PA. Reovirus as an oncolytic agent against experimental human malignant gliomas. J Natl Cancer Inst 2001; 93: 903–912.

44. Hirasawa K, Nishikawa SG, Norman KL, Alain T, Kossakowska A, Lee PW. Oncolytic reovirus against ovarian and colon cancer. Cancer Res 2002; 62:1696–1701.

45. Norman KL, Lee PW. Reovirus as a novel oncolytic agent. J Clin Invest 2000; 105:1035–1038.

46. Alain T, Hirasawa K, Pon KJ, Nishikawa SG, Urbanski SJ, Auer Y, Luider J, Martin A, Johnston RN, Janowska-Wieczorek A, Lee PW, Kossakowska AE. Reovirus therapy of lymphoid malignancies. Blood 2002; 100:4146–4153.

47. Hirasawa K, Nishikawa SG, Norman KL, Coffey MC, Thompson BG, Yoon CS, Waisman DM, Lee PW. Systemic reovirus therapy of metastatic cancer in immune-competent mice. Cancer Res 2003; 63:348–353.

7

Vesicular Stomatitis: an Oncolytic Virus that Exploits Tumor-Specific Defects in the Interferon Pathway

REBECCA ANN C. TAYLOR, JENNIFER M. PATERSON, and JOHN C. BELL

Ottawa Regional Cancer Centre
Research Laboratories
Ottawa, Ontario, Canada

ABSTRACT

Vesicular stomatitis virus (VSV) is part of a new generation of small RNA viruses being developed as replicating cancer therapeutics. Vesicular stomatitis virus replicates in and kills a wide variety of human cancer cell lines, and is highly effective in mouse cancer models. It is exquisitely sensitive to the antiviral effects of the interferon (IFN) family of cytokines and its oncolytic activity depends on the presence of inherent defects in the IFN signaling pathway in cancer cells. Recent research has shown that IFN-inducing VSV mutants are attenuated in normal cells, but not in cancer cells. These mutants can be administered i.v. to mice at high doses, and can effect durable cures in disseminated cancer models. Other research has highlighted the potential of VSV engineered to express transgenes that could enhance the antitumor immune response, cause bystander-cell killing, or retarget VSV to receptors expressed at high levels on cancer cells. Now that several first-generation oncolytic viruses have entered into clinical trials, the lessons learned from these viruses can be applied to second-generation viruses, such as VSV, to develop more effective, safe, replicating cancer therapeutics.

ONCOLYTIC VIRUSES ARE REPLICATING CANCER THERAPEUTICS

One of the expectations of the genomics revolution was that an understanding of the genetic differences between normal and diseased cells would lead to rationally designed therapeutics that could target those differences and thus avoid systemic toxicity. Although targeted drugs and gene therapy have had positive results in cancer therapy, they may be limited by inefficient delivery. Oncolytic viruses can take advantage of rational drug design principles, but as in situ replicating therapeutics, they can spread to and kill target cells that might otherwise be difficult to reach.

The idea of using replicating viruses to selectively kill cancer cells is not new; in fact, J. G. J. F. Sinkovics and J. C. Horvath reported the oncolytic effects of VSV in mouse cancer models was reported as early as 1969 [1]. However, the application of modern recombinant DNA technology to engineer better oncolytic viruses is new, and the field has exploded in the 10 years since the first report concerning Herpes simplex virus (HSV) in 1991 [2].

Adenoviruses were the first genetically engineered oncolytic viruses to be tested in humans [3]. Deletion of the E1B region resulted in a virus, ONYX-015, that could only replicate in cells with deficiencies in the p53 pathway, a characteristic shared by more than 50% of tumor cells [4]. Although this virus has proven to be very safe, its attenuation may have compromised its efficacy [5]. Since the early trials with ONYX-015 and HSV, there has been a shift toward developing viruses that replicate rapidly using their own polymerase in the cytoplasm, rather than the host polymerase in the nucleus. Examples include reovirus, vaccinia, measles, polio virus, Newcastle disease virus, and VSV (reviewed in [6] and [7]. Several characteristics of VSV suggest that it may be able to fulfill the full potential of oncolytic viruses. It replicates extensively in a wide variety of human cancer cells, it is not a natural human pathogen, and it can be easily genetically manipulated and purified.

THE BIOLOGY OF VSV

Vesicular Stomatitis Virus has been Studied Extensively

As a member of the genus *Vesiculovirus* of the Rhabdoviridae family, VSV is among the simplest of the RNA viruses in terms of its structure, genetics, and physiology and for this reason it is exploited as a model system for the study of viral replication and cytopathology. Vesicular stomatitis virus is an enveloped virus with a nonsegmented RNA genome of 11 kb encoding five mRNAs and give principle proteins, as well as several less well characterized internally initiated proteins. Structurally, it is divided into the ribonucleoprotein core and the viral membrane, as illustrated in Figure 1. The single-strand negative sense RNA, the N (nucleocapsid) protein, collectively with the P (phosphoprotein) and L (large) protein form the ribonucleoprotein core—a transcription unit that will synthesize, cap, and polyadenylate mRNAs in vitro. The L protein serves as the RNA-dependent RNA polymerase (RNP) but phosphorylated P is an absolute requirement for transcription [8]. The genome RNA, when tightly complexed with the N protein, is ribonuclease resistant and serves as the template for transcription [9]. The viral membrane is composed of the membrane spanning G (glycoprotein) and the M (matrix) protein, along with lipids and cholesterol derived from the host cell.

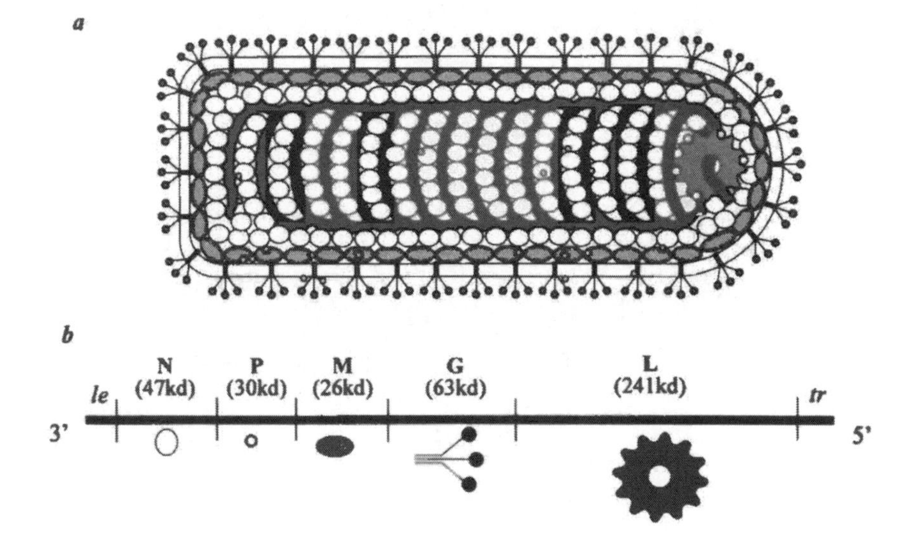

Figure 1 Diagram of Vesicular Stomatitis Virus (VSV), illustrating virion structure (*a*) and genome structure (*b*). This virus forms a bullet-shaped particle, enveloped in a host-derived plasma membrane, containing all five viral proteins and the single-stranded negative-sense RNA genome. The L (Large) protein forms the major component of the RNA-dependent RNA polymerase, along with P (Phosphoprotein) and N (Nucelocapsid) protein, which also has a role packaging the genome. The G (Glycoprotein) forms a glycosylated trimer that binds the host receptor, phosphatidylserine, and mediates pH-dependent membrane fusion. The M (Matrix) protein has a role in virus assembly and budding, cytopathic effects, and shutoff of host-cell defenses. The virion polymerase transcribes the five mRNAs in sequential order from the 3′ end, as shown in (*b*). The polymerase pauses at the end of each transcript, and reinitiates with imperfect efficiency, such that genes at the 3′ end (N) are transcribed more than genes at the 5′ end (L), providing a mechanism of regulation of gene expression. The leader (*le*) and trailer (*tr*) regions are transcribed, but not translated.

Binding of VSV to the cell surface and membrane fusion are mediated by the G protein. Not surprisingly, this viral surface protein is also the major antigenic determinant of VSV and stimulates a very strong antibody response [10]. The M protein is multifunctional in that it plays a crucial role in directing viral assembly and budding and has a number of cytopathogenic functions ascribed to it. These include inhibition of host transcription [11], disorganization of the cytoskeleton resulting in cell rounding and detachment [12], induction of apoptosis [13] and the inhibition of nucleocytoplasmic transport of mRNAs and proteins [14].

Infection with VSV begins with viral adsorption to the cell membrane by G protein binding to the ubiquitous phosphatidylserine and, possibly, other receptors. Receptor-mediated endocytosis and subsequent pH-dependent membrane fusion releases RNPs into the cytoplasm. The negative-strand RNA is transcribed sequentially into the five mRNAs: N, P, M, and L are translated by free cytoplasmic ribosomes, whereas G is translated on ribosomes bound to the endoplasmic reticulum, where the protein product becomes glycosylated and shuttled to the cell membrane. Viral RNA replication begins with synthesis of a full-length positive-strand copy of RNA, which serves as a template for the progeny negative-strand genomic RNA. These newly synthesized RNPs and the M protein are transported to the G protein-containing plasma membrane where viral assembly and budding is initiated [15]. Both the P and the M mRNAs encode additional proteins, the functions of which have not been fully elucidated [16,17].

An extensive understanding of the viral components of VSV and their interaction in the host cell will be invaluable for elucidating and manipulating its oncolytic properties. For example, the inhibition of mRNA nuclear export by the M protein of VSV is critical for its ability to evade the innate antiviral response and replicate in normal cells, whereas the strong antibody response induced by the G protein poses a significant barrier to in vivo viral replication during a multidose treatment regimen. Moreover, an appreciation of the life cycle of VSV is important because the relatively short replication will permit rapid viral spread, and its occurrence entirely within the cytoplasm ensures that VSV has no transforming ability by inadvertent incorporation into the host cell DNA.

Not a Human Pathogen

Vesicular Stomatitis Virus might at first appear to be an unlikely candidate for an oncolytic virus because its natural host is not human but livestock. In cattle and swine, the two main serotypes, New Jersey (NJ) and Indiana (IN), both cause a self-limiting disease clinically undistinguishable from foot-and-mouth disease. Vesicular stomatitis virus is endemic from northern South America to southern Mexico, where seasonal outbreaks infect roughly 10% of cattle every year, and sporadic outbreaks have been described in the southwestern United States. Although the natural life cycle of VSV is unknown, evidence suggests that insects, notably black flies and sand flies, carry the virus in endemic areas and are capable of infecting animals and, presumably, humans [10]. However, animal hosts with sustained viremia have never been found, strongly suggesting that they are dead-end hosts from which the virus does not return to its cycle in nature [11]. Under laboratory conditions, VSV cultures can be subjected to strong selective forces [9], however, extensive population studies among livestock have shown no evidence of immunological selection; VSV is genetically stable despite high levels of neutralizing antibodies among cattle populations [18]. In nonendemic areas, VSV infection in humans is rare and results in an influenza-like illness that is uniformly nonfatal [10].

These epidemiological data have important implications for the development of VSV as a therapeutic agent. First, they alleviate concern regarding the spread of VSV from the index treatment case to the population by establishing animals as dead-end hosts. Second, because VSV is a genetically stable virus in vivo, it is unlikely that VSV will mutate to a more virulent form following infection. Third, naturally occurring human VSV infection results in tolerable flu-like symptoms, and the virus is quickly and completely cleared. Moreover, because most humans are seronegative, the virus has an opportunity to replicate before neutralizing antibodies develop.

Exquisitely Sensitive to Interferon

Vesicular stomatitis virus is so well known for its interferon (IFN) sensitivity that an International Unit of IFN activity is defined by the ability to inhibit VSV replication. Interferon-α and -β are cytokines that belong to the family of Type I IFNs responsible for conferring an antiviral state in cells by the expression of IFN-stimulated genes (ISG). The IFN response results in an antiviral state consisting of cell-cycle arrest, inhibition of host translation, mRNA degradation, and apoptosis (Figure 2). Interferon-α and -β signal through Type I IFN receptors (IFNAR), which are associated with two Janus tyrosine kinases (Jak) and two signal transducer and activator of transcription molecules (STAT). The binding of Type I IFNs to their receptor results in transphosphorylation of Jaks and subsequent phosphorylation of STAT 1 and 2. The phosphorylated STATs form a heterodimer that translocates to the nucleus, where it associates with the DNA-binding protein IFN regulatory factor (IRF)-9 (previously termed *p48*). This heterotrimeric complex, known as ISGF3, binds to an IFN-stimulated response element to induce the transcription of more than 300 different ISGs [19].

The induction of the IFN-α and -β genes themselves is mediated by two IFN regulatory factors, IRF-3 and IRF-7, which define the first and the second wave of IFN production [20]. In the first wave following viral infection, IRF-3, which is constitutively expressed in all cell types, is activated by phosphorylation virus-activated kinases, recently reported to be members of the IKK family of kinases [21]. Interferon regulatory factor-3 translocates to the nucleus as a homodimer and binds to the IFN-β promoter as part of a multiprotein transcription-promoting complex called the *enhanceosome* that also contains nuclear factor-κB (NFκB), IRF-1 and ATF-2/c-Jun [19]. Once synthesized and secreted, IFN-β signals via the Type I IFN receptor to induce the synthesis of IRF-7, resulting in the second wave of the cascade. Interferon regulatory factor-7 is also a transcriptional activator that turns on the synthesis of IFN-α and many other ISGs. Interferon-α also signals via the Type 1 IFN receptor, resulting in a feedback loop that is

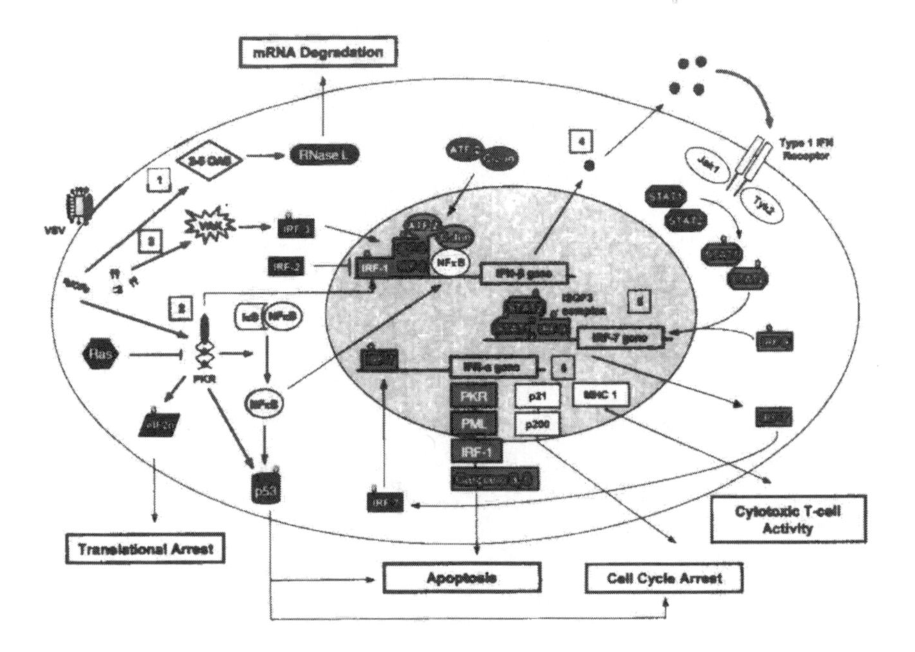

Figure 2 Cellular Response to Viral Infection. Infection with vesicular stomatitis virus (VSV) initiates the interferon (IFN) signaling cascade that results in a number of antiviral processes, including mRNA degradation, translational arrest, cytotoxic T-cell activity, cell-cycle arrest, and apoptosis. **1.** Double-stranded RNA activates 2–5 oligoadenylate synthetase, leading to the production of RNAse L and viral mRNA degradation. **2.** Protein kinase (PKR) is activated by autophosphorylation in response to dsRNA and, in turn, affects several pathways, including phosphorylation of eIF2α and subsequent translational arrest, activation of NFκB and IRF-1, components of the enhancesome complex formed on the IFN-β gene promoter, and the transcriptional upregulation and activation of p53, the downstream effects of which include cell-cycle arrest and apoptosis. **3.** Viral proteins result in the activation of the virus-activated kinase (VAK), leading to the phosphorylation of IRF-3 and its translocation to the nucleus where, along with NFκB and several other transcription factors, it forms the enhancesome and transcriptional upregulation of IFN-β. Interferon-β is synthesized and secreted from the cell where it binds to Type 1 IFN receptors on the cellular surface (itself and neighboring cells). This interaction results in the phosphorylation of signal transducer and activator of transcription molecules (STAT) 1 and 2 and, along with phosphorylated

essential for IFN priming, a phenomenon whereby pretreatment of cells with a low dose of IFN enhances IFN production during viral infection [20]. It is also essential for protective IFN signaling from infected cells to neighboring uninfected cells.

Of the many ISGs, the best-understood IFN-regulated pathways include the Mx family of proteins and the two double-stranded RNA (dsRNA)-dependent enzymes, the 2–5 oligoadenylate synthetase/RNase L system (2–5 OAS), and the protein kinase PKR. The Mx proteins belong to the dynamin family of guanosine triphosphatases. The human cytoplasmic MxA protein has been shown to inhibit viral replication at the level of transcription, as is the case with VSV [21], or at the posttranscriptional and translational levels, as seen with influenza [22] and measles [23]. The 2–5 oligoadenylate synthetases are transcriptionally induced by IFN but require dsRNA as a cofactor for activation. These enzymes polymerize adenosine triphosphate to produce 2′–5′ linked oligoadenylates whose only known function is to activate the latent ribonuclease RNase L. Activated RNase L mediates the antiviral activity through cleavage of single-stranded RNA, as demonstrated by the correlation between RNase L activation and IFN-induced inhibition of encephalomyocarditis virus [24], vaccinia virus [25], and reovirus infection [26].

The most extensively studied IFN-induced protein is PKR. Similar to 2–5 OAS, the transcription of PKR is upregulated by IFNs but the PKR protein needs to be activated by dsRNA. The PKR protein kinase is a serine-threonine kinase that undergoes autophosphorylation upon activation, which enables it to phosphorylate substrate targets. The translation initiation factor eIF2α is one such target, resulting in the effective shutdown of protein synthesis [27]. PKR upregulates the

◄————————————————

IRF-9, leads to the formation of the ISGF-3 complex on the IRF-7 promoter. **6.** Once synthesized, IRF-7 is phosphorylated by the VAK, resulting in its translocation to the nucleus and transcriptional pregulation of IFN-α and a number of other ISGs responsible for the variety of downstream effects of IFNs.

activity of several transcription factors, including NFκB, IRF-1 [28], both components of the enhanceosome complex critical for the induction of IFN-β. Protein kinase PKR is also found in a complex with STAT1, which dissociates upon PKR activation, allowing STAT1 to participate in the signaling cascade via the Type 1 IFN receptor [29]. In addition, PKR has been shown to regulate the activity of the tumor suppressor p53 and members of the stress-activated protein kinases, p38 MAP kinase and c-Jun NH2-terminal kinase, JNK [30]. The modulation of these varied targets by PKR confers a regulatory ability on diverse cellular processes, such as cell growth and differentiation, apoptotic, antitumor and, of course, antiviral activities. The biological importance of PKR in regulating virus infection is underscored by the existence of a multitude of viral inhibitors to PKR action, including adenovirus (inhibition of PKR activation), reovirus (sequestration of dsRNA), hepatitic C Virus (HCV) (inhibition of PKR dimerization), Herpesvirus (synthesis of PKR pseudosubstrates) and polio virus (PKR degradation) [27].

Interferons also signal through a number of alternative pathways include upregulation of caspases involved in the apoptotic cascade, cyclin-dependent kinase inhibitors, such as members of the p200 family and c-*myc*, exerting negative regulation of the cell cycle, and of major histocompatibility complex (MHC) Class I proteins for immune stimulation via cytotoxic T cells [19]. The relative importance of each component of the IFN signaling cascade remains to be elucidated.

In mouse experiments of VSV infection, treatment with Type I IFNs, even several days after inoculation, increases survival, whereas blocking the IFN response, either with antibodies or by using IFNα- or IFNβ-receptor knockout mice, decreases survival [31]. Similarly, mice lacking STAT1 [32] or p48/IRF-9 [33] genes are also remarkably susceptible to lethal infection with VSV. A clear role for PKR in IFN-mediated resistance to VSV has also been established by D. F. Stojdl [34]. In this study, fibroblasts derived from PKR-/- mice were shown to be more permissive to VSV infection than wild-type fibroblasts and to be deficient in IFN-mediated protection. In addition, PKR-/- mice were acutely susceptible to intranasal VSV infection and succumbed to an overwhelming respiratory infection, with an LD50 of less than 15 particle-forming-units (PFU) per mouse. By comparison, wild-type mice had an LD50 of 10^6 PFU and manifested central nervous system pathology, most notably hind limb paralysis. Moreover, pretreatment of wild-type mice with IFN-α or IFN-β completely alleviated symptoms of VSV infection, whereas pretreatment of PKR-/- mice resulted in no increase in survival or alleviation of symptoms. In an earlier study by Zhou et al, [35] VSV infection of fibroblasts from mice deficient in either PKR or triply deficient in PKR, Mx1, and RNAse L was 50- to 100-fold less inhibited by IFN than in RNAseL-/- or control mice. This suggests that PKR is largely responsible for the difference in the antiviral effect of IFN observed between the control and triply deficient fibroblasts following VSV infection. This difference notwithstanding, at the highest dose of IFN, more than 10^3 units of inhibition were observed in the PKR-/- and triply deficient mice, providing clear evidence for the existence of alternative antiviral pathways.

EXPLOITING DEFECTS IN CANCER CELLS

Tumor Cells Demonstrate Defects In Interferon Signaling

Interferons orchestrate a cascade of gene expression that leads to a number of fundamental changes in basic cell physiology, apart from induction of the antiviral state, are implicated in the antitumor response, including cell cycle arrest, differentiation, apoptosis, and immunomodulation. These cellular effects are the basis for their use in the treatment of neoplastic disease but, unfortunately, in the majority of cancers, IFNs have so far proven largely ineffective and patients who do respond to IFN therapy often have, at best, a short-lived remission [36]. This occurs because during tumor evolution, there is selective pressure for mutations in components of the IFN pathway, allowing cancer cells to acquire a growth and survival advantage over normal cells. Defects in IFN signaling may involve the inability to produce IFNs in response to viral infection, the inability to respond to IFN via the Type 1 IFNAR, or defects in specific ISGs.

Inability of cancer cells to produce IFNs may occur secondary to loss of heterozygosity at IFN α and β loci, as was demonstrated by deletions on chromosome 9p found in a number of human gliomas [37]. Alternatively, production of IFN may be impaired by defects in components of the enhanceosome [38], whereas submaximal induction of IFN is secondary to defects in IRF-1 and IRF-7 [39]. Interferon regulatory factor-3 appears to be involved in virus-induced oncogenesis given that the E6 oncoprotein of human papillomavirus (HPV) binds to and inhibits its transcriptional activation of IFN-β [40]. Loss of IRF-1, a demonstrated tumor-suppressor gene, is thought to be a critical event in the development of leukemia and has also been implicated in esophageal and gastric adenocarcinomas [41]. In addition, invasive human breast carcinomas [42] and aggressive malignant melanomas [43] are significantly less likely to express IRF-1 compared with normal or less invasive tumors. By contrast, overexpression of IRF-2, which blocks transcriptional activation of IRF-1 by competitive binding, causes oncogenic transformation of NIH 3T3 cells [44] and the ratio of IRF-1/IRF-2 mRNA levels are increased in proliferating cells [41]. Although PKR appears to be involved in the activation of IRF-1, NFκB, and ATF-2, PKR deficiency has not been demonstrated to impair virus- or dsRNA–dependent activation of IFN-α or IFN-β [45].

Tumor cells are also characterized by defects in their ability to respond to IFN signaling, including STAT phosphorylation and the formation of ISGF3. Human melanoma cells [46] and several transitional carcinoma cell lines [47] have been demonstrated to be deficient in the ISGF3 components, particularly STAT1, and the 'suppressor of cytokine signaling' SOCS-3, which is known to inhibit STAT1 activation, confers IFN resistance in cutaneous T-cell lymphoma [48]. In addition, the HPV oncogene E7 abrogates the antiviral effects of IFN-α in cervical carcinoma cells by preventing the translocation of p48/IRF-9 to the nucleus [40], whereas the oncogene E6 reduces Jak-STAT tyrosine phosphorylation in human fibrosarcoma cells [49].

Inhibition of ISGs has also been demonstrated in a number of human malignancies. Downregulation of IFN signaling is thought to be involved in prostate tumor progression because expression of 2–5 OAS and IRF-9 were found to be decreased in tumorigenic as compared with nontumorigenic prostatic hyperplasia cells and, moreover, 30% of human prostate adenocarcinoma samples exhibited downregulation of IFN-inducible genes [50].

What about the link between PKR and cancer? The role of PKR as a tumor-suppressor gene is still unclear. The expression and activity level of PKR have been shown to correlate with the degree of cellular differentiation in a variety of carcinomas, including lung, leukemia, breast, adenocarcinoma of the colon, and squamous cell carcinomas of the head and neck, but this has not been demonstrated in all cancer types [51]. In tissue culture cells, expression of either a dominant negative PKR or a PKR inhibitor, such as a variant of eIF2α that cannot undergo phosphorylation, leads to malignant transformation of NIH 3T3 cells [52]. In contrast, however, several studies of mice devoid of PKR protein or of PKR activity have not demonstrated increased tumorigenesis in these animals [45,52]. Mouse embryonic fibroblasts derived from mice with targeted disruption of the catalytic domain of PKR displayed normal virus- and stress-induced apoptosis and intact eIF2α phosphorylation, establishing the existence of a redundant pathway that can compensate for deficient PKR activity and maintain appropriate control of cell growth and apoptosis [52]. Although defective PKR alone may not be sufficient to promote unregulated cell growth, in combination with defects in other oncogenic pathways, PKR may be important for malignant transformation.

The protein kinase PKR may also be linked to oncogenesis by both its upstream and downstream pathways. Upstream of PKR, Ras activation, a hallmark of malignant transformation, downregulates PKR activity through induction of an endogenous PKR inhibitor [53]. Downstream, PKR activates the tumor suppressor p53 by an NFκB-dependent transcriptional upregulation and by phosphorylation. Defects in p53 or Ras activation are present in the vast majority of solid tumors, placing PKR at the center of a number of integrated and aberrant signaling pathways in cancer cells.

Selective Replication In Tumor Cells With Defects In The Interferon Pathway

Although tumor cells have gained a growth and survival advantage by acquiring defects in the IFN signaling pathway, they have simultaneously compromised their ability to defend themselves against certain pathogens. Because IFN is clearly essential for the defense against VSV it was hypothesized that VSV might replicate and spread more efficiently in cancer cells than in normal cells. Indeed, Several human tumor cell lines infected with VSV demonstrated earlier, had more extensive cytopathic effects, and yielded 10^1 to 10^4 more viral PFU/mL than normal human primary cell cultures [54]. Furthermore, IFN pretreatment completely protected normal primary cells and reduced viral production to less than 1,000 PFU, whereas tumor cell lines remained susceptible to VSV and continued to produce high viral titers. Moreover, in a mixed culture of leukemic cells from an Acute mylogenous Leukemin cell line and normal bone marrow cells, VSV demonstrated selective destruction of leukemic cells, even in the absence of IFN, illustrating the potential to use VSV in ex vivo bone-marrow purging. A similar study [55] also established the utility of VSV as an oncolytic agent by demonstrating efficient replication and cytolysis in a number of different human cancer cell lines with incomplete protection following IFN-β pretreatment. Interestingly, a second study by the same group [56] also demonstrated normal PKR activation and eIF2α phosphorylation in response to VSV infection in rat C6 glioblastoma cells, implying that other components of the IFN pathway and not PKR itself are responsible for the oncolytic activity of VSV in this cell line. This study also suggests that VSV kills infected tumor cells by inducing apoptosis.

Several groups investigated the in vivo oncolytic of VSV in human tumor venografts in a thymic mice [54–56]. These studies demonstrated significant tumor regression following intratumoral injection of VSV with residual virus in the tumor tissue but virtually no virus detectable in the lung, brain, kidney, spleen, or liver 21 days after infection [55]. Similar results were seen in athymic mice implanted with *c-myc* or *K-ras*–transformed BALB/3T3 cells, suggesting the potential of VSV against malignancies with a wide range of genetic lesions [56]. In terms of viral toxicity, whereas one group [56] used a fairly low dose of an VSV (2.5×10^7 PFU/mL) and observed no overt illness apparent in the treated animals during the study, another group [54] used a higher dose (10^8 PFU/mL) and found these immune-compromised mice began to die by day 10 of infection. In the latter study, however, mice that received IFN treatment during the infection were protected and survived symptom-free for more than 45 days. Finally, VSV is capable of reaching and replicating distal to the site of administration, and oncolytic viral replication has been demonstrated in tumors following i.v. injection and contralateral intratumoral injection [56]. As will be discussed below, results from our lab have established the in vivo efficacy of i.v.-administered VSV in a metastatic mouse model of colon cancer disseminated to the lung.

Naturally Occurring Interferon-Inducing Vesicular Stomatitis Mutants Have A Wider Therapeutic Index

We have shown that wild-type VSV can cure nude mice of human melanoma, although all animals eventually succumb to virus-induced mortality. These mice can, however, be completely protected from VSV by treatment with IFN, leaving tumor-killing ability intact [54]. As discussed above, VSV is exquisitely sensitive to exogenously administered IFN, but the Indiana strain of VSV itself induces very little IFN. It therefore was reasoned that a virus that could induce more IFN would retain its tumor-killing ability, yet be attenuated in normal cells.

Several IFN-inducing VSV mutants have been characterized in the literature [57,58]. These mutants have substitutions in the Matrix protein that result in decreased ability of the virus to inhibit host gene expression, including IFN induction. We have shown that the VSV mutants AV1 and AV2 induce 20 to 50 times more IFN than the parental wild-type virus on epithelial cell lines. Furthermore, these viruses retain their tumor-killing ability against 75% to 80% of cell lines in the NCI 60 panel of human cancer cell lines. As expected, most of the cell lines in this panel have defects in their response to IFN.

The mechanism by which some mutants induce more IFN than the wild-type VSV has been investigated in some detail. Expression of the wild-type M protein inhibited the transcription of transfected reporter genes [59] but mutant M proteins did not [60]. The physiological relevance of this phenomenon was confirmed by Ferran and Lucas-Lenard, who showed that the wild-type M protein also reduced expression of genes under the control of the human IFN-β promoter [61]. Although such studies favored a model in which M protein inhibits host transcription, other results provided evidence that the wild-type M protein actually inhibits nucleocytoplasmic trafficking, including mRNA export [14]. Further evidence has been provided to support both hypotheses (for example, Ref. 62,63), as well as another hypothesis, in which the VSV M protein inhibits STAT activation [64]. We have performed quantitative Reverse transcription (RT)-PCR to show that although AV1, AV2, and wild-type VSV Indiana induce equivalent levels of IFN-β mRNA in the nucleus, cytoplasmic IFN-β mRNA was reduced by almost 100-fold in wild-type virus-infected cells, providing further support for the nuclecytoplasmic transport hypothesis.

The attenuation of IFN-inducing VSV mutants has been confirmed in animals. For example, we have shown that the LD50 for AV1 is more than four logs greater than wild-type VSV administered intranasally into Balb/C mice. Importantly, the tumor-killing ability of AV1 and AV2 is maintained in-vivo. In a nude mouse model of intraperitoneally injected human ovarian cancer cells, three intraperitoneal doses of AV2 can durably cure 70% of mice, whereas a single dose of the wild-type virus is lethal. Furthermore, AV1 and AV2 can effect complete cures in immunocompetent models of subcutaneous and metastatic cancer. The development of these IFN-inducing mutants represents a major advance in VSV oncolytic therapy, because all previous experiments used viruses that were toxic when systemically administered to animals at high doses, unless exogenous IFN was applied. These IFN-inducing mutants, however, spread poorly in normal tissue, and therefore can be administered safely at very high doses, while tumor killing remains potent. Another safety advantage of the IFN-inducing viruses is that the IFN they induce also inhibits replication of the wild-type virus, so that if a revertant strain were to arise, it could not overwhelm the attenuated virus.

As discussed above, the selective oncolytic ability of VSV appears to depend on tumor specific defects in the IFN pathway. Because IFN is crucial to the innate defense against all viruses, one would expect that other viruses could also selectively target tumor cells, and indeed this appears to be the case. It has been reported that reovirus [65] and some HSV mutants [66] replicate selectively in cell with an activated Ras pathway because of the Ras-mediated upregulation of a PKR inhibitor. Furthermore, an influenza virus lacking the NS1 gene, the protein product of which inhibits host gene expression in general and the IFN pathway in particular, has also been shown to have selective oncolytic properties [67]. It will be interesting to see whether other viruses can also be rendered selectively oncolytic by manipulation of their IFN-antagonizing abilities.

ENGINEERING TO IMPROVE ONCOLYTIC EFFICACY

Another advantage of using VSV as an oncolytic platform is the relative ease with which it can be genetically engineered. A system has been established in which VSV can be rescued by cotransfecting plasmids encoding N, P, L, and the genome to be rescued [68]. Extra coding regions can be inserted with stability between the G and L genes. As discussed below, transgenes can be inserted that will increase toxicity to cancer cells, attenuate the virus in normal cells, induce an antitumor immune response, or provide a simple means of monitoring therapeutic progress. Furthermore, existing VSV genes can be specifically mutated, swapped, or rearranged in order [69].

Increasing Toxicity

Although VSV is a potent and rapidly cytolytic virus in vitro, multiple high doses of VSV seem to be required for durable cures, indicating that VSV is not killing all tumor cells in a single dose. Thus, it was reasoned that there may be room to increase the toxicity of VSV by engineering it to express thymidine kinase (TK) [70]. Thymidine kinase is an enzyme encoded by HSV that maintains a nucleotide pool, allowing infected cells to recycle nucleotides for DNA synthesis. It has often been used as a suicide gene in gene therapy because it converts the nontoxic prodrug Ganciclovir into a toxic metabolite and has a relatively strong bystander effect, being able to kill nearby nontransduced target cells, (reviewed in Ref. 70). G. Barber's group showed that combined treatment with VSV-TK and Ganciclovir was significantly better than wild-type VSV at inhibiting subcutaneous tumor growth in two different immunocompetent mouse models. The authors also noted that VSV-TK plus Ganciclovir was significantly better than either VSV-TK alone or Ganciclovir alone.

It has been suggested that arming oncolytic viruses with suicide genes such as TK could also provide a fail-safe mechanism for stopping viral replication in an undesirable context [71]. However, because VSV is highly sensitive to IFN, we believe that administration of exogenous IFN might be a faster and more effective way to halt viral replication.

Increasing Induced Antitumor Immunity

Although oncolytic viruses, because they are replicating therapeutics, have a better chance of targeting all cancer cells than do chemotherapeutics, the possibility still exists that a single cell will escape and proliferate. The most effective oncolytic viruses therefore need to induce an antitumor immune response. Tumors, although they may express foreign antigens, do not appear to stimulate an effective immune response, possibly because the antigens are not presented in the context of other "danger" signals. Viruses, however, are very efficient at stimulating the immune system, and may be able to supply these missing danger signals to tumor cells (reviewed in Ref. 72). Indeed, lysates of vaccinia virus-infected tumor cells are more efficient at stimulating an antitumor immune response than lysates of uninfected tumor cells [73]; and this therapy has been applied in clinical trials [74]. Another example is replicating HSV, which has been shown in several systems to induce a long-lasting protective tumor-specific immune response [75].

Vesicular stomatitis virus may also elicit this type of antitumor immune response. We have observed that five out of five immunocompetent mice cured of subcutaneous tumors for more than 7 months could not be reseeded with the same tumor cells. G. Barber's group hypothesized that the antitumor immune component of VSV oncolysis might be increased in a virus expressing IL-4 [70]. This cytokine has been shown to play a role in tumor rejection and is known to skew the immune response toward an antibody-mediated defense [76]. In two immunocompetent mouse models, intratumorally injected recombinant VSV–IL-4 was significantly better than the wild-type virus at inhibiting subcutaneous tumor growth. Histological examination of these subcutaneous tumors revealed significant inflammatory infiltration, especially in VSV–IL4 treated animals. However, no difference in tumor-specific cytotoxic T lymphocyte activity was observed in animals treated with the wild-type versus IL-4–expressing viruses.

Although VSV–IL4 may merit further investigation, viruses expressing other immunomodulatory transgenes may have greater potential. Oncolytic viruses expressing IL-2, tumor necrosis factor, B7.1, IL-10, IL-12, IFN-γ, and several others are currently under investigation [77].

Retargeting the Vesicular Stomatitis Virus Glycoprotein

Vesicular stomatitis virus has the ability to enter virtually all cell types, because it uses the ubiquitous phosphatidylserine as a receptor. Whereas all published strategies for enhancing the tumor selectivity of VSV have focused on events occurring after the virus has entered the cell, research into VSV as a therapeutic against human immunodeficiency virus (HIV) has demonstrated that VSV is amenable to glycoprotein engineering as a means of conferring selective killing [78]. In this work, the VSV glycoprotein gene was deleted and replaced with CD4 and CXCR4, the receptors expressed on T-cells that mediate HIV entry. This novel VSV could bind to and enter only CD4+ T-cells that were infected with HIV and was able to reduce HIV titers from infected cells by up to 10^4-fold. Clearly, the possibility exists that VSV, like other viruses and gene therapy vectors [79], could be specifically targeted to receptors that are overexpressed on cancer cells.

PROSPECTS FOR FUTURE RESEARCH

The areas in which currently used chemotherapies fail are those where oncolytic viruses hold the most promise. Traditional cancer therapeutics are limited by their toxicity to normal tissue because they exploit the proliferative capacity of malignant cells and therefore target all normal rapidly proliferating cells, leading to the most common dose-limiting side effects of chemotherapy, including neutropenia and enteral toxicity. Although newer, rationally designed chemotherapeutics generally target specific defects found in particular cancer cells, they are limited by the development of a resistant population of malignant cells within the tumor. Vesicular stomatitis virus and other oncolytic agents are postulated to be superior because they target multiple defects in the IFN pathway that combine to mediate susceptibility to VSV, so the chance of complete reversion to an IFN-sensitive VSV-resistant phenotype is less likely than for other targeted therapeutics. Furthermore, oncolytic viruses can amplify themselves by more than 1,000-fold in a single tumor cell with the potential to spread to all susceptible cells. Wild-type VSV has been shown to efficiently kill a wide range of human tumor cell lines, and can effect durable cures in a variety of cancer models with several different routes of administration. Recently, research in our lab has demonstrated that IFN-inducing VSV mutants are attenuated in normal cells, yet retain their tumor-killing ability, and therefore can be safely administered i.v. at high doses.

Certainly, future research with VSV will involve engineering recombinant viruses expressing novel transgenes to enhance the therapeutic index, but VSV is also an ideal virus with which to try more innovative tumor-specific regulatory mechanisms. DNA viruses, including adenovirus and HSV, have been extensively engineered with conditional promoters. For example, several adenoviruses were engineered with either the E1A or E4 regions under the control of tumor-specific promoters responsive to estrogen, hypoxia, telomerase, or a functional Rb pathway [80]. Because VSV transcription occurs in the cytoplasm using its own polymerase, host-specific promoters cannot be used to regulate VSV transcription. Translational and posttranslational targeting strategies could, however, be effectively exploited by an engineered VSV. For example, several translational elements, including IRES elements and 5'UTRs, have been shown to mediate enhanced translation in malignant cells [81,82], and, in addition, protein ubiquitination and degradation can be regulated in a tumor-specific manner [83]. Thus, although VSV cannot be targeted by tumor-specific promoters, posttranscriptional strategies exist that may be even more promising.

The development of VSV must also be furthered at the macroscopic level, including its delivery, spread, and biodistribution in animal models and humans. Work with Green-flourescent protein-expressing VSV in our lab has shown that i.v.-delivered VSV replicates extensively throughout lung tumors at very early time points (10 to 20 hours postinfection) whereas peak replication in subcutaneous tumors may occur slightly later. More extensive studies, perhaps using other reporter genes, such as luciferase and *lacZ* and in vivo detection methods, such as positron emission tomography scanning, will be useful in addressing some of these issues. Moreover, it will be increasingly important to elucidate the efficacy of VSV when combined with other cancer therapeutics, particularly radiation therapy, which has been demonstrated to increase oncolysis of other viruses. Although much work still needs to be done to establish the optimal dose, schedule, and tumor type for VSV, we believe that the IFN-inducing mutants currently under investigation in our lab possess all the requisites of a successful oncolytic virus, and could be evaluated in clinical trials in the future.

REFERENCES

1. Sinkovics JG, Howe CD. Superinfection of tumors with viruses Experientia 1969; 25:733–734.

2. Martuza RL, Malick A, Markert JM, Ruffner KL, Coen DM. Experimental therapy of human glioma by means of a genetically engineered virus mutant Science 1991; 252:854–856.

3. Kirn D. Oncolytic virotherapy for cancer with the adenovirus dl1520 (Onyx-015): results of phase I and II trials Expert Opin Biol Ther 2001; 1:525–538.

4. Bischoff JR, Kirn DH, Williams A, Heise C, Horn S, Muna M, Ng L, Nye JA, Sampson-Johannes A, Fattaey A, McCormick F. An adenovirus mutant that replicates selectively in p53-deficient human tumor cells Science 1996; 274:373–376.

5. Kirn D. Clinical research results with dl1520 (Onyx-015), a replication-selective adenovirus for the treatment of cancer: what have we learned? Gene Ther 2001; 8:89–98.

6. Russell SJ. RNA viruses as virotherapy agents Cancer Gene Ther 2002; 9:961–966.

7. Antonio Chiocca E. Oncolytic viruses Nat Rev Cancer 2002; 2:938–950.

8. Banerjee AK, Barik S. Gene expression of vesicular stomatitis virus genome R. N. A. Virology 1992; 188:417–428.

9. Wagner RR, Rhabdoviridae RJ. The Viruses and Their Replication. In: Fields BN KD, Ed. Fields Virology.. Philadelphia:: Lippincott-Raven, 1996:1121–1132.

10. Letchworth GJ, Rodriguez LL, Del cbarrera J. Vesicular stomatitis Vet J 1999; 157:239–260.

11. Kopecky SA, Lyles DS. Contrasting effects of matrix protein on apoptosis in HeLa and BHK cells infected with vesicular stomatitis virus are due to inhibition of host gene expression J Virol 2003; 77:4658–4669.

12. Kopecky SA, Lyles DS. The cell-rounding activity of the vesicular stomatitis virus matrix protein is due to the induction of cell death J Virol 2003; 77:5524–5528.

13. Kopecky SA, Willingham MC, Lyles DS. Matrix protein and another viral component contribute to induction of apoptosis in cells infected with vesicular stomatitis virus J Virol 2001; 75:12169–12181.

14. Her LS, Lund E, Dahlberg JE. Inhibition of Ran guanosine triphosphatase-dependent nuclear transport by the matrix protein of vesicular stomatitis virus Science 1997; 276: 1845–1848.

15. Flint SJEL, Krug RM, Racaniello VR, Skalka AM. Principles of Virology.. Washington:: ASM Press, 2000.

16. Jayakar HR, Whitt MA. Identification of two additional translation products from the matrix (M) gene that contribute to vesicular stomatitis virus cytopathology J Virol 2002; 76: 8011–8018.

17. Kretzschmar E, Peluso R, Schnell MJ, Whitt MA, Rose JK. Normal replication of vesicular stomatitis virus without C proteins Virology 1996; 216:309–316.

18. Rodriguez LL. Emergence and re-emergence of vesicular stomatitis in the United States Virus Res 2002; 85:211–219.

19. Goodbourn S, Didcock L, Randall RE. Interferons: cell signaling, immune modulation, antiviral response and virus countermeasures J Gen Virol 2000; 81:2341–2364.

20. Sen GC. Viruses and interferons Annu Rev Microbiol 2001; 55:255–281.

21. Sharma StB, Grandvaux N, Zhou GP, Lin R, Hiscott J. Triggering the interferon antiviral response through an IKK-related pathway Science 2003; 300:1148–1151.

22. Staeheli P, Pavlovic J. Inhibition of vesicular stomatitis virus mRNA synthesis by human MxA protein J Virol 1991; 65: 4498–4501.

23. Pavlovic J, Haller O, Staeheli P. Human and mouse Mx proteins inhibit different steps of the influenza virus multiplication cycle J Virol 1992; 66:2564–2569.

23a. Schnorr JJ, Schneider-Schaulies S, Simon-Jodicke A, Pavlovic J, Horisberger MA, ter Meulen V. MxA-dependent inhibition of measles virus glycoprotein synthesis in a stably transfected human monocytic cell line J Virol 1993; 67:4760–4768.

24. Li XL, Blackford JA, Hassel BA. RNase L mediates the antiviral effect of interferon through a selective reduction in viral RNA during encephalomyocarditis virus infection J Virol 1998; 72:2752–2759.

25. Diaz-Guerra M, Rivas C, Esteban M. Inducible expression of the 2–5A synthetase/RNase L system results in inhibition of vaccinia virus replication Virology 1997; 227:220–228.

26. Nilsen TW, Maroney PA, Baglioni C. Synthesis of (2′–5′)oligoadenylate and activation of an endoribonuclease in interferon-treated HeLa cells infected with reovirus J Virol 1982; 42:1039–1045.

27. Gil J, Esteban M. Induction of apoptosis by the dsRNA-dependent protein kinase (PKR): mechanism of action Apoptosis 2000; 5:107–114.

28. Kumar A, Yang YL, Flati V, Der S, Kadereit S, Deb A, Haque J, Reis L, Weissmann C, Williams BR. Deficient cytokine signaling in mouse embryo fibroblasts with a targeted deletion in the PKR gene: role of IRF-1 and NF-kappaB EMBO J 1997; 16:406–416.

29. Wong AH, Tam NW, Yang YL, Cuddihy AR, Li S, Kirchhoff S, Hauser H, Decker T, Koromilas AE. Physical association between STAT1 and the interferon-inducible protein kinase PKR and implications for interferon and double-stranded RNA signaling pathways EMBO J 1997; 16:1291–1304.

30. Williams B. Signal integration via P. K. R. SciSTKE 2001; 89: RE2.

31. van den Broek MF, Muller U, Huang S, Zinkernagel RM, Aguet M. Immune defence in mice lacking type I and/or type II interferon receptors Immunol Rev 1995; 148:5–18.

32. Durbin JE, Hackenmiller R, Simon MC, Levy DE. Targeted disruption of the mouse Stat1 gene results in compromised innate immunity to viral disease Cell 1996; 84:443–450.

33. Kimura T, Kadokawa Y, Harada H, Matsumoto M, Sato M, Kashiwazaki Y, Tarutani M, Tan RS, Takasugi T, Matsuyama

T, Mak TW, Noguchi S, Taniguchi T. Essential and non-redundant roles of p48 (ISGF3 gamma) and IRF-1 in both type I and type II interferon responses, as revealed by gene targeting studies Genes Cells 1996; 1:115–124.

34. Stojdl DF, Abraham N, Knowles S, Marius R, Brasey A, Lichty BD, Brown EG, Sonenberg N, Bell JC. The murine double-stranded RNA-dependent protein kinase PKR is required for resistance to vesicular stomatitis virus J Virol 2000; 74: 9580–9585.

35. Zhou A, Paranjape JM, Der SD, Williams BR, Silverman RH. Interferon action in triply deficient mice reveals the existence of alternative antiviral pathways Virology 1999; 258:435–440.

36. Grander D, Einhorn S. Interferon and malignant disease--how does it work and why doesn't it always? Acta Oncol 1998; 37: 331–338.

37. Bello MJ, de Campos JM, Kusak ME, Vaquero J, Sarasa JL, Pestana A, Rey JA. Molecular analysis of genomic abnormalities in human gliomas Cancer Genet Cytogenet 1994; 73: 122–129.

38. Mogensen TH, Paludan SR. Virus-cell interactions: impact on cytokine production, immune evasion and tumor growth Eur Cytokine Netw 2001; 12:382–390.

39. Barber GN. Host defense, viruses and apoptosis Cell Death Differ 2001; 8:113–126.

40. Bachmann A, Hanke B, Zawatzky R, Soto U, van Riggelen J, zur Hausen H, Rosl F. Disturbance of tumor necrosis factor alpha-mediated beta interferon signaling in cervical carcinoma cells J Virol 2002; 76:280–291.

41. Tanaka N, Taniguchi T. The interferon regulatory factors and oncogenesis Semin Cancer Biol 2000; 10:73–81.

42. Doherty GM, Boucher L, Sorenson K, Lowney J. Interferon regulatory factor expression in human breast cancer Ann Surg 2001; 233:623–629.

43. Lowney JK, Boucher LD, Swanson PE, Doherty GM. Interferon regulatory factor-1 and -2 expression in human melanoma specimens Ann Surg Oncol 1999; 6:604–608.

44. Harada H, Takahashi E, Itoh S, Harada K, Hori TA, Taniguchi T. Structure and regulation of the human interferon regulatory factor 1 (IRF-1) and IRF-2 genes: implications for a gene network in the interferon system Mol Cell Biol 1994; 14: 1500–1509.

45. Yang YL, Reis LF, Pavlovic J, Aguzzi A, Schafer R, Kumar A, Williams BR, Aguet M, Weissmann C. Deficient signaling in mice devoid of double-stranded RNA-dependent protein kinase EMBO J 1995; 14:6095–6106.

46. Wong LH, Krauer KG, Hatzinisiriou I, Estcourt MJ, Hersey P, Tam ND, Edmondson S, Devenish RJ, Ralph SJ. Interferon-resistant human melanoma cells are deficient in ISGF3 components, STAT1, STAT2, and p48-ISGF3gamma J Biol Chem 1997; 272:28779–28785.

47. Matin SF, Rackley RR, Sadhukhan PC, Kim MS, Novick AC, Bandyopadhyay SK. Impaired alpha-interferon signaling in transitional cell carcinoma: lack of p48 expression in 5637 cells Cancer Res 2001; 61:2261–2266.

48. Brender C, Nielsen M, Kaltoft K, Mikkelsen G, Zhang Q, Wasik M, Billestrup N, Odum N. STAT3-mediated constitutive expression of SOCS-3 in cutaneous T-cell lymphoma Blood 2001; 97:1056–1062.

49. Barnard P, McMillan NA. The human papillomavirus E7 oncoprotein abrogates signaling mediated by interferon-alpha Virology 1999; 259:305–313.

50. Shou J, Soriano R, Hayward SW, Cunha GR, Williams PM, Gao WQ. Expression profiling of a human cell line model of prostatic cancer reveals a direct involvement of interferon signaling in prostate tumor progression Proc Natl Acad Sci U S A 2002; 99:2830–2835.

51. Bell JC, Garson KA, Lichty BD, Stojdl DF. Oncolytic viruses: programmable tumor hunters Curr Gene Ther 2002; 2: 243–254.

52. Abraham N, Stojdl DF, Duncan PI, Methot N, Ishii T, Dube M, Vanderhyden BC, Atkins HL, Gray DA, McBurney MW, Koromilas AE, Brown EG, Sonenberg N, Bell JC. Characterization of transgenic mice with targeted disruption of the cata-

lytic domain of the double-stranded RNA-dependent protein kinase, PKR J Biol Chem 1999; 274:5953–5962.

53. Mundschau LJ, Faller DV. Endogenous inhibitors of the dsRNA-dependent eIF-2 alpha protein kinase PKR in normal and ras-transformed cells Biochimie 1994; 76:792–800.

54. Stojdl DF, Lichty B, Knowles S, Marius R, Atkins H, Sonenberg N, Bell JC. Exploiting tumor-specific defects in the interferon pathway with a previously unknown oncolytic virus Nat Med 2000; 6:821–825.

55. Balachandran S, Barber GN. Vesicular stomatitis virus (VSV) therapy of tumors IUBMB Life 2000; 50:135–138.

56. Balachandran S, Porosnicu M, Barber GN. Oncolytic activity of vesicular stomatitis virus is effective against tumors exhibiting aberrant p53, Ras, or myc function and involves the induction of apoptosis J Virol 2001; 75:3474–3479.

57. Francoeur AM, Poliquin L, Stanners CP. The isolation of interferon-inducing mutants of vesicular stomatitis virus with altered viral P function for the inhibition of total protein synthesis Virology 1987; 160:236–245.

58. Stanners CP, Francoeur AM, Lam T. Analysis of VSV mutant with attenuated cytopathogenicity: mutation in viral function, P, for inhibition of protein synthesis Cell 1977; 11:273–281.

59. Black BL, Lyles DS. Vesicular stomatitis virus matrix protein inhibits host cell-directed transcription of target genes in vivo J Virol 1992; 66:4058–4064.

60. Black BL, Rhodes RB, McKenzie M, Lyles DS. The role of vesicular stomatitis virus matrix protein in inhibition of host-directed gene expression is genetically separable.

61. Ferran MC, Lucas-Lenard JM. The vesicular stomatitis virus matrix protein inhibits transcription from the human beta interferon promoter J Virol 1997; 71:371–377.

62. Ahmed M, Lyles DS. Effect of vesicular stomatitis virus matrix protein on transcription directed by host RNA polymerases I., II, and I. I. I. J Virol 1998; 72:8413–8419.

63. von Kobbe C, van Deursen JM, Rodrigues JP, Sitterlin D, Bachi A, Wu X, Wilm M, Carmo-Fonseca M, Izaurralde E.

Vesicular stomatitis virus matrix protein inhibits host cell gene expression by targeting the nucleoporin Nup98 Mol Cell 2000; 6:1243–1252.

64. Terstegen L, Gatsios P, Ludwig S, Pleschka S, Jahnen-Dechent W, Heinrich PC, Graeve L. The vesicular stomatitis virus matrix protein inhibits glycoprotein 130-dependent STAT activation J Immunol 2001; 167:5209–5216.

65. Strong JE, Coffey MC, Tang D, Sabinin P, Lee PW. The molecular basis of viral oncolysis: usurpation of the Ras signaling pathway by reovirus EMBO J 1998; 17:3351–3362.

66. Farassati F, Yang AD, Lee PW. Oncogenes in Ras signaling pathway dictate host-cell permissiveness to herpes simplex virus 1 Nat Cell Biol 2001; 3:745–750.

67. Bergmann M, Romirer I, Sachet M, Fleischhacker R, Garcia-Sastre A, Palese P, Wolff K, Pehamberger H, Jakesz R, Muster T. A genetically engineered influenza A virus with ras-dependent oncolytic properties Cancer Res 2001; 61:8188–8193.

68. Lawson ND, Stillman EA, Whitt MA, Rose JK. Recombinant vesicular stomatitis viruses from D. N. A. Proc Natl Acad Sci U S A 1995; 92:4477–4481.

69. Ball LA, Pringle CR, Flanagan B, Perepelitsa VP, Wertz GW. Phenotypic consequences of rearranging the P., M, and G genes of vesicular stomatitis virus J Virol 1999; 73:4705–4712.

70. Fernandez M, Porosnicu M, Markovic D, Barber GN. Genetically engineered vesicular stomatitis virus in gene therapy: application for treatment of malignant disease J Virol 2002; 76:895–904.

71. Vile R, Ando D, Kirn D. The oncolytic virotherapy treatment platform for cancer: unique biological and biosafety points to consider Cancer Gene Ther 2002; 9:1062–1067.

72. Ochsenbein AF. Principles of tumor immunosurveillance and implications for immunotherapy Cancer Gene Ther 2002; 9: 1043–1055.

73. Iwaki H, Barnavon Y, Bash JA, Wallack MK. Vaccinia virus-infected C-C36 colon tumor cell lysates stimulate cellular responses in vitro and protect syngeneic Balb/c mice from tumor cell challenge J Surg Oncol 1989; 40:90–96.

74. Kim EM, Sivanandham M, Stavropoulos CI, Bartolucci AA, Wallack MK. Overview analysis of adjuvant therapies for melanoma—a special reference to results from vaccinia melanoma oncolysate adjuvant therapy trials Surg Oncol 2001; 10: 53–59.

75. Toda M, Iizuka Y, Kawase T, Uyemura K, Kawakami Y. Immuno-viral therapy of brain tumors by combination of viral therapy with cancer vaccination using a replication-conditional H. S. V. Cancer Gene Ther 2002; 9:356–364.

76. Mackensen A, Lindemann A, Mertelsmann R. Immunostimulatory cytokines in somatic cells and gene therapy of cancer Cytokine Growth Factor Rev 1997; 8:119–128.

77. Hermiston TW, Kuhn I. Armed therapeutic viruses: strategies and challenges to arming oncolytic viruses with therapeutic genes Cancer Gene Ther 2002; 9:1022–1035.

78. Johnson JE, Schnell MJ, Buonocore L, Rose JK. Specific targeting to CD4+ cells of recombinant vesicular stomatitis viruses encoding human immunodeficiency virus envelope proteins J Virol 1997; 71:5060–5068.

79. Peng KW, Donovan KA, Schneider U, Cattaneo R, Lust JA, Russell SJ. Oncolytic measles viruses displaying a single-chain antibody against CD38, a myeloma cell marker Blood 2003; 101:2557–2562.

80. Hernandez-Alcoceba R, Pihalja M, Qian D, Clarke MF. New oncolytic adenoviruses with hypoxia- and estrogen receptor-regulated replication Hum Gene Ther 2002; 13:1737–1750.

81. Graff JR, Zimmer SG. Translational control and metastatic progression: enhanced activity of the mRNA cap-binding protein eIF-4E selectively enhances translation of metastasis-related mRNAs Clin Exp Metastasis 2003; 20:265–273.

82. Clemens MJ, Bommer UA. Translational control: the cancer connection Int J Biochem Cell Biol 1999; 31:1–23.

83. Pugh CW, Ratcliffe PJ. The von Hippel-Lindau tumor suppressor, hypoxia-inducible factor-1 (HIF-1) degradation, and cancer pathogenesis Semin Cancer Biol 2003; 13:83–89.

8

Parvoviruses As Anti-Cancer Agents

JEAN ROMMELAERE, NATHALIA GIESE, CELINA CZIEPLUCH, and JAN J. CORNELIS

Applied Tumor Virology Program
(Abteilung F0100 and Institut National de la Santé et de la Recherche Médicale U375)
Deutsches Krebsforschungszentrum,
Heidelberg, Germany

OUTLINE

As a result of their oncotropism, oncolytic activity, and low capacity for danger signaling, some autonomous parvoviruses open new prospects to cancer therapy. Pilot Phase I clinical trials in humans using natural parvoviruses and pre-clinical studies in animals using both wild-type parvoviruses and vector derivatives transducing various cytokines, lend credit to the possible application of these agents to the fight against cancer. The oncosuppressive effects of natural and recombinant parvoviruses are reviewed, and possible mechanisms are discussed in the light of recent findings in the field of immunotherapy, which bear relevance to the development of more potent parvovirus-based anticancer agents.

VIRAL LIFE CYCLE

All members of the family Parvoviridae are nonenveloped nuclear-replicating single-stranded DNA viruses of only 20 nm to 25 nm in size. Members able to infect vertebrates can be placed into one of three genera. The genus *Dependovirus* comprises the so-called adeno-associated parvoviruses (AAV), which need assistance from some tumor viruses for their efficient replication, whereas the genera *Parvovirus* and *Erythrovirus* contain the autonomously replicating viruses, which do not require any helper viruses for productive infection. The only autonomous parvovirus known to be a human pathogen so far is the Erythrovirus B19 [1], yet further discoveries of new types cannot be excluded, as demonstrated by the recent description of a novel human skin parvovirus [2]. Despite growing evidence of a link between B19 and serious health problems (in particular the development of rheumatoid arthritis), attempts are being made to utilize the tropism of B19 toward erythropoietic progenitor cells for gene therapy [3]. This review focuses mainly on rodent viruses of the genus Parvovirus (shortly parvoviruses) such as MVM (minute virus of mice) and H-1 virus, which are for three reasons of main interest for applications in humans: (1) these agents can preferentially infect and kill a number of transformed human cells, (2) infection with H-1 virus seems to be apathogenic in adult humans, and (3) these viruses display oncosuppressive properties in vivo, for which implanted human tumors can be targets [4–7]. Relevant work performed with a number of closely related parvoviruses, such as the mouse virus MPV-1 (mouse parvovirus-1) and the rat viruses KRV (Kilham rat viruses), RPV-1 (rat parvovirus-1) and LuIII, is discussed as well.

Parvovirus infection is thought to proceed as follows: The virus adsorbs on to the host cell through binding to a membrane-associated receptor that is unknown for rodent parvoviruses. In case of the canine parvovirus (CPV), which binds the transferrin receptor, dynein- and microtubule-dependent passage through lysosomes has been demonstrated [8,9]. Minute virus of mice was recently shown to target late endosomes and the proteasome for cytoplasmatic trafficking [10]. Once decapsidated, single-stranded DNA genomes can be converted in the nucleus into a double-stranded form. This reaction has been demonstrated to depend on the S phase-specific factor cyclin A [11], banning parvovirus replication in resting cells. Conversion creates the template for expression of genes encoding the regulatory NS1 and NS2 proteins controlled by the P4 promoter (Figure 1A). Like the conversion reaction, the activity of this strong promoter is induced in the S phase of the cell cycle, mainly owing to E2F and ATF recognition elements [12,13]. The P38 promoter, which drives the expression of the genes encoding the structural capsid proteins, requires NS1 for transactivation [14]. Conversion is also required for subsequent genome amplification through a rolling circle-like mechanism and for the production of single-stranded progeny genomes, which proceeds concomitant to their packaging into preformed capsids [15,16]. In contrast to the related AAVs, autonomous parvoviruses showed no evidence so far of DNA integration in the host genome [17]. Whereas genome amplification and single-strand DNA synthesis require NS1 functions, the initial conversion reaction only relies on cellular factors.

NS1 seems to be the main cytopathic effector of autonomous parvoviruses, although in some cell lines, NS2 was shown to modulate cell killing [18–20]. The mechanism by which NS1 induces cell killing is not understood. However, NS1 is known to induce pleiotropic effects, which could well contribute to cell death because they may cause cell-cycle arrest, chromatin-nicking, and alteration of both the posttranslational modifications and localization of cellular proteins [21–26].

A number of in vitro transformed and tumor-derived cells appear to provide an internal milieu favoring parvovirus replication (discussed below). This oncotropic behavior is also reflected by the fact that several of these viruses have been isolated from tumor material [27]. First thought to be tumor

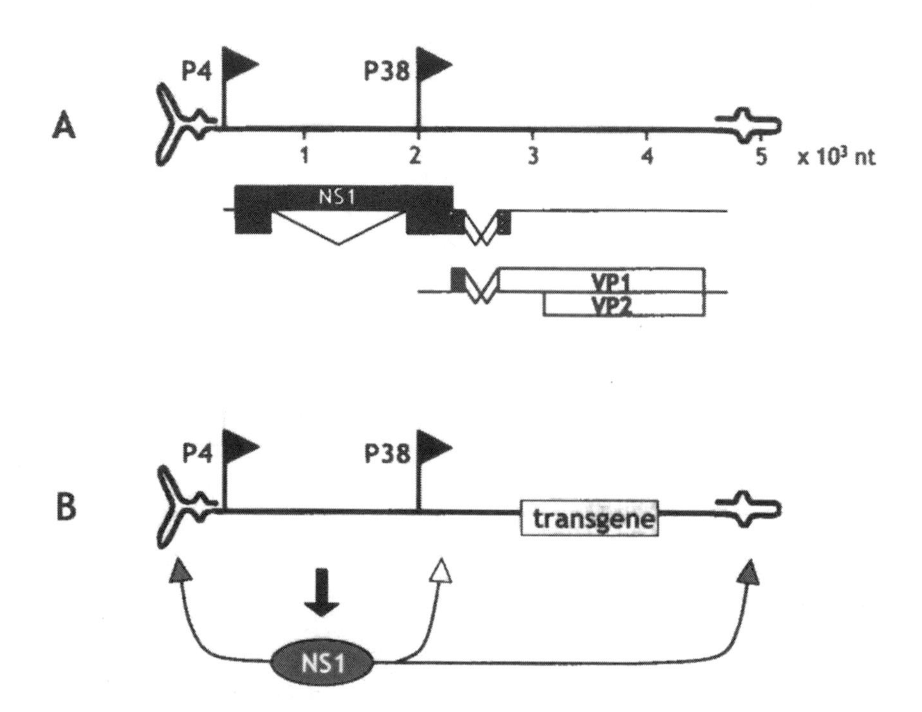

Figure 1 Schematic representation of the genome organization of autonomous rodent parvoviruses (A) and derived recombinant viruses as described in this chapter (B). A. The viral genome (thick line) flanked by palindromic sequences at both ends is depicted on top. Messenger RNAs (thin lines) are aligned underneath, with the positions of the transcription start sites for the P4 and P38 promoters indicated by flags. The NS- and VP-coding regions in the three reading frames are indicated by boxes marked in black, gray, and white. The genome terminal hairpins and RNA introns are not to scale. B. Inserted foreign cDNA is indicated by the open box labeled "transgene." The residual downstream VP sequence is not translated and varies in length depending on the transgene. The replicative and transactivating functions of the NS1 polypeptide are indicated by closed and open arrow heads, respectively.

viruses, on the contrary, parvoviruses were soon proved able to preferentially kill human tumor cells [28]. It is therefore not surprising that trials were initiated in humans with the hope of controlling neoplastic diseases by means of these lytic viruses.

OUTCOME OF PARVOVIRUS-BASED THERAPIES IN HUMANS

The first case of clinical application of parvoviruses dates back to 1965, when two girls with terminal osteosarcoma were treated with H-1 virus [29]. The choice of the tumor was based on the unique osteolytic effects of H-1 virus on developing bones of young infected hamsters [30]. Prior to application of the virus, both patients had undergone intensive yet unsuccessful courses of chemo- and radiation therapy. Extensive inoperable metastases and very poor prognosis left little hope for cure, and H-1 virus was administered in an attempt to control the neoplasms. Given a single time (intramuscularly [i.m.] and intratumorally [i.t.] for the first patient, i.m. for the second one), the H-1 virus inoculum was well tolerated, yet it failed to significantly affect tumor growth and to prevent the lethal outcome of the disease after about 1 month. From our recent pre-clinical studies using MVMp in mouse tumor models, it appears that this poor therapeutic efficiency may be ascribed, at least in part, to the inoculation of too-low doses of viruses in regard to the massive volumes of the tumors to be destroyed [31,32]. Another factor limiting parvovirus efficacy might be the rapid and strong humoral antiviral response of the host, as concluded from the analysis of blood samples taken at intervals [29]. Interestingly, the aforementioned patients underwent a viremia that showed a cyclic pattern: H-1 virus particles could be detected in the blood only from day 1 to 5 and 8 to 10 postinfection. The first wave of viremia most likely reflected the initial dissemination of virions from the inoculation site through the bloodstream [1] to the liver, where particulate materials (in particular, viruses, including MVM and AAV) are cleared [33] and [2] to the host cells, the majority of which are competent for virus uptake [31]. The second wave of viremia was probably attributable to the release of viruses from some of the latter cells as a result of the cytotoxicity of either the virus itself or specific immune effector cells [natural killer (NK) cells, cytotoxic T lymphocytes (CTL) or antibody-dependent cytotoxic cells]. Whatever the relative roles liver clearance, depletion of permissive cell reservoirs, intracellular degradation and innate or acquired immunity may have played in virus elimination, no virus could be detected in blood, tumor, liver, or kidney upon autopsy at 40 days postinfection. Furthermore, the strong neutralizing activity of circulating antibodies spoke against the usefulness of repeating the administration of the same virus over a prolonged period of time. A positive outcome of this trial lay in the absence of detectable untoward effects in virus-injected patients. Together with virus elimination, this tolerance argued for the safety of human treatment with H-1 virus, although no long-term follow-up could be done.

Conclusions from this initial case study were confirmed and extended in a Phase I clinical trial conducted in the early 1990s, again using parvovirus H-1 [34]. Limitations of this study lay in the strict enrollment criteria, preventing the selection of specific malignancies, and in the application of suboptimal virus doses. All 12 recruited patients had a history of previous unsuccessful treatments and were in a terminal stage of disease, carrying skin metastases from different kinds of solid tumors. The virus was inoculated into preselected neoplastic nodules through single or repeated injection. As in the previous clinical study, H-1 virus did not induce detectable toxicity, and the maximum tolerated dose was not attained in patients who received as much as 10^{10} PFU of virus (approximately 10^{13} physical particles). This dose, however, was not sufficient to reduce the size of the metastatic foci. Nevertheless, a response to the treatment, defined as a stabilization of injected nodules according to the World Health Organization criteria, was observed in three of seven patients with mammary and in one patient with bronchial carcinoma. None of the patients with kidney leiomyosarcoma (0/1), melanoma (0/2) or pancreatic carcinoma (0/1) showed clinical improvement. All patients developed neutralizing antibodies, indicating viral escape to the bloodstream and viremia similar to the situation observed during treatment of the osteosarcoma patients [29]. It is noteworthy that markers of virus replication and infectious virus could be detected in injected as well as noninjected lesions from some treated patients, but no correlation could so far be established between the stabilization of tumor size and the yield of recovered virus. Together with the fact that mammary tumor-derived epithelial cells showed a varying sensitivity to H-1 virus infection in in vitro studies [5,35], these data point to low-grade mammary carcinoma as a candidate target malignancy for future clinical trials.

In summary, these first clinical studies showed that parvovirus application can be considered rather safe. The lack of curative effect can probably be attributed, at least in part, to the fact that the tumor load was already very high at the time of virus injection. Hence, the development of recombinant parvoviruses capable of triggering antitumor bystander responses appears to be a very promising road to follow.

MOTIVATION FOR THE FURTHER DEVELOPMENT OF PARVOVIRUS-BASED THERAPEUTICS AGAINST CANCER

The development of parvoviruses as therapeutics for cancer treatment is founded on experimental observations discussed in the following paragraphs and demonstration that parvovirus infections interfere with tumor establishment and growth in various experimental settings. As outlined below, a further argument in favor of parvoviruses is their relative safety, which can be traced back to a low incidence of preimmunity and lack of danger signals. Major parameters influencing the outcome of parvovirus antitumor therapy are summarized in Figure 2.

Parvovirus Preimmunity is Low

To date, a major obstacle to the success of gene therapy protocols using viral vectors lies in the occurrence of antiviral immune responses. Despite limitations imposed by the emer-

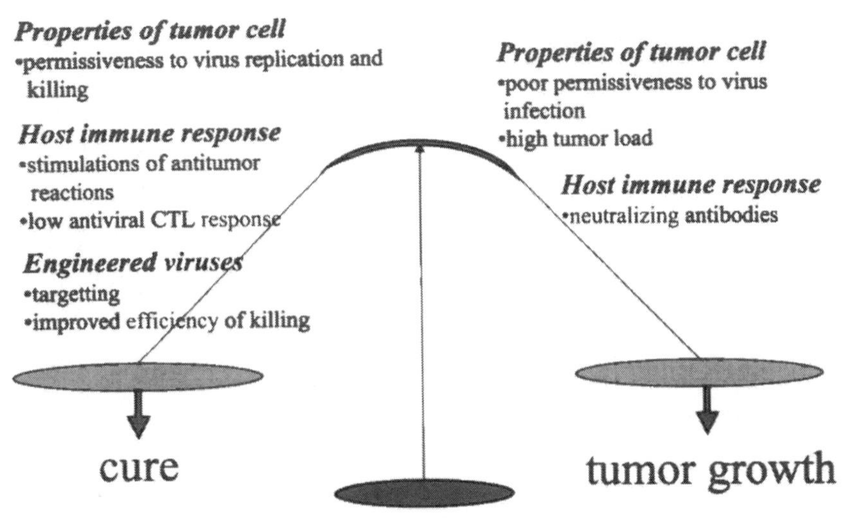

Figure 2 Parameters influencing the outcome of parvoviral antitumor therapy.

gence of neutralizing antibodies, a pitfall common to all virus-based therapies, MVMp and H-1 parvoviruses have, in our opinion, an immunological "profile" that makes them more suitable than other viruses for anti-cancer applications. In contrast to herpes-, adeno- and adeno-associated viruses, MVMp and H-1 are not common infectious agents in humans (27, and our own observations), resulting in the absence of pre-existing immunity against the latter viruses in the great majority of the human population. This is of clear benefit, regarding both efficiency and safety. In some studies, pre-existing antibodies were described as a positive factor in that they were able to limit the dissemination of adenovirus to peripheral organs and to restrict viral expression to the tumor site [36]. Yet the lethal outcome of a recent case of adenovirus-based therapy is currently questioned as a fatal consequence of extreme activation of the complement system by a high virus dose in the context of pre-existing adenoviral immunity

[37,38]. In contrast, the absence of pre-existing immunity against rodent parvoviruses should allow the initial inoculum of MVMp or H-1 viruses to be delivered efficiently to tumors without the risk of immediate side effects.

Parvoviruses Fail to Signal Danger

The immune system evolved to recognize and eliminate pathogenic invaders, including viruses, through both innate and adaptive immune reactions. Capsids and DNA of various viruses proved able to activate distinct defense cells through the recognition of Toll-like receptors (TLR) by pathogen-associated molecular patterns (PAMPs), leading to the production of endogenous danger signals that trigger local inflammation [39–41]. Furthermore, TLR-triggered signaling emerges as a link between innate and adaptive immunity in that it is essential for the maturation of dendritic cells (DC), their migration to draining lymph nodes, and the optimal antigen presentation, which enables the clonal expansion of effector lymphocytes. Results of in vivo studies involving the injection of H-1 or MVM viruses in young or adult humans (H-1), mice (MVMp), or rats (H-1) failed to show any deleterious side effect. In particular, no virus-related inflammatory histological changes could be detected in organs of autopsied animals, indicating that these viruses either failed to display relevant PAMPs or developed ways to control host antiviral immune reactions [32]. Most interestingly, the immune responses elicited by MVMp in in vivo-treated mice could be distinguished from those usually seen upon virus infection in that, on the innate side, NK cells were activated but type I interferon production was not, whereas on the adaptive side, neutralizing antibodies were produced but CTLs were not significantly induced [42]. Because the early presence of blood-derived type I IFNs and the later-generation CTLs both require DCs [43–46], these observations argue against the involvement of DCs in the process of parvovirus recognition and presentation. Knowing that antibody responses may be induced without DCs as a result of interactions between antigen-specific B cells and armed $CD4^{+}$T cells, we assume that activation of NK cells by the virus creates a milieu enabling presentation of processed virus components by major histocompatibility (MHC) class II on macrophages to naïve T cells. Interestingly enough and in contrast to many other viruses, one of the major known activators of DCs—TLR9-binding immunogenic CpG motifs—are statistically underrepresented in autonomous parvoviral DNA and, when present, are located within inhibitory sequences [47]. Although no information is available at present about the interaction of MVMp or H-1 virus with DCs, our results indicate that these viruses are poor direct inducers of danger signals, which is a potential advantage with regard to the safety of parvovirus-based therapies and the expression of transgenes harbored by parvoviral vectors. It should be stated, however, that parvoviruses are cytocidal agents and may still contribute to the formation of another important class of danger signals—namely, intracellular components released as a result of cell death [48]. Yet the oncolytic character of MVMp and H-1 viruses (discussed below) raises hope of these signals being preferentially generated at or in the vicinity of tumors, which would direct DC-mediated reactions against tumor cells and account for the observed lack of side effects in animal and human studies. It should not be overlooked, however, that other parvoviruses were implicated as causative agents in inflammatory and autoimmune disorders of animals [49–51] and possibly humans [52–55]. Furthermore, MVMp and H-1 viruses can be pathogenic under specific conditions in some animal species [56]. Therefore, the apathogenic behavior of MVMp in mice and H-1 in humans cannot be taken as a precedent for the general safety of autonomous parvoviruses.

In conclusion, the absence of pre-existing human immunity against H-1 and MVM viruses, the apparent inefficiency of these agents in eliciting inflammatory and cell-mediated immune reactions, together with their oncotropism and lack of evidence for integration into the host genome, place these and related parvoviruses among the safest oncolytic agents presently under consideration for anticancer applications.

Parvoviruses Interfere with Tumor Development

Antitumor effects of natural parvoviruses were observed under three different experimental conditions: (1) Animals infected with parvoviruses at birth were protected from spontaneous or induced cancer development in later life. (2) Parvovirus-bearing animals were refractory to tumor implantation. (3) Parvovirus treatment impeded the growth of pre-established tumors. These three categories will be discussed separately below.

Prevention Of Tumor Occurrence Due To Parvovirus-Infection At Birth

Soon after the discovery of parvoviruses, it was realized that life carriers were protected from the occurrence of cancer to a very significant extent, as determined through the survey of large cohorts of laboratory animals. In a pioneering work, neonatal hamsters were found to be sensitive to infection with H-1 virus, with a high incidence of dwarfed and deformed animals with mongoloid traits [30]. Later in life, these deformed animals proved to have a more than 20-fold lower incidence of spontaneous tumors than noninfected animals [57]. This observation was extended by showing that the mongoloid hamsters were also more resistant to tumor induction by adenovirus 12 or dimethylbenz anthracene [4,58]. Animals infected with H-1 virus that failed to become deformed were also protected from oncogenesis, albeit to a lesser extent. Neoplastic cell cultures prepared from the tumors of noninfected control hamsters and from the residual tumors of virus-infected mongoloid animals did not exhibit a differential sensitivity to H-1 virus, allowing no conclusion to be drawn regarding the role of viral oncolysis in the prevention of tumor appearance [4]. All animals developed virus-specific antibodies, with the highest titers found in the mongoloid hamsters. Similarly, Kilham and Margolis [59] reported that newborn hamsters infected with both the rat parvovirus KRV and Moloney leukemia virus (MoLV) did not develop tumors, whereas animals inoculated with MoLV alone did. This observation was confirmed by V. V. Bergs, who showed that the incidence of leukemia in two different strains of rats injected intraperitoneally at birth with a mixture of MoLV and KRV was approximately one-third that observed in control animals inoculated with the

MoLV tumor virus alone [60]. As in the aforementioned experiments with hamsters, the tumor-free rats had higher KRV-specific antibody titers than the leukemic animals.

It was shown in other studies that the anti-parvoviral immune response elicited in neonatally infected animals was not able to free them from the virus, which could be subsequently rescued by experimental manipulations, resulting more particularly in immunosuppression or tumor induction [61]. The protection of healthy carriers from oncogenesis may thus result from the reactivation of inapparent parvovirus infections at the site of nascent tumors, followed by oncolysis. However, this possibility remains to be proven, given that the persistence of parvovirus infection in seropositive animals protecting against tumor appearance was not demonstrated in any of the experiments outlined above. It is therefore an open question whether tumor prevention in animals inoculated at birth with parvoviruses required the reactivation of infectious viruses to kill emerging tumor cells, or resulted from the induction of a lifelong antitumor immune environment.

Refractoriness of Parvovirus Carriers to Tumor Implantation

The pre-infection of laboratory animals with parvoviruses was found to prevent various tumor implants from taking in these recipients. Beagle puppies that were unintentionally infected with the canine parvovirus (CPV) showed a rapid regression of transplanted venereal fibrosarcomas, in contrast to uninfected animals [62]. The related feline panleukopenia virus (FPV) was also able to prevent the transplanted fibrosarcomas from developing when the virus was inoculated concomitantly with the tumor cells but at a distant site [62]. Interestingly, FPV was inefficient as an anti-cancer agent if it was applied 3 weeks after sarcoma cell implantation, suggesting that the protective action of the virus got overwhelmed by too-large tumor burdens. Puppies that were cleared from tumors were fully protected because they did not produce neoplasias when challenged with tumor cells from the same line [62]. Similar observations were made for a mouse tumor model in which close to 90% of the animals that had been infected with the mouse parvovirus (MPV)-1 became refractory to the development of solid or ascitic tumors after implantation of SP2/0B myeloma cells [63].

Parvovirus Treatment Of Established Tumors

A number of attempts were made at arresting the growth or inducing the regression of established experimental tumors through their direct infection with parvoviruses. These approaches involved the use of transplantable tumors that were infected either prior to (ex vivo treatment) or after (in vivo treatment) their implantation in recipient animals. In 1986, E. Guetta showed that the development of tumors from Ehrlich ascites cells intraperitoneally implanted in mice could be drastically suppressed through the simultaneous subcutaneous injection of MVM [6]. This effect was obtained with the strain MVMp, for which Ehrlich ascites cells are permissive, but not with the related strain MVMi, to which these cells are resistant, arguing for the involvement of virus-induced oncolysis in the inhibition of tumor formation. The extent of tumor suppression dropped when the inoculation of virus was delayed relative to that of neoplastic cells. This again points to the critical impact of tumor load on the overall outcome after treatment with wild-type virus. Animals that rejected tumor implants as a result of the inoculation of tumor cells and the simultaneous subcutaneous injection with MVMp were fully protected against a challenge with naïve ascites cells [6]. Similar anti-neoplastic effects of parvoviruses were subsequently described in other tumor models. In mice, allogeneic P815 tumors growing as ascites were rejected more rapidly as a result of the infection of recipient animals with MVMp or MVMi, whereas the growth of S-194 solid tumors was not affected by these viruses [64]. This difference correlated with the varying susceptibility of these cells to virus infection, given that P815 cells were permissive for both MVM strains, whereas S-194 cells were not. As in the previous system, this correlation suggested that parvovirus-induced tumor suppression involved, at some point, a lytic interaction between virus and neoplastic target cells. On the other hand, the long-term protection of treated animals in the absence of firm evidence for persistent infection raised the possibility that this phenomenon also comprised an immune component. In keeping with this possibility, MPV-1 proved to be able to suppress grafts of allogeneic tumor cells that were not susceptible to virus infection in vitro [65]. It is worth noting that MPV-1 also accelerated the rejection of allogenic and, most surprisingly, syngeneic skin grafts [65,66], arguing for immunomodulating effects of parvovirus infections. Successful treatment of established tumors with parvoviruses was also reported for F344 rats in which tumor development from transplanted large granular lymphocytic (LGL) leukemia cells could be impeded by the later infection of recipient animals with the rat parvovirus 1 (RPV-1) [56]. Furthermore the intratumoral injection of H-1 virus into large allogeneic gliomas in rats led to a remarkable regression of these lesions (K. Geletneky, personal communication).

In several instances, parvovirus infections that proved efficient in suppressing tumor formation when carried out ex vivo prior to or concomitantly with the implantation of neoplastic cells failed to achieve a significant or durable protection when performed in vivo through intratumoral injection. This can be illustrated by the poor ability of MVMp to suppress MHC class I-positive immunogenic H5V hemangiosarcomas [32] or P815 mastocytomas (our unpublished results) through the in vivo treatment of tumor-bearing immunocompetent mice. MVMP had still an oncosuppressive potential in these systems because its inoculation to tumor cells prior to implantation prevented P815 cells from forming tumors and H5V cell implants from relapsing and generating metastases. Similarly, no significant antitumor effect was observed after MVM infection of mice bearing M-MSV or M-MLV–induced tumors in their footpads [64]. In another mouse model, 10 daily injections of MVMp into established tumors arising from subcutaneously implanted MHC class I-negative B78H1 melanoma cells caused tumor rejection only when the lesions were smaller than 2 mm in diameter (our unpublished results). With larger lesions (2 to 5 mm), only a transient tumor-growth arrest could be achieved, and no effect was observed when the initial tumor diameter exceeded 5 mm. These various cases exemplify the general difficulty in hitting high enough numbers of cells upon in vivo administration of virus-based therapeutics, and justify further efforts to improve the efficacy of such treatments (discussed below).

Tumors arising from human neoplastic or transformed cells implanted in recipient animals also proved to be targets for parvovirus-mediated oncosuppression in vivo. This was demonstrated using different models of human tumor xenogafts in nude and severe combined immune-deficient (SCID) mice. The direct injection of H 1 virus at the site of subcutaneous implantation of transformed human mammary epithelial cells (HBL100) in nude mice reduced the formation of palpable tumors by 80% [5]. Furthermore, a single i.v. injection of wild-type virus in animals bearing pre-established HBL100-cell-derived tumors slowed down the growth or even induced the regression of these tumors. More recently, S. Faisst [67] reported the dose-dependent regression of human cervical carcinoma cell (HeLa)-derived tumors in SCID mice as a result of a single injection of H-1 virus in the vicinity of the neoplasms. This regression was accompanied by evidence of parvovirus replication around tumor necrotic areas.

In conclusion, parvoviruses are endowed with the remarkable capacity for suppressing the development or causing the regression of certain tumors, although they cannot be considered magic bullets that are active under all circumstances. In a number of cases, the protective effect of these agents was only significant in vivo when the virus was already present at the onset of tumor development, and could be overwhelmed by increasing the tumor cell load. Therefore a race seems to exist between tumor growth and the generation of antitumor agents, whether the latter are the oncolytic viruses themselves or other defense effectors induced by infection. The occurrence of parvoviral anti-neoplastic effects in immunodeficient mice and the correlation observed in some systems between the extents of tumor suppression in vivo and tumor cell sensitivity to parvovirus infection in vitro argue for viral oncolysis as one contribution to oncosuppression. Thus, the varying sensitivity of tumor target cells to the cytotoxic activity of parvoviruses represents a first level of heterogeneity in the responsiveness of different tumors to parvovirus-based therapy. On the other hand, varying degrees of parvovirus-induced immune responses depending on the immunogenicity of tumors and the immune status of host organisms could account for another level of heterogeneity among tumors in their responsiveness to parvovirus treatments. Still another reason for the varying susceptibility of tumors to the suppressive effect of parvoviruses probably lies in the location of the lesions, which affects their accessibility to both viruses and defense cells and may account for the relatively high rate of success met with intraperitoneally growing tumors. One of the challenges for the future therefore lies in the identification of optimal target tumors for parvovirus-based therapies.

MECHANISMS UNDERLYING THE ANTITUMOR ACTIVITY OF NATURAL PARVOVIRUSES

Although parvoviruses may kill tumor cells in a direct way under in vivo conditions (viral oncolysis), the success of therapy will undoubtedly depend on at least two additional bystander effects: the immunological consequences of this tumor cell death, and the virus-induced perturbation of the immune system. These various aspects will be discussed separately in the following paragraphs.

Direct Evidence of Parvovirus Oncolytic Effects in Cell Cultures

Preferential Killing of (Pre) Neoplastic Cells in Culture

Cytocidal processes associated with tumor regression in infected animals are likely to be mediated, at least in part, by immune effector cells and do not constitute proof that the virus has a direct oncolytic activity. Therefore, in vitro cell cultures were used to determine whether parvoviruses could kill (pre) neoplastic cells in preference to the parental normal cells. Indeed, it was repeatedly observed that various normal human and rodent cells were more resistant to wild-type H-1 or MVMp infection than in vitro transformed derivatives or tumor-derived cells from the same tissue origin [68–75]. This holds true for cells of both mesenchymal and epithelial origins. Cultures of human tumor cells are more particularly sensitive to H-1 virus infection, albeit to various extents [5,69,75,76]. A striking example of the oncolytic effect of H-1 virus in vitro is illustrated in Figure 3. Several human leukemic cell lines could be killed

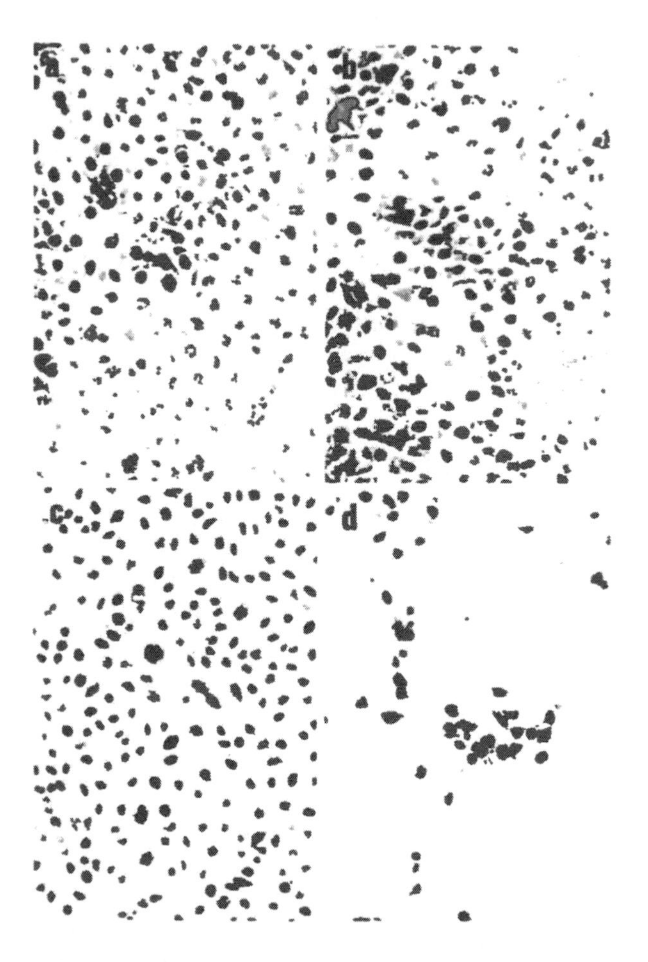

Figure 3 Hypersensitivity of a human squamous cell carcinoma cell line (c, d) to the cytopathic effects of wild-type parvovirus H-1, in comparison with normal human keatinocytes (a, c). a, c: mock-treated cultures; b, d: H1-virus infected cultures.

very efficiently by this virus [77–80], in keeping with the lymphotropism of the closely related KRV virus in its natural host, the rat [56]. Most of these studies were conducted with cell lines that had been established in vitro and may therefore have drifted from their original physiological condition. It should be stated, however, that short-term cultures of human mammary carcinoma [35], hepatoma [81], and glioma (K. Geletneky, personal communication) cells also proved to be preferential targets for the lytic activity of H-1 virus. In the systems analyzed so far, the greater susceptibility of neoplastic or transformed cells to parvovirus infection could not be ascribed to a more efficient virus uptake and correlated with an enhancement of viral DNA replication or gene expression [73,74,82]. A parallel could be drawn between the greater parvovirus-induced killing of transformed versus nontransformed human fibroblasts and the presence of a larger fraction of cells positive for viral gene expression in the former cultures [83,84].

In some cells, parvoviruses induce a typical apoptotic type of death. For example, human promonocytic tumor cells (U937) infected with H-1 virus were killed very efficiently through apoptosis, as monitored by the appearance of apoptotic bodies and the cleavage of poly(ADP-ribose)polymerase (PARP) [80]. Interestingly, variant cell clones surviving H-1 virus infection were also refractory to tumor necrosis factor (TNF)-α–mediated killing, suggesting that the parvovirus and TNF-α use at least partially overlapping signaling pathways to induce cell death. Similarly, rat glioma (C6) cells from infected cultures and cerebellum cells from infected rats were reported to be killed by H-1 virus through an apoptotic process [85]. Apoptotic death of primitive hemopoietic progenitor cells was also observed in the bone marrow of MVMi-infected mice [86]. On the other hand, HeLa cells that are also very susceptible to H-1 virus-induced death showed characteristics of both apoptosis and necrosis, depending on the conditions used [87]. In MVMp-infected rat fibroblasts, cell killing could not be attributed to apoptosis [88]. The mode of cell death is of importance when interactions with the immune system in infected organisms are considered (discussed below).

Viral Proteins Involved in Cytotoxicity

When the parvovirus gene coding for the nonstructural (NS) proteins was placed under control of an inducible promoter and stably introduced by transfection in SV40-transformed human fibroblasts, its expression correlated with a reduced number of cells able to form visible colonies [18]. Further site-directed mutagenesis studies led to the identification of the NS1 protein as a major determinant of parvovirus-induced cell disturbances [89–91]. The small parvoviral nonstructural proteins (NS2) may modulate the NS1 cytotoxic activity in a cell-dependent fashion [19]. Interestingly, when plasmids containing an inducible NS1 gene were expressed in normal rat fibroblasts or in transformed derivatives thereof, only the survival of the latter cells was impaired, suggesting that transformation sensitizes host cells to NS-induced killing [92]. The cytopathic function of NS1 is likely to be controlled by phosphorylation because distinct point mutations abrogating the proper phosphorylation of the viral polypeptid lead to the up or down modulation of its toxic activity in permissive cells [93]. It remains to be determined whether and how other viral proteins cooperate with NS1 to account for the overall cytotoxicity of parvovirus infections.

The mechanisms by which parvoviruses induce the death of permissive host cells are not yet fully understood even under in vitro conditions. This is a complex phenomenon involving both subversion of cellular defense processes and cell intoxication (L. Daeffler, personal communication). As stated above, the viral NS proteins are endowed with both cytostatic and cytotoxic properties, leading cells to undergo growth arrest, morphological alterations, and loss of adherence [93]. These phenotypic changes may be traced back to one or another of the multiple molecular disturbances induced by NS proteins in target cells [26], including the formation of specific NS–cellular polypeptide complexes (25, and references therein). In various systems, these alterations can culminate in cell lysis, in which the capsid proteins may be involved (our unpublished observations).

Indirect Evidence of Parvovirus Oncolytic Effects in Animal Models

The parvoviral oncolytic effects observed in cell cultures raise hopes of a similar activity being exerted against tumors under in vivo conditions. Yet evidence of parvoviral oncolysis in whole organisms is mostly indirect.

The expression of H-1 virus-cytotoxic NS proteins was detected in the necrotic zones of HeLa cell xenografts infected in vivo [67]. Furthermore, distinct necrotic foci were visible along MVM injection routes on histological sections of in vivo treated mouse melanoma implants (N. Giese, unpublished observations). Because some of these foci showed no infiltration with host neutrophils, monocytes, or lymphocytes, at least part of tumor necrosis can be assumed to be caused by the direct cytotoxic activity of the virus. This cannot be formally proven because defense cells were present in other necrotic areas of the tumor. The simultaneous occurrence of virus replication and leukocyte infiltration in regressing tumors is consistent with virus-induced oncolysis because this process is expected to be accompanied by a stimulation of local immune reactions, as discussed above. This, however, prevents a separate evaluation of the relative contributions of viral oncolysis and immune cytotoxicity through histological examination of regressing tumors. The same holds also for immunodeficient models given that innate defense mechanisms (and, in some models, antibody production) have to be taken into account. Indeed, A. Haag showed that NK cells were activated and infiltrated H-1 virus-infected HeLa xenografts in nude mice, raising the possibility that these cells may mediate, at least in part, the virus-induced suppression of tumor formation [94].

The occurrence of parvoviral oncolysis in vivo is also supported by the correlation generally observed between the sensitivity of target cells to virus killing under in vitro culture conditions, and the responsiveness of corresponding tumors to virus therapy, as determined after ex vivo or in vivo infection. Thus, the MVMp-sensitive Ehrlich ascites and P815 cells constitute tumor models that were efficiently inhibited by the parvovirus, whereas tumors derived from cells that were more resistant to MVMp either required higher virus doses to be suppressed

(H5V) or were refractory to the virus treatment (B78H1) (31,32, and unpublished results). Similarly, Ehrlich ascites tumors could be suppressed by MVMp but not by the related strain MVMi, paralleling the sensitivity of these cells to the former virus and their resistance to the latter [6]. It should be stated, however, that the sensitivity of a cell line to parvovirus infection correlates with its greater capacity for viral gene expression (see above). Therefore, it cannot be ruled out that under in vivo conditions, the virus-induced cellular disturbances do not progress up to the stage of cell death but result in phenotypic changes that stimulate innate cytotoxic reactions against (infected) neoplastic cells, leading to the cross presentation of tumor antigens by DC and the generation of antitumor CTL. Whether initiated by viral oncolysis or virus-activated innate killer cells, this scenario would account for the long-term protection of virus-treated animals against tumor-cell challenges in the absence of detectable viral imprints [6].

Tumor Cell Targets for Parvovirus Infection

From a therapeutic point of view, it would be useful to know whether specific oncogenes or tumor suppressor genes are involved in the sensitization of neoplastic cells to parvovirus attack because it would give clues to which tumor types are most appropriate for parvovirus-based treatments. On the one hand, it was shown that expression of a number of oncogene products, including MYC, SRC, RAS, and SV40 large T antigen, correlated with a greater susceptibility of rat fibroblast transformants to MVMp-induced killing, whereas the bovine papilloma virus transforming proteins were ineffective in this respect [73]. On the other hand, H-1 virus-resistant subclones derived from the H-1–sensitive human erythroleukemic cell line K562 were found to express wild-type p53, whereas this protein could not be detected by current methods in the original cell line [78]. Similarly, rat fibroblasts in which endogenous p53 was functionally inactivated by means of a dominant negative mutant thereof became more susceptible to H-1 virus infection and resultant killing compared with the parental cells expressing active p53 [78]. Furthermore, long-term cultures of Li-Fraumeni skin fibroblasts that lost their residual wild-type p53 allele proved to be more sensitive to H-1 virus-induced killing than earlier cell passages that still expressed functional p53 (our unpublished results). As a whole, these data indicate that the permissiveness of cells for parvovirus infection can be up- and down-modulated as a result of the expression of oncogenes and tumor suppressor genes, respectively. The mechanisms by which these genes affect the outcome of parvovirus infection are currently under investigation. One of the levels at which this modulation takes place was assigned to the activation of the viral early promoter P4 through specific signaling pathways [95].

An interesting parallel can be drawn between these data from autonomous parvoviruses and a recent report showing that AAV-2, a helper-dependent member of the Parvoviridae family, selectively killed human tumor cells lacking functional p53 through the induction of apoptosis. In contrast, cells with intact p53 survived the infection and became arrested in the G2 phase of the mitotic cycle [96]. These effects required very high multiplicities of infection with AAV (around $2.5 \times 10^4 - 5 \times 10^5$ particles per cell) and were apparently attributable to the incoming viral DNA. It was concluded from this work that AAV DNA, which is single-stranded, with hairpin structures at both ends, was sensed as DNA damage and elicited signals leading to death or growth arrest, depending on the p53 status of the host cell. Interestingly, the hairpin structures alone were able to trigger, at least in part, the damage response. It is noteworthy that the conditions under which AAV and autonomous parvoviruses caused cell death in a p53-dependent fashion were totally different. Indeed, the killing activity of autonomous parvoviruses was achieved using 50- to 200-fold lower multiplicity of infection (at which AAV had no effect) and required active virus replication and gene expression (in contrast to the involvement of input virus genomes in AAV toxicity) [20,80,97]. Because the cytopathogenicity of autonomous parvoviruses does not always correlate with the accumulation of viral DNA [72,74] and can result from the sole expression of NS proteins [18,92], H-1 virus and AAV appear to trigger cell death through different mechanisms. It should be stated, however, that the genomes of autonomous and helper-dependent parvoviruses share structural features, including terminal hairpins, which raises the possibility that at a higher multiplicity of infection, autonomous parvoviruses may provide death signals similar to those of AAVs. Most interestingly, the injection of high amounts of AAV virions (the genomes of which retained the hairpin ends) prevented the establishment of subcutaneous tumors, suppressed the growth of established tumors, or even caused their regression in various mouse models, according to the p53 status of the neoplastic cells grafted [96]. These observations give strong support to the development of drugs that mimic the DNA damage signals provided by parvoviral genomes. Indeed, one may expect that, like parvoviral DNA, these drugs will have an anti-cancer potential through their ability to kill cells devoid of functional p53 (which is the case of most human tumor cells) while sparing normal cells (with intact p53).

Contribution of the Immune System to Tumor Clearance In Vivo

Stimulation of Tumor Immunity

The manner in which tumor cells die after parvovirus infection is not only of academic interest but constitutes an important factor determining the susceptibility of tumor cell to macrophages and DC-mediated processing, which is decisive for the induction of acquired immunity [48,98]. Irrespective of the mechanism involved, parvovirus infections appear to stimulate cells to release distress signals such as heat shock proteins [99], which are known to generate a "danger" environment favorable to the activation of the host immune response [100,101]. Therefore, the interaction of parvoviruses with tumors may contribute not only to the destruction of a fraction of the neoplastic cells but also to the stimulation of antitumor immune reactions. This possibility is supported by recent in vitro cross-presentation experiments showing that H-1 virus infection and ensuing killing of human melanoma cells led to DC loading with tumor antigens, whereas mock-treated melanoma cells were ineffective in this respect (M. Möhler, personal communication).

Parvoviruses as Immune Trouble Makers

Treatments with MVMp and H-1 virus, including repeated injection regimens, were not associated with any detectable deleterious side effects in mouse and rat models used to demonstrate oncosuppression, nor in the human Phase I clinical trials conducted to assess tolerance (discussed above; Refs. 29,31,32,34; and K. Geletneky, personal communication). Yet pathological signs can be induced by other parvoviral infections and may shed light on at least some facets of the tumor-suppressing activity of these agents. In particular, certain parvoviruses have been implicated as causative agents in inflammatory and autoimmune diseases, including endothelialitis and rheumatoid arthritis (human B19), vasculitis, glomerulonephritis, polyarthritis and anti-DNA immunity (aleutian disease virus [ADV]), enteritis and myocarditis (CPV), and diabetes (KRV) (49–55,102,103). In some experimental settings, a strain-dependent mortality caused by intestinal and renal hemorrhages was described for neonatal mice infected with MVMi (a parvovirus with a high degree of homology to MVMp), whereas MVMp was apathogenic [64,104,105]. Infection of endothelial cells, presence of circulating antibodies, and formation of immune complexes are thought to play a major role in these pathologies. These observations raise the intriguing possibility that some aspects of parvovirus oncosuppression may be traced back to a virus-induced state of immune hyperactivity for which tumors would be preferential targets.

The inter-twinning of parvovirus-mediated oncosuppression and immunomodulation can be exemplified by the effects of the lymphocytotropic MPV-1. This virus was found to potentiate rejection not only of tumor allografts but also of allogeneic and—most surprisingly—syngeneic skin grafts [63,65,66]. These reactions did not result from virus infection of the cellular or vascular components of the various grafts and were mediated by T cells. In this system, the parvovirus appears to be devoid of a direct oncolytic activity and to act by exerting immunomodulatory effects, including the disruption of peripheral tolerance and development of autoimmune responses. The immune cell targets and mode of action of MPV-1 remain to be determined. Similarly, the destruction of pancreatic β cells that is responsible for the induction of autoimmune diabetes by KRV in DR-BB rats did not appear to result from direct viral cytopathic effects, but correlated with immunological changes (induction of specific T_H1-type cytokines and $CD4^+$ cells), which may disrupt the immune balance and lead to the generation of autoreactive T lymphocytes [49,103]. In keeping with these observations, infections of rats with KRV were recently reported to trigger the release of the T_H1-type cytokine IFN-γ, but not that of the T_H2-type cytokine interleukin (IL)-4 [106]. Besides the involvement of T_H1 cytokines in the activation of autoimmune cells [107], there is more and more evidence to suggest that T_H1–type responses generate a physiological milieu detrimental to oncogenesis. For instance, the antimetastatic effect of the chemical adjuvant cyclophosphamide has been ascribed to its ability to generate a T_H1 cell response [108]. Furthermore, successful rejection of tumors using various immunotherapeutic approaches was shown to be accompanied with autoimmune reactions, as exemplified by skin depigmentation due to melanocyte destruction in the case of melanoma treatment in humans and mice (109,110). Therefore, conditions promoting autoimmunity appear to potentiate tumor rejection, arguing for the possibility that the competence of parvoviruses for inducing such conditions may well play an important role in the anti-neoplastic activity of these agents.

A direct targeting of immune cells is observed after infection with several rodent parvoviruses that have lymphocytotropic properties [56] and therefore may hit the appropriate cells to induce changes in the immune environment. It cannot be ruled out that this scenario also applies to a so-called fibrotropic parvovirus such as MVMp because this agent was reported to infect mesenteric lymph nodes in mice [105]. Indeed, we found in our pre-clinical studies that besides neoplastic lesions, lymphoid organs were a major site of MVMp mRNA accumulation in virus-infected tumor-bearing mice [32]. Target cells for MVMp transcription in lymph nodes and spleen could be assigned to a minute subpopulation of lymphocytes rather than metastasizing tumor cells. This extra-tumoral MVMp gene expression in lymphoid tissues remains to be evaluated for its impact on tumor growth. Interestingly, a progressive disappearance of activated or memory $CD8^+$ $CD44^+$ $CD26L^-$ lymphocytes (N. Giese, unpublished results) was observed, which may contribute to the poor induction of virus-specific CTLs in MVMp-infected C57Bl6 mice [42]. Although MVMp gene expression proved to be restricted in various in vitro cell cultures of lymphoid origin [111], aforementioned observations suggest that immunomodulating effects of this virus still need to be considered. Together, these data also indicate that advantage may be taken of the unique lymphotropism of parvoviruses to deliver appropriate transgenes to lymphoid tissues through parvoviral vectors in immune gene therapy protocols (discussed below).

PRE-CLINICAL STUDIES USING RECOMBINANT PARVOVIRUSES

A most effective antitumor protection was achieved when parvoviruses were inoculated to the animals prior to the appearance of neoplasms (in particular, at birth) or to the tumor cells prior to their implantation into recipient hosts. In contrast, variable results were obtained when parvoviruses were administered to animals harboring established tumors, which may be ascribed at least in part to excessive tumor burdens. To overcome this restraint, efforts are being made to increase efficiency of parvoviruses through the inclusion of therapeutic transgenes.

Construction of Recombinant Parvovirus-Based Vectors

The design, production, and purification of recombinant autonomous parvoviruses have recently been reviewed [112,113] and are not discussed in detail here. It is sufficient to say that MVM and H-1 virus-derived recombinant parvoviruses designed so far for cancer therapy purposes are so-called capsid-replacement vectors, in which the P4 and P38 promoters and the NS genes from the parental virus are kept, while part or all of the capsid (VP) genes are replaced by a therapeutic transgene, as illustrated in Figure 1B [76,114–116]. These vectors are expected to be endowed with the oncotropic and oncolytic properties of the wild-type viruses because they retain the *cis*- and *trans*-acting elements thought to be responsible for these features. Upon cell infection, the viral regulatory protein NS-1 is expressed from the genuine P4 promoter, leading to the amplification of the vector genomes and the potent transactivation of the P38 promoter. The resulting transgene expression proved to be very efficient but transient in vitro as well as in vivo, and to depend on cell oncogenic transformation [94,114], which seems especially suitable for the production of toxic principles in the context of cancer gene therapy application. The transience of transgene expression was ascribed to the cytopathic effects of the vectors, although the recombinants were found to be less toxic for some cells than the corresponding parental viruses in single-infection experiments, pointing to a possible contribution of newly synthesized capsid (proteins) to parvovirus-induced cell disturbances (unpublished results). Limitations of parvovirus-based vectors lie in the limited size of the transgenes that they can accommodate (less than 1800 bp in the aforementioned constructs) and the present difficulties in producing high-titer recombinant virus stocks. Furthermore, presently available vectors can only achieve one-hit infections because they are unable to yield progeny particles attributable to their defect in VP genes. The main features of recombinant and wild-type viruses are compared in Table 1.

Improved Antitumor Capacity of Parvoviral Vectors Harboring Immunomodulating Transgenes

The ultimate goal of cancer immunotherapy is to break down tolerance to tumor-associated antigens (TAAs). There is clear evidence to suggest that CTLs are major effector cells that can attack and kill tumor cells in vivo through the recognition of TAAs presented by MHC class I molecules on the neoplastic cell surface. The capture and presentation of TAAs by DCs is central to the induction of effective antitumor CTL responses. This process is dictated by the initial death of some tumor cells to promote the uptake of antigens by DCs, and by the local release of cytokines, chemokines, and heat shock proteins to

Table 1 Properties of Natural Parvoviruses and Derived Vectors

	Virus			
Properties	Wild-type		Vector[(1)]	Refs
Stability		High		135
Preparation	Infection		Cotransfection	112, 113
Infectious particles/10^6 cells	$> 10^9$		$2-10\times10^6$	113
Maximum size of DNA insert			1.8 kb	115, 116
DNA integration		No Evidence		17
Anti-virus immune responses:				
Humoral	Strong		NR	42, 56, 106
Cellular	Weak		NR	42
Pre-existing immunity in man		No or low		2
Safety in humans	No adverse reaction reported		NR	29, 34
Competence for progeny virus production	Yes		No	115, 116
Preferential binding to tumor cells	No		No	69, 72, 73
Oncolysis	Yes		Yes	80

(1) Only for the vectors described in this contribution.
NR = not reported.

induce DC maturation. Innate immune cells, including NK cells, macrophages, and several types of leukocytes, are thought to play an important role in this initiation of the tumor rejection process [48,117,118]. These were the grounds for supplying parvoviruses with transgenes encoding various immunoregulatory molecules.

Parvoviral vectors transducing distinct co-stimulatory molecules or cytokines and chemokines were shown to constitute valuable antitumor therapeutics in various tumor models, as summarized in Table 2. These recombinants reduce tumor growth and prolong the survival of the animals by enhancing the immunogenicity of neoplastic cells or recruiting and acti-

Table 2 Anti-Tumor Effects of Recombinant Parvoviruses Transducing Costimulatory Factors or Cytokines/Chemokines

Tumor models	Mice	Protocol	Virus	Transgene	Wild-type	Antitumor Effect: Recombinant	Antitumor Effect: Recombinant/wt	Ref
HeLa human carcinoma	Nude	Ex vivo	H-1	IL-2	Yes	Yes	>	94
HeLa human carcinoma	Nude	Ex vivo	H-1	MCP-1	Yes	Yes	~	94
K1735 mouse melanoma	Syng	Ex vivo	MVMp	IL-2	NR	Yes		123
H5V mouse endothelioma	Syng	In vivo	MVMp	IP-10	No	Yes	>	32
B78/H1 mouse melanoma	Syng	Ex vivo	MVMp	MCP-3	Yes	Yes	>	122
	Syng	In vivo	MVMp	MCP-3	Yes	Yes	>	122
HeLa human carcinoma	Nude	Ex vivo	H-1	MCP-3	NR	Yes		120
EL4 mouse thymoma	Syng	In vivo	MVMi	B7.1	NR	Yes		119

Syng = syngeneic; NR = not reported; No = no antitumor effect was recorded; > = antitumor activity of recombinant greater than that of corresponding wild-type virus; ~ = antitumor activity of recombinant and wild-type viruses were similar.

vating immune effector cells. Thus, the treatment of mice bearing EL4 thymomas with a single dose of recombinant MVMi expressing the co-stimulatory molecule B7.1 elicited a protective immune response generating tumor-specific CTL memory cells [119]. Furthermore, MVMp or H-1 virus-based vectors transducing the genes encoding human interleukin (IL)-2, monocyte chemotactic protein 3 (MCP-3), or interferon (IFN)-γ–inducible protein 10 (IP-10) proved to have a significant anti-neoplastic activity in various animal tumor models in which their wild-type counterparts were inefficient (31,32,94,120,121, and unpublished results). The ability of these vectors to support a cytokine or chemokine production was demonstrated in treated animals and obviously overtook the defectiveness and reduced toxicity of the recombinants to endow them with a globally enhanced oncosuppressive capacity. In nude mice, parvovirus-mediated MCP-3 and IL-2 expression caused an intratumoral infiltration and activation of NK cells and macrophages that was associated with increased neoplastic cell killing and reduced growth of human carcinoma cell xenografts [94,120]. In immunocompetent animals, MCP-3-transducing MVMp fully suppressed the formation of neoplasias from ex vivo-infected B78H1 melanoma cells, and drastically reduced the growth of in vivo-treated established tumors [122]. Regressing or stabilized tumors were found to become infiltrated with T lymphocytes and NK cells in which the death program was activated, as revealed by positive transcription signals for IFN-γ, IL-2, Fas L, granzymes, and perforin. In another mouse melanoma model (K1735), ex vivo infection with MVMp and IL-2 was found to prevent tumors from appearing in about 70% of implanted animals under conditions in which the wild-type virus only delayed the onset of tumor growth [123]. Interestingly, tumor-free animals became resistant to a later challenge with live melanoma cells, indicating that the recombinant parvovirus elicited a long-term antitumor protection. Promising results were also obtained using parvoviral vectors to express the chemokine IP-10, which binds to the CXCR3 receptor present on activated T lymphocytes, NK cells, and vascular cells. The development of syngeneic hemangiosarcomas in immunocompetent mice

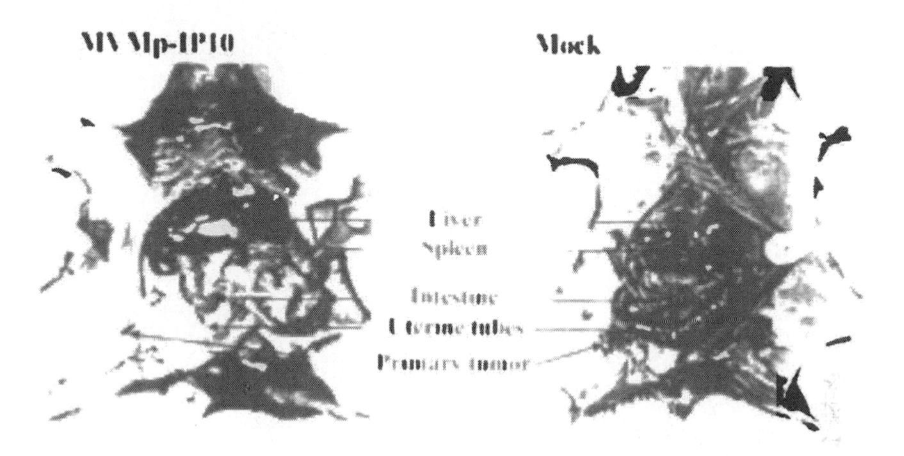

Figure 4 Suppression of primary tumor relapse and metastases formation (open arrow heads) through infection of hemangiosarcoma-bearing mice with recombinant MVMp expressing the chemokine IP-10. (From Ref. 32.)

could be efficiently suppressed through multiple intratumoral injections of IP-10–transducing MVMp recombinants [32]. As illustrated in Figure 4, besides the recurrence of primary tumors, the formation of multiple metastases in the liver, spleen, uterine tubes, peritoneal cavity, and ovaries was drastically reduced in animals treated with MVMp and IP-10.

In summary, these data show that parvovirus vectors are efficient transducers of cytokines and chemokines. Therefore, they deserve to be assessed for applications in the immune gene therapy of cancer.

PERSPECTIVES

Besides inclusion of immunomodulating transgenes, as outlined above, further attempts are currently being made to improve the potential of autonomous parvoviruses for cancer therapy through virus engineering. Possible strategies are the insertion of cytotoxic transgenes, alteration of the pattern of virus-driven gene expression through modification of the P4 promoter, or modulation of NS1-mediated toxicity through site-directed mutagenesis. In this context, the possible application of parvoviruses in vaccination strategies is elaborated.

Transduction of Cytotoxic Transgenes

Besides their direct contribution to the reduction of the tumor load, suicide genes encoding conditional toxins may also stimulate antitumor immunity by promoting the uptake of dying tumor cells by DCs and the presentation of tumor antigens to T cells [98]. A recombinant MVMp expressing the conditionally toxic Herpes simplex virus thymidine kinase gene, proved able, in vitro, to cause efficient killing of a number of tumor cell types in the presence of the prodrug ganciclovir, while having no or little effect on normal proliferating cells [83]. Similar attempts were recently made using H-1 virus-based vectors that transduced and expressed the chicken anemia virus product Apoptin [124]. As a result of the ability of this polypeptide to induce death specifically in transformed cells [125], the H-1/Apoptin recombinants were capable of killing in vitro a greater proportion of human tumor cells than transgene-free viruses, without becoming toxic for nontransformed cells. In vivo experiments are awaited to assess the oncosuppressive potential of this class of recombinant parvoviruses.

Vaccination

A number of clinical trials lend credit to the use of DNA vectors for vaccination purposes [126]. The suitability of recombinant autonomous parvoviruses to achieve immunization was recently demonstrated by successfully protecting mice against *Borrelia burgdorferi* through intranasal or i.v. administration of a MVMi-based vector expressing the bacterial outer surface protein A (OspA) [127]. Treated mice developed high-titer anti-OspA–specific antibodies that provided sustained protection from live spirochete challenge over the lifetime of the animals.

By analogy, parvovirus vectors expressing tumor-associated antigens deserve to be considered as potential vaccines to elicit antitumor immunity, given their aforementioned capacity for preferentially transducing both neoplastic and lymphoid tissues. In addition, recombinant parvoviruses harboring immunomodulating transgenes might be used to facilitate the recruitment, maturation, or functioning of effector cells participating in the antitumor immune response. Besides their transduction ability, parvoviruses are endowed with cytotoxic properties that promote not only the death of tumor cells but also the release of cellular danger signals such as heat shock proteins [99]. This represents a potential advantage regarding the induction of antitumor immunity, given the capacity of tumor cell lysates and, in particular, heat shock proteins for promoting tumor antigen presentation by DCs [98,117]. Parvovirus-mediated vaccination may be carried out through direct administration of the recombinant viruses in vivo, or by means of adoptive protocols in which the vectors are inoculated to tumor or immune cells in vitro.

These developments are still in their infancy, but are supported by a recent attempt to suppress B78H1 mouse melanoma tumors in immunocompetent animals using wild-type MVMp and a vector derivative expressing the chemokine MCP-3 [122]. The ex vivo infection of B78H1 cells with MVMp/ and MCP-3 (but not MVMp) did not only prevent these cells from forming tumor upon implantation in recipient mice, but protected more than half of treated animals from naïve tumor cell challenge over their lifetime. This system exemplifies tumors against which the wild-type parvovirus is poorly protective (correlating with a restriction on virus-driven gene expression in a large fraction of the cells), unless it is supplemented with a transgene that has an immunostimulating bystander effect. Altogether, these data suggest that the use of parvoviral vectors for vaccination purposes deserves to be further explored.

Targeting Specific Tumors Through P4 Promoter Modifications

The host cell range of autonomous parvoviruses can be altered by modifying the pivotal P4 promoter. This was initially demonstrated by substituting the human immunodeficiency virus (HIV) TAT protein-response (tar) element for the proximal region of the P4 promoter of parvovirus MVMi [112]. Infectious MVMi viruses harboring this substitution could be produced in a cell line constitutively expressing the HIV TAT transactivator and proved to depend on TAT for their multiplication and cytotoxic effects. Although this study was aimed at killing HIV-infected cells by means of modified parvoviruses, a similar strategy may conceivably be used to target specific tumors with these agents. Indeed, tumors of a certain type may be distinguished from other tissues by patterns of oncogene and tumor-suppressor gene alterations that result in the activation of a signaling pathway involving a specific transcription factor. Advantage may be taken of this situation to design parvoviruses that depend on such a transcription factor for their replication and can be used to specifically destroy the tumor cells in which it is activated. This situation can be exemplified by human colon tumors, which are characterized by the frequent deregulation of the wnt signaling pathway [128], resulting in the sustained induction of promoters that contain binding sites for the heterodimeric Tcf/β-catenin transcription factor [129]. This prompted us to produce hybrid H-1/MVMp viruses in the genome of which Tcf consensus binding motifs were substituted for distinct sites of the parvoviral P4 promoter [130]. These substitutions proved to make virus replication Tcf dependent, resulting in the induction of cytopathic effects in colon cell lines with high Tcf activity but not in cells that have inactive wnt signaling pathways. This is proof of principle that parvoviruses can be rendered specific for certain tumor cells through P4 promoter engineering, although the oncotropism and anti-neoplastic efficiency of viruses modified in this way remain to be assessed in vivo.

Modulation of Parvovirus Cytotoxicity Through Site-Directed Mutagenesis

Recent developments showed that phosphorylation events driven by protein kinases involved in the regulation of cell growth, differentiation, and transformation are essential to tune the MVM NS1 protein for induction of toxic effects in host cells. Using site-directed mutagenesis techniques to alter the phosphorylation pattern of the NS1 polypeptide, mutant MVMp viruses were generated that displayed increased or reduced toxic potentials (our unpublished observations). This gives hope of producing MVMp mutant viruses or vectors that are endowed with an enhanced oncolytic activity or a greater capacity for transgene expression compared with their wild-type equivalents.

CONCLUSION

Ongoing clinical trials, with modified adenoviruses, in particular, gave compelling evidence to suggest that virus therapy deserves to be considered among novel nonconventional approaches to cancer treatment [131]. This is especially true for tumor types against which surgical and radio- or chemotherapeutical interventions have only limited or transient efficacy and metastatic stages of cancers. In the same way the use of antimitotics is restrained by their nonspecific toxicity, a number of viruses endowed with anti-neoplastic potential can have undesirable side effects, even if they are devoid of intrinsic pathogenic activity. This can be exemplified by the life-threatening inflammatory reactions triggered by some adenoviral treatments [38]. In this respect, the rodent autonomous parvoviruses considered in this review are distinguishable in that they do not elicit overt immune reactions in animal models, even after repeated injections, and are not associated with known diseases in humans or adverse reactions in treated patients. This safety results, at least in part, from the fact that these viruses have a restricted host cell range, in spite of their ability to enter most cells in infected organisms. Thus, few of the infected cells support parvovirus replication and expression, and still fewer are capable of progeny virus production, which limits the effective virus targets to distinct tissues including in particular some tumors (hence the quotation of these agents are oncolytic). In addition, the development of a neutralizing antibody response to incoming virions certainly contributes to the containment of parvovirus infections.

Yet these advantages of parvoviruses regarding safety can also represent limitations with respect to their efficiency in tumor suppression. Indeed, pre-clinical studies have shown that the protective effect of parvoviruses against cancer is significant but can often be overtaken by too-large tumor burdens. Attempts are therefore being made to improve the antineoplastic potential of parvoviruses by optimizing their oncolytic activity and taking advantage of their preferential expression in tumors and lymphoid tissues.

Besides the various strategies outlined in previous sections, recent findings that still await anti-cancer applications are worth mentioning. On the one hand, parvovirus pseudotyping is feasible [132], which should increase the efficiency of multiple virus infections by minimizing the neutralization of a given inoculum by the antibodies raised against a previously administered virus batch. On the other hand, the cell-surface-targeting specificity of members of the Parvoviridae family can be altered through capsid modifications that prevent the viruses from binding to their genuine ubiquitously expressed receptors and redirect them to distinct new cellular receptors [133]. As stated above, parvoviruses naturally possess oncotropic properties that manifest themselves at the intracellular level. Targeting of the neoplastic cell surface through capsid modifications would constitute an additional safeguard to ensure the tumor specificity of parvoviruses and, even more importantly, should increase the efficiency of parvovirus-based treatments by preventing a large fraction of the inoculated viruses from getting lost as a result of their dead-end uptake by normal tissues [31]. Besides virus engineering to improve oncosuppressive capacity, efforts are certainly worth being made to identify tumors that are most susceptible to parvovirus infection, and to combine parvoviruses with other conventional or nonconventional therapies. Particular consideration may be given to the application of parvoviruses as postoperative treatment or as a complement to immunomodulating drugs that bring neoplastic cells into contact with T cells in a lymphoid tissue environment favorable to the induction of anti-tumor immunity [134]. Recombinant parvoviruses may allow combining the viral and immune therapeutic principles in the form of a single agent.

ACKNOWLEDGMENTS

Relevant work in the authors' laboratory was supported by grants from the European Union (Programme "Quality of Life and Management of Living Resources").

We are grateful to E. Burkard for help with the preparation of this manuscript and S. Lang and U. Ackermann for assistance with the artwork. All members of the laboratory and collaborators are acknowledged for discussions and sharing unpublished observations.

REFERENCES

1. Young NS. Parvoviruses. In: Field BN, Ed. Virology. Philadelphia: Lipincott-Raven, 1996:2199–2220.

2. Hokynar K, Soderlund-Venermo M, Pesonen M, et al. A new parvovirus genotype persistent in human skin Virology 2002; 302:224–228.

3. Wang XS, Yoder MC, Zhou SZ, Srivastava A. Parvovirus B19 promoter at map unit 6 confers autonomous replication competence and erythroid specificity to adeno-associated virus 2 in primary human hematopoietic progenitor cells Proc Natl Acad Sci U S A 1995; 92:12416–12420.

4. Toolan HW, Rhode SL, Gierthy JF. Inhibition of 7,12-dimethylbenz(a)anthracene-induced tumors in Syrian hamsters by prior infection with H-1 parvovirus Cancer Res 1982; 42: 2552–2552.

5. Dupressoir T, Vanacker JM, Cornelis JJ, Duponchel N, Rommelaere J. Inhibition by parvovirus H-1 of the formation of tumors in nude mice and colonies in vitro by transformed human mammary epithelial cells Cancer Res 1989; 49: 3203–3208.

6. Guetta E, Graziani Y, Tal J. Suppression of Ehrlich ascites tumors in mice by minute virus of mice J Natl Cancer Inst 1986; 76:1177–1180.

7. Rommelaere J, Cornelis JJ. Antineoplastic activity of parvoviruses J Virol Methods 1991; 33:233–251.

8. Parker JS, Murphy WJ, Wang D, O'Brien SJ, Parrish CR. Canine and feline parvoviruses can use human or feline transferrin receptors to bind, enter, and infect cells J Virol 2001; 75:3896–3902.

9. Suikkanen S, Saajarvi K, Hirsimaki J, et al. Role of recycling endosomes and lysosomes in dynein-dependent entry of canine parvovirus J Virol 2002; 76:4401–4411.

10. Ros C, Burckhardt CJ, Kempf C. Cytoplasmic trafficking of minute virus of mice: low-pH requirement, routing to late endosomes, and proteasome interaction J Virol 2002; 76: 12634–12645.

11. Bashir T, Horlein R, Rommelaere J, Willwand K. Cyclin A activates the DNA polymerase delta -dependent elongation machinery in vitro: A parvovirus DNA replication model Proc Natl Acad Sci U S A 2000; 97:5522–5527.

12. Deleu L, Fuks F, Spitkovsky D, Horlein R, Faisst S, Rommelaere J. Opposite transcriptional effects of cyclic AMP-responsive elements in confluent or p27KIP-overexpressing cells versus serum-starved or growing cells Mol Cell Biol 1998; 18: 409–419.

13. Deleu L, Pujol A, Faisst S, Rommelaere J. Activation of promoter P4 of the autonomous parvovirus minute virus of mice at early S phase is required for productive infection J Virol 1999; 73:3877–3885.

14. Rhode SL. Trans-Activation of parvovirus P38 promoter by the 76K noncapsid protein. J-Virol 1985; 55:886–889.

15. Richards R, Linser P, Armentrout RW. Kinetics of assembly of a parvovirus, minute virus of mice, in synchronized rat brain cells J Virol 1977; 22:778–793.

16. Tullis GE, Burger LR, Pintel DJ. The minor capsid protein VP1 of the autonomous parvovirus minute virus of mice is dispensable for encapsidation of progeny single-stranded DNA but is required for infectivity J Virol 1993; 67:131–141.

17. Richards RG, Armentrout RW. Early events in parvovirus replication: lack of integration by minute virus of mice into host cell D. N. A. J Virol 1979; 30:397–399.

18. Caillet-Fauquet P, Perros M, Brandenburger A, Spegelaere P, Rommelaere J. Programmed killing of human cells by means of an inducible clone of parvoviral genes encoding non-structural proteins EMBO J 1990; 9:2989–2995.

19. Brandenburger A, Legendre D, Avalosse B, Rommelaere J. NS-1 and NS-2 proteins may act synergistically in the cyto-

pathogenicity of parvovirus MVMp Virology 1990; 174: 576–584.

20. Li X, Rhode SL. Mutation of lysine 405 to serine in the parvovirus H-1 NS1 abolishes its functions for viral DNA replication, late promoter trans activation, and cytotoxicity J Virol 1990; 64:4654–4660.
21. Op de Beek A, Caillet-Fauquet P. The NS1 protein of the autonomous parvovirus minute virus of mice blocks cellular DNA replication: a consequence of lesions to the chromatin? J Virol 1997; 71:5323–5329.
22. Anouja F, Wattiez R, Mousset S, Caillet Fauquet P. The cytotoxicity of the parvovirus minute virus of mice nonstructural protein NS1 is related to changes in the synthesis and phosphorylation of cell proteins J Virol 1997; 71:4671–4678.
23. Cziepluch C, Lampel S, Grewenig A, Grund C, Lichter P, Rommelaere J. H-1 parvovirus-associated replication bodies: a distinct virus-induced nuclear structure J Virol 2000; 74: 4807–4815.
24. Bashir T, Rommelaere J, Cziepluch C. In vivo accumulation of cyclin A and cellular replication factors in autonomous parvovirus minute virus of mice-associated replication bodies J Virol 2001; 75:4394–4398.
25. Young PJ, Jensen KT, Burger LR, Pintel DJ, Lorson CL. Minute virus of mice NS1 interacts with the SMN protein, and they colocalize in novel nuclear bodies induced by parvovirus infection J Virol 2002; 76:3892–3904.
26. Vanacker JM, Rommelaere J. Non-structural proteins of autonomous parvoviruses : from cellular effects to molecular mechanisms Semin Virol 1995; 6:291–297.
27. Siegl G. Biology and pathogenicity of autonomous parvoviruses. In: Berns KI, Ed. The Parvoviruses.. New York:: Plenum Press, 1984:297–362.
28. Toolan H, Ledinko N. Growth and cytopathogenicity of H-viruses in human and simian cell cultures Nature 1965; 208: 812–813.
29. Toolan HW, Saunders EL, Southam CM, Moore AE, Levin AG. H-1 virus viremia in the human Proc Soc Exp Biol Med 1965; 119:711–715.

30. Toolan HW. Experimental production of mongoloid hamsters Science 1960; 131:1446.

31. Giese N, DeMartino L, Haag A, Dinsart C, Cornelis JJ, Rommelaere J. Autonomous parvovirus-based vectors: potential for gene therapy of cancer. In: Gregoriades G, Ed. Proceeding of NATO ASI on targeting of drugs: strategies of gene constructs and delivery: IOS Press, 2000:34–52.

32. Giese NA, Raykov Z, DeMartino L, et al. Suppression of metastatic hemangiosarcoma by a parvovirus MVMp vector transducing the IP-10 chemokine into immunocompetent mice Cancer Gene Ther 2002; 9:432–442.

33. Alemany R, Suzuki K, Curiel DT. Blood clearance rates of adenovirus type 5 in mice J Gen Virol 2000; 81:2605–2609.

34. Le Cesne A, Dupressior T, Janin N, et al. Intra-lesional administration of a live virus, parvovirus H-1 (PHV-1) in cancer patients: feasibility study Proc Ann Meet Am Soc Clin Oncol 1993; 12:297.

35. Van Pachterbeke C, Tuynder M, Cosyn JP, Lespagnard L, Larsimont D, Rommelaere J. Parvovirus H-1 inhibits growth of short-term tumor-derived but not normal mammary tissue cultures Int J Cancer 1993; 55:672–677.

36. Bramson JL, Hitt M, Gauldie J, Graham FL. Pre-existing immunity to adenovirus does not prevent tumor regression following intratumoral administration of a vector expressing IL-12 but inhibits virus dissemination Gene Ther 1997; 4: 1069–1076.

37. Cichon G, Boeckh-Herwig S, Schmidt HH, et al. Complement activation by recombinant adenoviruses Gene Ther 2001; 8: 1794–1800.

38. Bostanci A. Gene therapy. Blood test flags agent in death of Penn subject Science 2002; 295:604–605.

39. Matzinger P. Tolerance, danger, and the extended family Annu Rev Immunol 1994; 12:991–1045.

40. Brown BD, Lillicrap D. Dangerous liaisons: the role of "danger" signals in the immune response to gene therapy Blood 2002; 100:1133–1140.

41. Medzhitov R, Janeway C. Innate immune recognition: mechanisms and pathways Immunol Rev 2000; 173:89–97.
42. Lang S, Raykov Z, Giese N, et al. Strong humoral but poor cellular immune response to MVMp in C57BL/6 mice.. Bologna. Italy: IX Parvovirus workshop, 2002.
43. Kim KD, Kim JK, Kim SJ, et al. Protective antitumor activity through dendritic cell immunization is mediated by NK cell as well as CTL activation Arch Pharm Res 1999; 22:340–347.
44. Fitzgerald-Bocarsly P. Natural interferon-alpha producing cells: the plasmacytoid dendritic cells Biotechniques 2002(Suppl:16–20):22, 24–29.
45. Barchet W, Cella M, Odermatt B, Asselin-Paturel C, Colonna M, Kalinke U. Virus-induced interferon alpha production by a dendritic cell subset in the absence of feedback signaling in vivo J Exp Med 2002; 195:507–516.
46. Schlecht G, Leclerc C, Dadaglio G. Induction of CTL and nonpolarized Th cell responses by CD8alpha(+) and CD8alpha(−) dendritic cells J Immunol 2001; 167:4215–4221.
47. Karlin S, Doerfler W, Cardon LR. Why is CpG suppressed in the genomes of virtually all small eukaryotic viruses but not in those of large eukaryotic viruses? J Virol 1994; 68:2889–2897.
48. Melcher A, Todryk S, Hardwick N, Ford M, Jacobson M, Vile RG. Tumor immunogenicity is determined by the mechanism of cell death via induction of heat shock protein expression Nat Med 1998; 4:581–587.
49. Chung YH, Jun HS, Kang Y, et al. Role of macrophages and macrophage-derived cytokines in the pathogenesis of Kilham rat virus-induced autoimmune diabetes in diabetes-resistant BioBreeding rats J Immunol 1997; 159:466–471.
50. Henson JB, Gorham JR, Padgett GA, Davis WC. Pathogenesis of the glomerular lesions in aleutian disease of mink. Immunofluorescent studies Arch Pathol 1969; 87:21–28.
51. Porter DD, Larsen AE, Porter HG. The pathogenesis of Aleutian disease of mink. 3. Immune complex arteritis Am J Pathol 1973; 71:331–344.
52. Magro CM, Crowson AN, Dawood M, Nuovo GJ. Parvoviral infection of endothelial cells and its possible role in vasculitis and autoimmune diseases J Rheumatol 2002; 29:1227–1235.

53. Crowson AN, Magro CM, Dawood MR. A causal role for parvovirus B19 infection in adult dermatomyositis and other autoimmune syndromes J Cutan Pathol 2000; 27:505–515.

54. Cioc AM, Sedmak DD, Nuovo GJ, Dawood MR, Smart G, Magro CM. Parvovirus B19 associated adult Henoch Schonlein purpura J Cutan Pathol 2002; 29:602–607.

55. Takahashi Y, Murai C, Shibata S, et al. Human parvovirus B19 as a causative agent for rheumatoid arthritis Proc Natl Acad Sci U S A 1998; 95:8227–8232.

56. Jacoby RO, Ball-Goodrich LJ. Parvovirus infections of mice and rats Sem Virol 1995; 6:329–337.

57. Toolan HW. Lack of oncogenic effect of the H-viruses for hamsters Nature 1967; 214::1036.

58. Toolan HW, Ledinko N. Inhibition by H-1 virus of the incidence of tumors produced by adenovirus 12 in hamsters Virology 1968; 35:475–478.

59. Kilham L, Margolis G. Transplacental infection of rats and hamsters induced by oral and parenteral inoculations of H-l and rat viruses (RV) Teratology 1969; 2:111–123.

60. Bergs VV. Rat virus-mediated suppression of leukemia induction by Moloney virus in rats Cancer Res 1969; 29:1669–1672.

61. Rommelaere J, Tattersall P. Oncosuppression by Parvoviruses. In: Tijssen P, Ed. CRC Handbook of Parvoviruses.. Vol. Vol. II.. Boca Raton. Fl: CRC Press, 1990:41–74.

62. Yang WC, Schultz RD, Spano JS. Isolation and characterization of porcine natural killer (NK) cells. Vet Immunol Immunopathol 1987; 14:345–356.

63. McKisic MD, Lancki DW, Otto G, et al. Identification and propagation of a putative immunosuppressive orphan parvovirus in cloned T cells J Immunol 1993; 150:419–428.

64. Kimsey PB, Engers HD, Hirt B, Jongeneel CV. Pathogenicity of fibroblast- and lymphocyte-specific variants of minute virus of mice J Virol 1986; 59:8–13.

65. McKisic MD, Paturzo FX, Smith AL. Mouse parvovirus infection potentiates rejection of tumor allografts and modulates T cell effector functions Transplantation 1996; 61:292–299.

66. McKisic MD, Macy JD, Delano ML, Jacoby RO, Paturzo FX, Smith AL. Mouse parvovirus infection potentiates allogeneic skin graft rejection and induces syngeneic graft rejection Transplantation 1998; 65:1436–1446.

67. Faisst S, Guittard D, Benner A, et al. Dose-dependent regression of HeLa cell-derived tumours in SCID mice after parvovirus H-1 infection Int J Cancer 1998; 75:584–589.

68. Mousset S, Rommelaere J. Minute virus of mice inhibits cell transformation by simian virus 40 Nature 1982; 300:537–539.

69. Chen YQ, de Foresta F, Hertoghs J, Avalosse BL, Cornelis JJ, Rommelaere J. Selective killing of simian virus 40-transformed human fibroblasts by parvovirus H-1 Cancer Res 1986; 46:3574–3579.

70. Chen YQ, Tuynder MC, Cornelis JJ, Boukamp P, Fusenig NE, Rommelaere J. Sensitization of human keratinocytes to killing by parvovirus H-1 takes place during their malignant transformation but does not require them to be tumorigenic Carcinogenesis 1989; 10:163–167.

71. Cornelis JJ, Becquart P, Duponchel N, Salome N, Avalosse BL, Namba M, Rommelaere J. Transformation of human fibroblasts by ionizing radiation, a chemical carcinogen, or simian virus 40 correlates with an increase in susceptibility to the autonomous parvovirus H-1 and minute virus of mice J Virol 1988; 62:1679–1686.

72. Cornelis JJ, Chen YQ, Spruyt N, Duponchel N, Cotmore SF, Tattersall P, Rommelaere J. Susceptibility of human cells to killing by the parvoviruses H-1 and minute virus of mice correlates with viral transcription J Virol 1990; 64:2537–2544.

73. Salome N, van Hille B, Duponchel N, et al. Sensitization of transformed rat cells to parvovirus MVMp is restricted to specific oncogenes Oncogene 1990; 5:123–130.

74. Van Hille B, Duponchel N, Salome N, et al. Limitations to the expression of parvoviral nonstructural proteins may determine the extent of sensitization of EJ-ras-transformed rat cells to minute virus of mice Virology 1989; 171:89–97.

75. Moehler M, Blechacz B, Weiskopf N, et al. Effective infection, apoptotic cell killing and gene transfer of human hepatoma

cells but not primary hepatocytes by parvovirus H1 and derived vectors Cancer Gene Ther 2001; 8:158–167.

76. Dupont F, Tenenbaum L, Guo LP, Spegelaere P, Zeicher M, Rommelaere J. Use of an autonomous parvovirus vector for selective transfer of a foreign gene into transformed human cells of different tissue origins and its expression therein J Virol 1994; 68:1397–1406.

77. Bass LR, Hetrick FM. Human lymphoblastoid cells as hosts for parvoviruses H-1 and rat virus J Virol 1978; 25:486–490.

78. Telerman A, Tuynder M, Dupressoir T, et al. A model for tumor suppression using H-1 parvovirus Proc Natl Acad Sci U S A 1993; 90:8702–8706.

79. Faisst S, Schlehofer JR, zur Hausen H. Transformation of human cells by oncogenic viruses supports permissiveness for parvovirus H-1 propagation J Virol 1989; 63:2152–2158.

80. Rayet B, Lopez-Guerrero JA, Rommelaere J, Dinsart C. Induction of programmed cell death by parvovirus H-1 in U937 cells: connection with the tumor necrosis factor alpha signalling pathway J Virol 1998; 72:8893–8903.

81. Lin W, Cui Y, Yu B, Luo Z, Lin Z. Preliminary comparison of sensitivity toward parvovirus H-1 of human hepatoma cells and parahepatoma tissue Chin J Cancer Res 1989; 1:15–20.

82. Cornelis JJ, Spruyt N, Spegelaere P, Guetta E, Darawshi T, Cotmore SF, Tal J, Rommelaere J. Sensitatization of transformed rat fibroblasts to killing by parvovirus minute virus of mice correlates with an increase in viral gene expression J Virol 1988; 62:3438–3444.

83. Dupont F, Avalosse B, Karim A, et al. Tumor-selective gene transduction and cell killing with an oncotropic autonomous parvovirus-based vector Gene Ther 2000; 7:790–796.

84. Dinsart C, Cornelis JJ, Rommelaere J. Recombinant autonomous parvoviruses: new tools for the gene therapy of cancer? Chemistry Today 1996; 9:32–38.

85. Ohshima T, Iwama M, Ueno Y, et al. Induction of apoptosis in vitro and in vivo by H-1 parvovirus infection J Gen Virol 1998; 79:3067–3071.

86. Segovia JC, Gallego JM, Bueren JA, Almendral JM. Severe leukopenia and dysregulated erythropoiesis in SCID mice persistently infected with the parvovirus minute virus of mice J Virol 1999; 73:1774–1784.

87. Ran Z, Rayet B, Rommelaere J, Faisst S. Parvovirus H-1-induced cell death: influence of intracellular NAD consumption on the regulation of necrosis and apoptosis Virus Res 1999; 65:161–174.

88. Op De Beeck A, Caillet Fauquet P. The NS1 protein of the autonomous parvovirus minute virus of mice blocks cellular DNA replication: a consequence of lesions to the chromatin? J Virol 1997; 71:5323–5329.

89. Li X, Rhode SL. Nonstructural protein NS2 of parvovirus H-1 is required for efficient viral protein synthesis and virus production in rat cells in vivo and in vitro Virology 1991; 184: 117–130.

90. Rhode SL. Construction of a genetic switch for inducible trans-activation of gene expression in eucaryotic cells J Virol 1987; 61:1448–1456.

91. Legendre D, Rommelaere J. Terminal regions of the NS-1 protein of the parvovirus minute virus of mice are involved in cytotoxicity and promoter trans inhibition J Virol 1992; 66: 5705–5713.

92. Mousset S, Ouadrhiri Y, Caillet-Fauquet P, Rommelaere J. The cytotoxicity of the autonomous parvovirus minute virus of mice nonstructural proteins in FR3T3 rat cells depends on oncogene expression J Virol 1994; 68:6446–6453.

93. Corbau R, Duverger V, Rommelaere J, Nuesch JP. Regulation of MVM NS1 by protein kinase C: impact of mutagenesis at consensus phosphorylation sites on replicative functions and cytopathic effects Virology 2000; 278:151–167.

94. Haag A, Menten P, Van Damme J, Dinsart C, Rommelaere J, Cornelis JJ. Highly efficient transduction and expression of cytokine genes in human tumor cells by means of autonomous parvovirus vectors; generation of antitumor responses in recipient mice Hum Gene Ther 2000; 11:597–609.

95. Perros M, Deleu L, Vanacker JM, et al. Upstream CREs participate in the basal activity of minute virus of mice promoter

P4 and in its stimulation in ras-transformed cells J Virol 1995; 69:5506–5515.

96. Raj K, Ogston P, Beard P. Virus-mediated killing of cells that lack p53 activity Nature 2001; 412:914–917.
97. Rhode SL, Richard SM. Characterization of the trans-activation-responsive element of the parvovirus H-1 P38 promoter J Virol 1987; 61:2807–2815.
98. Somersan S, Larsson M, Fonteneau JF, Basu S, Srivastava P, Bhardwaj N. Primary tumor tissue lysates are enriched in heat shock proteins and induce the maturation of human dendritic cells J Immunol 2001; 167:4844–4852.
99. Moehler M, Zeidler M, Schede J, et al. The oncolytic parvovirus H1 induces release of heat shock protein HSP72 in susceptible human tumor cells but may not affect primary immune cells Cancer Gene Ther 2003; 10:477–480.
100. Srivastava PK. Immunotherapy of human cancer: lessons from mice Nat Immunol 2000; 1:363–366.
101. Blachere NE, Li Z, Chandawarkar RY, et al. Heat shock protein-peptide complexes, reconstituted in vitro, elicit peptide-specific cytotoxic T lymphocyte response and tumor immunity J Exp Med 1997; 186:1315–1322.
102. Appel MJ, Scott FW, Carmichael LE. Isolation and immunisation studies of a canine parco-like virus from dogs with haemorrhagic enteritis Vet Rec 1979; 105:156–159.
103. Chung YH, Jun HS, Son M, et al. Cellular and molecular mechanism for Kilham rat virus-induced autoimmune diabetes in DR-BB rats J Immunol 2000; 165:2866–2876.
104. Brownstein DG, Smith AL, Jacoby RO, Johnson EA, Hansen G, Tattersall P. Pathogenesis of infection with a virulent allotropic variant of minute virus of mice and regulation by host genotype Lab Invest 1991; 65:357–364.
105. Brownstein DG, Smith AL, Johnson EA, Pintel DJ, Naeger LK, Tattersall P. The pathogenesis of infection with minute virus of mice depends on expression of the small nonstructural protein NS2 and on the genotype of the allotropic determinants VP1 and VP2 J Virol 1992; 66:3118–3124.

106. Ball-Goodrich LJ, Paturzo FX, Johnson EA, Steger K, Jacoby RO. Immune responses to the major capsid protein during parvovirus infection of rats J Virol 2002; 76:10044–10049.

107. Druet P, Sheela R, Pelletier L. Th1 and Th2 cells in autoimmunity Chem Immunol 1996; 63:138–170.

108. Matar P, Rozados VR, Gervasoni SI, Scharovsky GO. Th2/Th1 switch induced by a single low dose of cyclophosphamide in a rat metastatic lymphoma model Cancer Immunol Immunother 2002; 50:588–596.

109. van Elsas A, Hurwitz AA, Allison JP. Combination immunotherapy of B16 melanoma using anti-cytotoxic T lymphocyte-associated antigen 4 (CTLA-4) and granulocyte/macrophage colony-stimulating factor (GM-CSF)-producing vaccines induces rejection of subcutaneous and metastatic tumors accompanied by autoimmune depigmentation J Exp Med 1999; 190: 355–366.

110. Dudley ME, Wunderlich JR, Robbins PF, et al. Cancer regression and autoimmunity in patients after clonal repopulation with antitumor lymphocytes Science 2002; 298:850–854.

111. Tattersall P, Bratton J. Reciprocal productive and restrictive virus-cell interactions of immunosuppressive and prototype strains of minute virus of mice J Virol. 1983; 46:944–955.

112. Palmer GA, Tattersall P. Autonomous parvoviruses as gene transfer vehicles. In: Faisst S, Rommelaere J, Eds. Contributions to Microbiology., Karger. 2000; Vol. 4.:178–202.

113. Cornelis JJ, Haag A, Kornfeld C, et al. Autonomous parvovirus vectors. In: Cid-Arregui A, Garcia-Carranca A, Eds. Contributions to Microbiology., Eaton Publishing. 2000:97–115.

114. Russell SJ, Brandenburger A, Flemming CL, Collins MK, Rommelaere J. Transformation-dependent expression of interleukin genes delivered by a recombinant parvovirus J Virol 1992; 66:2821–2828.

115. Brandenburger A, Coessens E, El Bakkouri K, Velu T. Influence of sequence and size of DNA on packaging efficiency of parvovirus MVM-based vectors Hum Gene Ther 1999; 10: 1229–1238.

116. Kestler J, Neeb B, Struyf S, et al. Cis requirements for the efficient production of recombinant DNA vectors based on autonomous parvoviruses Hum Gene Ther 1999; 10:1619–1632.

117. Todryk S, Melcher AA, Hardwick N, et al. Heat shock protein 70 induced during tumor cell killing induces Th1 cytokines and targets immature dendritic cell precursors to enhance antigen uptake J Immunol 1999; 163:1398–1408.

118. Dredge K, Marriott JB, Todryk SM, et al. Protective antitumor immunity induced by a costimulatory thalidomide analog in conjunction with whole tumor cell vaccination is mediated by increased Th1-type immunity J Immunol 2002; 168: 4914–4919.

119. Palmer GA, Tattersall P. MVM-based vectors for transducing human and murine T cells.. VIII Parvovirus Workshop.. Canada: Mont-Tremblant, 2000.

120. Wetzel K, Menten P, Opdenakker G, et al. Transduction of human MCP-3 by a parvoviral vector induces leukocyte infiltration and reduces growth of human cervical carcinoma cell xenografts J Gene Med 2001; 3:326–337.

121. Clement N, Velu T, Brandenburger A. Construction and production of oncotropic vectors, derived from MVM(p), that share reduced sequence homology with helper plasmids Cancer Gene Ther 2002; 9:762–770.

122. Wetzel K. Untersuchung MCP-3-rekombinanter Parvoviren zur Gentherapie von Krebs: Vorklinische Studien in zwei Tiermodellen University of Heidelberg, 2000.

123. Clement N, Bakkouri KE, Velu T, Brandenburger A. Production of oncotropic vectors derived from the autonomous parvovirus MVM(p).. IX Parvovirus Workshop.. Bologna. Italy, 2002.

124. Olijslagers S, Dege AY, Dinsart C, et al. Potentiation of a recombinant oncolytic parvovirus by expression of Apoptin Cancer Gene Ther 2001; 8:958–965.

125. Danen-Van Oorschot AA, Fischer DF, Grimbergen JM, et al. Apoptin induces apoptosis in human transformed and malignant cells but not in normal cells Proc Natl Acad Sci U S A 1997; 94:5843–5847.

126. Restifo NP, Ying H, Hwang L, Leitner WW. The promise of nucleic acid vaccines Gene Ther 2000; 7:89–92.

127. Palmer GA, Brogdon JL, Constant SL, Tattersall P. A nonproliferating parvovirus vaccine vector elicits sustained protective humoral immunity following a single intravenous or intranasal inoculation J Virol 2004; 78:1101–1108.

128. Polakis P. Wnt signaling and cancer Genes Dev 2000; 14: 1837–1851.

129. Korinek V, Barker N, Morin PJ, et al. Constitutive transcriptional activation by a beta-catenin-Tcf complex in APC-/- colon carcinoma Science 1997; 275:1784–1787.

130. Malerba M, Daeffler L, Rommelaere J, Iggo RD. Replicating parvoviruses that target colon cancer cells J Virol 2003; 77: 6683–6691.

131. Kirn D, Martuza RL, Zwiebel J. Replication-selective virotherapy for cancer: Biological principles, risk management and future directions Nat Med 2001; 7:781–787.

132. Wrzesinski C, Tesfay L, Salome N, et al. Chimeric and pseudotyped parvoviruses minimize the contamination of recombinant stocks with replication-competent viruses and identify a DNA sequence that restricts parvovirus H-1 in mouse cells J Virol 2003; J Virol 2003; 77:3851–3858.

133. Girod A, Ried M, Wobus C, et al. Genetic capsid modifications allow efficient re-targeting of adeno-associated virus type 2 Nat Med 1999; 5:1052–1056.

134. Ochsenbein AF, Sierro S, Odermatt B, et al. Roles of tumour localization, second signals and cross priming in cytotoxic T-cell induction Nature 2001; 411:1058–1064.

135. Arella M, Garzon S, Bergeron J, Tijssen P. Physicochemical properties, production, and purification of parvoviruses. In: Tijssen P, Ed. CRC Handbook of Parvoviruses., CRC Press. 1990; Vol. I.:11–30.

9

Newcastle Disease Virus Oncolysate in the Management of Stage III Malignant Melanoma

WILLIAM A. CASSEL, DOUGLAS R. MURRAY,
and ZBIGNIEW L. OLKOWSKI†

Emory University School of Medicine
Atlanta, Georgia

INTRODUCTION

Awareness of tumor regression during viral infection in humans dates back to 1912, and the first virus-induced tumor lysis in a laboratory animal was observed with vaccinia virus in 1922. In clinical trials, the West Nile arbovirus was the first

† deceased

to show beneficial effects. It soon became evident that some tumor-destroying viruses had a predilection for neural tissues. Unfortunately, several treated patients developed encephalitis, and by 1970 interest in direct action by oncolytic viruses had essentially run its course, although the effectiveness of mumps virus, even against brain tumors, was reported in 1978. The subject is reviewed by H. E. Webb and C. E. G. Smith [1], and more recently by J. Nemunaitis [2].

Evidence that virally lysed tumors had immunizing properties was presented by G. R. Sharpless et al in 1950 [3], and H. Koprowski et al in 1957 [4]. Finally, viral oncolysates per sé were employed to immunize laboratory animals, as described by J. Lindenmann and P. A. Klein in 1957 [5]. A virus budding from the cell membrane was an essential feature in the process. The role of the virus was elucidated by the fact that, in animals made immunologically tolerant to the virus, the oncolysates were ineffective in immunization [6,7,8]. The evidence therefore suggests that the virus acts as a helper antigen in assisting the immune system to recognize the poorly antigenic tumor cells. Application of oncolysates to humans was introduced by J. G. Sinkovics et al [9] and W. A. Cassel et al [10] in patients with sarcomas and malignant melanoma. For further orientation on viral oncolysates, the reader is referred to an excellent review by F. C. Austin and C. W. Boone [11].

RATIONALE

Newcastle Disease virus (NDV) employed in our studies was acquired from the American Cyanamid Company, through the courtesy of Dr. Robert Love. The virus had extensive passage in Ehrlich ascites tumor cells in vitro and in vivo and was capable of curing mice bearing the Ehrlich ascites tumor that was allowed to progress to 41% of its potential development [12]. When cured animals were challenged with a dose of 2×10^7 tumor cells, most animals showed a solid immunity to the tumor [13]. The virus also was active against a human adenocarcinoma carried in the hamster cheekpouch, and human cervical cancer cells (HeLa). It would not multiply in the adult brains of six animal species [12], suggesting that NDV was safe for human application. A patient with uterine cervical cancer received a heavy dose of the virus injected into the tumor. Extensive sloughing occurred, with cessation of bleeding, shrinkage of involved lymph nodes, and considerable subjective improvement in the patient. Several years later, T. C. Merigan [14] observed this strain of NDV to be the best endogenous interferon inducer of all agents examined in humans, following intravenous administration, and no neuropathogenicity was encountered. Another cytokine induced by the virus is tumor necrosis factor, as demonstrated by R. M. Lorence et al. [15].

We considered the particulate nature of the lysate to be an asset in stimulating an immune response, as suggested by the studies of G. J. V. Nossal [16], A. L. Notkins et al [17], and D. M. Weir [18]. V. Schirrmacher et al [19], however, employing a non-lytic strain of NDV, found intact, infected cells were preferred for immunization. In our study, infectious virus was maintained in the oncolysate, for the purpose of providing a maximal impact on the patient's immune system, and possibly to destroy directly any tumor cells remaining after surgery. All evidence indicated the virus would not be harmful to patients. A nullifying effect by antibody to the virus, developing from repeated oncolysate application, appeared minimal. J. Lindenmann and P. A. Klein [20] noted that mice immunized to influenza virus, then administered an oncolysate prepared with the same virus, were protected from challenge by the tumor employed in the lysate. A. Horak et al [21] found that in the NDV-oncolysate system, antiviral immunity induced in humans does not interfere with the development of antimelanoma immune cell responses. The presence of antibody might actually play a role in oncolysis, should the patient's tumor cells become infected by the virus in the oncolysate and succumb to complement, NK cells, or macrophage-mediated lysis [1]. The wide spectrum of human tumor cells susceptible to infection by NDV suggested that oncolysates probably could be prepared for application in a variety of malignancies.

Considering the above information, and the long-existing awareness of the antigenic nature of malignant melanoma, this malignancy was selected for study. We focused on Stage III (AJCC) cases: patients with regional lymph node metastases that were palpable and required therapeutic dissection.

In summary, some reasons for favoring NDV as a lytic agent include: (1) The virus has not shown neuropathogenicity in humans. (2) It induces lymphokines. (3) It shows a wide range of infectivity for tumor cells. (4) It has a high neuraminidase activity [12], which could be involved in uncovering concealed antigenic sites on the tumor cell surface. (5) The virus has a stable genome, reducing the likelihood of it mutating to a harmful variant.

PROCEDURES

The methods employed in our studies are presented in depth elsewhere [10,22]. Oncolysates were prepared from two existing malignant melanoma cell lines and 21 cell lines established from the ongoing patients. Highly concentrated oncolysate was administered subcutaneously on the anterior thigh as a 2.5-mL dose, consisting of a systematically rotated composition of autogenous and allogeneic cell lines, with three different cell lines per dose. An additional dose of 0.2 mL was injected intracutaneously on the forearm. The oncolysates were shown to be devoid of viable tumor cells. Patients were entered into the study as they presented and were confirmed as Stage III (AJCC) cases of malignant melanoma, requiring a therapeutic lymph node dissection. We introduced the practice of continuing immunization for a period of years. Treatment began 4 weeks after surgery. Oncolysate then was administered weekly for 4 weeks, followed by a 2-week schedule until week 52, at which time the interval was increased to 6 weeks. After 10 years, the patients went to a 3-month schedule. Those treated included 81 on-study, Stage III (AJCC) patients, and 80 patients with widely disseminated disease who were treated as a desperate measure. Over 11,000 doses of oncolysate were administered without any significant adverse effect, thus attesting to the relative safety of the oncolysate and its live virus component.

CLINICAL FINDINGS

With approval of the Emory University Clinical Trials Committee, the first patient was treated in 1971, and our first Phase II study reports appeared in 1977 [10,22]. A 10-year follow-up report was presented in 1992 [23]. The earliest recurrence of disease in a patient was seen after 36 weeks of treatment. At 10 years, 63% of the patients remained free of detectable disease. Currently, surviving patients have been under study for 20 to 28 years. Thirty-six (71%) of patients who survived 10 years remained free of malignant melanoma for the next 10 years, resulting in 44% of the patients remaining clinically free from disease at 20 years (Table1). Of the 15 patients lost between 10 and 20 years, six died from malignant melanoma. Each of the six patients had followed faithfully his treatment schedule up to the time of recurrent disease. The other nine patients were clinically free of malignant melanoma when they died from other causes. Although these nine patients cannot be included in the survival data, if some of them had survived to 20 years, the tumor-free survival figure would have risen. We are not aware of any other 20-year hard data on the survival of a group of Stage III (AJCC) cases.

DISCUSSION

We compared our survival findings in this Phase II study with those in seven independent reports concurrent with our patient accrual times [23]. These historical controls included 832 patients, and survival figures in the different studies ranged from zero to 15% at 10 years following therapeutic dissections. More recent reports describe 33% survival rates at 10 years [24,25], but 20% and 24% of the patients had elective lymph node dissections. According to D. L. Morton et al [26], there is a highly significant improved survival in patients receiving elective lymphadenectomy versus those requiring therapeutic dissections. To compare therapeutic dissections with a mixed group is tenuous. Arguably, there appears to be a consensus that the natural history of malignant melanoma has not changed in recent decades [27]. Historical data, therefore, should be considered acceptable for a Phase II study. Furthermore, in accord with regulations codified by the U.S. Food and Drug Administration [(21 C.F.R. 314. 111(a) (5) (ii) (a) (4)], in a disease with high risk and predictable mortality, the results of the use of a treatment procedure may be compared quantitatively to prior historical experience in the natural history of the disease.

Aside from overall survival, two subgroups were noted in this study. Malignant melanoma showing regional lymph node disease of the head and neck is particularly serious. As pointed out by D. F. Roses [28], such involvement is almost a certain indication of systemic disease. We observed improved survival of head and neck cases in our treated patient group [29]. Because of time limits imposed by other reports, 29 months after surgery necessarily was taken as the time for comparisons. At this point, 71% of our patients evidenced disease-free survival, in contrast to 3% to 25% to at other institutions (Dr. Robert M. Byers, personal communication), and D. F. Roses et al [28]. The second group that drew our attention was patients with brain metastases. Surgical specimens from all six patients in this category showed striking inflammatory infiltration (lymphocytes, plasma cells) into the brain metastases [30], even though the immune reaction thus induced failed to eradicate the brain tumors. Any cellular immune response in a cerebral metastasis, particularly in malignant melanoma, is a rarity. Our uniform findings are statistically significant, and indicate an immune response induced by NDV oncolysate. Three of these patients died within 3 years after brain surgery, two from malignant melanoma and one from complications relating to surgery. The other three patients remained free of malignant melanoma relapse for 17 years, at which point two died from causes other than malignant melanoma, and the remaining patient continues to be free of detectable disease. These certainly are extraordinary survival times. Postsurgical life expectancy in patients with cerebral metastases usually is 2 to 7 months, with an occasional extension to 20 months [31,32] or, rarely, longer. The longest postoperative survival report in the English literature may be 14 years [33].

Evidence of favorable clinical progress in the patients prompted us to perform various immunologic tests, beginning in the early phase of this study, particularly to assess cellular responses. Marked increases in lymphocyte cytotoxicity against autogenous and allogeneic target malignant melanoma cells were observed in all treated patients [10]. Support for a critical role by activated CD-8 T-lymphocytes in the eradication of malignant melanoma came from a study by Z. L. Olkowski [34], employing human malignant melanoma transplants in nude mice. T-lymphocytes from five patients, stimulated by recombinant interleukin-2, were incubated with a human melanoma cell line (M40) for several weeks. This treatment resulted in increased lymphocyte cytotoxic activity when tested against each patient's melanoma cell line, as determined by a Cr51 release assay. Injection of the activated T-lymphocytes into nude mice bearing melanoma M40 nodules resulted in selective accumulation of the technetium 99m-labeled T-lymphocytes in the malignant melanoma nodules of the mice, providing evidence for a role of CD-8 cells in the eradication of malignant melanoma. More recently, F. M. Batliwalla et al [35] observed a marked oligoclonality of CD-8 T-lymphocytes in our patients.

Cursory examination of our 20-year data (Table 1) might suggest that because almost half the surviving patients initially had a single diseased lymph node, we were dealing with a low-risk group. If so, one must explain why 56% of the patients we lost to malignant melanoma in the control group of this study also had only a solitary diseased lymph node. Furthermore, among the patients not on study, but treated for humanitarian reasons, three had failed surgery and chemotherapy, and were regarded as hopeless cases. One patient had multiple nodules of the head and neck, with metastases to the chest wall. The other two patients had 17 and 32 diseased lymph nodes. These individuals remain free of detectable disease at 29, 21, and 20 years respectively, following surgery and oncolysate treatment.

Several staging systems exist for malignant melanoma, the primary purpose of which is to predict survival, and hopefully act as a guide in deciding on treatment procedure. Based

Table 1 Clinical Data on Stage III (AJCC) Melanoma Patients Who Survived Disease Free

Patient	Gender/ age	Location of primary lesion	Primary level/ thickness	Diseased lymph nodes	DF survival* (years)
1	F/18	Scapula	III/1.3 mm	4 (axilla)	28
2	M/26	Shoulder	IV/1.5 mm	1 (axilla)	27
3	F/30	Shoulder	IV/2.5 mm	1 (axilla)**	26
4	M/27	Interscapular	III/1.8 mm	3 (axilla)	26
5	F/36	Thigh	III/1.1 mm	1 (groin)	26
6	F/29	Calf	III/2.0 mm	1 (groin)	25
7	M/22	Scapula	IV/2.1 mm	1 (axilla)**	25
8	M/37	Unknown	Unknown	2 (axilla)	25
9	M/60	Scapula	Unknown	2 (axilla)	24
10	M/54	Subpateller	II/1.0mm	1 (groin)	24
11	M/38	Chest	III/1.5 mm	3 (axilla)**	24
12	F/22	Scapula	IV/5.5 mm	1 (axilla)**	24
13	M/43	Chest	IV/2.5 mm	1 (axilla)	24
				2 (in transit)	
14	F/41	Scapula	II/0.7 mm	2 (axilla)**	24
				2 (in transit)	
15	F/39	Unknown	Unknown	2 (neck)	24
16	M/36	Shoulder	Unknown	1 (axilla)	23
17	M/32	Subscapular	III/1.7 mm	3 (axilla)	23
18	F/52	Wrist	IV/2.6 mm	1 (axilla)	23
				7 (in transit)	
19	M/29	Unknown	Unknown	15 (axilla)	23
20	F/21	Gluteal fold	IV/5.0 mm	1 (groin)	22
21	F/27	Calf	III/1.1 mm	2 (groin)**	22
		Shoulder	II/2.3 mm	0	
22	F/45	Interscapular	Unknown	4 (axilla)	21
23	F/32	Knee	V/5.0 mm	2 (groin)**	21
24	M/54	Ear	IV/1.7mm	1 (neck)	21
25	F/49	Upper arm	IV/3.8 mm	1 (axilla)**	21
26	M/29	Shoulder	III/0.5 mm	6 (axilla)**	21
27	F/30	Calf	III/Unknown	1 (inguinal)	20
28	M/35	Thigh	IV/1.7 mm	3 (groin)	20
				Plus 8-cm mass	
29	F/25	Axilla	Unknown	4 (axilla)**	20
30	F/70	Interscapular	III/2.5 mm	1 (axilla)	20
31	M/37	Chest	III/1.1 mm	3 (axilla)	20
32	M/35	Upper arm	III/0.9 mm	1 (axilla)	20
33	M/28	Shoulder	IV/1.7 mm	1 (axilla)	20
34	M/58	Lumbar	IV/1.9 mm	5 (axilla)	20
35	M/45	Elbow	V/6.0 mm	1 (axilla)**	20
36	M/43	Forearm	III/1.2 mm	2 (axilla)	20

* Disease-free survival: Figures rounded off to the nearest year.
** Palpable lymph nodes occurring concomitantly with the primary skin lesion.

on empirical clinical observations, such systems neglect a major part of the picture, namely, the variance in virulence of the invading tumor. Historically, the six clinical variables given in Table 1 are those most frequently seen in the literature. Other criteria exist, and as the list lengthens practical application becomes increasingly unmanageable. A current report [36] addresses the consideration of 13 variables. It is difficult to imagine doing a randomized, stratified study and finding enough patients to fill all categories. Whereas large numbers of diseased lymph nodes would portend for a poor prognosis, this is not always the case. Conversely, the presence of a solitary, palpable lymph node can lead to serious consequences. The need to complement clinical observations with scientific, laboratory methods is evident. Routine information on chromosome aberrations, oncogene amplification, tumor invasiveness in animal models and tissue cultures, gene expression profiling, and measures relating to the host's immune reactions to the tumor could help in prognostication. Admittedly, these are sophisticated methodologies, but until a scientific base is coupled with clinical empiricism, prognosis will not be as precise as is claimed by staging systems currently in use. Methods for measuring the status of the patient's immune system have long been available. Some of the immune reactions are not tumor inhibiting, but can be tumor enhancing instead. For a full evaluation of the host–tumor relationship, a combination of immunity status, clinical criteria, and tumor virulence is required. Although this is not beyond possibility, it is not available or routinely practiced, currently.

CONCLUSION

In the past five years there has been an outpouring of approximately 700 publications dealing with a vaccine or other forms of adjunctive immunotherapy for the management of malignant melanoma. Innovative approaches have included heat shock proteins, anti-idiotypic antibodies, hybrid cells, bacteria or viruses as gene vectors, oncolytic viruses, peptides, epitopes, gangliosides, dendritic cells, liposome encapsulated agents, cytokines, macrophage-stimulating factor, various forms of modified tumor cells, DNA vaccines, antisense oligonucleotides, and gene therapy, to mention some of the evolving methods. We believe NDV oncolysate has provided benefits for malignant melanoma Stage III (AJCC) patients, and that it should remain in place for the management of the most rapidly increasing malignancy worldwide.

ACKNOWLEDGMENTS

Supported by the General Assembly of the Georgia State Legislature and the Goldhirsch Foundation.

This report is dedicated to the memory of Zbigniew L. Olkowski, M.D., Sc.D.

REFERENCES

1. Webb HE, Smith CEG. Viruses in the treatment of cancer Lancet 1970; 1:1206–1209.
2. Nemunaitis J. Live viruses in cancer treatment Oncology 2002; 16:1483–1492.
3. Sharpless GR, Davies MC, Cox HR. Antagonistic action of certain neurotropic viruses toward a lymphoid tumor in chickens with resulting immunity Proc Soc Exp Biol Med 1950; 78: 270–275.
4. Koprowski H, Love R, Koprowska I. Enhancement of susceptibility to viruses in neoplastic tissues Tex Rep Biol Med 1957; 15:559–576.
5. Lindenmann J, Klein PA. Viral oncolysis: Increased immunogenicity of host cell antigens associated with influenza virus J Exp Med 1967; 126:93–108.
6. Svet-Moldavsky GJ, Hamburg VP. An approach to the immunological treatment of tumors by artificial hetrogenization. In: Harris JC, Ed. Specific Tumor Antigens (UICC Monograph Series, vol 2). Copenhegen: Munksgaard, 1967:323–327.

7. Kobayashi H, Kodama T, Gotha E. Xenogenization of tumor cells Hokkaido Univ Med Libr Ser 1977; 9:1–24.
8. Boone CW, Paranjpe M, Orme T, Gillette R. Virus-augmented tumor transplantation antigens: Evidence for a helper antigen mechanism Int J Cancer 1974; 13:543–551.
9. Sinkovics JG, Gonzales F, Campos LT, Kay HD, Shirato E, Gyorkey F. Immunology of human sarcomas Abstracts of the Annual Meeting of the American Society of Microbiology 1974:206.
10. Cassel WA, Murray DR, Torbin AH, Olkowski ZL, Moore ME. Viral oncolysate in the management of malignant melanoma. I. Preparation of the oncolysate and measurement of immunologic responses Cancer 1977; 40:672–679.
11. Austin FC, Boone CW. Virus augmentation of the antigenicity of cell extracts Adv Cancer Res 1979; 30:301–345.
12. Cassel WA, Garrett RE. Newcastle disease virus as an antineoplastic agent Cancer 1965; 18:863–868.
13. Cassel WA, Garrett RE. Tumor immunity after viral oncolysis J Bacteriol 1966; 92:792.
14. Merigan TC, De Clereq E, Finkelstein MS, Clever L, Walker S, Waddell DJ. Clinical studies employing interferon inducers in man and animals Ann NY Acad Sci 1970; 173:746.
15. Lorence RM, Rood PA, Kelly KW. Newcastle disease virus as an antineoplastic agent: Induction of tumor necrosis factor and augmentation of its cytotoxicity J Natl Cancer Inst 1988; 80: 1305.
16. Nossal GJV. The mechanism of action of antigens Australas Ann Med 1965; 14:321.
17. Notkins AL, Mergenhagen SE, Rizzo AA, Scheele C, Waldman TA. Elevated gamma-globulin and increased antibody production in mice infected with lactic dehyrrogenase virus J Exp Med 1966; 123:349–364.
18. Weir DM. The immunological consequences of cell death Lancet 1967; 2:1071–1073.
19. Schirrmacher V, Ahlert T, Heicappell R, Appelhaus B, Hoegen P. Successful application of non-oncogenic viruses for antimetastatic cancer immunotherapy Cancer Rev 1986; 5:19–49.

20. Lindenmann J, Klein PA. Immunological aspects of viral oncolysis Recent Results Cancer Res 1967; 9:1–84.

21. Horak A, Horvath J, Sinkovics JG, Pendleton S, Pritchard M. Antiviral and antimelanoma immune reactions in patients actively immunized with melanoma-NDV oncolysates, St. Joseph's Cancer Institute and University of South Florida Melanoma Conference, 1998, Tampa, FL.

22. Murray DR, Cassel WA, Torbin AH, Olkowski ZL, Moore ME. Viral oncolysate in the management of malignant melanoma. II. Clinical studies Cancer 1977; 40:680–686.

23. Cassel WA, Murray DR. A ten-year follow-up of stage II malignant melanoma patients treated postsurgically with Newcastle disease virus oncolysate Med Oncol 1992; 9:169–171.

24. Callery C, Cochran AJ, Roe DL, Rees W, Nathanson SD, Benedetti JK, Elashoff RM, Morton DL. Factors prognostic for survival in patients with malignant melanoma spread to regional lymph nodes Ann Surg 1982; 196:69–75.

25. Bevilacqua RG, Coit DG, Rogatko A, Younes RN, Brennan MF. Auxillary dissection in melanoma. Prognostic variables in node-positive patients Ann Surg 1990; 212:125–131.

26. Morton DL, Wanek L, Nizze JA, Elashoff RM, Wong JH. Improved long-term survival after lymphadenectomy of melanoma metastatic to regional nodes Ann Surg 1991; 214:491–499.

27. Morton DL, Foshag LJ, Hoon DSB, Nizze JA, Wanek LA, Chang C, Davtyan DG. Prolongation of survival in metastatic melanoma after active specific immunotherapy with a new polyvalent melanoma vaccine Ann Surg 1992; 215:463–482.

28. Roses DF, Harris MN, Grunberger I. Selective surgical management of cutaneous melanoma of the head and neck Ann Surg 1980; 192:629–632.

29. Cassel WA, Murray DR. Viral oncolysate in the treatment of regional metastases of melanoma. In: Larson DL, Ballantyne AJ, Guillamondegui OM, Eds. Cancer in the Neck. New York: Macmillan Publishing Company, 1986:235–242.

30. Cassel WA, Weidenheim KM, Campbell WG, Murray DR. Inflammatory mononuclear cell infiltrates in cerebral metastases

during concurrent therapy with viral oncolysate Cancer 1986; 57:1302–1312.

31. Madajewicz S, Karakousis C, West CR, Caracandas J, Avellanosa AM. Malignant melanoma brain metastases. Review of Roswell Park Memorial Institute experience Cancer 1984; 53: 2550–2552.
32. Balch CM, Milton GW, Shaw HM, Soong S. Cutaneous Melanoma. Philadelphia: JB Lippincott, 1985.
33. McCann WP, Weir BKA, Elvidge AR. Long-term survival after removal of metastatic malignant melanoma of the brain J Neurosurg 1968; 28:483–487.
34. Olkowski ZL. The role of human 99_mTC labeled cytotoxic lymphocytes in radioimmunodetection and potential radioimmunotherapy of malignant melanoma Immunoconjugates Radiopharmaceuticals 1991; 4:55.
35. Batliwalla FM, Bateman BA, Serrano D, Murray DR, Macphail S, Maino VC, Ansel JC, Gregersen PK, Armstrong C. A 15-year follow-up of AJCC stage III malignant melanoma patients treated with Newcastle disease virus (NDV) oncolysate and determination of alterations in the CD8 T cell repertoire Mol Med 1998; 4:783–794.
36. Sprakel B, Stenschke F, Unnewehr M, Ladas A, Senninger N. Prognosis of malignant melanoma following dissection of regional lymph node metastases Der Chirurg 2003; 74:55–60.

10

Vaccinia Viral Lysates in Immunotherapy Of Melanoma

PETER HERSEY
Oncology & Immunology Unit
Newcastle, NSW, Australia,
Peter.Hersey@newcastle.edu.au

ABSTRACT

Melanoma continues to be difficult to treat once it has spread from its primary site in the skin. There is much evidence that immune responses against the tumor play an important role in control of the disease. Experimental evidence suggested that viral lysates of tumor cells may be effective vaccines for treatment of tumors. In view of this, we developed a vaccine from vaccinia viral lysates of a single-cultured melanoma line and tested its effectiveness in treatment of American Joint Committee on Cancer (AJCC) Stage IIB and III melanoma in a series of Phase II studies and a randomized control trial. We verified the immunogenicity of the vaccine and its safety in patients with melanoma. The Phase II studies showed impressive gains in survival compared with historical controls. The Phase III study was confounded by much longer survival of all patients in the study, which meant that the number of events was relatively low and the trial was consequently underpowered to detect the relatively small differences seen between treated and control groups. There was an improvement in the hazard of dying from melanoma of 20% but the confidence limits included 1, indicating the result may have been attributable to chance. Median survivals of control and treated groups were 88 and 151 months, respectively. There was an approximate 10% and 20% improvement in survival at 5 and 10 years, respectively, but the best P value was 0.068. The apparent benefit from this treatment may therefore have been attributable to chance but, if true, the results from the trial are superior to those from other randomized vaccine trials. Treatment was not associated with any significant side effects. Approaches to improve treatment results include co-treatment with agents that may sensitize melanoma to apoptosis and control of regulatory T cells in patients with melanoma.

INTRODUCTION

Patients with thick primary melanoma or regional lymph node (LN) metastases are known to be at high risk of disease recurrence and death. Previously published estimates of survival over a 5-year period approximate 63% to 67% in patients with primary melanoma 4 mm or greater in thickness (AJCC Stage IIB) and 46% to 69% in those with regional LN metastases (AJCC Stage III disease) [1–3]. Median survivals for patients with Stage IIB and III disease in three recent trials were 34 [4], 41.3 [5], and 72 months [6]. A number of different adjuvant therapies given after surgery have not been shown to reduce the risk of recurrence or death in such patients. These include chemotherapy with Dimethyl, triazero, imidazole carboxamide (DTIC) with or without immunotherapy with bacillus Calmette-Guerin [7] and recombinant interferon-α2a (IFN-α2a) given at relatively low doses (3 million units subcutaneously three times a week) for 3 years [8,9]. Trials of IFN-α2b at higher doses conducted by the Eastern Cooperative Oncology Group (ECOG) in the United States have given inconclusive results, with one trial (ECOG 1684) showing clear benefit of relapse-free (RFS) and overall survival (OS) [4], whereas a subsequent trial (ECOG 1690) showed improved RFS but no improvement in OS [6]. The most recent trial (ECOG 1694) showed benefit in terms of RFS and OS in comparison with immunotherapy with a ganglioside GM2/keyhole limport hemocyanin (GM-2/KLH) vaccine. An untreated control group was not included in the latter study.[10].

Following the disappointing results from adjuvant therapy trials of nonspecific immunotherapy with agents such as Bacillus Calmette Guérin (BCG) and *Corynebacterium parvum* [11], much interest was generated in specific immunotherapy using autologous or allogeneic melanoma cells, and particularly in the use of viral lysates. The basis for using an immunological approach in therapy was the frequent occurrence of regression of primary melanoma [12] associated with lymphocytic infiltration into the tumor [13,14] and many in vitro studies showing antibody and T-cell responses to melanoma [15,16]. The use of viral lysates as a melanoma vaccine was well founded by prior studies in animal tumor models, where it was shown that viral infection of cancer cells increased their immunogenicity [17,18]. Augmentation of immunity required that the tumor cells be infected with the virus, and separate administration of virus and tumor cells was not effective. H. Fujiwara [19] and Y. Shimizu [20] demonstrated that vaccinia virus generated T cell help against the hapten, trinitrophenyl and against syngeneic murine tumors.

VACCINIA VIRUS IN IMMUNOTHERAPY OF MELANOMA

Vaccinia virus is a member of the poxviruses with a nucleoprotein core and coat material containing at least seven major antigens. Infection with the virus induces both antibody and T cell responses, and life-long immunity. On entry into the cell, the viral coat is digested and early and late mRNA are transcribed within 7 hours. Production of host proteins in the cell is stopped during synthesis of the virus and transcription ceases by approximately 7 hours. The host cells usually undergo lysis with release of viral particles, but some viral particles can leave intact infected cells [21]. Viral proteins are expressed on the cell surface by 1 to 2 hours after infection and may induce killing of the target cell by cytotoxic T cells [22,23]. The presence of viral antigens in the host cell membranes and organelles is believed responsible for induction of strong helper activity for induction of immune responses to weak tumor antigens [24].

Studies in melanoma patients by J. C. Belisario and G. W. Milton [25] showed that injection of vaccinia virus into cutaneous metastases in seven patients resulted in regression of the metastases. Systemic responses against noninjected nodules were also seen in five of the patients. K. H. Burdick and W. A. Hawk [26] reported that intralesional injection of vaccinia virus was associated with regression of both cutaneous and regional LN metastases. Another three patients showed temporary improvement and five showed no response. Similar studies were reported by H. H. Roenijk et al [27]. J. D. Everall et al [28] injected vaccinia virus into primary melanoma 14 days prior to surgical excision and reported that 23 patients so treated had improved RFS compared with 25 patients treated by surgical excision alone. M. K. Wallack et al [29] used vaccinia viral lysates of cancer cells to immunize patients. Studies in 29 patients with various cancers showed the procedure to be safe and to be associated with lack of disease progression in nine patients. Antibodies against melanoma lines were induced in melanoma patients undergoing this treatment [30]. Phase I and II studies by the Southeastern Oncology Group were conducted in 48 patients with resected high-risk primary and lymph node metastases using a vaccine prepared from vaccinia virus-induced lysates of four melanoma cells [31]. These studies led to a randomized controlled trial on 217 patients that had its final report in 1998 [5]. There was no significant increase in overall ($p = 0.79$) or disease-free survival ($p = 0.61$) in the patients treated with the vaccine.

In view of the promising experimental and clinical results from earlier studies on vaccinia virus, we initiated in 1984 a series of Phase II studies in patients with AJCC Stage IIB and III melanoma to determine whether immunization with vaccinia melanoma cell lysates (VMCL) over a 2-year period may have therapeutic efficacy. The first of these trials indicated a highly significant difference in survival for 80 patients treated with VMCL compared with the survival of 151 historical and 56 concurrent nonrandomized controls [32]. The second Phase II study examined whether pretreatment of patients with cyclophosphamide improved results any further. This was prompted by previous reports that pretreatment with cyclophosphamide increased cell-mediated immunity to an autologous melanoma vaccine [33]. Again, the survival of 102 patients treated with VMCL plus cyclophosphamide was significantly above that of the historical controls but was no better than the group of patients treated with VMCL alone [34]. The vaccine appeared highly immunogenic in terms of antibody production in the patients [35,36] and electron microscopy confirmed the particulate nature of the vaccine [35] and the presence of viral material in the membrane fragments used.

The results from the Phase II studies prompted a randomized, controlled trial, which was initiated in 1988 within Australia to test the efficacy of the vaccine against a control group treated only by surgery. An interim analysis of this trial [37] after accrual of 569 patients in 1996 indicated a difference in favor of the VMCL-treated group that was not significant.

RESULTS FROM THE RANDOMIZED TRIAL ON VACCINIA MELANOMA CELL LYSATES VACCINE IN TREATMENT OF STAGE IIB AND III MELANOMA

Preparation of VMCL for vaccine production was carried out by methods similar to those described by M. K. Wallack et al [31] except that one melanoma line was used instead of four melanoma lines as used by Wallack. The vaccine contained the centrifuged material from the vaccinia viral lysate of MM200 melanoma cells. The particulate nature of the vaccine is shown in Figure 1. The line chosen contained the main melanoma differentiation antigens known to induce responses against melanoma. Prior studies on antibody responses to melanoma had shown that sera from melanoma patients frequently reacted to this particular melanoma line [15,38]. Further details of vaccine production, patient selection, study design, and statistical analysis of the trial are described elsewhere [39].

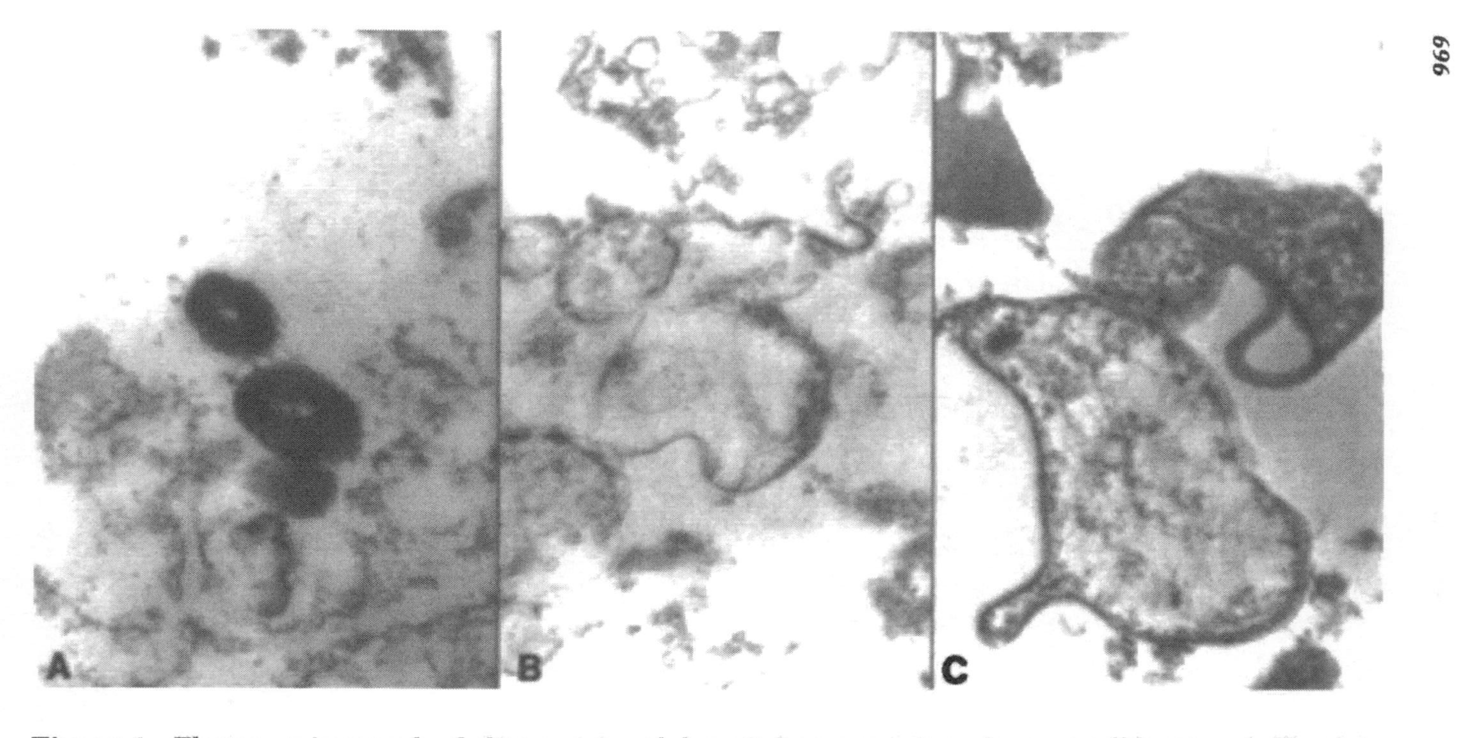

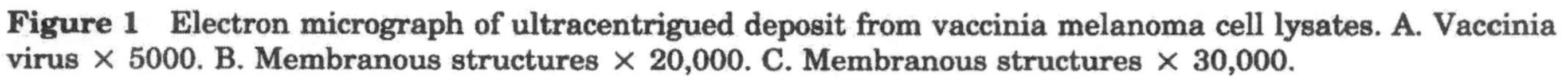

Figure 1 Electron micrograph of ultracentrigued deposit from vaccinia melanoma cell lysates. A. Vaccinia virus × 5000. B. Membranous structures × 20,000. C. Membranous structures × 30,000.

Seven hundred patients were accrued into the study from four main centers in Australia. Median follow-up at the time of analysis was 8 years. Accrual into the different prognostic strata is shown in Figure 2 and patient characteristics are summarized in Table 1. Twenty-five patients were considered ineligible for various reasons. These were evenly distributed between the groups.

Relapse-free survival and OS were as shown in Table 2. Overall survival by Kaplan-Meier estimates is shown for treated and control groups in Figure 3. Median OS for the control group was 88 months (Figure 3A) and approximately 151 months for the treated group. The univariate stratified analysis of OS after exclusion of the 25 ineligible and two untreated patients showed a P value of 0.068. No multivariate analysis model indicated statistical significance for treatment allocation. The point estimate of the hazard ratio (HR) (0.81) indicated approximately a 20% reduction in risk of dying from

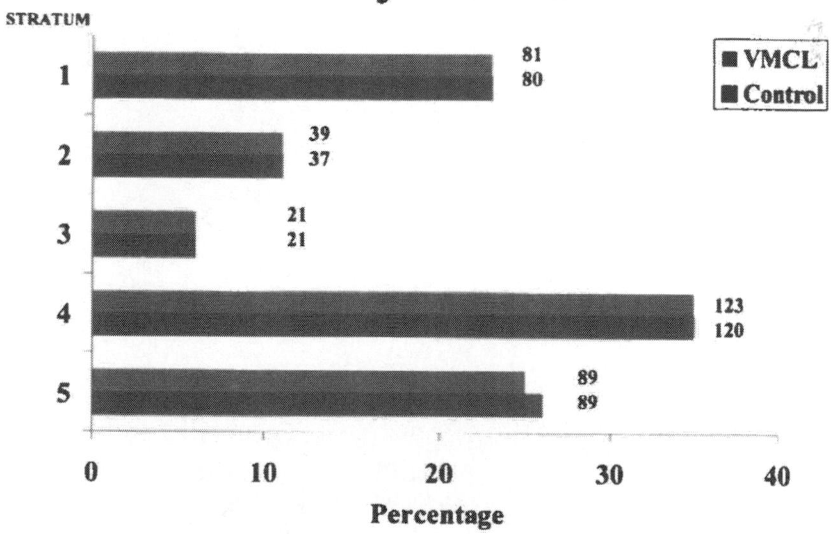

Figure 2 The number and percentage of patients accrued into each stratum. Sixty percent of the patients (stratum 4 and 5) had detection of nodal metastases 8 weeks or more after removal of the primary.

Table 1 Patient Characteristics

	n	Median age (years)	Gender		Median tumor thickness (mm)	Ulceration*
			M	F		
All patients (48%)	700	51	471 (67%)	217 (33%)	2.7	335/70
VMCL (48%)	353	51	222 (63%)	125 (37%)	2.7	168/35
Control (48%)	347	50	249 (71%)	92 (29%)	2.7	167/34

* Ulceration was not recorded in 78 and 83 of the patients in vaccinia melanoma cell lysate (VMCL) and Control Group, respectively. These were recorded as *not* ulcerated in the table.

Table 2 Relapse-Free and Overall Survival

	Total	Alive	Dead	Median RFS* (years)	Median OS** (years)	5 Years		10 Years	
						RFS %	OS %	RFS %	OS %
Eligible Patients									
All Patients 48.4	673	379	294	4.4	9.3	49.1	57.7	43.9	
VMCL 53.4	338	201	137	6.9	12.6	50.9	60.6	45.4	
Control	335	178	157	3.6	7.3	46.8	54.8	42.5	41

Stratified Cox regression univariate analysis of treatment.
Eligible Patients
* P = 0.17; HR = 0.86; CI = 0.70–1.07.
** P = 0.068; HR = 0.81; CI = 0.64–1.02.
VMCL = vaccinia melanoma cell lysate; RFS and OS = relapse-free and overall survival; HR = hazard ratio; CI = 95% confidence intervals.

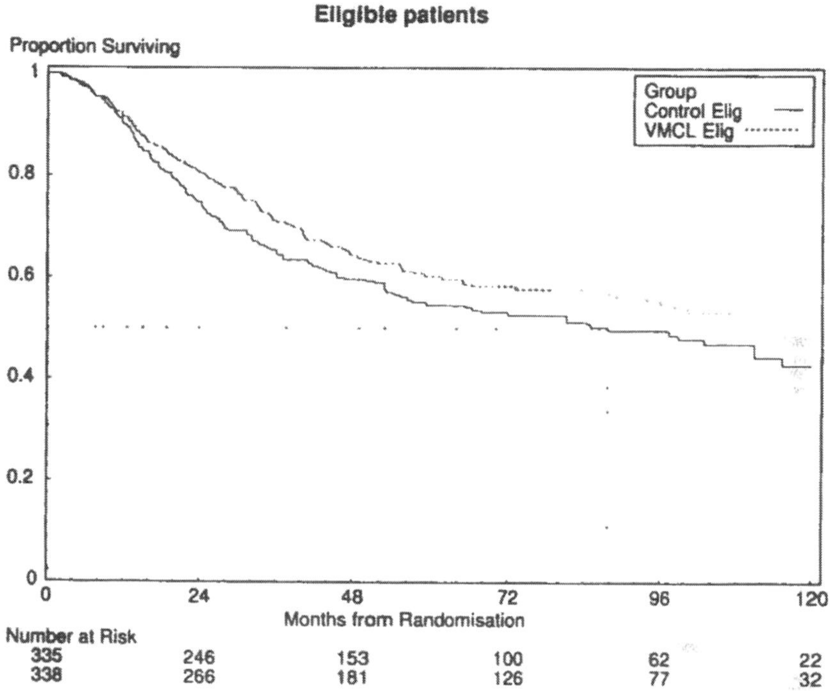

Figure 3 Kaplan Meier estimates of overall survival of eligible patients. The numbers at risk are indicated for control (top line) and vaccinia melanoma cell lysate-treated patients.

melanoma in the treated group. The 95% confidence limits, however, included 1.0 (Table 2). At 5 years and 10 years, there were 5.8% and 12.4% survival difference respectively in favor of the treated group. The median RFS were 43 and 83 months in the control and treated groups respectively ($p = 0.17$). The treatment was not associated with significant toxicity.

DISCUSSION OF THE RESULTS

The results indicate that immunotherapy with VMCL was associated with a nonsignificant trend to improvement in OS that amounted to a 20% reduction in the risk of dying when assessed by hazard ratios. The confidence limits, however, included 1.0, indicating this may have been a chance finding. The lack of any significant effect on RFS makes this a possible explanation of the results. Nevertheless, the size of the estimated effects on survival (as illustrated by an increase in estimated median survival from 88 months to 151 months or a 10% increase in absolute survival at 5 years) would, if true, be important. Similarly, the treated group had a 14% reduction in the HR for development of recurrences that, if carried through to a similar reduction in the risk of death, would represent worthwhile clinical benefit. When viewed in these terms and given the negligible toxicity of the vaccine, a useful clinical benefit from the vaccine remains a possibility. Even though the trial included approximately 700 patients and had a median follow-up of 8 years, the number of deaths (n = 294) was relatively low and the results need to be reviewed after a longer follow-up. Subset analysis did not reveal benefit from treatment in any particular group of patients. The best interpretation of the results is that immunization with the vaccine produced a true increase in survival of approximately 20%. The relatively low statistical significance of the difference (p = .068) could be because of the wide variation in survival of individual patients and the resulting wide confidence limits.

Nevertheless, the therapeutic results were much less than expected based on the Phase II studies, in which the median survival of the historical and nonrandomized control group was 32 months and the median survival of the treated group had not been reached with a minimum follow-up period of 5 years [32]. The main reason for this may be the improved survival of all patients in the study compared with the historical controls. This improvement in survival of patients occurred throughout the duration of the trial and Cox regression analysis showed that this was a more powerful factor than the effect of treatment. It is not uncommon for the control group in Phase III studies to have much better survivals than the control groups in Phase II studies. Nevertheless, the improved survival of the control patients on this study appears remarkable, given that at the commencement of this study, the published 5-year survival rates for patients with AJCC Stage IIB and Ill disease ranged from 23 to 42 months and median survivals

Table 3 Median Overall Survival of Patients in Control and Treated Groups and Comparison of Key Prognostic Factors in Recent Melanoma Vaccine and Interferon Adjuvant Trials

			Median survivals (months)		T4 N0	Median tumor	*Ulceration*	1LN	≥2LNs
	Stage	n	Controls	Treated	(%)	thickness (mm)	(%)	(%)	%
VMCL Phase II (1987)	IIB,III	80		>60	10	2.0	26	45	45 (1.0)*
Historical Controls (SMU) (1980–84)	IIB,III	151	32		2	1.98	31	38	60 (.63)
Wallack et al (1995)	III	217	41	50**	0	—	37	58	44 (1.3)
VMCL Phase III (2000)	IIB,III	700	88	151**	23	2.7	48	45	32 (1.4)
WHO No 16 (1996)	III	427	31	36**	0	—			
ECOG 1684 (1996)	IIB,III	280	34	46**	11		16.8	89	
ECOG 1690 (2000)	IIB,III	642	72	6**	26		38	33	41 (0.8)

* Figures in brackets indicate proportion of patients with l involved LN.
** P values for OS were 0.79, 0.068, 0.50, 0.024, and 0.74 for studies by Wallack et al, vaccinia melanoma cell lysates (VMCL) Phase III. Eastern Cooperative Oncology Group (ECOG) 1684 and ECOG 1690, respectively.
??? WHO = World Health Organization.

ranged from 32 to 36 months (7,40–42). In more recent studies, listed in Table 3, the median survivals of patients in control groups in the ECOG 1684 [4] and World Health Organization 16 [8] IFNα-2a trials were 34 and 31 months, respectively, and 41 months in the immunotherapy trial of M. K. Wallack et al [5]. The most recent ECOG 1690 IFNcx-2a trial, however, reported a median survival of 72 months, which was closer to the 88 months recorded in the present study [6]. One of the practical consequences of the improved survival of the patients in the study was the impact it had on the trial design, in that many more patients were needed to detect a difference in survival. The study was therefore underpowered to detect small differences in survival.

Despite the relatively small benefits from the treatment with VMCL, it is worth noting that the results (if not attributable to chance) are the best yet reported from randomized trials of melanoma vaccines in the adjuvant setting. In the H. K. Wallack study, the differences in median survivals of treated (50.2 months) and control (41 .3 months) groups were quite small and not significant [5]. The vaccine and the protocol used by Wallack et al were different from those used in our studies and the control group was given vaccinations against vaccinia alone. Whether the latter may have had some beneficial effects is unknown but cannot be discounted. The GM2/KLH vaccine developed by Livingston et al was found to be less effective than high-dose IFN [10]. An allogeneic vaccine, "Melacine," given with an adjuvant "Detox" to patients with AJCC Stage IIA melanoma also failed to have any effect on disease-free survival [43]. Subgroups of patients in the latter study who were human leukocyte antigen-A2 or -C3 had highly significant prolongation of disease-free survival [44].

WHY WERE THE VACCINIA VIRAL LYSATE VACCINES NOT MORE EFFECTIVE?

Much more is now known about melanoma antigens and design of vaccines than when these studies were started in 1984. Viral lysates, however, continue to have many properties that make them attractive as vaccines; e.g., the antigens are particulate rather than soluble so they are more likely to be taken up by antigen-presenting cells, such as dendritic cells (DCs). The vaccines were given intradermally, so it is likely that DCs took up the lysates and transported them to the regional LNs. Increased size and tenderness of regional LNs draining the vaccination site were common in the patients, especially in the first 6 to 8 weeks, suggesting that stimulation of regional LNs occurred. Studies since 1984 have also shown that T cell responses may be directed to antibody (Th2) rather than representing cell-mediated (Th1) responses. Antibody-related responses may be more frequent following repeated antigenic stimulation, as may occur in tumor-bearing patients. Cell-mediated responses are more likely in response to viruses such as vaccinia, which could be expected to preferentially favor TH1 IC responses to the vaccine. Unfortunately, these studies were initiated before assays to measure Th1 and Th2 responses were available, so whether TH1 responses were initiated by VMCL injections is not known. It is clear, however, that the vaccine induced immunoglobulin-G lymphocyte-dependent antibody responses against melanoma in the patients [24].

Studies on antibody responses showed that induction of the response was rapid, with peak titers at approximately 4 weeks, but following this, antibody titers declined (Figure 3). The reason for this is not clear but raises the possibility that vaccine administration induced regulatory T cells, as reported by others [45]. Induction of the latter is poorly understood but we have shown that heavy-chain ferritin from melanoma cells may induce regulatory T cells [46,47]. They may also be induced during immunization with self-antigens and during frequent vaccine administration, as would have occurred in this study. Future studies may need to pay more attention to the activity of these cells and methods to inhibit their activity, particularly their production of interleukin-10.

Another question is whether the vaccinia MM200 lysate contained melanoma antigens known to be associated with cytotoxic T lymphocytes (CTL) cell responses against melanoma. Studies in animal models suggest that antigens specific to individual tumors may be the most effective for rejection of tumors and some recent vaccine trials have used autologous vaccines in the hope of generating such responses. These antigens would not have been present in the VMCL vaccine or in other allogeneic melanoma cell vaccines. Nevertheless, the vaccine did contain the common melanoma antigens known to be frequently recognized by T cells. These included the differentiation antigens tyrosinase, gpl00, and MART-I, and the tumor-specific antigens MAGE-A3, MAGE-Al0, BAGE, GAGE, and XAGE. It did not contain MAGE-l, CT7, NA17-l, or NY-ESO-l (unpublished data). Studies on a limited number of patients treated with VMCL off study have shown induction of

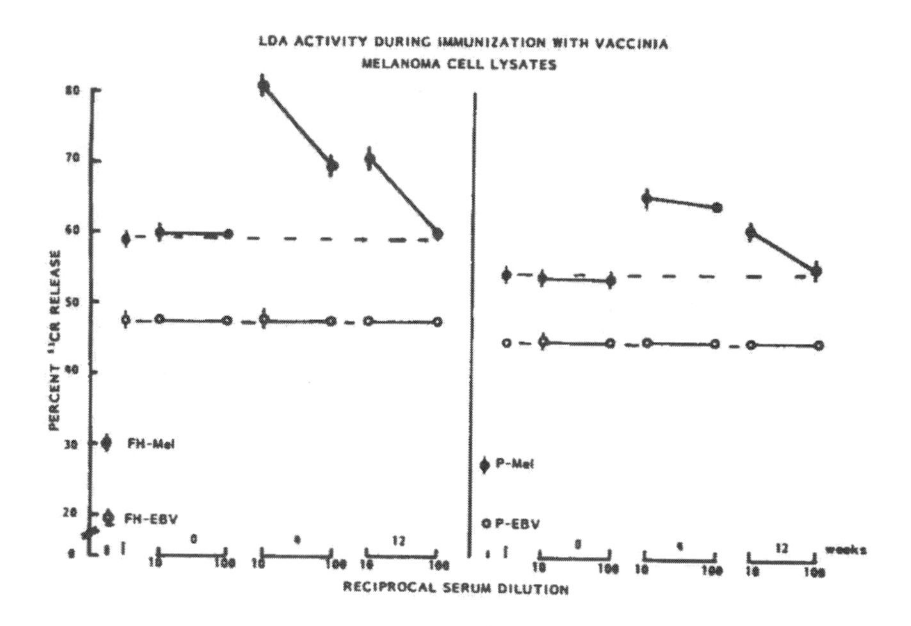

Figure 4 LDA activity of sera taken from patient J.Bl, before, and at 4 and 12 weeks during immunizations with vaccinia melanoma cell lysate. FH-Mel and P(MP)-Mel are two allogeneic melanoma cells and FH-EBV and P-EBV are Epstein-Barr virus (EBV)-transformed B cells from these patients. Baseline release from the target cells alone are shown by the target cell designations and the ^{51}Cr release in the presence of the effector cells caused by their natural killer activity is shown by the dashed lines. ^{51}Cr release above these lines in the presence of (heat-activated) sera is attibutable to LDA activity. Human leukocyte antigen phenotype of Mel-MP was A.24. −: B 7.60: DR 2.4 and EBV-FHA 2.24: B 35.60: Dr 4. −.

CTL activity against gp100 280, MART-1, tyrosinase, and MAGE-3A2 (as shown in Figure 5). Vaccinia melanoma cell lysates were at least as effective as the human leukocyte antigen-A2 epitope peptides from these antigens in inducing responses against these melanoma epitopes. The figure shows, however, that IFN-γ production was not induced in all patients. Unfortunately, measurement of T cell responses against these antigens was not possible during the course of the VMCL trial, so we were unable to correlate CTL responses or lack of them with the clinical course of individual patients.

MORE GENERAL ISSUES IN USE OF MELANOMA VACCINES

Recent studies suggest that the number of T cells against melanoma may be increased by immunizing patients after lymphocyte depletion [48,49]. Our phase II study after relatively low doses of oral cyclophosphamide did not show any benefit in patient survival, but this issue needs to be studied more carefully. Adoptive transfer studies with cloned T cells or tumor-infiltrating lymphocytes have shown that migration of T cells into the tumor is limited and fewer than 0.005% of transferred lymphocytes actually reached the tumor. Methods to increase T cell infiltration are being explored. Lymphocyte migration is a two-step process, with the first step being adhesion of lymphocytes to ligands on endothelial cells induced by inflammation, such as intercellula adhesion molecule (ICAM-1) binding to selectins [50]. The second step is migration in response to chemokines. W. H. Kershaw et al [51] suggested that growth-regulated oncogene alpha [52] is produced selectively by melanoma and that transfection of its chemokine receptor CXCR2 into T cells may aid localization in melanoma.

Apart from issues related to vaccine design and administration, we have been increasingly interested in the hypothesis that melanoma in patients has been selected by the immune system to be resistant to host defenses. The escape mechanisms may be multiple and vary among patients and different tumors in patients [53,54]. The immune system appears to kill tumor cells by induction of apoptosis and induction of resistance to apoptosis in melanoma cells may be a fundamental

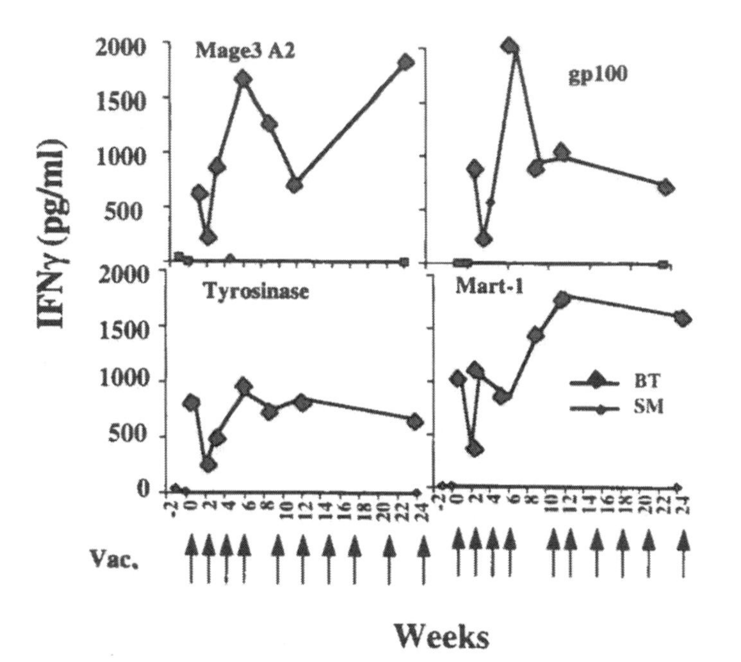

Figure 5 Interferon (IFN)-γ production from blood lymphocytes of two patients during immunotherapy with vaccinia melanoma cell lysates vaccine. PBL from BT but not SM showed good responses to peptide epitopes from MAGE3-A2, gp100 280-9V and MART-1. PBL (5×10^5) were incubated for 20 hours with 5×10^5 T2 target cells that had been pretreated with the peptides at 10 μg/mL for 2 hours then washed two times. Supernatants were assayed for IFN-γ using the IFN-γ Eliza kit from Pharmingen (Catalog No. 554548). Peptides used were MAGE3-A2 FLWGPRALV, gp100 280-9V YLEPGPVTV, MART-1 ELAGIGILTV, and Tyrosinase YMDGTMSQV (unpublished data).

escape mechanism [55]. Our studies on the mechanism of resistance of melanoma cells to apoptosis induced by tumor necrosis factor-related apoptosis-inducing ligand have revealed several different mechanisms, such as loss of death receptors or inhibition by intracellular Bcl-2 family proteins and inhibitor-of-apoptosis family proteins [54–56]. Therapeutic approaches to overcome these apoptosis-resistance mechanisms are becoming available and may be valuable adjuncts to treatment with melanoma vaccines.

In conclusion, the use of viral lysates to treat patients with melanoma may be as effective as more recent vaccine approaches to induce immune responses against melanoma. Limitations of the vaccines may be those that apply to the efficacy of tumor vaccines in general. New approaches might include lymphocyte depletion to increase the number of specific memory T cells and reduce or control regulatory T cells. In addition, imaginative approaches may be needed to facilitate homing of T cells to the site of small, extravascular, "immunologically silent" foci of melanoma cells. Combination of these approaches with methods to sensitize melanoma cells to apoptosis may result in much-needed improvement in treatment of the disease.

REFERENCES

1. Balch CM, Buzaid AC, Soong SJ, Atkins MB, Cascinelli N, Coit DG, Fleming ID, Gershenwald JR, Houghton A, Kirkwood JM, McMasters KM, Mihm MF, Morton DL, Reintgen DS, Ross MI, Sober A, Thompson JA, Thompson JF. Final version of the American Joint Committee on Cancer staging system for cutaneous melanoma J Clin Oncol 2001; 19:3635–3648.
2. Balch CM, Soong S-J, Gershenwald JE, Thompson JF, Reintgen DS, Cascinelli N, Urist M, McMasters KM, Ross ML, Kirkwood JM, Atkins MB, Thompson JA, Coit DG, Byrd D, Desmond R, Zhang Y, Liu P-Y, Lyman GH, Morabito A. Prognostic factors analysis of 17,600 melanoma patients: validation of the American Joint Committee on Cancer Melanoma Staging System J Clin Oncol 2001; 19:3622–3634.
3. Dickler MN, Coit DG, Meyers ML. Adjuvant therapy of malignant melanoma Surg Oncol Clin N Am 1997; 6:793–812.
4. Kirkwood JM, Strawderman MH, Ernstoff MS. Interferon-alpha-2b adjuvant therapy of high-risk resected cutaneous melanoma: the Eastern Cooperative Oncology Group Trial EST 168 J Clin Oncol 1996; 14:7–17.
5. Wallack MK, Sivanandham M, Balch CM, Urist MM, Bland KI, Murray D, Robinson WA, Flaherty L, Richards JM, Bartolucci

AA, Rosen L. Surgical adjuvant active specific immunotherapy for patients with stage III melanoma: the final analysis of data from a phase III. randomized, double-blind, multicenter vaccinia melanoma oncolysate trial J Am Coll Surg 1998; 187: 69–77.

6. Kirkwood JM, Ibrahim JG, Sondak VK, Richards J, Flaherty LE, Ernstoff MS, Smith TJ, Rao U, Steele M, Blum RH. High- and low-dose Interferon alpha-2b in high-risk melanoma: first analysis of intergroup trial E1690/S91 I 1/C9190 J Clin Oncol 2000; 18:2444–2458.
7. Veronesi U, Adamus J, Aubert C, Bajetta E, Beretta G, Bonadonna G, Bufalino R, Cascinelli N, Cocconi G, Durand J, De Marsillac J, Ikonopisov RL, Kiss B, Lejeune F, MacKie R, Madej G, Mulder H, Mechl Z, Milton GW, Morabito A, Peter H, Priario J, Paul E, Rumke P, Sertoli R, Tomin R. A randomized trial of adjuvant chemotherapy and immunotherapy in cutaneous melanoma N Engl J Med 1982; 307:913–916.
8. Cascinelli N. Evaluation of efficacy of adjuvant rIFN 2A in melanoma patients with regional node metastases Proc American Society for Clinical Oncology (ASCO) 1995; 14:A129.
9. Cascinelli N, Bufalino R, Morabito A, MacKie R. Results of adjuvant interferon study in WHO melanoma programme Lancet 1994; 343:913–914.
10. Kirkwood JM, Ibrahim JG, Sosman JA, Sondak VK, Agarwala SS, Ernstoff MS, Rao U. High-dose interferon alfa-2b significantly prolongs relapse-free and overall survival compared with the GM2-KLH/QS-21 vaccine in patients with resected stage jib-iii melanoma: results of intergroup trial e1694/s9512/c509801 J Clin Oncol 2001; 19:2370–2380.
11. Hersey P, Balch CM. Current status and future prospects for adjuvant therapy of melanoma Aust N Z J Surg 1984; 54: 303–315.
12. McGovern VJ. Spontaneous regression of melanoma. Pathology 1975; 7:91–9.
13. Hersey P, Murray E, Grace J, McCarthy WH. Current research on immunopathology of melanoma: analysis of lymphocyte populations in relation to antigen expression and histological features of melanoma Pathology 1985; 17:385–391.

14. Brocker EB, Zwadlo G, Holzmann B, Macher E, Sorg C. Inflammatory cell infiltrates in human melanoma at different stages of tumor progression Int J Cancer 1988; 41:562–567.

15. Hersey P, Edwards A, Murray E, McCarthy WH, Milton GW. Prognostic significance of leukocyte-dependent antibody activity in melanoma patients Journal of the National Cancer Inst. 1983; 71:45–53.

16. Platsoucas CD. Human autologous tumor-specific T cells in malignant melanoma Cancer Met Rev 1991; 10:151–176.

17. Lindenmann J. Viruses as immunological adjuvants in cancer Biochim Biophys Acta 1974; 355:49–75.

18. Austin FC, Boone CW. Virus augmentation of the antigenicity of tumor cell extracts Adv Cancer Res 1979; 30:301–345.

19. Fujiwara H, Shimizu Y, Takai Y, Wakamiya N, Ueda S, Kato S, Hamaoka T. The augmentation of tumor-specific immunity by virus help. II. Demonstration of vaccinia virus-reactive helper T cell activity involved in enhanced induction of cytotoxic T lymphocyte and antibody responses Eur J Immunol 1984; 14: 171–175.

20. Shimizu Y, Fujiwara H, Ueda S, Wakamiya N, Kato S, Hamaoka T. The augmentation of tumor-specific immunity by virus help. II. Enhanced induction of cytotoxic T lymphocyte and antibody responses to tumor antigens by vaccinia virus-reactive helper T cells Eur J Immunol 1984; 14:839–843.

21. Davis BD. Variola (Smallpox) and vaccinia. In: Davis BD, Dulbecco R, Eisen HN, Ginsberg HS, Eds. Microbiology. Maryland: Harper & Row, 1980:1079–1091.

22. Mallon V, Domber EA, Holowczak JA. Vaccinia virus proteins on the plasma membranes of infected cells. II. Expression of viral antigens and killing of infected cells by vaccinia virus-specific cytotoxic T cells Virology 1985; 145:1–23.

23. Wilton S, Gordon I, Dales S. Identification of antigenic determinants by polyclonal and hybridoma antibodies induced during the course of infection by vaccinia virus Virology 1986; 148: 84–96.

24. Mitchison NA. Immunologic approach to cancer Transplant Proc 1970; 2:92–103.

25. Belisario JC, Milton GW. The experimental local therapy of cutaneous metastases of malignant melanoblastomas with cow pox vaccine or colcemid (demecolcine or omaine) Aust J Dermatol 1961; 6:113–118.

26. Burdick KH, Hawk WA. Vitiligo in a case of vaccinia virus-treated melanoma Cancer 1964; 17:708–712.

27. Roenigk HH, Deodhar S, St. Jacques R, Burdick K. Immunotherapy of malignant melanoma with vaccinia virus Arch Dermatol 1974; 109:668–673.

28. Everall JD, O'Doherty CJ, Wand J, Dowd PM. Treatment of primary melanoma by intralesional vaccinia before excision Lancet 1975; 27:583–587.

29. Wallack MK, Steplewski Z, Koprowski H, Rosato E, George J, Hulihan B, Johnson J. A new approach in specific, active immunotherapy Cancer 1977; 39:560–564.

30. Wallack MK, Michaelides M. Serologic response to human melanoma lines from patients with melanoma undergoing treatment with vaccinia melanoma oncolysates Surgery 1984; 96: 791–800.

31. Wallack MK, McNally KR, Leftheriotis E, Seigler H, Balch C, Wanebo H, Bartolucci AA, Bash JA. A southeastern cancer study group phase I/II trial with vaccinia melanoma oncolysates Cancer 1986; 57:649–655.

32. Hersey P, Edwards A, Coates A, Shaw H, McCarthy W, Milton G. Evidence that treatment with vaccinia melanoma cell lysates (VMCL) may improve survival of patients with stage II melanomia Cancer Immunol Immunother 1987[??]; 25:257–265.

33. Berd D, Maquire HC, Mastrangelo MJ. Induction of cell-mediated immunity to autologous melanoma cells and regression of metastases after treatment with a melanoma cell vaccine preceded by cyclophosphamide Cancer Res 1986; 46: 2572–2577.

34. Hersey P. Active immunotherapy with viral lysates of micrometastases following surgical removal of high risk melanoma World J Surg 1992; 16:251–260.

35. Hersey P, Edwards A, D'Alessandro G, MacDonald M. Phase II study of vaccinia melanoma cell lysates (VMCL) as adjuvant

to surgical treatment of stage II melanoma Cancer Immunol Immunother 1986; 22:221–231.

36. Hersey P, Werkman H, Edwards AE. Western blot analysis of serological responses following immunization with vaccinia viral lysates of melanoma cells Int J Cancer 1990; 46:612–617.
37. Hersey P, Coates A, McCarthy WH. Interim analysis of a randomized trial of immunotherapy with vaccinia melanoma cell lysates (VMCL) following surgical removal of high risk melanoma. Abstract Proc Am Assoc Cancer Res 1996; 37:489.
38. Hersey P, Edwards AE, Murray E, McCarthy WH, Milton GW. Sequential studies of melanoma leukocyte-dependent antibody activity in melanoma patients Eur J Cancer 1978; 14:629–637.
39. Hersev P, Coates AS, McCarthy WH, Thompson JF, Sillar RW, McLeod R, et al. Adjuvant immunotherapy of patients with high risk melanoma with vaccinia viral lysates of melanoma. Results of a randomized trial J Clin Oncol 2002; 20:4181–4190.
40. Balch CM, Soong SI, Murad TM, Ingalls AL, Maddox WA. A multifactorial analysis of melanoma: III. Prognostic factors in melanoma patients with lymph node metastases (stage II) Ann Surg 1981; 193:377–388.
41. Balch CM, Soong SJ, Shaw HM, Milton GW. An analysis of prognostic factors in 4000 patients with cutaneous melanoma. Cutaneous Melanonia. Editor Balch C.M., Milton G.W. Philadelphia: Lippincott, 1985:338.
42. Cascinelli N, Vaglini M, Nava M. Prognosis of skin melanoma with regional node metastases J Surg Oncol 1984; 25:240–247.
43. Sondak VK, Liu P-Y, Tuthill RJ. Adjuvant therapy of intermediate-thickness node-negative melanoma with an allogeneic tumor vaccine Ann Surg 2002; 20:2067–2075.
44. Sosman JA, Unger JM, Liu P-Y, Flaherty LE, Park MS, Kempf RA, Thompson JA, Terasaki PI, Sondak VK. Adjuvant immunotherapy of resected, intermediate-thickness, node-negative melanoma with an allogeneic tumor vaccine: II. Impact of HLA class I antigen expression on outcome J Clin Oncol 2002; 20: 2067–2075.
45. Javia LR, Rosenberg SA. CD4+CD25+ suppressor lymphocytes in the circulation of patients immunized against melanoma antigens J Immunother 2003; 26:85–93.

46. Gray CP, Franco AV, Arosio P, Hersey P. Immunosuppressive effects of melanoma-derived heavy-chain ferritin are dependent on stimulation of IL-10 production Int J Cancer 2001; 92: 843–850.
47. Gray CP, Arosio P, Hersey P. Heavy chain ferritin activates regulatory T cells by induction of changes in dendritic cells Blood 2002; 99:3326–3334.
48. Dudley ME, Wunderlich JR, Robbins PF, Yang JC, Hwu P, Schwartzentruber D, Topalian SL, Sherry R, Restifo NP, Hubicki AM, Robinson MR, Raffeld M, Duray P, Seipp CA, Rogers-Freezer L, Morton KE, Mavroukakis SA, White DE, Rosenberg SA. Cancer regression and autoimmunity in patients after clonal repopulation with antitumor lymphocytes Science 2002; 298:850–854.
49. Asavaroengchai W, Kotera Y, Mule JJ. Tumor lysate-pulsed dendritic cells can elicit an effective antitumor immune response during early lymphoid recovery Proc Natl Acad Sci U S A 2002; 99:931–936.
50. Schadendorf D, Heidel J, Gawlik C, Suter L, Czarnetzki BM. Association with clinical outcome of expression of VLA-4 in primary cutaneous malignant melanoma as well as P-selectin and Eselectin on intratumoral vessels J Natl Cancer Inst 1995; 87: 366–371.
51. Kershaw MH, Wang G, Westwood JA, Pachynski RK, Tiffany HL, Marincola FM, Wang E, Young H, Murphy PM, Hwu P. Redirecting migration of T cells to chemokine secreted from tumors by genetic modification with CXCR2 Hum Gene Ther 2002; 13:1971–1980.
52. Haghnegahdar H, Du J, Wang D, Strieter RM, Burdick MD, Nanney LB, Cardwell N, Luan J, Shattuck-Brandt R, Richmond A. The tumorigenic and angiogenic effects of MGSA/GRO proteins in melanoma J Leukoc Biol 2000; 67:53–62.
53. Hersey P. Impediments to successful immunotherapy Pharmacol Ther 1999; 81:111–119.
54. Hersey P. Advances in the non-surgical treatment of melanoma Expert Opin Investig Drugs 2002; 11:75–85.
55. Hersey P, Zhang XD. How melanoma cells evade TRAIL-induced apoptosis Nature Rev 2001; 1:142–150.
56. Hersey P, Zhang XD. Overcoming resistance of cancer cells to apoptosis J Cell Physiol 2003; 196:9–18.

11

Fusogenic Oncolytic Herpes Simplex Viruses for Therapy of Solid Tumors

XINPING FU, MIKIHITO NAKAMORI, and XIAOLIU ZHANG
Center for Cell and Gene Therapy, Departments of Pediatrics, Molecular Virology and Microbiology, Baylor College of Medicine
Houston, Texas

HERPES SIMPLEX VIRUS INFECTION AND REPLICATION: AN OVERVIEW

Herpes simplex virus type 1 (HSV-1) is a member of the family Herpesviridae, which includes other well-known human herpesviruses, such as HSV type 2, cytomegalovirus, varicella zoster virus, and Epstein-Barr virus. Among them, HSV-1 is the most extensively studied and therefore the more thoroughly understood human herpesvirus [1]. During natural infection, HSV-1 usually enters its host through skin or mucosal surfaces. The HSV envelope, a lipid bilayer derived from host cell membrane, contains at least 11 viral glycoproteins [2]. For most herpesviruses, less than half of the viral glycoproteins are thought to participate in viral entry. The initial attachment of HSV to cells is mediated by interaction of glycoprotein C (gC) and/or glycoprotein B (gB) with cell-surface glycosaminoglycans, primarily heparan sulfate [3]. This initial virus-cell contact is not sufficient to trigger virus entry, but apparently is required to position the virus for interaction with receptors recognized by glycoprotein D (gD), including HVEM (a member of the tumor necrosis factor-receptor family) and nectin-1 or nectin-2 (two related members of the immunoglobulin superfamily). The binding of gD to one of its receptors, with the participation of gB and gH-gL heterodimers, activates pH-independent viral membrane fusion, leading directly to internalization of the viral nucleocapsid and tegument into the cell cytoplasm.

After productive infection and local replication at the mucosal surfaces, the virus enters sensory nerve endings and then migrates by retrograde axonal transportation to the neuronal cell bodies [4–6]. Here, a more restricted replication cycle occurs, most often culminating in latent infection of these neurons [7,8]. Once latency is established, viral gene expression becomes severely restricted and viral replication ceases, with only the latency-associated transcripts being abundantly transcribed [9]. During latency, the viral genome exists predominantly in an episomal form within the nuclei of neurons [10], and the ends of the virion genomic DNA are covalently joined [11–13]. A variety of stimuli can reactivate the latent virus, which then travels via the axons to the mucosal surfaces, where it produces a new round of infection.

Like that of many DNA viruses, the transcriptional program of HSV-1 is a regulated cascade in which early and late phases of gene expression are separated by viral DNA synthesis [14]. The immediate early genes are mainly involved in gene regulation, whereas the products of early genes are essentially required for HSV DNA replication. Both immediate and early genes are transcribed before the initiation of viral DNA replication. Late genes, which largely function as structural proteins, are expressed at high levels only after viral DNA replication has taken place. Late transcripts can be further categorized as leaky-late, which are readily detectable before the onset of viral genome replication, or strict late, which are reliably detectable only after the onset of viral DNA replication [15–17].

The HSV genome is a 152-kilobase (kb) double-stranded DNA molecule with approximately 80 open reading frames [18]. It comprises two covalently linked segments, designated as long (L) and short (S). Each segment consists of unique sequences (U_L and U_S), each bracketed by inverted repeats [19]. There are three internal origins of replication along the viral genome, one located in the long segment (OriL) and a diploid origin (OriS) in the repeated region bracketing the short segment [1]. Seven of the gene products encoded by the virus are necessary and sufficient for origin-dependent viral DNA replication [20–23]. They include a DNA polymerase (UL30), a protein designated ICP8 (UL29) that binds to single-stranded DNA, an origin-binding protein (UL9) that binds specifically to multiple sites at or near the replication origins of the viral genome, a double-strand DNA-binding protein (UL42) that also forms complexes with the DNA polymerase, and three additional proteins (UL5, UL9, and UL52) that form a complex in equimolar ratios and function as a primase or helicase in the presence of UL29. During HSV replication, the termini of the input liner viral genome fuse soon after infection. The circularized DNA molecules then serve as templates for the synthesis of concatemeric DNA, by a rolling-circle mechanism. Viral assembly begins in the nuclei of infected cells, where unit-length progeny genomes are thought to be cleaved from the concatemeric intermediates and packaged into viral capsids. Mature viruses are formed when these capsids traffic through the cytoplasm and bud into an organelle, such as the Golgi network, and acquire their final envelope [24].

DEVELOPMENT OF ONCOLYTIC HERPES SIMPLEX VIRUS FOR TUMOR THERAPY

Human viruses, including HSV, have the natural ability to efficiently infect and kill target cells. Because viruses can be genetically engineered to kill tumor cells while sparing normal cells, they afford an attractive approach to antitumor therapy. Oncolytic viruses have two principal advantages. First, unlike conventional chemotherapy and radiotherapy, they specifically target cancer cells because of their restricted ability to replicate in normal cells [25–27]. Second, compared with replication-incompetent vectors, oncolytic viruses can propagate from initially infected tumor cells to surrounding tumor cells, thereby achieving a large volume of distribution and enhanced antitumor effects.

HSV has been modified for oncolytic purposes, most commonly by deleting viral genes necessary for efficient replication in normal (nondividing) cells but not tumor cells. Deletion of the viral *r34.5* gene, which functions as a neurovirulence factor during HSV infection [28], blocks viral replication in nondividing cells [29–31]. The viral *ICP6* gene encodes the large subunit of ribonucleotide reductase, which generates sufficient deoxynucleoside 5'-triphosphate (dNTP) pools for efficient viral DNA replication [32–34], and is abundantly expressed in tumor cells but not in nondividing cells. Consequently, viruses with a mutation in this gene can preferentially replicate in and kill tumor cells. The oncolytic HSV G207, which has been extensively tested in animal studies and is currently in clinical trials, harbors deletions in both copies of the *r34.5* locus and an insertional mutation in the *ICP6* gene by the *Escherichia coli lacZ* gene [35–37]. Alternatively, an oncolytic HSV can be constructed by using a tumor-specific promoter to drive *r34.5* or other genes essential for HSV replication [38].

Compared with other viruses that have been investigated for oncolytic purposes, HSVs possess several unique features that enhance their potential as antitumor agents. First, antiherpetic medications such as acyclovir and gancyclovir are available as safety measures in the event of undesired infection or toxicity from the HSV. Second, productive infection by

HSV usually kills target cells much more rapidly than infection by other viruses. For example, a HSV can form visible plaques in cultured cells in only 2 days, in contrast to 7 to 9 days for an adenovirus. In vitro studies have also shown that, at a multiplicity of infection of 0.01, a HSV can kill almost 100% of cultured cancer cells in 2 days [39], whereas a much higher dose or a longer infection time is required to achieve equivalent cell killing with an adenovirus [40]. Rapid replication and spreading among target cells may be vital properties allowing a virus to execute its full oncolytic potential in vivo because the body's immune mechanism may be more likely to restrict the spread of slower-growing viruses. Third, HSV seems to be able to replicate and spread even in the presence of anti-HSV immunity. This feature has been most clearly demonstrated during recurrent HSV infection, in which the anterogradely transported virus can still grow and spread extensively in the local skin area of infection, despite obvious antiviral immunity. Moreover, pre-existence of anti-HSV immunity in experimental animals has no significant effect on the therapeutic potency of oncolytic HSVs administered either intratumorally or systemically [41–43]. Fourth, HSVs have very wide cell tropism, infecting almost every type of human cell tested so far. Thus, oncolytic viruses derived from HSV would likely have wide applicability among cancer patients. Finally, the risk of introducing an insertional mutation during HSV oncolytic therapy appears minimal because HSVs generally do not integrate into cellular DNA.

Oncolytic HSVs were initially designed and constructed for the treatment of brain tumors—glioblastomas in particular [37,44]. Subsequently, they have proved to be effective in treating a variety of other human solid tumors, including breast cancer [39,45]. The safety of the oncolytic virus G207 has been extensively tested in mice [46] and a primate species (*Aotus*) that is extremely sensitive to HSV infection [47,48]. These studies have confirmed that oncolytic HSVs are safe for in vivo administration. These encouraging results in animals have prompted clinical trials of these viruses in patients with malignant gliomas [49,50].

STRATEGIES TO ENHANCE THE POTENCY OF ONCOLYTIC HERPES SIMPLEX VIRUSES

Despite encouraging preclinical results, the available evidence indicates that current oncolytic viruses, although safe, may have only limited antitumor activity on their own. For example, although such viruses inhibited tumor growth and improved the survival of experimental animals bearing human tumor xenografts [35,37,43,45,51], they effected cure in only a fraction of the animals. In ongoing Phase I and II clinical trails, several oncolytic viruses have been well-tolerated and have produced some evidence of antitumor effect. However, it is evident that the therapeutic efficacy of these viruses must be improved before significant changes in the tumor growth curve, and hence prognosis, are seen [49,50,52]. The suboptimal efficacy of oncolytic viruses may be attributable to viral gene deletions, which can reduce the replicative potential of the virus in tumors. Deletion of the *γ34.5* gene, for example, significantly reduces virus replication even in rapidly dividing cells [31,53–55]. Thus, strategies to enhance the potency of current oncolytic viruses appear warranted.

Attempts to improve the antitumor activity of oncolytic HSVs have included their combined administration with radiotherapy and chemotherapy [33,56–59] and insertion of immune modulator genes such as *IL-4, IL-12*, and *B7.1* into the viral genome [60–62]. Less successful have been efforts to directly increase oncolytic potency by combining the virus with genes encoding prodrug converting enzymes, such as HSV thymidine kinase (TK) [63,64], most likely because the antitumor effect of prodrug activation is offset by its inhibitory action on viral replication.

RATIONALES FOR DEVELOPING FUSOGENIC ONCOLYTIC HERPES SIMPLEX VIRUSES FOR TUMOR THERAPY

Several viruses kill their target cells through the formation of multinucleated syncytia, a process that involves membrane fusion between the infected and uninfected cells. The viral components contributing to syncytia formation are mainly the fusogenic membrane glycoproteins (FMGs). An attractive antitumor strategy therefore would be to incorporate syncytia-forming capability into an oncolytic HSV, so the virus would kill tumor cells by two efficient and complementary mechanisms: direct cytolysis (through virus replication) and cell membrane fusion. This approach might even yield a synergistic antitumor effect, given that syncytia formation in the tumor tissue can facilitate the spread of virus, leading, in turn, to widespread syncytia formation. Although wild-type HSV isolated from patients does not cause significant cell fusion in vitro, cells infected with certain spontaneously occurring syncytial mutants fuse extensively, either with each other or with uninfected cells [65,66]. Analysis of these syncytial mutants has shown that the syncytia formation results from aberrant expression of several viral glycoproteins, such as gB and gK [66–69]. It is therefore possible that an oncolytic virus capable of conferring a syncytial phenotype could be selected from a well-characterized oncolytic HSV. Furthermore, because certain viral glycoproteins (including gB and gK) are encoded by late genes (γ), the expression of which depends upon viral DNA replication, an oncolytic virus expressing these proteins might retain the safety of the original virus because syncytia formation occurs in tumor cells (where the virus can undergo a full infectious cycle), but not in normal nondividing cells (where viral replication is restricted and little glycoprotein is expressed).

Alternatively, a fusogenic oncolytic HSV can be constructed by inserting a fusogenic membrane glycoprotein from another virus into an oncolytic HSV genome, preferably under the control of a tumor-selective promoter. The resultant virus may have advantages over fusogenic viruses directly selected from an HSV. That is, syncytia formation mediated by FMGs relies on the initial binding of these glycoproteins with their specific receptors on target cells, which then induces ordered structural changes of the membrane lipid bilayers, leading to lipid mixing and eventually to membrane fusion either between viral and cellular membranes or among cellular membranes [70]. Because the molecular requirements for viral–cellular membrane fusion (a prerequisite for virus entry into target cells) and cell–cell membrane fusion (for syncytia formation) are essentially the same, tumor cells that become resistant to virus infection by receptor downregulation or mutation may also be refractory to syncytia formation mediated by the same virus. However, when a FMG from a different virus is incorporated into an oncolytic HSV, it uses a receptor that differs entirely from those normally recognized by HSV glycoproteins. This acquired property should reduce the emergence of therapy-resistant tumor cells because any tumor cells resistant to infection or syncytia formation mediated by HSV would still be destroyed by syncytia formation through the expression of a FMG from a different virus. This is an important consideration because, unlike xenografted tumors established from cultured tumor cells, which are relatively homogenous, the malignant cells in naturally occurring tumors may be heterogeneous in their sensitivity to HSV infection owing to variation among their cell surface receptors.

CONSTRUCTION OF FUSOGENIC ONCOLYTIC HERPES SIMPLEX VIRUSES THROUGH RANDOM MUTAGENESIS

Previous studies have shown that certain spontaneously occurring HSV syncytial mutants can be isolated in vitro, and are able to fuse extensively either with each other or with uninfected cells [65,66]. To learn whether syncytial mutants can also be selected for oncolytic purpose, X. Fu et al [39] subjected the well-characterized oncolytic HSV G207 to random mutagenesis through incorporation of the thymidine analog bromodeoxyuridine during virus replication in Vero cells. The mutagenized virus stock was then screened for the ability to confer a syncytial phenotype upon infection of Vero cells. Plaques resulting predominantly formed from syncytia formation were collected, and one isolate, designated Fu-10, consistently showed a strong syncytial phenotype after a few consecutive passages.

Phenotypic characterization showed that infection with Fu-10 caused widespread syncytia formation in a variety of tumor cells of different tissue origins, including A-549 (lung), DU-145 (prostate), HepG2 (liver), MDA-MB-435 (breast), and U-87 MG (glioblastoma). In vitro analysis also showed that Fu-10 had a significantly stronger ability to kill these tumor cells than did G207. Notably, there was a strong correlation between the intensity of syncytia formation and the degree of toxicity toward individual tumor cells during Fu-10 infection. For example, Fu-10 displayed the strongest killing effect in A-549 and HepG2 cells, which also showed the most extensive syncytia formation. In MDA-MB-435 and DU-145 cells, characterized by only modest syncytia formation, twice as many cells were killed by Fu-10 than by G207, when the viruses were used at a dose of 0.1 pfu/cell. Fu-10 showed marginally better killing than G207 in the U-87 MG cell line (28% vs. 44% survival) when the cells were infected at 0.1 pfu/cell, but did not produce any detectable cytotoxicity at a lower dose (0.01 pfu/cell) [39].

In vivo studies showed that Fu-10 is significantly more effective than the nonfusogenic G207 against established metastatic breast cancer in lung. Systemic administration of Fu-10 resulted in a significant reduction of lung tumor nodules: The average number of tumor nodules in Fu-10–injected mice was only 2.6 per animal, compared with 22.2 and 69.2 nodules in G207- and PBS- treated groups, respectively. There were no visible tumor nodules in the lungs from two of five mice in the group treated with Fu-10 [39]. These findings demonstrate that Fu-10 has a significantly enhanced oncolytic effect on disseminated metastatic breast cancer after being systemically injected into experimental animals, an outcome predicted by in vitro studies of this fusogenic virus.

CONSTRUCTION OF FUSOGENIC ONCOLYTIC HERPES SIMPLEX VIRUS BY INSERTION OF A HYPERFUSOGENIC ENVELOPE PROTEIN OF GIBBON APE LEUKEMIA VIRUS INTO THE VIRAL GENOME

Earlier investigations of FMGs for use in cancer treatment relied on viruses whose infections naturally induce syncytia formation in target cells. These studies demonstrated that, in principle, administration of these viruses could lead to significant oncolysis; in some cases, tumor destruction by the viruses subsequently elicited potent antitumor immune responses [71]. More recently, A. Bateman et al [72] reported that FMGs delivered either as plasmid DNA or by viral vectors are also extremely cytotoxic to tumor cells. An example is provided by a C-terminal truncated form of the gibbon ape leukemia virus envelope glycoprotein (GALV.fus), which lacks the 16-amino-acid R-peptide of the wild-type protein [72,73], that restricts fusion of the envelope until it is cleaved during viral infection [74]. Hence, truncation of the R-peptide from this glycoprotein leads to a constitutive and hyperfusogenic version of GALV.fus [75]. The cellular receptor for GALV.fus has been identified as Pit-1, a type III sodium-dependent phosphate transporter, and is abundantly distributed on the surface of most mammalian cells [76,77]. Transduction of the *GALV.fus* gene into a range of human tumor cells efficiently kills the cells through syncytia formation [78], which destroys neighboring cells through a "bystander" effect at least 10-fold greater than that observed with HSV-TK or cytosine deaminase-mediated therapy [78,79]. The full therapeutic potential of *GALV.fus* will probably not be realized until means are devised for its efficient delivery and specific expression in tumor cells.

We hypothesized that this feat might be achieved by inserting the *GALV.fus* gene into an oncolytic HSV, because oncolytic HSV can specifically replicate and spread in tumor cells. However, because expression of *GALV.fus* can also cause syncytia formation in normal cells, uncontrolled expression of the gene, even in the context of a tumor-restricted oncolytic virus, still poses a safety concern. This is particularly true when systemic administration is required in cases of metastatic diseases, for example. One way to minimize this risk is to use a tumor- or tissue-specific promoter to control *GALV.fus* expression. Although many tumor- or tissue-specific promoters have been described, they generally have much lower activity than viral promoters and may lose their tissue specificity once they are cloned into viral vectors [80]. Herpes simplex virus genome transcription is a tightly regulated molecular cascade in which early and late phases of gene expression are separated by viral DNA replication [81]. In particular, some of the late transcripts can be characterized as strict-late, the expression of which depends rigorously on the initiation of viral DNA replication. It is therefore possible that directing *GALV.fus* expression with such a strict-late viral promoter, in the context of an oncolytic HSV, might lead to high-level and selective *GALV.fus* gene expression in tumor cells. It is expected that such a strict-late viral promoter will be extremely active in tumor tissue, where the oncolytic virus can fully replicate, but silent in normal cells if these are nondividing or postmitotic, because viral replication would be limited [82].

Earlier attempts to clone the *GALV.fus* gene into an oncolytic adenovirus were not successful [79]. Two reasons may have caused this failure. First, the rapid and widespread formation of syncytia after *GALV.fus* gene transfection may have interfered with the homologous recombination needed for the genome recombinant of virus. Second, expression of *GALV.fus* in the context of an adenovirus may have interfered with the infectious process of the virus. Unlike adenoviruses, which are nonenveloped, HSVs are enveloped viruses, infection with which naturally involves membrane fusion. We therefore anticipated that HSVs would be able to withstand the membrane-fusion effect from genes such as *GALV.fus* and continue to grow. To avoid potential interference from the initial membrane fusion of *GALV.fus* gene transfection, we used an enforced direct-ligation strategy, which does not require homologous recombination for the cloning of *GALV.fus* gene into the genome of an oncolytic HSV. As illustrated in Figure 1, fHSV-delta-pac is a bacterial artificial chromosome (BAC)-based construct that contains a mutated HSV genome in which the diploid gene encoding r34.5 was partly deleted, and both copies of HSV packaging signal (pac) were completely deleted [83]. Hence, transfection of this construct into HSV-permissive cells will not form any infectious virus unless an intact HSV pac is provided in *cis*. Also, any virus generated from this construct will be replication selective because of partial deletion of the *r34.5* gene. The gene cassettes were initially linked with a HSV packaging sequence and were then inserted into the

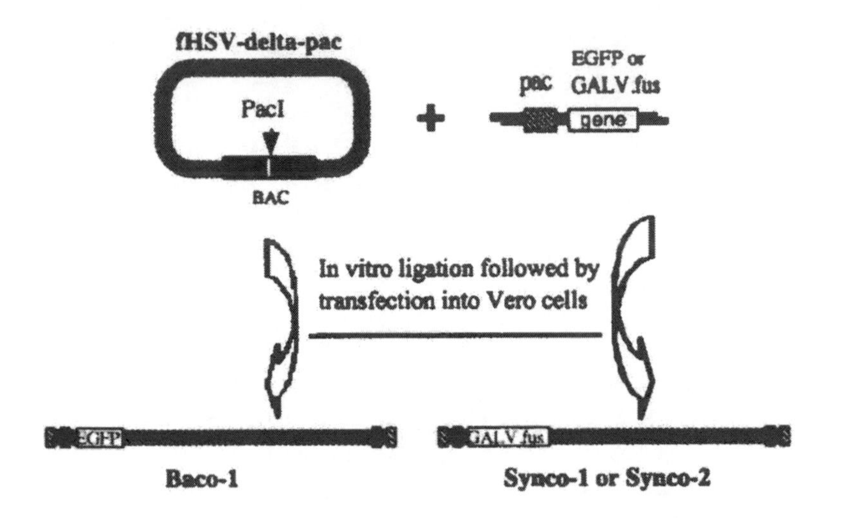

Figure 1 Schematic illustration of the enforced ligation strategy for insertion of foreign DNA into an oncolytic Herpes simplex virus (HSV). The HSV DNA sequence is represented by hatched areas and the bacterial artificial chromosome (BAC) sequence by black areas (not proportional) in the fHSV-delta-pac. The HSV packaging signal (pac) and the gene cassette (either *EGFP* or *GALV.fus*) in a DNA fragment excised from shuttle plasmids are each individually labeled. The gene cassettes together with the pac were ligated onto the BAC–HSV construct in vitro before the ligation mixture was transfected into Vero cells for the generation of infectious viruses.

unique PacI site located in the BAC sequence of fHSV-delta-pac. The ligated DNA was directly transfected into Vero cells, to generate Baco-1 (containing *EGFP*), Synco-1 (containing *GALV.fus* driven by the IE promoter of CMV), and Synco-2 (containing *GALV.fus* driven by the UL38 promoter of HSV).

Phenotypic characterization of these fusogenic oncolytic HSVs showed clear syncytia formation in all the tumor cells tested, including Hep 3B (liver), DU-145 (prostate), and U87 (breast). The extent of tumor cell fusion did not differ appreciably between Synco-1 and Synco-2. In contrast, the same types of tumor cells infected with Baco-1 lacked this phenotype. In vitro cell killing assay showed that *GALV.fus*-mediated syncytia formation in the context of oncolytic HSV replication increased tumor cell killing in these tumor cells [84].

Two experiments were then conducted to determine whether *GALV.fus*-mediated syncytia formation after Synco-2 infection truly depends on viral DNA replication. First, human tumor cells were infected with either Synco-1 or Synco-2 in the presence or absence of acyclovir, which inhibits HSV DNA replication. Syncytia formation by Synco-1 in Hep 3B cells was largely unaffected by the presence of acyclovir, whereas cell membrane fusion by Synco-2 in the same cells was completely blocked by this drug. Second, to directly test whether Synco-2 loses its ability to cause syncytia formation in normal nondividing human cells, primary human fibroblasts in either a quiescent or cycling state were infected with one of the viruses. The results demonstrated that infection with either Synco-1 or Synco-2 caused syncytia formation in these normal human cells when they were in cycle. Synco-1–mediated cell fusion was only marginally affected in cells whose cycling was slowed by serum starvation or completely arrested by lovastatin. Synco-2–mediated cell fusion, on the other hand, was absent in cells whose cycling was decreased or arrested [84]. These results demonstrate that syncytia formation induced by *UL38* promoter-controlled *GALV.fus* expression in the context of an oncolytic HSV is cell-cycle-dependent.

Subsequent animal experiments indicated that the syncytial phenotype and the enhanced ex vivo tumor cell killing shown by Synco-1 and Synco-2 translate into an enhanced antitumor effect in vivo. Both Synco-1 and Synco-2 have significantly better antitumor effects than Baco-1, starting from week 2 after virus administration. However, there was no significant difference in the tumor growth ratio between Synco-1– and Synco-2–treated mice [84]. These results indicate that Synco-2 has equivalent antitumor potency to Synco-1, even though its *GALV.fus* expression is driven by a conditional viral promoter. Histological examination revealed that syncytia comprising a variable number of cell nuclei were frequently encountered across the tissue section of tumors injected with either Synco-1 or Synco-2 virus. Such syncytia were not seen in tissue sections of tumors injected with either Baco-1 or PBS.

The envelope glycoprotein of human immunodeficiency virus type 1 (HIV-1), which can induce syncytia in the presence of CD4+ target cells, has recently been cloned into a conditionally replicating adenovirus. In vitro characterization of this virus (Ad5HIVenv) in permissive (CD4-positive) cells showed that syncytium formation induced by this FMG significantly increased the dispersion of viral gene products within the cytoplasm of the syncytial as well as viral particles in the nuclei of the syncytial mass. In addition, progeny virions were released more efficiently from syncytia compared with nonsyncytial cells, facilitating cell–cell spread of the virus particles [85]. These observations are consistent with results obtained from characterization of fusogenic oncolytic HSVs [39], indicating that an oncolytic adenovirus containing a FMG with a wider tropism, if successfully constructed, may also have enhanced antitumor potency in vivo.

CONSTRUCTION OF A FUSOGENIC ONCOLYTIC HERPES SIMPLEX VIRUS THAT CONTAINS BOTH FUSION MECHANISMS

To further increase the potency of the oncolytic virus, we recently constructed a newer version of fusogenic HSV by incorporating both membrane fusion mechanisms into a single oncolytic HSV, generating Synco-2D. This was achieved through a two-step procedure. Initially, membrane fusion capability was introduced into Baco-1, the nonfusogenic oncolytic HSV, through random mutagenesis [39]. The virus was phenotypically identified and purified to homogeneity. Then the hyperfusogenic *GALV.fus* gene, driven by the strict-late promoter of the *UL38* gene, was cloned into the BAC-based viral genome through an enforced ligation strategy as described earlier, to replace the enhanced green fluorescent protein gene of Baco-1 [86].

The antitumor potency of Synco-2D has been assessed in two tumor models: (1) metastatic ovarian cancer in the abdominal cavity [86] and (2) prostate cancer in both the orthotopic site and in the lung [89]. Peritoneal invasion of ovarian cancer is a common and serious clinical problem. It has been reported that about 70% of late-stage ovarian cancer patients have metastatic disease in the peritoneal cavity [87]. We therefore chose a peritoneal metastasis model (xenografted Hey-8 cells) as a means to test the efficacy of Synco-2D against human ovarian cancer. Our results demonstrated that intraperitoneal administration of Synco-2D at a moderate dose at a site distant from that of tumor cell implantation has a dramatic therapeutic effect on established ovarian cancer, rendering 75% of animals tumor free. By contrast, none of the animals treated with Baco-1 was tumor-free, and most of the mice in the phosphate-buffered saline (PBS) control group died or became ill and were euthanized during the experiment.

The objective of evaluating Synco-2D in a prostate cancer model was twofold. First, we wished to determine whether this doubly fusogenic oncolytic HSV could infect and kill metastatic prostate tumor, which is incurable with conventional therapies. Second, the model afforded the opportunity to compare doubly fusogenic and singly fusogenic HSVs in terms of antitumor efficacy. Both primary and metastatic prostate cancer xenografts were initially established in severe combined immune-deficient mice through orthotopic and systemic injection of PC-3M-Pro4 cells, which were selected from PC-3M cells through repeated cycles of orthotopic inoculation and harvest in athymic mice, and shown to efficiently establish lung metastases after i.v. injection into immune-deficient mice [88]. This tumor xenograft model, representing both primary and metastatic prostate cancer, is more relevant than subcutaneously implanted tumor cells to advanced disease in patients. Our results showed that systemic delivery of Synco-2D had a significantly stronger therapeutic effect than Synco-2 on growth of prostate cancer confined to the primary site. Systemic administration of Synco-2D also produced strikingly better therapeutic effects on lung metastases, indicating that the doubly fusogenic oncolytic HSV is more potent than the singly fusogenic virus, as predicted by its increased activity against cultured tumor cells.

SAFETY OF FUSOGENIC ONCOLYTIC HERPES SIMPLEX VIRUSES

Although oncolytic HSVs replicate only in tumor cells, the syncytia formation by fusogenic oncolytic HSVs is potentially toxic to normal cells and, if uncontrolled, can pose a safety concern, especially the virus is given systemically. During construction of Synco-2 and Synco-2D, several strategies were employed to restrict the syncytia formation to tumor tissues. First, a strict-late viral promoter, UL38p, was used to direct *GALV.fus* gene expression in both Synco-2 and Synco-2D. The activity of UL38p remains confined to the tumor tissue after systemic administration of an oncolytic HSV [82]. Thus, it is expected that GALV.fus-mediated syncytia formation by these two viruses would be conditionally linked to virus replication in tumor cells. Second, syncytia formation from mutagenized HSV is mainly attributable to aberrant expression of several viral glycoproteins, such as gB and gK [66–69]. Because these glycoproteins are encoded by late genes, the expression of which depends upon viral DNA replication, the cell-membrane fusion mediated by this mechanism would occur in tumor cells (where virus can undergo a full infection cycle) but not in normal nondividing cells (where the virus replication is restricted and very low levels of glycoproteins are expressed). Our findings in previous studies, that blocking viral DNA replication completely abolishes the syncytia-forming ability of Fu-10 (the selected syncytial mutant from G207) [39] and that Synco-2 (containing *GALV.fus* driven by UL38p) cannot induce syncytia formation in nondividing cells [84], strongly suggest that Synco-2D retains the safety profile of a conventional oncolytic HSV. More comprehensive toxicity studies of these fusogenic oncolytic HSVs in sensitive animals are needed to confirm their safety for clinical use.

SUMMARY

Oncolytic viruses have shown considerable promise in the treatment of solid tumors, but their potency needs to be improved if their full clinical potential is to be realized. Recent studies by us and others have shown that the addition of cell membrane fusion activity to an oncolytic virus represents an effective strategy for enhancing the antitumor potency of the virus. In particular, doubly fusogenic oncolytic HSVs have shown increased potency in the treatment of both primary and metastatic tumors after local or systemic administration. Compelling demonstration of the safety of these fusogenic viruses in sensitive animals, including nonhuman primates, is needed to ensure their approval for clinical testing. Most intriguing, perhaps, is the strong likelihood that successful treatment of one type of human solid tumor using fusogenic oncolytic HSVs will translate into successful therapeutic strategies for an array of other human cancers and warrant their future human clinical trails.

REFERENCES

1. Roizman B, Sears AE. Herpes simplex viruses and their replication. In: Fields BN, Ed. Virology. New York: Raven Press Ltd, 1996:2231–2295.
2. Shukla D, Liu J, Blaiklock P, Shworak NW, Bai X, Esko JD, Cohen GH, Eisenberg RJ, Rosenberg RD, Spear PG. A novel role for 3-O-sulfated heparan sulfate in herpes simplex virus 1 entry Cell 1999; 99:13–22.
3. Spear PG, Shieh MT, Herold BC, WuDunn D, Koshy TI. Heparan sulfate glycosaminoglycans as primary cell surface receptors for herpes simplex virus Adv Exp Med Biol 1992; 313: 341–353.
4. Kristensson K, Lycke E, Roytta M, Svennerholm B, Vahlne A. Neuritic transport of herpes simplex virus in rat sensory neurons in vitro. Effects of substances interacting with microtubular function and axonal flow [nocodazole, taxol and erythro-9-3-(2-hydroxynonyl)adenine] J Gen Virol 1986; 67:2023–2028.
5. Openshaw H, Stampalia L, Asher LS. Retrograde axoplasmic transport of herpes simplex virus Trans Am Neurol Assoc 1978; 103:238–239.

6. Topp KS, Meade LB, LaVail JH. Microtubule polarity in the peripheral processes of trigeminal ganglion cells: relevance for the retrograde transport of herpes simplex virus J Neurosci 1994; 14:318–325.
7. Leib DA, Coen DM, Bogard CL, Hicks KA, Yager DR, Knipe DM, Tyler KL, Schaffer PA. Immediate-early regulatory gene mutants define different stages in the establishment and reactivation of herpes simplex virus latency J Virol 1989; 63:759–768.
8. Sawtell NM, Thompson RL. Herpes simplex virus type 1 latency-associated transcription unit promotes anatomical site-dependent establishment and reactivation from latency J Virol 1992; 66:2157–2169.
9. Stevens JG, Wagner EK, Devi-Rao GB, Cook ML, Feldman LT. RNA complementary to a herpesvirus alpha gene mRNA is prominent in latently infected neurons Science 1987; 235: 1056–1059.
10. Mellerick DM, Fraser NW. Physical state of the latent herpes simplex virus genome in a mouse model system: evidence suggesting an episomal state Virology 1987; 158:265–275.
11. Rock DL, Fraser NW. Detection of HSV-1 genome in central nervous system of latently infected mice Nature 1983; 302: 523–525.
12. Rock DL, Fraser NW. Latent herpes simplex virus type 1 DNA contains two copies of the virion DNA joint region J Virol 1985; 55:849–852.
13. Efstathiou S, Minson AC, Field HJ, Anderson JR, Wildy P. Detection of herpes simplex virus-specific DNA sequences in latently infected mice and in humans J Virol 1986; 57:446–455.
14. Wagner EK, Guzowski JF, Singh J. Transcription of the herpes simplex virus genome during productive and latent infection Prog Nucleic Acid Res Mol Biol 1995; 51:123–165.
15. Holland LE, Anderson KP, Shipman C, Wagner EK. Viral DNA synthesis is required for the efficient expression of specific herpes simplex virus type 1 mRNA species Virology 1980; 101: 10–24.
16. Flanagan WM, Papavassiliou AG, Rice M, Hecht LB, Silverstein S, Wagner EK. Analysis of the herpes simplex virus

type 1 promoter controlling the expression of UL38, a true late gene involved in capsid assembly J Virol 1991; 65:769–786.

17. Johnson PA, Everett RD. DNA replication is required for abundant expression of a plasmid-borne late US11 gene of herpes simplex virus type 1 Nucleic Acids Res 1986; 14:3609–3625.
18. McGeoch DJ, Dalrymple MA, Davidson AJ, Dolan A, Frame MC, McNab D, Perry LJ, Scott JE, Taylor P. The complete DNA sequence of the long unique region in the genome of herpes simplex virus type 1 J Gen Virol 1988; 69:1531–1574.
19. Sheldrick P, Berthelot N. Inverted repetitions in the chromosome of herpes simplex virus Cold Spring Harb Symp Quant Biol 1975; 39:667–678.
20. Challberg MD. A method for identifying the viral genes required for herpesvirus DNA replication Proc Natl Acad Sci U S A 1986; 83:9094–9098.
21. Wu CA, Nelson NJ, McGeoch DJ, Challberg MD. Identification of herpes simplex virus type 1 genes required for origin- dependent DNA synthesis J Virol 1988; 62:435–443.
22. Stow ND. Herpes simplex virus type 1 origin-dependent DNA replication in insect cells using recombinant baculoviruses J Gen Virol 1992; 73:313–321.
23. McGeoch DJ, Dalrymple MA, Dolan A, McNab D, Perry LJ, Taylor P, Challberg MD. Structures of herpes simplex virus type 1 genes required for replication of virus DNA J Virol 1988; 62:444–453.
24. Griffiths G, Rottier P. Cell biology of viruses that assemble along the biosynthetic pathway Semin Cell Biol 1992; 3: 367–381.
25. Martuza RL. Act locally, think globally. News, comment Nat Med 1997; 3:1323.
26. Alemany R, Gomez-Manzano C, Balague C, Yung WK, Curiel DT, Kyritsis AP, Fueyo J. Gene therapy for gliomas: molecular targets, adenoviral vectors, and oncolytic adenoviruses Exp Cell Res 1999; 252:1–12.
27. Pennisi E. Will a twist of viral fate lead to a new cancer treatment? News, comment Science 1996; 274:342–343.

28. Chou J, Kein ER, Whitley RJ, Roizman B. Mapping of herpes simplex virus-1 neurovirulence to gamma, 34.5, a gene nonessential for growth in culture Science 1990; 250:1262–1266.

29. Bolovan CA, Sawtell NM, Thompson RL. ICP34.5 mutants of herpes simplex virus type 1 strain 17syn+ are attenuated for neurovirulence in mice and for replication in confluent primary mouse embryo cell cultures J Virol 1994; 68:48–55.

30. Chou J, Roizman B. The gamma 1(34.5) gene of herpes simplex virus 1 precludes neuroblastoma cells from triggering total shutoff of protein synthesis characteristic of programed cell death in neuronal cells Proc Natl Acad Sci U S A 1992; 89:3266–3270.

31. McKie EA, MacLean AR, Lewis AD, Cruickshank G, Rampling R, Barnett SC, Kennedy PG, Brown SM. Selective in vitro replication of herpes simplex virus type 1 (HSV-1) ICP34.5 null mutants in primary human CNS tumours—evaluation of a potentially effective clinical therapy Br J Cancer 1996; 74:745–752.

32. Mineta T, Rabkin SD, Martuza RL. Treatment of malignant gliomas using ganciclovir-hypersensitive, ribonucleotide reductase-deficient herpes simplex viral mutant Cancer Res 1994; 54:3963–3966.

33. Chase M, Chung RY, Chiocca EA. An oncolytic viral mutant that delivers the CYP2B1 transgene and augments cyclophosphamide chemotherapy Nat Biotechnol 1998; 16:444–448.

34. Boviatsis EJ, Scharf JM, Chase M, Harrington K, Kowall NW, Breakefield XO, Chiocca EA. Antitumor activity and reporter gene transfer into rat brain neoplasms inoculated with herpes simplex virus vectors defective in thymidine kinase or ribonucleotide reductase Gene Ther 1994; 1:323–331.

35. Walker JR, McGeagh KG, Sundaresan P, Jorgensen TJ, Rabkin SD, Martuza RL. Local and systemic therapy of human prostate adenocarcinoma with the conditionally replicating herpes simplex virus vector G207 Hum Gene Ther 1999; 10:2237–2243.

36. Todo T, Rabkin SD, Sundaresan P, Wu A, Meehan KR, Herscowitz HB, Martuza RL. Systemic antitumor immunity in experimental brain tumor therapy using a multimutated, replication-competent herpes simplex virus Hum Gene Ther 1999; 10: 2741–2755.

37. Mineta T, Rabkin SD, Yazaki T, Hunter WD, Martuza RL. Attenuated multi-mutated herpes simplex virus-1 for the treatment of malignant gliomas Nat Med 1995; 1:938–943.

38. Chung RY, Saeki Y, Chiocca EA. B-myb promoter retargeting of herpes simplex virus gamma34.5 gene-mediated virulence toward tumor and cycling cells J Virol 1999; 73:7556–7564.

39. Fu X, Zhang X. Potent systemic antitumor activity from an oncolytic herpes simplex virus of syncytial phenotype Cancer Res 2002; 62:2306–2312.

40. Yu DC, Sakamoto GT, Henderson DR. Identification of the transcriptional regulatory sequences of human kallikrein 2 and their use in the construction of calydon virus 764, an attenuated replication competent adenovirus for prostate cancer therapy Cancer Res 1999; 59:1498–1504.

41. Chahlavi A, Rabkin S, Todo T, Sundaresan P, Martuza R. Effect of prior exposure to herpes simplex virus 1 on viral vector-mediated tumor therapy in immunocompetent mice Gene Ther 1999; 6:1751–1758.

42. Lambright ES, Kang EH, Force S, Lanuti M, Caparrelli D, Kaiser LR, Albelda SM, Molnar-Kimber KL. Effect of preexisting anti-herpes immunity on the efficacy of herpes simplex viral therapy in a murine intraperitoneal tumor model Mol Ther 2000; 2:387–393.

43. Yoon SS, Nakamura H, Carroll NM, Bode BP, Chiocca EA, Tanabe KK. An oncolytic herpes simplex virus type 1 selectively destroys diffuse liver metastases from colon carcinoma FASEB J 2000; 14:301–311.

44. Martuza RL, Malick A, Markert JM, Ruffner KL, Coen DM. Experimental therapy of human glioma by means of a genetically engineered virus mutant Science 1991; 252:854–856.

45. Toda M, Rabkin SD, Martuza RL. Treatment of human breast cancer in a brain metastatic model by G207, a replication-competent multimutated herpes simplex virus 1 Hum Gene Ther 1998; 9:2177–2185.

46. Sundaresan P, Hunter WD, Martuza RL, Rabkin SD. Attenuated, replication-competent herpes simplex virus type 1 mutant G207: safety evaluation in mice J Virol 2000; 74:3832–3841.

47. Hunter WD, Martuza RL, Feigenbaum F, Todo T, Mineta T, Yazaki T, Toda M, Newsome JT, Platenberg RC, Manz HJ, Rabkin SD. Attenuated, replication-competent herpes simplex virus type 1 mutant G207: safety evaluation of intracerebral injection in nonhuman primates J Virol 1999; 73:6319–6326.

48. Todo T, Feigenbaum F, Rabkin SD, Lakeman F, Newsome JT, Johnson PA, Mitchell E, Belliveau D, Ostrove JM, Martuza RL. Viral shedding and biodistribution of G207, a multimutated, conditionally replicating herpes simplex virus type 1, after intracerebral inoculation in aotus Mol Ther 2000; 2:588–595.

49. Rampling R, Cruickshank G, Papanastassiou V, Nicoll J, Hadley D, Brennan D, Petty R, MacLean A, Harland J, McKie E, Mabbs R, Brown M. Toxicity evaluation of replication-competent herpes simplex virus (ICP 34.5 null mutant 1716) in patients with recurrent malignant glioma Gene Ther 2000; 7: 859–866.

50. Markert JM, Medlock MD, Rabkin SD, Gillespie GY, Todo T, Hunter WD, Palmer CA, Feigenbaum F, Tornatore C, Tufaro F, Martuza RL. Conditionally replicating herpes simplex virus mutant, G207, for the treatment of malignant glioma: results of a phase I trial Gene Ther 2000; 7:867–874.

51. Kooby DA, Carew JF, Halterman MW, Mack JE, Bertino JR, Blumgart LH, Federoff HJ, Fong Y. Oncolytic viral therapy for human colorectal cancer and liver metastases using a multimutated herpes simplex virus type-1 (G207) FASEB J 1999; 13:1325–1334.

52. Vasey PA, Shulman LN, Campos S, Davis J, Gore M, Johnston S, Kirn DH, O'Neill V, Siddiqui N, Seiden MV, Kaye SB. Phase I trial of intraperitoneal injection of the E1B-55-kd-gene- deleted adenovirus ONYX-015 (dl1520) given on days 1 through 5 every 3 weeks in patients with recurrent/refractory epithelial ovarian cancer J Clin Oncol 2002; 20:1562–1569.

53. Kramm CM, Chase M, Herrlinger U, Jacobs A, Pechan PA, Rainov NG, Sena-Esteves M, Aghi M, Barnett FH, Chiocca EA, Breakefield XO. Therapeutic efficiency and safety of a second-generation replication-conditional HSV1 vector for brain tumor gene therapy Hum Gene Ther 1997; 8:2057–2068.

54. Todo T, Martuza RL, Rabkin SD, Johnson PA. Oncolytic herpes simplex virus vector with enhanced MHC class I presentation

and tumor cell killing Proc Natl Acad Sci U S A 2001; 98: 6396–6401.

55. Andreansky S, Soroceanu L, Flotte ER, Chou J, Markert JM, Gillespie GY, Roizman B, Whitley RJ. Evaluation of genetically engineered herpes simplex viruses as oncolytic agents for human malignant brain tumors Cancer Res 1997; 57: 1502–1509.
56. Advani SJ, Sibley GS, Song PY, Hallahan DE, Kataoka Y, Roizman B, Weichselbaum RR. Enhancement of replication of genetically engineered herpes simplex viruses by ionizing radiation: a new paradigm for destruction of therapeutically intractable tumors Gene Ther 1998; 5:160–165.
57. Spear MA, Sun F, Eling DJ, Gilpin E, Kipps TJ, Chiocca EA, Bouvet M. Cytotoxicity, apoptosis, and viral replication in tumor cells treated with oncolytic ribonucleotide reductase-defective herpes simplex type 1 virus (hrR3) combined with ionizing radiation Cancer Gene Ther 2000; 7:1051–1059.
58. Chahlavi A, Todo T, Martuza RL, Rabkin SD. Replication-competent herpes simplex virus vector G207 and cisplatin combination therapy for head and neck squamous cell carcinoma Neoplasia 1999; 1:162–169.
59. Toyoizumi T, Mick R, Abbas AE, Kang EH, Kaiser LR, Molnar-Kimber KL. Combined therapy with chemotherapeutic agents and herpes simplex virus type 1 ICP34.5 mutant (HSV-1716) in human non-small cell lung cancer Hum Gene Ther 1999; 10: 3013–3029.
60. Andreansky S, He B, van Cott J, McGhee J, Markert JM, Gillespie GY, Roizman B, Whitley RJ. Treatment of intracranial gliomas in immunocompetent mice using herpes simplex viruses that express murine interleukins Gene Ther 1998; 5:121–130.
61. Parker JN, Gillespie GY, Love CE, Randall S, Whitley RJ, Markert JM. Engineered herpes simplex virus expressing IL-12 in the treatment of experimental murine brain tumors Proc Natl Acad Sci U S A 2000; 97:2208–2213.
62. Todo T, Martuza RL, Dallman MJ, Rabkin SD. In situ expression of soluble B7-1 in the context of oncolytic herpes simplex virus induces potent antitumor immunity Cancer Res 2001; 61: 153–161.

63. Todo T, Rabkin SD, Martuza RL. Evaluation of ganciclovir-mediated enhancement of the antitumoral effect in oncolytic, multimutated herpes simplex virus type 1 (G207) therapy of brain tumors Cancer Gene Ther 2000; 7:939–946.

64. Yoon SS, Carroll NM, Chiocca EA, Tanabe KK. Cancer gene therapy using a replication-competent herpes simplex virus type 1 vector Ann Surg 1998; 228:366–374.

65. Hoggan MD, Roizman B. The isolation and properties of a variant of herpes simplex producing multinucleated giant cells in nomolayer cultures in the presence of antibody Am J Hyg 1959; 70:208–219.

66. Read GS, Person S, Keller PM. Genetic studies of cell fusion induced by herpes simplex virus type 1 J Virol 1980; 35: 105–113.

67. Bond VC, Person S, Warner SC. The isolation and characterization of mutants of herpes simplex virus type 1 that induce cell fusion J Gen Virol 1982; 61:245–254.

68. Pogue-Geile KL, Lee GT, Shapira SK, Spear PG. Fine mapping of mutations in the fusion-inducing MP strain of herpes simplex virus type 1 Virology 1984; 136:100–109.

69. Person S, Kousoulas KG, Knowles RW, Read GS, Holland TC, Keller PM, Warner SC. Glycoprotein processing in mutants of HSV-1 that induce cell fusion Virology 1982; 117:293–306.

70. Lentz BR, Malinin V, Haque ME, Evans K. Protein machines and lipid assemblies: current views of cell membrane fusion Curr Opin Struct Biol 2000; 10:607–615.

71. Sinkovics J, Horvath J. New developments in the virus therapy of cancer: a historical review Intervirology 1993; 36:193–214.

72. Bateman A, Bullough F, Murphy S, Emiliusen L, Lavillette D, Cosset FL, Cattaneo R, Russell SJ, Vile RG. Fusogenic membrane glycoproteins as a novel class of genes for the local and immune-mediated control of tumor growth Cancer Res 2000; 60:1492–1497.

73. Fielding AK, Chapel-Fernandes S, Chadwick MP, Bullough FJ, Cosset FL, Russell SJ. A hyperfusogenic gibbon ape leukemia envelope glycoprotein: targeting of a cytotoxic gene by ligand display Hum Gene Ther 2000; 11:817–826.

74. Januszeski MM, Cannon PM, Chen D, Rozenberg Y, Anderson WF. Functional analysis of the cytoplasmic tail of Moloney murine leukemia virus envelope protein J Virol 1997; 71: 3613–3619.

75. Forestell SP, Dando JS, Chen J, de Vries P, Bohnlein E, Rigg RJ. Novel retroviral packaging cell lines: complementary tropisms and improved vector production for efficient gene transfer Gene Ther 1997; 4:600–610.

76. Johann SV, van Zeijl M, Cekleniak J, O'Hara B. Definition of a domain of GLVR1 which is necessary for infection by gibbon ape leukemia virus and which is highly polymorphic between species J Virol 1993; 67:6733–6736.

77. Kavanaugh MP, Miller DG, Zhang W, Law W, Kozak SL, Kabat D, Miller AD. Cell-surface receptors for gibbon ape leukemia virus and amphotropic murine retrovirus are inducible sodium-dependent phosphate symporters Proc Natl Acad Sci U S A 1994; 91:7071–7075.

78. Higuchi H, Bronk SF, Bateman A, Harrington K, Vile RG, Gores GJ. Viral fusogenic membrane glycoprotein expression causes syncytia formation with bioenergetic cell death: implications for gene therapy Cancer Res 2000; 60:6396–6402.

79. Diaz RM, Bateman A, Emiliusen L, Fielding A, Trono D, Russell SJ, Vile RG. A lentiviral vector expressing a fusogenic glycoprotein for cancer gene therapy Gene Ther 2000; 7:1656–1663.

80. Babiss LE, Friedman JM, Darnell JE. Cellular promoters incorporated into the adenovirus genome. Effect of viral DNA replication on endogenous and exogenous gene transcription J Mol Biol 1987; 193:643–650.

81. Weir JP. Regulation of herpes simplex virus gene expression Gene 2001; 271:117–130.

82. Fu X, Meng F, Tao L, Jin A, Zhang X. A strict late viral promoter is a strong tumor-specific promoter in the context of an oncolytic herpes simplex virus Gene Ther 2003; 10:1458–1464.

83. Saeki Y, Ichikawa T, Saeki A, Chiocca EA, Tobler K, Ackermann M, Breakefield XO, Fraefel C. Herpes simplex virus type 1 DNA amplified as bacterial artificial chromosome in Escherichia coli: rescue of replication-competent virus progeny and

packaging of amplicon vectors Hum Gene Ther 1998; 9: 2787–2794.

84. Fu X, Tao L, Jin A, Vile R, Brenner M, Zhang X. Expression of a fusogenic membrane glycoprotein by an oncolytic herpes simplex virus provides potent synergistic anti-tumor effect Mol Ther 2003; 7:748–754.

85. Li H, Haviv YS, Derdeyn CA, Lam J, Coolidge C, Hunter E, Curiel DT, Blackwell JL. Human immunodeficiency virus type 1-mediated syncytium formation is compatible with adenovirus replication and facilitates efficient dispersion of viral gene products and de novo-synthesized virus particles Hum Gene Ther 2001; 12:2155–2165.

86. Nakamori M, Fu X, Meng F, Jin A, Tao L, Bast RCJ, Zhang X. Effective therapy of metastatic ovarian cancer with an oncolytic herpes simplex virus incorporating two membrane-fusion mechanisms Clin Cancer Res 2003; 9:2727–2733.

87. Buy JN, Moss AA, Ghossain MA, Sciot C, Malbec L, Vadrot D, Paniel BJ, Decroix Y. Peritoneal implants from ovarian tumors: CT findings Radiology 1988; 169:691–694.

88. Pettaway CA, Pathak S, Greene G, Ramirez E, Wilson MR, Killion JJ, Fidler IJ. Selection of highly metastatic variants of different human prostatic carcinomas using orthotopic implantation in nude mice Clin Cancer Res 1996; 2:1627–1636.

89. Nakamori M, Fu X, Pettaway CA, Zhang X. Potent antitumor activity after systemic delivery of a doubly fusogenic oncolytic herpes simples virus against metastatic prostate cancer Prostate 2004; 60:53–60.

12

Poliovirus Recombinants Against Malignant Glioma

MELINDA MERRILL

Department of Molecular Genetics & Microbiology, Duke University Medical Center
Durham, NC

DAVID SOLECKI

Department of Developmental Neurobiology, Rockefeller University
New York, NY

MATTHIAS GROMEIER

Department of Molecular Genetics & Microbiology, Duke University Medical Center
Durham, NC

INTRODUCTION

Poliovirus has recently been added to the rapidly expanding list of potential viral oncolytic agents. Its inherent tropism for certain malignant cell types, highly lytic growth cycle, and permissiveness for genetic manipulation qualify poliovirus as a prime target for therapeutic use against cancer. On the other hand, the danger of serious neurological damage following administration of poliovirus-based oncolytic agents restricts its use. We have developed an innovative strategy to harness poliovirus for therapeutic use against central nervous system (CNS) malignancy. Our approach is the result of thorough analyses of the molecular pathogenesis of poliomyelitis and takes advantage of the natural targeting and invasive properties of poliovirus. We have exploited an unconventional mechanism of translation initiation upon which poliovirus relies to achieve a tumor-specific replication phenotype and to exclude viral cytopathogenicity in normal tissues.

A prototype oncolytic poliovirus recombinant recently entered production according to "good manufacturing practice" and will undergo toxicology testing in non-human primates. Phase-I clinical trials with our agent against malignant glioma are expected to begin within the near future.

POLIOVIRUS MOLECULAR BIOLOGY AND PATHOGENESIS

The Poliovirus Life Cycle

Poliovirus, a nonenveloped plus-stranded RNA virus, is the prototype member of the Picornaviridae family of viruses. The level of understanding of picornavirus molecular biology is second to none and many aspects of the poliovirus life cycle, schematically represented in Figure 1, have been elucidated.

Susceptibility to poliovirus is determined by its sole cellular receptor, the immunoglobulin superfamily (IGSF) molecule CD155 [1]. Binding to CD155 alone is sufficient to induce conformational alteration of the viral particle [2] and results in passage of the virion across the cell membrane and uncoating of the viral RNA genome. Once released from the capsid, the viral RNA serves as a template for viral translation and for negative-strand RNA synthesis (Figure 1). The plus-stranded RNA genome of poliovirus lacks a 5′ cap structure [3] and instead features a 5-terminal covalently linked protein, VPg (4;

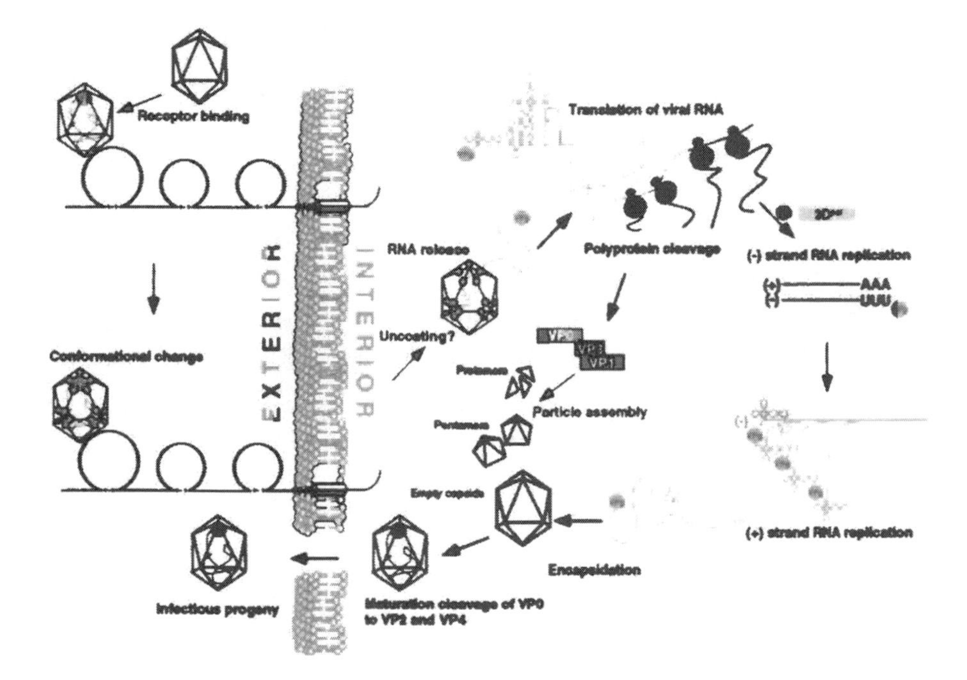

Figure 1 Schematic representation of the life cycle of poliovirus (see text for details).

Figure 1). The viral genome contains an unusually large and highly structured 5′ nontranslated region (NTR) studded with noninitiating AUG codons (Figure 2B). Characterization of the uncommon features of picornaviral 5′ NTRs resulted in the discovery of the internal ribosomal entry site (IRES), that mediates translation initiation in a 5′ end-, cap-independent manner [5,6].

During poliovirus infection, critical determinants of cap-dependent translation initiation, e.g., the eukaryotic initiation factor 4G (eIF4G) [7] and the poly(A) binding protein (PABP) [8,9] are cleaved by the viral proteinase $2A^{pro}$(Figure 2B). Eukaryotic cap-dependent translation relies upon the interaction between PABP and eIF4G (10,11; Figure 2A). The interaction of eIF4G with eIF4E (binding to the cap structure) and PABP results in circularization of mRNAs (12; Figure 2A). Bridging

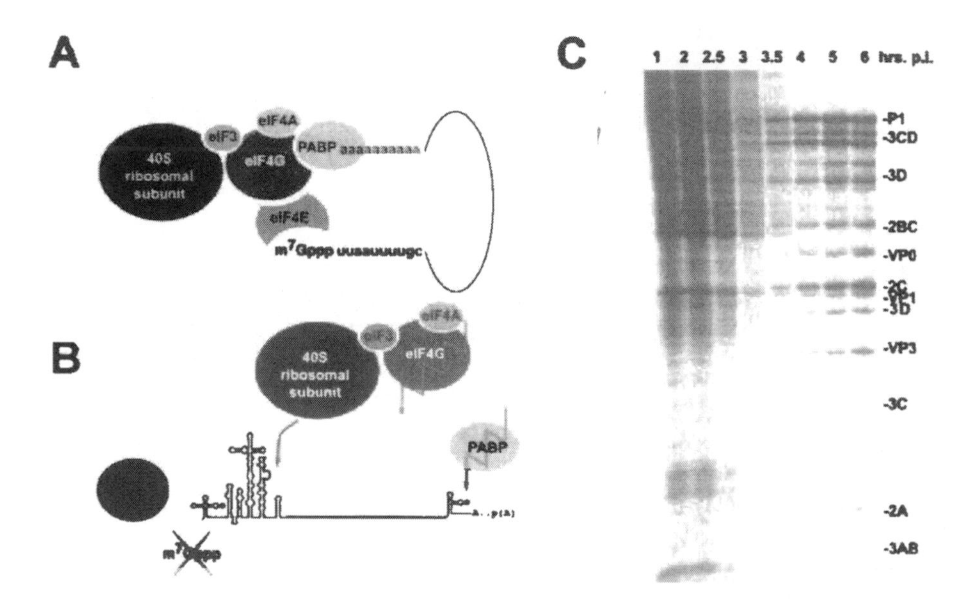

Figure 2 **A.** Proposed model for the circularization of eukaryotic mRNAs. eIF4G, through its affinity for PABP [10,11] and the cap-binding factor eIF4E, links the cap-binding complex and the poly(A) tail. **B.** During poliovirus infection, eIF4G [7] and PABP [8,9] are proteolytically degraded. The absence of a 5′ terminal cap structure and cleavage of translation initiation factors prevents formation of a circularized template. The position and predicted secondary structure of the IRES element is outlined. **C.** Decline of host-cell protein synthesis and emerging viral translation in poliovirus-infected HeLa cells. Pulse-chase metabolic labeling of infected HeLa cells with $[S_{35}]$-methionine. By 2.5 to 3 hours postinfection, $2A^{pro}$-mediated degradation of eIF4G and PABP result in gradual shut-off of host translation. In parallel, IRES-mediated translation of viral gene products eclipses cellular cap-dependent translation already by 4 hours postinfection. The individual viral gene products released by proteolytic processing of the polyprotein are indicated.

of poly(A) and the initiation complex stimulates translation, possibly by favoring 3′ to 5′ shunting of ribosomes [13].

Cleavage of eIF4G and PABP selectively inhibits cap-dependent cellular translation without impairing cap-independent viral protein synthesis mediated by the viral IRES (Figure 2C). As a result, host-cell protein synthesis dramatically declines as early as 2 to 3 hours postinfection (p.i.) while IRES-mediated translation usurps the cellular protein synthesis machinery (Figure 2C).

The IRES drives expression of a single large polyprotein proteolytically processed by virus-encoded proteinases to yield four structural and seven nonstructural gene products. Using viral genomic RNA as a template, the newly synthesized viral RNA-dependent RNA polymerase produces negative-strand intermediates that serve as templates for the synthesis of plus-strand RNA. Encapsidation of plus-strand RNA copies yields virus progeny (Figure 1). The poliovirus life cycle is highly productive and results in the release of infectious progeny through lytic destruction of the host cell.

Poliovirus infection interferes with cellular function at multiple levels and induces profound and lethal disruption of host-cell biology. Shut-off of host-cell protein synthesis, cleavage of eukaryotic translation and transcription factors, disruption of intracellular vesicular traffic, disturbance of exonuclear transport, expression of cytotoxic viral gene products (e.g., the viral proteinase $2A^{pro}$), and overwhelming virus particle reproduction ultimately result in lytic destruction of the host cell. The lethal effects of poliovirus gene expression on the host cell occur early (2 to 4 hours p.i.) and are irreversible. In contrast to many other oncolytic virus candidates, the poliovirus life cycle and its lethal effects are cell-cycle independent.

Poliovirus Pathogenesis

Poliovirus infection occurs through fecal–oral transmission and results in primary replication within the gastrointestinal tract [14]. Gastrointestinal infection generally remains asymptomatic, but 1% to 2% of infected individuals develop neurological symptoms. Poliovirus pathogenicity is strictly limited to the primate CNS, where selective viral tropism for anterior horn spinal cord motor neurons produces a highly characteristic neurological syndrome dominated by flaccid paralysis (14; Figure 3).

Despite major research efforts for almost a century, the critical determinants of poliovirus neuropathogenesis are incompletely understood. Because the inherent neuropathogenic properties of poliovirus may affect its clinical use as an oncolytic agent, continuing efforts will be required to elucidate the molecular mechanisms determining poliovirus neuroinvasion and its peculiarly restricted tropism for anterior horn motor neurons.

Enteroviruses face severe obstacles on their route from the primary site of replication in the gut to the spinal cord. Central nervous system invasion may require viremia and blood-borne seeding to the CNS across the blood–brain barrier [14]. Alternatively, peripheral nerves may serve as a portal of entry for poliovirus invasion of the CNS. A variety of potential invasion routes have been demonstrated to be operational in experimental animals; e.g., direct penetration of the blood–brain barrier [15] or retrograde axonal transport [16,17]. Although the mechanism of poliovirus neuroinvasion remains under debate, available data clearly suggest that this process is highly inefficient. The low proportion of neurological

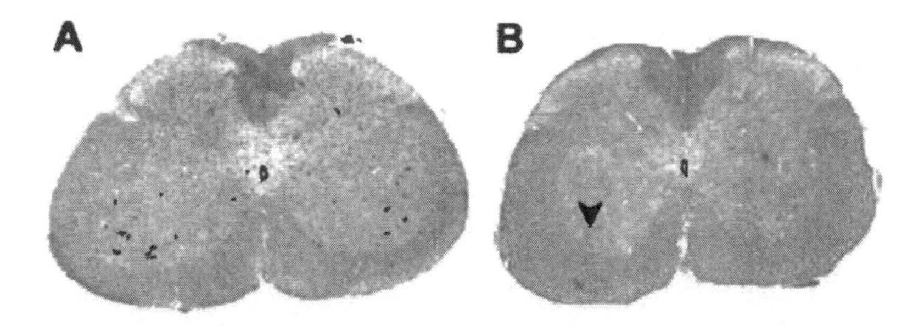

Figure 3 Histopathology of paralytic poliomyelitis in poliovirus-infected mice transgenic for the poliovirus receptor CD155. **A.** The normal murine lumbar spinal cord. **B.** Poliovirus infection results in selective destruction of anterior horn motor neurons (arrowhead).

syndromes among individuals infected with poliovirus (~1% to 2%) is believed to reflect the difficulty for the virus to reach and invade the CNS. These observations indicate that peripheral administration of poliovirus-based oncolytic agents may not be suitable for efficient targeting of intracerebral malignant lesions (discussed below).

The reason for the peculiar predilection of poliovirus for a highly specialized and minute population of CNS neurons has been a contentious issue since the first description of polio histopathology. The destruction of motor neurons is the result of direct targeting of these cells by poliovirus, rather than a bystander effect of infectious or inflammatory processes in their vicinity (as had been suggested in early reports; discussed in Ref. 18). However, the molecular mechanisms that constitute the basis for poliovirus specificity for motor neurons are not entirely clear. Virus susceptibility may be exclusively controlled by the expression patterns of cell surface receptors, or by the combined effect of cell external (e.g., receptors) as well as internal (e.g., host replication factors) determinants. For poliovirus, evidence for both cell-external as well as -internal determinants of host-cell specificity exists (discussed below).

Poliovirus infection induces an acute neurological syndrome that may vary in extent from minor monopareses to fatal paralysis. Generally, the virus is unable to persist or induce chronic infection or latency. Most infected individuals clear the virus several weeks after infection and cease to shed infectious particles with their stool. It has become apparent, recently, that persistent enteric infection can occur in patients with inherent or acquired immunodeficiency [19]. Available evidence suggests that a depressed immune system may fail to clear the virus and permit prolonged enteric replication.

ONCOLYTIC AGENTS BASED ON POLIOVIRUS

Viral Target Tropism

Targeting of oncolytic viral agents to tumor cells may be the most obvious and basic requirement to achieve oncolytic activity. Target tropism of oncolytic viruses is determined by interaction of the virus particle surface with specific receptors expressed on the host-cell surface. This is followed by particle uptake and uncoating that will lead to the delivery of the viral genome into the host cell.

Empirical results from clinical investigations of oncolytic viruses have suggested tumor cell targeting and virus spread as one of the main obstacles to achieve efficient oncolysis in vivo (discussed below). It therefore is imperative, prior to clinical application of oncolytic viruses, to determine the conditions for effective tumor cell targeting and to clarify expression patterns of their cellular receptor(s) in the target tumor type.

Expression analysis of viral receptors in tumors is hindered by the fact that established tumor cell lines, the mainstay of pre-clinical investigations of oncolytic agents in tissue culture and xenotransplantation models, are poor representations of actual tumor tissue. Expression patterns of molecular determinants of virus attachment and entry in tissue culture cell lines commonly do not correspond to those observed in actual tumors. Consequently, experimental data of viral oncolysis in cultured tumor cell lines may yield results that poorly relate to the oncolytic effects in heterogeneous tumors. These difficulties demonstrate the need for more thorough investigation of the determinants for virus host-cell attachment, entry, and tissue invasion; their relationship with cancer; and their role in viral oncolysis.

This fact has become evident with oncolytic adenoviruses, which use the IGSF member coxsackievirus/adenovirus receptor (CAR) for cell entry [20,21]. Oncolytic adenoviruses were applied therapeutically in clinical trials against malignant glioma prior to analysis of CAR expression profiles in actual tumors [22]. Recent investigations demonstrated that CAR expression levels in glioma cell lines selected for exquisite adenovirus sensitivity for pre-clinical investigations did not correspond to expression levels in patients' tumors [23–26].

Given the cardinal importance of virus-receptor expression for tumor targeting and oncolytic activity, we have undertaken considerable efforts to elucidate the role of the poliovirus receptor CD155 in the pathogenesis of poliomyelitis and its potential role as a mediator of poliovirus tropism for malignant cell types. To this end, we have focused on the assessment of transcriptional regulation and expression patterns of the *CD155* gene in the normal human CNS and in cancer. Detection of CD155 in the adult human CNS has been notoriously difficult because of exceedingly low expression levels in neural tissues and even in cell lines commonly used for poliovirus propagation (e.g., HeLa cervical carcinoma cells; 27). Structural similarities with insect members of the IGSF expressed during larval neurogenesis [28] suggested that CD155 expression may occur primarily during embryonic development.

Developmental Expression of the CD155 Gene

CD155 is the founding member of the nectin subfamily of the IGSF, structurally characterized by a V-C2-C2 domain arrangement (Table 1). Four poliovirus receptor-related proteins (PRR1–4; recently renamed *nectins 1–4*) have been identified in humans [29–32]. Apart from their pathogenic roles as viral receptors, nectins have been reported to mediate homophilic cell adhesion [33,34] and, through interaction with afadin, form part of the E-cadherin system [35–37].

Table 1 Members of the Nectin Family of Genes

Gene	Species	Gene products	Associated function(s)	Reference
CD155, PVR	Human	CD155 (PVR) (soluble)	poliovirus binding "	1
CD111, PRR-1, Nectin-1	Human	Nectin-1	-herpesvirus binding	30, 38
CD112, PRR-2, Nectin-2	Human	Nectin-2	homophilic adhesion	29, 33
PRR-3, Nectin-3	Human	Nectin-3	afadin binding	31
PRR-4, Nectin-4	Human	Nectin-4	homophilic adhesion	32
AGM 1, AGM 2	Simian	sCD155 (sPVR)	poliovirus binding	39
TagE4	Murine	TagE4	tumor antigen	40

Given that intense searches for CD155 homologues in non-primates did not produce any obvious candidates, we evaluated the developmental regulation of the *CD155* gene in a transgenic context [28]. Transgenic mice expressing the entire *CD155* gene develop a neurological syndrome upon poliovirus infection indistinguishable from primate poliomyelitis [41,42]. These findings suggested that upstream regulatory regions of the *CD155* gene faithfully control expression in a transgenic context.

We constructed transgenic mice that express the reporter gene *β-galactosidase* (*lacZ*) under control of *CD155* upstream regulatory regions [28]. Analysis of offspring embryos revealed that the CD155 promoter is active during midgestation in a triad of CNS structures known to induce motor neuron differentiation and maturation (28; Figure 4). Localized expression of CD155 in the floor plate, notochord, and spinal cord anterior horn of the developing CNS may explain why poliovirus neuropathogenicity is limited to the motoneuronal system. Activity of the CD155 promoter diminished considerably toward par-

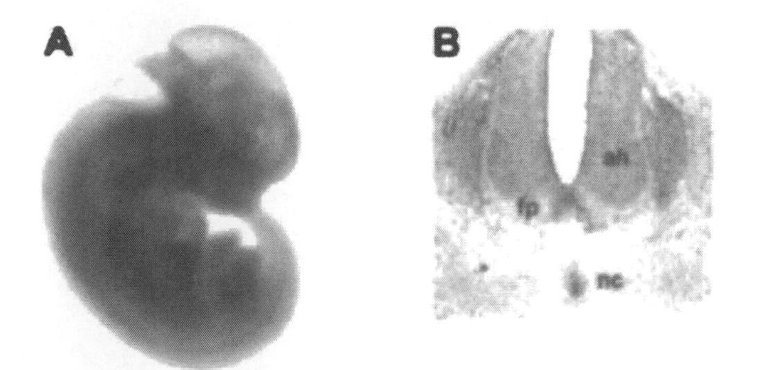

Figure 4 Embryonic reportergene expression in mice transgenic for β-galactosidase (*lacZ*) controlled by *CD155* upstream regulatory regions. **A.** As shown in this E12.0 embryo, reporter gene expression in the spinal cord and brain peaked during midgestation. **B.** LacZ expression was limited to notochord (nc), floor plate (fp) and the anterior horn (ah) of the developing spinal cord. The horizontal section shown corresponds to the upper cervical portion of the spinal cord.

tum, and reporter gene expression became undetectable as early as 16 days post conception [28].

Our observations of highly restricted CD155 promoter activity in the motoneuronal system implicate the poliovirus receptor as a major determinant of specific vulnerability to poliovirus infection. Furthermore, its temporo-spatial expression pattern suggests a physiological role in cell adhesion, migration, and axonal guidance during embryonic CNS morphogenesis [28]. Prepartal downregulation of expression of the *CD155* gene may explain low expression rates in adult tissues.

Transcriptional Regulation of the CD155 Gene

After the developmental profile of *CD155* promoter activity had been determined in reporter transgenic mice, subsequent cell culture studies focused on dissecting the *cis*-acting elements and *trans*-acting factors that determine the expression pattern observed in the developing CNS. A combination of deletion studies, DNaseI footprinting and linker scanning mutagenesis were used to identify a 280-base-pair core promoter fragment of the CD155 gene that harbored four functional *cis*-acting elements (termed *FPI-IV*) and a region of multiple transcriptional start sites (43–46; Figure 5A).

The CD155 core promoter identified by these studies possessed all the information needed to direct cell type-specific transcription of a reporter gene; therefore much effort was subsequently dedicated to the identification of the transcription factors that interacted with FPI-IV.

FPI and FPII are both bound by the developmentally expressed protein AP-2, a transcription factor that has been described to participate in the regulation of gene expression in the developing lens, retinal ganglion cell layer, and neural tube [47,48]. The expression of AP-2 in the developing CNS partially overlaps temporally and spatially with that of CD155, suggesting that the AP-2 transcription factors may influence the CD155 expression profile during embryogenesis.

FPIV is bound by the transcription factor (NRF-1) nuclear respiratory factor, a potent activator of the CD155 core promoter in vitro [45]. Interestingly, vertebrate NRF-1 related

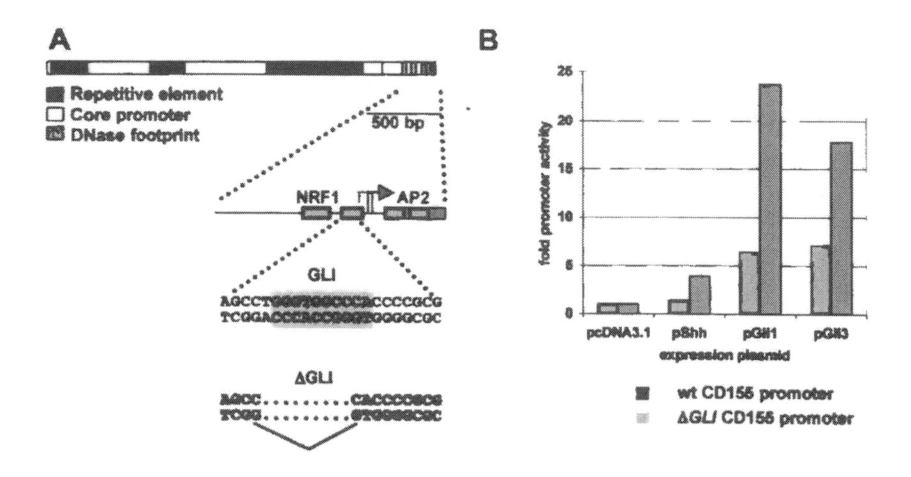

Figure 5 **A.** Genetic structure of the upstream region of the *CD155* gene. Footprints (FP) I/II were identified to interact with AP-2; FPIV with NRF-1; and FPIII contains a putative GLI binding site. The lower panel shows a GLI binding site deletion variant of the *CD155* core promoter. **B.** Transient transfection assays. HeLa cells were co-transfected with either a wild-type *CD155* promoter-driven reporter construct or the ΔGLI variant and the indicated expression vector. Both Sonic hedgehog and GLI transcription factors induced *CD155* promoter-mediated reporter expression.

factors have been shown to regulate the expression of genes required for proper retinal development [49]. The *CD155* promoter is active in the developing retina [28] suggesting that regulation of the CD155 core promoter by NRF-1 could play a role in directing promoter activity to this location.

The secreted morphogen *Sonic hedgehog* (Shh) plays an essential role in patterning many structures during development, including the notochord and ventral CNS, the developing limb, lung, and foregut [50–52]. The most thoroughly characterized site of Shh action is the anterior neural tube, associated with notochord, floor plate, and the motoneuronal system [50]. Secreted Shh from the notochord induces development of the floor plate, which, in turn, becomes a source of Shh that patterns many ventral neuron types along the entire anterior–posterior axis of the developing CNS [50].

Given the overlapping expression patterns of Shh and CD155 in the developing CNS (46; Figure 6), we tested whether CD155 expression was modulated by Shh signaling [46]. Interestingly, *CD155* mRNA was induced when a neuroepithelial cell line was treated with Shh protein [46]. Further studies showed that transient overexpression of Shh activated reporter-gene expression driven by the *CD155* core promoter (46; Figure 5B). Upregulation of reporter expression was dependent on an intact GLI binding site within the *CD155* core promoter (46; Figure 5B). In accordance with their observed function as downstream effectors of the Shh signal [52,53], GLI proto-oncogene transcription factors (*Gli1* and *Gli3*) were also potent activators of the *CD155* core promoter (46; Figure 5B).

The Shh-GLI pathway has been recognized to be actively involved in the oncogenesis of neuroectodermal tumors and skin cancers [54–57]. Transcriptional regulation of the *CD155* gene through Shh-GLI signaling during embryonic CNS development therefore may also affect its expression in tumors.

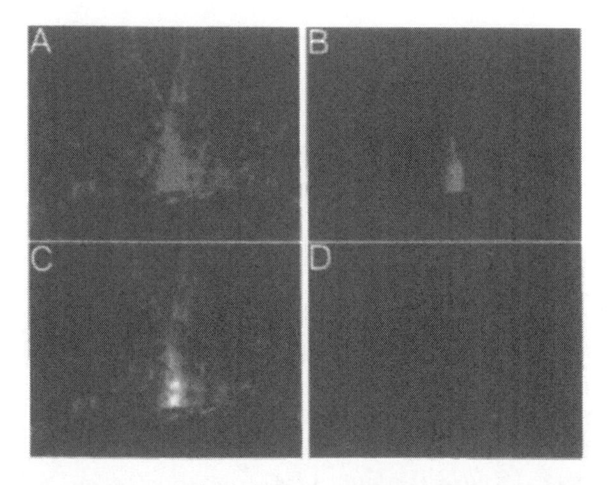

Figure 6 Expression of the poliovirus receptor CD155 (**A**) and Sonic hedgehog (**B**) overlaps (**C,** merged signal from A and B) in the floor plate of the developing spinal cord of CD155 transgenic mouse embryos [46]. Nonspecific rabbit immunoglobulin G yielded no signal (**D**).

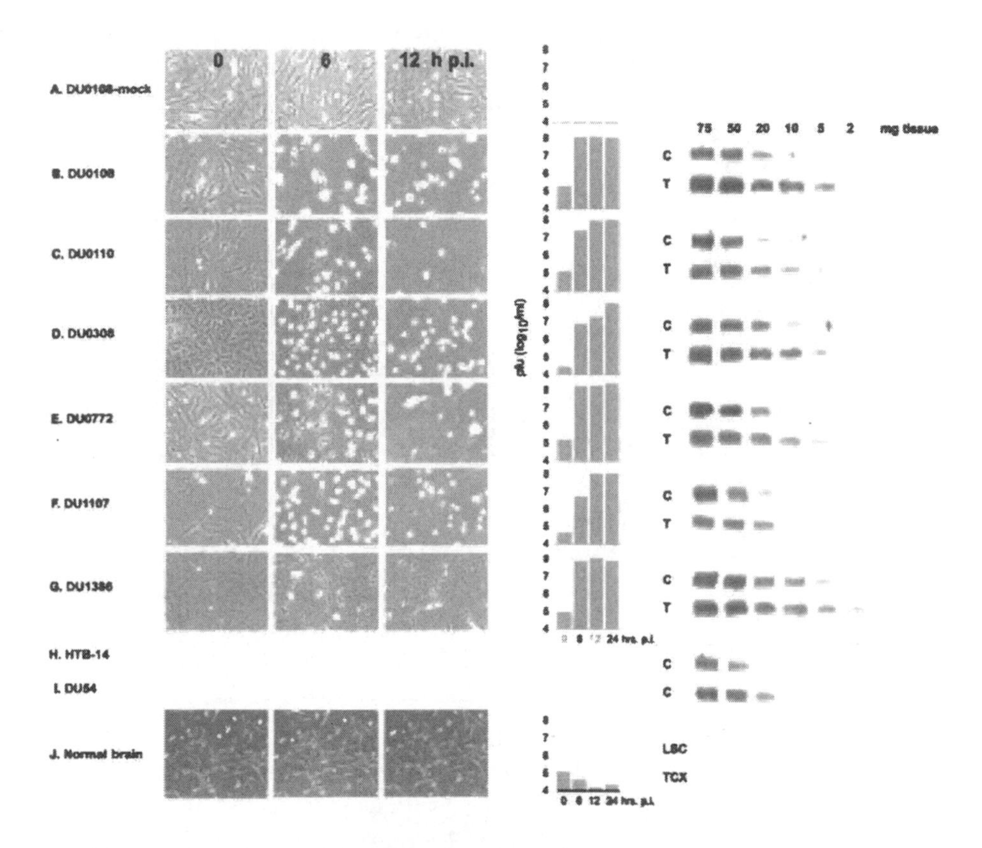

Figure 7 Cytopathic changes after PVS-RIPO infection and combined antigen capture and Western blot analysis of CD155 expression in cell and tissue homogenates of primary explant glioma cell lines (**B-G**), established glioma cell lines (**H, I**), and primary explant normal brain cultures (**J**). **Left panel:** The incubation intervals following infection are indicated atop. **A.** Mock-infected DU0108 cells did not show signs of cytopathic change stemming from the infection procedure. **B–G.** Infection with PVS-RIPO uniformly produced prominent cytopathic effects and lysis of primary explant glioma cultures already 6 hours postinfection. **Middle panel:** All cultures were freeze-thawed after documentation of their morphology to determine the virus yield by plaque assay. Efficient virus propagation rates were observed in all primary explant glioma cultures in accordance with the observed progression of cytopathic changes (B–G). In contrast, viral propagation failed in normal brain explant cultures (H, I).

Association of CD155 Expression with Neuroectodermal Malignancy

Increasing evidence links ectopic expression of cell adhesion molecules (CAM) of the IGSF to cancer and malignant glioma in particular [58–61]. Immunoglobulin superfamily-CAMs have been implicated in tumor progression [62,63], invasion, and metastasis [62,64,65] and have become prominent targets for anti-neoplastic therapy. Most oncolytic viruses in clinical trials against malignant glioma target IGSF-CAM receptors—adenoviruses through their affinity for CAR [20,21]; herpesviruses through their affinity for PRR-1 (nectin-1; 38).

Four observations prompted us to evaluate the association of the IGSF member CD155 with neuroectodermal malignancy: (1) Other members of the nectin family are known tumor antigens (e.g., TagE4) (40; Table 1). (2) A large number of IGSF family members with structures or developmental expression patterns related to CD155 are ectopically expressed in cancer [62]. (3) Analyses of transcriptional regulation of the *CD155* gene revealed that the morphogenic factor Shh and its downstream effectors, transcription factors of the *gli* family, potently activate CD155 transcription and protein expression [46]. (4) All established malignant glioma cell lines tested exhibit high levels of susceptibility to poliovirus.

Given the known discrepancies of cell-surface-marker expression of established glioma cell lines and tumors occurring in patients, we focused our analyses on patient material and primary explant cultures derived therefrom. We analyzed six cases of glioblastoma multiforme (GBM) for their association with CD155 (27; Figure 7). Tumor biopsy material was obtained at the time of surgery and subjected to further analy-

Right panel: A–J. The amount of total protein analyzed is indicated atop.Antigen capture/Western blot assays of cell and tissue homogenates are labeled "C" and "T", respectively.The case origin of the homogenate is indicated on the left. HTB-14 and DU54 are widely used established glioma cell lines.Tissue homogenates of lumbar spinal cord (LSC) and temporal cortex (TC) were analyzed in parallel to lysatesproduced from tumor tissues.

sis of virus susceptibility and CD155 expression levels. In accordance with universal upregulation of *CD155* gene expression in established glioma cell lines [27,83], CD155 was detected in all GBM biopsy specimens (Figure 7). Subcultivation of tumor tissue obtained from patients did not alter receptor expression, given that Western blot analysis revealed parallel amounts in tumor tissue and primary explant cultures (27; Figure 7). CD155 expression in the normal human brain, including the spinal cord, is exceedingly low (27; Figure 7). In accordance with our receptor expression data, susceptibility to PVS-RIPO was universal among the panel of GBM primary cell lines (Figure 7).

The proposal to consider PVS-RIPO for use as an oncolytic agent is not based on properties of the virus alone, but motivated mainly by natural target tropism for malignant glioma cells mediated by ectopic expression of CD155. Our pre-clinical evaluations strongly suggest that, at least in high-grade malignant glioma, universal receptor expression leads to exquisite sensitivity to the oncolytic effects of PVS-RIPO infection. Most importantly, our empirical results are not based on studies of established glioma cell lines but rather on patients' tumor material.

Studies of poliovirus susceptibility suggest that CD155 expression may also occur in non-CNS malignancies. Accordingly, an association of CD155 with colorectal cancer has recently been reported [66].

Physiological Functions of CD155

CD155 was identified because of its role as a cellular poliovirus receptor [1]. To date, apart from hypothetical functions suggested by its predicted structure, its relation to other members of the IGSF, or its expression pattern during CNS embryogenesis, CD155 has not been implicated functionally in any physiological process. Because it is likely that CD155 exerts its physiological function(s) through interaction with specific ligands, experimental approaches have focused on the identification of CD155 binding partners.

Recently, a search for a physiological binding partner for CD155 has identified the extracellular matrix (ECM) protein vitronectin [67]. Although vitronectin is widely distributed in a number of organs, its presence coincides with known sites of CD155 expression—e.g., the floor plate of the developing neuraxis and retina [28,67,68,69]. Moreover, vitronectin has been implicated in motor neuron induction and differentiation [69,70]. Most significantly, vitronectin is the major ECM component in malignant gliomas and is particularly abundant in high-grade malignant gliomas, where it has been reported to promote glioma cell survival and invasion [70,71].

In general, IGSF-CAMs are believed to confer invasive properties to cancer cells through their interactions with ECM proteins [62]. CD155 affinity for vitronectin establishes a further link between IGSF-CAMs and the ECM in malignant glioma. Because local invasion is a standard feature of malignant glioma, posing severe limits on surgical and radiological treatment regimens, the functional significance of IGSF-CAM expression in malignant glioma has been under intense investigation. The physiological significance of CD155–vitronectin interactions in the developing CNS and in malignant CNS tumors remains to be evaluated.

Tumor-Specific Virulence

Once the viral genome has reached the interior of the host tumor cell, oncolytic activity depends on the virus' ability to replicate and induce cytopathic changes and tumor cell killing. To render oncolytic viruses clinically applicable, viral replication and cytopathogenicity need to be selective, killing off tumor cells while sparing healthy cells. Thus, pathogenic viruses considered for use as oncolytic agents—e.g., poliovirus—need to be manipulated to prevent virus propagation and associated toxicity in nonmalignant tissues without affecting growth in transformed cells. This property has been termed a *conditional replication* phenotype [73–75].

A number of mechanisms have been targeted to achieve conditional virus replication in cancerous cells. These include deletion of endogenous viral genes coding for products required for replication in nonmalignant cell types but expendable in tumor cells, and insertion of tumor-specific transcriptional control elements into the viral genome, promoting viral gene expression selectively in cancer cells.

Despite its inherent ability to infect and destroy cancer cells, poliovirus' neuropathogenic properties prevent its use as an oncolytic agent. Because poliovirus pathogenicity is limited to infection of spinal cord and brain stem motor neurons, a conditional replication phenotype in tumor cells would be achieved by eliminating the virus' ability to propagate in neuronal cell types.

In contrast to large DNA viruses under consideration for use as cancer therapeutics—e.g., adeno- and herpesviruses—poliovirus genomes are characterized by extreme genetic austerity (the poliovirus genome is ~7.5 kB in length, compared with ~150 kB for α-herpesviruses; 76). To adapt to the extraordinarily high mutation rate of their replication machinery, picornaviruses have limited the size of their genome and number of their gene products to an essential minimum [76]. Therefore, any manipulations to delete viral coding sequences or add extraneous regulatory sequences would either severely impair virus viability or trigger deletion events to remove foreign inserts.

To confer a conditional replication phenotype to poliovirus, we manipulated the viral IRES element, which is critically involved in translation control of the poliovirus genome (discussed above). Our approach is the first to achieve a conditional replication phenotype through manipulation of non-coding sequences regulating translation and is the first treatment targeting tumor cells at the level of translation control.

Swapping the Poliovirus Internal Ribosomal Entry Site Element

Poliovirus has demonstrated a surprising level of flexibility with regard to accommodating heterologous IRES elements of variable origin within its genome. Insertion of closely related enteroviral IRESes (e.g. Coxsackievirus B4, Coxsackievirus A9, or Enterovirus 71; M. Gromeier et al, unpublished observations), more distantly related cardioviral IRESes (e.g., encephalomyocarditis virus) [77] or even the unrelated hepatitis C virus IRES [78] produces viable polioviruses with wild-type growth properties. Furthermore, these chimeric polioviruses maintained their pathogenic potential and caused paralytic poliomyelitis in CD155 transgenic mice [79].

The most remarkable effect of IRES exchange was observed with a chimeric poliovirus, termed *PVS-RIPO*, carrying the IRES element of human rhinovirus type 2 (HRV2), a fellow picornavirus of the *rhinovirus* genus (79; Figure 8). The chimeric PVS-RIPO, while replicating with wild-type kinetics in HeLa cells, had acquired a dramatic reduction of neuropathogenicity (79; Figure 8). This recombinant was unable to replicate within the spinal cord of CD155 transgenic mice (Figure 8) and failed to cause poliomyelitis in these animals, even after intracerebral administration of doses exceeding the wild-type LD_{50} by seven orders of magnitude [79]. Accordingly, polioviruses containing the HRV2 IRES element failed to induce spinal cord pathology and neurological disease after intraspinal inoculation in *Cynomolgus* macaques (80; Table 2).

Rhinoviruses cause the common cold and do not exhibit tropism for neuronal tissues. During infection, rhinovirus rep-

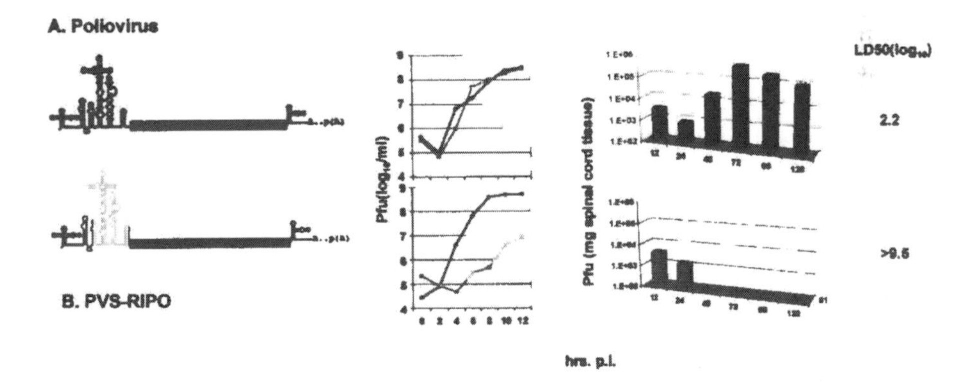

Figure 8 The cell type-specific replication deficit of PVS-RIPO. Genetic structure (left panel), replication profile in DU54 glioblastoma (blue diamonds) and Sk-N-Mc neuroblastoma (red squares) (left middle panel), replication in the spinal cord of CD155 transgenic mice (right middle panel), and LD50 in CD155 transgenic mice (right panel) of wild-type poliovirus (**A**) and PVS-RIPO (**B**). The HRV2 IRES is outlined in red.

Table 2 Monkey Neurovirulence Testing Results

Virus	No. Cynomolgus macaques	Mean lesion score	Clinical outcome
PV1(MAHONEY)*	4*	2.48±0.34*	75% fatal*
NA-4**	14	0.87±0.38	No clinical symptoms
PV1(RIPO)	14	0.92±0.42	No clinical symptoms
PVS-RIPO¶	ND	ND	

* The data shown on wild-type PV1(Mahoney) are derived from a separate study (103). **NA-4 is a reference PV1(S) strain. NA-4 and PV1(RIPO) were studied in parallel by K. Chumakov (U. S. Food and Drug Administration, Center for Biologics Evaluation and Research). †¶PVS-RIPO is scheduled to undergo primate testing at the same facility in the near future.

lication is restricted to the upper respiratory tract. Thus, neuronal dysfunction of its IRES element would not affect rhinovirus pathogenicity.

The nonpathogenic phenotype of PVS-RIPO corresponded to a pronounced defect of this virus to propagate in cultured cell lines of neuronal origin,—e.g., Sk-N-Mc neuroblastoma cells (79; Figure 8). Neuroblastoma cell lines, albeit their malignant origin, have served as a convenient tissue culture model to study poliovirus replication in a neuronal environment [81,82]. Remarkably, despite a major impediment to virus replication in neuronal cell types, PVS-RIPO retained wild-type growth kinetics and superb oncolytic potential for malignant glioma cell lines—both established cell lines [83] and primary explant cultures (27; Figures 7 and Figures 8).

Our observations suggested that insertion of a heterologous HRV2 IRES into the poliovirus genome suppressed neuronal replication and, hence, pathogenesis. Because IRES-mediated selective replication deficits provide the basic rationale to use polioviruses as oncolytic agents, we intend to unravel the molecular mechanisms responsible for the observed phenotype.

The Genetic Basis for the Tumor-Specific Replication Phenotype

To elucidate the molecular mechanism for tumor-specific function of the HRV2 IRES element, we are pursuing genetic and biochemical approaches—identifying the genetic locus for the non-neuropathogenic phenotype of PVS-RIPO and characterizing the interaction of the IRES with the cytoplasmic milieu of glioma vs. normal neuronal cells.

Following the simplest strategy of genetic analyses of poliovirus, we carried out adaptation experiments of PVS-RIPO to identify genetic alterations that might restore the neurovirulent phenotype upon serial passages in neuronal cell lines or CNS tissues. Interestingly, passage in Sk-N-Mc neuroblastoma cells or in CD155 transgenic mice did not produce any neurovirulent revertants [83]. This finding indicated a high degree of genetic stability of the nonpathogenic phenotype of PVS-RIPO (a desirable safety feature for oncolytic viruses; discussed below), which prevented isolation and characterization of neurovirulent revertants.

We therefore are limited to reverse genetics to delineate the genetic basis of PVS-RIPO neuro-attenuation. Because of the high level of conservation of the predicted secondary structure of the HRV2 and poliovirus IRESes, individual stem-loop structures of either IRES element can be recombined without affecting the ability of the virus to grow in malignant glioma cells [80]. We used such a "mix-and-match" recombination strategy to determine the genetic locus for neuronal IRES defects (80; Figure 9).

Our experiments implicated IRES domains V/VI, in the neurovirulent phenotype (80; Figure 9). Insertion of HRV2 IRES domains V/VI was sufficient to reduce neurovirulence in CD155 transgenic mice to the level of PVS-RIPO (80; Figure 9). Conversely, HRV2 IRES domains V/VI alone conferred attenuation of neurovirulence to poliovirus (80; Figure 9).

Our findings implicating IRES domains V/VI in the neurovirulent phenotype of poliovirus are in accordance with a proposed role of the IRES in attenuation of the live attenuated vaccine strains of poliovirus (the Sabin strains). All three Sabin serotypes feature point mutations within a confined region of IRES domain V that have been implicated in the attenuation phenotype [14].

Sequence differences mapping to IRES domain V have been proposed to cause functional deficits through their effect

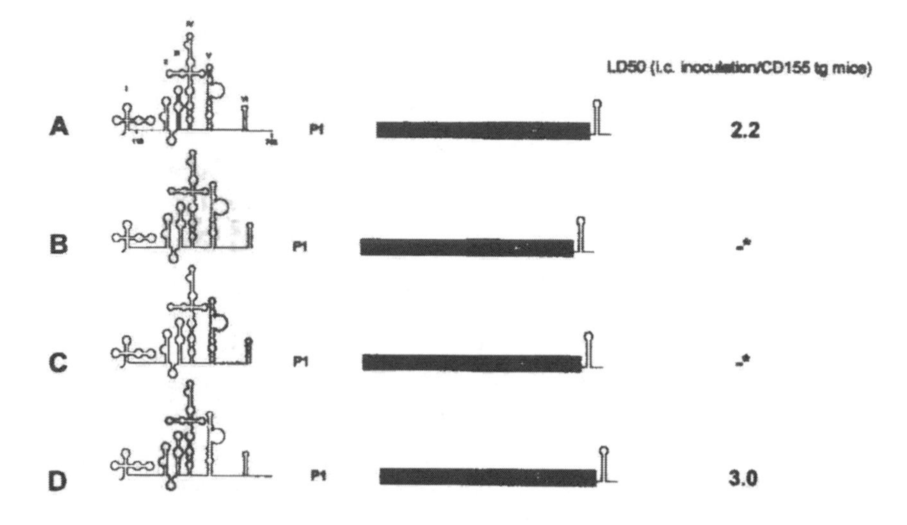

Figure 9 The genetic locus for neuronal internal ribosomal entry site (IRES) function. Poliovirus (**A**), PVS-RIPO (**B**), and mix-and-match IRES hybrids (**C, D**) were tested for their neuropathogenic potential in CD155 transgenic mice. *No neurological symptoms were observed after administration of 10^9 plaque forming units.

on the predicted secondary structure of the IRES element [84,85]. This notion is reinforced by serial passaging experiments of poliovirus type 1 (Sabin) in cultured neuronal cells. These experiments yielded neurovirulent variants in which base pairing and, hence, stem loop stability were restored to correspond to the wild-type structure [86]. The evaluation of the influence of predicted secondary structure on HRV2 IRES function in neuronal vs. malignant glioma cells is in progress.

Apart from *cis* effects of IRES structure on cell type-specific function, *trans*-acting factors supplied by the infected host cell have been proposed as a major determinant of IRES function and its cell type-specific restrictions.

Cellular trans-Acting Factors

The mechanism of translation initiation at the IRES is fundamentally different from that of eukaryotic mRNAs (discussed above, Figure 2). Like the latter, translation mediated by internal ribosomal entry appears to depend on canonical translation initiation factors required for ribosomal scanning, with the exception of the cap-binding initiation factor eIF4E [87]. In addition, noncanonical *trans*-activating eukaryotic factors may be needed for efficient IRES function.

A number of *trans*-activating factors have been suggested to interact with picornavirus IRESes. These include the polypyrimidine tract binding protein (PTB) [88], the La autoantigen (the target for autoantibodies in systemic lupus erythematosus) [89], the poly(rC) binding protein 2 (PCBP2) [90,91], and the RNA chaperone 'upstream of N-ras' (unr) [92].

The identification of *trans*-acting factors followed procedures tracking polypeptide binding to IRES sequences. Isolation of factors that bind to IRES sequences does not necessarily imply this binding activity to be functionally significant. Recently, we have provided the first evidence for PTB to be functionally required for poliovirus replication *in vivo* [93].

The requirements for IRES function appear to differ individually, meaning that distinct IRESes require different sets of interacting *trans*-activating factors. Polypyrimidine tract binding protein may stimulate all picornavirus IRESes—e.g., cardioviruses [94], aphthoviruses [95], enteroviruses [88], rhinoviruses [92], and hepatoviruses [96]—and has been reported to *trans*-activate the IRES of HCV [97] and to bind even to putative eukaryotic IRESes [98]. In contrast, requirement for unr seems, at this point, to be limited to the IRESes of polio- and rhinoviruses [92,99].

The role of cellular noncanonical *trans* activities in IRES translation suggests that IRES function and, thus, pathogenic properties, could co-vary with the set of stimulating factors available in the infected cell. For example, neuron-specific deficits of the HRV2 IRES could be a reflection of the differences of the cytoplasmic make-up of neuronal vs. glioma cells. Such a scenario has been proposed to be a determinant of the neuropathogenic properties of another picornavirus, Theiler's murine encephalitis virus [100].

We are investigating a number of candidate eukaryotic RNA-binding proteins with affinity for the HRV2 IRES and their role in the conditional replication phenotype of PVS-RIPO. All investigations of the interaction of eukaryotic IRES *trans* activities were based on the premise that IRES activity is determined by the availability of activating factors alone. Our experimental results strongly suggest that not only is IRES function subject to the presence of stimulating factors, but repression of IRES function can occur through IRES-binding proteins with inhibitory function (M. Merrill et al, unpublished results).

PVS-RIPO as an Oncolytic Agent

Inherent tropism for CD155, a tumor antigen ectopically expressed in neuroectodermal malignancy, and a nonpathogenic profile in primates, raise the possibility of using PVS-RIPO as an oncolytic agent. We have carried out extensive studies in animal xenotransplantation models to characterize the oncolytic potential of PVS-RIPO in vivo. Treatment of athymic mice harboring subcutaneous glioma xenografts with PVS-RIPO resulted in dramatic tumor regression and, eventually, elimination (83; Figure 10). Intratumoral virus replication corresponded with gradual shrinkage of the xenograft, indicating that virus propagation within tumor cells reduced the xenotransplant (Figure 10B). Similar results were obtained after i.v. or intramuscular administration of PVS-RIPO [83]. Treatment of athymic mice carrying bilateral xenografts by intratumoral inoculation of PVS-RIPO produced shrinkage of the contralateral mass, indicating efficient targeting of tumor cells by the agent [83]. Histopathological evaluation of the diminishing xenografts revealed intense infiltration of the neoplastic lesion followed by replacement with fibrous scar tissue (Figure 10A).

Intratumoral inoculation of PVS-RIPO into intracerebral glioma xenografts similarly produced tumor regression (Figure 11A). However, peripheral forms of administration of the virus had little effect on tumor progress and survival (Figure 11B). These observations confirm the notion of inefficient CNS invasion of circulating poliovirus (discussed above) and exclude the possibility of treating intracerebral neoplastic processes via peripheral administration of PVS-RIPO.

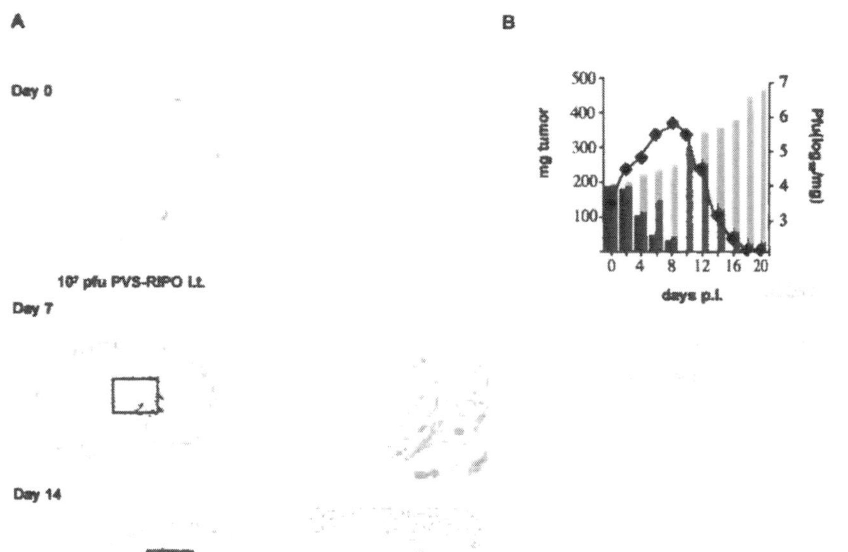

Figure 10 Treatment of subcutaneous glioma xenografts with PVS-RIPO. **A.** Subcutaneous HTB15 glioma xenotransplants (Day 0 = 10 mm diameter) were treated with a single intratumoral inoculation of PVS-RIPO. Seven days after virus treatment, the viable tumor had shrunk considerably. Approximately 90% of the tumor mass had been destroyed and replaced by fibrous scar tissue. The detail images (right panel) show the remaining tumor nodus surrounded by infiltrating cells and scar tissue. At day 14 after treatment, only a minuscule remnant of tumor can be distinguished (arrowhead in the detail image). At this time, the scar had replaced almost the entire tumor mass. **B.** The kinetics of tumor regression in PVS-RIPO–treated xenografted animals (blue bars) and their untreated peers (red bars). Tumor regresson corresponded to the kinetics of intratumoral virus propagation (black diamonds).

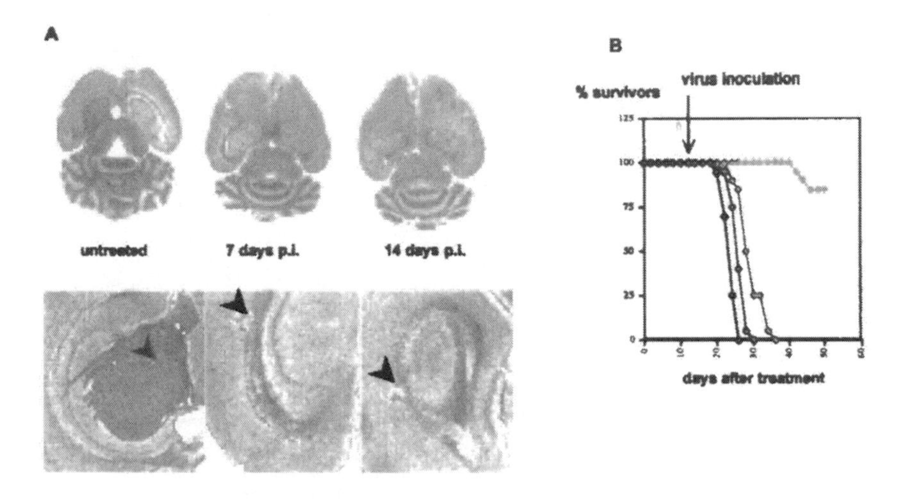

Figure 11 Treatment of intracerebral glioma xenografts with PVS-RIPO. **A.** Tumor regression and elimination after intratumoral inoculation of PVS-RIPO into intracerebral HTB14 glioma xenotransplants [83]. The lower panel shows the tumor implantation site in detail. The injection locale is indicated by arrowheads. **B.** Survival of athymic mice harboring intracerebral HTB14 xenotransplants after mock treatment (dark blue diamonds), intramuscular (green diamonds), intravenous (red diamonds), and intratumoral (light blue diamonds) administration of PVS-RIPO. Virus treatment was implemented at day 12 after xenografting.

Further evaluations of the microdiffusion of PVS-RIPO from the site of inoculation and the evaluation of alternative forms of intratumoral administration (stereotactic bolus injection vs. convection-enhanced delivery) are in progress.

SAFETY

Attenuation of Neuropathogenic Properties

The most imminent and serious safety concern associated with therapeutic use of polioviruses is their neuropathogenic potential. We have identified a powerful new way to achieve superb levels of poliovirus neuroattenuation without affecting growth in malignant glioma cell types. Insertion of a heterologous

HRV2 IRES into the poliovirus genome drastically reduced the virus' ability to propagate in cells of neuronal origin and eliminated neuropathogenicity in a transgenic mouse model for poliomyelitis and in *Cynomolgus* macaques [80].

The phenomenon of poliovirus neurovirulence and its attenuation has been the subject of numerous investigations ever since epidemic paralytic poliomyelitis emerged as a public health concern. The first thoroughly characterized attenuated poliovirus strains were generated by A. Sabin through serial passaging of wild-type neurovirulent polioviruses in a variety of host cell types [101]. Sequencing of the poliovirus serotype 1 (S) [PV1(S)] genome in 1981 [102] and comparison with the sequence of its wild-type progenitor [103] provided the first glimpse into molecular mechanisms of the attenuation of poliovirus neurovirulence.

Analysis of the molecular basis for attenuation of the Sabin vaccine strains of poliovirus has implicated major attenuating determinants within the capsid as well as point mutations within stem loop domain V of the IRES (discussed above; 14). The coding regions for the capsid proteins of the Sabin vaccine strains, compared with their wild-type progenitors, contain mutations (encoding 11 amino acid exchanges in serotype 1 and 3 in serotype 3; serotype 2 is a naturally occurring attenuated variant neurovirulent progenitor of which is unknown). Furthermore, reversions to the neurovirulent phenotype in vaccinees are commonly accompanied by reversion to the wild-type capsid or IRES sequences [14,76].

Because the attenuating effects of the Sabin capsids are well documented, we opted to construct a prototype oncolytic poliovirus in a PV1(S) background. PV1(S) is considered to have the highest level of neuroattenuation because it features the largest number of capsid mutations associated with attenuated neurovirulence. This consideration is supported by the fact that reversion to neurovirulence in immunized children is far more common with the type 2 and 3 (S) strains than PV1(S) [14,76].

Preliminary investigation of poliovirus/rhinovirus chimeras in monkey neurovirulence assays demonstrated that insertion of the HRV2 IRES alone, yielding PV1(RIPO), reduced poliovirus neurovirulence to the level of PV1(S) (Table 2). It is reasonable to assume that combination of the HRV2 IRES and the coding region of PV1(S) containing the major attenuation determinants mapping to the capsid will further reduce neurovirulence of PVS-RIPO in primates.

Prior to clinical application, the oncolytic PVS-RIPO will be submitted to comprehensive neurovirulence evaluation after intraspinal and intrathalamic inoculation in *Cynomolgus* macaques following standard World Health Organization procedures.

Genetic Stability and Spontaneous Reversion to Neurovirulence

The live attenuated Sabin vaccine strains are known to spontaneously revert to a neurovirulent phenotype. This event has been reported to occur upon prolonged passage in tissue culture, in the gastrointestinal tract of vaccinees without clinical signs of paralytic polio, and in cases of vaccine-associated poliomyelitis [14,104]. Despite this tendency, vaccine-related incidences of paralytic poliomyelitis are extremely rare (2 to 3 cases per million vaccinated).

The inherent genetic variability of polioviruses is cause for the concern that genetic manipulations carried out to achieve attenuation of neurovirulence may revert upon virus replication in vivo. In contrast to the Sabin vaccine strains, which are attenuated by a series of point mutations scattered throughout the genome (e.g., only four nucleotide substitutions for serotype 3), PVS-RIPO contains an entire genetic element (the IRES) with severe functional deficits in neuronal cells. Serial passaging of PVS-RIPO in Sk-N-Mc neuroblastoma cells or CD155 transgenic mice failed to produce neurovirulent revertants (discussed above; 83), indicating that the genetic basis for attenuation of PVS-RIPO must be comparatively broad. Our current insight suggests that the neuronal deficit of the HRV2 IRES rests on structural features determined by complex sequence elements (stem loop domains V/VI) rather than single nucleotide exchanges, as has been proposed for the Sabin strains [14].

Neurovirulence per se is not an indicator of poliovirus fitness or replication potential. Therefore, to explain the occurrence of adaptation events with the Sabin strains, the question arises how reversion to neurovirulence benefits their growth rate. Notably, reversion to neurovirulence of the Sabin vaccine strains commonly occurs in non-neuronal cell types—e.g., VERO cells [105] or the human gut [106]. Because the Sabin strains are characterized by reduced growth compared with their wild-type progenitors in many cell lines—e.g. VERO monkey kidney, HeLa adenocarcinoma, DU54 glioma—adaptation events may simply restore optimal replication rates. This would mean that reversion to neurovirulence is a side effect of adaptation to optimal growth rates.

Despite its superior attenuation profile, PVS-RIPO, in contrast to the Sabin strains, grows with wild-type kinetics in all permissive cell lines [79,83]. Prolonged passaging of PVS-RIPO in VERO, HeLa, or DU54 cells and even neuronal cell types did not alter the PVS-RIPO genotype (Dobrikovaey et al., unpublished observations). Our observations may indicate that the observed genetic stability of PVS-RIPO is a reflection of unimpeded growth in permissive cell lines and the broad genetic basis of its neuron-specific replication defect.

Intergenomic Recombination

It is well established that during picornavirus replication, extensive intergenomic recombination can occur. These events have been observed in tissue culture systems [76] and in cell-free poliovirus replication systems [107], as well as in vivo [108]. Although there is substantial evidence for intergenomic recombination among strains of the same species (e.g., among the three live attenuated vaccine strains of poliovirus) [108], the occurrence of recombination events involving different enterovirus species has never been documented. Given the extensive sequence homology among enteroviruses [14], intergenomic recombination events involving multiple enterovirus species are conceivable.

The propensity for intergenomic recombination entails the possibility that genetically engineered attenuated poliovirus strains may acquire pathogenic properties through recombination with other poliovirus strains or non-polio enteroviruses. Because it is highly improbable that cancer patients may harbor active poliovirus infection (only patients with documented anti-poliovirus immunization will be enrolled in clinical trials), recombination events involving circulating wild-type or live attenuated poliovirus strains are unlikely to occur. This leaves the possibility of recombination with non-polio enteroviruses in patients harboring concomitant enterovirus infections at the time of oncolytic virus treatment.

Using established tissue culture and cell-free replication systems, we have analyzed the frequency of recombination events involving PVS-RIPO and non-polio enteroviruses. Because of their very close genetic kinship to poliovirus [109], our experiments have focused on 'C' cluster Coxsackie A viruses (CAV17, 21, and 24). Experimental conditions that favored the occurrence of recombination events involving different poliovirus strains failed to produce recombinant hybrid genomes containing CAV sequences. Our tissue culture experiments clearly demonstrate that intergenomic recombination events involving polio- and non-polio enteroviruses are very unlikely to occur. However, we cannot exclude the possibility that specific conditions prevailing in the human gastrointestinal tract or CNS may favor recombination events that fail to materialize in vitro.

REFERENCES

1. Mendelsohn CL, Wimmer E, Racaniello VR. Cellular receptor for poliovirus: molecular cloning, nucleotide sequence, and expression of a new member of the immunoglobulin superfamily Cell 1989; 56:855–865.
2. He Y, Bowman VD, Mueller S, Bator CM, Bella J, Peng X, Baker TS, Wimmer E, Kuhn RJ, Rossmann MG. Interaction of the poliovirus receptor with poliovirus Proc Natl Acad Sci U S A 2000; 97:79–84.
3. Nomoto A, Lee YF, Wimmer E. The 5′ end of poliovirus mRNA is not capped with m7G(5′)ppp(5′)Np Proc Natl Acad Sci U S A 1976; 73:375–380.

4. Lee YF, Nomoto A, Detjen BM, Wimmer E. A protein covalently linked to poliovirus genome RNA Proc Natl Acad Sci U S A 1977; 74:59–63.

5. Jang SK, Kraeusslich HG, Nicklin MJ, Duke GM, Palmenberg AC, Wimmer E. A segment of the 5′ nontranslated region of encephalomyocarditis virus RNA directs internal entry of ribosomes during in vitro translation J Virol 1988; 62: 2636–2643.

6. Pelletier J, Sonenberg N. Internal initiation of translation of eukaryotic mRNA directed by a sequence derived from poliovirus RNA Nature 1988; 334:320–325.

7. Etchison D, Milburn SC, Edery I, Sonenberg N, Hershey JW. Inhibition of HeLa cell protein synthesis following poliovirus infection correlates with the proteolysis of a 220,000-dalton polypeptide associated with eukaryotic initiation factor 3 and a cap binding protein complex J Biol Chem 1982; 257: 14806–14810.

8. Joachims M, Van Breugel PC, Lloyd RE. Cleavage of poly(A)-binding protein by enterovirus proteases concurrent with inhibition of translation in vitro J Virol 1999; 73:718–727.

9. Kerekatte V, Keiper BD, Badorff C, Cai A, Knowlton KU, Rhoads RE. Cleavage of Poly(A)-binding protein by coxsackievirus 2A protease in vitro and in vivo: another mechanism for host protein synthesis shutoff? J Virol 1999; 73:709–717.

10. Le H, Tanguay RL, Balasta ML, Wei CC, Browning KS, Metz AM, Goss DJ, Gallie DR. Translation initiation factors eIF-iso4G and eIF-4B interact with the poly(A)-binding protein and increase its RNA binding activity J Biol Chem 1997; 272: 16247–16255.

11. Tarun SZ, Sachs AB. Association of the yeast poly(A) tail binding protein with translation initiation factor eIF-4G EMBO J 1996; 15:7168--7177.

12. Wells SE, Hillner PE, Vale RD, Sachs AB. Circularization of mRNA by eukaryotic translation initiation factors Mol Cell 1998; 2:135–140.

13. Sachs AB. Physical and functional interactions between the mRNA cap structure and the poly(A) tail. In: Sonenberg N,

Hershey JWB, Matthews MB, Eds. Translational Control of Gene Expression. NY: Cold Spring Harbor Laboratory Press, 2000:447–466.

14. Gromeier M, Nomoto A. Determinants of poliovirus pathogenesis. In: Semler B, Wimmer E, Eds. The Molecular Biology of Picornaviruses. Washington, DC: ASM Press, 2002:426–443..
15. Yang WX, Terasaki T, Shiroki K, Ohka S, Aoki J, Tanabe S, Nomura T, Terada E, Sugiyama Y, Nomoto A. Efficient delivery of circulating poliovirus to the central nervous system independently of poliovirus receptor Virology 1997; 229: 421–428.
16. Ren R, Racaniello VR. Poliovirus spreads from muscle to the central nervous system by neural pathways J Infect Dis 1992; 166:747–752.
17. Gromeier M, Wimmer E. Mechanism of injury-provoked poliomyelitis J Virol 1998; 72:5056–5060.
18. Bodian D. Poliomyelitis. In: Minckler J, Ed. Pathology of the Nervous System. Vol. 3. New York: McGraw-Hill, 1972: 2323–2344.
19. Bellmunt A, May G, Zell R, Pring-Akerblom P, Verhagen W, Heim A. Evolution of poliovirus type I during 5.5 years of prolonged enteral replication in an immunodeficient patient Virology 1999; 265:178–184.
20. Bergelson JM, Cunningham JA, Droguett G, Kurt-Jones EA, Krithivas A, Hong JS, Horwitz MS, Crowell RL, Finberg RW. Isolation of a common receptor for coxsackie B viruses and adenoviruses 2 and 5 Science 1997; 275:1320–1323.
21. Tomko RP, Xu R, Philipson L. HCAR and MCAR: The human and mouse cellular receptors for subgroup C adenoviruses and group B coxsackieviruses Proc Natl Acad Sci U S A 1997; 94: 3352–3356.
22. Heise C, Sampson-Johannes A, Williams A, McCormick F, Von Hoff DD, Kirn DH. ONYX-015, an E1B gene-attenuated adenovirus, causes tumor-specific cytolysis and antitumoral efficacy that can be augmented by standard chemotherapeutic agents Nat Med 1997; 3:639–645.
23. Asaoka K, Tada M, Sawamura Y, Ikeda J, Abe H. Dependence of efficient adenoviral gene delivery in malignant glioma cells

on the expression levels of the coxsackievirus and adenovirus receptor J Neurosurg 2000; 92:1002–1008.

24. Douglas JT, Kim M, Sumerel LA, Carey DE, Curiel DT. Efficient oncolysis by a replicating adenovirus (ad) in vivo is critically dependent on tumor expression of primary ad receptors Cancer Res 2001; 61:813–817.
25. Li D, Duan L, Freimuth P, O'Malley BW. Variability of adenovirus receptor density influences gene tranfer efficiency and therapeutic response in head and neck cancer Clin Cancer Res 1999; 5:4175–4181.
26. Li Y, Pong RC, Bergelson JM, Hall MC, Sagalowsky AI, Tseng CP, Wang Z, Hsieh JT. Loss of adenoviral receptor expression in human bladder cancer cells: a potential impact on the efficacy of gene therapy Cancer Res 1999; 59:325–330.
27. Merrill M, Sampson JH, Bernhardt G, Wikstrand C, Bigner DD, Gromeier M. Molecular targeting of the human poliovirus receptor CD155 in malignant glioma Neuro Oncology 2004; 6: 208–217.
28. Gromeier M, Solecki D, Patel D, Wimmer E. Expression of the human poliovirus receptor/CD155 gene during development of the CNS: implications for the pathogenesis of poliomyelitis Virology 2000; 273:259–268.
29. Eberle F, Duybreuil P, Mattei MG, Devilard E, Lopez M. The human PRR2 gene, related to the human poliovirus receptor gene (PVR) is the true homolog of the murine MPH gene Gene 1995; 159:267–272.
30. Lopez M, Eberle R, Mattei MG, Gabert J, Birg F, Bardin F, Maroc C, Dubreuil P. Complementary DNA characterization and chromosomal localization of a human gene related to the poliovirus receptor-encoding gene Gene 1995; 155:261–265.
31. Reymond N, Borg J, Lecocq E, Adelaide J, Campadelli-Fiume G, Dubreuil P, Lopez M. Human nectin3/PRR3: A novel member of the PVR/PRR/nectin family that interacts with afadin Gene 2000; 255:347–355.
32. Reymond N, Fabre S, Lecocq E, Adelaide J, Dubreuil P, Lopez M. Nectin4/PRR4, a new afadin-associated member of the nectin family that trans-interacts with nectin1/PRR1 through V domain interaction J Biol Chem 2001; 276:43205–43215.

33. Lopez M, Aoubala M, Jordier F, Isnardon D, Gomez S, Dubreuil P. The human poliovirus receptor related 2 protein is a new hematopoietic/endothelial homophilic adhesion molecule Blood 1998; 92:4602–4611.

34. Aoki J, Koike S, Asou H, Ise I, Suwa H, Tanaka T, Miyasaki M, Nomoto A. Mouse homolog of poliovirus receptor-related gene 2 product, mPRR2, mediates homophilic cell aggregation Exp Cell Res 1997; 235:374–384.

35. Tachibana K, Nakanishi H, Mandai K, Ozaki K, Ikeda W, Yamamoto Y, Nagafuchi A, Tsukita S, Takai Y. Two cell adhesion molecules, nectin and cadherin, interact through their cytoplasmic domain-associated protein J Cell Biol 2000; 150: 1161–1176.

36. Takahashi K, Nakanishi H, Mihahara M, Mandai K, Satoh K, Satoh A, Nishioka H, Aoki J, Nomoto A, Mizoguchi A, Takai Y. Nectin/PRR: An immunoglobulin-like cell adhesion molecule recruited to cadherin-based adherens junctions through interaction with afadin, a PDZ domain-containing protein J Cell Biol 1999; 145:539–549.

37. Miyahara M, Nakanishi H, Takahashi K, Satoh-Horikawa K, Tachibana K, Takai Y. Interaction of nectin with afadin is necessary for its clustering at cell-cell contact sites but not for its *cis* dimerization or *trans* interaction J Biol Chem 2000; 275:613–618.

38. Geraghty RJ, Krummenacher C, Cohen GH, Eisenberg RJ, Spear PG. Entry of alphaherpesviruses mediated by poliovirus receptor-related protein 1 and poliovirus receptor Science 1998; 280:1618–1620.

39. Koike S, Ise I, Sato Y, Yonekawa H, Gotoh O, Nomoto A. A second gene for the African green monkey poliovirus receptor that has no putative N-glycosylation site in the functional N-terminal immunoglobulin-like domain J Virol 1992; 66: 7059–7066.

40. Chadeneau C, LeCabellec M, LeMoullac B, Meflah K, Denis MG. Over-expression of a novel member of the immunoglobulin superfamily in Min mouse intestinal adenomas Int J Cancer 1996; 68:817–821.

41. Ren RB, Costantini F, Gorgacz EJ, Lee JJ, Racaniello VR. Transgenic mice expressing a human poliovirus receptor: a new model for poliomyelitis Cell 1990; 63:353–362.

42. Koike S, Taya C, Kurata T, Abe S, Ise I, Yonekawa H, Nomoto A. Transgenic mice susceptible to poliovirus Proc Natl Acad Sci U S A 1991; 88:951–955.

43. Solecki D, Schwarz S, Wimmer E, Lipp M, Bernhardt G. The promoters from human and monkey poliovirus receptors J Biol Chem 1997; 272:5579–5586.

44. Solecki DJ, Wimmer E, Lipp M, Bernhardt G. Identification and characterization of the *cis*-acting elements of the human *CD155* gene core promoter J Biol Chem 1999; 274:1791–1800.

45. Solecki DJ, Bernhardt G, Lipp M, Wimmer E. Identification of a nuclear respiratory factor-1 binding site within the core promoter of the human poliovirus receptor/CD155 gene J Biol Chem 2000; 275:12453–12462.

46. Solecki DJ, Gromeier M, Mueller S, Wimmer E, Bernhardt G. Expression of the human poliovirus receptor/CD155 gene is activated by Sonic Hedgehog J Biol Chem 2002; 277: 25697–25702.

47. Zhang J, Hagopian-Donaldson S, Serbedzija G, Elsemore J, Plehn-Dujowich D, McMahon AP, Flavell RA, Williams T. Neural tube, skeletal and body wall defects in mice lacking transcription factor AP-2 Nature 1996; 381:238–241.

48. Schorle H, Meier P, Buchert M, Jaenisch R, Mitchell PJ. Transcription factor AP-2 essential for cranial closure and craniofacial development Nature 1996; 381:235–238.

49. Becker TS, Burgess SM, Amsterdam AH, Allende ML, Hopkins N. not really finished is crucial for development of the zebrafish outer retina and encodes a transcription factor highly homologous to human Nuclear Respiratory Factor-1 and avian Initiation Binding Repressor Development 1998; 125:4369–4378.

50. Tanabe Y, Jessell TM. Diversity and pattern in the developing spinal cord Science 1996; 274:1115–1123.

51. Hammerschmidt M, Brook A, McMahon AP. The world according to hedgehog Trends Genet 1997; 13:14–21.

52. Kalderon D. Transducing the hedgehog signal Cell 2000; 103: 371–374.

53. Ingham PW, McMahon AP. Hedgehog signaling in animal development: paradigms and principles Genes Dev 2001; 15: 3059–3087.

54. Goodrich LV, Milenkovic L, Higgins KM, Scott MP. Altered neural cell fates and medulloblastoma in mouse patched mutants Science 1997; 277:1109–1113.

55. Ruiz i Altaba A, Sanchez P, Dahmane N. Gli and hedgehog in cancer: tumours, embryos and stem cells Nat Rev Cancer 2002; 2:361–372.

56. Dahmane N, Sanchez P, Gitton Y, Palma V, Sun T, Beyna M, Weiner H, Ruiz i Altaba A. The Sonic Hedgehog-Gli pathway regulates dorsal brain growth and tumorigenesis Development 2001; 128:5201–5212.

57. Kinzler KW, Bigner SH, Bigner DD, Trent JM, Law ML, O'Brien SJ, Wong AJ, Vogelstein B. Identification of an amplified, highly expressed gene in a human glioma Science 1987; 236:70–73.

58. Izumoto S, Ohnishi T, Arita N, Hiraga S, Taki T, Hayakawa T. Gene expression of neural cell adhesion molecule L1 in malignant gliomas and biological significance of L1 in glioma invasion Cancer Res 1996; 56:1440–1444.

59. Sasaki H, Yoshida K, Ikeda E, Asou H, Inaba M, Otani M, Kawase T. Expression of the neural cell adhesion molecule in astrocytic tumors: an inverse correlation with malignancy Cancer 82:1921–1931.

60. Rickman DS, Rachana T, Xiao-Xiang Z, Bobek MP, Song S, Blaivas M, Misek DE, Israel MA, Kurnit DM, Ross DA, Kish PE, Hanash SM. The gene for the axonal cell adhesion molecule TAX-1 is amplified and aberrantly expressed in malignant glioma Cancer Res 2001; 61:2162–2168.

61. Sehgal A, Boynton AL, Young RF, Vermeulen SS, Yonemura KS, Koehler EP, Aldape HC, Simrell CR, Murphy GP. Cell adhesion molecule Nr-CAM is over-expressed in human brain tumors Int J Cancer 1998; 76:451–458.

62. Goldbrunner RH, Bernstein JJ, Tonn JC. Cell-extracellular matrix interaction in glioma invasion Acta Neurochir 1999; 141:295–305.

63. Johnson JP. Cell adhesion molecules in the development and progression of malignant melanoma Cancer Metastasis Rev 1999; 18:345–357.

64. Ohene-Abuakwa Y, Pignatelli M. Adhesion molecules in cancer biology Adv Exp Med Biol 2000; 465:115–126.

65. Taguchi A, Blood DC, Del Toro G, Canet A, Lee DC, Qu W, Tanji N, Lu Y, Lalla E, Fu C, Hofmann MA, Kislinger T, Ingram M, Lu A, Tanak H, Hori O, Ogawa S, Stern DM, Schmidt AM. Blockade of RAGE-amphoterin signaling suppresses tumor growth and metastases Nature 2000; 405:354–360.

66. Masson D, Jarry A, Baury B, Blanchardie P, Laboisse C, Lustenberger P, Denis MG. Overexpression of the CD155 gene in human colorectal carcinoma Gut 2001; 49:236–240.

67. Lange R, Peng X, Wimmer E, Lipp M, Bernhardt G. The poliovirus receptor CD155 mediates cell-to-matrix contacts by specifically binding to vitronectin Virology 2001; 285:218–227.

68. Martinez-Morales JR, Barbas JA, Marti E, Bovolenta P, Edgar D, Rodriguez-Tebar A. Vitronectin is expressed in the ventral region of the neural tube and promotes the differentiation of motor neurons Development 1997; 124:5139–5147.

69. Martinez-Morales JR, Marti E, Frade JM, Rodriguez-Tebar A. Developmentally regulated vitronectin influences cell differentiation, neuron survival and process outgrowth in the developing chicken retina Neuroscience 1995; 68:245–253.

70. Pons S, Marti E. Sonic hedgehog synergizes with the extracellular matrix protein vitronectin to induce spinal motor neuron differentiation Development 2000; 127:333–342.

71. Gladson CL, Wilcox JN, Sanders L, Gillespie GY, Cheresh DA. Cerebral microenvironment influences expression of the vitronectin gene in astrocytic tumors J Cell Sci 1995; 108: 947–956.

72. Uhm JH, Dooley NP, Kyritsis AP, Rao JS, Gladson CL. Vitronectin, a glioma-derived extracellular matrix protein, pro-

tects tumor cells from apoptotic death Clin Cancer Res 1999; 5:1587–1594.

73. Gromeier M. Viruses as therapeutic agents against malignant disease of the central nervous system J Natl Cancer Inst 2001; 93:889–890.
74. Gromeier M, Wimmer E. Viruses for the treatment of malignant glioma Curr Opin Mol Ther 2001; 3:503–508.
75. Gromeier M. Oncolytic viruses for cancer therapy Am J Cancer 2004:In press.
76. Wimmer E, Hellen CUT, Cao XM. Genetics of poliovirus Annu Rev Genet 1993; 27:353–436.
77. Alexander L, Lu HH, Wimmer E. Polioviruses containing picornavirus type 1 and/or type 2 internal ribosomal entry site elements: genetic hybrids and the expression of a foreign gene Proc Natl Acad Sci U S A 1994; 91:1406–1410.
78. Lu HH, Wimmer E. Poliovirus chimeras replicating under the translational control of genetic elements of hepatitis C virus reveal unusual properties of the internal ribosomal entry site of hepatitis C virus Proc Natl Acad Sci U S A 1996; 93: 1412–1417.
79. Gromeier M, Alexander L, Wimmer E. Internal ribosomal entry site substitution eliminates neurovirulence in intergeneric poliovirus recombinants Proc Natl Acad Sci U S A 1996; 93:2370–2375.
80. Gromeier M, Bossert B, Arita M, Nomoto A, Wimmer E. Dual stem loops within the poliovirus internal ribosomal entry site control neurovirulence J Virol 1999; 73:958–964.
81. La Monica N, Racaniello VR. Differences in replication of attenuated and neurovirulent poliovirus in human neuroblastoma cell line SH-SY5Y J Virol 1989; 63:2357–2360.
82. Agol VI, Drozdov SG, Ivannikova TA, Kolesnikova MS, Korolev MB, Tolskaya EA. Restricted growth of attenuated poliovirus strains in cultured cells of a human neuroblastoma J Virol 1989; 63:4034–4038.
83. Gromeier M, Lachmann S, Rosenfeld M, Gutin P, Wimmer E. Nonpathogenic intergeneric poliovirus recombinants for the

treatment of glioma Proc Natl Acad Sci U S A 2000; 97: 6803–6808.

84. Macadam AJ, Ferguson G, Burlison J, Stone D, Almond JW. Correlation of RNA secondary structure and attenuation of Sabin vaccine strains of poliovirus in tissue culture Virology 1992; 189:415–422.

85. Skinner MA, Racaniello VR, Dunn G, Cooper J, Minor PD, Almond JW. New model for the secondary structure of the 5′ non-coding RNA of poliovirus is supported by biochemical and genetic data that also shows that RNA secondary structure is important in neurovirulence J Mol Biol 1989; 207:379–392.

86. Christodoulou C, Colbere-Garapin F, Macadam A, Taffs LF, Marsden S, Horaud F. Mapping of mutations associated with neurovirulence in monkeys infected with Sabin 1 poliovirus revertants selected at high temperature J Virol 1990; 64: 4922–4929.

87. Pestova TV, Hellen CUT, Shatsky IN. Canonical eukaryotic initation-factors determine initation of tranlation by internal ribosomal entry Mol Cell Biol 1996; 16:6859–6869.

88. Hellen CUT, Witherell GW, Schmid M, Shin SH, Pestova TV, Gil A, Wimmer E. A cytoplasmic 57 kDa protein that is required for translation of picornavirus RNA by internal ribosomal entry is identical to the nuclear pyrimidine tract-binding protein Proc Natl Acad Sci U S A 1993; 90:7642–7646.

89. Meerovitch K, Svitkin YV, Lee HS, Lejbkowicz F, Kenan DJ, Chan EKL, Agol VI, Keene JD, Sonenberg N. La autoantigen enhances and corrects aberrant translation of poliovirus RNA in reticulocyte lysate J Virol 1993; 67:3798–3807.

90. Blyn LB, Swiderek KM, Richards O, Stahl DC, Semler BL, Ehrenfeld E. Poly(rC) binding protein 2 binds to stem-loop IV of the poliovirus RNA 5′ noncoding region: identification by automated liquid chromatography-tandem mass spectrometry Proc Natl Acad Sci U S A 1996; 93:11115–11120.

91. Blyn LB, Towner J, Semler BL, Ehrenfeld E. Requirement of poly(rC) binding protein 2 for translation of poliovirus RNA J Virol 1997; 71:6243–6246.

92. Hunt SL, Hsuan JJ, Totty N, Jackson RJ. Unr, a cellular cytoplasmic RNA-binding protein with five cold-shock domains, is

required for internal initiation of translation of human rhinovirus RNA Genes & Dev 1999; 13:437–448.

93. Florez PM, Sessions O, Wagner E, Gromeier M, Garcia-Blanco M. RNAi-mediated PTB depletion reveals a requirement for picornavirus IRES-dependent translation and propagation J. Virology 2004 Submitted.

94. Witherell G, Gil A, Wimmer E. Interaction of polypyrimdine tract binding protein with the encephalomyocarditis virus mRNA internal ribosomal entry site Biochemistry 1993; 32: 8268–8275.

95. Niepmann M, Petersen A, Meyer K, Beck E. Functional involvement of polypyrimidine tract-binding protein in translation initiation complexes with the internal ribosome entry site of foot-and-mouth disease virus J Virol 1997; 71:8330–8339.

96. Chang KH, Brown EA, Lemon SM. Cell type-specific proteins which interact with the 5′ nontranslated region of hepatitis A virus RNA J Virol 1993; 67:6716–6725.

97. Anwar A, Ali N, Tanveer R, Siddiqui A. Demonstration of functional requirement of polypyrimidine tract-binding protein by SELEX RNA during hepatitis C virus internal ribosome entry site-mediated translation initiation J Biol Chem 2000; 275: 34231–34235.

98. Huez I, Creancier L, Audigier S, Gensac MC, Prats AC, Prats H. Two independent internal ribosome entry sites are involved in translation initiation of vascular endothelial growth factor mRNA Mol Cell Biol 1998; 18:6178–6190.

99. Boussadia O, Niepmann M, Creancier L, Prats AC, Dautry F, Jacquemin-Sablon H. Unr is required in vivo for efficient initiation of translation from the internal ribosome entry site of both rhinovirus and poliovirus J Virol 2003; 77:3353–3359.

100. Pilipenko EV, Viktorova EG, Guest ST, Agol VI, Roos RP. Cell-specific proteins regulate viral RNA translation and virus-induced disease EMBO J 2001; 20:6899–6908.

101. Sabin AB. Oral poliovirus vaccine. History of its development and prospects for eradication of poliomyelitis JAMA 1965; 194: 872–876.

102. Nomoto A, Omata T, Toyoda H, Kuge S, Horie H, Kataoka Y, Genba Y, Nakano Y, Imura N. Complete nucleotide sequence

of the attenuated poliovirus Sabin 1 strain genome Proc Natl Acad Sci U S A 1982; 79:5793–5797.

103. Omata T, Kohara M, Kuge S, Komatsu T, Abe S, Semler BL, Kameda A, Itoh H, Arita M, Wimmer E, Nomoto A. Genetic analysis of the attenuation phenotype of poliovirus type 1 J Virol 1986; 58:348–358.

104. Minor PD. Attenuation and reversion of the Sabin vaccine strains of poliovirus Dev Biol Stand 1993; 78:17–26.

105. Taffs RE, Chumakov KM, Rezapkin GV, Lu Z, Douthitt M, Dragunsky EM, Levenbook IS. Genetic stability and mutant selection in Sabin 2 strain of oral poliovirus vaccine grown under different cell culture conditions Virology 1995; 209: 366–373.

106. Dunn G, Begg NT, Cammack N, Minor PD. Virus excretion and mutation by infants following primary vaccination with live oral poliovaccine from two sources J Med Virol 1990; 32: 92–95.

107. Duggal R, Cuconati A, Gromeier M, Wimmer E. Genetic recombination of poliovirus in a cell-free system Proc Natl Acad Sci U S A 1997; 94:13786–13791.

108. Georgescu MM, Balanant J, Macadam A, Otelea D, Combiescu M, Combiescu AA, Crainic R, Delpeyroux F. Evolution of the Sabin type 1 poliovirus in humans: characterization of strains isolated from patients with vaccine-associated paralytic poliomyelitis J Virol 1997; 71:7758–7768.

109. Gromeier M, Wimmer E, Gorbalenya A. Genetics, pathogenesis and evolution of picornaviruses. In: Domingo E, Webster R, Holland JJ, Eds. Origin and Evolution of Viruses, Academic Press. 1999:287–344.

13

Oncolytic Viruses that Depend on Loss of RB or p53

FRANK MCCORMICK
University of California-San Francisco Cancer Research Institute
San Francisco, CA

INTRODUCTION

Loss of the RB checkpoint is a fundamental step in the development of cancer [1,2]. Loss may occur through direct inactivation of the *RB* gene itself, or through hyperactivation of CDK4, causing inactivation of RB through uncontrolled phosphorylation. In many human cancers, the HPV protein E6 inactivates RB directly. One way or another, tumor cells lose RB function and upregulate activity of the transcription factor complex referred to as *E2F* [3] (Figure 1; 4). This allows unscheduled entry into S-phase, and suppression of differentiation and senescence, because E2F directs transcription of genes involved directly in replicating DNA and synthesizing pools of DNA precursors. Dihydrofolate reductase and thymidylate synthase, for example, are targets of E2F activity, along with cyclins A and E, MCM proteins, and many other key regulators of S-phase (Figure 2).

Another consequence of E2F activation through loss of RB is induction of p14ARF, an inhibitor of MDM2 (Figure 1). In the absence of MDM2 activity, p53 accumulates and can turn on target genes involved in growth suppression or apoptosis. To tolerate high levels of E2F, tumor cells must therefore suppress p53 activity. This is achieved through mutation of p53 itself, through suppression of p14ARF expression via promoter methylation or mutation, or through increased MDM2 activity. Expression of HPV E6 is another way of degrading p53 that occurs in human tumors of squamous cell origin. In general terms, tumor cells can therefore be characterized as cells in which E2F is de-repressed and p53 is sup-

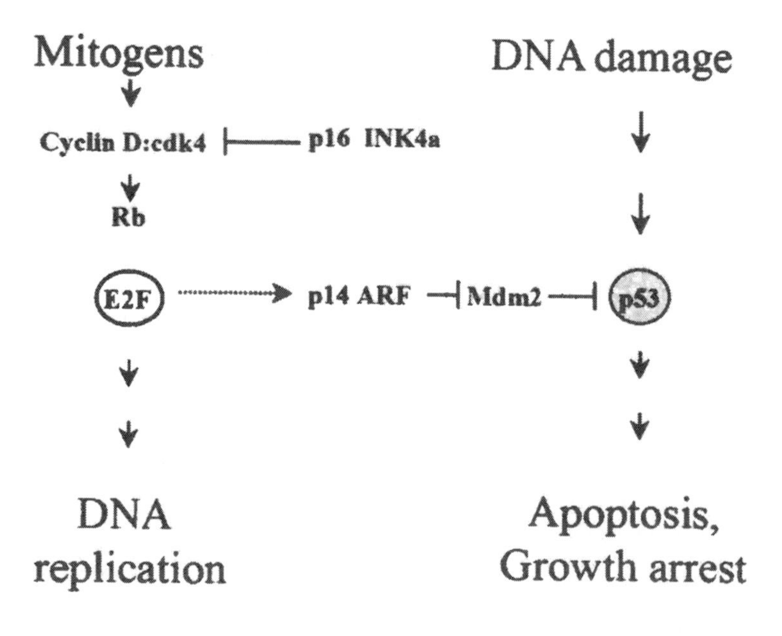

Figure 1

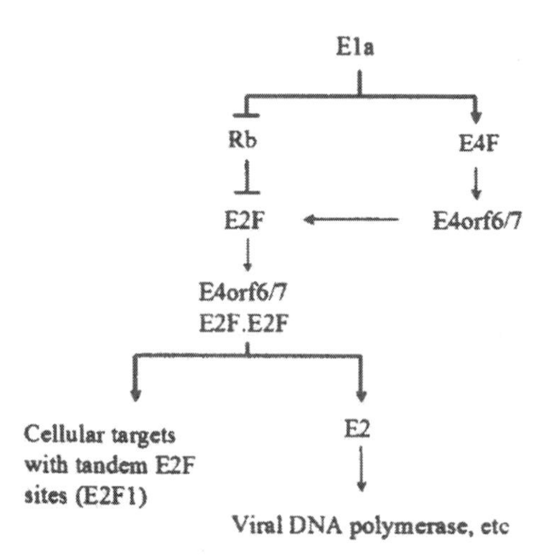

Figure 2

pressed. These cells are characterized by uncontrolled entry into S-phase and reduced susceptibility to apoptosis.

High levels of E2F and low levels of p53 are conditions that are induced by infection of epithelial cells by adenoviruses. E1a binds RB and liberates E2F [5–9]. Products of these genes are used to synthesize viral DNA in infected cells. In addition, E2F transcribes the E2 region of the viral genome, a region that encodes viral DNA polymerase and DNA-binding proteins. This, of course, is where the name E2F comes from: a cellular factor that transcribes E2 in an E1a-dependent manner.

Uncontrolled activation of E2F, however, provokes accumulation of p53, as described above. p53 could attenuate replication by causing apoptosis, or by other means. For example, p53 could, in principle, block entry into S-phase by induction of p21. However, E1a is known to overcome p21-induced arrest and, indeed, p21 is not induced significantly during infection. It is well established, however, that adenoviruses, like other DNA viruses, need to inactivate p53 to achieve efficient replication, although the precise reason for this has not been formally established [10]. Adenoviruses eliminate p53 by production of E1B 55k, which binds p53 directly and blocks its function [11–16], and by E4orf6, which forms a complex with E1B 55k and p53, promoting p53 degradation [17–19].

Based on the remarkable parallel between molecular events that cause cancer and those involved in initiating virus replication, we have developed adenoviruses that replicate selectively in cancer cells. Specificity is based on the following hypotheses: (1) Viruses that lack E1a are defective for replication in normal cells, because RB blocks replication, and are permissive in cancer cells, in which RB is defective. (2) Viruses that lack E1B 55k are defective in normal cells because p53 blocks replication, but permissive in cancer cells that lack functional p53.

Both of these hypotheses are based on the supposition that the sole function of E1a and E1B 55k relate to inhibition of RB and p53, respectively. Clearly, this is not the case, and we discuss below how the general concept embodied in the hypothesis has been translated into reality through analysis of other functions of these viral proteins.

VIRUSES THAT DEPEND ON LOSS OF RB

The interaction between E1a and RB was described in 1988, and set the stage for understanding how viral proteins transform cells. E1a mutants that fail to bind RB were identified [20], and it was realized that RB binding is one of many functions of E1a that contribute to transformation. In 1992, we speculated that an adenovirus encoding a mutant form of E1a that fails to bind RB should replicate efficiently in RB-deficient cells, but fail to grow in normal cells. One such mutant virus is dl922/947: It contains a deletion in the CR2 region of E1a that prevents interaction with RB and related proteins We expected that rapidly proliferating cells would be somewhat permissive for this mutant virus because normal cells inactivate RB during normal transition through the cell cycle. To our surprise, we found that even quiescent cells could, in fact, support replication by this mutant virus [21]. Proliferating cells, as expected, were more permissive and tumor cells, as expected, were fully permissive for replication.

Replication of dl922/947 in primary cells must be attributable to alternative mechanisms for activating E2F. The E4 region expresses a protein referred to as *E4orf6/7*, the function of which is to bind to E2F and dimerize the transcription factor in an orientation that directs E2F activity to promoters with two head-to-tail E2F binding sites (Figure 3) [22–27]. The E2 region of the viral genome has such an orientation, as does the cellular promoter for E2F-1. Binding of E4orf6/7 to E2F can displace RB binding. We therefore speculated that leaky expression of E4orf6/7 in primary cells in culture can lead to E2F activation and thereby initiate entry into S-phase and activation of E2 transcription in the absence of E1a/RB complexes. We, therefore, constructed a virus in which the E4 region was placed under the control of E2F, so that E4 expression would be selective for tumor cells in which E2F levels are high. We also engineered E1a with a deletion in the RB binding site, under the E2F promoter. The resulting virus, ONYX-411, has the properties of being highly restricted for replication in normal cells, even while they proliferate, but being fully permissive for replication in tumor cells (Figure 4) [21].

Whereas ONYX-411 contains three elements of selectivity (E2F drives E1a, E1a is mutated for RB binding, and E2F

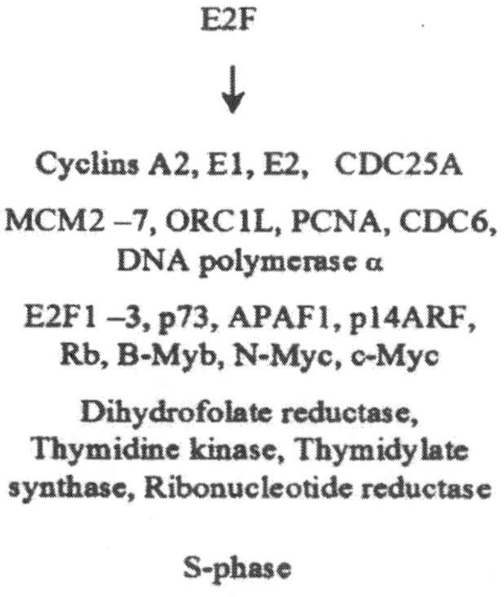

E2F

↓

Cyclins A2, E1, E2, CDC25A

MCM2 –7, ORC1L, PCNA, CDC6, DNA polymerase α

E2F1 –3, p73, APAF1, p14ARF, Rb, B-Myb, N-Myc, c-Myc

Dihydrofolate reductase, Thymidine kinase, Thymidylate synthase, Ribonucleotide reductase

S-phase

Figure 3

ONYX-411

Rb Rb

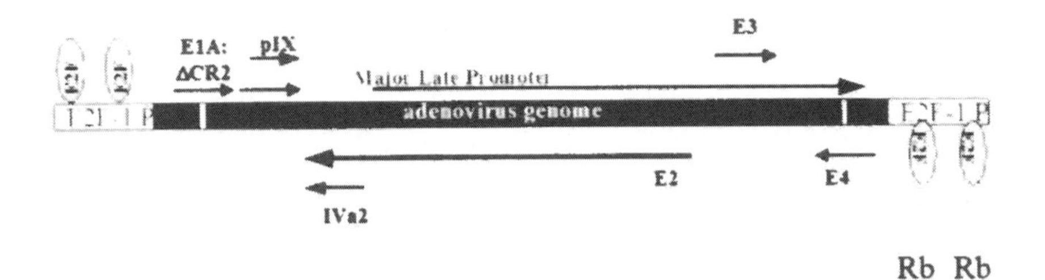

Figure 4

drives E4), the simpler virus, which contains only an E1a deletion (dl922/947) for selectivity [28,29], has shown promise as a cancer-selective virus, and an improved version of this virus that includes an RGD motif to improve infectivity recently entered pre-clinical evaluation. (Fueyo et al, in press)

ONYX-015 AND THE P53 PATHWAY

ONYX-015 is dl1520, an adenovirus made by A. J. Berk and colleagues in 1987 [11]. This virus lacks the E1B 55K gene and is therefore unable to block p53. In 1992, we proposed that such a virus should replicate selectively in cancer cells, since these cells typically lack p53 and should therefore not need E1B 55K. In normal cells, p53 would block replication. ONYX-015 [30] entered clinical trials for head and neck cancer in 1996 [31] and is still under clinical investigation.

p53 is induced during infection of normal cells by adenoviruses, as a result of expression of E1a and activation of E2F. p14ARF is induced under these conditions, and MDM2 is inhibited, allowing p53 to accumulate as described above (Figure 1). Primary cells infected with ONYX-015 are indeed attenuated for replication, and p53 is induced to very high levels in these cells. Suppression of p53 by dominant negative p53 mutants increases ONYX-015 replication, showing that p53 is responsible, in part, for ONYX-015 attenuation.

In tumor cells, replication of ONYX-015 varies tremendously relative to wild-type adenovirus. Originally, we thought that this might be because of different states of p53 activity in these cells. The most permissive cells we examined, cervical cancer C33a cells, replicated ONYX-015 at wild-type levels of efficiency. These cells harbor a mutation in p53 that renders the protein inactive. On the other hand, U2OS osteosarcoma cells, which express wild-type p53 activity, are severely attenuated for ONYX-015 replication. Later, it was shown by Zur Hausen and colleagues that attenuation of ONYX-015 in U2OS cells is unrelated to p53 function: They eliminated p53 activity using a dominant negative construct and failed to rescue ONYX-015 replication [32]. Likewise, we and others observed a lack of correlation between p53 status and ONYX-015 replication [33–36].

Two issues needed to be addressed: First, how is ONYX-015 able to replicate in tumor cells that retain wild-type p53? Second, why does ONYX-015 fail to replicate in some cancer cells in which p53 function is missing?

Replication of ONYX-015 in Tumor Cells with Wild-Type p53

The first of these issues was quite easy to address. It was discovered in 1999 that tumor-derived cell lines that retain wild-type p53 lose expression of p14ARF. This protein connects E2F to p53, as shown in Figure 1. We, therefore, tested whether tumor cells retaining p53 support replication of ONYX-015 because they are defective for p53 function. Indeed, infection of such cells by ONYX-015 fails to activate p53. Forced expression of p14ARF induces high levels of p53 and blocks ONYX-015 replication in a p65-dependent manner. In such tumor cells, p53 can be activated by DNA damaging agents because these insults do not depend on p14ARF. This led to the misleading impression that tumor cells with wild-type p53 are normal with respect to the p53 pathway. We concluded that loss of p14ARF permits replication of ONYX-015 in tumor cells with wild-type p53 [37].

Failure of ONYX-015 to Replicate in p53-Defective Tumor Cells

E1B 55K has additional functions: Failure to replicate in p53-defective cells must reflect ONYX-015's requirement for these other functions in nonpermissive tumor cells such as U2OS. Equally, permissive cells, such as HCT116 colon cancer cells and C33a cells, must supply all functions of E1B 55K to allow replication of ONYX-015 at wild-type levels in these cells. Other functions of E1B 55K include export and translation of viral late mRNAs, shutoff host protein synthesis. The molecular basis of these functions and the relationship to one another are under investigation.

A mutant form of E1B 55K that fails to bind p53 but retains other functions might seem to provide a solution to ONYX-015's replication defect in certain cancer cells, while retaining attenuation in primary cells. Such a mutant has been constructed: Shen and workers characterized a panel of virus-containing point mutations in E1B 55K for replication in U2OS cells and for p53 binding [38]. Mutants were obtained that had the desired properties—specifically, mutants R240A and H260A failed to bind and degrade p53. The H260A of E1B 55k also failed to interact with E4orf6, whereas R240A remained wild-type in this function. R240A replicated efficiently in cancer cells in which ONYX-015 is defective. However, this mutant replicates quite efficiently in normal cells, despite induction of high levels of p53. This is because the mutant restores shutoff of host protein synthesis, in contrast to ONYX-015. As a result, p53 that is induced early in infection cannot induce expression of target genes, which, of course, depend on host protein synthesis. It is therefore apparent that ONYX-015's attenuation of replication in normal cells depends on lack of both functions of E1B 55K—p53 binding and host shutoff. Replication in tumor cells depends on the degree to which tumor cells supply E1B 55K functions that are unrelated to p53.

CLINICAL EXPERIENCE WITH ONYX-015

ONYX-015 entered clinical trials for head and neck cancer in 1996. Response rates of over 60% were reported in tumors injected with high doses of virus in a Phase II clinical trial. More than 20% of injected tumors responded completely [31]. Based on these encouraging data, a Phase III trial was initiated, but this trial was not pursued, largely because of financial and corporate policies (development of ONYX-015 fell to Pfizer, New York, NY, as part of a corporate merger in 1997). Other clinical studies have continued, however. For example, T. Reid and coworkers developed a protocol for treating metastatic colon cancer via hepatic artery infusion [39,40]. This approach results in high doses of virus delivery directly to tumor beds in the liver, which receive most of their blood supply from the hepatic artery. In theory, this circumvents the problem of virus dilution encountered by i.v. infusion, and has two potential benefits: high concentration of virus at the site of delivery is expected to increase efficiency of infection and reduces the impact of neutralizing antibodies that develop during repeated dosing. In this study, about half of treated patients became viremic, producing an estimated 10e11 genomes of virus over a period of days after infusion. Significant clinical responses were observed, but this was in a Phase II setting and was not randomized or compared directly with a control group. Further studies are required to evaluate the effectiveness of this agent more thoroughly.

One variable that may affect the outcome of such a trial is the expression of the receptor for adenovirus infection [41]. This protein, referred to as *Coxsackie adenovirus receptor* (CAR) is expressed primarily in epithelial cells and, like other epithelial cell markers, is downregulated in some tumors [42]. About 50% of metastatic colon cancers downregulate CAR, through the action of the Ras pathway, which is hyperactivated in such cells [43–45]. It is therefore possible that ONYX-015 is more effective in patients in whose tumors CAR remains at high levels. Alternatively, CAR could be upregulated by pharmacological intervention; we recently reported that MEK inhibitors upregulate CAR expression, resulting in increased infectivity and more than 10-fold increase in virus production.

Although many tumors show decreased CAR expression as they become progressively more malignant, we were surprised to discover that metastatic prostate cancer cells have high levels of CAR—far higher than high-grade local tumors [46]. This suggests that metastatic prostate cancer may be a suitable target for adenoviral therapy, at least from the perspective of infectivity. Clearly, this disease will need to be treated by systemic administration of virus, and production of neutralizing antibodies will need to be suppressed. To this end, a clinical study has been initiated to investigate the effect of pretreating patients with Rituxan to prevent expansion of B-cells and suppress production of neutralizing antibodies. If successful, this approach could open a number of possibilities for future use of adenoviruses and other oncolytic agents.

CONCLUSION AND FUTURE PROSPECTS

We and others have created viruses whose replication depends on loss of the tumor suppressors RB and p53. These viruses kill cancer cells selectively and have shown promising preclinical and clinical evidence of efficacy. Future development of this technology will depend on a clear understanding of factors that limit viral spread in vivo and effective ways of preventing accumulation of neutralizing antibodies that must limit systemic efficacy. These are issues that need to be addressed in the clinical setting, to a large degree, because suitable animal models are not available. The agents that have been developed so far have proven safe and provide a solid foundation for the future development of this relatively new field. New agents that express transgenes to increase potency may be the next wave of this technology. In the meantime, I anticipate that existing agents such as ONYX-015 can be improved tremendously through combination therapy, without compromising the excellent safety profile that has been demonstrated so far. There is therefore good reason to believe that oncolytic viruses will become useful clinical agents that help decrease the morbidity and mortality associated with human cancer.

ACKNOWLEDGMENTS AND AFFILIATIONS

The author is a share holder in ONYX Pharmaceuticals and is supported, in part, by ONYX. I wish to thank my colleagues, Ali Fattaey, Leisa Johnson, Jerry Shen, and Aleida Perez, from ONYX, and Clodagh O'Shea, Allan Balmain, and Michael Korn from University of California–San Francisco for their excellent contributions to this program.

REFERENCES

1. Sherr CJ. Cancer cell cycles Science 1996; 274:1672–1677.
2. Sherr CJ, McCormick F. The RB and p53 pathways in cancer Cancer Cell 2002; 2:103–112.
3. Dyson N. The regulation of E2F by pRB-family proteins Genes Dev 1998; 12:2245–2262.
4. Sherr CJ. Tumor surveillance via the ARF-p53 pathway Genes Dev 1998; 12:2984–2991.
5. Mymryk JS, Bayley ST. Multiple pathways for activation of E2A expression in human KB cells by the 243R E1A protein of adenovirus type 5 Virus Res 1994; 33:89–97.
6. Shepherd SE, Howe JA, Mymryk JS, Bayley ST. Induction of the cell cycle in baby rat kidney cells by adenovirus type 5 E1A in the absence of E1B and a possible influence of p53 J Virol 1993; 67:2944–2949.
7. Howe JA, Bayley ST. Effects of Ad5 E1A mutant viruses on the cell cycle in relation to the binding of cellular proteins including the retinoblastoma protein and cyclin A Virology 1992; 186: 15–24.
8. Wang HG, Draetta G, Moran E. E1A induces phosphorylation of the retinoblastoma protein independently of direct physical association between the E1A and retinoblastoma products Mol Cell Biol 1991; 11:4253–4265.
9. Wang HG, Rikitake Y, Carter MC, et al. Identification of specific adenovirus E1A N-terminal residues critical to the binding

of cellular proteins and to the control of cell growth J Virol 1993; 67:476–488.

10. McCormick F. ONYX-015 selectivity and the p14ARF pathway Oncogene 2000; 19:6670–6672.

11. Barker DD, Berk AJ. Adenovirus proteins from both E1B reading frames are required for transformation of rodent cells by viral infection and DNA transfection [published erratum appears in Virology 1987 May;158(1):263] Virology 1987; 156: 107–121.

12. Yew PR, Berk AJ. Inhibition of p53 transactivation required for transformation by adenovirus early 1B protein Nature 1992; 357:82–85.

13. Yew PR, Liu X, Berk AJ. Adenovirus E1B oncoprotein tethers a transcriptional repression domain to p53 Genes Dev 1994; 8: 190–202.

14. Teodoro JG, Branton PE. Regulation of p53-dependent apoptosis, transcriptional repression, and cell transformation by phosphorylation of the 55-kilodalton E1B protein of human adenovirus type 5 J Virol 1997; 71:3620–3627.

15. Teodoro JG, Branton PE. Regulation of apoptosis by viral gene products J Virol 1997; 71:1739–1746.

16. Teodoro JG, Halliday T, Whalen SG, Takayesu D, Graham FL, Branton PE. Phosphorylation at the carboxy terminus of the 55-kilodalton adenovirus type 5 E1B protein regulates transforming activity J Virol 1994; 68:776–786.

17. Querido E, Marcellus RC, Lai A, et al. Regulation of p53 levels by the E1B 55-kilodalton protein and E4orf6 in adenovirus-infected cells J Virol 1997; 71:3788–3798.

18. Nevels M, Rubenwolf S, Spruss T, Wolf H, Dobner T. The adenovirus E4orf6 protein can promote E1A/E1B-induced focus formation by interfering with p53 tumor suppressor function Proc Natl Acad Sci U S A 1997; 94:1206–1211.

19. Steegenga WT, Riteco N, Jochemsen AG, Fallaux FJ, Bos JL. The large E1B protein together with the E4orf6 protein target p53 for active degradation in adenovirus infected cells Oncogene 1998; 16:349–357.

20. Whyte P, Buchkovich KJ, Horowitz JM, et al. Association between an oncogene and an anti-oncogene: the adenovirus E1A proteins bind to the retinoblastoma gene product Nature 1988; 334:124–129.

21. Johnson L, Shen A, Boyle L, et al. Selectively replicating adenoviruses targeting deregulated E2F activity are potent, systemic antitumor agents Cancer Cell 2002; 1:325–337.

22. Helin K, Harlow E. Heterodimerization of the transcription factors E2F-1 and DP-1 is required for binding to the adenovirus E4 (ORF6/7) protein J Virol 1994; 68:5027–5035.

23. Neill SD, Nevins JR. Genetic analysis of the adenovirus E4 6/7 trans activator: interaction with E2F and induction of a stable DNA-protein complex are critical for activity J Virol 1991; 65: 5364–5373.

24. Raychaudhuri P, Bagchi S, Neill SD, Nevins JR. Activation of the E2F transcription factor in adenovirus-infected cells involves E1A-dependent stimulation of DNA-binding activity and induction of cooperative binding mediated by an E4 gene product J Virol 1990; 64:2702–2710.

25. Neill SD, Hemstrom C, Virtanen A, Nevins JR. An adenovirus E4 gene product trans-activates E2 transcription and stimulates stable E2F binding through a direct association with E2F Proc Natl Acad Sci U S A 1990; 87:2008–2012.

26. Reichel R, Neill SD, Kovesdi I, Simon MC, Raychaudhuri P, Nevins JR. The adenovirus E4 gene, in addition to the E1A gene, is important for trans-activation of E2 transcription and for E2F activation J Virol 1989; 63:3643–3650.

27. O'Connor RJ, Hearing P. Mutually exclusive interaction of the adenovirus E4-6/7 protein and the retinoblastoma gene product with internal domains of E2F-1 and DP-1 J Virol 1994; 68: 6848–6862.

28. Fueyo J, Gomez-Manzano C, Alemany R, et al. A mutant oncolytic adenovirus targeting the Rb pathway produces anti- glioma effect in vivo Oncogene 2000; 19:2–12.

29. Heise C, Hermiston T, Johnson L, et al. An adenovirus E1A mutant that demonstrates potent and selective systemic antitumoral efficacy Nat Med 2000; 6:1134–9.

30. Bischoff JR, Kirn DH, Williams A, et al. An adenovirus mutant that replicates selectively in p53-deficient human tumor cells. Comments Science 1996; 274:373–376.

31. Khuri FR, Nemunaitis J, Ganly I, et al. a controlled trial of intratumoral ONYX-015, a selectively-replicating adenovirus, in combination with cisplatin and 5-fluorouracil in patients with recurrent head and neck cancer Nat Med 2000; 6:879–885.

32. Rothmann T, Hengstermann A, Whitaker NJ, Scheffner M, zur Hausen H. Replication of ONYX-015, a potential anticancer adenovirus, is independent of p53 status in tumor cells J Virol 1998; 72:9470–9478.

33. Heise C, Sampson-Johannes A, Williams A, McCormick F, Von Hoff DD, Kirn DH. ONYX-015, an E1B gene-attenuated adenovirus, causes tumor-specific cytolysis and antitumoral efficacy that can be augmented by standard chemotherapeutic agents. Comments Nat Med 1997; 3:639–645.

34. Turnell AS, Grand RJA, Gallimore PH. The replicative capacities of large E1B-null group A and group C adenoviruses are independent of host cell p53 status. In process citation J Virol 1999; 73:2074–2083.

35. Goodrum FD, Ornelles DA. p53 status does not determine outcome of E1B 55-kilodalton mutant adenovirus lytic infection J Virol 1998; 72:9479–9490.

36. Harada JN, Berk AJ. p53-Independent and -dependent requirements for E1B-55K in adenovirus type 5 replication J Virol 1999; 73:5333–5344.

37. Ries SJ, Brandts CH, Chung AS, et al. Loss of p14ARF in tumor cells facilitates replication of the adenovirus mutant dl1520 (ONYX-015) Nat Med 2000; 6:1128–1133.

38. Shen Y, Kitzes G, Nye JA, Fattaey A, Hermiston T. Analyses of single-amino-acid substitution mutants of adenovirus type 5 E1B-55K protein J Virol 2001; 75:4297–4307.

39. Reid T, Galanis E, Abbruzzese J, et al. Hepatic arterial infusion of a replication-selective oncolytic adenovirus (dl1520): phase II viral, immunologic, and clinical endpoints Cancer Res 2002; 62:6070–6079.

40. Reid T, Warren R, Kirn D. Intravascular adenoviral agents in cancer patients: lessons from clinical trials Cancer Gene Ther 2002; 9:979–986.
41. Bergelson JM, Cunningham JA, Droguett G, et al. Isolation of a common receptor for Coxsackie B viruses and adenoviruses 2 and 5 Science 1997; 275:1320–1323.
42. Li Y, Pong RC, Bergelson JM, et al. Loss of adenoviral receptor expression in human bladder cancer cells: a potential impact on the efficacy of gene therapy Cancer Res 1999; 59:325–330.
43. Anders M, Christian C, McMahon M, McCormick F, Korn WM. Inhibition of the Raf/MEK/ERK pathway up-regulates expression of the Coxsackie virus and adenovirus receptor in cancer cells Cancer Res 2003; 63:2088–2095.
44. Anders U, Korn O. Model selection in neural networks Neural Netw 1999; 12:309–323.
45. Anders M, Hansen R, Ding RX, Rauen KA, Bissell MJ, Korn WM. Disruption of 3D tissue integrity facilitates adenovirus infection by deregulating the coxsackievirus and adenovirus receptor Proc Natl Acad Sci U S A 2003; 100:1943–1948.
46. Rauen KA, Sudilovsky D, Le JL, et al. Expression of the Coxsackie adenovirus receptor in normal prostate and in primary and metastatic prostate carcinoma: potential relevance to gene therapy Cancer Res 2002; 62:3812–3818.

Epilogue: "Damn the Torpedoes! Full Speed Ahead!"*

JOSEPH G. SINKOVICS
JOSEPH C. HORVATH

The concise summaries of their recent work that the invited authors submitted for this volume, reflect well the state of virus therapy for human cancers. After a protracted and difficult beginning many decades ago, now major advances are being made both in the selection of naturally oncolytic viruses, but above all, in the creation of genetically engineered viruses

* After Admiral David Glasgow Farragut (1801–1870) exclaiming over the mine fields of Mobile Bay, AL, on August 5, 1864 (and getting through!).

that attack tumors highly selectively. Either the promoters that drive these viruses are restricted to certain malignantly transformed tissues, or the viral genome is altered in such a way that normal cells would eliminate the virus, the task that tumor cells due to their own genetic defects can not accomplish. Another generation of genetically engineered viruses insert genes into the genome of tumor cells. These are tumor-suppressor genes, cytokine genes, genes of immunological co-stimulators, and genes that trigger cell suicide. A set of viral vectors deliver antisense oligonucleotides (As-ODN) that antagonize the replication of genomic sequences of oncogenes and fusion oncoproteins, or those of genes that promote neoangiogenesis. The tumor cell can now be attacked at many levels of its operation. What are the most prominent intracellular targets at which virotherapy should be aimed, and can multiple targets be attacked at the same time simultaneously or sequentially?

Naturally oncolytic viruses encounter little, if any opposition, to their full replicative cycles in tumor cells with down-regulated interferon production machinery. Certain oncoproteins, especially those encoded by the c-*ras* family of oncogenes, inactivate the protein kinase (PKR) system, which is essential for the initiation of interferon production within the normal but virally infected cell. If farnesyltransferase inhibitors of the c-*ras*→Ras pathway restore cell metabolism, shifting it toward normalization, these tumor cells may regain PKR activity, produce interferon, and eliminate a "naturally oncolytic" virus, thereby escaping tumor cell lysis. This would be a system in which a biologically active molecular mediator and a naturally oncolytic virus could not cooperate. However, genetically engineered viruses depositing the herpesviral *tk* gene in these tumor cells still could be tested in combination with farnesyltransferase inhibitors. R115777 reduces hypoxia, induces oxygenation, and reduces neovascularization of human glioblastoma xenografts [1]. It should be tested in combination with a genetically engineered oncolytic virus for which excellent candidates are the polio-rhinoviral recombinants developed at Duke University, Durham, NC.

Other designated targets for oncolytic virotherapy are the insulinlike growth factor receptor-I and ligand (IGF-I and R) systems; the survivins; the proteasomes; the telomeres; the notorious epidermal growth factor (EGF)-R; and the neoangiogenesis inducer gene product proteins, from vascular endothol growth factor (VEGF) to the hepatocyte growth factor/scatter factor (HGF/SF), reacting with its receptor c-Met [2]. Insulin-like growth factor and its receptor form para- and autocrine circuits to promote tumor cell growth, whereas IGF-binding protein-related protein-1 (IGFBP-rP1) antagonizes the system and induces apoptosis in those tumor cells that may overexpress it [3]; however, its gene is silenced by hypermethylation in tumor cells. After demethylating it with 5-aza-dC, the suppressed gene of IGFBP-rP1 in prostate cancer cells or in stromal cells of the tumor bed regains its function. Prostate cancer cells transfected with the IGFBP-rP1 gene lost a great deal of their malignant behavior: Their growth rate was reduced, they formed no colonies in soft agar, and they succumbed to apoptosis [3]. This is one of the candidate genes for reinsertion into the genome of tumor cells by adeno- or adeno-associated viruses. Insulin-like growth factor-IR emits anti-apoptotic signals through the Ras pathway, such as the activation of PI3K/Akt and the translocation of Raf-1 protein into mitochondria. However, soluble sIGF-IR, acting as a decoy, blocks the activity of the real receptor and thereby inhibits the growth of ovarian carcinoma cells [4]. Overexpression of the growth-inhibitory insulin receptor-related receptor and the nerve growth factor receptor (NGF-R/TrkA) by neuroblastoma cells overrules the tumor-promoting effects of the amplified N-Myc and that of the ligand-activated IGF-IR [5]. These are the inhibitory genes that need to be reinserted into neuroblastoma cells by viral vectors. There is a fully human monoclonal antibody specific to IGF-IR [6]. Its administration should be combined with the adenoviral vectors that encode decoy receptors antigenically different from the real receptor or the proteins that are antagonistic to IGF-IR.

Survivin emerges as one of the most powerful anti-apoptotic proteins operational in neoplastic cells. The survivin protein is recognized by T lymphocyte clones of the tumor-bearing host. These T cell clones may be expanded and may exert cytotoxicity on survivin-expressing tumor cells in a HLA-A2–restricted fashion [7,8]. Autologous $CD34^+$ stem cells in the bone marrow are spared by these clones of immune T cells. Anti-survivin plasmids containing a reverse primer (pcDNA3-survivin) inhibited tumorigenicity of gastric carcinoma cells [9]; so does the Thr34Ala survivin mutant gene-carrier adenovirus. In a sequential fashion, immune T cell therapy followed by adenoviral therapy to eliminate the remaining survivin-expressing tumor cells that survived the lymphocyte attack appears to be feasible.

The short-living survivin protein is usually destroyed in proteasomes [10], an event that would favor the tumor-bearing host. Retroviral vector pMSCV-Fas achieves the same effect [11]. Destruction of survivin in proteasomes is promoted by interferon (IFN)-γ. Interferon-γ–treated colon carcinoma cells also overexpress the CD95 Fas receptor, rendering these tumor cells more apoptosis-prone when binding FasL [11]. However, the multicatalytic proteasome system often destroys pro-apoptotic proteins (p53; Bax) within the tumor cell. Proteasome inhibitors such as bortezomib or lactacystin can preserve the intracellular pro-apoptotic proteins in tumor (multiple myeloma and many other) cells. Would the Mayo Clinic's measles virus act additively or even synergistically with bortezomib in the treatment of multiple myeloma? If another viral vector delivery of proteasome inhibitors at a critical phase of the cell cycle could antagonize MDM and thereby rescue p53 protein from destruction, the tumor cell would remain susceptible to programmed death.

Regarding sarcomas, activin receptor-like kinases, especially Alk1 and Alk5, act as signal receptors for transforming growth factor (TGF)-β, a growth factor for many of these tumors. The Alk4/5/7 inhibitor SB-431542, in combination with imatinib, suppressed proliferation of osteosarcoma cells. Imatinib acted against TGF-β–induced platelet-derived growth factor (PDGF) production in these cells, which is another growth factor for many sarcoma cells [12]. Imatinib suppresses the PDGF circuits in gastrointestinal stromal tumor (GIST) and in pancreatic carcinoma cells [13,14]. The same circuit could be targeted by As-ODN carrier adenoviruses. Rhabdomyosarcoma cells are driven by HGF/SF, the ligand of the receptor c-Met. In addition, these tumor cells also express the CXC chemokine receptor-4 and respond to its ligand, SDF-1 [15]. Histone deacetylase inhibitors (suberoylanilide hydroxamide acid or SAHA; pyroxamide) induce apoptotic death of rhabdomyosarcoma cells [16]. Cytomegalovirus and herpesvirus G207 kill rhabdomyosarcoma cells (Cincinnati Children's Hospital, Cincinnati, OH, and MediGene, San Diego, CA). G207 and SAHA should be administered in combination first in xenografts and then in a clinical trial. Ewing sarcoma cells express the chimeric EWS/ETS oncoprotein resulting from chromosome breaks and translocations forming t(11;22)(q24;q12) [17]. The fused oncoprotein activates the telomerase system. These tumor cells also overexpress the adhesion molecule MUC18, which was discovered in melanoma cells but is overexpressed in many other tumors, including osteosarcomas [18]. A fully human monoclonal antibody, ABX-MA1, suppresses the metastatic spread of xenografts of these tumors [18]. The E1A adenoviral type 5 gene inserted into tumor cells by many different means inhibits tumor growth and induces apoptosis in tumor cells. In Ewing's sarcoma cells, AdE1A downregulated HER2/neu expression and suppressed VEGF production [19]. Trastuzumab (Herceptin, Genentech, South San Francisco, CA) and AdE1A administered in combination is a therapeutic modality for this tumor and is to be tested soon in clinical trials. Mesothelioma cells overexpress cyclooxygenase-2 and this pathway is blocked by rofecoxib. Cyclooxygenase-2 inhibition combined with the administration of Ad.IFNβ induced tumor-attacking $CD8^+$ T cell clones [20]. Roscovitine, the inhibitor of cycline-dependent kinases, induces apoptotic death in HTLV-1–transformed leukemic T cells [21]. The H. L. Moffitt Cancer Center, Tampa, FL, will use it against sarcoma cell lines either alone or in combination with other agents in preparation for a clinical trial. Antibodies aimed at the PDGF-R and Ad.E1A are good candidates for possible additive action with cycline-dependent kinase inhibitors.

It has been discovered that the ansamycin 17-N-allylamino-17-demethyl geldanamycin synergizes with radio-

and chemotherapy for prostatic carcinoma [22]. Geldanamycins inhibit heat shock protein 90; and ZD1839 (gefitinib) suppresses signal transduction (Akt activation) from ErbB2 in breast cancer cells, thereby removing a major blockade of apoptotic death [23]. Because suppressors of oncogene signal pathways work best in combinations, naturally oncolytic and genetically engineered viruses should be applied in the same manner. For a combination to be tested, a good candidate is GGTI-2154, a geranylgeranyl transferase inhibitor that readily induces apoptotic death of H-Ras–dependent murine breast cancers [24]. A mouse mammary virus-related retroviral agent exists in some human breast cancers; this association may render these tumor cells more susceptible to an external viral attack, inasmuch as viruses frequently interfere with one another. Geldanamycins and certain inhibitors of oncogenic kinase autophosphorylation and viral vectors antagonizing the oncogene itself appear to be reasonable candidates for combined protocols.

Further services that genetically engineered adenoviruses could render to the tumor-bearing host are transduction of the interleukin (IL)-12 gene into macrophages of the tumor bed (AdmIL-12) [25] for instigating an antitumor attack by the transduced macrophages. Further, an adenoviral construct (Ad5-TERT-E1) replicating exclusively under the control of human telomerase reverse transcriptase selectively lyses prostate cancer cells [26]. A replication-defective adenovirus (Introgen, Houston, TX) replaces the wild-type p53 in tumor cells. ONYX-015 targeting p53-deficient tumor cells now performs well in the oral cavity as a mouthwash in resolving premalignant dysplastic lesions, predecessors of squamous cell carcinomas [27,28], proving that simplistic technology of administration of an oncolytic virus may not be second in therapeutic value to a sophisticated mode of administration. The question immediately arises: Could the new live attenuated influenza A viral vaccine (Flumist, Wyeth/MedImmune, Gaithersburg, MD; *Physicians Desk Reference*, 2004; pp 3424–3427) have any role in the treatment of squamous cell carcinomas of the oronasal cavities?

The long list of agents targeting the most notorious EGF-R (ZD1839, gefitinib; OSI-774, erlotinib; CI-1033; C225, cetuximab; ABX-EGF; EGF-P64K vaccine; DAB389 EGF immunotoxin) contains no viral agent [29]. Is there a major obstacle that would doom to failure any attack on EGF-R by naturally oncolytic or genetically engineered viruses? We could recognize none. Cytomegalovirus (CMV) enters cells through the EGF-R [30]. If tumor cells over expressing EGF-R were increasingly accessible to CMV entry, would apathogenic oncolytic variants of CMV selectively attack these cells? Amphiregulin (AR) activates EGF-R which, in turn, induces overexpression of the malignant phenotype-associated cell-surface protein, the extracellular matrix metalloprotease receptor-inducer. Antisense cDNA aimed at the interaction of AR and EGF-R reduces tumorigenicity of breast cancer cells [31]. Adenoviral vectors could deliver the AR antisense sequences and break the circuit.

In addition to delivering suicide and cytokine genes into tumor cells, some viruses deliver genes of enzymes that render prodrugs to be effective chemotherapeutic agents within the affected cells. Of the long list of these viral vectors, the latest is vesicular stomatitis virus (VSV), delivering the gene of cytosine deaminase/uracil phosphoribosyl transferase into tumor (mouse breast cancer) cells, rendering the tumor cells highly susceptible to 5-FU and inducing extensive bystander effect [32]. Viruses may render chemotherapy-resistant human tumor cells sensitive to selected chemotherapeutical agents.

The latest results with high dose IV virotherapy of various human tumors using a newly applied strain of NDV and resulting in partial remissions and stabilizations of disease have recently been reviewed [33]. Most recent reports now in print suggest that some (not all) chemoradiotherapy-resistant glioblastoma multiforme tumor cells remain susceptible to oncolysis by NDV [34,35]. It is most remarkable that different NDV strains (73T, Ulster, PV701, LaSota, OV001, and the veterinary attenuated vaccine strain renamed MTH-68H) all exert oncostatic-oncolytic effects either by apoptosis induction and/or by mobilizing host immune reactions against NDV-infected tumor cells, and that antiviral immune globulin production does not inhibit oncolysis, and that no immune complex disease ensues (it has not been reported) after repeated administration through different routes of small to very large doses of the virus. The Proceedings of the American Society of Clinical Oncology (ASCO) 40th Annual Meeting held in New Orleans, Louisiana, in June 2004 contains an abstracted description of successful treatment for cutaneous lymphomas with an adenoviral vector of the IFN-γ cDNA. The Proceedings of the American Association for Cancer Research (AACR) 95th Annual Meeting held in Orlando, Florida, in March 2004 contains several abstracted reports on oncolysis of human cancer cells by fusogenic and other oncolytic herpes viruses (Synco-2D; NV1023; NV1066; OncoVEX $^{GM\text{-}CSF}$; G207; MGH1), a triple-mutated vaccinia virus (vSPT), and genetically engineered adenoviruses (CG5757, CG8840, adenoviruses driven by VEGF or survivin specifically attacking malignant glioma cells, and the hTERT promoter-driven and endostatin gene containing CNHK300-mE adenoviral construct).

As to immune T cells, not only the amplified antiapoptotic survivins but also the amplified or mutated telomere systems engender the rise of such autologous clones [36–38]. Some mixed clones of NK and T cells may be implicated to induce tumor cell (chondrosarcoma) differentiation, senescence, and death [39]. When rituximab-coated human lymphoma cells are phagocytized and digested by autologous dendritic cells (DC), lymphoma-associated antigens expressed by these DC induce tumor rejection-strength immune reactions mediated by T cell clones [40]. Patients with advanced arteriosclerosis/atheromatosis circulate CD4 T cells that kill smooth muscle cells in the arterial cell walls; these lymphocyte clones are cytotoxic to allogenic leiomyosarcoma cells [41]. Expanded clones of immune T cells and/or NK cells (LAK cells) are not to be discounted, they remain strong agents of adoptive immunotherapy.

The human multiple-drug-resistance-1 gene is transcriptionally activated by the YB-1 protein, when this protein translocalizes into the nucleus. Further, YB-1 interacts with p53 and suppresses the *fas* gene, thus exerting anti-apoptotic effects. Intranuclear overexpression of YB-1 protein occurs in breast, colon, lung, gastric, ovarian, and pancreatic carcinomas and in osteosarcoma cells, and YB-1 mRNA is overexpressed in glioblastoma and melanoma cells. E1A-mutated (Ad520; dl520) or E1A-deleted (Ad312) adenoviruses are replication-defective in normal cells, but replicate to full viral particle maturation and cytolysis in malignantly transformed MDR cells, in which nuclear YB-1 is overexpressed. Cell stress by irradiation, chemotherapy, or hyperthermia promotes YB-1 translocation into the nucleus and thereby facilitates full oncolytic replication of Ad520 in such cells. Whereas P-glycoprotein–overexpressing MDR tumor cells resist enveloped (myxo- and paramyxo-) oncolytic viruses [42], these very same tumor cells are susceptible to, and succumb to, full replicative cycles of Ad520 [43].

Tumor growth and its acceleration or inhibition remains a most treacherous territory, when it comes to scientifically designed interventions. No single biological agent emerges as being able to render full and lasting reversal of the oncogenic pathways that are operational in malignantly transformed cells. However, we begin to recognize two- or three-agent combinations that appear to approach this effect, whether applied synchronously or sequentially. It is not the xenograft assays, but the clinical trials that identify the most effective combinations. When biological agents are powerful constituents of these combined treatment regimens, lower and less toxic doses of chemotherapeuticals can exert their full efficacy. In some fully biologically combined regimens, the effects match those of the highest dosages of chemotherapy, especially in notoriously chemotherapy-resistant tumors (glioblastoma, melanoma, kidney carcinoma). One member of biologically combined regimens should be an oncolytic virus. Even when the virus is replication-deficient and works by apoptosis induction, apoptosomes of tumor cells within incompletely lysed tumor masses could immunize the host through dendritic cell (DC)→T cell interaction or by evoking ADCC reactions involving natural killer (NK) cells or macrophages for further antitumor effects. Some sort of a special co-expression of tumor and viral antigens occurs in viral oncolysates (VO). It should be studied further, whether VO are more efficient means of tumor lysis induction and active tumor-specific immunization than the postoncolytic immunity elicited by a naturally oncolytic virus; and if viruses other than influenza A or Newcastle disease virus (NDV) (such as reovirus or VSV) could produce oncolysates efficient in tumor lysis and antitumor immunization. Indeed, viral oncolysates contain tumor cell apoptosomes, solubilized tumor-associated antigens, and viral antigens budding from tumor cell membranes. Viral oncolysates are expected to co-stimulate DC→T cell interactions as well as mobilize NK/NKT cell populations. Viral oncolysates may act efficiently against micrometastases, but seldom score as therapeutic vaccines working against established metastatic tumors. Some DC vaccines may exert this high level of efficacy. When a DC vaccine consists of an inactivated tumor cell fused with a live mature, autologous DC, the fusogenic protein may be the contribution of a viral agent (herpesvirus; NDV; VSV). Virotherapy could re-orient the stroma of the tumor bed, rendering it from a supportive network in collusion with the tumor to an indifferent bystander or, better, an army of warriors launching an attack on the tumor at many levels—denying blood supply to it or actually expressing hostility to it with the production of molecular mediators, inducers of apoptosis in the tumor cells. Thus, virotherapy should find its integrated place within the combination regimens that forecast the new biological therapy of cancer.

REFERENCES

1. Delmas C, End D, Rochaix P, Favre G, Toulas C, Cohen-Jonathan E. The farnesyltransferase inhibitor R115777 reduces hypoxia and matrix metalloproteinase 2 expression in human glioma xenografts Clin Cancer Res 2003; 9:6062–6068.
2. Sengupta S, Sellers LA, Cindrova T, Skepper J, Gherardi E, Sasisekharan R, Fan T-PD. Cyclooxygenase-2-selective nonsteroidal anti-inflammatory drugs inhibit hepatocyte growth factor/scatter factor-induced angiogenesis Cancer Res 2003; 63: 8351–8359.
3. Mutaguchi K, Yasumoto H, Mita K, Matsubara A, Shiina H, Igawa M, Dahiya R, Usui T. Restoration of insulin-like growth

factor binding protein-related protein 1 has a tumor suppressive activity through induction of apoptosis in human prostate cancer Cancer Res 2003; 53:7717–7723.

4. Hongo A, Kuramoto H, Nakamura Y, Hasegawa K, Nakamura K, Kodama J, Hiramatsu Y. Antitumor effects of a soluble insulin-like growth factor I receptor in human ovarian cancer cells: advantage of recombinant protein administration in vivo Cancer Res 2003; 63:7834–7839.
5. Weber A, Huesken C, Bergmann E, Kiess W, Christiansen NM, Christiansen H. Coexpression of insulin receptor-related receptor and insulin-like growth factor 1 receptor correlates with enhanced apoptosis and dedifferentiation in human neuroblastomas Clin Cancer Res 2003; 9:5683–5692.
6. Burtrum D, Zhu Z, Lu D, Anderson DM, Prewett M, Pereira DS, Bassi R, Abdullah R, Hooper AT, Koo H, Jimenez X, Johnson D, Apblett R, Kussie P, Bohlen P, Witte L, Hicklin DJ, Ludwig DL. A fully human monoclonal antibody to the insulin-like growth factor I receptor blocks ligand-dependent signaling and inhibits human tumor growth in vivo Cancer Res 2003; 63:8912–8921.
7. Reed JC, Wilson DB. Cancer immunotherapy targeting survivin Clin Cancer Res 2003; 9:6310–6315.
8. Pisarev V, Yu B, Salup R, Sherman S, Altieri DC, Gabrilovich DI. Full-length dominant-negative survivin for cancer immunotherapy Clin Cancer Res 2003; 9:6523–6533.
9. Tu SP, Jiang XH, Lin MC, Cui JT, Yang Y, Lum CT, Zou B, Zhu YB, Jiang SH, Wong WM, Chan AO-O, Yuen MF, Lam SK, Kung HF, Wong C-Y. Suppression of survivin expression inhibits in vivo tumorigenicity and angiogenesis in gastric cancer Cancer Res 2003; 63:7724–7732.
10. Voorhees PM, Dees EC, O'Neil B, Orlowski RZ. The proteasome as a target for cancer therapy Clin Cancer Res 2003; 9: 6316–6325.
11. Geller J, Petak I, Szekely Szucs K, Nagy K, Tillman DM, Houghton JA. Interferon-γ-induced sensitization of colon carcinomas to ZD9331 targets caspases, downstream of Fas, independent of mitochondrial signaling and the inhibitor of apoptosis survivin Clin Cancer Res 2003; 9:6504–6515.

12. Matsuyama S, Iwadate M, Kondo M, Saitoh M, Hanyu A, Shimizu K, Aburatani H, Mishima HK, Imamura T, Miyazono K, Muiyazawa K. DSB-431542 and Gleevec inhibit transforming growth factor-β-induced proliferation of human osteosarcoma cells Cancer Res 2003; 63:7791–7798.

13. Heinrich MC, Corless CL, Demetri GD, Blanke CD, von Mehren M, Joensuu H, McGreevey LS, Chen C-J, Van den Abbeele AD, Druker BJ, Kiese B, Eisenberg B, Roberts PJ, Singer S, Fletcher CDM, Silberman S, Dimitrijevic S, Fletcher JA. Kinase mutations and imatinib response in patients with metastatic gastrointestinal stromal tumor J Clin Oncol 2003; 21: 4342–4349.

14. Hwang RF, Yokoi K, Bucana CD, Tsan R, Killion JJ, Evans DB, Fidler IJ. Inhibition of platelet-derived growth factor receptor phosphorylation by STI571 (Gleevec) reduces growth and metastasis of human pancreatic carcinoma in an orthotopic nude mouse model Clin Cancer Res 2003; 9:6534–6544.

15. Jankowski K, Kucia M, Wysoczynski M, Reca R, Zhan D, Tryzna E, Trent J, Peiper S, Zembala M, Ratajczak J, Houghton P, Janowska-Wieczorek A, Ratajczak MZ. Both hepatocyte growth factor (HGF) and stromal-derived factor-1 regulate the metastatic behavior of human rhabdomyosarcoma cells, but only HGF enhances their resistance to radiotherapy Cancer Res 2003; 53:7926–7935.

16. Kutko MC, Glick RD, Butler LM, Coffey DC, Rifkind RA, Marks PA, Richon VM, LaQuaglia MP. Histone deacetylase inhibitors induce growth suppression and cell death in human rhabdomyosarcoma in vitro Clin Cancer res 2003; 9:5749–5755.

17. Takahashi A, Higashino F, Aoyagi M, Yoshida K, Itoh M, Kyo S, Ohno T, Taira T, Ariga H, Nakajima K, Hatta M, Kobayashi M, Sano H, Kohgo T, Shindoh M. EWS/ETS fusions activate telomerase in Ewing's tumors Cancer Res 2003; 63:8338–8344.

18. McGary EC, Heimberger A, Mills L, Weber K, Thomas GW, Shtivelband M, Chelouche Lev D, Bar-Eli M. A fully human antimelanoma cellular adhesion molecule/MUC18 antibody inhibits spontaneous pulmonary metastasis of osteosarcoma cells in vivo Clin Cancer Res 2003; 9:6560–6566.

19. Zhou Z, Zhou R-R, Guan H, Bucana CD, Kleinerman ES. E1A gene therapy inhibits angiogenesis in a Ewing's sarcoma animal model Mol Cancer Ther 2003; 21:1313–1319..
20. DeLond P, Tanaka T, Kruklitis R, Henry AC, Kapoor V, Kaisre LR, Sterman DH, Albelda SM. Use of cyclooxygenase-2 inhibition to enhance the effect of immunotherapy Cancer Res 2003; 63:7845–7852.
21. Mohapatra S, Chu B, Wei S, Djeu J, Epling-Burnette PK, Loughran T, Jove R, Pledger WJ. Roscovitine inhibits STAT5 activity and induces apoptosis in the human leukemia virus type 1-transformed cell line MT-2 Cancer Res 2003; 63:8523–8530.
22. Enmon R, Yang W-H, Ballangrud ÅM, Solit DB, Hiller G, Rosen N, Scher HI, Sgouros G. Combination treatment with 17-N-allylamino-17-demethoxy geldanamycin and acute irradiation produces supra-additive growth suppression in human prostate carcinoma spheroids Cancer Res 2003; 63:8393–8399.
23. Xu W, Yuan X, Jung YJ, Yang Y, Basso A, Rosen N, Chung EJ, Treprl J, Neckers L. The heat shock protein 90 inhibitor geldanamycin and the ErbB inhibitor ZD1839 promote rapid PP1 phophatase-dependent inactivation of AKT in Erb2 overexpressing breast cancer cells Cancer Res 2003; 63:7777–7784.
24. Sun J, Okhanda J, Coippola D, Yin H, Kothare M, Busciglio B, Hamilton AD, Sebti SM. Geranylgeranyltransferase I inhibitor GGTI-2154 induces breast carcinoma apoptosis and tumor regression in H-Ras transgenic mice Cancer Res 2003; 63: 8922–8929.
25. Satoh T, Saika T, Ebara S, Kusaka N, Timme TL, Yang G, Wang J, Mouraviev V, Cao G, Abdel Fattah EM, Thompson TC. Macrophages transduced with an adenoviral vector expressing interleukin 12 suppress tumor growth and metastasis in a preclinical metastatic prostate cancer model Cancer Res 2003; 63: 7853–7860.
26. Lanson NA, Friedlander PL, Schwarzenberger P, Kolls JK, Wang G. Replication of an adenoviral vector controlled by the human telomerase reverse transcriptase promoter causes tumor-selective tumor lysis Cancer Res 2003; 63:7936–7941.
27. Ganly I, Singh B. Topical ONYX-015 in the treatment of premalignant oral dysplasia: another role for the cold virus? J Clin Oncol 2003; 21:4476–4478.

28. Rudin CM, Cohen EEW, Papadimitrakopoulou VA, Silverman S Jr, Recant W, El-Nagger AK, Stenson K, Lippman SM, Hong WK, Vokes EE. An attenuated adenovirus, ONYX-015, as mouthwash therapy for premalignant oral dysplasia J Clin Oncol 2003; 21:4546–4552.

29. Herbst RS, Bunn PA. Targeting the epidermal growth factor receptor in non-small cell lung cancer Clin Cancer Res 2003; 9: 5813–5824.

30. Wang X, Huang SM, Chin ML, Raab-Traub N, Huang ES. Epidermal growth factor receptor is a cellular receptor for human cytomegalovirus Nature 2003; 424:456–461.

31. Menashi S, Serova M, Ma L, Vignot S, Mourah S, Calvo F. Regulation of extracellular matrix metalloproteinase inducer and matrix metalloproteinase expression by amphiregulin in transformed human breast epithelial cells Cancer Res 2003; 63: 7575–7580.

32. Porosnicu M, Mian A, Barber GN. The oncolytic effect of recombinant vesicular stomatitis virus is enhanced by expression of the fusion cytosine deaminase/uracil phosphoribosyltransferase suicide gene Cancer Res 2003; 63:8366–8376.

33. Lorence RM, Pecora AL, Major PP, Hotte SJ, Laurie SA, Roberts MS, Groene WS, Bamat MK. Overview of phase I studies of intravenous administration of PV701, an oncologic virus. Curr Opin Molec Therap 2003; 5:618–624.

34. Csatary LK, Gosztonyi G, Szeberenyi J, Fabian Z, Liszka V, Bodey B, Csatary CM. MTH-68/H oncolytic viral treatment in human high-grade gliomas. J Neurooncol 2004; 67:83–93.

35. Freeman AI, Gomori JM, Linetsky E, Zakay-Rones Z, Panet A, Libson E, Irving CS, Galun E, Siegal T. Phase I/II trial intravenous OV001 oncolytic virus in resistant glioblastoma multiforme. Am Soc Clin Oncol 40th Annual Meeting, New Orleans, June 5–8. Proceedings 2004; 23:110 #1515.

36. Minev B, Hipp J, Firat H, Schmidt JD, Langlade-Demoyen P, Zanetti M. Cytotoxic T cell immunity against telomerase reverse transcriptase in humans. Proc Nat Acad Sci USA 2000; 97:4796–4801.

37. Hernández J, Garcia-Pons F, Lone YC, Firat H, Schmidt JD, Langlade-Demoyen P, Zanetti M. Identification of a human te-

lomerase reverse transcriptase of low affinity for HLA A2.1 that induces cytotoxic T lymphocytes and mediates lysis of tumor cells. Proc Nat Acad Sci USA 2002; 99:12275–12280.

38. Vonderheide RH, Schultze JL, Anderson KS, Maecker B, Butler MO, Xia Z, Kuroda MJ, von Bergwelt-Baildon MS, Bedor MM, Hoar KM, Schnipper DR, Brooks MW, Letvin NL, Stephans KF, Wucherpfenning KW, Hanh WC, Nadler LM. Equivalent induction of telomerase-specific cytotoxic T lymphocytes from tumor-bearing patients and healthy individuals. Cancer Res 2001; 61:8366–8370.

39. Sinkovics JG. Chondrosarcoma cell differentiation. Experimental data and possible molecular mechanisms. Pathology Oncology Research 2004 (in print).

40. Franki SN, Levy R, Timmerman JM. Vaccination with dendritic cells co-cultivated with antibody-coated tumor cells provides protective immunity against B cell lymphoma in vivo. 12th Immunology, Montreal, Canada. Clin Investig Medicine 2004; 27:212D Th53.319.

41. Sato K, Kopecky SI, Frye RI, Goronzy JJ, Weyand CM, T cells kill vascular smooth muscle cells via the death pathway. A role for TRAIL-producing T cells in the rupture of the atherosclerotic plaque. 12th Immunology, Montreal, Canada. Clin Investig Medicine 2004; 27:159A M37.45.

42. Raviv Y, Puri A, Blumenthal R. P-glycoprotein-overexpressing multidrug-resistant cells are resistant to infection by enveloped viruses that enter via the plasma membrane FASEB 2000; 14: 511–515.

43. Holm PS, Lage H, Bergmann S, Jürchott K, Glockzin G, Bernshausen A, Mantwill K, Landhoff A, Wichert A, Mymryk JS, Ritter T, Dietel M, Gänsbacher B, Royer H-D. Multidrug-resistant cancer cells facilitate E1-independent adenoviral replication: impact for cancer gene therapy Cancer Res 2004; 64: 322–328.

Index